Introduction to Reticular Chemistry

Introduction to Reticular Chemistry

Metal-Organic Frameworks and Covalent Organic Frameworks

Omar M. Yaghi
Markus J. Kalmutzki
Christian S. Diercks

WILEY-VCH

Authors

Prof. Omar M. Yaghi
University of California, Berkeley
Department of Chemistry
602 Latimer Hall
94720 Berkeley, CA
United States

Dr. Markus J. Kalmutzki
University of California, Berkeley
Department of Chemistry
618 Latimer Hall
94720 Berkeley, CA
United States

Christian S. Diercks
University of California, Berkeley
Department of Chemistry
618 Latimer Hall
94720 Berkeley, CA
United States

Library of Congress Card No.:
applied for

British Library Cataloguing-in-Publication Data
A catalogue record for this book is available from the British Library.

Bibliographic information published by the Deutsche Nationalbibliothek
The Deutsche Nationalbibliothek lists this publication in the Deutsche Nationalbibliografie; detailed bibliographic data are available on the Internet at http://dnb.d‐nb.de.

Print ISBN: 978-3-527-34502-1
ePDF ISBN: 978-3-527-82108-2
ePub ISBN: 978-3-527-82110-5
oBook ISBN: 978-3-527-82109-9

Cover Design Formgeber, Mannheim, Germany
Typesetting SPi Global, Chennai, India
Printing and Binding CPI Group (UK) Ltd, Croydon, CR0 4YY

Printed on acid-free paper

C9783527345021_271125

To emerging scholars whose curiosity and power of observation make Nature reveal itself

Contents

About the Companion Website

This book is accompanied by a companion website:

http://www.wiley.com/go/yaghi/reticular

The Instructor Companion Site includes:

1) **Figures**
2) **Diamond files**

The provided *.diamdoc files can only be viewed using the crystal structure visualization software DIAMOND: http://www.crystalimpact.de/diamond/Default.htm.

Foreword

Our knowledge of how atoms are linked in space to make molecules and how such molecules react has now reached a sophisticated level leading not only to the formation of useful crystalline materials but also in deciphering important disciplines (e.g. chemical biology, materials chemistry), where chemistry plays an indispensable role in understanding matter. In contrast, the science of making and studying extended chemical structures has remained relatively untouched by the tremendous progress being made in molecular chemistry. This is because solid-state compounds are usually made at high temperatures where the structures of organics and metal complexes do not survive and where their molecular reactivity is not retained. Although this has led to useful inorganic solids being made and studied, the need for translating organic and inorganic complex chemistry with all its subtleties and intricacies into the realm of solid state continued until the end of the twentieth century. At that time, it became clear that the successful synthesis and crystallization of metal-organic frameworks (MOFs) and later covalent organic frameworks (COFs) constituted an important step in developing strong covalent bond and metal–ligand bond chemistry beyond the molecular state. MOFs of organic carboxylates linked to multi-metallic clusters were shown to be architecturally robust and proven to have permanent porosity. Both are critical factors for carrying out precision organic reactions and metal complexations within solid-state structures. With COFs, their successful synthesis and crystallization ushered in a new era for they extended organic chemistry beyond molecules (0D) and polymers (1D) to layered (2D) and framework (3D) structures. The fact that both MOFs and COFs are made under mild conditions, which preserve the structure and reactivity of their building blocks, and that their building blocks are made entirely from strong bonds and are also linked to each other by strong bonds to make crystals of porous frameworks, gave rise to a new thinking in chemistry. By knowing the geometry of the building blocks it became possible to design specific MOF and COF structures, and by knowing the conditions under which such structures formed it became possible to expand their metrics and functionalize their pores without affecting their crystallinity or underlying topology. This is completely new in solid-state chemistry. On the fundamental level, MOFs and COFs represent whole new classes of materials and the intellectual aspects of their chemistry provided a new thinking for the practicing scientist. One might go as far as to say that this new chemistry, termed reticular chemistry, gave credence to the notion of materials on demand. At present, reticular chemistry is being practiced and researched in over a thousand laboratories around the world in academia, industry, and government. The utility of reticular materials in many fields such as gas adsorption, water harvesting, and energy storage, to mention a few, makes this new field all the more interesting to explore and teach since it covers aspects from basic science to real world applications. Accordingly, we have endeavored in this book to provide an introductory entry into this vast field. The book is divided roughly into four parts, which are seamlessly joined in their presentation. The first part (Chapters 1–6) focuses on MOF chemistry and presents their synthesis, building blocks, characterization, structures, and porosity. The second part (Chapters 7–11) presents COF chemistry in a sequence similar to that of MOFs but with emphasis on the organic chemistry used to produce their linkers and linkages. The third part (Chapters 12–17) is dedicated to the applications of MOFs with some mention of those pertaining to COFs. Here, we have endeavored to give a basic description of the physical principles for each application and how reticular materials are deployed. The fourth part (Chapters 18–21) is what we have referred to as special topics that are related to reticular chemistry thinking and analysis. The book is written to allow instructors to use each part independently from the others, and for most chapters, they can also be taught out of sequence or even separately. We hope the students and instructors will appreciate through this textbook that reticular chemistry as a field of study is rooted in organic, inorganic, and physical chemistry, and that it has merged these traditional disciplines into one to produce useful crystalline materials without losing the precision of molecular chemistry. The book is unique in its coverage of the basic science leading to the synthesis, structure, and properties as well as to the applied science of using these materials in addressing societal challenges. Reticular chemistry extends molecular chemistry and its precision in making and breaking bonds to solid-state framework structures being linked by strong bonds. It is now realistic to think in the following way: what the atom is to the molecule, the molecule is to the framework. The molecule fixes the atom in a specific orientation and spatial arrangement, while the framework fixes the molecule into specific orientation and spatial arrangement; except that the framework also encompasses space within which matter can be further manipulated and controlled. It is a new field that combines the beauty of chemical structures, chemistry of building units and their frameworks, and relevance to societal challenges. We have sought to communicate these aspects in our book to provide a rich and stimulating arena for learning.

Berkeley
March 2018

Markus J. Kalmutzki
Christian S. Diercks
Omar M. Yaghi

Acknowledgment

The authors wish to thank the following scholars from the Yaghi research group at the University of California, Berkeley, who contributed selflessly to proofreading of the manuscript: Dr. Eugene Kapustin, Mr. Kyle Cordova, Mr. Robinson Flaig, Mr. Peter Waller, Mr. Steven Lyle, and Dr. Bunyarat Rungtaweevoranit.

We also wish to express our gratitude for the commitment and extensive efforts of Ms. Paulina Kalmutzki, who lent her precious time to the Yaghi group, and Dr. Yuzhong Liu (Yaghi group) for help with the preparation of illustrations. We want to acknowledge Prof. Adam Matzger (University of Michigan), Dr. Bunyarat Rungtaweevoranit, and Yingbo Zhao (Yaghi group) for providing some of the microscopy images found in this text.

Finally, we would like to thank our publisher, Wiley VCH Weinheim, especially Anne Brennführer and Sujisha Karunakaran, for the understanding and assistance provided throughout all stages of the elaborate and laborious task of producing this book.

Introduction

Reticular Chemistry is concerned with making and breaking bonds in molecules and how this can be done in a controlled fashion. When a new molecule is discovered, the need and desire to build it up from simple starting materials using logical means becomes a central objective. Thus, chemists first and foremost are architects and builders: generally, a "blueprint" for a target molecule is designed and a reaction pathway is determined for making it. Often, this blueprint also includes a strategy for achieving the desired molecular geometry and spatial arrangement of atoms, as these dramatically impact the properties of molecules. This sequence of operations is so well developed in organic chemistry that virtually any reasonable target can be designed and made with high precision. The deliberate chemical synthesis approach thus employed is less developed for metal complexes because a metal ion can adopt different geometries and coordination numbers thereby introducing uncertainty into the outcome of the synthesis. Furthermore, unlike organic molecules, where multiple chemical reactions can be carried out to functionalize them, metal complexes are modified largely by substitution–addition reactions. This is because of the limitations imposed by the chemical stability of metal complexes. Thus, the step-by-step approach to the synthesis of organic compounds is severely limited in the synthesis of metal complexes, and this adds a significant component of trial-and-error to metal ion chemistry. It should be noted that the uncertainty in metal-complex chemistry is sometimes obviated by sophisticated design of multi-dentate organic ligands, whereby a metal ion can be locked into a specific geometry and coordination mode. It remains, however, that although immense diversity can be created, the ability to control the geometry around the metal ion and spatial arrangement of ligands is an ongoing challenge.

A new level of precision and control in chemical synthesis is achieved when linking molecules together to make larger discrete and extended structures. There are two basic aspects to consider in linking molecules: the first pertains to the type of interactions used in such linkages and the directionality they impart to the formation of the resulting structure, and the second is concerned with the geometry of the molecular building units and how their metric characteristics such as length, size, and angles guide the synthesis to a specific structure. These aspects are at the core of reticular chemistry, which is concerned with linking molecular building units by strong bonds to make crystalline large and extended structures.

Reticular chemistry started by linking metal ions through strong bonds using charged organic linkers such as carboxylates leading to metal-organic frameworks (MOFs) and related materials. These frameworks in effect expanded the scope of inorganic complex chemistry to include extended structures in which the building units are fixed in precise geometrical and spatial arrangements. Another development was to extend organic chemistry beyond molecules and polymers by using reticular chemistry to link organic building blocks into crystalline two- and three-dimensional covalent organic frameworks (COFs).

The subject of reticular chemistry is also concerned with providing a logical framework for using molecular building units to make structures with useful properties. The concept of node and link that was introduced by Alexander F. Wells to describe a net (collection of nodes and links) has become central to the "grammar" and "taxonomy" of reticular structures, which we discuss in this book. They encompass both, large discrete entities such as metal-organic polyhedra (MOPs) and covalent organic polyhedra (COPs) and extended frameworks such as MOFs, zeolitic imidazolate frameworks (ZIFs), and COFs. This field expanded dramatically and has come to represent a significant segment of the larger field of chemistry.

Among the extensive body of knowledge produced from linking building units using reticular chemistry there are a number of challenges that have been addressed: First, the propensity of metal ions to have variable coordination number and geometries, as mentioned above, is detrimental to controlling the outcome of linking metal ions with organic linkers into MOFs or MOPs. Although exceptions may be found where a metal ion prefers a specific arrangement such as square planar for divalent platinum, in general the use of single metal ions as nodes detracts from the needed control in producing a specific structure. The use of poly-nuclear complexes named secondary building units (SBUs), as in metal carboxylate clusters, locks the metal ions into position and thereby the coordination geometry of the entire SBU is the determining factor in the reticulation process. Second, since the SBUs are clusters by necessity and the organic linkers are multi-atomic, reticular synthesis inevitably yields open structures. The fact that the SBUs are rigid and directional provides for the possibility of design and control of the resulting material. Since the SBUs are made of strong bonds, when joined by organic linkers, they ensure architectural stability and permanent porosity of the framework when the molecules filling its pores are removed. The strong bonds also impart thermal stability and, when they are kinetically inert, chemical stability of the overall porous structure. Third, the ability to determine the conditions under which a specific SBU forms has led to isoreticular synthesis where the same SBU can be joined by a variety of linkers having the same linkage modality but with different size, length, and functional groups attached to them. Fourth, the discovery of the conditions to crystallize the products of these reticular syntheses has enabled the definitive characterization of the outcome of the structures by X-ray diffraction and has facilitated structure–property relationships. Ultimately, this aspect has vastly contributed to the design of structures with specific functionality and pore metrics. Fifth, the permanent porosity, thermal and chemical stability, and crystallinity of these frameworks allow for chemical modification to be carried out on their interior with full preservation of porosity and crystallinity. This meant that large and extended structures can be transformed post-synthetically, and that the incorporation of a specific functionality can be achieved either before or after formation of the product. Sixth, the precision with which such frameworks can be made and their interior modified coupled to the flexibility in deploying a variety of SBUs and organic linkers to make metal-organic and organic reticular materials have given rise to a vast number of properties and applications.

Reticular chemistry has advanced to the point where flexibility and dynamics can be incorporated into large and extended structures. This is accomplished by using flexible constituents or by introducing mechanically interlocking rings within the organic linker. More recently, mechanical entanglement was successfully used in interlacing organic threads to make woven extended structures. In principle, this strategy is also applicable to the interlocking of large discrete rings.

To fully appreciate reticular chemistry and its potential, it is instructive to view reticular structures as being composed of backbone, functionality attached to the backbone, and space encompassed by this construct. The backbone provides the overall structural integrity while the functionality provides for optimal pore environment. The pores can be adjusted to allow for molecules of various sizes, shapes, and character to be incorporated and potentially transformed. In cases when multiple functionalities are used to decorate the pores, the possibility of having unique sequences of chemical entities becomes a reality and the potential for such sequences to code for specific properties exists. The diffusion of molecules within such pore space will undoubtedly be influenced by the specific sequence. This ushers a new era in chemistry where it becomes possible to design and make sequence-dependent materials. The recent advance in "editing" reticular structures by linker or metal substitution without changing the overall porosity and order within the structure is a very promising direction for being able to deliberately alter such chemical sequences. It follows from this discussion that reticular structures are amenable to the introduction of heterogeneity such as defects and functionality by design making it possible to target specific reactivity in ways not possible otherwise.

By linking molecules together into large and extended structures, reticular chemistry has in effect endowed the molecule with additional properties inaccessible without it being linked. Specifically, since the molecule in the reticular structure is fixed in position, it becomes more directly addressable, and depending on where it is linked, the units surrounding it can be considered effectively as "protecting groups." The fact that molecules are repeated throughout the structure provides opportunities for that molecule to be part of a whole that could function above and beyond the sum of its parts. The interface between the molecules making up the structure and other molecules freely residing in the pores as guests is a well-defined region of the overall structure. This interface is also endowed with the same precision of design and definition that is so characteristic to reticular structures. Accordingly, the interface can be varied and tailored in ways the molecule cannot experience outside this intricate environment. In essence, what reticular chemistry has done is to provide means of controlling matter beyond molecules, in large and extended structures, and to also provide the space within which molecules can be further controlled and manipulated.

Abbreviations

1,2-H_2DACH	1,2-diaminocyclohexane
(V)MIL-47	V(O)(BDC)
^{13}C CP-MAS	^{13}C cross-polarization magic angle spinning
2,6-H_2NDC	naphthalene-2,6-dicarboxylic acid
1,4-H_2NDC	4,4′-(naphthalene-2,6-diyl)dibenzoic acid
2-mBIM	2-methylbenzimidazolate
4,4′-H_2DMEDBA	4,4′-(1,2-dimethoxyethane-1,2-diyl)dibenzoic acid
4-nIM	4-nitroimidazolate
5-BBDC	5-*tert*-butyl-1,3-benzenedicarboxylate
AB	4-aminobenzoate
acac	acetylacetonate
AD	adeninate
ADHP	adsorption-driven heat pumps
ADI	adiponitrile
AFM	atomic force microscope/atomic force microscopy
aIM	2-carbaldehyde imidazolate
Al-PMOF-1	$Al_2(OH)_2(TCPP\text{-}H_2)$
Al-soc-MOF-1	$[In_3O(H_2O)_3]_2(TCPT)_3(NO_3)$
ANH	aromatic *N*-heterocycle
APTES	3-aminopropyltriethoxysilane
ASA	*p*-arsanilic acid
ATZ	5-amino-triazolate
BASF	Badische Anilin und Soda Fabrik
BBC	4,4′,4″-(benzene-1,3,5-triyl-tris(benzene-4,1-diyl))tribenzoate
BBCDC	9*H*-carbazole-3,6-dicarboxylate
bBIM	5-bromo-1H-benzo[d]imidazole
bBIM	6-bromobenzimidazolate
BBO-COF-1	$[(TFB)_2(PDA\text{-}(OH)_2)_3]_{benzoxazole}$
BBO-COF-2	$[(TFPB)_2(PDA\text{-}(OH)_2)_3]_{benzoxazole}$
BDA	terephthaldehyde
BDA-(F)	2-fluoroterephthaldehyde
BDA-$(F)_4$	2,3,5,6-tetrafluoroterephthaldehyde

BDA-(H_2C—C≡CH)	2,5-bis(2-propynyloxy)terephthalaldehyde
BDA-$(OH)_2$	2,5-dihydroxy-1,4-benzenedialdehyde
BDA-$(OMe)_2$	2,5-dimethoxyterephthaldehyde
BDBA	1,4-phenylenediboronic acid
BDH-$(OEt)_2$	2,5-diethoxyterephtalohydrazide
BET model	Brunauer–Emmett–Teller model
BIM	benzimidazolate
bio-MOF-100	$[Zn_6O_2(AD)_4(BPDC)_6](NO_3)_4$
bio-MOF-101	$[Zn_6O_2(AD)_4(NDC)_6](NO_3)_4$
bio-MOF-102	$[Zn_6O_2(AD)_4(ABDC)_6](NO_3)_4$
bio-MOF-103	$[Zn_6O_2(AD)_4(NH_2\text{-}TDC)_6](NO_3)_4$
BIPY	4,4′-bpyridine
BLP	1,3,5-(*p*-aminophenyl)-benzene-borane
Boc	*tert*-butyloxycarbonyl
BPDA	4,4′-biphenyldialdehyde
BPEE	(*E*)-1,2-di(pyridin-4-yl)ethene
Br-H_2BDC	2-bromoterephthalic acid
BTB	4,4′,4″-benzene-1,3,5-triyltribenzoate
BTBA	benzene-1,3,5-triyltriboronic acid
BTCTB	4,4′,4″-[benzene-1,3,5-triyltris (carbonylimino)]tris-benzoate)
BTDD	bis(1*H*-1,2,3-triazolo[4,5-*b*],[4′,5′-*i*]) dibenzo[1,4]dioxin
BTE	4,4′,4″-(benzene-1,3,5-triyl-tris(benzene-4,1-diyl))tribenzoate
BTEB	4′,5′-bis(4-carboxyphenyl)-[1,1′:2′,1″-terphenyl]-4,4″-dicarboxylic acid
Bu	butyl
BZD-$(NO_2)_2$	2,2′-dinitrobenzidine
CAL	coordinative alignment
CAU-10	Al(OH)(*m*-BDC)
cBIM	5-chloro-1*H*-benzo[*d*]imidazole
cBIM	6-chlorobenzimidazole
CBP	Cu(I)bis-4,4′-(1,10-phenanthroline-2,9-diyl) diphenol
CCS	CO_2 capture and sequestration
CdIF-4	$Cd(eIM)_2$
CdIF-9	$Cd(nIM)_2$
cIM	2-chloro imidazolate
Cl_2-H_2BDC	2,5-dichloroterephthalic acid
CNG	compression of natural gas
Co(TAP)	tetra(4-aminophenyl)porphinato cobalt
COD	1,5-cyclooctadiene
COF	covalent organic framework
COF-1	$[BDBA]_{boroxine}$
COF-102	$[TBPM]_{boroxine}$
COF-103	$[TBPS]_{boroxine}$

COF-105	$[(TBPS)_3(HHTP)_4]_{boronate\ ester}$
COF-108	$[(TBPM)_3(HHTP)_4]_{boronate\ ester}$
COF-202	$[(TBPM)_3(\textit{tert}\text{-butylsilane triol})_4]_{borosilicate}$
COF-300	$[(TAM)(BDA)_2]_{imine}$
COF-320	$[(TAM)(BPDA)_2]_{imine}$
COF-366	$[(H_2TAP)(BDA)_2]_{imine}$
COF-366-Co	$[(Co(TAP))(BDA)_2]_{imine}$
COF-367-Co	$[(Co(TAP))(BPDA)_2]_{imine}$
COF-42	$[(TFB)_2(BDH\text{-}(OEt)_2)_2]_{hydrazone}$
COF-43	$[(TFP)_2(BDH\text{-}(OEt)_2)_2]_{hydrazone}$
COF-5	$[(HHTP)_2(BDBA)_3]_{boronate\ ester}$
COF-505-Cu	$(Cu)(BF_4)[(PDB)(BZD)_2]_{imine}$
COP	covalent organic polyhedron
CP-MAS	cross-polarization magic angle spinning
CP-MAS NMR	cross-polarization magic angle spinning NMR
CS-COF	$[(HATP)_2(PT)_3]_{phenazine}$
CTF-1	$[DCyB]_{triazine}$
Cu(TAP)	[5,10,15,20-tetrakis(4-aminophenyl) porphinato]-copper
CuBTTri	$H_3[(Cu_4Cl)_3(BTTri)_8]$
DAA	2,6-diaminoanthracene
DAB	(([2,2′-bipyridine]-5,5′-diylbis(oxy)) bis(4,1-phenylene))dimethanamine
DABCO	1,4-Diazabicyclo[2.2.2]octan
DABCO	1,4-Diazabicyclo[2.2.2]octane
DBA	hexahydroxy-dehydrobenzoannulene
DBS	4-(dodecycloxy)benzoic acid
dcIM	4,5-dichloroimidazolate
DCyB	1,4-dicyanobenzene
DEA	diethylamine
DFP	2,6-pyridinedicarboxaldehyde
DIT	1,14-di-iodo-3,6,9,12-tetraoxy-tetradecane
DLS	dynamic light scattering
DMA	dimethylamine
dmBIM	5,6-dimethylbenzimidazole
DMF	*N*,*N*-dimethylformamide
DMOF	$Zn(BDC)(DABCO)_{0.5}$
DMOF-1(NH_2)	$Zn_2(NH_2\text{-}BDC)_2(DABCO)$
DOBPDC	4,4′-dioxidobiphenyl-3,3′-dicarboxylate
DOE	US Department of Energy
DOX	doxorubicin
DSC	differential scanning calorimetry
DUT	Dresden University of Technology
DUT-32	$Zn_4O(BPDC)(BTCTB)_{4/3}$
DUT-51	$Zr_6O_6(OH)_2(DTTDC)_4(CH_3COO)_2$
DUT-67	$Zr_6O_6(OH)_2(TDC)_4(CH_3COO)_2$
DUT-69	$Zr_6O_4(OH)_4(TDC)_5(CH_3COO)_2$

EDDB	4,40-(ethyne-1,2-diyl)dibenzoic acid
EDX	energy dispersive X-ray spectroscopy
eIM	2-ethyleimidazolate
ElAPO	metal-aluminophosphate with additional Li, Be, B, Ga, Ge, As, Ti
ElAPSO	metal-silicoaluminophosphate with additional Li, Be, B, Ga, Ge, As, Ti
en	1,2-ethylene diamine
Et	ethyl
ETTA	1,1,2,2-tetrakis(4-aminophenyl)ethane
FDM	Fudan Materials
FDM-3	$[(Zn_4O)_5(Cu_3OH)_6(PyC)_{22.5}(OH)_{18}(H_2O)_6][Zn(OH)(H_2O)_3]_3$
FT-IR	Fourier-transform infrared spectroscopy
GCMC	grand canonical Monte Carlo
gea-MOF-1	$Y_9(\mu_3\text{-}OH)_8(\mu_2\text{-}OH)_3(BTB)_6$
GIWAXS	grazing incidence wide angle X-ray scattering
GLU	glutaronitrile
H_2ABDC	(*E*)-4,4′-(diazene-1,2-diyl)dibenzoic acid
H_2ADC	anthracene-9,10-dicarboxylic acid
H_2BATZ	bis(5-amino-1H-1,2,4-triazol-3-yl)methane
H_2BBTA	1*H*,5*H*-benzo(1,2-d:4,5-d′)bistriazole
H_2BDC	terephthalic acid (benzene-1,4-dicarboxylic acid)
H_2BPCu	Cu^{2+}-4,7,10,13,16,19,22,25-octaoxa-2(2,9)-phenanthrolina-1,3(1,4)-dibenzenacyclohexacosaphane @ 4,4′-(1,10-phenanthroline-3,8-diyl)dibenzoic acid
H_2BPDC	[1,1′-biphenyl]-4,4′-dicarboxylic acid
H_2BPyDC	[2,2′-bipyridine]-5,5′-dicarboxylic acid
H_2CBDA	4,4′-carbonyldibenzoic acid
$H_2CONQDA$	4,4′-(5,6,12,13-tetrachloro-1,3,8,10-tetraoxo-1,3,8,10-tetrahydroanthra[2,1,9-def:6,5,10-d′e′f′]diisoquinoline-2,9-diyl)dibenzoic acid
H_2DMBDA	4,4′-((2,5-dimethoxy-1,4-phenylene) bis(ethyne-2,1-diyl))dibenzoic acid
H_2DTTDC	dithieno[3,2-b:2′,3′-d]thiophene-2,6-dicarboxylic acid
H_2EDBA	(*E*)-4,4′-(ethene-1,2-diyl)dibenzoic acid
H_2HPDC	4,5,9,10-tetrahydropyrene-2,7-dicarboxylic acid
H_2MPBA	4-(3,5-dimethylpyrazol-4-yl)benzoic acid
H_2MPDA	4,4′-(2,9-dimethyl-1,10-phenanthroline-3,8-diyl)dibenzoic acid
H_2NDC	naphthalene-2,6-dicarboxylic acid
H_2OBA	4,4′-oxybis(benzoic acid)

H_2PDC	pyrene-2,7-dicarboxylic acid
H_2TAP	5,10,15,20-tetrakis(4-amino-phenyl)porphyrin
H_2TDC	thiophene-2,5-dicarboxylic acid
H_2TPDC	[1,1′:4′,1″-terphenyl]-4,4″-dicarboxylic acid
H_2TTC	2,2′:5′,2″-terthiophene-5,5″-dicarboxylic acid
H_3BBC	5″-(4′-carboxy-[1,1′-biphenyl]-4-yl)-[1,1′:4′,1″:3″,1‴:4‴,1⁗-quinquephenyl]-4,4⁗-dicarboxylic acid
H_3BHTC	[1,1′-biphenyl]-3,4′,5-tricarboxylic acid
H_3BTB	5′-(4-carboxyphenyl)-[1,1′:3′,1″-terphenyl]-4,4″-dicarboxylic acid
H_3BTC	benzene-1,3,5-tricarboxylate
H_3BTE	4,4′,4″-(benzene-1,3,5-triyltris(ethyne-2,1-diyl)tribenzoic acid
H_3BTN	6,6′,6″-(benzene-1,3,5-triyl)tris(2-naphthoic acid)
H_3BTT	1,3,5-benzetristetrazole
H_3BTTC	benzo[1,2-b:3,4-b′:5,6-b″]trithiophene-2,5,8-tricarboxylic acid
H_3BTTri	1,3,5-tris(1H-1,2,3-triazol-5-yl)benzene
H_3HTB	4-[7,11-bis(4-carboxyphenyl)-2,4,6,8,10,12,13-heptaazatricyclo[7.3.1.05,13]trideca-1,3,5,7,9,11-hexaen-3-yl]benzoic acid
H_3IMDC	1*H*-imidazole-4,5-dicarboxylic acid
H_3TAPB	4′,4‴,4⁗′-(1,3,5-triazine-2,4,6-triyl)tris(([1⁗,1⁗′-biphenyl]-4-carboxylic acid))
H_3TATAB	4,4′,4″-((1,3,5-triazine-2,4,6-triyl)tris(azanediyl))tribenzoic acid
H_3TATAB	4,4′,4″-((1,3,5-triazine-2,4,6-triyl)tris(azanediyl))tribenzoic acid
H_3TATB	4,4′,4″-(1,3,5-triazine-2,4,6-triyl)tribenzoic acid
H_3TCA	4,4′,4″-nitrilotribenzoic acid
H_3TCPBA	4′,4‴,4⁗′-nitrilotris(([1⁗,1⁗′-biphenyl]-4-carboxylic acid))
H_3TTCA	triphenylene-2,6,10-tricarboxylic acid
H_3TZI	5-tetrazolylisophthalic acid
H_4ABTC	(*E*)-5,5′-(diazene-1,2-diyl)diisophthalic acid
H_4ADBTD	5′,5⁗-(anthracene-9,10-diyl)bis(([1,1′:3′,1″-terphenyl]-4,4″-dicarboxylic acid))
H_4ADIP	4,4′-(anthracene-9,10-diyl)dibenzoic acid
H_4ATB	4,4′,4″,4‴-(adamantane-1,3,5,7-tetrayl)tetrabenzoic acid
H_4BBDC	5-boronobenzene-1,3-dicarboxylate

H_4BITC	18-crown-6 @ 4,4′,4″,4‴-(1,4-phenylenebis(1*H*-benzo[d]imidazole-2,4,7-triyl))tetrabenzoic acid
H_4BNETBA-$(OEt)_2$	4,4′,4″,4‴-((1*E*,1′*E*,1″*E*,1‴*E*)-(2,2′-diethoxy-[1,1′-binaphthalene]-4,4′,6,6′-tetrayl) tetrakis(ethene-2,1-diyl))tetrabenzoic acid
H_4BPDCD	9,9′-([1,1′-biphenyl]-4,4′-diyl)bis(9*H*-carbazole-3,6-dicarboxylic acid)
H_4BPTC	[1,1′-biphenyl]-3,3′,5,5′-tetracarboxylic acid
H_4CBI	1,12-Bis(3′,5′-bis(hydroxycarbonyl) phen-1-yl)-1,12-dicarba-closododecaborane
H_4CQDA$(OEt)_2$	5′,5″-bis(4-carboxyphenyl)-2′,2″-diethoxy-[1,1′:3′,1″:3″,1‴-quaterphenyl]-4,4‴-dicarboxylic acid
H_4DH_{11}PhDC/DOT-XI	4′-[4′-(4′-{4′-[4′-(4-carboxy-3-hydroxyphenyl)-2,2′,5,5′-tetramethyl-[1,1′-biphenyl]-4-yl]-5′-hexyl-2,5-dimethyl-2′-pentyl-[1,1′-biphenyl]-4-yl}-2,2′,5,5′-tetramethyl-[1,1′-biphenyl]-4-yl)-2,5-dimethyl-2′,5′-dipentyl-[1,1′-biphenyl]-4-yl]-3-hydroxy-2′,5′-dimethyl-[1,1′-biphenyl]-4-carboxylic acid
H_4DOT	2,5-dihydroxyterephthalic acid
H_4DOT-III	3,3″-dihydroxy-2′,5′-dimethyl-(1,1′:4′,1″-terphenyl)-4,4″-dicarboxylic acid
H_4ETTC	4′,4‴,4′′′′′,4′′′′′′′-(ethene-1,1,2,2-tetrayl) tetrakis(([1,1′-biphenyl]-4-carboxylic acid))
H_4MTB	4,4′,4″,4‴-methanetetrayltetrabenzoic acid
H_4MTPA	4,4′,4″,4‴-((methanetetrayltetrakis (benzene-4,1-diyl)tetrakis(ethyne-2,1-diyl)) tetrabenzoic acid
H_4MTPB	4′,4‴,4′′′′′,4′′′′′′′-methanetetrayltetrakis (([1,1′-biphenyl]-4-carboxylic acid))
H_4PyrDI	5,5′-(pyrimidine-2,5-diyl)diisophthalic acid
H_4QPTCA	[1,1′:4′,1″:4″,1‴:4‴,1⁗-quinquephenyl]-3,3⁗,5,5⁗-tetracarboxylic acid
H_4SFTT	4,4′,4″,4‴-(9,9′-spirobi[fluorene]-2,2′,7,7′-tetrayl)tetrabenzoic acid
H_4STBA	4,4′,4″,4‴-silanetetrayltetrabenzoic acid
H_4TBADB-18Cr6	4,4′,4″,4‴-(6,7,9,10,17,18,20,21-octahydrodibenzo[b,k][1,4,7,10,13,16] hexaoxacyclooctadecine-2,3,13,14-tetrayl) tetrabenzoic acid
H_4TBAPy	4,4′,4″,4‴-(1,8-dihydropyrene-1,3,6,8-tetrayl) tetrabenzoic acid

H_4TCBPP-H_2	4′,4‴,4⁗′,4″‴″-(porphyrin-5,10,15,20-tetrayl)tetrakis(([1,1′-biphenyl]-4-carboxylic acid))
H_4TCPP-H_2	4,4′,4″,4‴-(porphyrin-5,10,15,20-tetrayl) tetrabenzoic acid
H_4TPTC	terphenyl-3,3′,5,5′-tetracarboxylaic acid
H_5PTPCA	5′-(4-carboxyphenyl)-[1,1′:3′,1″-terphenyl]-3,3″,5,5″-tetracarboxylic acid
$H_6BHEHPI$	1,3,5-tris[(1,3-carboxylic acid-5-(4-(ethynyl) phenyl))butadiynyl]-benzene
H_6BHEI	5,5′,5″-(benzene-1,3,5-triyltris(buta-1,3-diyne-4,1-diyl))triisophthalic acid
H_6HTTEI	5,5′,5″-(((benzene-1,3,5-triyltris(ethyne-2,1-diyl))tris(benzene-4,1-diyl))tris(ethyne-2,1-diyl))triisophthalic acid
H_6PTEI	4,4′-((5′-(4-((4-((oxo-λ^3-methyl)-λ^3-oxidaneyl) phenyl)ethynyl)phenyl)-[1,1′:3′,1″-terphenyl]-4,4″-diyl)bis(ethyne-2,1-diyl))dibenzoic acid
H_6TDCPB	4,4′,4″,4‴,4⁗,4⁗′-((nitrilotris(benzene-4,1-diyl))tris(azanetriyl))hexabenzoic acid
H_6TPBTM	5,5′,5″-((benzene-1,3,5-tricarbonyl) tris(azanediyl))triisophthalic acid
H_6TTA	5′,5‴-bis(4-carboxyphenyl)-5″-(4,4″-dicarboxy-[1,1′:3′,1″-terphenyl]-5′-yl)-[1,1′:3′,1″:3″,1‴:3‴,1⁗-quinquephenyl]-4,4⁗-dicarboxylic acid
H_6TTATP	5,5′,5″-((1,3,5-triazine-2,4,6-triyl) tris(azanediyl)triisophthalic acid
H_8BPTCD	9,9′,9″,9‴-([1,1′-biphenyl]-3,3′,5,5′-tetrayl) tetrakis(9H-carbazole-3,6-dicarboxylic acid)
H_8MTBDA	4′,4‴,4⁗′,4″‴″-methanetetrayltetrakis (([1,1′-biphenyl]-3,5-dicarboxylic acid))
$H_8TBCPPP$-H_2	5′,5⁗,5″‴″,5″″‴″-(porphyrin-5,10,15,20-tetrayl)tetrakis(([1,1′:3′,1″-terphenyl]-4,4″-dicarboxylic acid))
$H_8TDPEPE$	4′,4‴,4⁗′,4″‴″-(ethene-1,1,2,2-tetrayl) tetrakis(([1,1′-biphenyl]-3,5-dicarboxylic acid))
HATP	2,3,6,7,10,11-hexaaminoterphenylene
HDN	hydrodenitrogenation
HEIMIM	(*E*)-2-(((2-hydroxyethyl)imino)methyl) imidazolate
HHTP	2,3,6,7,10,11-hexahydroxyterphenylene
HKUST	Hong Kong University of Science and Technology

HKUST-1	$Cu_3(BTC)_2$
HPP	1,3,4,6,7,8-hexahydro-2*H*-pyrimido [1,2-a]pyrimidine
HR-PXRD	High resolution X-ray diffraction
HSAB	hard-soft acid–base
IAST	ideal adsorbed solution theory
ICOF-1	$[(OHM)(TMB)_2]_{spiroborate}$
ICP	inductively coupled plasma
In-soc-MOF	$[In_3O(H_2O)_3]_2(ABDC)_3(NO_3)$
*i*Pr	iso-propyl
IRMOF-74-III	Mg(DOT-III)
IRMOF-74-III(CH_2NH_2)	Mg(CH_2NH_2-DOT-III)
IRMOF-74-III(CH_2NHMe)	Mg(CH_2NHMe-DOT-III)
IRMOF-993	$Zn_4O(ADC)_3$ pcu topology (theoretical)
IUPAC	International Union of Pure and Applied Chemistry
IZA	Structure Commission of the International Zeolite Association
JUC-77	In(OH)(OBA)
KAUST-7 or NbOFFIVE-1-Ni	$Ni(Pyr)_2(NbOF_5$
Keggin Type POM	$(NH_4)_3[(XO_4)Mo_{12}O_{36}])$, X = P, Si, S among others and M = Mo, W
L-Asp	L-aspartate
LD_{50}	The median lethal dose in toxicology. LD_{50} = lethal dose, 50%
LMCT	ligand-to-metal charge transfer
LNG	liquefied natural gas
LZU-1	$[(TFP)_2(PDA)_3]_{imine}$
MAF-25	$Mn_2^{2+}Cl_2(BBTA)$
MAF-25-ox	$Mn^{2+}Mn^{3+}(OH)Cl_2(BBTA)$
MAF-27	$Co_2^{2+}Cl_2(BBTA)$
MAF-27-ox	$Co^{2+}Co^{3+}(OH)Cl_2(BBTA)$
MAF-49	$[Zn(BATZ)](H_2O)_{0.5}$
MAF-X8	Zn(MPBA)
MAMS-1	$Ni_8(5\text{-}BBDC)_6(\mu_3\text{-}OH)_4$
m-BDC	isophthalic acid
mBIM	5-methyl-1*H*-benzo[*d*]imidazole
mBIM	6-methylbenzimidazolate
Me	methyl
Me_2-H_2TPDC	2′,5′-dimethyl-[1,1′:4′,1″-terphenyl]-4,4″-dicarboxylic acid
Me_4-BPDC	2,2′,6,6′-tetramethylbiphenyl-4,4′-dicarboxylic acid
Me_4-DMOF	$Zn(Me_4\text{-}BDC)(DABCO)_{0.5}$
MeAPO	metal-aluminophosphate
MeAPSO	metal-silicoaluminophosphate
MeOH	methanol

MeOHIM	2-hydroxymethylimidazolate
MIL	Materials Institute Lavoisier
MIL-100	$[M_3O(H_2O)_2L](BTC)_2/[M_3OL_3](BTC)_2$
MIL-100(Fe_BTB)	$[Fe_3O(H_2O)_2(L)](BTB)_2$
MIL-101	$[M_3OL_3](BDC)_3$
MIL-125	$Ti_8O_8(OH)_4(BDC)_6$
MIL-125(NH_2)	$Ti_8O_8(OH)_4(NH_2\text{-}BDC)_6$
MIL-53	M(OH)(BDC)
MIL-88	$Fe_3O(OH)(H_2O)_2(BDC)_3$
mIM	2-methyl-1*H*-imidazole
mmen	*N,N′*-dimethylethylenediamine
MMM	mixed-matrix membranes
MOF	metal-organic framework
MOF-177	$Zn_4O(BTB)_2$
MOF-180	$Zn_4O(BTE)_2$
MOF-2	$Zn(BDC)(H_2O)$
MOF-200	$Zn_4O(BBC)_2$
MOF-205	$Zn_4O(BTB)_{4/3}(NDC)$
MOF-210	$(Zn_4O)_3(BPDC)_4(BTE)_3$
MOF-325	$Cu_3(H_2O)_3[(Cu_3O)(PyC)_3(NO_3)L_2]_2$
MOF-5	$Zn_4O(BDC)_3$
MOF-520	$Al_8(OH)_8(HCOO)_4(BTB)_4$
MOF-520-BPDC	$Al_8(OH)_8\ (BTB)_4(BPDC)_2$
MOF-525	$Zr_6O_4(OH)_4(TCPP\text{-}H_2)_3$ (ftw topology)
MOF-545	$Zr_6O_4(OH)_4(TCPP\text{-}H_2)_2(H_2O)_8$ (scsq topology)
MOF-74	$M_2(DOT)$
MOF-801	$Zr_6O_4(OH)_4(fumarate)_{12}$
MOF-808	$Zr_6O_4(OH)_4(HCOO)_6(BTC)_2$
MOF-812	$Zr_6O_4(OH)_4(MTB)_3(H_2O)_2$
MOF-841	$Zr_6O_4(OH)_4(MTB)_2(HCOO)_4(H_2O)_4$
MOF-901	$Ti_6O_6(OCH_3)_6(AB)_6$
MOP	metal-organic polyhedron
MS	mass spectroscopy
MTV	multivariate
MUF	Massey University Metal–Organic Frameworks
MUF-7a	$(Zn_4O)_3(BTB)_{4/3}(BDC)_{1/2}\ (BPDC)_{1/2}$
MWC	maximum working capacity
NASA	National Aeronautics and Space Administration
nBIM	6-nitrobenzimidazolate
NBPDA	4-(*tert*-butoxycarbonylamino)-aniline
n-BuLi	*n*-butyllithium
NG	natural gas
$NH_2\text{-}H_2BDC$	2-aminoterephthalic acid

NH_2-H_2TPDC	2′-amino-[1,1′:4′,1″-terphenyl]-4,4″-dicarboxylic acid
nIM	2-nitroimidazolate
NLDFT	nonlinear density functional theory
NLDFT	nonlinear density functional theory
NMP	*N*-methyl-2-pyrrolidone
NMR	nuclear magnetic resonance
NOTT	Nottingham
NOTT-101	$Cu_2(H_2O)_2$(TPTC)
NOTT-103	$Cu_2(H_2O)_2$(2,6-NDI)
NOTT-109	$Cu_2(H_2O)_2$(1,4-NDI)
NU	Northwestern University
NU-100	$Cu_3(H_2O)$(HTTEI)
NU-1000	$Zr_6(\mu_3\text{-OH/O})_8(H_2O,OH)_8(TBAPy)_2$
NU-110	$Cu_3(H_2O)$(BHEHPI)
NU-902	$Zr_6O_4(OH)_4(TCPP\text{-}H_2)_2(H_2O)_4(OH)_4$ (scu topology)
OAc	acetate, CH_3COO^-
OHM	octa-hydroxy functionalized macrocycle
O_h-nano-Ag	octahedral silver nanocrystal
OTf	triflate
OX	oxalate
PCN	porous coordination network
PCN-125	$[Cu_2(H_2O)_2]$(TPDC)
PCN-13	$Zn_4O(H_2O)_3(ADC)_3$
PCN-14	$Cu_2(H_2O)_2$(ADIP)
PCN-223	$Zr_6O_4(OH)_4(TCPP\text{-}H_2)_3$ (shp topology)
PCN225	$Zr_6O_4(OH)_4(TCPP\text{-}H_2)_2(H_2O)_4(OH)_4$ (sqc topology)
PCN-332	$[M_3O(H_2O)_2(L)](BTTC)_2$
PCN333	$[M_3O(H_2O)_2(L)](TATB)_2$
PCN-6	$Cu_3(TATB)_2$ interpenetrated
PCN-6′	$Cu_3(TATB)_2$
PCN-61	$Cu_3(H_2O)_3$(BTEI)
PCN-610	$Cu_3(H_2O)$(HTTEI)
PCN-68	$Cu_3(H_2O)_3$(PTEI)
PCN-700	$Zr_6O_4(OH)_4(Me_2\text{-}BPDC)_4(OH)_4(H_2O)_4$
PCN-700	$Zr_6O_4(OH)_8(H_2O)_4(Me_2\text{-}BPDC)_{8/2}$
PCN-701	$Zr_6O_4(OH)_6(H_2O)_2(Me_2\text{-}BPDC)_{8/2}(BDC)_{2/2}$
PCN-702	$Zr_6O_4(OH)_6(H_2O)_2(Me_2\text{-}BPDC)_{8/2}(Me_2\text{-}TPDC)_{1/2}$
PCN-703	$Zr_6O_4(OH)_6(H_2O)_2(Me_2\text{-}BPDC)_{8/2}(BDC)_{2/2}(Me_2\text{-}TPDC)_{1/2}$
PCN-777	$Zr_6O_4(OH)_4(HCOO)_6(TATB)_2$
PCN-9	$Cu_3(HTB)_2$ interpenetrated
PCN-9′	$Cu_3(HTB)_2$
PDA	1,4-phenylenediamine

PDA-$(OH)_2$	2,5-dihydroxy-1,4-phenylenediamine
PDAN	2,2′-(1,4-phenylene)diacetonitrile
PDB	0,13,16,19,22,25-octaoxa-2(2,9)-phenanthrolina-1,3(1,4)-dibenzenacyclohexacosaphane
PDH	1,4-dicarbonyl-phenyl-dihydrazide
PET	polyethyleneterephthalate
PIC	γ-picoline
PI-COF-1	$[(TAPA)_2(PMDA)_3]_{imide}$
PI-COF-2	$[(TAPB)_2(PMDA)_3]_{imide}$
PI-COF-3	$[(TABPB)_2(PMDA)_3]_{imide}$
PI-COF-4	$[(TAA)(PMDA)_2]_{imide}$
PI-COF-5	$[(TAM)(PMDA)_2]_{imide}$
PMDA	pyromellitic dianhydride
PMOF-1	$Cu_3(H_2O)(TPBTM)$
POM	polyoxometallate
PSA	pressure swing adsorption
PSE	post-synthetic linker exchange
PSM	post-synthetic modification
PT	*tert*-butyl pyrenetetraone
PTA	phosphotungstic acid
PTO	2,7-di-*tert*-butyl-pyrene-4,5,9,10-tetraone
Pur	purine
PVP	polyvinylpyrolidone
PX	*p*-xylene
PXRD	powder X-ray diffraction
Py	pyridine
PyC	4-pyrazolecarboxylic acid
Pyr	pyrazine
PyTA	4,4′,4″,4‴(Pyrene-1,3,6,8-tetrayl)tetraaniline
QCM	quartz crystal microbalance
Q_{st}	isoteric heat of adsorption
RCSR	reticular Chemistry Structure Resource
RED	3D rotation electron diffraction
RH	relative humidity
rho Z-MOF	$In(HIMDC)_2(HPP)$ with **rho** topology
rht-MOF-1	$[Cu_3(TZI)_2(H_2O)_2]_{12}[Cu_3O(OH)(H_2O)_2]_8$
RON	research octane number
ROX	roxorsone
RPM3-Zn	$Zn_2(BPDC)_2(BPEE)$
SALE	solvent assisted linker exchange
SALEM-1	$Cd(mIM)_2$
SALI	solvent assisted ligand incorporation
SAPO	silicoaluminophosphate
SBU	secondary building unit
SDA	structure directing agent
SEM	scanning electron microscope

SIFSIX-2-Cu	$Cu(DPA)_2(SiF_6)$
SIFSIX-2-Cu-i	$Cu(DPA)_2(SiF_6)$ interpenetrated
SIFSIX-3-Ni	$Ni(Pyr)_2(SiF_6)$
SLG	single layer graphene
SLI	sequential linker installation
S-MOF-808	$Zr_6O_5(OH)_3(BTC)_2(SO_4)_{2.5}(H_2O)_{2.5}$
sod Z-MOF	$In(HIMDC)_2(HIM)$ with **sod** topology
sp^2C-COF	$[(TFPPy)(PDAN)_2]_{acrylonitrile}$
SQ	squaric acid
ST-1	$(Zn_4O)_3(TATAB)_4(BDC)_3$
ST-2	$(Zn_4O)_3(TATAB)_4(NDC)_3$
ST-3	$(Zn_4O)_3(TATAB)_4(BPDC)_2(BDC)$
ST-4	$(Zn_4O)_5(TATAB)_4(BPDC)_6$
STM	scanning tunneling microscopy
SUC	succinonitrile
TAA	1,3,5,7-tetraaminoadamantane
TABPB	1,3,5-tris[4-amino(1,1biphenyl-4-yl)]benzene
TAM	tetra-(4-aminophenyl)methane
TAPA	tris(4-aminophenyl)amine
TAPB	1,3,5-tris(4-aminophenyl)benzene
TAPB	1,3,5-tris(4-aminophenyl)benzene
TBPM	tetra(4-dihydroxyborylphenyl)methane
TBPS	tetra(4-dihydroxyborylphenyl)silane
TBPy	5,5′-bis(2-(5′-methyl-[2,2′-bipyridin]-5-yl)ethyl)-2,2′-bipyridine
TBU	tertiary building unit
TCA	(1,1′,3′,1″-Terphenyl)-3,3″,5,5″-tetracarbaldehyde
TCAT	4-(*tert*-butyl)benzene-1,2-diol
TCP	4,4′,4″,4‴-(porphyrin-5,10,15,20-tetrayl) tetrabenzonitrile
TCTPM	4,4′,4″,4‴-tetracyanotetraphenylmethane
TEM	transmission electron microscope
TEM	transmission electron microscopy
TEMPO	4-azido-2,2,6,6-tetramethyl-1-piperidinyloxy
TEOA	triethanolamine
TFA	trifluoro-acetic acid
TFB	1,3,5-triformyl-benzene
TFP	1,3,5-tris-(4-formylphenyl)-benzene
TFP	triformylphloroglucinol
TFPPy	4,4′,4″,4‴-(pyrene-1,3,6,8-tetrayl) tetrabenzaldehyde
TGA	thermogravimetric analysis
THF	tetrahydrofuran
TMB	trismethoxy borate
TMTPDC	2′,3′,5′,6′-tertramethylterphenyl-4,4″-dicarboxylic acid

TPa-1	$[(TFP)_2(PDA)_3]_{\beta\text{-}ketoenamine}$
TPP	5,10,15,20-tetra(pyridin-4-yl)porphyrin
TSA	temperature swing adsorption
TTH	(9s,10s)-13,16-diethyl-9,10-dihydro-9,10-[1,2]benzenoanthracene-2,3,6,7-tetraol
UiO	University of Oslo
UiO-66	$Zr_6O_4(OH)_4(BDC)_{12}$
UiO-67	$Zr_6O_4(OH)_4(BPDC)_{12}$
UiO-68	$Zr_6O_4(OH)_4(TPDC)_{12}$
UMCM	University of Michigan Crystalline Material
UMCM-1	$Zn_4O(BDC)(BTB)_{4/3}$
UMCM-1(NH_2)	$Zn_4O(BDC)(BTB)_{4/3}$
UMCM-10	$Zn_4O(BDC)_{0.75}(Me_4\text{-}BPDC)_{0.75}(TCA)$
UMCM-11	$Zn_4O(BDC)_{0.75}(EDDC)_{0.75}(TCA)$
UMCM-12	$Zn_4O(BDC)_{0.75}(MTMTPDC)_{0.75}(TCA)$
UMCM-150	$Cu_3(BHTC)_2(H_2O)_3$
UMCM-2	$Zn_4O(T^2DC)(BTB)_{4/3}$
UMCM-309a	$Zr_6O_4(OH)_4(BTB)_6(OH)_6(H_2O)_6$
UMCM-4	$Zn_4O(BDC)_{1.5}(TCA)$
usf-Z-MOF	$In_5(HIMDC)_{10}(1,2\text{-}H_2DACH)_{2.5}$ with **med** topology
UTSA-76	$Cu_3(H_2O)_3(PyrDI)$
UV–Vis	ultraviolet–visible spectroscopy
VED	volumetric energy density
VSA	vacuum swing adsorption
XAS	X-ray absorption spectroscopy
XPS	X-ray photoelectron spectroscopy
ZABU SBU	$Zn_8O_2(AD)_4(-COO)_{12}$
ZIF	zeolitic imidazolate framework
ZIF-20	$Zn(Pur)_2$
ZIF-300	$Zn(2\text{-}mIM)_{0.86}(bBIM)_{1.14}$
ZIF-301	$Zn(2\text{-}mIM)_{0.94}(cBIM)_{1.06}$
ZIF302	$Zn(2\text{-}mIM)_{0.67}(mBIM)_{1.33}$
ZIF-376	$Zn(nbIM)_{0.25}(mIM)_{0.25}(IM)_{1.5}$
ZIF-412	$Zn(BIM)_{1.13}(nIM)_{0.62}(IM)_{0.25}$
ZIF-414	$Zn(nbIM)_{0.91}(mIM)_{0.62}(IM)_{0.47}$
ZIF-486	$Zn(nbIM)_{0.20}(mIM)_{0.65}(IM)_{1.15}$
ZIF-68	Zn(BIM)(nIM)
ZIF-7	$Zn(BIM)_2$
ZIF-8	$Zn(mIM)_2$
ZIF-90	$Zn(aIM)_2$
ZIF-91	$Zn(MeOHIM,aIM)_2$
ZIF-92	$Zn(HEIMIM,aIM)_2$
Z-MOF	metal-organic framework with zeolitic topology

Part I

Metal-Organic Frameworks

1

Emergence of Metal-Organic Frameworks

1.1 Introduction

Reticular chemistry[1] is the study of linking discrete chemical entities (molecules and clusters) by strong bonds to make extended structures such as metal-organic frameworks (MOFs). In MOFs, polynuclear metal clusters are joined together by organic linkers to make crystalline porous frameworks. MOFs combine the synthetic control exercised in making organic molecules with the vast geometric and compositional variations possible by using inorganic units. The reticular chemistry of MOFs has combined two fields of chemistry that have been practiced and taught separately, into one. Accordingly, the synthesis of MOFs requires the well-honed skills of both organic and inorganic chemists to make extended solids with precisely designed structures and properties. These are imparted by the constituents yet go beyond what would be possible by the individual molecular building units. One such property is the open space encompassed by the framework into which molecules can be introduced and transformed in a manner not possible otherwise. Given the potential of reticular synthesis and the place it is beginning to occupy in the larger context of chemistry, it is instructive to provide a historical perspective on how this new field has emerged. Since MOFs were the first class of crystalline solids to be developed in the realm of reticular chemistry, their history figures prominently in its initial development.

1.2 Early Examples of Coordination Solids

The field of synthetic metal-organic chemistry as it is practiced today has emerged from coordination chemistry. Early examples of transition metal complexes were discovered by serendipity centuries ago and at that time only little was known about their structure and composition. The first reported example of a synthetic coordination compound can be traced back to the discovery of the pigment "Prussian blue" in Berlin, Germany, in the beginning of the eighteenth century [1]. The story of this finding is captured in a book by Georg E. Stahl [2]. According to him, the discovery of Prussian blue took place

1 The term "reticular" is derived from Latin "*rēticulum*" meaning "*having the form of a net*" or "netlike."

Introduction to Reticular Chemistry: Metal-Organic Frameworks and Covalent Organic Frameworks,
First Edition. Omar M. Yaghi, Markus J. Kalmutzki, and Christian S. Diercks.

Companion website: www.wiley.com/go/yaghi/reticular

in the laboratories of Johann K. Dippel who was preparing a so-called "animal oil" by distillation of animal materials. This was then repeatedly distilled from potash (K_2CO_3) to remove undesired impurities. This procedure promotes the decomposition of organic components to form cyanide, which subsequently reacts with residual iron from the animal blood to form hexacyanoferrate ions $[M_2Fe(CN)_6]$ ($M = Na^+, K^+$), which stays behind as an impurity in the potash. At that time, a color maker named Johann J. Diesbach worked in Dippel's laboratory synthesizing "Florentine lake," an organic red pigment based on cochineal red. Usually, he accomplished this by precipitation of an extract of cochineal with potash and the addition of alum $[KAl(SO_4)_2 \cdot 12H_2O]$ and iron sulfate ($FeSO_4$) to enhance both the color and the processing of the resulting pigment. At one point, Diesbach had run out of potash so he borrowed some of the potash that had been used in the production of Dippel's animal oil. To his surprise, upon addition of this contaminated potash he observed an unexpected rich blue precipitate, later termed Prussian blue, $Fe^{3+}_4[Fe^{2+}(CN)_6]_3 \cdot H_2O$.

Owing to their intense colors, a variety of coordination compounds have had widespread practical use throughout history as pigments (e.g. Prussian blue) and dyes (e.g. alizarin) without knowledge of their chemical composition or structure [1c, 3]. As illustrated with this representative example, the serendipitous discoveries of coordination compounds at that time severely limited the number of accessible materials and hence conclusions about their behavior were exclusively based on phenomenological observations.

1.3 Werner Complexes

The conceptual foundation of coordination chemistry was laid by the Swiss chemist Alfred Werner, who was ultimately awarded the Nobel Prize in chemistry in 1913 for his efforts [4]. When he started his career in 1890 he tried to elucidate and conceptualize the spatial arrangement of atoms in coordination complexes [5]. In 1857, F. August Kekulé proposed the model of constant valence, which was based on the general assumption that every element only exists in one valence and therefore only has one fixed coordination number [6]. Chemical formulae were consequently given using the dot notation, as in $CoCl_3 \cdot 6NH_3$, which gave a correct description of the chemical composition but, as Werner later determined, did not represent the actual molecular structure (Figure 1.1).

A key observation that led to this conclusion was that addition of hydrochloric acid to a solution of $CoCl_3 \cdot 6NH_3$ did not result in the quantitative liberation of all six ammonia molecules per complex. The fact that some ammonia was

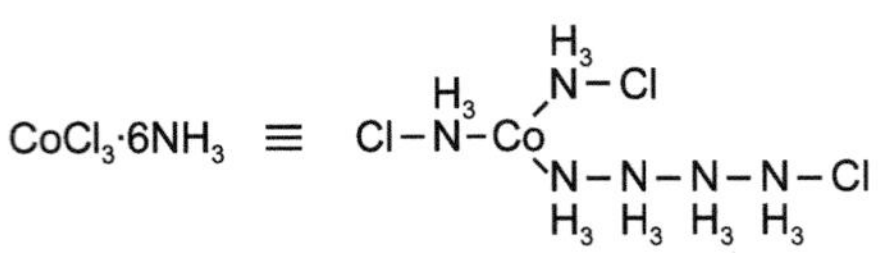

Figure 1.1 Chemical structure of $CoCl_3 \cdot 6NH_3$ based on the theory of constant valence. According to this theory cobalt has a valence of three and therefore has three ligands attached (trigonal arrangement) with the remaining ligands forming chains.

not released led Werner to deduce that it must be bound tightly to the central cobalt atom. In contrast, upon addition of aqueous silver nitrate, all the chloride ions were precipitated as silver chloride. Furthermore, in experiments conducted on a series of compounds of general formula $CoCl_3{\cdot}nNH_3$ (n = 1–6) containing various amounts of ammonia, the amount of silver chloride formed by addition of silver nitrate was shown to be directly proportional to the number of ammonia molecules bound to the Co^{3+} center (Figure 1.2)[2] [7c]. Werner carried out conductivity measurements on solutions containing these different complexes, where he observed a trend in conductivity that could be directly correlated to the number of free chloride ions [8]. Based on these findings, Werner concluded that an attractive force must exert uniformly from the central metal ion toward all parts of its surface and that six ligands arrange around this center of attraction in order to minimize the interactions between themselves but maximize their interactions with the metal ion. According to this new concept the aforementioned complexes were denoted as $[Co(NH_3)_6]Cl_3$, $[Co(NH_3)_5Cl]Cl_2$, and $[Co(NH_3)_4Cl_2]Cl$, illustrating that they are in fact built from six ligands surrounding one central Co^{3+} ion.

The coordination number 6 found for this complex can adopt three different geometries: hexagonal planar, trigonal-prismatic, and octahedral. These geometries can be distinguished by the number of their possible isomers. In order to determine the geometry of $CoCl_3{\cdot}nNH_3$ complexes (i.e. which one of these conformations is in fact favored) Werner conducted detailed studies on $[Co(NH_3)_4Cl_2]Cl$. For this complex, a hexagonal planar or trigonal prismatic coordination affords three different stereoisomers, whereas the octahedral coordination can only result in two such isomers (Figure 1.3). Werner verified

$$CoCl_3{\cdot}6NH_3 \xrightarrow{Ag(NO_3)} Co(NO_3)_3{\cdot}NH_3 + 3AgCl \implies [Co(NH_3)_6]Cl_3$$

$$CoCl_3{\cdot}5NH_3 \xrightarrow{Ag(NO_3)} Co(NO_3)_2Cl{\cdot}NH_3 + 2AgCl \implies [Co(NH_3)_4Cl]Cl_2$$

$$CoCl_3{\cdot}4NH_3 \xrightarrow{Ag(NO_3)} Co(NO_3)Cl_2{\cdot}NH_3 + 1AgCl \implies [Co(NH_3)_5Cl_2]Cl$$

Figure 1.2 Precipitation of silver(I) chloride by addition of silver(I) nitrate to solutions of different ionization isomers of $CoCl_3{\cdot}nNH_3$. The amount of silver(I) chloride precipitated was found to be different for each isomer. The chemical formulae shown on the right indicate a coordination number of 6 for the Co^{3+} center.

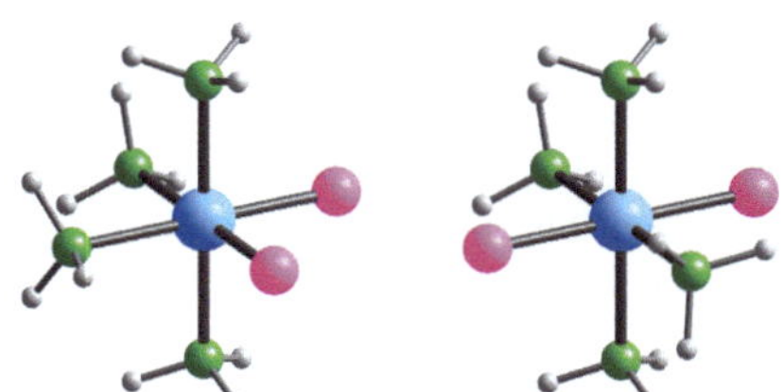

cis-isomer *trans*-isomer

Figure 1.3 Possible isomers for the octahedral complex of formula $Co(NH_3)_4Cl_2$. The violet *cis*-isomer ("violeo" complex) is shown on the left, the green *trans*-isomer ("praseo" complex) is shown on the right. The two isomers can be distinguished by their vivid red and green colors, respectively. Color code: Co, blue; N, green; Cl, pink; H, light gray.

2 These findings could also be explained by the chain theory developed by Christian Blomstrand, which was later further developed by Sophus Jørgensen [7].

the latter by isolating two, not three, isomers. This work laid the foundation for the subsequent development of coordination chemistry [9].

1.4 Hofmann Clathrates

The newly gained insight into the precise molecular structure provided by Werner's work served as an inspiration to extend the practice of coordination chemistry from the molecular (0D) regime into higher dimensions, especially 2D and 3D extended structures. An early example of a coordination compound with an extended 2D structure was published by Karl A. Hofmann in 1897 [10]. Slow diffusion of C_6H_6 into an NH_3 solution of $Ni(CN)_2$ yielded a crystalline material of the general formula $[Ni(CN)_2(L)](C_6H_6)$ ($L = NH_3$), commonly referred to as Hofmann clathrate (Figure 1.4).[3] This compound was first speculated to be a molecular solid composed of $Ni(CN)_3(\eta^6\text{-}C_6H_6)$ molecules, but when its crystal structure was solved by single crystal X-ray diffraction, this material was found to be an extended coordination compound, built from 2D layers of alternating octahedral and square planar Ni^{2+} ions linked by CN^- ions [12]. Terminal ammonia ligands on the octahedral nickel centers pointing toward adjacent layers facilitate the formation of cavities, rendering the compound capable of encapsulating benzene as guests. These guest molecules are, as in many cases, solvent molecules trapped during the synthesis of the material that function as templates and hence play an important role in the

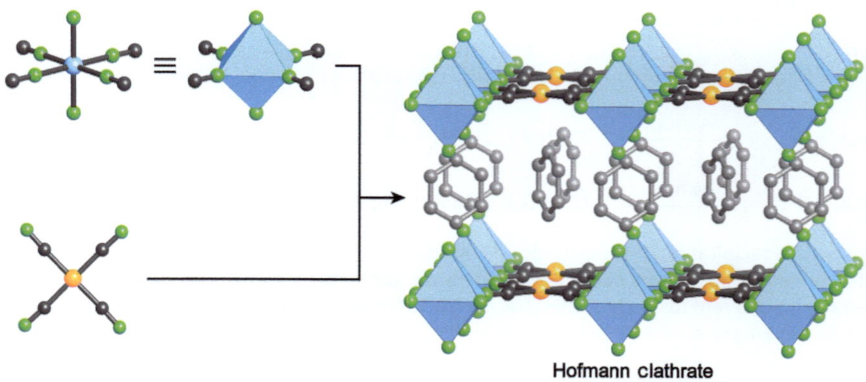

Figure 1.4 Representation of the crystal structure of the original Hofmann clathrate as determined by Herbert M. Powell and coworkers in 1952. Octahedral and square planar nickel moieties are linked by CN^- ions into stacked layers of composition $Ni(CN)_2(NH_3)$ that are separated by benzene guests. The two different coordination geometries for Ni^{2+} (d^8) can be explained by the strength of the ligand field. While strong ligands ($-NH_3$ and $-NC$) result in an octahedral splitting, a square planar splitting is more favorable for weaker ligands ($-CN$). All hydrogen atoms are omitted for clarity. Color code: Ni, blue and orange spheres; C, gray; N, green; benzene guest, light gray.

3 The term clathrate was first coined by Herbert M. Powell [11].

formation of the clathrate material. Structural collapse of Hofmann clathrates and related materials upon removal of the guest molecules from the structures is commonly observed.

The structural elucidation of this material sparked an interest in extended coordination compounds and consequently a variety of Hofmann clathrates have been reported. Iwamoto et al. focused on a more systematic approach for the synthesis of Hofmann-type compounds and discovered that in general this type of material is built from two different units, namely $[M_a^{2+}(CN)_4]^{2-}$ and $[M_b^{2+}(NH_3)_2]^{2+}$ (where a and b indicate different divalent metals such as Cd^{2+} or Ni^{2+}) and that the terminal ammonium ligands can be replaced by alkylamines [13]. They employed precursors of these complex ions in a reaction mixture involving neutral aromatic solvents to build structures of the general formula $[M_a(NH_3)_2M_b(CN)_4]G$ (G = benzene, aniline, pyrrole, or thiophene guest molecules) following Eq. (1.1).

$$[Ma^{2+}(CN)_4]2^- + [M_b^{2+}(NH_3)_2]^{2+} \xrightarrow{G} [M_a(NH_3)_2M_b(CN)_4] \tag{1.1}$$

After the successful substitution of the ammonia ligands by alkylamines, the next logical step was to employ bifunctional amino-linkers to connect adjacent layers (Figure 1.5) [14]. Iwamoto and coworkers demonstrated that the terminal ammonia ligands can be replaced with α,ω-diaminoalkanes that link adjacent layers and thereby create space for encapsulation of guests. The length of the organic spacer can be systematically varied to allow for size-selective inclusion of guest molecules [15].

The introduction of organic linkers between adjacent layers facilitates the adjustment of the interlayer distance and thus has a strong impact on the properties of the extended coordination compound. To increase the control

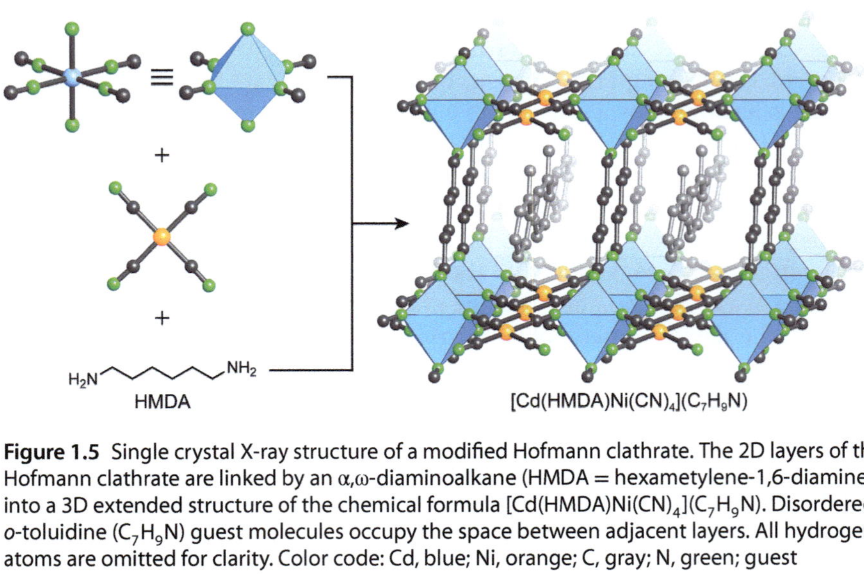

Figure 1.5 Single crystal X-ray structure of a modified Hofmann clathrate. The 2D layers of the Hofmann clathrate are linked by an α,ω-diaminoalkane (HMDA = hexametylene-1,6-diamine) into a 3D extended structure of the chemical formula $[Cd(HMDA)Ni(CN)_4](C_7H_9N)$. Disordered *o*-toluidine (C_7H_9N) guest molecules occupy the space between adjacent layers. All hydrogen atoms are omitted for clarity. Color code: Cd, blue; Ni, orange; C, gray; N, green; guest molecules, light gray.

that can be exercised over the metrics of extended structures, the next logical progression was to link metal ions entirely through organic linkers to form what have come to be known as coordination networks (also referred to as coordination polymers, although we prefer the use of the term networks as such compounds are crystalline extended structures).

1.5 Coordination Networks

The first members of this new class of materials were reported by Saito and coworkers who made use of the well-established chemistry of Cu^+ ions and linked them through bis(alkylnitrilo) units of different lengths to yield a series of crystalline materials with structures of varying dimensionality [16]. While the use of a short linker such as succinonitrile (SUC) results in a 1D structure, slightly longer linkers favor the formation of layers, as was shown for glutaronitrile (GLU), and further elongation leads to the formation of an interpenetrated 3D structure, as in the example of adiponitrile (ADI). The key compound in this series is $[Cu(ADI)_2](NO_3)$, which adopts a 3D structure based on the diamond net (**dia**) (Figure 1.6). The "open" architecture of this structure, owing to the length of the organic linker, leads to sixfold interpenetration, leaving enough space for the nitrate ions balancing the charge on the cationic framework.

The topological classification of $[Cu(ADI)_2](NO_3)$ is based on the geometric principles of crystal chemistry established by Alexander F. Wells, who developed a system to simplify crystal structures by describing them in terms of nets constructed from nodes and links [17].

Since this concept is frequently used to describe extended structures, especially those of MOFs, it is instructive to briefly illustrate the basics underlying this concept. Here, topology refers to a simplified representation of a crystal structure considering only the connectivity and not the chemical information or metrics of its constituents. It is invariant to bending, stretching, and collapsing, but not to the making and breaking of connections (see Chapter 18). This principle is illustrated by a fisherman's net representing a square grid similar to that of $[Cu(ADI)_2](NO_3)$ (Figure 1.6). The net retains its square grid structure whether it is folded or distorted, but loses it if one or more threads are cut in half. This principle is useful in simplifying and classifying the crystal structures of solids [18]. The nomenclature for net topologies uses three letter codes (small bolded letters) that are compiled in the reticular chemistry structure resource (RCSR) database. These names may be assigned arbitrarily but often they are related to the names of naturally occurring minerals of that specific topology (e.g. diamond, **dia**; quartz, **qtz**). The topology of the net underlying a crystal structure is derived by deconstructing it into vertices and edges (nodes and links). These are distinguished based on their number of points of extension: the number of connections to other building units within the structure. An edge has two points of extension, such as the ditopic linker adiponitrile (Figure 1.6), and a vertex is defined as a building unit with three or more points of extension, such as a metal ion with coordination number 4 or a cluster of atoms making 4 connections. These two definitions will enable us to simplify any given crystal

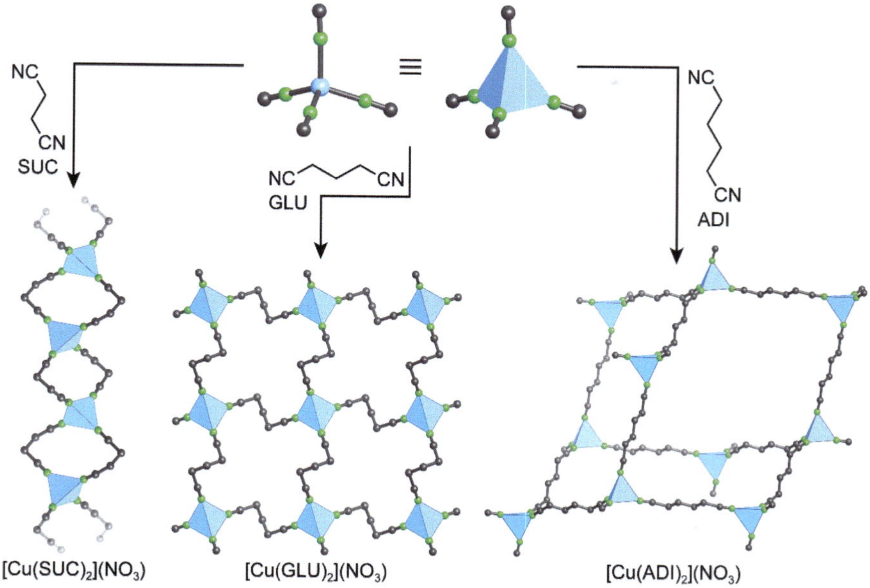

Figure 1.6 Structures of a series of bis(alkylnitrilo) linked Cu^+ coordination networks. Short linkers such as succinonitrile (SUC) yield 1D chains of the kind shown on the left. 2D layers (one is shown) are obtained from longer glutaronitile (GLU) linkers (center), and a 3D network with **dia** topology is formed with adiponitrile (ADI) linkers (right). All hydrogen atoms are omitted and only one framework of the sixfold interpenetrated framework in the **dia** structure of $[Cu(ADI)_2](NO_3)$ is shown for clarity. Color code: Ni, blue; C, gray; N, green.

structure to a net of vertices that are linked by edges. We exercise this for the structure of $[Cu(ADI)_2](NO_3)$ with **dia** topology. Figure 1.7a shows a fragment of the $[Cu(ADI)_2](NO_3)$ structure [16c]. ADI units are 2-connected linkers while the copper atoms are 4-connected nodes as shown in Figure 1.7b in the simplified net. An even clearer representation can be achieved when adding the corresponding polyhedra or vertex figures to give the augmented net **dia-a** (Figure 1.7c). Linking metal centers through organic struts leads to the formation of frameworks encompassing open space. Within such structures this open space is sometimes filled with additional frameworks that are identical in both composition and topology. These are mechanically entangled rather than chemically linked, a phenomenon referred to as interpenetration [18]. A more detailed discussion on the topic of topology can be found in Chapter 18.

In an attempt to synthesize a radical anion salt of 2,5-dimethyl-*N*,*N*-dicyanoquinonediimine, Siegfried F. Hünig and coworkers prepared another coordination network of **dia** topology [19]. Despite the fact that its crystal structure was not discussed in detail, Akiko Kobayashi and coworkers synthesized isostructural forms using functionalized linkers bearing methoxy-, chloro-, and bromo-substituents, which have the same sevenfold interpenetrated structure [20]. Adding functionality onto the backbone of such networks, without changing the overall metrics and underlying topology, brought the molecular precision of organic chemistry into the realm of extended solids.

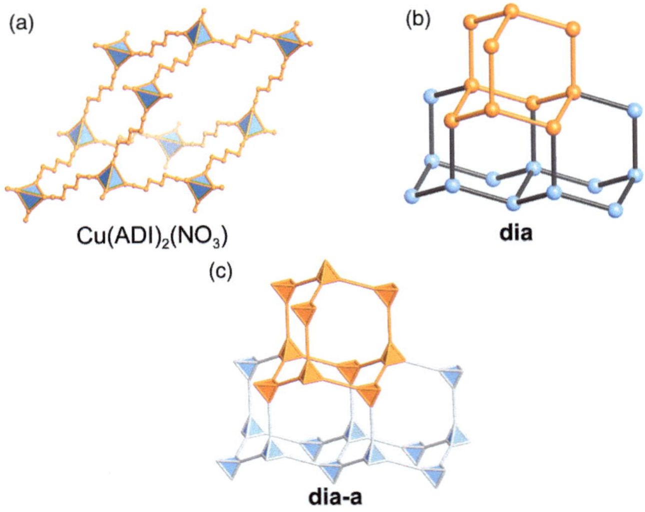

Figure 1.7 (a) Simplification of the crystal structure of [Cu(ADI)$_2$](NO$_3$) adopting a diamond-like structure. (b) Representation of building units with two points of extension as edges and building units with four points of extensions as nodes yields the underlying **dia** topology. (c) Representing the vertices as their corresponding vertex figures (polyhedra) yields the augmented **dia-a** net in its highest symmetry embedding. Tetrahedral nodes are shown in blue, edges in gray. One adamantane cage is shown in (a) and highlighted in orange in (b) and (c).

The immense diversity of theoretically accessible coordination network structures made in a manner akin to the methods reported by Saito et al. inevitably led to the necessity of deploying generally applicable design principles for this class of materials. Such principles were already well developed in the field of crystal engineering, where chemists seek to understand weak interactions (C—H· · ·A, hydrogen bonds, halogen bonds, π-interactions, and van der Waals forces) between individual molecules in molecular solids in order to engineer their arrangement within the crystal [21]. Since coordination networks are also held together by rather weak non-covalent interactions (Metal–N–donor interactions), the deliberate design of coordination networks is often considered to fall under the rubric of crystal engineering [22]. In this context, Richard Robson and Bernard Hoskins recognized that Wells principles of nodes and links as outlined earlier can be applied to predict structures that will result from linking of molecular building units of a given geometry and connectivity[4] [24]. They demonstrated that this approach facilitates the deliberate design of coordination networks with predetermined structures. For example, linking tetrahedral Cu^+ single metal nodes and 4,4′,4″,4‴-tetracyanotetraphenylmethane (TCTPM) results in a non-interpenetrated coordination network of the chemical formula

4 In this paper Hoskins and Robson also report the designed synthesis of Zn(CN)$_2$ and Cd(CN)$_2$, which previously had been synthesized and described (1941 and 1945, respectively) by Zhdanov et al. and whose ability to form clathrates was reported by Iwamoto et al. in 1988 [23].

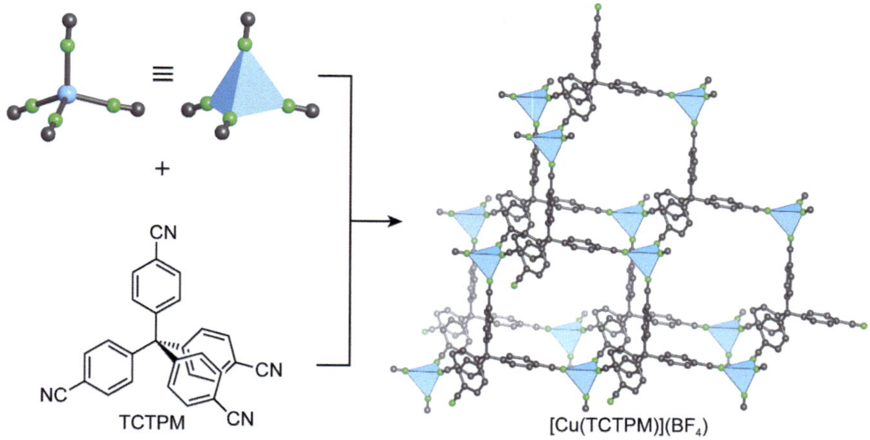

Figure 1.8 Crystal structure of the cationic coordination network $[Cu(TCTPM)](BF_4)$ (TCTPM = 4,4′,4″,4‴ tetracyanotetraphenylmethane). The network has a **dia** topology and is composed of tetrahedral Cu^+ single metal nodes and tetrahedral TCTPM linkers. All counter ions, solvent molecules, and hydrogen atoms are omitted for clarity. Color code: Cu, blue; C, gray; N, green.

$[Cu(TCTPM)](BF_4)$ and **dia** topology (Figure 1.8). The adamantane cages of this structure have an estimated pore volume of 700 Å^3 and are occupied by BF_4^- ions that can be exchanged with PF_6^-, as evidenced by infrared spectroscopy, while the crystallinity of the material is retained.

It was shown that the use of elongated linkers such as 1,4-dicyanobenzene, 4,4′-dipyridyl, and 2,5-dimethylpyrazine yields isostructural analogs with different degrees of interpenetration due to the different pore sizes of the resulting networks [25]. In addition to changing the metrics of the building units their general geometry and number of points of extension can be altered to yield networks of different structure types.

The combination of tetrahedral and square planar building units leads to structures based on the platinum sulfide (**pts**) net. In the first such example, Cu^+ ions were linked with $Pt(CN)_4^{2-}$ units. Here, the Cu^+ and the $Pt(CN)_4^{2-}$ units replace the tetrahedral S^{2-} and square planar Pt^{2+} ions in the structure of the PtS mineral, respectively [26]. The resulting anionic framework has the chemical formula $[CuPt(CN)_4](NMe_4)$ and the pores are filled with $(NMe_4)^+$ counter ions. Control over the metrics of the system was demonstrated by deliberate expansion of the pore size by replacing the inorganic $Pt(CN)_4^{2-}$ units with porphyrin-based square building units (Figure 1.9). Here, a cyanophenyl-functionalized porphyrin (TCP) was used as the square planar unit to give a twofold interpenetrated structure of the chemical formula $[Cu(Cu\text{-}TCP)](BF_4)$ [27]. It was then shown that interpenetration can be avoided by using a pyridyl-functionalized porphyrin linker (TPP). Linking TPP with tetrahedral Cu^+ single metal nodes gives a non-interpenetrated structure of the formula $[Cu(Cu\text{-}TPP)](BF_4)$. This finding is rationalized by the smaller internal pore space of the network constructed from TPP compared to that constructed from TCP linkers [27].

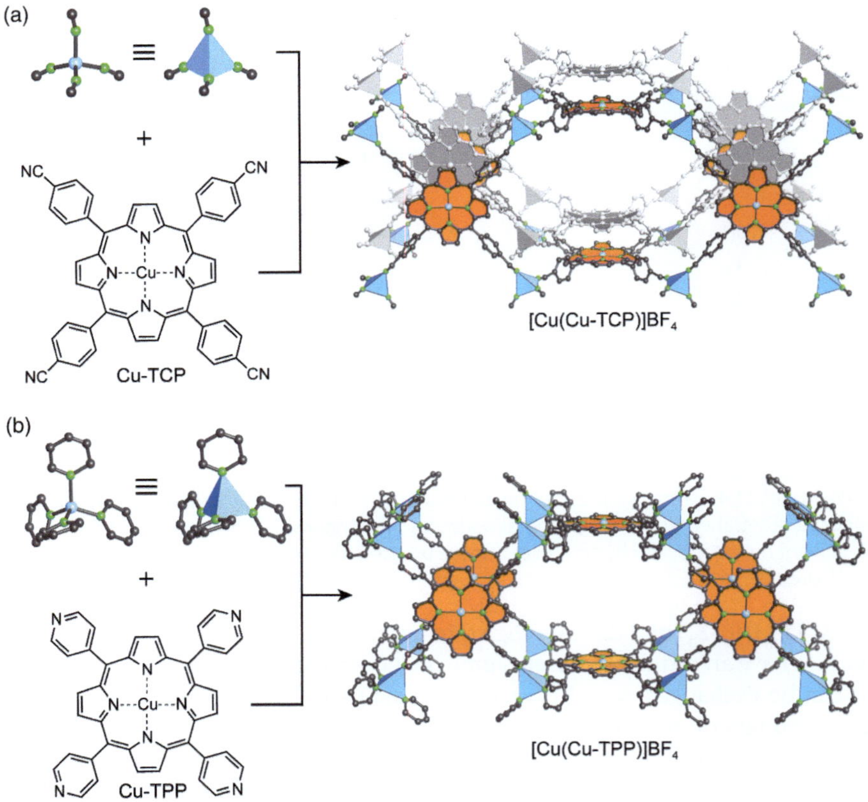

Figure 1.9 Comparison of two coordination networks built from tetrahedral Cu^+ and square planar porphyrin-based linkers, crystallizing in the **pts** topology. (a) A twofold interpenetrated framework [Cu(Cu-TCP)](BF_4) is obtained from cyanophenyl-functionalized porphyrin (TCP) and Cu^+ ions. (b) Replacing the terminal benzonitrile coordinating groups by pyridine groups (TPP = tetrapyridyl-functionalized porphyrin) prevents interpenetration and gives rise to the non-interpenetrated framework [Cu(Cu-TPP)](BF_4). All hydrogen atoms, counter ions, and solvent molecules are omitted for clarity. The interpenetrating net in (a) is shown in gray. Color code: Cu^+/Cu^{2+}, blue; C, gray; N, green; square planar porphyrin building units are highlighted as orange polygons. The crystal structure drawings are based on modified datasets where the porphyrin rings are fixed in a planar shape.

The use of geometric design principles for coordination networks and the molecular building unit approach signified an important evolution in the synthesis of extended structures. The resulting level of synthetic control was largely unknown prior to coordination networks. It is however worthy of note that at this point only a hand full of structure types was reported, most of which suffered from interpenetration and lack of accessibility of their internal pore space.

In 1990, Makoto Fujita used ethylenediamine-capped Pd^{2+} units to make a square-shaped polynuclear macrocyclic complex of composition $[(en)Pd(BIPY)(NO_3)_8]_4$ (en = ethylendiamine, BIPY = 4,4′-bipyridine) [28].

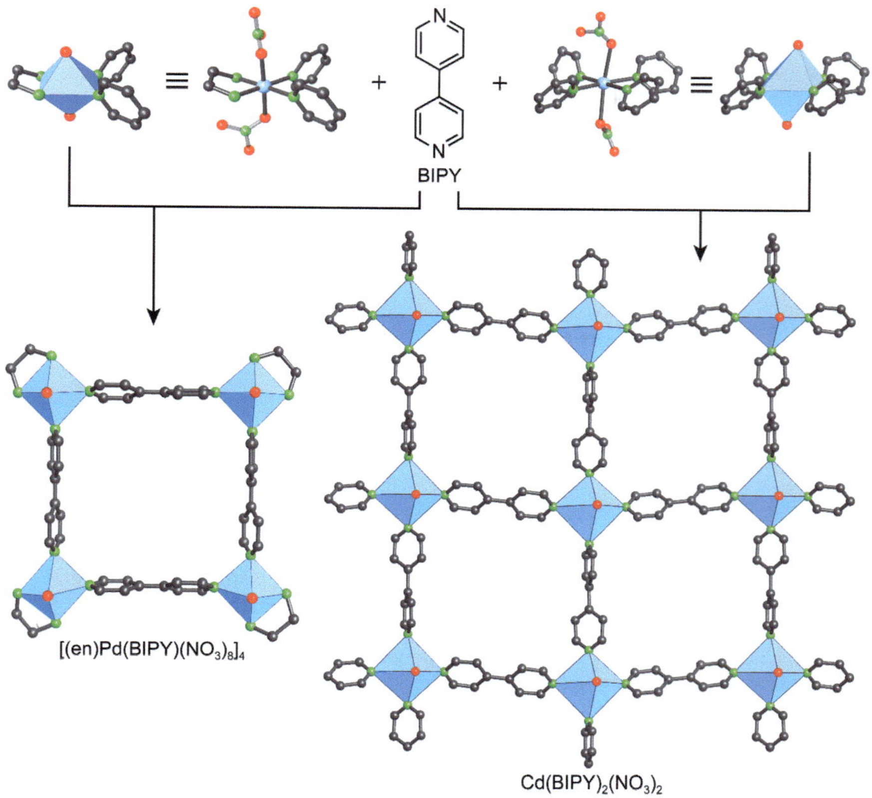

Figure 1.10 Molecular square synthesized by reacting a capped Pd^{2+} complex with BIPY. Using Cd^{2+} ions results in the formation of an extended square grid (**sql**) structure of formula $Cd(BIPY)_2(NO_3)_2$. Dichlorobenzene guest molecules reside in the square channels formed by the eclipsed stacking of the **sql** layers of the network. All guest molecules and hydrogens are omitted for clarity. Color code: Pd and Cd, blue; C, gray; N, green; O, red.

When the capped Pd^{2+} units in the synthesis of this macrocycle are replaced by uncapped Cd^{2+} ions an extended 2D square grid (**sql**) is formed (Figure 1.10) [29].

In 1995, two extended coordination networks related to $M(BIPY)_2$ were published, both of which are essential in the development of the field of MOFs. In fact, the term metal-organic framework was first coined in one of these contributions, in which Omar M. Yaghi and coworkers reported the solvothermal synthesis of $[Cu(BIPY)_{1.5}](NO_3)$ (Figure 1.11) [30]. The term metal-organic framework was originally used to describe the overall composition (metal ion and organic) and character of the structure (framework). Later on, the term MOF was more meaningfully used to describe additional structural attributes (rigidity) and properties (porosity).[5] The structure of $[Cu(BIPY)_{1.5}](NO_3)$ is built from trigonal planar Cu^+

5 Currently, the IUPAC definition of a MOF is: "A coordination network with organic ligands containing potential voids."

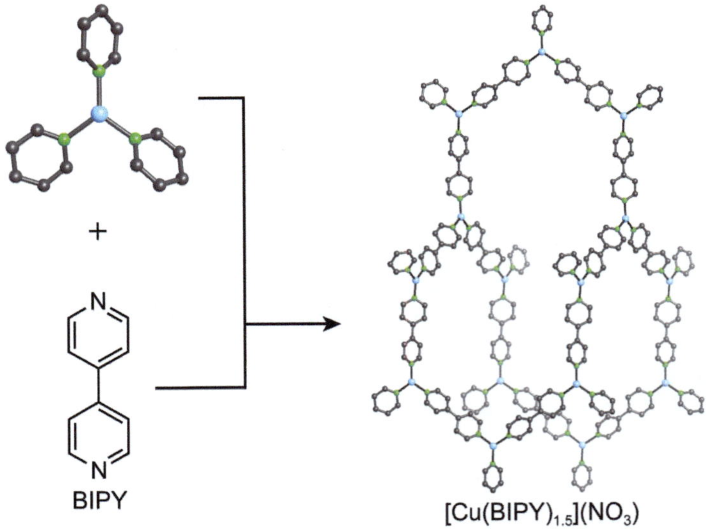

Figure 1.11 3D framework of $[Cu(BIPY)_{1.5}](NO_3)$ based on trigonal planar Cu^+ single metal nodes connected by linear BIPY linkers. The twofold interpenetrated structure has a **ths** topology. Only one cage is shown to illustrate the connectivity and orientation of the individual building units within the **ths** net. Interpenetrating frameworks, solvent molecules, counter ions residing in the channels, and all hydrogen atoms are omitted for clarity. Color code: Cu, blue; C, gray; N, green.

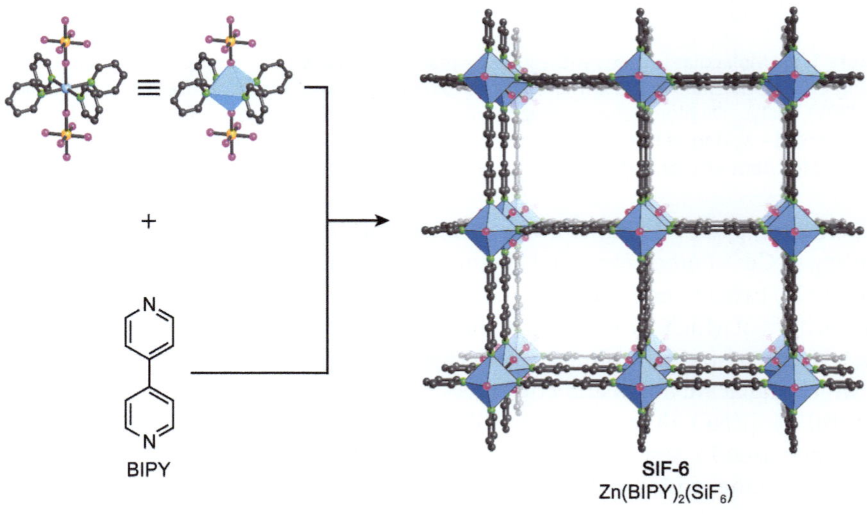

Figure 1.12 Single crystal structure of $Zn(BIPY)_2(SiF_6)$ with view along the *c*-direction. Octahedrally coordinated Zn^{2+} ions are joined by BIPY linkers to form 2D **sql** layers. These layers are pillared by SiF_6^{2-} resulting in the assembly of a charge neutral 3D **pcu** network, with channels of 8 × 8 Å running along the crystallographic *c*-axis. All hydrogen atoms and solvent molecules are omitted for clarity. Color code: Zn, blue; Si, orange; F, purple; C, gray; N, green.

centers connected by linear BIPY linkers to form an interpenetrated 3D network with an underlying $ThSi_2$ (**ths**) topology. The NO_3^- counter ions reside in the 8 × 6 Å and 4 × 5 Å channels of the structure and they can be readily exchanged for simple inorganic anions such as BF_4^- or SO_4^{2-} with full retention of the overall structure. The solvothermal synthesis of this material resembles the synthetic routes used in zeolite chemistry and this approach has since proven fruitful for the synthesis of many MOFs.

That same year, Michael J. Zaworotko and coworker reported a coordination network of formula $Zn(BIPY)_2SiF_6$ having a square grid of octahedral Zn^{2+} ions linked by BIPY (Figure 1.12) [31]. These layers are pillared by SiF_6^- to form a charge neutral non-interpenetrated cubic primitive structure with 8 × 8 Å channels running along the crystallographic *c*-direction. The potential empty space in this network represents 50% of the unit cell volume. However, the structure of $Zn(BIPY)_2SiF_6$ collapses when the guest molecules are removed from its pores.

1.6 Coordination Networks with Charged Linkers

While the aforementioned design principles can be used to construct a wide variety of coordination networks through the judicious choice of metal ions and organic linkers, the resulting materials generally suffer from inherent architectural and chemical instability. To overcome these limitations, charged chelating linkers were introduced. The use of such linkers has two important advantages: increased bond strength results in higher thermal and chemical stability and the charge on the linker can balance the charge of the cationic metal centers to circumvent the formation of ionic networks and avoid the need for counter ions filling the pores. This was first illustrated in 1995 with the synthesis of $Co(BTC)(Py)_2$ (BTC, benzene-tricarboxylate). The structure of $Co(BTC)(Py)_2$ consists of alternating stacked layers of pyridine and Co-BTC [32]. within the Co-BTC layers, each Co^{3+} ion is coordinated by three carboxylates of neighboring BTC linkers (Figure 1.13). One of the BTCs is coordinated to three metal centers in a bidentate fashion, while the other BTCs coordinate to three metal centers in a monodentate fashion. The pyridine ligands between adjacent layers provide for an interlayer distance of 7 Å. $Co(BTC)(Py)_2$ is exceptionally stable for an extended network material, decomposing only at temperatures above 350 °C. As expected, owing to the strong bonds between the metal centers and the charged BTC linkers, removal of the pyridine molecules does not lead to the collapse of the structure. The Co-BTC layers remain intact and after the pyridine guest molecules have been removed thermally, they can selectively be re-inserted between the layers, thereby regenerating the original structure as evidenced by powder X-ray diffraction.

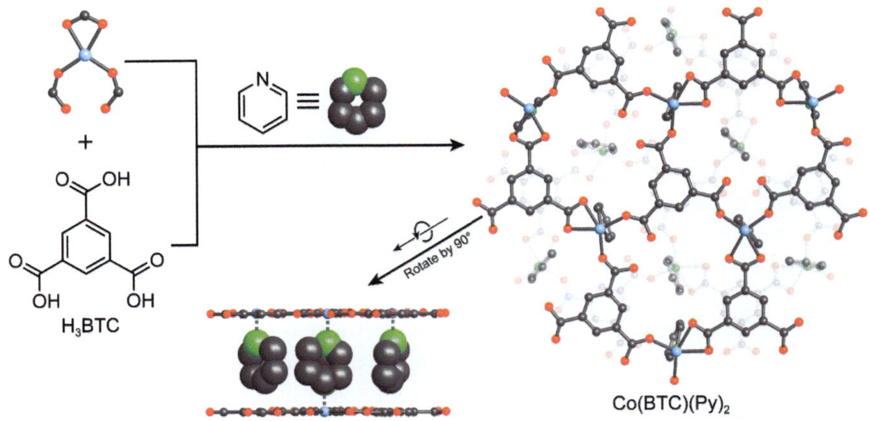

Figure 1.13 Linking Co^{3+} ions and BTC results in the formation of a layered 2D structure of formula $Co(BTC)(Py)_2$. The layers are constructed from square planar Co^{3+} and trigonal planar BTC linkers and are stacked along the crystallographic *c*-axis. The individual layers are separated by pyridine ligands coordinated to give Co^{3+} centers to given an overall octahedral coordination geometry. The pyridine guest molecules can be removed thermally and reinserted, regenerating the original structure of MOF-1 of the original structure of MOF-1. Color code: Co, blue; C, gray; N, green; O, red.

1.7 Introduction of Secondary Building Units and Permanent Porosity

To further increase the stability of metal-organic extended structures, polynuclear clusters, commonly referred to as secondary building units (SBUs), were sought as nodes to replace the single metal-ion nodes in coordination networks. The SBUs offered several advantages toward realizing more robust structures: the chelation of metal ions to make polynuclear clusters provided for rigidity and directionality while the charge on the linker led to increased bond strength and the formation of neutral frameworks. In combination, these factors were expected to contribute greatly to the overall stability of the resulting material. This concept was realized in 1998 when the synthesis and gas sorption properties of the first metal-organic framework, MOF-2 $Zn(BDC)(H_2O)$ were reported (Figure 1.14). MOF-2 has a neutral framework structure and is synthesized by slow vapor diffusion of a mixture of trimethylamine/toluene into a DMF/toluene solution of $Zn(NO_3)_2{\cdot}6H_2O$ and benzenedicarboxylic acid (H_2BDC) [33]. The layered structure of MOF-2 is built from dimeric $Zn_2(\text{—}COO)_4$ paddle wheel SBUs (rather than single metal nodes) that are linked by BDC struts to form a square grid (**sql**).

The increased stability imparted by the paddle wheel SBUs made it possible to remove all solvent molecules from the pores without collapsing the structure of MOF-2, leading to permanent microporosity as evidenced by reversible nitrogen gas adsorption at 77 K. The proof of permanent porosity in this MOF signaled a turning point in the chemistry of extended metal-organic solids and led to the use of the term MOF to emphasize their distinct stability and porosity.

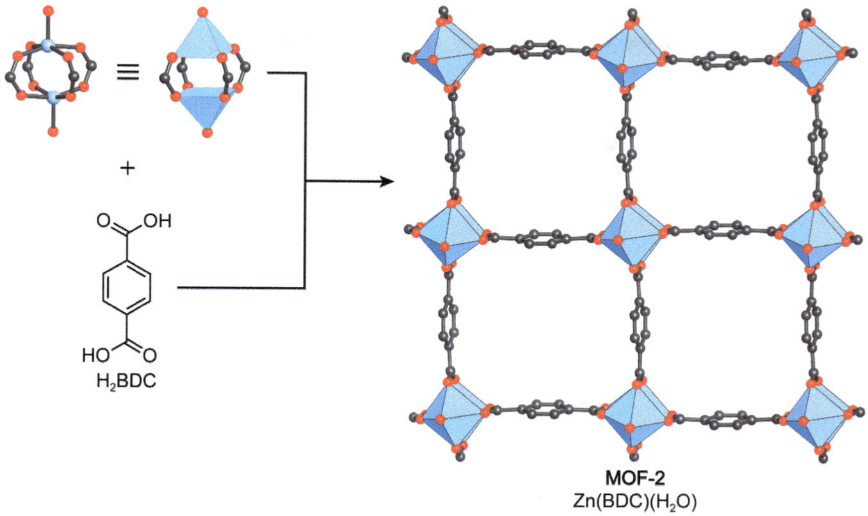

Figure 1.14 Crystal structure of MOF-2 viewed along the crystallographic *a*-axis, emphasizing the trapezoidal channels. Dinuclear Cu^{2+} paddle wheel SBUs are connected by ditopic BDC linkers to form layers of **sql** topology. The architecturally stable combination of paddle wheel SBUs and charged chelating linkers endow MOF-2 with permanent porosity. All hydrogen atoms and guest molecules are omitted for clarity. Color code: Cu, blue; C, gray; N, green; O, red.

Furthermore, this development led to extensive work on combining metals with carboxylates and other charged chelating linkers to give crystalline frameworks with SBUs as nodes. The term MOF has been overwhelmingly applied to distinguish such structures and henceforth we will adopt this terminology. The discovery of permanent porosity in MOF-2 generated interest in the further development of MOFs as it indicated that it is possible to make a wide range of 2D and 3D MOFs by combining different inorganic SBUs and organic linkers.

1.8 Extending MOF Chemistry to 3D Structures

The inorganic SBUs are polynuclear clusters in which the positions of the metal ions are locked in place by the binding groups of the linkers (in this book mainly carboxylates) as exemplified by the di-nuclear $M_2(CH_3COO)_4$ (M^{2+} = Cu, Zn) paddle wheel complex [34]. Their geometry and connectivity can be varied in order to allow for the formation of a variety of different MOF structures. These features, along with rigidity, and definitive directionality and connectivity facilitate the possibility for reticular synthesis and for the design of new, rigid, and permanently porous frameworks adopting a targeted structure. The synthetic and structural chemistry of polynuclear metal carboxylate clusters was well developed early on and many of their structures were solved soon after the discovery of X-ray diffraction by crystals [35]. As a matter of fact, the structure of the acetate capped paddle wheel clusters, as is found in the structure of MOF-2, was determined as early as 1953 [35g]. Based on the presumption that the replacement of the capping acetate ligands with multifunctional organic molecules promotes the formation of open extended framework structures, the idea of employing other carboxylate clusters as SBUs in the formation of MOFs emerged. First attempts to extend the chemistry of MOFs into 3D involved the use of the basic zinc acetate, a tetra-nuclear carboxylate cluster coordinated by six acetates in an octahedral fashion, as an SBU [35f].

1.8.1 Targeted Synthesis of MOF-5

It was known by that time that basic zinc acetate $Zn_4O(CH_3COO)_6$ can be prepared by adding small amounts of hydrogen peroxide to a solution of a zinc salt in acetic acid [36]. This facilitates the formation of O^{2-}, which lies at the center of the resulting polynuclear cluster [37]. The knowledge of both, the synthesis route affording the molecular $Zn_4O(CH_3COO)_6$ cluster as well as that employed in the preparation of MOF-2, allowed for the deduction of a synthetic procedure targeting a 3D MOF based on octahedral $Zn_4O(—COO)_6$ SBUs and ditopic linear linkers.

One of the lessons learned from the synthesis of MOF-2 was that precise synthetic control is required in order to avoid the rapid precipitation of ill-defined amorphous powders as a result of the low reversibility of the formation of strong metal–carboxylate bonds. This is in stark contrast to structures held together by relatively weak metal–N–donor bonds (e.g. bipyridines and dinitriles) whose crystallization is relatively straightforward owing to the high reversibility and facile error correction during crystallization. In the case of MOF-2, the formation of a crystalline material was achieved by slow diffusion of a base (trimethylamine) into a solution of a mixture of the metal salt ($Zn(NO_3)_3{\cdot}6H_2O$) and the organic linker H_2BDC (benzenedicarboxylic acid). Slow deprotonation of the carboxylic acid groups of the linker slowed down the formation of MOF-2 and allowed for error correction and consequently the crystallization of MOF-2. This strategy was largely retained in the synthesis of MOF-5 and only modified by adding a small amount of hydrogen peroxide to a mixture of $Zn(NO_3)_2{\cdot}4H_2O$ and H_2BDC in analogy to the synthesis of the molecular $Zn_4O(CH_3COO)_6$ cluster, to favor the formation $Zn_4O(—COO)_6$ SBUs over the previously obtained Zn_2 $(—COO)_4$ paddle wheel units.

Despite the rational approach to the synthesis of MOF-5, the bulk material that collected on the bottom of the vial turned out to be MOF-2.[6] One of the authors recalls that following this procedure, his student observed a small amount of cube-shaped crystals, having a morphology different from the main phase collecting at the bottom of the reaction vessel. These cubic crystals were floating at the meniscus of the mother liquor and adhered to the sides of the flask in the

6 Solvothermal methods to prepare MOF-5 in high yield were established in the following years where the slow diffusion of base into the reaction mixture was replaced by using DMF (dimethylformamide) or DEF (diethylformamide), which slowly decompose upon heating to release small amounts of dimethyl- or diethylamine base. It was also shown that the use of hydrogen peroxide is not needed since O^{2-} ions can be formed from trace amounts of water in the reaction mixture. Typical reaction temperatures of 80–100 °C as well as the applicability of this route to different metal salts were reported [38].

same vicinity. The comparison of the powder X-ray diffraction pattern of MOF-2 and that of these cubic crystals confirmed the presence of two structurally distinct compounds. However, when attempting to mount these cubic crystals on a single crystal X-ray diffractometer, the formation of cracks and the loss of transparency were observed, indicating the loss of mono-crystallinity and thus initially precluding their structural characterization. It proved difficult to handle this material because the crystals degraded upon loss of solvent by evaporation after they were removed from the mother liquor. Eventually, the structure of MOF-5 was determined by keeping the crystals in the mother liquor and sealing them in a capillary prior to examination by single crystal X-ray diffraction.

1.8.2 Structure of MOF-5

The synthesis, characterization, and structure of MOF-5, $[Zn_4O(BDC)_3](DMF)_x$ was reported in 1999 by Yaghi and coworkers.[7] It was shown that the structure of MOF-5 is indeed composed of octahedral $Zn_4O(—COO)_6$ SBUs, consisting of four tetrahedral ZnO_4 units sharing a common vertex, joined by ditopic BDC linkers to give a 3D framework structure of **pcu** topology (Figure 1.15). The large size (8.9 Å) and high connectivity of the SBUs in combination with the long BDC linker (6.9 Å) provide for an open porous structure with alternating interconnected pores of 15.1 and 11.0 Å in diameter, and a pore aperture of 8.0 Å.

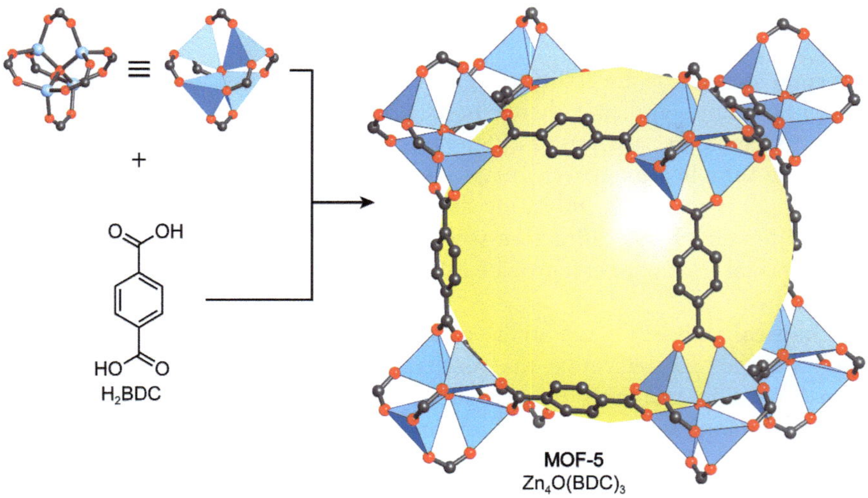

Figure 1.15 Crystal structure of MOF-5, constructed from octahedral $Zn_4O(—COO)_6$ SBUs and linear ditopic BDC linkers. The resulting primitive cubic net (**pcu**) has alternating large (15.1 Å diameter) and small (11.0 Å diameter) pores whose different size is a result of the orientation of the phenyl units of the BDC linkers with respect to the center of the pore. Only the large pore is shown for clarity. The yellow sphere indicates the largest sphere that can be placed inside the pore without coming within the van der Waals radius of any framework atom. All hydrogen atoms are omitted for clarity. Color code: Zn, blue; C, gray; O, red.

---

7 The name MOF-5 was chosen in analogy to the well-known zeolite ZMS-5.

These large cavities make up 61% of the unit cell volume and are filled with solvent molecules (DMF) in the as-synthesized material. One of the most striking features of the MOF-5 structure is that the pores have no walls. This provides for an unprecedented openness of the structure that allows guest molecules to move with great facility without clogging the pores. In contrast, the pores in more traditional porous solids such as zeolites have walls and diffusion can be subject to complications related to blocked pores. The structure of MOF-5 is shown in Figure 1.15 and the open space within this structure is illustrated by a yellow sphere that represents the largest sphere that can occupy the pore without penetrating the van der Waals radius of any framework atom. We will use these spheres to highlight the accessible open space within the structures of all porous frameworks discussed throughout this book.

Among the very first questions to be addressed about MOF-5 was whether the guests filling the pores could be removed without collapsing the overall structure and whether, like MOF-2, MOF-5 is stable enough to support permanent porosity. Before addressing this issue, we digress slightly to enumerate the different types of stability relevant to this and other MOFs that follow.

1.8.3 Stability of Framework Structures

Chemical stability is the ability of a given material to withstand chemical treatment without any significant change in its structure. This can be evaluated by subjecting a material to different liquid or gaseous chemicals, followed by X-ray diffraction analysis to verify that the structure of the material has not been altered or degraded.

Thermal stability is the ability of a given material to withstand thermal treatment without any significant change in its structure. This can often be assessed by thermogravimetric analysis or differential-scanning-calorimetry where, upon heating the sample, an apparent mass loss or a thermal effect (exothermic or endothermic) is recorded, indicating decomposition and changes in the structure. Additionally, X-ray diffraction studies performed on the material after or during thermal treatment can provide information on whether the structure has been retained.

Mechanical stability is the ability of a given material to withstand external forces. Methods to determine the mechanical stability of MOFs are similar to those used in materials science such as pressurization (compressibility), nano-indentation (Young's modulus) or determination of the tensile strength to name a few.

Architectural stability is the ability of a framework material to retain its structural integrity in the absence of guest molecules. It can be proven by evacuating the solvent from the pores of a MOF and subsequent confirmation of its crystal structure and porosity.

1.8.4 Activation of MOF-5

To realize the full potential of MOF-5, the challenge of removing guest molecules to yield an open framework was addressed. Initial attempts to evaporate the solvent guest molecules from the crystal caused cracking and a concomitant partial loss of porosity that were ascribed to the strong mechanical forces acting on the framework upon solvent removal. These forces are proportional to the surface tension of the solvent in the pores and the extent of the "adhesive forces" between the guest molecules and the inner surface of the MOF. To facilitate the evacuation of the material, the highly mobile guest molecules present in the pores of the as-synthesized material were fully exchanged with chloroform ($CHCl_3$), which upon removal "puts less stress on the framework." The complete removal of all guest molecules from the pores of MOF-5 was eventually achieved by evacuation of the solvent exchanged material at 5×10^{-5} Torr and room temperature for three hours with full retention of the crystallinity of the architecturally stable framework [37]. The process of removing volatile guest molecules from the pores of MOFs is commonly referred to as "activation."

Since no change in morphology or transparency was observed upon activation of MOF-5, single crystal X-ray diffraction studies of the activated material were carried out. This is usually difficult because porous solid-state materials often lose their monocrystallinity upon removal of guest molecules. However, in this case the unit cell parameters and atomic positions determined from these measurements were shown to be almost identical to those of the as-synthesized material. In fact, the remaining electron density within the pores was significantly lower than for the as-synthesized material, providing further proof that all guest molecules had been removed and that MOF-5 is indeed permanently porous[8] [37].

1.8.5 Permanent Porosity of MOF-5

The next step in proving the permanent porosity of MOF-5 was the determination of its internal surface area. For this purpose, nitrogen adsorption experiments at 77 K (as recommended by IUPAC) were performed (Figure 1.16). These measurements allow for the determination of both pore size and surface area. The pore volume calculated from these measurements ($0.54–0.61\,cm^3\,cm^{-3}$) was higher than those reported for the best performing zeolites at that time (up to $0.47\,cm^3\,cm^{-3}$) [37]. With a value of $2900\,m^2\,g^{-1}$, the Langmuir surface area reported in this contribution surpassed by far that of all zeolites, activated carbons, and other porous materials.[9] In later contributions, even higher surface areas up to $3800\,m^2\,g^{-1}$ were reported as better methods for the activation of MOFs were developed [38a].

The combination of a 6-connected $Zn_4O(—COO)_6$ cluster and charged bridging carboxylate linkers suggest that the resulting framework should exhibit high thermal stability, and indeed, neither the morphology nor the crystallinity of the fully activated MOF-5 was affected by heating the material in dry air at 300 °C for 24 hours. This was further evidenced by subsequent single crystal X-ray diffraction studies carried out on MOF-5 samples that underwent this procedure [37]. Furthermore, MOF-5 was shown to be stable at temperatures up to 400 °C under vacuum. The structural degradation of MOF-5 under atmospheric conditions can

8 A material is defined as permanently porous if it is proven to be stable upon removal of the guests from the pores without collapsing. This is measured by nitrogen gas adsorption experiments (at 77 K relative pressures between 0 and 1), the gold standard for evaluation of porosity.

9 Ulrich Müller, a research director at BASF SE, recalls his reaction when he came across this study on MOF-5 and stated: "That number was so unbelievably high, I thought it had to be a misprint." Only after having repeated the measurement himself was he convinced [39].

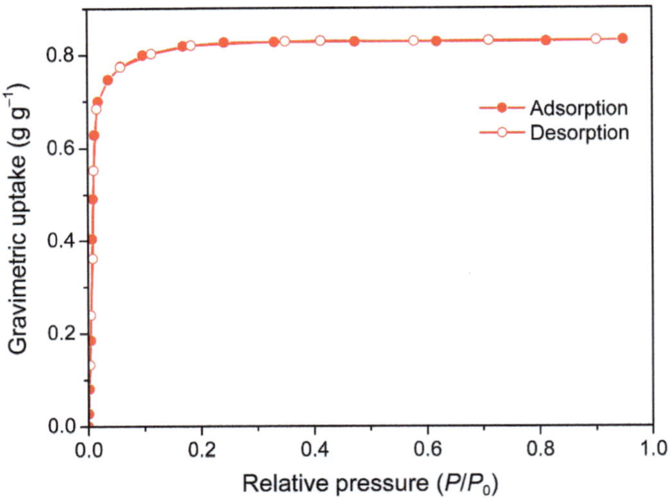

Figure 1.16 Nitrogen adsorption isotherm measured at 77 K. A pore volume of 0.54–0.61 cm^3 cm^{-3} and a Langmuir surface area of 2900 m^2 g^{-1} have been calculated from this measurement. The fact that the desorption branch perfectly traces the adsorption branch highlights the outstanding architectural and mechanical stability of MOF-5 and gives further evidence of its permanent porosity.

therefore be ascribed to humidity in the air rather than to oxygen. This is further supported by the fact that treating MOF-5 with dry solvents or dry air has no effect on its crystallinity and surface area, whereas treatment with humid air or moist solvents results in the slow decomposition of MOF-5 and the formation of a nonporous product [38a].

1.8.6 Architectural Stability of MOF-5

It is worthy of note, that when MOF-5 was first reported, there were many doubters as no one expected such an open structure, composed of largely open space, to be architecturally and thermally stable. Many expected the framework to collapse onto itself once the solvent guests are removed. To gain a deeper understanding of the key factors rendering MOF-5 architecturally stable, it is helpful to take a closer look at its structure. The cubic structure of MOF-5 (Figure 1.17a) can be deconstructed into the basic **pcu** net, that is, a framework built from single atom vertices connected by edges (Figure 1.17b). When a shear force is applied to this basic **pcu** net little resistance is expected. This however does not hold true for the actual crystal structure of MOF-5. In its crystal structure, the vertices of the basic **pcu** net are cationic zinc-oxide clusters that have an envelope[10] of truncated tetrahedral shape. These vertices are joined together by the rigid planar BDC linkers, which can be represented by a planar flat envelope (Figure 1.17c). Each set of linkers located on opposing sides of the truncated

10 The envelope representation of individual building units in carboxylate MOFs are geometrical shapes identical to those obtained when wrapping the respective building units in paper (thus envelope) while making sure, that all oxygen atoms of the carboxylate groups are touching the paper.

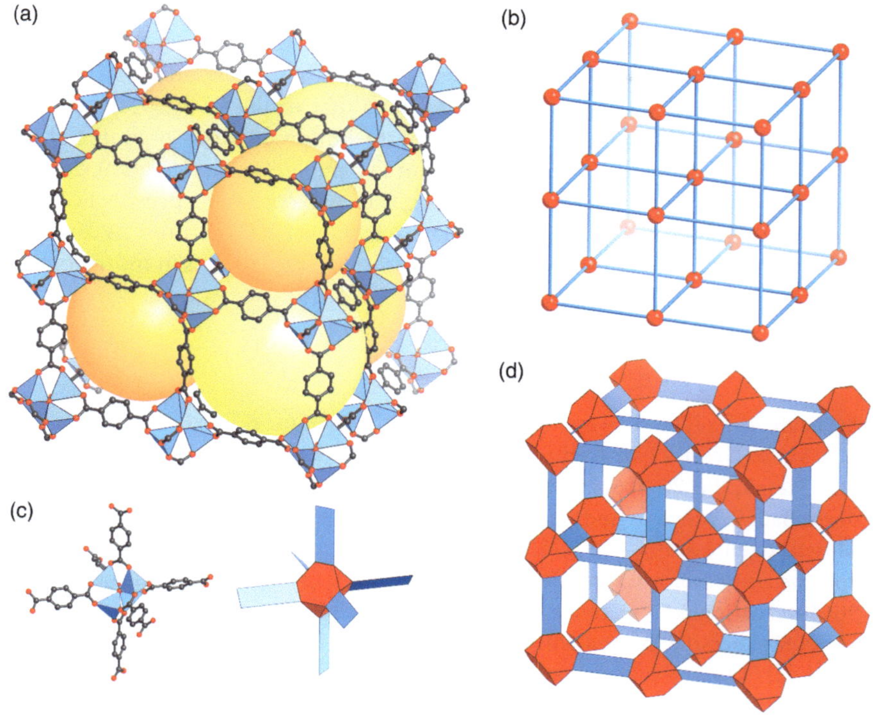

Figure 1.17 (a) Crystal structure of MOF-5, the two differently sized pores are highlighted by yellow (large pore, 15.1 Å diameter) and orange spheres (small pore, 11.0 Å), respectively. (b) Simplified representation of the basic **pcu** net of MOF-5. SBUs are replaced by single atom vertices and the BDC linkers are replaced by edges. (c) Envelope representation of the octahedral $Zn_4O(—COO)_6$ SBUs and the BDC linker as truncated tetrahedra and rectangles, respectively. (d) Envelope representation of the extended framework structure of MOF-5, highlighting its architectural stability that originates from the mutually perpendicular arrangement of BDC linkers around the SBUs. Color code: Zn, blue tetrahedra; C, gray; O, red. In the topology and envelope representation, nodes are shown in red, linkers in blue.

tetrahedron has a dihedral angle of 90°; i.e. they are rotated by 90° with respect to each other. Linking these two building units into an extended 3D framework results in an inherently rigid structure, held together by mutually perpendicular hinges (Figure 1.17d). This arrangement provides for the high architectural stability needed to allow for the activation and support of permanent porosity. The high thermal stability of MOF-5 on the other hand is attributed to the fact that the backbone of MOF-5 is composed entirely of strong bonds (Zn—O, C—O, and C—C), all of which are significantly stronger and therefore thermodynamically more stable than those in coordination networks (M–N–donor) [40].

1.9 Summary

In this chapter we have outlined the history of the development of MOFs. We showed the transition from 0D amine and nitrile-based coordination compounds into 2D and 3D coordination networks and highlighted the key points in making robust, chemically, mechanically, and architecturally stable compounds that support permanent porosity: (i) The use of charged chelating linker and (ii) the SBU approach. In this way, the need for counter ions that reside in the pores of the framework can be avoided, and the rigidity of the building units – organic linker and SBU – renders the framework architecturally stable. We showed that different SBUs can be targeted in a rational manner, thus presenting the prospect of the designed synthesis of a vast variety of possible framework structures. In the following chapters we will consider the porosity of such frameworks in more detail.

References

1 (a) Kraft, D.A. (2012). *Wege des Wissens: Berliner Blau, 1706–1726.*; Frankfurt/Main: Gesellschaft Deutscher Chemiker/Fachgruppe Geschichte der Chemie. Bd 22. ISSN 0934-8506. https://www.gdch.de/fileadmin/downloads/Netzwerk_und_Strukturen/Fachgruppen/Geschichte_der_Chemie/Mitteilungen_Band_22/2012-22-02.pdf (b) Ball, P. (2003). *Bright Earth: Art and the Invention of Color*. Penguin. (c) Bartoll, J. (2008). Proceedings of the 9th International Conference on NDT of Art https://www.ndt.net/article/art2008/papers/029bartoll.pdf.

2 Stahl, G. (1731). Experimenta, observationes, animadversiones. *Chymicae et Physicae (Berlin)* 300: 281–283.

3 (a) Orna, M.V., Kozlowski, A.W., Baskinger, A., and Adams, T. (1994). *Coordination Chemistry: A Century of Progress*, American Chemical Society Symposium Series 565, 165–176. Washington, DC: American Chemical Society. (b) Wunderlich, C.-H. and Bergerhoff, G. (1994). Konstitution und Farbe von Alizarin- und Purpurin-Farblacken. *Chemische Berichte* 127 (7): 1185–1190.

4 (a) Kauffman, G.B. (2013). *Alfred Werner: Founder of Coordination Chemistry*. Springer Science & Business Media. (b) Constable, E.C. and Housecroft, C.E. (2013). Coordination chemistry: the scientific legacy of Alfred Werner. *Chemical Society Reviews* 42 (4): 1429–1439.

5 Werner, A. (1893). Beitrag zur konstitution anorganischer verbindungen. *Zeitschrift für Anorganische Chemie* 3 (1): 267–330.

6 (a) Kekulé, A. (1857). Ueber die sg gepaarten Verbindungen und die Theorie der mehratomigen Radicale. *European Journal of Organic Chemistry* 104 (2): 129–150. (b) Kekulé, A. (1858). Über die Constitution und die Metamorphosen der chemischen Verbindungen und über die chemische Natur des Kohlenstoffs. *European Journal of Organic Chemistry* 106 (2): 129–159.

7 (a) Blomstrand, C.W. (1869). *Chemie der Jetztzeit*. Heidelberg: C. Winter. (b) Jörgensen, S. (1894). Zur Konstitution der Kobalt-, Chrom-und Rhodiumbasen. *Zeitschrift für Anorganische und Allgemeine Chemie* 5 (1): 147–196. (c) Kauffman, G.B. (1959). Sophus Mads Jorgensen (1837–1914): a chapter in coordination chemistry history. *Journal of Chemical Education* 36 (10): 521–527.

8 Werner, A. and Miolati, A. (1894). *Zeitschrift für Physik Chem Leipzig* 14: 506–511.

9 Werner, A. (1907). Über 1.2-Dichloro-tetrammin-kobaltisalze. (Ammoniak-violeosalze). *European Journal of Inorganic Chemistry* 40 (4): 4817–4825.

10 Hofmann, K. and Küspert, F. (1897). Verbindungen von kohlenwasserstoffen mit metallsalzen. *Zeitschrift für Anorganische Chemie* 15 (1): 204–207.

11 Powell, H.M. (1948). 15. The structure of molecular compounds. Part IV. Clathrate compounds. *Journal of the Chemical Society (Resumed)* 61–73.

12 Rayner, J. and Powell, H.M. (1952). 67. Structure of molecular compounds. Part X. Crystal structure of the compound of benzene with an ammonia–nickel cyanide complex. *Journal of the Chemical Society (Resumed)* 319–328.

13 (a) Iwamoto, T., Miyoshi, T., Miyamoto, T. et al. (1967). The metal ammine cyanide aromatics clathrates. I. The preparation and stoichiometry of the diamminemetal(II) tetracyanoniccolate(II) dibenzene and sianiline. *Bulletin of the Chemical Society of Japan* 40 (5): 1174–1178. (b) Iwamoto, T., Nakano, T., Morita, M. et al. (1968). The Hofman-type clathrate: $M(NH_3)_2M'(CN)_4 . 2G$. *Inorganica Chimica Acta* 2: 313–316. (c) Miyoshi, T., Iwamoto, T., and Sasaki, Y. (1972). The structure of catena-μ-ethylenediaminecadmium(II)tetracyanoniccolate(II)dibenzene clathrate: $Cd(en)Ni(CN)_4{\cdot}2C_6H_6$. *Inorganica Chimica Acta* 6: 59–64. (d) Walker, G. and Hawthorne, D. (1967). Complexes between n-alkylamines and nickel cyanide. *Transactions of the Faraday Society* 63: 166–174.

14 Nishikiori, S.-I. and Iwamoto, T. (1984). Crystal structure of Hofmann-dma-type benzene clathrate bis(dimethylamine)cadmium(II) tetracyanonickelate(II) benzene(2/1). *Chemistry Letters* 13 (3): 319–322.

15 (a) Hasegawa, T., Nishikiori, S.-I., and Iwamoto, T. (1984). *Clathrate Compounds, Molecular Inclusion Phenomena, and Cyclodextrins*, 351–357. Springer. (b) Hasegawa, T., Nishikiori, S.-I., and Iwamoto, T. (1985). Isomer selection of 1,6-diaminohexanecadmium(II) tetracyznonickelate(II) for *m*- and *p*-toluidine. Formation of 1,6-diaminohexanecadmium(II) tetracyznonickelate(II) *m*-toluidine (1/1) inclusion compound and bis(*p*-toluidine)-1,6-diaminohexanecadmium(II)tetracyznonickelate(II) complex. *Chemistry Letters* 14 (11): 1659–1662. (c) Nishikiori, S.-I., Hasegawa, T., and Iwamoto, T. (1991). The crystal structures of α,ω-diaminoalkanecadmium(II) tetracyanonickelate(II) aromatic molecule inclusion compounds. V. Toluidine clathrates of the hosts built of the diamines, 1,4-diaminobutane, 1,5-diaminonentane, and 1,8-diaminooctane. *Journal of Inclusion Phenomena and Molecular Recognition in Chemistry* 11 (2): 137–152.

16 (a) Kinoshita, Y., Matsubara, I., and Saito, Y. (1959). The crystal structure of bis(succinonitrilo)copper(I) nitrate. *Bulletin of the Chemical Society of Japan* 32 (7): 741–747. (b) Kinoshita, Y., Matsubara, I., and Saito, Y. (1959). The crystal structure of bis(glutaronitrilo)copper(I) nitrate. *Bulletin of the Chemical Society of Japan* 32 (11): 1216–1221. (c) Kinoshita, Y., Matsubara, I., Higuchi, T., and Saito, Y. (1959). The crystal structure of

bis(adiponitrilo)copper(I) nitrate. *Bulletin of the Chemical Society of Japan* 32 (11): 1221–1226.

17 Wells, A. (1954). The geometrical basis of crystal chemistry. Part 1. *Acta Crystallographica* 7 (8–9): 535–544.

18 Ockwig, N.W., Delgado-Friedrichs, O., O'Keeffe, M., and Yaghi, O.M. (2005). Reticular chemistry: occurrence and taxonomy of nets and grammar for the design of frameworks. *Accounts of Chemical Research* 38 (3): 176–182.

19 Aumüller, A., Erk, P., Klebe, G. et al. (1986). A radical anion salt of 2,5-dimethyl-*N,N'*-dicyanoquinonediimine with extremely high electrical conductivity. *Angewandte Chemie International Edition in English* 25 (8): 740–741.

20 Kato, R., Kobayashi, H., and Kobayashi, A. (1989). Crystal and electronic structures of conductive anion-radical salts,(2,5-R_1R_2-DCNQI)$_2$Cu (DCNQI = *N,N'*-dicyanoquinonediimine; R_1, R_2 =CH_3, CH_3O, Cl, Br). *Journal of the American Chemical Society* 111 (14): 5224–5232.

21 Desiraju, G.R. and Parshall, G.W. (1989). *Crystal Engineering: The Design of Organic Solids*, Materials Science Monographs, vol. 54. Elsevier.

22 Moulton, B. and Zaworotko, M.J. (2001). From molecules to crystal engineering: supramolecular isomerism and polymorphism in network solids. *Chemical Reviews* 101 (6): 1629–1658.

23 (a) Zhdanov, H. (1941). The crystalline structure of $Zn(CN)_2$. *Comptes Rendus de l'Académie des Sciences de l'URSS* 31: 352–354. (b) Shugam, E. and Zhdanov, H. (1945). The crystal structure of cyanides. II. The structure of $Cd(CN)_2$. *Acta Physicochim. URSS* 20: 247–252. (c) Takafumi, K., Shin-ichi, N., Reiko, K., and Toschitake, I. (1988). Novel clathrate compound of cadmium cyanide host with an adamantane-like cavity. Cadmium cyanide–carbon tetrachloride(1/1). *Chemistry Letters* 17 (10): 1729–1732.

24 (a) Hoskins, B.F. and Robson, R. (1989). Infinite polymeric frameworks consisting of three dimensionally linked rod-like segments. *Journal of the American Chemical Society* 111 (15): 5962–5964. (b) Hoskins, B. and Robson, R. (1990). Design and construction of a new class of scaffolding-like materials comprising infinite polymeric frameworks of 3D-linked molecular rods. A reappraisal of the zinc cyanide and cadmium cyanide structures and the synthesis and structure of the diamond-related frameworks $[N(CH_3)_4][Cu^IZn^{II}(CN)_4]$ and Cu^I[4,4',4'',4'''-tetracyanotetraphenylmethane]$BF_4.xC_6H_5NO_2$. *Journal of the American Chemical Society* 112 (4): 1546–1554.

25 Zaworotko, M.J. (1994). Crystal engineering of diamondoid networks. *Chemical Society Reviews* 23 (4): 283–288.

26 Gable, R.W., Hoskins, B.F., and Robson, R. (1990). Synthesis and structure of $[NMe_4][CuPt(CN)_4]$: an infinite three-dimensional framework related to PtS which generates intersecting hexagonal channels of large cross section. *Journal of the Chemical Society, Chemical Communications* (10): 762–763.

27 Abrahams, B.F., Hoskins, B.F., Michail, D.M., and Robson, R. (1994). Assembly of porphyrin building blocks into network structures with large channels. *Nature* 369 (6483): 727–729.

28 Fujita, M., Yazaki, J., and Ogura, K. (1990). Preparation of a macrocyclic polynuclear complex, $[(en)Pd(4,4'\text{-}bpy)]_4(NO_3)_8$ (en = ethylenediamine, bpy = bipyridine), which recognizes an organic molecule in aqueous media. *Journal of the American Chemical Society* 112 (14): 5645–5647.

29 Fujita, M., Kwon, Y.J., Washizu, S., and Ogura, K. (1994). Preparation, clathration ability, and catalysis of a two-dimensional square network material composed of cadmium(II) and 4,4′-bipyridine. *Journal of the American Chemical Society* 116 (3): 1151–1152.

30 Yaghi, O. and Li, H. (1995). Hydrothermal synthesis of a metal-organic framework containing large rectangular channels. *Journal of the American Chemical Society* 117 (41): 10401–10402.

31 Subramanian, S. and Zaworotko, M.J. (1995). Porous solids by design: $[Zn(4,4'\text{-}bpy)_2(SiF_6)]_n \cdot xDMF$, a single framework octahedral coordination polymer with large square channels. *Angewandte Chemie International Edition in English* 34 (19): 2127–2129.

32 Yaghi, O.M., Li, G., and Li, H. (1995). Selective binding and removal of guests in a microporous metal-organic framework. *Nature* 378 (6558): 703.

33 Li, H., Eddaoudi, M., Groy, T.L., and Yaghi, O.M. (1998). Establishing microporosity in open metal-organic frameworks: gas sorption isotherms for Zn(BDC) (BDC = 1,4-Benzenedicarboxylate). *Journal of the American Chemical Society* 120 (33): 8571–8572.

34 Tranchemontagne, D.J., Mendoza-Cortes, J.L., O'Keeffe, M., and Yaghi, O.M. (2009). Secondary building units, nets and bonding in the chemistry of metal-organic frameworks. *Chemical Society Reviews* 38 (5): 1257–1283.

35 (a) Bragg, W.L. (1914). Die Beugung kurzer elektromagnetischer Wellen durch einen Kristall. *Zeitschrift für Anorganische Chemie* 90 (1): 153–168. (b) Friedrich, W., Knipping, P., and Laue, M. (1913). Interferenzerscheinungen bei Röntgenstrahlen. *Annalen der Physik* 346 (10): 971–988. (c) Komissarova, L.N., Simanov, Y.P., Plyushchev, Z.N., and Spitsyn, V.I. (1966). Zirconium and hafnium oxoacetates. *Russian Journal of Inorganic Chemistry* 11 (9): 2035–2040. (d) Komissarova, L.N.K., Prozorovskaya, S.V., Plyushchev, Z.N., and Plyushchev, V.E. (1966). Zirconium and hafnium oxoacetates. *Russian Journal of Inorganic Chemistry* 11 (2): 266–271. (e) Koyama, H. and Saito, Y. (1954). The crystal structure of zinc oxyacetate, $Zn_4O(CH_3COO)_6$. *Bulletin of the Chemical Society of Japan* 27 (2): 112–114. (f) van Niekerk, J.N., Schoening, F.R.L., and Talbot, J.H. (1953). The crystal structure of zinc acetate dihydrate, $Zn(CH_3COO)_2 \cdot 2H_2O$. *Acta Crystallographica* 6 (8–9): 720–723. (g) van Niekerk, J.N. and Schoening, F.R.L. (1953). X-ray evidence for metal-to-metal bonds in cupric and chromous acetate. *Nature* 171 (4340): 36–37.

36 Lionelle, J.E. and Staffa, J.A. (1983). Metal oxycarboxylates and method of making same. US Patent US10596310.

37 Li, H., Eddaoudi, M., O'Keeffe, M., and Yaghi, O.M. (1999). Design and synthesis of an exceptionally stable and highly porous metal-organic framework. *Nature* 402 (6759): 276–279.

38 (a) Kaye, S.S., Dailly, A., Yaghi, O.M., and Long, J.R. (2007). Impact of preparation and handling on the hydrogen storage properties of

Zn_4O(1,4-benzenedicarboxylate)$_3$ (MOF-5). *Journal of the American Chemical Society* 129 (46): 14176–14177. (b) Tranchemontagne, D.J., Hunt, J.R., and Yaghi, O.M. (2008). Room temperature synthesis of metal-organic frameworks: MOF-5, MOF-74, MOF-177, MOF-199, and IRMOF-0. *Tetrahedron* 64 (36): 8553–8557.

39 Jacoby, M. (2008). Heading to market with MOFs. *Chemical and Engineering News* 86 (34): 13–16.

40 (a) Yaghi, O.M., O'Keeffe, M., Ockwig, N.W. et al. (2003). Reticular synthesis and the design of new materials. *Nature* 423 (6941): 705–714. (b) Kiang, Y.-H., Gardner, G.B., Lee, S. et al. (1999). Variable pore size, variable chemical functionality, and an example of reactivity within porous phenylacetylene silver salts. *Journal of the American Chemical Society* 121 (36): 8204–8215. (c) Jiang, J., Zhao, Y., and Yaghi, O.M. (2016). Covalent chemistry beyond molecules. *Journal of the American Chemical Society* 138 (10): 3255–3265.

2
Determination and Design of Porosity

2.1 Introduction

The porosity of a material is defined as the ratio of the volume of the pores to the total volume occupied by the solid. Since porous materials are partially composed of empty space that can be accessed by guest molecules, they are commonly described in terms of their gas uptake properties. In Chapter 1, we showed that MOF-5 exhibits permanent porosity with a specific surface area surpassing those of well-known zeolites, porous silicates, and porous carbons. Indeed, many metal-organic frameworks (MOFs) developed later exhibit ultra-high porosity surpassing that of all other classes of porous solids. In this chapter, we provide the basics of gas adsorption and present the unique aspects of MOF chemistry that endow them with their unusual porosity.

2.2 Porosity in Crystalline Solids

The concept of porosity was proposed as early as 1896 in inorganic zeolites and was proven experimentally less than 50 years later [1]. Since then, the chemistry of zeolites (from the Greek ζέω (*zéō*), meaning "to boil" and λίθος (*líthos*), meaning "stone") has been further developed, and principles for their synthesis and characterization have been established [2]. Although the chemistry and structural diversity of zeolites is extensive, these materials are inherently limited by the number of accessible building units and the difficulty with which they can be functionalized. This is largely due to the chemical nature of their inorganic structures that are constructed from linked tetrahedral MO_4 (M = Si, Ge, Al, …) units [2a]. The incorporation of transition metal ions and organics into the backbone of zeolites has always been a desirable goal, thus giving an impetus to the discovery and development of porous metal-organic frameworks.

Porosity in coordination compounds was first observed in a series of molecular Werner-type complexes of the general formula β-[M(PIC)$_4$(SCN)$_2$] (M = Ni^{2+}, Co^{2+}; PIC = γ-picoline; SCN = thiocyanate) [3]. The crystal structures of these compounds have empty voids between the individual complexes as a result of the

Introduction to Reticular Chemistry: Metal-Organic Frameworks and Covalent Organic Frameworks,
First Edition. Omar M. Yaghi, Markus J. Kalmutzki, and Christian S. Diercks.

Companion website: www.wiley.com/go/yaghi/reticular

sterically demanding PIC ligands [4]. Richard Barrer and coworkers showed that these materials are indeed capable of reversible uptake of gases such as nitrogen, oxygen, argon, and hydrocarbons in a zeolite-like manner. While these molecular crystals show gas uptake, cycling of uptake and release causes a decrease in porosity attributable to a phase transition of the structure into a dense nonporous form. This is not surprising, considering that the metal complexes constituting the crystal structure are held together only by weak non-covalent interactions [5]. This example brought forward the need that in order to achieve permanent porosity it is crucial to link the molecular entities by strong bonds to afford mechanical and architectural stability.

The Hofmann clathrate and related compounds discussed in Chapter 1 are capable of absorbing specific molecules from liquid phase, but no reversible uptake of gases was reported. In contrast, the adsorption of gases by Prussian blue compounds was measured to prove their permanent porosity [6]. In 1995, Dorai Ramprasad et al. demonstrated that lithium pentacyanocobaltate $Li_3[Co(CN)_5](DMF)_2$ (DMF = *N,N*-dimethylformamide), a crystalline ionic material, is capable of reversible uptake of oxygen at room temperature by solid-state diffusion and binding of dioxygen to cobalt in a chemisorptive fashion, and as such this material does not exhibit permanent porosity [7].

In 1997, Susumu Kitagawa and coworkers reported the high pressure (up to 36 atm) uptake of methane, nitrogen, and oxygen by a coordination network of formula $[M_2(BIPY)_3(NO_3)_4]$ [8]. Wasuke Mori and coworkers reported dicarboxlyates of copper, molybdenum, and rhodium, and it was speculated that their structures consist of 2D layers, analogous to MOF-2 [9]. The gas occlusion of Cu(BDC)(Py) (Py = pyridine) and Rh(BDC)(Py) at low temperatures and ambient pressure were investigated but no gas adsorption isotherms were recorded [9b, c]. The conditions under which these gas uptake measurements were performed do not provide the means to quantify parameters such as pore volume or surface area and therefore a proof of permanent porosity in coordination networks remained outstanding.

According to the International Union of Pure and Applied Chemistry (IUPAC) requirements, the evaluation of permanent porosity demands the measurement of at least three and preferably five or more points in the appropriate pressure range on the N_2 adsorption isotherm at the normal boiling point of liquid nitrogen. This is because all models used to describe gas adsorption in porous solids and calculate the surface areas and other measures of porosity are only valid at the boiling point of the adsorbate and for relatively low pressure. The unambiguous proof of permanent porosity in a metal-organic extended solid was first achieved with the report of reversible nitrogen adsorption measurements carried out at 77 K for MOF-2 in 1998. Since then the gas adsorption in MOFs has been studied according to the MOF-2 example, and the chemistry of MOFs was further developed to routinely provide for mechanical and architectural stability [10]. The discovery of MOF-5 ushered in the era of 3D MOFs with large accessible pore volumes and surface areas [11]. In Sections 2.3 and 2.4, we will introduce the terminology and basic principles used to describe and evaluate porosity as well as design principles for MOFs with maximized pore size, pore volume, and ultrahigh surface areas.

2.3 Theory of Gas Adsorption

2.3.1 Terms and Definitions

Adsorption phenomena and porosity have been extensively studied and their theoretical principles and experimental measurements are well established. Here, the basics of gas adsorption by porous solids are introduced and the four parameters that are used as a measure of porosity are defined. These are specific surface area, specific pore volume, porosity, and pore size (Figure 2.1).

The term "adsorption" originally denoted the condensation of gas on a flat free surface as opposed to its entry into the bulk of a solid, as in absorption [12]. However, this distinction is frequently not observed, and the uptake of fluids (gaseous or liquid) by porous materials is often referred to as adsorption or simply sorption. The latter term is used to describe both adsorption and negative adsorption (desorption) regardless of the physical mechanism involved [13]. The exposure of a porous solid to a gas in a closed space at a definite pressure results in the adsorption of the gas (adsorbate) by the solid (adsorbent). This leads to an increase of the weight of the solid and consequently, a decrease in the gas pressure. An overview of the terminology and nomenclature used with respect to gas sorption in porous solids is given in Table 2.1.

According to IUPAC, porous materials are divided into three classes based on their pore size: (i) the term "microporous" refers to pore diameters smaller than 2 nm, (ii) "mesoporous" to pore sizes between 2 and 50 nm, and (iii) "macroporous" to pores larger than 50 nm in diameter (Figure 2.2) [14]. MOFs typically have pores within the micro- and mesoporous regimes.

2.3.2 Physisorption and Chemisorption

The interactions between the adsorbate and the adsorbent can be of physical (physisorption) or chemical (chemisorption) nature. Even though physisorption

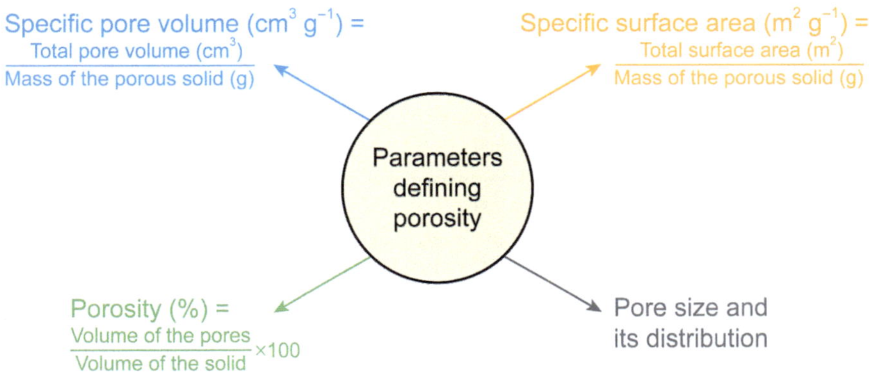

Figure 2.1 Parameters used to define porosity. The specific pore volume and the porosity are calculated from crystal structure data or density measurements. The specific surface area and the pore size distribution are estimated from adsorption data collected at the boiling point of the respective adsorbent.

Table 2.1 Terminology and nomenclature for gas adsorption.

Term	Definition
Adsorption	Enrichment of one or more components at or near an interface
Desorption	Reverse of adsorption
Adsorbent	Solid material onto which adsorption occurs
Adsorption	Absorbable substance in the fluid phase (i.e. gas or liquid)
Adsorbate	Substance in the adsorbed state
Chemisorption	Adsorption involving strong chemical interactions (e.g. chemical bonding)
Physisorption	Adsorption involving weak physical interactions
Porous solid	Solid with cavities which are deeper than wide
Open pores	Cavity or channel that can be accessed from the surface
Closed pores	Cavity or channel that cannot be accessed from the surface
Interconnected pore	Pore which is connected to neighboring pores
Blind pore/dead end pore	Pore with only one opening that connects to the surface
Interparticle space	Space between individual microscopic particles
Interparticle condensation	Condensation of adsorbent in the interparticle space
Pore size	Largest geometrical pore width, without penetrating the van der Waals radius of the atoms framing the pore
Micropore/microporous	Pores with an internal diameter < 2 nm
Mesopore/mesoporous	Pores with an internal diameter between 2 and 50 nm
Macropore/macroporous	Pores with an internal diameter > 50 nm
Nanopore/nanoporous	Pores with an internal diameter ~ 100 nm
Pore volume	Experimentally determined volume of the pores
Porosity	Pore volume/sample volume ($cm^3\ cm^{-3}$)
Total porosity	(Void volume + pore volume)/sample volume ($cm^3\ cm^{-3}$)
Open porosity	(Void volume + open pore volume)/sample volume ($cm^3\ cm^{-3}$)
Surface area	Surface accessible by an adsorbate under steady-state conditions
External surface area	Area of the surface excluding the pores (often only excluding micropores)
Internal surface area	Area of the surface of all pore walls (often only of micropores)
Monolayer capacity	Amount of physisorbed adsorbent needed to cover the surface with a monolayer or amount of chemisorbed adsorbent needed to occupy all surface sites
Surface coverage	Adsorbed substance/monolayer capacity
True density	Density of the solid excluding the pores and voids
Apparent density	Density including closed or inaccessible pores

Figure 2.2 Classification of pore sizes according to IUPAC in microporous (<2 nm), mesoporous (2–50 nm), and macroporous materials (>50 nm) along with examples listed for the respective sizes.

is an exothermic process, the interactions involved are generally weak and of a similar magnitude as those responsible for the condensation of vapors and deviations from the ideal gas behavior in real gases. Owing to the relatively low degree of specificity and directionality, physisorbed molecules retain their identity (minor changes in bond angles due to partial orbital overlap can occur) and multilayers can be formed at high relative pressures. The equilibrium in physisorption systems is generally attained rapidly but may be limited by a slow rate-determining transport process.

Chemisorption involves interactions of the same order of magnitude as those occurring during chemical reactions and the formation of chemical bonds. The highly directional nature of these interactions results in strong binding of adsorbate molecules to reactive sites on the surface that is often accompanied by surface reactions (e.g. dissociation) and a concomitant loss of identity of the adsorbate. Accordingly, chemisorbed molecules are confined to a monolayer, in contrast to the multilayer adsorption commonly observed for physisorption. Chemisorption is an activated process and hence at low temperature the system may not have sufficient thermal energy to initiate this process, resulting in slow equilibration.

2.3.3 Gas Adsorption Isotherms

Tens of thousands of gas adsorption isotherms measured on a large variety of solids are found in the literature of porous materials. Most isotherms of porous solids are due to physical adsorption and IUPAC has defined six types of isotherm: Types I to VI as depicted in Figure 2.3[1] [15], In this classification, the isotherms are distinguished based on adsorbents that are microporous (Type I), nonporous or macroporous (Types II, III, and VI), or mesoporous (Types IV and V) [14a, 16]. The relatively rare Type VI isotherm is also called stepped isotherm and is often found for nonporous materials with very uniform surfaces and is therefore mainly of theoretical interest.

The desorption process does not always proceed in exactly the reverse pathway of the adsorption, and oftentimes hysteresis (lower branch represents the

1 These isotherms have been originally classified by Brunauer, Deming, Deming, and Teller, but they are most often referred to as BET isotherms after Brunauer, Emmett, and Teller. The multistep isotherm was not included in this classification.

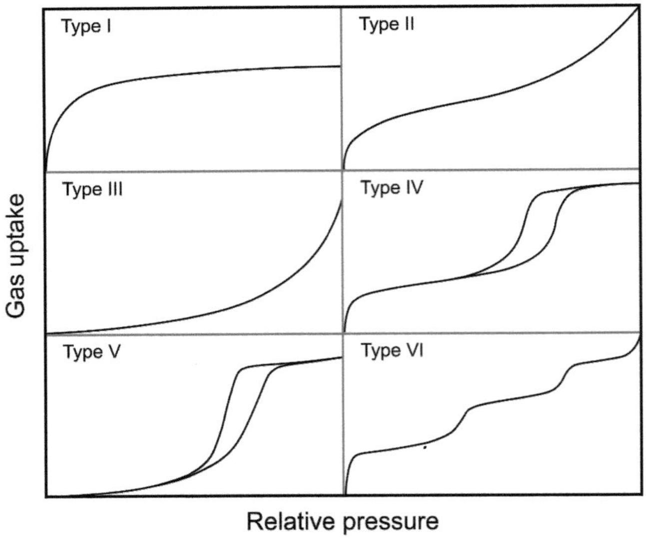

Figure 2.3 IUPAC classification of adsorption isotherms for microporous (Type I), nonporous or macroporous (Types II, III, and VI), or mesoporous (Types IV and V) materials [14a, 16].

adsorption, the upper branch the desorption process) is observed. Isotherms of Types IV and V show hysteresis and it is widely accepted that the shape of the hysteresis loop and the texture of a porous material (i.e. pore size distribution, pore geometry, and connectivity) can be correlated. IUPAC has defined four types of hysteresis that are typically observed for Type IV and V isotherms (Figure 2.4). A hysteresis of Type H1 is typically observed for materials with well-defined cylindrical pores or agglomerates of approximately uniform spheres. Materials with a wide distribution of pore sizes show Type H2 hysteresis, which can also be indicative of bottleneck constrictions. Narrow slit pores result in Type H3 hysteresis similar to that observed for nonrigid aggregates of plate-like particles, and Type H4 hysteresis is associated with wide slit pores within the material. The low-pressure hysteresis, represented by the dashed lines, can be ascribed to a change in volume of the adsorbent (the swelling of nonrigid pores) or irreversible gas uptake, which is observed when the pores are of about the same width as that of the adsorbate molecules.

In Figure 2.5, the process of gas adsorption is schematically illustrated for a porous material whose structure contains three types of pore (micro-, meso-, and macropores), correlating the gas uptake with the appearance of the gas adsorption isotherm. In the low-pressure region only little gas is adsorbed on the inner surface and this is the case for all three pore types independent of their size, as well as for the outer surface of the adsorbent (Figure 2.5a). As the pressure is increased, a monolayer is formed, covering the whole surface of the adsorbent. Simultaneously, the micropore is filled completely, resulting in a plateau in the corresponding isotherm (Figure 2.5b). A similarly shaped isotherm, indicating "mircopore filling," can thus be observed in the absence of micropores, unless

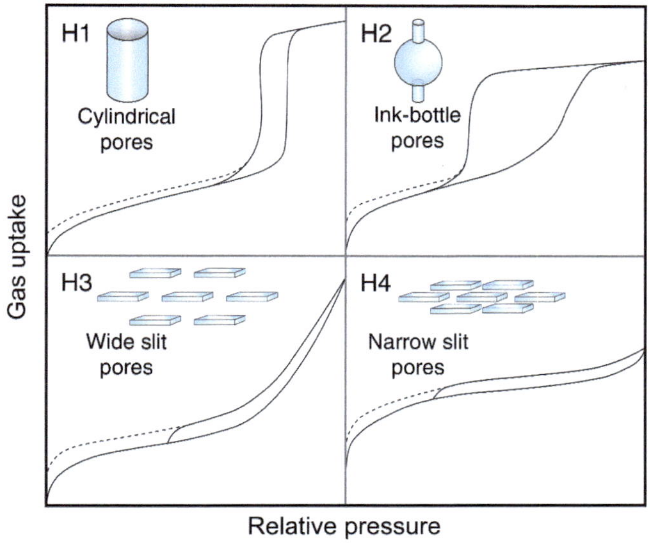

Figure 2.4 Definition of hysteresis for Type IV and V isotherms according to IUPAC. A certain shape of the hysteresis loop is commonly associated with the texture of the porous material. H1 is observed for materials containing cylindrical pores, H2 is indicative of bottleneck restrictions, and H3 and H4 are commonly observed for materials with narrow and wide slit pores, respectively.

repulsive forces between the adsorbent and the adsorbate dominate. A further increase in pressure is necessary to achieve multilayer adsorption and "mesopore filling," which leads to a second steep ascent in the adsorption isotherm (Figure 2.5c). Similarly, a third ascent is observed, when the appropriate pressure is reached to facilitate filling of the macropores (Figure 2.5d). It should be noted, that an adsorption behavior like that of a meso- or macropore can also result from interparticle condensation, hence one must be careful when interpreting isotherm data. From the preceding discussion it has become clear that at lower pressures the uptake of gas is mainly governed by the surface area because only then the surface of the adsorbent is covered with adsorbate molecules (monolayer adsorption) while the rest of the free pore volume does not significantly contribute to the uptake.

2.3.4 Models Describing Gas Adsorption in Porous Solids

Two models are commonly used to describe adsorption phenomena in porous solids (Figure 2.6). The Langmuir model describes the surface specific adsorption involving strong forces and (sometimes) surface reactions such as the dissociation of H_2 on a Pt surface and subsequent binding as two H atoms. The Brunauer–Emmett–Teller (BET) model is applicable to cases where weak nondirectional interactions result in the non-surface-specific adsorption in the absence of surface reactions, as exemplified by H_2 molecules being adsorbed on the surface of a MOF and bound by only van der Waals interactions. The gas adsorption

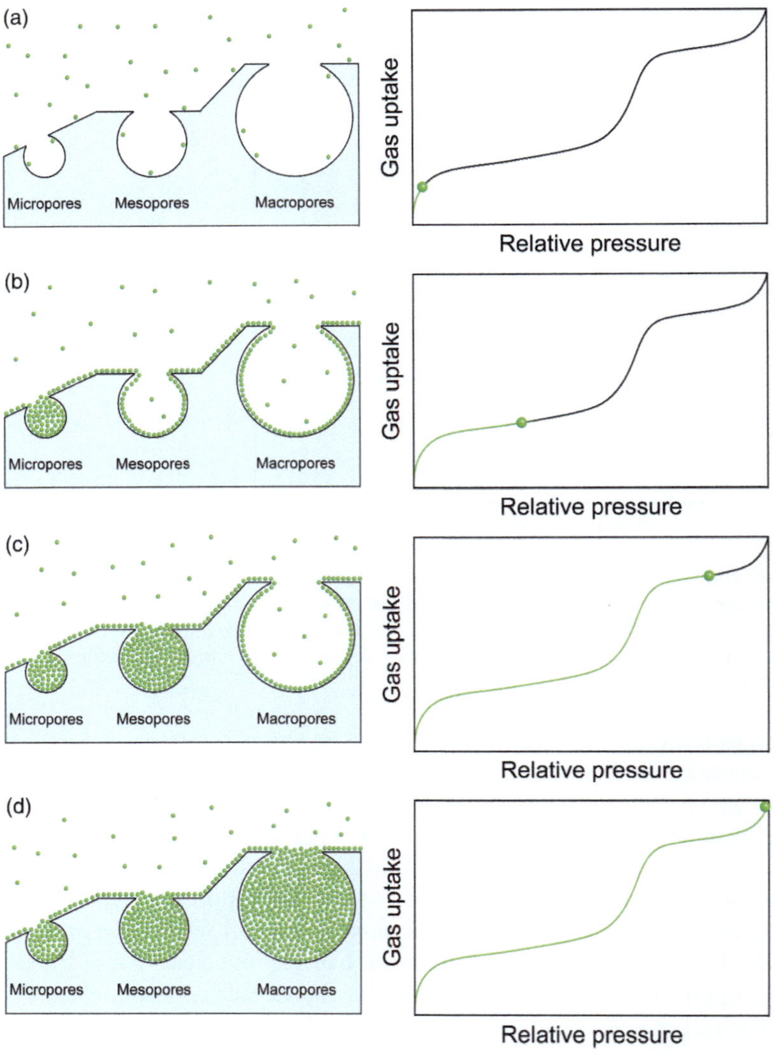

Figure 2.5 Adsorption process for a material containing micro-, meso-, and macropores. (a) In the low-pressure region, the adsorption is mainly governed by the surface area, and (b) at slightly higher pressures micropore filling is observed. (c) A further increase in pressure affords the filling of the mesopores, and (d) ultimately, when pressures close to the bulk saturation pressure are reached, to macropore filling.

in MOFs is typically governed by weak interactions (physisorption) and can, with some limitations, be described by both models. It should be noted that the main difference between them is that unlike the Langmuir model, the BET model takes into account multilayer adsorption, which is particularly important when dealing with larger pores (>2 nm).

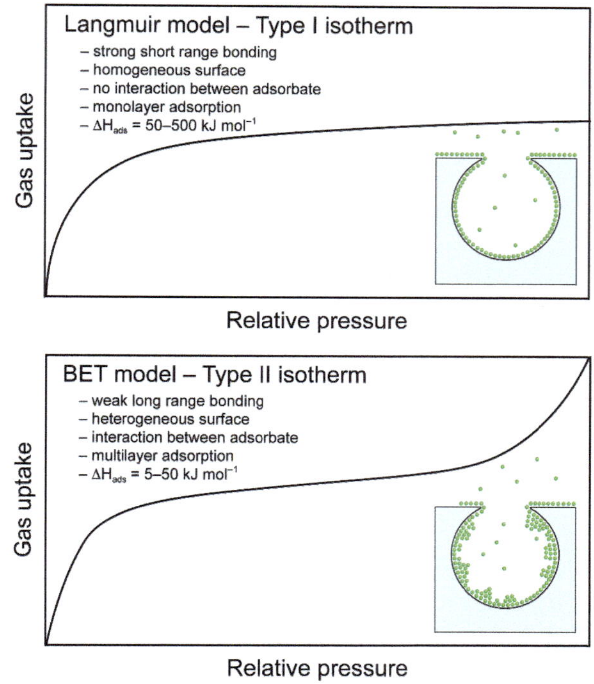

Figure 2.6 Comparison of the Langmuir and BET models for adsorption. The Langmuir model assumes strong directional adsorption (chemisorption) on surfaces in the form of monolayers, whereas the BET model assumes weak nondirectional adsorption (physisorption) and can be used to describe multilayer adsorption.

2.3.4.1 Langmuir Model

The Langmuir model is based on the idea of a limited number of adsorption sites [17]. There are multiple possible scenarios with this model: the presence of (i) only one kind of adsorption site, (ii) more than one kind of adsorption site, (iii) an amorphous adsorbing surface providing a continuum of (different) adsorption sites, (iv) sites that can accommodate more than one adsorbate molecule, (v) dissociative adsorption, and (vi) multilayer adsorption. Commonly, the Langmuir model refers to the first case: adsorption on a plane surface with only one kind of adsorption site, where each adsorption site can hold only one adsorbate molecule. This assumption highlights that this model initially did not allow for either porosity or physisorption, both present in the adsorption of gas in porous solids. Nevertheless, this model can be applied (with some restrictions) to describe gas adsorption in porous solids. Additionally, it supplies the basis on which more refined models, such as the BET model, were developed and it is therefore instructive to briefly discuss how the Langmuir model is derived [17, 18].

The Langmuir equation is derived from kinetic considerations of the adsorption process. The surface of the adsorbent is represented by an array of N^s equivalent and independent adsorption sites that can only adsorb one molecule per site. The fraction θ of sites occupied by a number N^a of adsorbed molecules is calculated following Eq. (2.1):

$$\theta = \frac{N^{\mathrm{s}}}{N^{\mathrm{a}}} \tag{2.1}$$

From the kinetic theory of gases, we know that the rate of adsorption is dependent on the pressure and the number of unoccupied adsorption sites $(1 - \theta)$, and the rate of desorption is dependent on the number of occupied sites (θ) and the activation energy E_A (positive energy of adsorption). The equilibrium is obtained when both the rate of adsorption and desorption are equal as described by Eq. (2.2):

$$\theta = \frac{dN^a}{dt} = \alpha p(1 - \theta) - \beta\theta e^{\left(\frac{-E_A}{RT}\right)} = 0 \tag{2.2}$$

where α and β are constants for a given gas–solid system. Under the presumption, that in the ideal case the probability of adsorption and desorption are independent of the surface coverage (no adsorbate–adsorbate interactions) and the activation energy E_A (heat of adsorption) is constant for a given set of adsorbate and adsorbent, we can rearrange and simplify to obtain the well-known Langmuir isotherm equation

$$\theta = \frac{bp}{1 + bp} \quad \text{with} \quad bp = Ke^{\left(\frac{-E_A}{RT}\right)} \tag{2.3}$$

where b is the adsorption coefficient and the pre-exponential factor K is equal to α/β, the ratio of the adsorption and desorption coefficients. For low coverage $(\theta \to 0)$, Eq. (2.3) can be reduced to Henry's law, and at high coverage $(\theta \sim 1)$ a plateau is reached, corresponding to monolayer adsorption. Considering the assumptions used to derive the Langmuir model, it is only strictly applicable to chemisorption. Three prerequisites must be fulfilled in order to render the Langmuir model applicable: (i) the isotherm should have a linear region at low pressures and show a plateau for $\theta \to 1$, (ii) the energy of adsorption should be independent of surface coverage, and (iii) the differential entropy of adsorption should vary in accordance with the ideal localized model [19]. These considerations indicate that the Langmuir model is not applicable to real physisorption systems even though it is commonly used to determine the surface area of porous materials. This finding is attributed to the fact that many adsorption isotherms show a shape similar to that expected from the Langmuir model (Type I). It should however be pointed out that most Type I isotherms reported in the literature are associated with micropore filling rather than monolayer coverage, and therefore the surface areas calculated using the Langmuir model are often overestimated and should be confirmed using more applicable methods such as the BET method.

2.3.4.2 Brunauer–Emmett–Teller (BET) Model

The BET model is based on the observation that physisorption is not restricted to monolayer coverage if the relative pressure P/P_0 exceeds a certain value. Accordingly, the beginning of the almost linear section of a Type II isotherm (B in Figure 2.7) was concluded to be the point likely to correspond to the monolayer completion. This is supported by experimentally determined surface areas calculated at different points (A, B, C, and D in Figure 2.7) from isotherms measured for the adsorption of several different gases, including N_2 and Ar, at or near their respective boiling points. The best agreement between the various values is found for point B indicating that at this point the monolayer adsorption

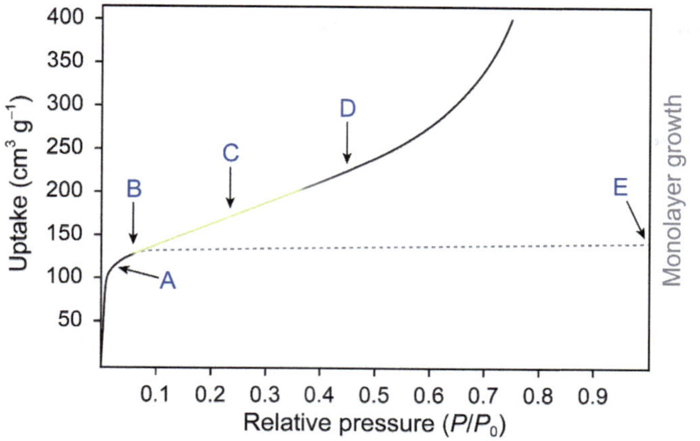

Figure 2.7 Schematic representation of the adsorption of N_2 on an Fe/Al_2O_3 surface at 77 K (black). The gray line represents an isotherm showing only monolayer adsorption. The beginning of the linear part of the isotherm (point B, linear part is highlighted in green) indicates the completion of a monolayer, as evidenced by surface areas calculated at point B from isotherms of different gases.

is complete. These findings are supported by a notable decrease of the heat of adsorption near point B.

The introduction of a few additional assumptions to the Langmuir model yields an extension of the adsorption mechanism that allows for multilayer adsorption. The resulting BET equation describes a Type II isotherm. The original BET model is an extension of the Langmuir kinetic theory that considers the formation of an infinite number of adsorbed layers at the saturation pressure (P_0). This implies that adsorbed molecules themselves can act as adsorption sites at any pressure $P < P_0$. Accordingly, the adsorbed layer is envisaged to be made up of random stacks of molecules rather than a layer of uniform thickness, where fractions of the surface are covered ($\theta_0, \theta_1, \theta_2, \theta_3, \ldots, \theta_i$, where θ_0 is the fraction of unoccupied sites) by 0, 1, 2, …, i layers. Under the assumption that in equilibrium at pressure P the fractions of both bare (θ_0) and covered (θ_i) surface are constant, the rate of adsorption and desorption for a single layer can be given by Eq. (2.4)

$$a_1 p \theta_0 = b_1 \theta_1 \mathrm{e}^{\left(\frac{-E_{\mathrm{A1}}}{\mathrm{RT}}\right)} \tag{2.4}$$

where a_1 and b_1 are the adsorption and desorption constants, and E_{A1} is the heat of adsorption for the first layer. This only holds true for the absence of lateral adsorbate–adsorbate interactions. Treating all possible layers the same way and assuming that (i) the sum of all θ_i equals 1, (ii) the energy of adsorption $E_{\mathrm{A}i}$ equals the liquefaction energy for all layers with $i \geq 2$, and (iii) the multilayer has infinite thickness at $P = P_0$ we can derive Eq. (2.5)

$$\frac{n}{n_{\mathrm{m}}} = \frac{c\left(\frac{P}{P_0}\right)}{\left(1 - \frac{P}{P_0}\right)\left(1 - \frac{P}{P_0} + c\left(\frac{P}{P_0}\right)\right)} \tag{2.5}$$

where n is the total amount of adsorbed gas, n_{m} is the amount of gas adsorbed within the monolayer, and c is a constant. A plot of n/n_{m} against P/P_0 at constant

c ($c > 2$), gives a curve of Type II shape, where the shape of the knee near $n/n_m = 1$ is dependent on the value of c. Larger c values ($c \sim 80$) result in a sharper knee, whereas positive values of $c < 2$ result in the disappearance of the inflection point and a Type III isotherm. Rearrangement of Eq. (2.5) gives the "linear transformed BET equation" (Eq. (2.6)), which provides the basis for the analysis of experimental isotherm data in the form of the "$P/n(P_0-P)$ versus P/P_0 BET plot":

$$\frac{P}{n\,(P_0 - P)} = \frac{1}{n_m c} + \frac{c-1}{n_m c} \times \frac{P}{P_0} \tag{2.6}$$

As discussed above there is more than one model for the determination of the surface area from experimental isotherms. The most commonly used model is the BET model, originally developed to describe the multilayer adsorption on free flat surfaces, which is not the case for MOFs whose large inner surfaces are far from flat. Therefore, it was long argued whether it was valid to use this method for the determination of the surface areas of MOFs and only recently theoretical calculations made it clear that the BET model yields truly meaningful data [20]. Similar discussions were launched regarding the nonlocal density functional theory (NLDFT) models used to calculate pore sizes and their distribution. These models were also developed for materials that are substantially different from MOFs; however, the fact that good agreement between geometrically determined pore sizes and those determined from isotherms using NLDFT methods is often observed has helped to overcome these concerns. In conclusion, the choice of a suitable model to evaluate experimental gas adsorption data is crucial to obtaining meaningful values of all parameters calculated therefrom.

2.3.5 Gravimetric Versus Volumetric Uptake

A quantitative description of gas sorption is given by an adsorption isotherm: the amount of gas adsorbed by a material at a constant temperature as a function of pressure. This is expressed in either volumetric ($cm^3\,cm^{-3}$, $g\,cm^{-3}$) or gravimetric ($g\,g^{-1}$, $cm^3\,g^{-1}$, or wt%) uptake, the former being the more commonly used in the literature. The adsorption behavior is strongly dependent on the size and shape of the pores, and therefore IUPAC conventions for classifying pore sizes and gas sorption isotherms that reflect this relationship have been proposed [14]. Having introduced the terminology and the principles underlying the adsorption of gases by porous solids, we will elucidate methods to deliberately maximize the pore size and surface area of MOFs by design.

2.4 Porosity in Metal-Organic Frameworks

2.4.1 Deliberate Design of Pore Metrics

Section 2.3 illustrated that the pore shape and especially the pore size have a direct impact on the sorption properties. It has been an outstanding challenge in the synthesis of solid-state materials to alter the metrics and functionality of a given structure without changing its underlying topology. This works in MOF chemistry, where elongation or functionalization of the linker without modifying its general shape and connectivity results in the formation of isoreticular frameworks: frameworks with different pore size or functionality but the same underlying topology as the parent structure. Isoreticular structures are achieved because of *a priori* knowledge of the synthetic conditions under which the targeted secondary building unit (SBU) is formed. The successful implementation of this strategy was first demonstrated with the report of an isoreticular series of structures based on MOF-5. Here, a variety of ditopic carboxylate linkers is reticulated with Zn^{2+} ions under conditions closely related or identical to those used in the synthesis of the parent MOF-5, also termed IRMOF-1 [21]. This yields frameworks of the same general structure but with either added substituents or altered pore sizes (Figure 2.8). The expansion of linkers in the primitive cubic system may however result in interpenetrated frameworks, and in fact, for the isoreticular series of MOF-5 only the functionalized H_2BDC and 2,6-H_2NDC linkers form non-interpenetrated frameworks, whereas structures based on further expanded linkers (H_2BPDC, H_2HPDC, H_2PDC, and H_2TPDC; IRMOF-9, IRMOF-11, IRMOF-13, and IRMOF-15, respectively) form twofold interpenetrated structures [21, 22]. The non-interpenetrated forms of members of the isoreticular metal-organic framework (IRMOF) series with pore diameters of up to 28.8 Å (IRMOF-16) are prepared from more dilute solutions of starting materials. Another way to avoid interpenetration when aiming for larger pore sizes is to target structures whose topologies do not allow it, as we will discuss below.

Interpenetration of frameworks is experimentally observed for certain topologies but not for others. The potential of a structure to exhibit the phenomenon of interpenetration is assessed by considering the type of net and the open space encompassed by a structure. To describe the open space within a structure we use an approach based on nets as tilings. Tilings are polyhedra that fill the internal open space within a net (see Chapter 18) [23]. By inserting new vertices in the centers of each tile and connecting them with new edges passing through the faces of the tiles, we create a new net within the first net, and we call this the dual net. A net is termed self-dual when the interpenetrating nets are identical and a hetero-dual when they are of different topology (Figure 2.9).

It is important to note, that interpenetrating frameworks are only linked by mechanical not chemical bonds, they are entangled. Consequently, in the first step in the design of MOFs with exceptionally large pores, it is helpful to select hetero-dual nets as the product of the reticular synthesis, as these are less likely to form interpenetrated structures.[2]

For the design of MOFs with large pore sizes simple geometric consideration can be used to derive the ideal pore symmetry. Equation (2.7) describes the

---

2 Information on whether interpenetrated nets of a certain topology exist or not can be found in the RCSR database (http://rcsr.anu.edu.au/).

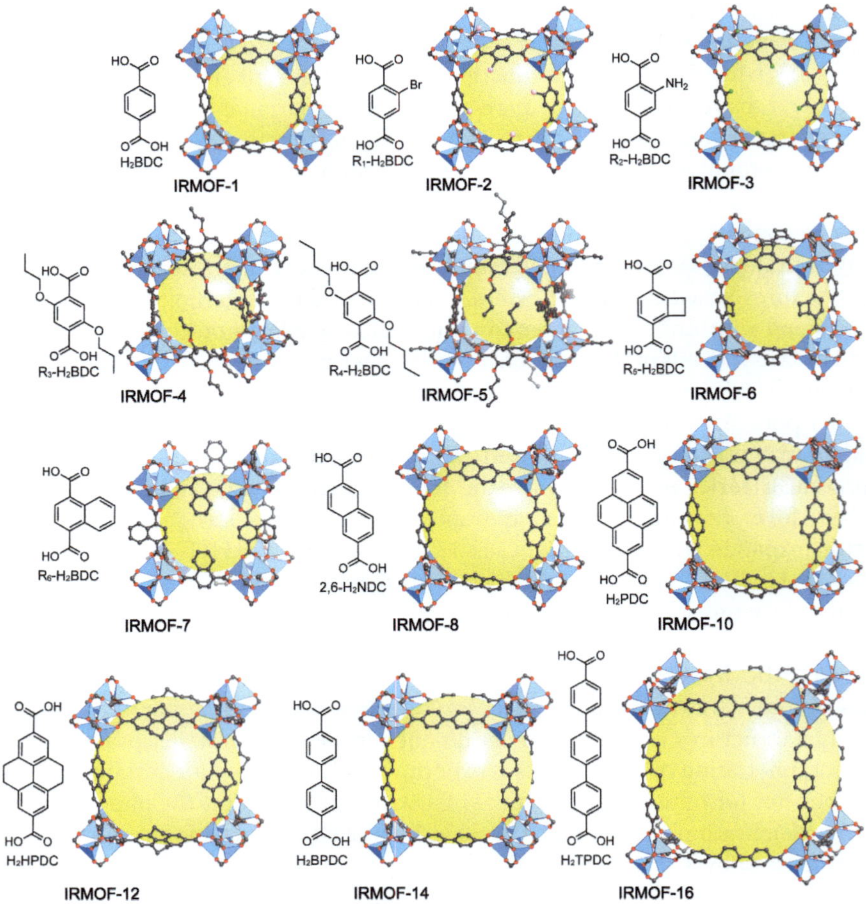

Figure 2.8 Isoreticular metal-organic framework (IRMOF) series based on MOF-5 (also termed IRMOF-1). H_2BDC derivatives decorated with various substituents yield isoreticular functionalized frameworks, and employing expanded linkers gives isoreticular expanded frameworks, both of **pcu** topology. The linkers used in the synthesis of these frameworks are shown next to the respective IRMOF structure. Only the non-interpenetrated structures are shown and all hydrogen atoms are omitted for clarity. Color code: Zn, blue; O, red; C, gray.

relation between the radius of the inscribed circle of a polygon (r) and the length of its edge:

$$r = \frac{a}{2} \cot\left(\frac{180^\circ}{n}\right) \tag{2.7}$$

where a is the length of the edge (link) and n is the number of edges. Accordingly, for a given length a of the linker the inscribed circle is larger the more corners the polygon has (Figure 2.10). By selecting a MOF with hexagonal channels that crystallizes in a net that is not prone to interpenetration, it should therefore be possible to achieve very large pore diameters while using comparatively short linker molecules.

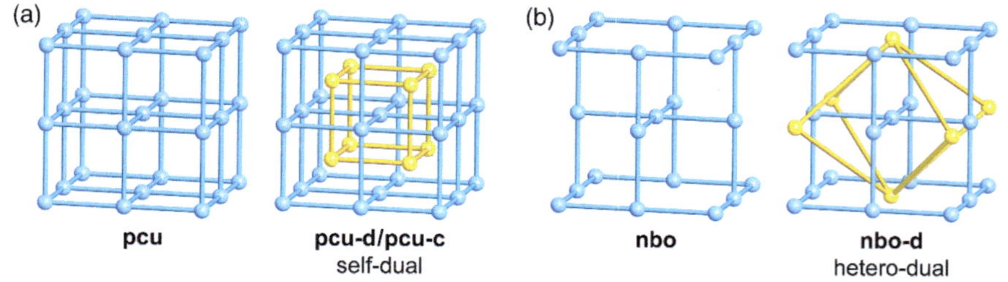

Figure 2.9 Dual nets of the **pcu** (a) and **nbo** (b) net. (a) The **pcu** net forms a "self-dual" net, meaning the dual net (yellow) has the same topology as the original one (blue). (b) In contrast, the dual (yellow) of the **nbo** net is not of the same topology as the original one (blue) and is hence termed "hetero-dual." From this consideration, it is clear that a given set of building units with well-defined geometry are less likely to form interpenetrated structures in topologies that only allow for the formation of hetero-dual nets. The **nbo** net can however interpenetrate and form a self-dual net, if the origin of the second framework is not positioned at the center of the tile of the original net. Thus, a careful analysis of the net topology is required to determine whether interpenetration is likely to occur, or not.

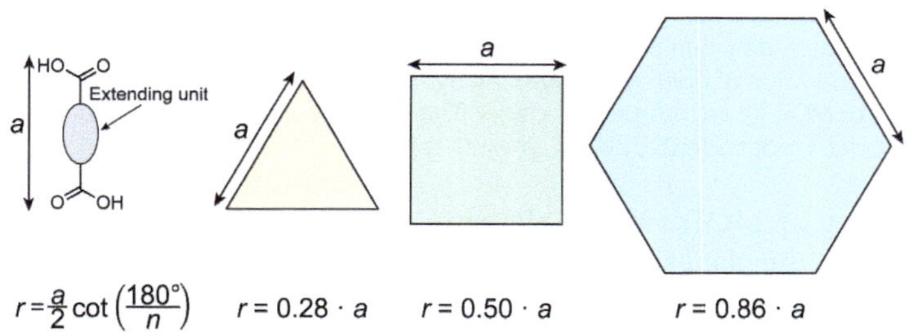

Figure 2.10 Geometric consideration of the relation between the number of edges (n) of an n-sided polygon, the length a of the edges, and the radius r of the inscribed circle. The radius of the inscribe circle of polygons with n sides of a given length a becomes larger the more edges the polygon has. In the context of reticular chemistry this means that the same linker can afford the formation of various pore sizes depending on the geometry of the pore.

Toward this end efforts were made to expand the pore size of many MOFs into the mesoporous regime. The isoreticular expansion of MOF-74, which is a MOF built from 1D rod SBUs connected by linear H_4DOT (DOT = 2,5-dioxidoterephthalate) linkers forming a structure with 1D hexagonal channels is promising, especially since its **etb** topology is unlikely to interpenetrate because of its hetero-dual nature (Figure 2.11).

The preparation and characterization of an isoreticular MOF-74 series constructed from linkers ranging from 7 to 50 Å and having pore diameters ranging from 14×10 to 85×98 Å is illustrated in Figure 2.12a [24]. To achieve pore sizes of this magnitude many challenges have to be overcome: (i) the design of the organic linkers, (ii) their synthesis in pure form, and (iii) ultimately the crystallization of the resulting isoreticular expanded MOFs. Long chains of p-linked aryl units, however, are generally poorly soluble due to strong π–π stacking interactions. The solubility of the starting materials is a necessary requirement

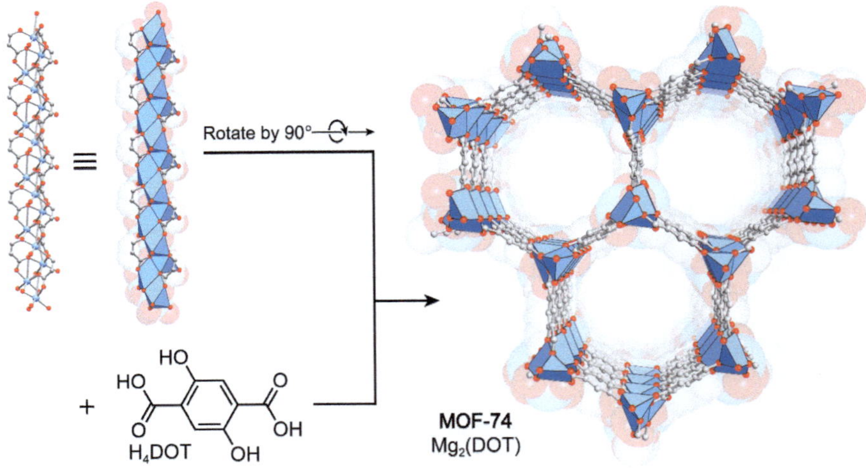

Figure 2.11 Crystal structure of MOF-74. Linking 1D rod SBUs by linear DOT linkers affords the formation of a framework of **etb** topology. The structure resembles that of a honeycomb with 1D channels running along the crystallographic *c*-axis. The coordination sphere of the metal centers in the SBUs is completed by one neutral water ligand per metal center that can be removed to afford the generation of open metal sites. All hydrogen atoms and terminal water ligands in the structure of MOF-74 are omitted for clarity. The structure is superimposed onto its own space filling model. Color code: Zn, blue; C, gray; O, red.

for the crystallization of MOFs, hence aliphatic side chains engendering a torsional twist between the planes of the rings are introduced, thereby reducing π–π stacking interactions. The series of palindromic oligophenylene derivatives terminated with α-hydroxy-carboxylic acid functionalities based on the H_4DOT linker of MOF-74 used to prepare members of the IRMOF-74 series are given in Figure 2.12b.

In structures with 3D spherical pores, the cross-section of the cages confining these pores often consists of polygons with a large number of edges (n), which allows for the formation of large 3D pores from relatively short linkers. MIL-100 is an ideal starting point for the isoreticular expansion of a MOF with 3D pores following the principles outlined earlier. The structure of this MOF comprises large cages with pore diameters of 25 and 29 Å, built from supertetrahedral tertiary building units (TBUs), that are joined to form an extended structure based on the **mtn** topology, a hetero-dual net [25]. Replacing the 1,3,5-benzenetricarboxylate (BTC) linker in MIL-100 by an expanded analog (i.e. BTB) results in the formation of MIL-100(Fe_BTB). Theoretical pore sizes of 55 and 68 Å were calculated from crystal structure data since the nonplanar BTB linker induces stress into the framework rendering the activation of this difficult [26]. Thus, two completely planar linkers, BTTC and TATB, are used to prepare another two isoreticular expanded analogs of MIL-100, termed PCN-332 and PCN-333, respectively (Figure 2.13) [27]. These materials are architecturally stable and their pore sizes were determined experimentally after activation. In these MOFs, the cages are expanded from 25 and 29 Å in MIL-100, to 34 and 45 Å in PCN-332 and 55 and 68 Å in PCN-333. PCN-333 has the largest

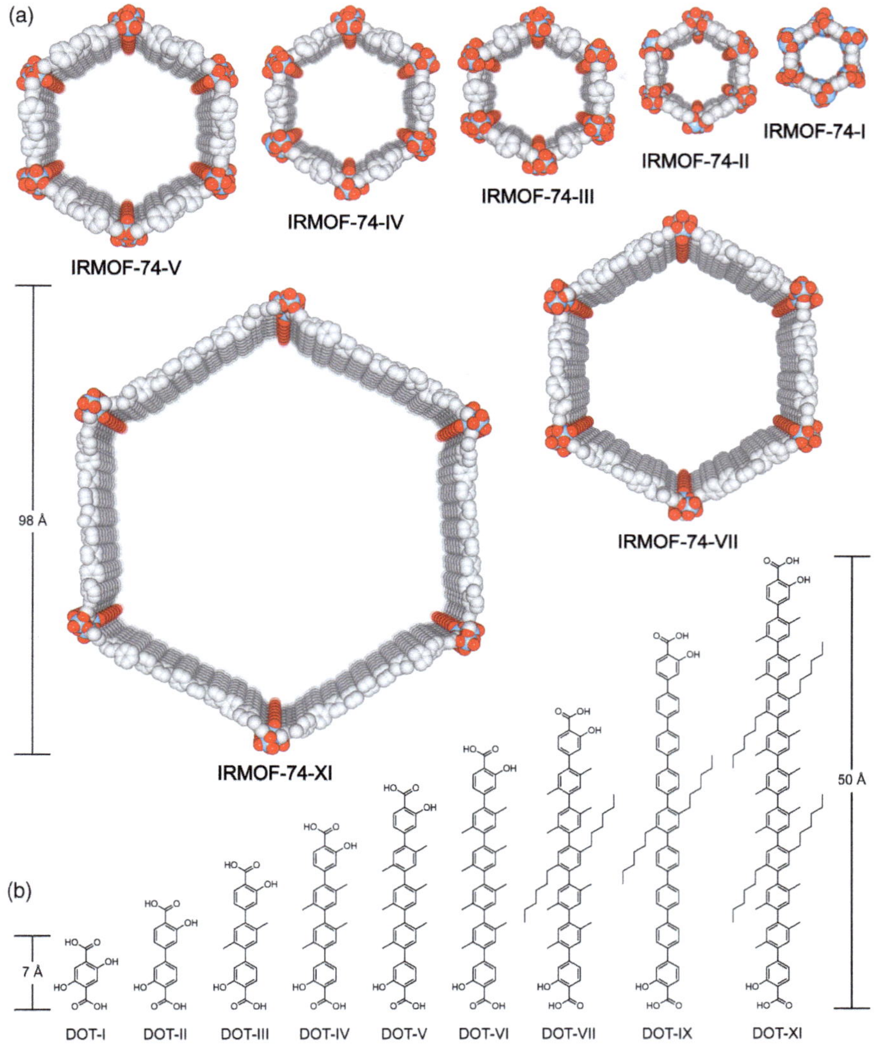

Figure 2.12 (a) Isoreticular series of MOFs with **etb** topology. By successive elongation of the palindromic DOT linker pore sizes of up to 98 Å are achieved. The space filling model of selected members of the IRMOF-74 is shown. (b) Palindromic linkers with terminal hydroxybenzoic acid groups used to prepare members of the IRMOF-74 series. The aliphatic side chains result in a torsional twist of the linkers, rendering them more soluble due to reduced π–π stacking interactions. In the crystal structure drawings all alkyl side chains and hydrogen atoms are omitted for clarity. Color code: Zn, blue; C, gray; O, red.

accessible cages in permanently porous carboxylate-based MOFs reported to date, making it an interesting candidate for trapping large molecules such as enzymes. This example highlights the power of the isoreticular principle for tailoring pore metrics [27]. It is important to mention that the presence of large pores does not necessarily translate into high surface areas as discussed in Section 2.4.2.

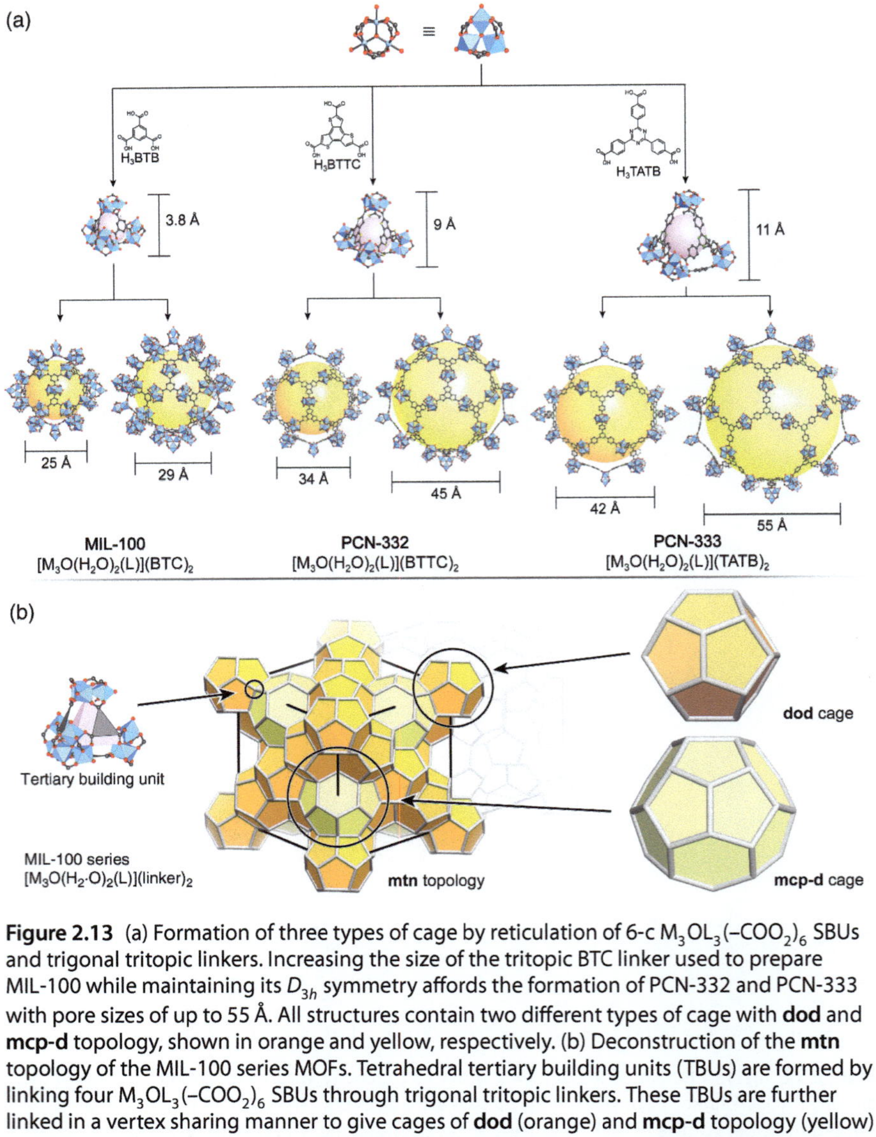

Figure 2.13 (a) Formation of three types of cage by reticulation of 6-c $M_3OL_3(-COO_2)_6$ SBUs and trigonal tritopic linkers. Increasing the size of the tritopic BTC linker used to prepare MIL-100 while maintaining its D_{3h} symmetry affords the formation of PCN-332 and PCN-333 with pore sizes of up to 55 Å. All structures contain two different types of cage with **dod** and **mcp-d** topology, shown in orange and yellow, respectively. (b) Deconstruction of the **mtn** topology of the MIL-100 series MOFs. Tetrahedral tertiary building units (TBUs) are formed by linking four $M_3OL_3(-COO_2)_6$ SBUs through trigonal tritopic linkers. These TBUs are further linked in a vertex sharing manner to give cages of **dod** (orange) and **mcp-d** topology (yellow) that eventually are fused into an extended structure of **mtn** topology. All hydrogen atoms are omitted for clarity. Color code: Cr, blue; C, gray; N, green; O, red.

2.4.2 Ultrahigh Surface Area

Even though PCN-333 has the largest cages in carboxylate-based MOFs reported to date, it has a comparatively low surface area of 4000 m^2 g^{-1}. For comparison, MOF-5 has pores with a diameter of only 15.1 Å but a surface area of 3800 m^2 g^{-1}. This illustrates that there is no direct correlation between the pore size and the surface area. As discussed earlier, the gas uptake in the low-pressure region, which is more relevant for practical applications, is mostly determined by the accessible surface area and not the pore size. Therefore, with respect to the gas adsorption-based applications, the development of materials with ultrahigh surface areas is of interest (see Chapters 14–17). In the following text we illustrate parameters that determine the surface area with respect to gas adsorption, and outline principles that specifically help design MOFs with ultrahigh surface areas.

The first step in understanding how to achieve high surface areas is to identify the adsorption sites in MOFs and to maximize their numbers (Figure 2.14). In general, the major adsorption sites are found in the vicinity of the SBU. They are preferred due to the polar nature of the SBU and the resulting strong interactions with the adsorbate. Additional adsorption sites are located close to the organic linker, more precisely on the faces and the edges of the aryl units, highlighting the role of the linker in the design of materials with large accessible surface areas. The adsorption sites for argon and nitrogen in MOF-5 shown in Figure 2.14 were determined by X-ray diffraction at 30 K and later corroborated by inelastic neutron diffraction, Raman measurements, and theoretical studies [28].

The binding of adsorbed gas molecules to the inner surface of MOFs is stronger than in carbon-based materials. Experiments and calculations suggest that in MOFs, the strength of interactions with the framework is further enhanced by the use of more polarizing centers or by the presence of open metal sites [28c]. The strength of an interaction of a gas molecule adsorbed on a surface is approximated

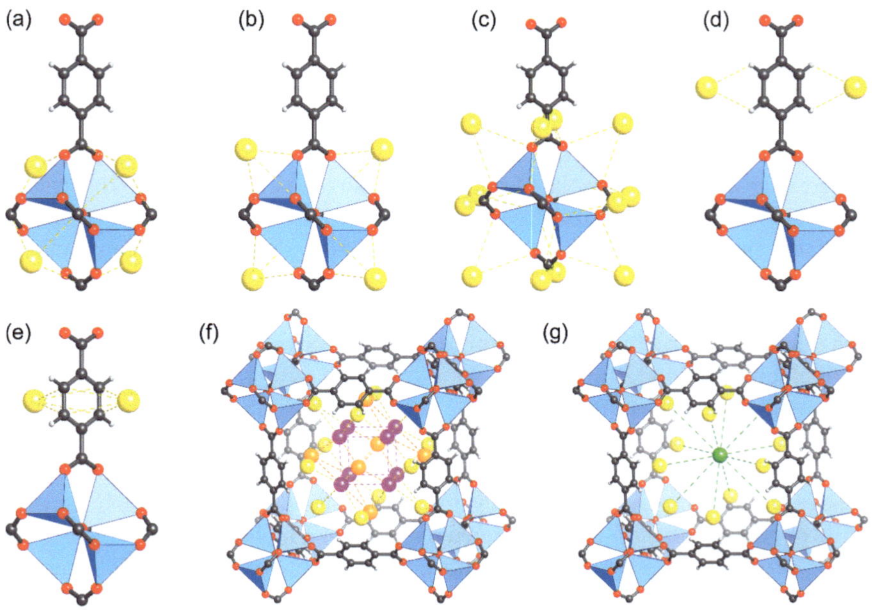

Figure 2.14 Adsorption sites of argon in MOF-5 determined by X-ray diffraction at 30 K, eight symmetry-independent sites were crystallographically identified. (a–c) Gas molecules are preferentially adsorbed close to the polar SBU. (d, e) Additional adsorption sites are located on the edges and faces of the organic linker. (f) A second layer is formed in the large pore of MOF-5 (pink and orange spheres), and (g) one additional adsorption site is found in the center of the small pore (green sphere). Color code: Zn, blue; C, gray; O, red; adsorbed Ar, yellow/orange/pink/green.

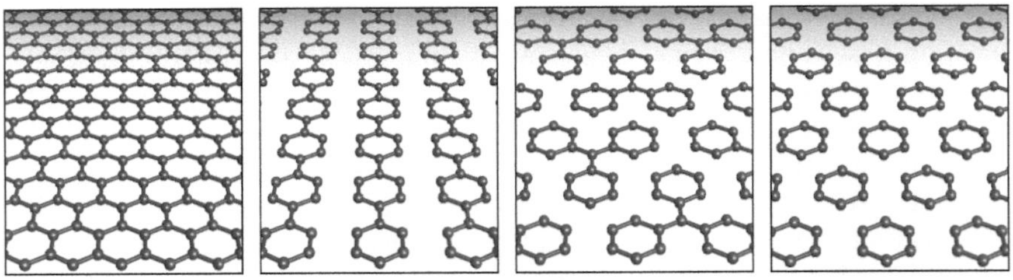

Figure 2.15 A strategy to maximize the number of adsorption sites and surface area of MOFs by designing linkers with a maximized number of exposed faces and edges. Trigonal 1,3,5-triphenyl benzene-based linkers provide a high number of exposed adsorption sites while still allowing for both, isoreticular expansion and functionalization.

by measuring its rotational barrier near that surface using inelastic neutron scattering. Such measurements were carried out on MOF-5 as a function of hydrogen pressure, revealing that hydrogen is bound strongly to its SBU, whereas the interaction with additional adsorption sites on the faces and edges of the aryl units of the linkers is comparatively weak. Nevertheless, these adsorption sites still play a key role in enhancing the gas uptake in MOFs because their numbers can be deliberately maximized by optimizing the shape and metrics of the linker.

This is illustrated by the following theoretical exercise (Figure 2.15). Starting from a simple graphene sheet the number of exposed edges is increased by forming smaller polymeric or molecular subunits. A single infinite graphene sheet has a theoretical surface area of $2965\,m^2\,g^{-1}$, which is increased to $5683\,m^2\,g^{-1}$ for the case of infinite chains of poly-*p*-linked benzene, as more edges are exposed. Dividing the graphene sheet into 1,3,5-triphenylbenzene units increases the theoretical surface area even further ($6200\,m^2\,g^{-1}$) and maximizing the number of exposed edges by forming isolated benzene rings results in a theoretical surface area of $7745\,m^2\,g^{-1}$. Since in the case of isolated benzene rings attaching binding groups significantly lowers the number of adsorption sites, it seems most promising to employ trigonal 1,3,5-triphenylbenzene based linkers when targeting ultrahigh surface areas, as they provide the maximum number of faces and edges while still allowing for the synthesis of framework structures and their isoreticular expansion [29].

This strategy was implemented by reticulating 1,3,5-benzenetribenzoate (H_3BTB) and basic zinc carboxylate clusters ($Zn_4O(COO)_6$) to form MOF-177, a MOF of **qom**[3] topology. In contrast to other net topologies constructed from nodes of a similar general geometry (octahedron and triangle), such as the default **pyr** net (self-dual net), frameworks of **qom** topology (hetero-dual net) are not prone to interpenetration. Thus, the structure of MOF-177 is non-interpenetrated and its BET surface area is $4500\,m^2\,g^{-1}$ despite the relatively small pores (~12 Å) [29].

Calculations suggest that linkers containing cumulene (=C=C=) and polyyne (—C≡C—) units have even higher specific surface areas (up to $14\,937\,m^2\,g^{-1}$, presumably the highest possible specific surface area) than those built only from

3 **qom** = Queen of MOF. This name was chosen since it was thought that this topology will allow for the highest surface area possible.

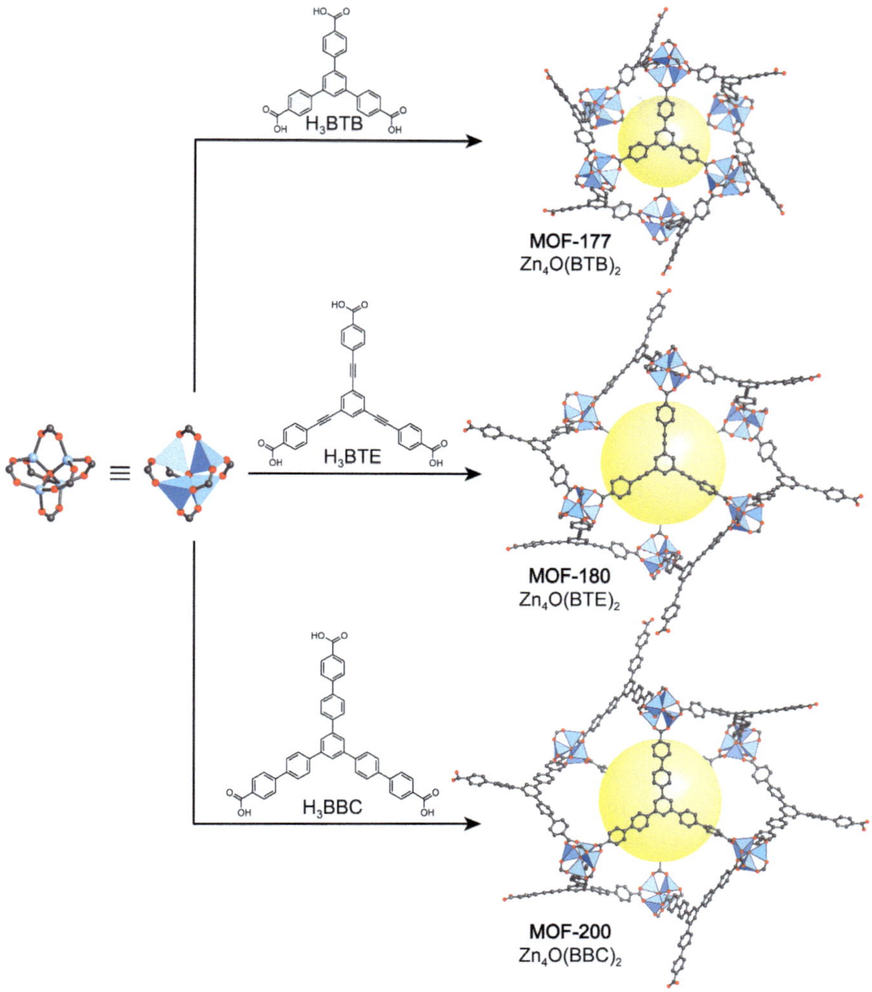

Figure 2.16 Isoreticular series of MOFs with **qom** topology formed by reticulation of trigonal tritopic linkers (H_3BTB, H_3BTE, and H_3BBC) and octahedral $Zn_4O(-COO)_6$ SBUs. (a) MOF-177, (b) MOF-180, and (c) MOF-200 have geometric surface areas of 4740, 6080, and 6400 $m^2\ g^{-1}$, respectively. The **qom** net is a hetero-dual net, hence no interpenetration is observed when expanded linkers are used. All hydrogen atoms are omitted for clarity. Color code: Zn, blue; C, gray; O, red.

aryl units. Therefore, elongated linkers derived from BTB with additional alkynyl (BTE) or aryl units (BBC) were designed to synthesize an isoreticular series of MOFs based on MOF-177 (Figure 2.16). The resulting frameworks, MOF-180 and MOF-200, have calculated porosities of 89% and 90%, respectively. The highly porous structures of these MOFs made it difficult or, as in the case of MOF-180, even impossible to activate them, which is the reason why the reported BET surface area for MOF-200 (4530 $m^2\ g^{-1}$) is far lower than what is expected from geometrical calculations (6400 $m^2\ g^{-1}$) [30].

These principles of linker design allow to conceive MOFs with even higher surface areas. We will illustrate this for an isoreticular series of MOFs with **ntt** topology (equivalent to **rht** topology, see Section 5.2.3). The parent MOF to this series is PMOF-1 ($Cu_3(H_2O)(TPBTM)$), which has a surface area of 3160 $m^2\ g^{-1}$ [31]. Isoreticular expansion using HTTEI linkers affords $Cu_3(H_2O)(TTEI)$, which is referred to as both, PCN-610 and NU-100 [32].[4] The activation of this MOF is challenging but eventually affords a material with a surface are of 6143 $m^2\ g^{-1}$. To reach even higher surface areas the linker used to prepare NU-100 is expanded even further (BHEHPI) yielding NU-110, which after activation has a surface area of 7140 $m^2\ g^{-1}$ [33]. The isoreticular series of **ntt** MOFs alongside the linkers used to prepare them is given in Figure 2.17. Note that the linker design is based on the principles discussed earlier.

The expansion of the octahedral pore (19 Å) in PMOF-1 by a factor of 39% (26 Å for PCN-610) results in a 94% larger surface area whereas further increase in pore size by another 26% (33 Å in NU-110) only leads to a 16% higher surface area. This highlights that there is no direct correlation between pore size and surface area and that above a certain diameter, large pores partially consist of "dead volume" where the adsorbate does not interact with the surface of the pore. Another disadvantage of large pores when targeting MOFs with good sorption properties is the fact that while the gravimetric gas uptake ($m^2\ g^{-1}$) might be high a low volumetric uptake ($m^2\ cm^{-3}$) is inevitable. The gravimetric uptake increases with the pore size until the theoretical limit is reached. In contrast, the volumetric uptake will only increase until a critical pore size is reached and will decrease upon further expansion.

We illustrate this based on an isoreticular expansion of a hypothetical MOF of **pcu** topology (Figure 2.18). In structures of this topology, the geometric surface area can be estimated as a sphere where all parts of the spherical surface that are not covered by the framework structure are subtracted, as they do not contribute to the surface area. Upon expansion, these areas occupy a continuously growing part of the spherical surface, hence the effective surface area does not grow at a constant rate. The infinitely expanded structure can be approximated to consist only of chains of poly-*p*-linked benzene rings and its molecular mass can therefore be approximated by that of poly(*p*-phenylene). From this point on no further gain in surface area per weight will be observed according to this model, and the theoretical limit of 6200 $m^2\ g^{-1}$ is reached (green line in Figure 2.18) [34]. This is different when analyzing the surface area per volume, which will reach a maximum before the decrease in density outweighs the increase in surface area resulting in a decrease of the surface area per volume upon further expansion of the framework (blue line in Figure 2.18).

This behavior is also important in considering practical applications and it must be pointed out that all effects mentioned above are strongly dependent on the nature and size of the adsorbent in question. Therefore, the structures of MOFs have to be carefully designed, examined, and modified in order to create

4 This MOF was reported by two groups simultaneously and therefore has two different names.

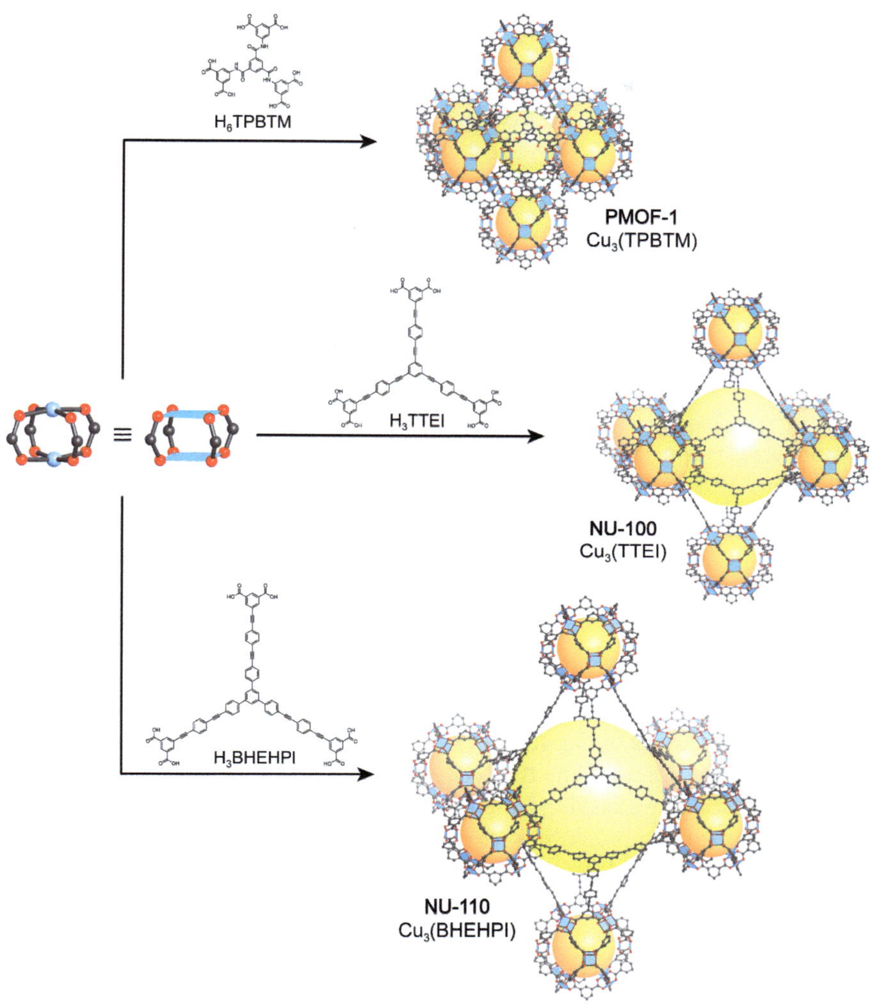

Figure 2.17 Isoreticular series of MOFs with **ntt** topology formed by reticulation of trigonal hexatopic linkers (H_6TPBTM, H_6TTEI, and $H_6BHEHPI$) and square 4-c $Cu_2(-COO)_4$ paddle wheel SBUs. (a) PMOF-1, (b) NU-100, and (c) NU-110 have surface area of 3160, 6143, and 7140 $m^2\ g^{-1}$, respectively. The **ntt** net is a hetero-dual net and therefore unlikely to interpenetrate. The terminal *m*-BDC moieties of the hexatopic linker connect the SBUs to form 24-c tertiary building units of **rco** topology. These cages maintain their size upon isoreticular expansion and only the large pore located in the octahedral hole formed by the TBUs increases in size. All hydrogen atoms as well as all terminal water ligands on the paddle wheel SBUs are omitted for clarity. Color code: Zn, blue; C, gray; N, green; O, red.

high-performance materials with outstanding properties. Over the past years the number of reported MOF structures increased exponentially, and efforts were made to simplify and systemize their structures to understand how to use reticular chemistry as a means to construct extended structures by design and this will be the focus of the Chapters 3–6.

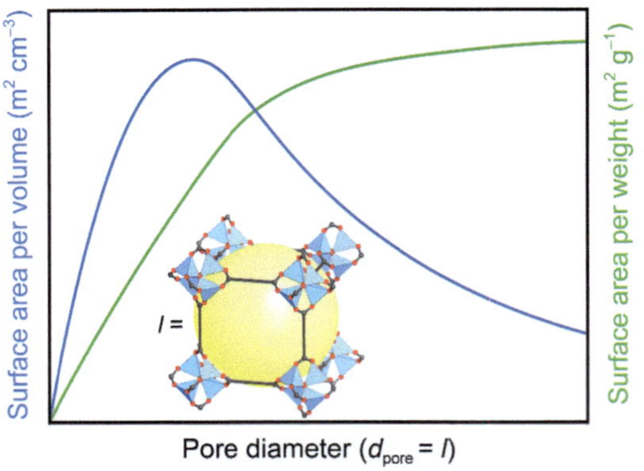

Figure 2.18 Schematic plot of the surface area per volume (blue) and per weight (green) versus the pore diameter. The change in surface area is based on a simplified cubic model system with a linker of variable length *l*. For very long linkers, the molecular mass of the compound is approximately that of poly(*p*-phenylene) since the contribution of the SBUs is considered to be negligible. The surface area per mass can therefore also be approximated by that of poly(*p*-phenylene). In contrast, the surface per volume passes through a maximum before the decrease in density outweighs the increase in surface area.

2.5 Summary

In this chapter we introduced the terminology used to describe porous materials. We showed the process of gas adsorption, the resulting types of isotherms and hysteresis, and the interactions that can be involved in this process. Based on these considerations we outlined the theory of gas adsorption in porous solids and discussed the two models most frequently used to describe it: the Langmuir and the BET-model. The different assumptions used to derive these models allow to understand their applicability to MOFs and specific types of measurements. With the isoreticular principle we provided a tool for the design MOFs with large pores and predefined topology and examples making use of this principle were given. We showed that there is no direct correlation between the pore size and the surface area and consequently introduced design principles for the targeted synthesis of new MOFs with ultrahigh surface areas alongside illustrative examples. In Chapter 3 we will introduce the building units of MOFs – organic linkers and inorganic SBUs – and general synthetic approaches to their synthesis.

References

1 (a) Cronstedt, A.F. (1756). Rön och beskrifning om en obekant bärg art, som kallas Zeolites. *Svenska Vetenskaps akademiens Handlingar* 17: 120. A translation can be found in Schlenker, J.L. and Kühl, G.H. (1993). *Proc. 9th Intl. Conference on Zeolites*. (b) Claire-Deville, H.d.S. (1862). Reproduction de la levyne. *Comptes Rendus* 54 (1862): 324–327. (c) Friedel, G. (1896). New experiments on zeolites. *Bulletin de la Societe Francaise de Mineralogie* 19: 363–390. (d) Friedel, G. (1896). Sur quelques propriétés nouvelles

des zéolithes. *Bulletin de la Société Française de Minéralogie* 19: 94–118. (e) Barrer, R.M. (1938). The sorption of polar and non-polar gases by zeolites. *Proceedings of the Royal Society of London Series A: Mathematical and Physical Sciences* 167 (A930): 0392–0420. (f) Barrer, R.M. (1949). Transient flow of gases in sorbents providing uniform capillary networks of molecular dimensions. *Transactions of the Faraday Society* 45 (4): 358–373. (g) Barrer, R.M. and Macleod, D.M. (1954). Intercalation and sorption by montmorillonite. *Transactions of the Faraday Society* 50 (9): 980–989.

2 (a) Baerlocher, C., McCusker, L.B., and Olson, D.H. (2007). *Atlas of Zeolite Framework Types*. Elsevier. (b) Corma, A., Díaz-Cabañas, M.J., Jiang, J. et al. (2010). Extra-large pore zeolite (ITQ-40) with the lowest framework density containing double four- and double three-rings. *Proceedings of the National Academy of Sciences of the United States of America* 107 (32): 13997–14002. (c) Kulprathipanja, S. (ed.) (2010). *Zeolites in Industrial Separation and Catalysis*, 1–26. Wiley.

3 Schaeffer, W.D., Dorsey, W.S., Skinner, D.A., and Christian, C.G. (1957). Separation of xylenes, cymenes, methylnaphthalenes and other isomers by clathration with inorganic complexes. *Journal of the American Chemical Society* 79 (22): 5870–5876.

4 Soldatov, D.V., Enright, G.D., and Ripmeester, J.A. (2004). Polymorphism and pseudopolymorphism of the $[Ni(4\text{-methylpyridine})_4(NCS)_2]$ Werner complex, the compound that led to the concept of "organic zeolites". *Crystal Growth & Design* 4 (6): 1185–1194.

5 Allison, S.A. and Barrer, R.M. (1969). Sorption in the β-phases of transition metal(II) tetra-(4-methylpyridine)thiocyanates and related compounds. *Journal of the Chemical Society A: Inorganic, Physical, Theoretical* 1717–1723.

6 Krap, C.P., Balmaseda, J., Zamora, B., and Reguera, E. (2010). Hydrogen storage in the iron series of porous Prussian blue analogues. *International Journal of Hydrogen Energy* 35 (19): 10381–10386.

7 Ramprasad, D., Pez, G.P., Toby, B.H. et al. (1995). Solid state lithium cyanocobaltates with a high capacity for reversible dioxygen binding: synthesis, reactivity, and structures. *Journal of the American Chemical Society* 117 (43): 10694–10701.

8 Kondo, M., Yoshitomi, T., Matsuzaka, H. et al. (1997). Three-dimensional framework with channeling cavities for small molecules: $\{[M_2(4,4'\text{-bpy})_3(NO_3)_4]\cdot xH_2O\}_n$ (M = Co, Ni, Zn). *Angewandte Chemie International Edition in English* 36 (16): 1725–1727.

9 (a) Takamizawa, S., Mori, W., Furihata, M. et al. (1998). Synthesis and gas-occlusion properties of dinuclear molybdenum(II) dicarboxylates (fumarate, terephthalate, trans-trans-muconate, pyridine-2,5-dicarboxylate, and trans-1,4-cyclohexanedicarboxylate). *Inorganica Chimica Acta* 283 (1): 268–274. (b) Mori, W., Hoshino, H., Nishimoto, Y., and Takamizawa, S. (1999). Synthesis and gas occlusion of new micropore substance rhodium(II) carboxylates bridged by pyrazine. *Chemistry Letters* 28 (4): 331–332. (c) Wasuke, M., Fumie, I., Keiko, Y. et al. (1997). Synthesis of new adsorbent copper(II) terephthalate. *Chemistry Letters* 26 (12): 1219–1220.

10 Li, H., Eddaoudi, M., Groy, T.L., and Yaghi, O.M. (1998). Establishing microporosity in open metal-organic frameworks: gas sorption isotherms for Zn(BDC) (BDC = 1,4-benzenedicarboxylate). *Journal of the American Chemical Society* 120 (33): 8571–8572.

11 Li, H., Eddaoudi, M., O'Keeffe, M., and Yaghi, O.M. (1999). Design and synthesis of an exceptionally stable and highly porous metal-organic framework. *Nature* 402 (6759): 276–279.

12 (a) Kayser, H. (1881). Wiederman's. *Annals of Physical Chemistry* 14: 451. (b) Kayser, H. (1881). Über die verdichtung von gasen an oberflächen in ihrer abhängigkeit von druck und temperatur. *Annalen der Physik* 250 (11): 450–468. (c) Everett, D. (1972). Manual of symbols and terminology for physicochemical quantities and units, appendix II: definitions, terminology and symbols in colloid and surface chemistry. *Pure and Applied Chemistry* 31 (4): 577–638.

13 McBain, J.W. (1909). XCIX. The mechanism of the adsorption ("sorption") of hydrogen by carbon. *The London, Edinburgh, and Dublin Philosophical Magazine and Journal of Science* 18 (108): 916–935.

14 (a) Sing, K.S. (1985). Reporting physisorption data for gas/solid systems with special reference to the determination of surface area and porosity (Recommendations 1984). *Pure and Applied Chemistry* 57 (4): 603–619. (b) Wilkinson, A. and McNaught, A. (1997). *IUPAC Compendium of Chemical Terminology, (The "Gold Book")*. International Union of Pure and Applied Chemistry.

15 (a) Brunauer, S., Emmett, P.H., and Teller, E. (1938). Adsorption of gases in multimolecular layers. *Journal of the American Chemical Society* 60 (2): 309–319. (b) Brunauer, S., Deming, L.S., Deming, W.E., and Teller, E. (1940). On a theory of the van der Waals adsorption of gases. *Journal of the American Chemical Society* 62 (7): 1723–1732. (c) Brunauer, S. (1945). *The Adsorption of Gases and Vapours*, vol. 1. Oxford: Oxford University Press.

16 (a) Broekhoff, J. (1979). Mesopore determination from nitrogen sorption isotherms: fundamentals, scope, limitations. *Studies in Surface Science and Catalysis* 3: 663–684. (b) Lowell, S., Shields, J.E., Thomas, M.A., and Thommes, M. (2012). *Characterization of Porous Solids and Powders: Surface Area, Pore Size and Density*, vol. 16. Springer Science & Business Media. (c) Thommes, M., Kaneko, K., Neimark, A.V. et al. (2015). Physisorption of gases, with special reference to the evaluation of surface area and pore size distribution (IUPAC Technical Report). *Pure and Applied Chemistry* 87 (9–10): 1051–1069.

17 Langmuir, I. (1918). The adsorption of gases on plane surfaces of glass, mica and platinum. *Journal of the American Chemical Society* 40 (9): 1361–1403.

18 Langmuir, I. (1916). The evaporation, condensation and reflection of molecules and the mechanism of adsorption. *Physical Review* 8 (2): 149–176.

19 (a) Everett, D.H. (1950). Thermodynamics of adsorption. Part I. General considerations. *Transactions of the Faraday Society* 46: 453–459. (b) Everett, D.H. (1950). The thermodynamics of adsorption. Part II. Thermodynamics of monolayers on solids. *Transactions of the Faraday Society* 46: 942–957.

20 Gomez-Gualdron, D.A., Moghadam, P.Z., Hupp, J.T. et al. (2016). Application of consistency criteria to calculate BET areas of micro- and mesoporous metal-organic frameworks. *Journal of the American Chemical Society* 138 (1): 215–224.

21 Eddaoudi, M., Kim, J., Rosi, N. et al. (2002). Systematic design of pore size and functionality in isoreticular MOFs and their application in methane storage. *Science* 295 (5554): 469–472.

22 Reineke, T.M., Eddaoudi, M., Moler, D. et al. (2000). Large free volume in maximally interpenetrating networks: the role of secondary building units exemplified by $Tb_2(ADB)_3[(CH_3)_2SO]_4 \cdot 16[(CH_3)_2SO]^1$. *Journal of the American Chemical Society* 122 (19): 4843–4844.

23 Delgado-Friedrichs, O. and O'Keeffe, M. (2005). Crystal nets as graphs: terminology and definitions. *Journal of Solid State Chemistry* 178 (8): 2480–2485.

24 Deng, H., Grunder, S., Cordova, K.E. et al. (2012). Large-pore apertures in a series of metal-organic frameworks. *Science* 336 (6084): 1018–1023.

25 Ferey, G., Serre, C., Mellot-Draznieks, C. et al. (2004). A hybrid solid with giant pores prepared by a combination of targeted chemistry, simulation, and powder diffraction. *Angewandte Chemie International Edition* 43 (46): 6296–6301.

26 Horcajada, P., Chevreau, H., Heurtaux, D. et al. (2014). Extended and functionalized porous iron(iii)tri-or dicarboxylates with MIL-100/101 topologies. *Chemical Communications* 50 (52): 6872–6874.

27 Feng, D., Liu, T.-F., Su, J. et al. (2015). Stable metal-organic frameworks containing single-molecule traps for enzyme encapsulation. *Nature Communications* 6: 5979.

28 (a) Rowsell, J.L.C., Spencer, E.C., Eckert, J. et al. (2005). Gas adsorption sites in a large-pore metal-organic framework. *Science* 309 (5739): 1350–1354. (b) Rowsell, J.L., Eckert, J., and Yaghi, O.M. (2005). Characterization of H_2 binding sites in prototypical metal-organic frameworks by inelastic neutron scattering. *Journal of the American Chemical Society* 127 (42): 14904–14910. (c) Centrone, A., Siberio-Perez, D.Y., Millward, A.R. et al. (2005). Raman spectra of hydrogen and deuterium adsorbed on a metal-organic framework. *Chemical Physics Letters* 411 (4–6): 516–519. (d) Dubbeldam, D., Frost, H., Walton, K.S., and Snurr, R.Q. (2007). Molecular simulation of adsorption sites of light gases in the metal-organic framework IRMOF-1. *Fluid Phase Equilibria* 261 (1): 152–161. (e) Greathouse, J.A., Kinnibrugh, T.L., and Allendorf, M.D. (2009). Adsorption and separation of noble gases by IRMOF-1: grand canonical Monte Carlo simulations. *Industrial and Engineering Chemistry Research* 48 (7): 3425–3431.

29 Chae, H.K., Siberio-Pérez, D.Y., Kim, J. et al. (2004). A route to high surface area, porosity and inclusion of large molecules in crystals. *Nature* 427 (6974): 523–527.

30 Furukawa, H., Ko, N., Go, Y.B. et al. (2010). Ultrahigh porosity in metal-organic frameworks. *Science* 329 (5990): 424–428.

31 Zheng, B., Bai, J., Duan, J. et al. (2010). Enhanced CO_2 binding affinity of a high-uptake rht-type metal-organic framework decorated with acylamide groups. *Journal of the American Chemical Society* 133 (4): 748–751.

32 (a) Yuan, D., Zhao, D., Sun, D., and Zhou, H.C. (2010). An isoreticular series of metal-organic frameworks with dendritic hexacarboxylate ligands and exceptionally high gas-uptake capacity. *Angewandte Chemie International Edition* 49 (31): 5357–5361. (b) Farha, O.K., Yazaydin, A.O., Eryazici, I. et al. (2010). De novo synthesis of a metal-organic framework material featuring ultrahigh surface area and gas storage capacities. *Nature Chemistry* 2 (11): 944–948.

33 Farha, O.K., Eryazici, I., Jeong, N.C. et al. (2012). Metal-organic framework materials with ultrahigh surface areas: is the sky the limit? *Journal of the American Chemical Society* 134 (36): 15016–15021.

34 Schnobrich, J.K., Koh, K., Sura, K.N., and Matzger, A.J. (2010). A framework for predicting surface areas in microporous coordination polymers. *Langmuir* 26 (8): 5808–5814.

3

Building Units of MOFs

3.1 Introduction

In the last decade the number of reported metal-organic framework (MOF) structures has increased exponentially, and the principle of topology used to simplify and systematize them based on nets constructed of linked vertices and edges, has already been introduced [1]. To determine the underlying topology of a given framework, its structure has to be deconstructed into individual building units (see chapter 18). The building units of MOFs are the inorganic and organic components, commonly referred to as secondary building units (SBUs) and linkers, respectively. Linkers bearing binding groups such as carboxylates, phosphates, pyrazolates, tetrazolates, catecholates, and imidazolates (see Chapter 21) have been used to stitch these building units together into extended framework structures. MOFs linked by carboxylate functionalities comprise by far the largest group, and thus these will be the focus of this section. Owing to their chelating nature, carboxylate groups favor the formation of polynuclear metal carboxylate SBUs over single metal nodes. These SBUs provide directionality, and, in combination with the strong bonding between the constituents, endow the resulting MOFs with high mechanical, architectural, and chemical stability.[1] The exact geometrical structure of the individual building units defines the outcome of reticular synthesis and thus enables the *a priori* design of frameworks with targeted structures.[2] In this chapter we will provide an overview of organic linkers and inorganic SBUs frequently encountered in MOF chemistry, and classify them based on their geometry and number of points of extension.

3.2 Organic Linkers

Up to the mid-1990s neutral donor linkers such as bipyridines and nitriles were used to prepare coordination networks. As highlighted in Chapter 1, such materials suffer from many drawbacks and thus the linkers used to prepare

1 In carboxylate-based MOFs, such as MOF-5, the Zn—O bond has a formal valence of ½. The bond energy of the two Zn—O bonds is expected to be close to that of a typical C—C bond [2].
2 We will focus on SBUs containing transition and main group metals only and will not discuss rare earth SBUs since their formation is less studied.

Introduction to Reticular Chemistry: Metal-Organic Frameworks and Covalent Organic Frameworks,
First Edition. Omar M. Yaghi, Markus J. Kalmutzki, and Christian S. Diercks.

Companion website: www.wiley.com/go/yaghi/reticular

them were later replaced by charged chelating linkers with binding groups such as carboxylates to make more robust frameworks. Carboxylate-based linkers provide four advantages over neutral donor linkers: (i) the charged carboxylates neutralize the positive charges of the metal nodes, thus allowing for the formation of neutral frameworks and obviating the need for counter ions; (ii) their chelating ability provides more structural rigidity and directionality; (iii) they favor the formation of polynuclear clusters (SBUs) with a fixed overall coordination geometry and connectivity; (iv) the strong bonding between the linkers and the metal centers of the SBUs results in MOFs with high thermal, mechanical, and chemical stability. As illustrated in Figure 3.1, progress in making carboxylate-based MOFs advanced reticular chemistry into the strong bond regime and helped to overcome the crystallization challenge that is commensurate with stronger interactions.

Most linkers employed in the synthesis of MOFs are of relatively high symmetry and are commonly built from rigid unsaturated hydrocarbon fragments. These attributes help to provide chemically and mechanically stable frameworks. Organic linkers typically have rigid backbones that are synthesized by coupling reactions such as carbon–carbon, carbon–heteroatom, or heteroatom–heteroatom couplings. Cross-coupling reactions are used to join aryl units either directly (Suzuki coupling) by alkynyle units (Sonogashira coupling), or by alkenyl units (Heck reaction). The synthetic procedures for these reactions are well established and allow for the facile synthesis of linkers in

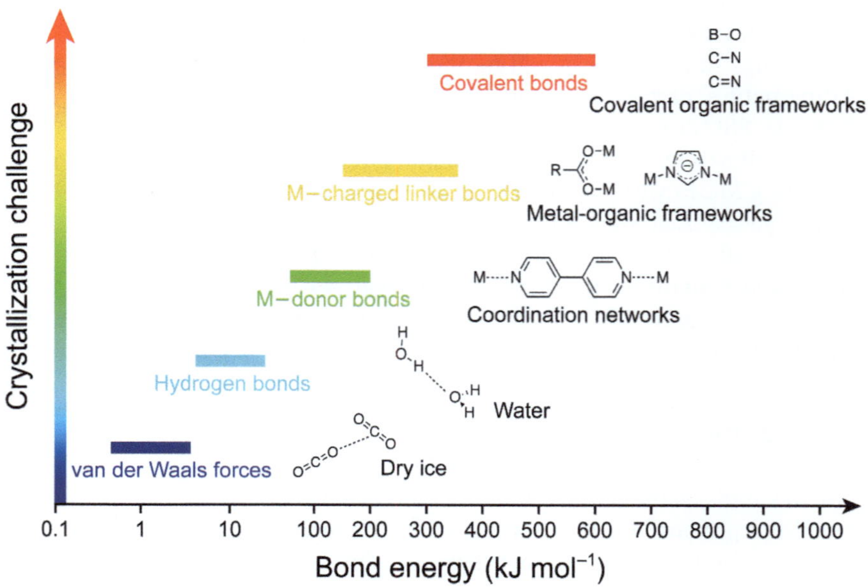

Figure 3.1 Comparison of the strength of different types of interactions ranging from weak (blue and cyan, van der Waals forces, and hydrogen bonding), to intermediate (green, metal–donor bonds), to strong (yellow and red, metal-charged linker and covalent bonds, respectively). The challenge in crystallizing extended structures based on these interactions increases with increasing bond strength.

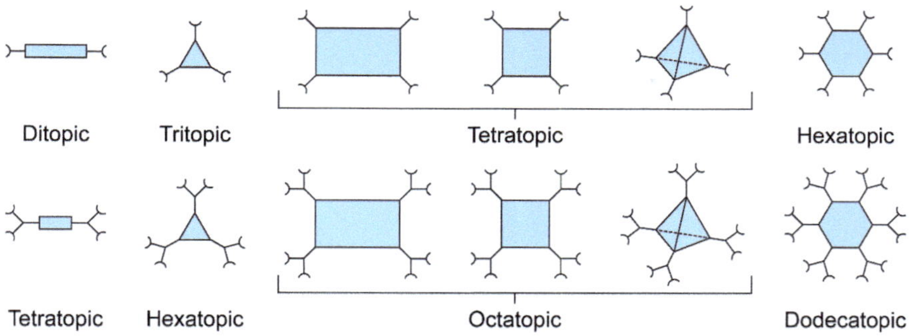

Figure 3.2 Basic geometries of organic linkers used in the synthesis of MOFs typically ranging from 2 to 12 points of extension. The term topicity is also used to describe the number of points of extension. Lower symmetry forms of these geometries are possible but not shown.

high purity and yields. Further organic transformations can be used to introduce additional functionality onto such linkers as outlined in Chapter 6.

In the context of the isoreticular expansion of MOFs we highlighted the influence of the linker length on the metrics of the framework. We also saw that differently shaped linkers afford the formation of frameworks of different topologies. The number of accessible linker shapes is large; however, since organic chemistry is based on carbon, angular constraints are introduced leading to certain shapes being easier to access. The organic linkers connecting the SBUs typically have two, three, four, six, eight, or twelve points of extension and the terminologies ditopic, tritopic, tetratopic, and so on are used to describe them (Figure 3.2).

As we will see in the following section, there is a variety of possibilities to synthesize a linker of a given geometrical shape and specific number of points of extension from different combinations of organic fragments.

3.2.1 Synthetic Methods for Linker Design

The retrosynthetic analysis of a given molecule allows to dissect it into the corresponding synthetic precursors. In general, organic linkers are built from three fragments: (i) the "core unit" of the linker that often defines the geometry of the backbone, (ii) the "binding groups" that connect the linker molecules to the SBUs, and (iii) the "extending units" that define the size of the linker and thereby the metrics of MOFs built therefrom (Figure 3.3).

The binding groups, such as carboxylic acids, are commonly introduced at the very end of the linker synthesis, in most cases along with an extending unit. Therefore, the linker is synthesized by coupling the synthetic precursors of the core and the extending units by a selection of chemical transformations and subsequent coupling of the resulting fragment to the synthetic precursors of the binding groups. The core unit often codes the geometrical information of the linker (e.g. tribromobenzene units impart threefold symmetry), which can be retained or altered by judicious choice of the extending units. Accordingly, the extending units are grouped into three classes: (i) linear units that allow for the extension of the linker without changing its overall geometry and

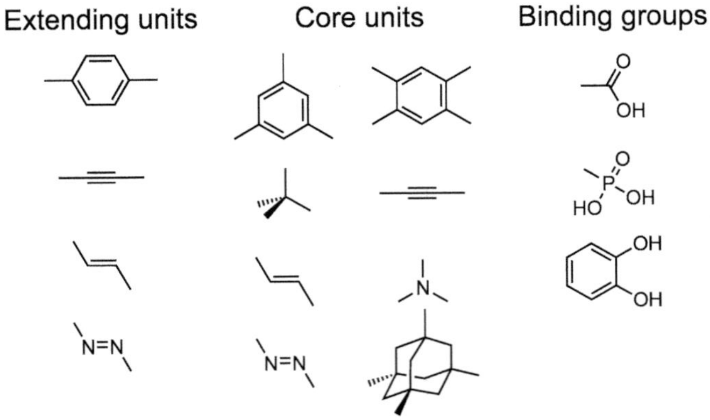

Figure 3.3 Organic units used in the construction of linkers. Note that most units are unsaturated hydrocarbons and/or possess rigid geometries allowing for the controlled construction of geometrically well-defined organic building units.

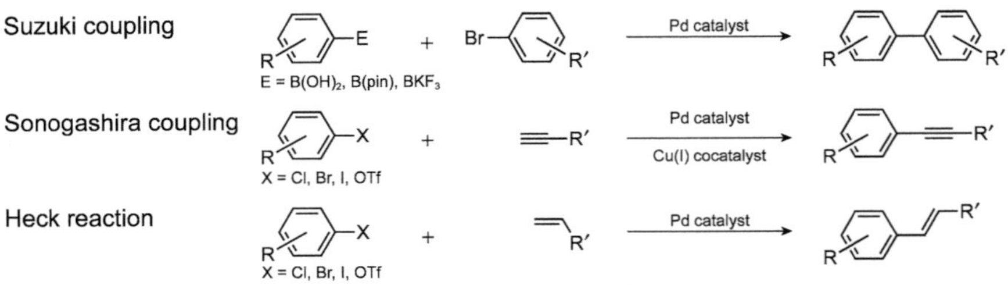

Figure 3.4 Coupling reactions used in the synthesis of organic linkers. Three reactions are frequently used to join organic units such as aryls, alkynes, or alkenes. Standard procedures for the shown reactions are well established allowing for the synthesis of linkers in pure form and high yields.

connectivity, (ii) angled or offset units that alter the overall symmetry but not the connectivity, and (iii) branched units that affect both the geometry and the connectivity of the core unit. Once the appropriate synthetic precursors for the fragments constituting the targeted linker have been selected, the backbone of the linker is synthesized by the formation of covalent bonds between them. This is typically accomplished using a variety of coupling reactions depending on the units between which the new bonds are formed. Figure 3.4 shows the three types of palladium-catalyzed carbon–carbon coupling reaction routinely used to construct the backbone of linker molecules.

Suzuki coupling reactions are employed to form carbon–carbon bonds between two aryl units, where the synthetic equivalents of the corresponding units in the linker are an aryl boronic acid or -ester and an aryl halide [3]. In a Sonogashira coupling reaction, an aryl and an alkyne unit are joined [4]. Aside from the palladium catalyst used to catalyze the carbon–carbon bond formation, a copper cocatalyst is required for the metalation of the alkyne and subsequent transmetalation. The synthetic precursors for this reaction are aryl halides and alkynes. The third important carbon–carbon coupling reaction is the Heck reaction that is used to form a carbon–carbon bond between an aryl and an alkene that are employed as their synthetic equivalents, aryl halide or -triflate and alkene, respectively [5]. Other coupling reactions such as the metal-catalyzed Buchwald–Hartwig amination, Gilman coupling, and Glaser coupling as well as coupling reactions that do not require a precious metal catalyst including diazo coupling, amide coupling, Friedel–Crafts alkylation/acylation, nucleophilic substitution, and imine condensation are frequently used to construct the backbone of linker molecules. The products of such reactions often exhibit higher degrees of flexibility compared to those synthesized using one of the three carbon–carbon cross-coupling reactions mentioned before (Suzuki, Sonogashira, Heck), since both, the structures of the synthetic precursor and the newly formed carbon–heteroatom or heteroatom–heteroatom bonds tend to be less rigid.

Figure 3.5 shows two examples of typical linker syntheses. Figure 3.5a illustrates the synthesis of H_3BTE **(a7)** (4,4′,4″-[benzene-1,3,5-triyltris(ethyne-2,1-diyl)]

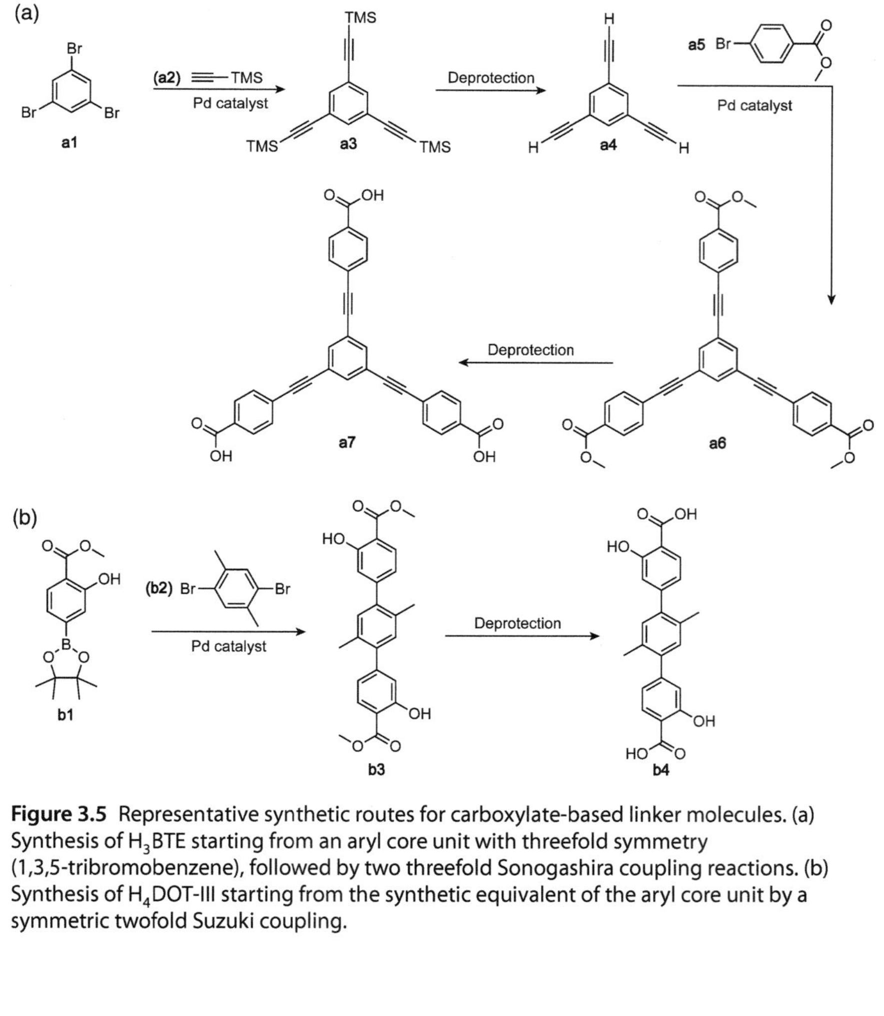

Figure 3.5 Representative synthetic routes for carboxylate-based linker molecules. (a) Synthesis of H_3BTE starting from an aryl core unit with threefold symmetry (1,3,5-tribromobenzene), followed by two threefold Sonogashira coupling reactions. (b) Synthesis of H_4DOT-III starting from the synthetic equivalent of the aryl core unit by a symmetric twofold Suzuki coupling.

tribenzoic acid), the linker molecule employed in the synthesis of MOF-200 (see Figure 2.16). The brominated core unit of the linker **(a1)** (1,3,5-tribromobenzene) is reacted with the synthetic precursor of the ethyne extension units **(a2)** (ethynyltrimethylsilane) in a threefold Sonogashira coupling. After deprotection the crude product **(a4)** is purified by sublimation. Another threefold Sonogashira coupling with the terminal protected benzoic acid binding group affords **(a5)**. After purification through liquid column chromatography H_3BTE is formed by saponification of the corresponding ester **(a6)**.

The synthesis of H_4DOT-III **(b4)** (3,3″-dihydroxy-2′,5′-dimethyl-(1,1′:4′,1″-terphenyl)-4,4″-dicarboxylic acid), the linker employed in the synthesis of IRMOF-74-III, is shown in Figure 3.5b. A twofold Suzuki coupling between the brominated core unit **(b2)** (1,4-dibromo-2,5-dimethylbenzene) and boronic acid pinacol ester of the protected 2-hydroxy benzoic acid binding groups **(b1)** affords the ester **(b3)** of the targeted linker. Saponification and subsequent purification by liquid column chromatography yields H_4DOT-III **(b4)**.

With the tools for the design and synthesis of linkers at hand it is important to gain an overview of the linker geometries that are frequently used in MOF chemistry. In the next section, we will sketch out the different groups of linkers classified based on their number of points of extension and discuss possible geometries and ways to alter them.

3.2.2 Linker Geometries

For a given number of points of extension, a variety of different linker geometries is possible. The number of points of extensions defines the number of connections a linker can make to adjacent SBUs within a framework structure. Therefore, it is instructive to categorize linkers according to the number of points of extension, rather than only by their geometries. The following section will provide an overview of linkers with numbers of points of extension ranging from two to eight and give an insight into their molecular structures.

3.2.2.1 Two Points of Extension

Ditopic linkers are of linear, offset, or angled geometry (Figure 3.6), and their carboxylate binding groups can be either coplanar or twisted with respect to each other. Despite the consideration of all ditopic linkers as edges in terms of topology, regardless of their true molecular structure, their precise geometry has a profound impact on the outcome of reticular synthesis involving these linkers.

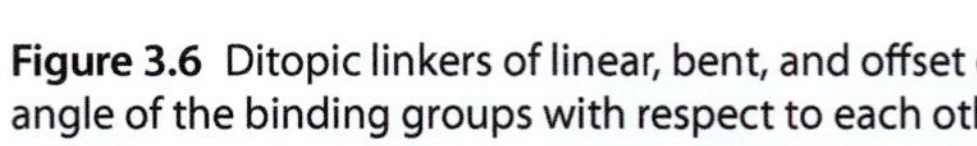

Figure 3.6 Ditopic linkers of linear, bent, and offset geometry. In addition to the position and angle of the binding groups with respect to each other, the dihedral angle between them has a profound influence on the structure of the product of their reticulation with metal ions (not shown).

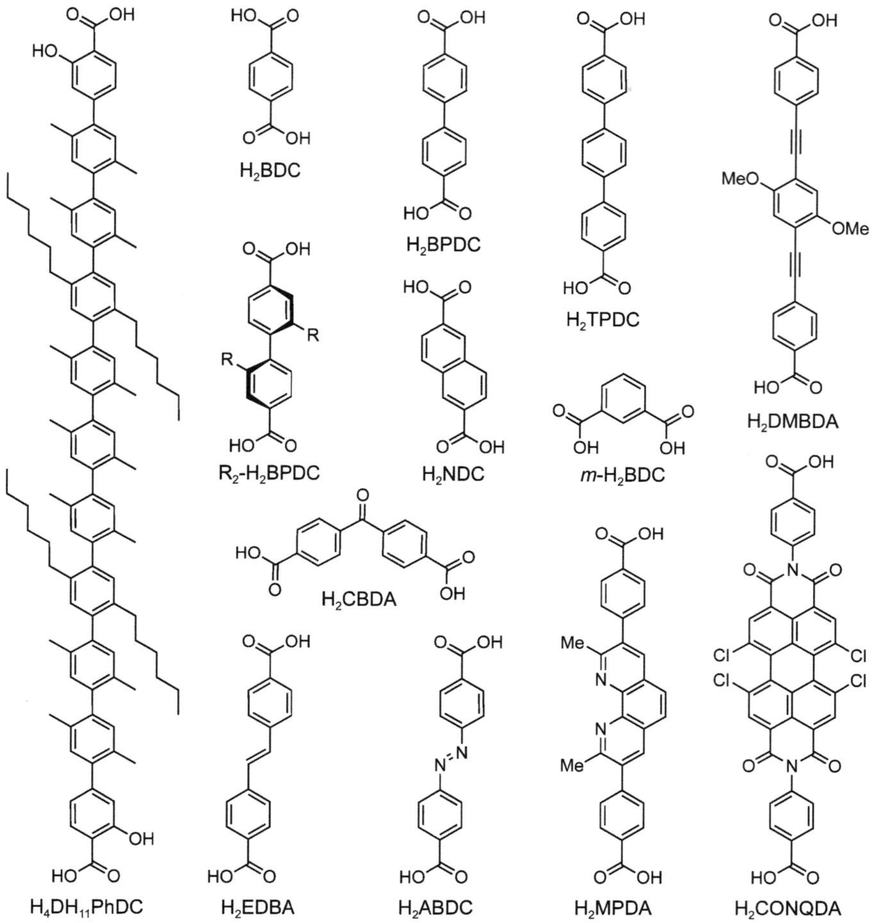

Figure 3.7 Examples of ditopic linkers used in the synthesis of carboxylate-based MOFs. Both the length and geometry can be controlled and functionalities such as additional coordination sites for further functionalization by post-synthetic modification (see Chapter 6) can be incorporated into the backbone of the linker molecules.

The shortest and most common linear ditopic linker used in MOF chemistry is H_2BDC (terephthalic acid) (Figure 3.7). It can be elongated by either additional aryl or by acetylene units with retention of its linear shape. In contrast, insertion of alkenyl units or diazo bridges introduces an offset between the carboxylic acid groups that may lead to the formation of structures with topologies that are different from those obtained from reticular synthesis involving linear linkers. Furthermore, a bending angle between the carboxylic acid groups is introduced by using *m*-substituted aryl units (*m*-H_2BDC, isophthalic acid) or ketones (H_2CBDA, benzophenonedicarboxylic acid) as the core. In many cases,

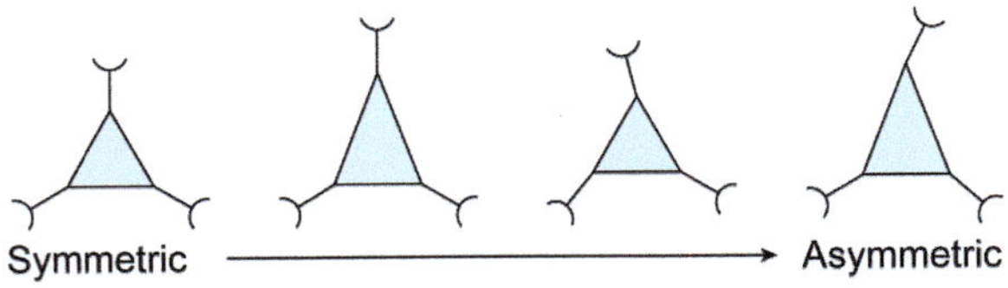

Figure 3.8 De-symmetrization of a trigonal linker. The triangle can be de-symmetrized by elongating one side of the triangle, changing the angle of the binding groups with respect to the center of the triangle using asymmetric extending units, or both (left to right). Alterations of the dihedral angle between the core unit and the binding groups are not shown.

additional substituents are added onto such linkers without effecting the metrics or structure of a given framework. The effects on frameworks resulting from the reticulation of non-linear ditopic linkers with 4-c square SBUs will be discussed in Chapter 4 in detail.

3.2.2.2 Three Points of Extension

Tritopic linkers are always of trigonal shape but can be either equilateral or de-symmetrized. Ways to lower the symmetry of an equilateral trigonal linker with point group D_{3h} are shown in Figure 3.8. Additionally, the dihedral angle between the core unit and the binding groups has to be considered. A compilation of different trigonal linkers is given in Figure 3.9.

The smallest symmetric trigonal tritopic linker commonly used in the synthesis of MOFs is H_3BTC (1,3,5-benzenetricarboxylic acid, trimesic acid). It can be expanded by introducing additional aryl units to give H_3BTB 4,4′,4″-benzene-1,3,5-triyltribenzoate and H_3BBC 4,4′,4″-(benzene-1,3,5-triyl-tris(benzene-4,1-diyl))tribenzoate or by introducing acetylene units to yield H_3BTE 4,4′,4″-(benzene-1,3,5-triyl-tris(ethyne-2,1-diyl))tribenzoate. The D_{3h} symmetry of H_3BTC may be lost upon expansion due to the frequently observed dihedral angles between the aryl core unit and the terminal benzoic acid binding groups resulting in "propeller shape" linker molecules. This is avoided by selecting a triazine core, thereby eliminating repulsive H···H interactions of the aromatic protons leading to a perfect planar arrangement. Using this strategy, expanded versions of H_3BTC with D_{3h} symmetry such as H_3TATB (4,4′,4″-(1,3,5-triazine-2,4,6-triyl) tribenzoic acid) are prepared.

Asymmetry can be introduced in different ways. Frequently observed strategies are the replacement of one of the carboxylic acid binding groups on H_3BTC by a benzoic acid group as in H_3TTCA ([1,1′-biphenyl]-3,4′,5-tricarboxylic acid) or the expansion using asymmetric extending units as exemplified by H_3BTN (6,6′,6″-(benzene-1,3,5-triyl) tris (2-naphthoic acid)).

3.2.2.3 Four Points of Extension

Tetratopic linkers can have various shapes and are grouped into (i) tetrahedral, (ii) square, and (iii) irregular shapes (Figure 3.10). A compilation of tetratopic linkers is shown in Figures 3.11 and 3.12.

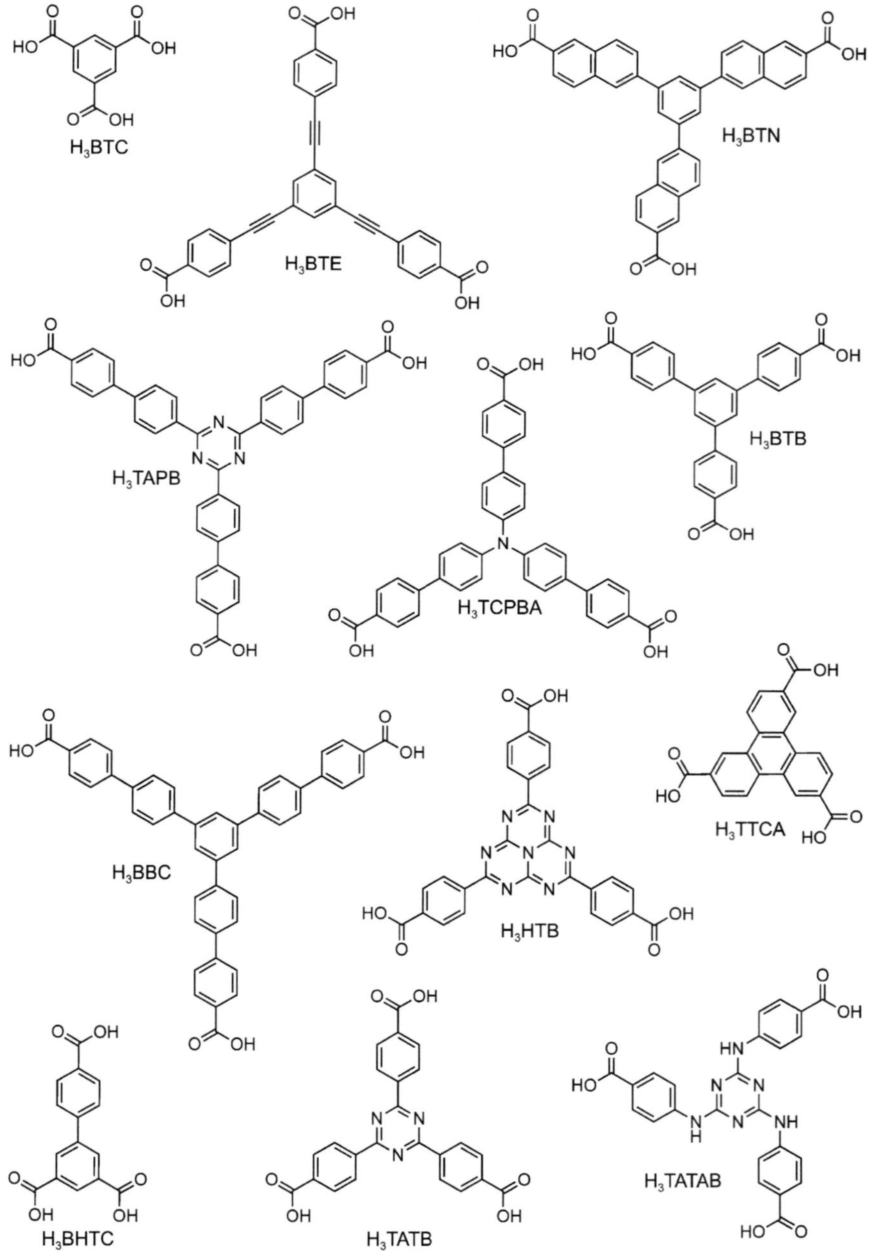

Figure 3.9 Examples of tritopic linkers used in the synthesis of carboxylate-based MOFs. Linkers of different metrics and geometry ranging from ideal D_{3h} to C_{2h} symmetry have been reported. The precise 3D structure of the linker can be a useful guide in the design of new MOFs.

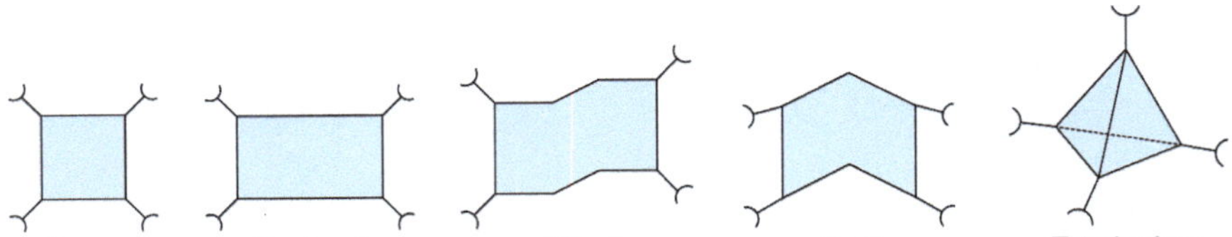

Figure 3.10 Possible geometries for tetratopic linkers. Linkers with D_{4h} symmetry are de-symmetrized by elongation along one direction, or introducing an offset or an angle (left to right). Linker molecules with T_d symmetry are typically based on either a central carbon or silicon atom, or an adamantane cage. In addition, the angle of the binding groups with respect to each other the dihedral angle between them has a profound influence on the outcome of their reticulation with metal ions (not shown).

Square linkers are often based on porphyrin units, where the simplest one is H_4TCPP-H_2 ((4,4′,4″,4‴-(porphyrin-5,10,15,20-tetrayl) tetrabenzoic acid)). Depending on the number of benzene rings between the porphyrin core unit and the carboxylic acid binding groups, the acid groups will either be in-plane with the porphyrin core or perpendicular to it. As we will see in Chapters 3–5, such subtle changes in the geometry of the linker can have a profound impact on what structure will form by their reticulation with metal ions.

The smallest rectangular linker commonly used is H_4BPTC ([1,1′-biphenyl]-3,3′,5,5′-tetracarboxylic acid) and there are two known ways of altering its geometry and metrics. Extending units can be added either at the center of the linker between the two terminal isophthalic acids, as exemplified by H_4QPTCA ([1,1′:4′,1″:4″,1‴:4‴,1⁗-quinquephenyl]-3,3⁗,5,5⁗-tetracarboxylic acid), or next to the carboxylic acid binding groups (H_4CQDA, 5′,5″-bis (4-carboxyphenyl)-[1,1′:3′,1″:3″,1‴-quaterphenyl]-4,4‴-dicarboxylic acid) to elongate one or both sides of the rectangle, respectively. By insertion of C=C double bonds or diazo bridges an offset is introduced. Another approach to the design of rectangular linkers is to use an alkenyle core unit as in H_4ETTC (4′,4‴,4′′′′′,4′′′′′′′-(ethene-1,1,2,2-tetrayl)tetrakis([1,1′-biphenyl]-4-carboxylic acid)). It should be noted that in this case less control over the ratio of the length of the two sides of the rectangle is exercised since they cannot be modified independently.

Tetrahedral linkers are typically based on an sp^3 hybridized carbon or silicon atom, as in H_4ATB (4,4′,4″,4‴-(adamantane-1,3,5,7-tetrayl)tetrabenzoic acid), or an adamantane core, as in H_4MTB (4,4′,4″,4‴-methanetetrayltetrabenzoic acid). The use of asymmetric core units allows for the synthesis of tetratopic linkers with irregular shapes that do not fit any of the aforementioned categories, highlighting the possible structural variety of the frameworks formed by their reticulation.

Figure 3.11 Examples of tetratopic linkers of square and rectangular geometry used in the synthesis of carboxylate-based MOFs. Porphyrin core units are often used to construct square linkers, core units of a symmetry lower than C_4 translate in a change of the precise geometry upon addition of extending units.

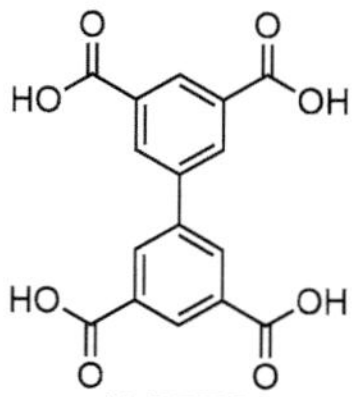

H_4BPTC

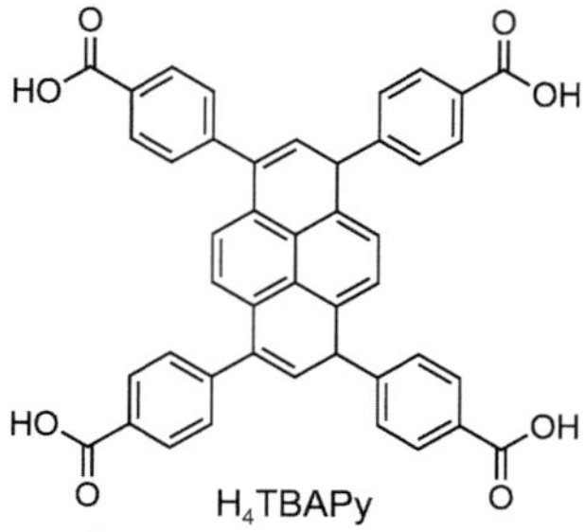

H_4TBAPy

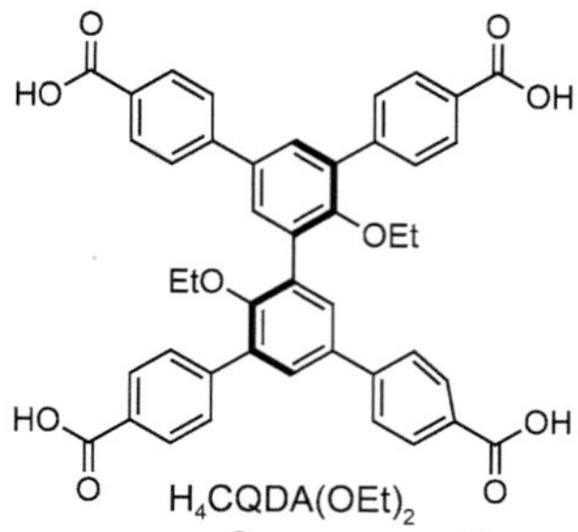

$H_4CQDA(OEt)_2$

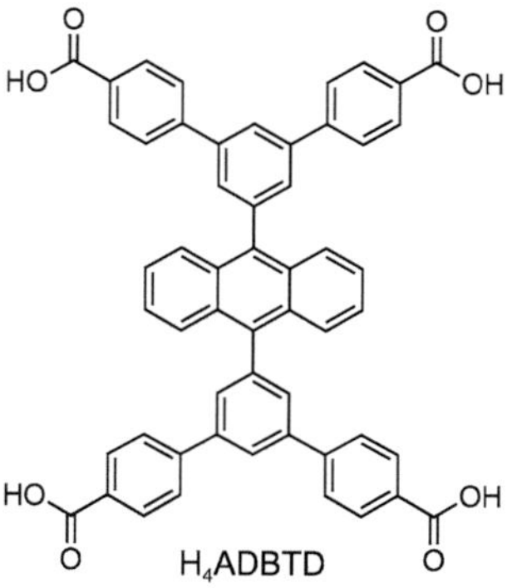

H_4ADBTD

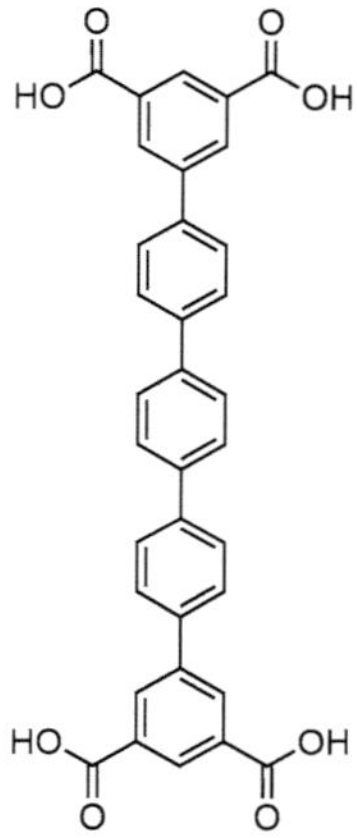

H_4QPTCA

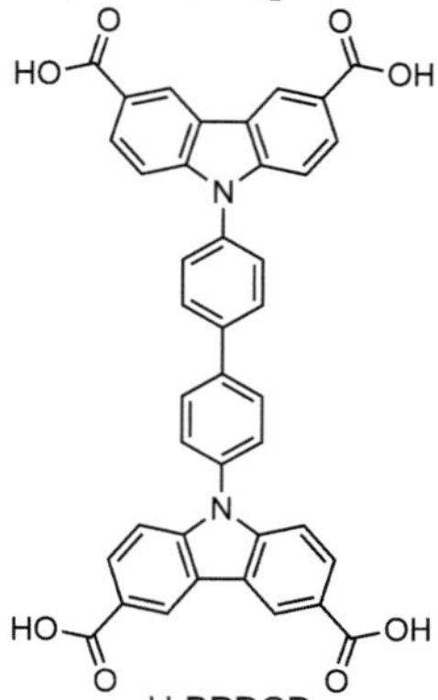

H_4BPDCD

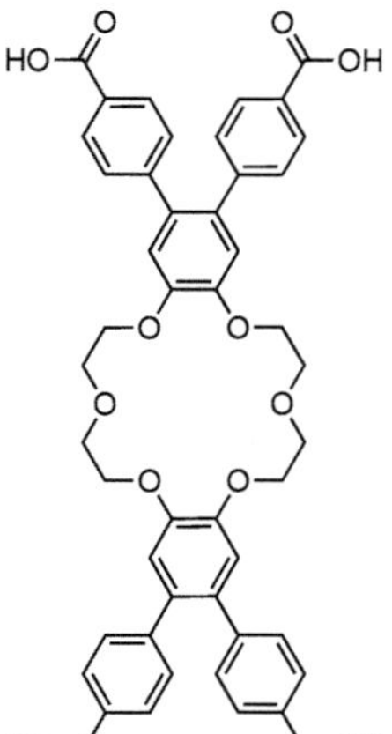

H_4TBADB-18Cr6

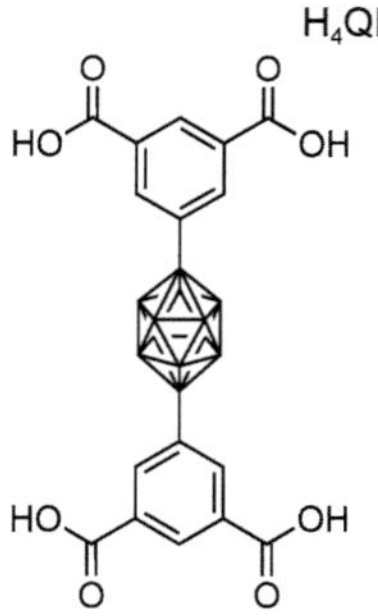

H_4CBI

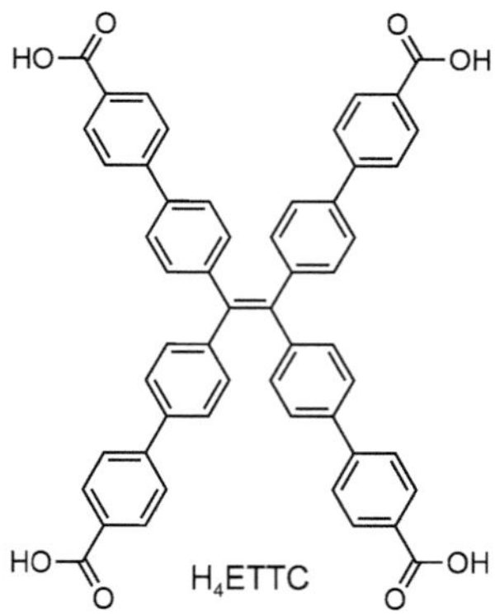

H_4ETTC

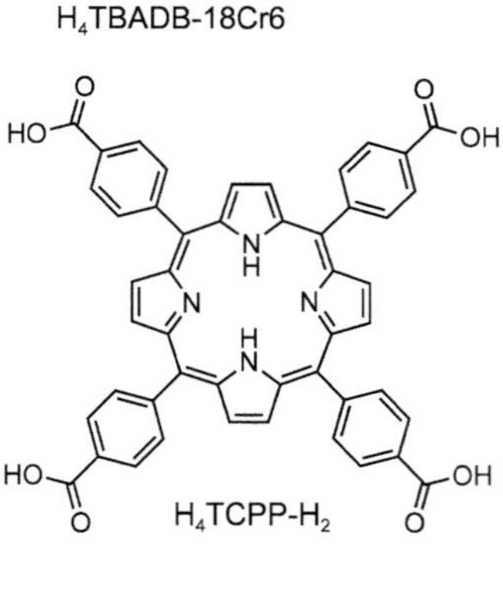

H_4TCPP-H_2

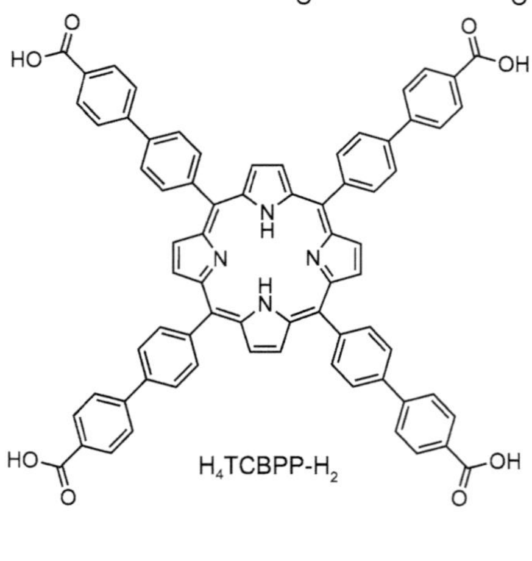

H_4TCBPP-H_2

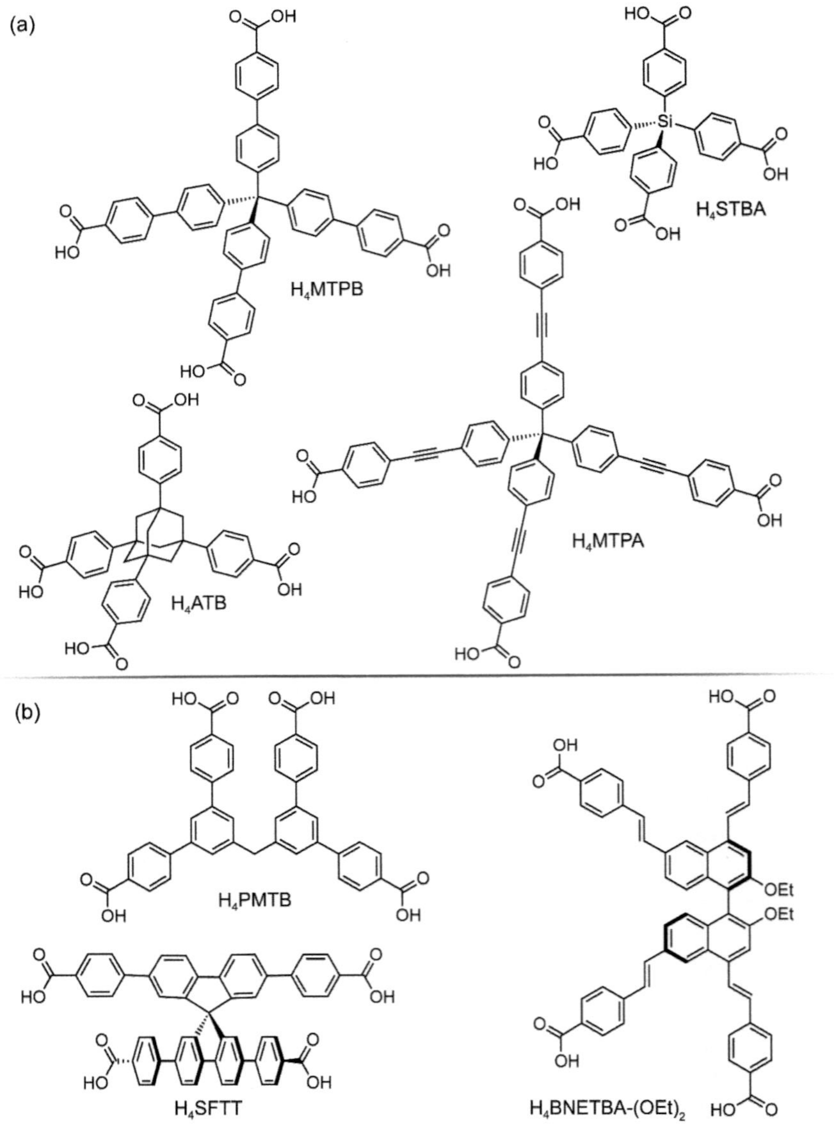

Figure 3.12 Examples of tetratopic linkers of tetrahedral and irregular shape. (a) Tetrahedral linkers are commonly based on a central sp^3 hybridized carbon or silicon atom, or an adamantane core. (b) Irregular shapes are achieved by employing asymmetric core and/or extending units.

3.2.2.4 Five Points of Extension

In the synthesis of MOFs, the use of linkers with high symmetry is common practice. Nonetheless, there are rare examples of MOFs constructed from lower symmetry linkers such as the pentatopic linker H_5PTPC (5′-(4-carboxyphenyl)-[1,1′ : 3′,1″-terphenyl]-3,3″,5,5″-tetracarboxylic acid) used in the construction of a zinc-based MOF (Figure 3.13) [6]. This example illustrates that there are very few limits to the design of new linkers as constituents of framework structures.

3.2.2.5 Six Points of Extension

Even though first examples of hexatopic linkers were reported as early as 2001, the wide variety known today was not systematically developed until the discovery of **ntt**-topology MOFs (Figure 3.14) [7]. Most hexatopic linkers share the same trigonal core with many of the tritopic linkers discussed above but are terminated with isophthalic rather than benzoic acid groups, thus resulting in six instead of three points of extension. The points of extension for carboxylate-based linkers are typically defined as the carboxylate carbon atom. Trigonal linkers bearing six binding groups can also act as trigonal "tritopic" units, where the gravimetric center of two carboxylic acids is chosen as the point of extension, i.e. the C-5 carbon of the *m*-BDC subunits. This concept will be discussed in more detail in Chapter 5 in the context of MOFs built from tertiary building units (TBUs).

3.2.2.6 Eight Points of Extension

Although they are rare, frameworks constructed from octatopic linkers are known. Octatopic linkers typically share the same general core with most tetratopic linkers listed earlier but are decorated with double the number of carboxylic acid functionalities; an aspect that is achieved by terminating the linker with isophthalic instead of benzoic acid groups (Figure 3.15). Analogous to hexatopic linkers, some octatopic linkers can also act as "tetratopic" units, where again the gravimetric center of two carboxylic acids is chosen as the point of extension.

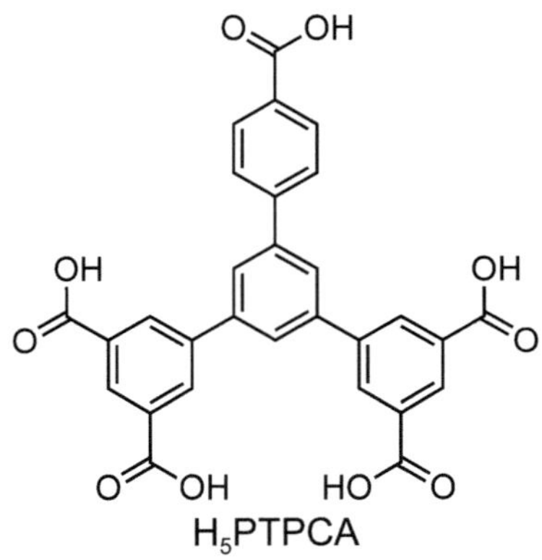

H_5PTPCA

Figure 3.13 Pentatopic linkers are rare but not unknown in the chemistry of MOFs. The example above was used to synthesize highly porous zink-based MOFs.

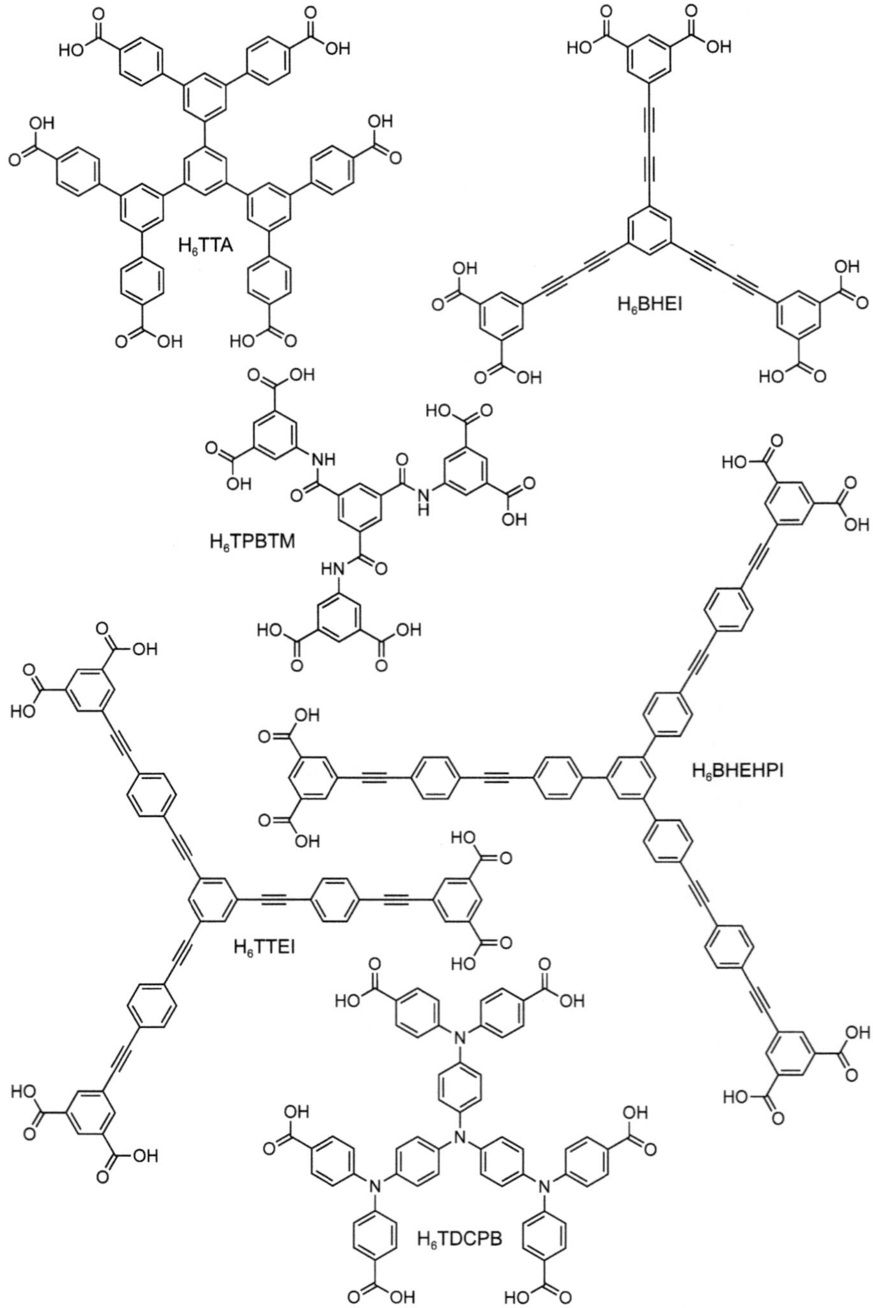

Figure 3.14 Examples of hexatopic linkers. The core and extension units are similar to those used in the construction of tritopic linkers where the monofunctional benzoic acid binding groups are replaced by bifunctional isophthalic acid groups.

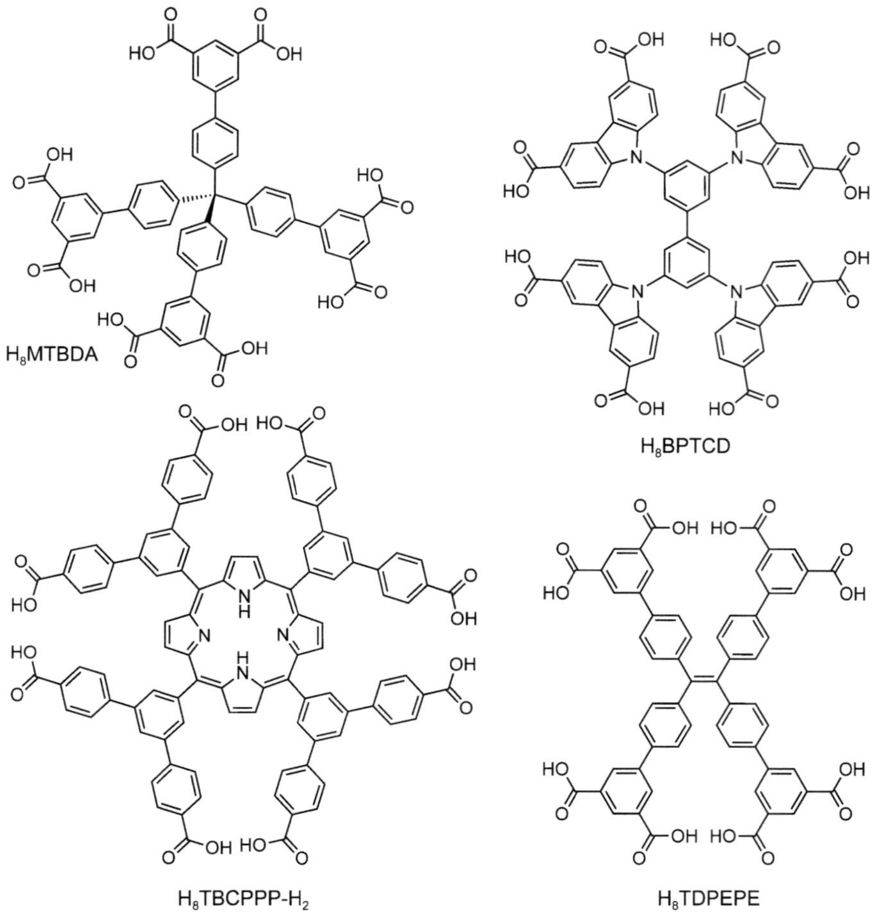

Figure 3.15 Examples of octatopic linkers. The core and extension units are similar or identical to those used to construct tetratopic linkers where the monofunctional benzoic acid groups are replaced by bifunctional isophthalic acid groups.

3.3 Secondary Building Units

The term "secondary building unit" or SBU was originally coined to classify and describe the rich structural chemistry of zeolites by deconstructing them into finite or infinite structural subunits. This strategy allows for the description of highly complex structures in terms of simple nets by defining arrangements of atoms as building units. This concept has been transferred to MOF chemistry, where the term SBU is used to describe the inorganic building units in MOF structures. These units are typically polynuclear clusters of metal ions, where the polydentate binding groups of the linkers are an integral part of these units. SBUs are commonly formed *in situ*, thus allowing for the slow and reversible assembly of the overall structure, error correction, and consequently the formation of highly crystalline products.

The structural richness of MOFs is largely attributed to the diverse geometries and the wide range of connectivity, typically ranging from three to twelve, found in inorganic SBUs. Molecular carboxylate clusters with up to 66 points of extension and the potential to be reticulated into MOF structures are known in the literature [8]. The high connectivity and diverse geometries of such SBUs, in contrast to single metal nodes, make them ideal building units for accessing a large variety of framework structures. As illustrated in Figure 3.16, most metals and metalloid elements have been incorporated into MOF structures, most of them into their SBUs.

Combining the rigid and geometrically well-defined structures of the SBUs with deliberately designed organic linkers facilitates the targeted synthesis of extended frameworks. This is possible due to the geometric constraints inherent to these rigid building units, which significantly limits the number of possible resulting net topologies [9]. Additionally, MOF structures with predictable pore shape and metrics can be designed by making use of the isoreticular principle or linker exchange reactions (see Chapter 6) [10].

When all following prerequisites are taken into consideration, the molecular building unit approach is a powerful tool for the *a priori* design of MOFs: (i) it is essential that the chemistry of the metal ion used to form the specific SBUs is well understood since subtle variations in the synthetic conditions may result in the formation of different SBUs,[3] (ii) the geometry of the organic linker and its specific functionalities must be precisely determined and remain unaltered during the synthesis to impart rigidity upon framework formation and therefore introduce a degree of predictability, (iii) the given reaction conditions must allow for the formation of an ordered crystalline material, and (iv) no structure directing agent is added so that the assembly of the building units will lead to their default structure guided by their inherent directionality and rigidity.

Before we analyze specific synthesis conditions and MOF structures, it is instructive to understand the coordination chemistry of metal carboxylates. In the structures of carboxylate-based MOFs, the carboxylates can bind to the metal centers in different ways: ionic bonds (e.g. sodium formate), unidentate (e.g. lithium acetate), symmetric chelating binding modes (e.g. indium acetate), and a variety of bridging binding modes (e.g. copper acetate) (Figure 3.17), where the nature of the R—COO—M bond ranges from purely ionic to partially covalent.

The observation that all six binding modes are found in the structures of binary metal carboxylates of different metals highlights the importance of knowing the reaction conditions to target the formation of a specific SBU. In the following section examples of general synthesis conditions for the formation of SBUs built from divalent, trivalent, and tetravalent metals are given.

3 Therefore, in targeting SBUs of metal ions previously not used in MOF chemistry imposes a degree of uncertainty about the geometry of the resulting SBU.

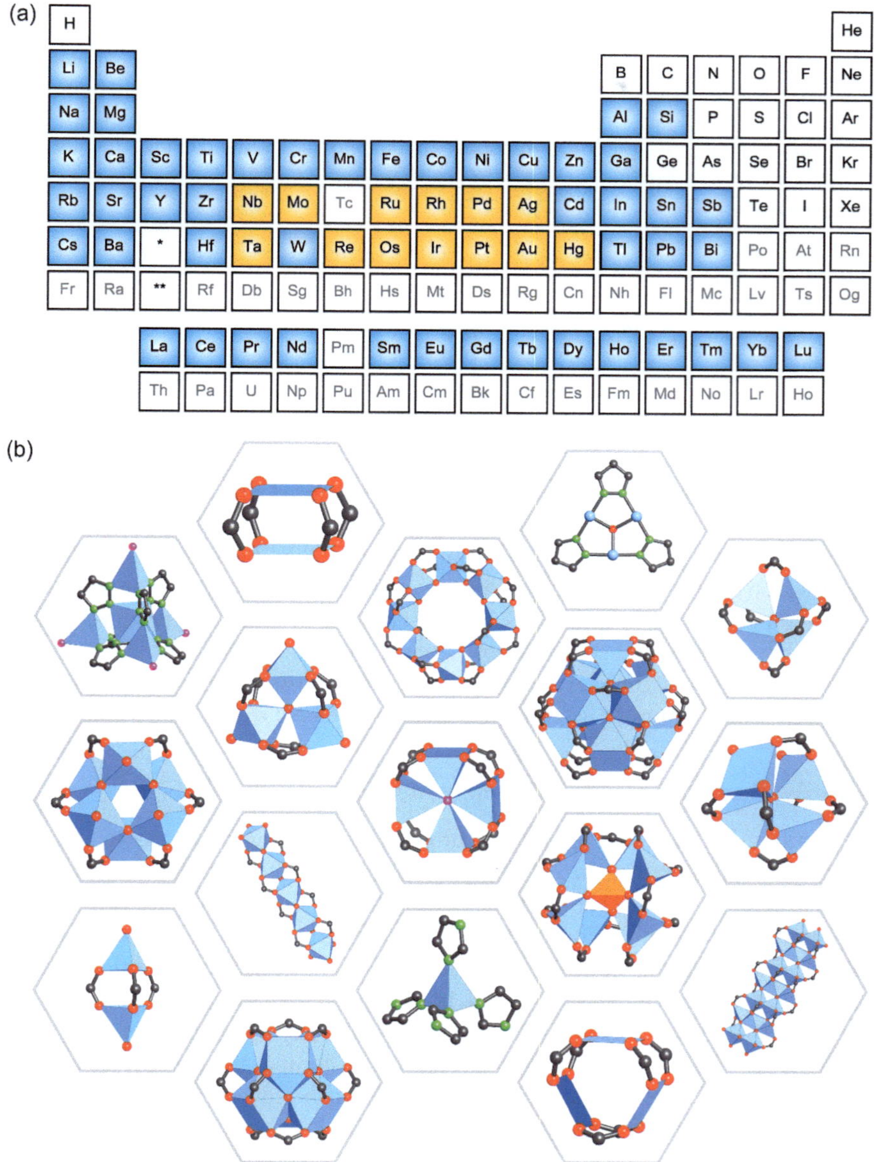

Figure 3.16 (a) Periodic table showing metals that have been incorporated into MOF structures. The elements highlighted in blue can form SBUs or are part of the organic linker (metallo-linker), elements highlighted in orange have been incorporated into the structure of MOFs by metalation of the linker in the form of complexes. (b) Some examples of 0D and 1D SBUs with points of extension ranging from 3 to 12.

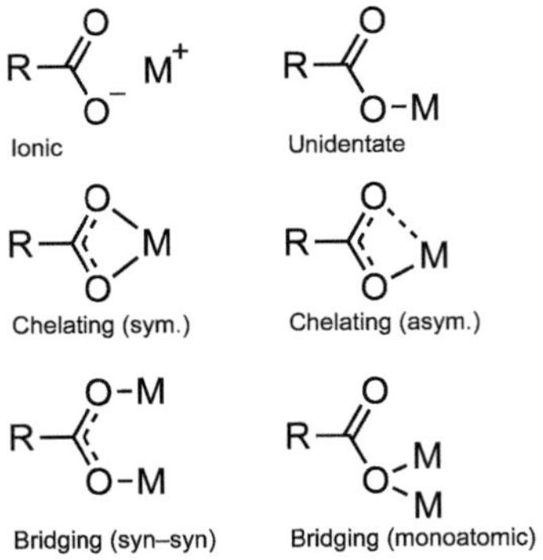

Figure 3.17 Different binding modes of R-COO$^-$ to metal centers. The most favorable binding mode is strongly dependent on the nature of the metal. Since chelating and bridging binding modes afford the formation of polynuclear clusters, these are the predominant binding modes found in the structures of MOFs.

3.4 Synthetic Routes to Crystalline MOFs

The synthesis of MOF-2 and MOF-5 were discussed in Chapter 1 and the fact that precise control over the reversibility and rate of the formation reaction is decisive for the formation of crystalline materials has been highlighted. The way in which such reversibility is achieved differs for every metal that is used to construct an SBU. Since the discussion of synthetic procedures for all reported MOFs is beyond the scope of this book, we will illustrate three different synthetic paradigms for metals that carry two, three, or four positive charges and provide information on the proper activation of the resulting MOFs.

3.4.1 Synthesis of MOFs from Divalent Metals

The bonding between the organic carboxylates and divalent *s*-block and transition metals has partial ionic character. Figure 3.18 illustrates the binding modes found in molecular copper(II) acetates in different hydration states: monodentate bridging in $Cu(OAc)_2$, syn–syn bridging mode in $Cu(OAc)_2 \cdot H_2O$, and anti-syn bridging mode in $Cu(OAc)_2 \cdot 2H_2O$. A decreasing covalent character of the metal–carboxylate bond is observed with decreasing ionization energy of the metal. This in turn results in a more reversible formation of the metal–carboxylate bond and therefore it is comparatively easy to adjust the reaction conditions in such a way as to form highly crystalline products. A general procedure for the synthesis of MOFs built from SBUs containing divalent metals such as Ca^{2+}, Be^{2+}, Zn^{2+}, and Cu^{2+} (note that Be, unlike other members of group 2 prefers a tetrahedral coordination due to its comparatively high charge to radius ratio) involves the following steps: (i) dissolution of the organic linker in DMF or other amide-based solvents, (ii) dissolution of the metal

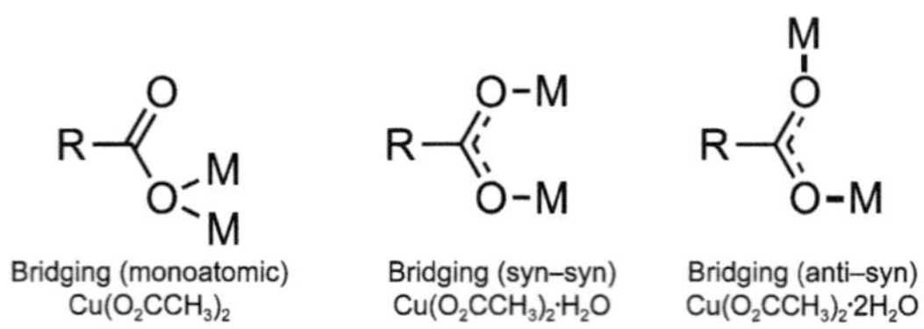

Figure 3.18 Binding modes found in copper(II) acetate complexes depending on their hydration state ranging from monoatomic bridging to different bidentate bridging modes.

salt, most commonly metal nitrates, in an appropriate solvent, (iii) combining both solutions in a scintillation vial or in special cases, sealed borosilicate tubes, (iv) addition of a cosolvent is sometimes required to aid in solubilizing the larger fragments formed during the reticulation, and (v) heating of the reaction mixture at temperatures ranging from room temperature to 140 °C.

The appropriate choice of solvent, concentration, and reaction temperature is important. To elucidate the role of the solvent further, investigation of the chemical reactions occurring during isothermal heating of DMF (*N*,*N*′-dimethylformamide) solutions containing linkers and metal ions is necessary. It is well known that amides slowly decompose at elevated temperatures to form an amine base and a carboxylic acid. In the case of DMF a certain amount of dimethylamine (DMA), a strong base with $pK_b = 3.29$, and formic acid, an acid with $pK_a = 3.77$ acting as a modulator, are released. The rate of release is controlled by adjusting the reaction temperature. DMA affords the deprotonation of the linker whereas formic acid acts as a modulator (Figure 3.19). A modulator is an organic molecule bearing the same binding group as the linker but having a lower topicity (formic acid is a "monotopic" ligand) and a different pK_a. The deprotonated anionic form of the linker can directly react with the metal ions leading to the formation of the desired SBUs and subsequently the targeted framework structure. In contrast, the lower topicity of the modulator means that it cannot facilitate a linkage between SBUs in the same way the linker does, thereby introducing reversibility and modulating the rate of framework formation. In conclusion, the solvent determines the strength of the base

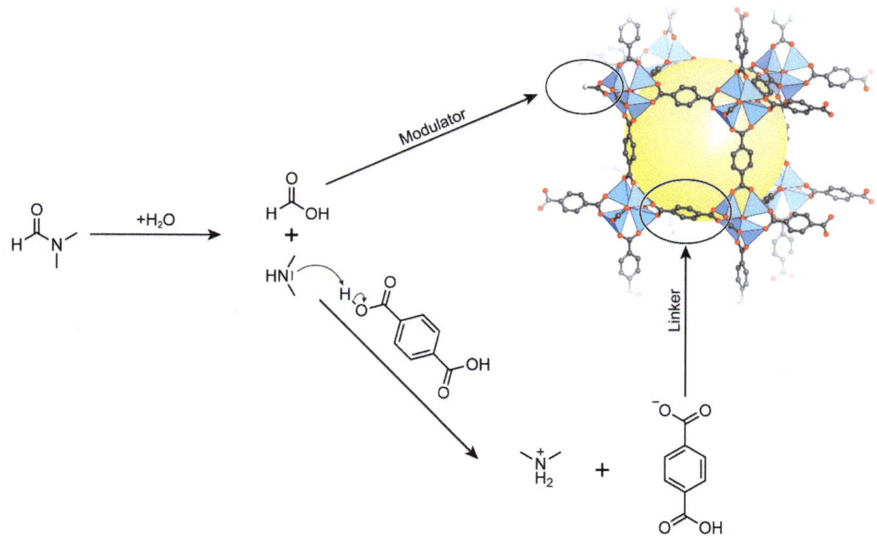

Figure 3.19 Thermal decomposition of DMF and the reaction pathway of its decomposition products, DMA and formic acid, illustrated for the formation of MOF-5. DMA facilitates the deprotonation of the H_2BDC linker, while formic acid acts as a modulator, rendering the reticulation process reversible. This allows for error correction and the formation of a highly crystalline product.

liberated upon heating, whereas the rate of release of base is determined by the reaction temperature. Additional to finding the optimal reaction temperature, precise adjustment of the concentration of both, the linker and metal source, is required to afford the formation of crystalline materials.

3.4.2 Synthesis of MOFs from Trivalent Metals

3.4.2.1 Trivalent Group 3 Elements

The solution chemistries of aluminum, gallium, and indium is fundamentally different as illustrated by the fact that aluminum and gallium hydroxide are both amphoteric, whereas indium hydroxide is a basic metal hydroxide [11]. It is therefore not surprising that the reaction conditions affording crystalline MOFs built from these three metals are often very different. The hydrolysis of aluminum in aqueous solution typically results in the formation of clusters of condensed tetrahedrally and octahedrally coordinated Al^{3+} ions such as the polyoxoaluminates $[(AlO_4)_{12}[Al_{12}(OH)_{24}(H_2O)_{12}]^{7+}$ and $[Al_2O_8Al_{28}(OH)_{56}(H_2O)_{26}]^{18+}$ [12]. This is in stark contrast to gallium and indium, which preferentially form smaller condensed clusters and isolated hexaquo-complexes, respectively [13]. The increasing ionic radius form Al^{3+} (0.675 Å) to Ga^{3+} (0.760 Å) and In^{3+} (0.940 Å) is associated with a decrease of the surface charge, equating to a sixfold increase of the kinetic lability from aluminum to indium in $[M(H_2O)_6]^{3+}$ (M = Al or In) complexes and therefore a much higher ligand exchange rate [14]. This is manifested in the formation of larger crystals of indium MOFs compared to aluminum or gallium MOFs. The M—O bond length decreases, and the bond strength increases from indium to aluminum, which is in good agreement with the stability of the corresponding MOFs. In contrast to In-MOFs that are commonly prepared from organic solutions (DMF, DMA, NMP) at comparatively low temperatures ($T \ll 150\,°C$), Al- and Ga-MOFs are typically synthesized from basic aqueous solutions under solvothermal conditions at higher temperatures (120–220 °C). Additionally, for the formation of crystalline products, the presence of strong acidic modulators such as inorganic mineral acids (HF, HNO_3, HCl) is required. The comparatively harsh reaction conditions are necessary to render the metal–carboxylate bond formation and thereby the reticulation process reversible and allow for error correction and consequently the formation of crystalline products.

3.4.2.2 Trivalent Transition Metals

As for the trivalent main group metals discussed earlier, the high reactivity and strong metal–carboxylate bonds of trivalent transition metals such as Fe^{3+}, Cr^{3+}, and V^{3+} makes it difficult to establish reaction conditions allowing for reversibility and the formation of crystalline products. The fact that metal oxides and hydroxides of these trivalent cations are formed over a wide pH range leaves only a narrow set of reaction conditions for the formation of crystalline MOFs. The pH domain allowing for a soluble cationic iron species lies below a pH value of 2, hence a successful synthesis of iron MOFs from neutral or basic solutions can only be achieved through the addition of strongly acidic modulators such as inorganic mineral acids (HF, HCl) or strong monocarboxylic acids (CF_3COOH, HCOOH). The kinetic lability of the metal ion dictates the exchange rate of the linker. The low exchange rate of kinetically inert Cr^{3+} compared to Fe^{3+} means that even harsher reaction conditions are needed for the formation of chromium MOFs ($T > 180\,°C$) than for iron MOFs ($T < 150\,°C$) [15]. Both iron and chromium MOFs are often prepared from aqueous solutions under solvothermal conditions and in the presence of strongly acidic modulators. The tendency of vanadium to form polyoxovanadates in its tetravalent state makes it necessary to adjust the reaction parameters in such a way so as to prevent the oxidation of V^{3+} to V^{4+} when targeting the formation of vanadium(III) MOFs.

3.4.3 Synthesis of MOFs from Tetravalent Metals

Zirconium and hafnium have a d^0 electron configuration in their tetravalent state with almost identical ionic radii due to lanthanide contraction [16]. Polynuclear metal carboxylate clusters of zirconium, especially those of the general formula $Zr_6(OH)_4O_4(R{-}COO)_{12}$, are well known in molecular inorganic chemistry. The synthesis of crystalline MOFs based on these clusters is however challenging. This is due to the high charge of the metal ions and the concomitant strong $R{-}COO{-}M^{4+}$ bonds and the strong oxophilicity of zirconium. This challenge can be addressed and reversibility in the metal–carboxylate bond formation is introduced in a manner akin to that described in the previous section. The formation of zirconium and hafnium MOFs is commonly achieved by reacting an appropriate metal source ($MOCl_2$, $M(SO_4)_2$, or MX_4 where X = Cl, Br, I and M = Zr^{4+}, Hf^{4+}) with an organic carboxylate-based linker, in an organic amide solvent, in the presence of a modulator (typically monocarboxylic acids such as formic, acetic, or benzoic acid) at temperatures ranging from 50 to 140 °C [17].

The synthesis of titanium MOFs is synthetically even more challenging than that of zirconium and hafnium MOFs due to a higher charge to radius ratio rendering the resulting bond even stronger and thus the bond formation even less reversible. Additionally, the symmetry and connectivity of most titanium clusters known in molecular inorganic chemistry cannot lead to the formation of well-ordered crystalline extended framework structures, an aspect that will be discussed in more detail in the following chapters. The synthesis of titanium MOFs is commonly achieved by reacting the organic linker with an appropriate highly reactive titanium source ($Ti(OR)_4$, where R = Me, Et, *i*Pr, Bu, etc.) dissolved in an organic solvent (DMF or similar) at elevated temperatures (100–160 °C) for a prolonged period of time (2–7 days). Titanium MOFs are typically formed as microcrystalline powders and their synthesis in single crystalline form is challenging. Recently, the synthesis of a titanium MOF by linking of pre-synthesized decorated clusters through imine condensation, thereby avoiding the crystallization challenge, has been reported [18].

3.5 Activation of MOFs

The removal of guest molecules from the pores of MOFs is commonly achieved by performing multiple solvent exchange steps and subsequent activation under dynamic vacuum (Figure 3.20). First, unreacted materials and decomposition

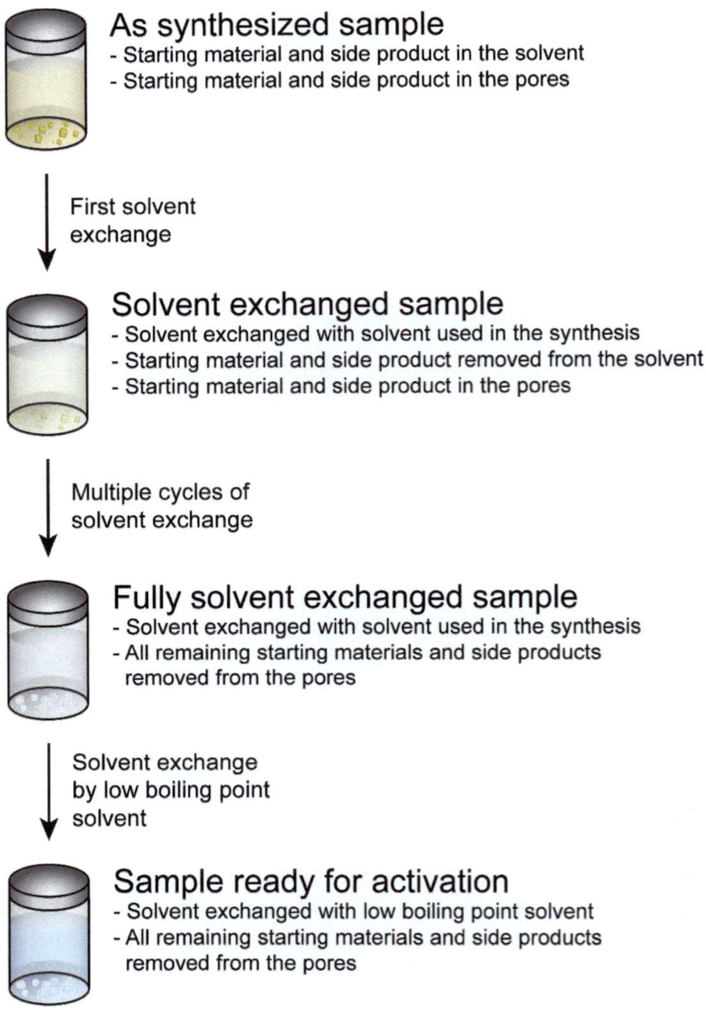

Figure 3.20 Washing and solvent exchange steps prior to activation. The solvent of the reaction mixture is exchanged by the solvent used in the synthesis or a chemically similar solvent to remove unreacted starting materials and byproducts from the pores of the MOF. This procedure is repeated multiple times. Subsequently, the solvent is exchanged by a lower boiling point solvent that is appropriate for further activation by evacuation (dichloromethane, chloroform, or acetone) or supercritical CO_2 exchange (ethanol or methanol).

products residing within the pores of the MOF are washed out using the solvent employed in the synthesis of the MOF. Once the pores are free from these byproducts, the solvent is exchanged by one that has a lower surface tension and therefore "puts less stress on the framework" during the evacuation step. Solvents commonly used in this step include chloroform, dichloromethane, acetone, ethanol, or methanol.

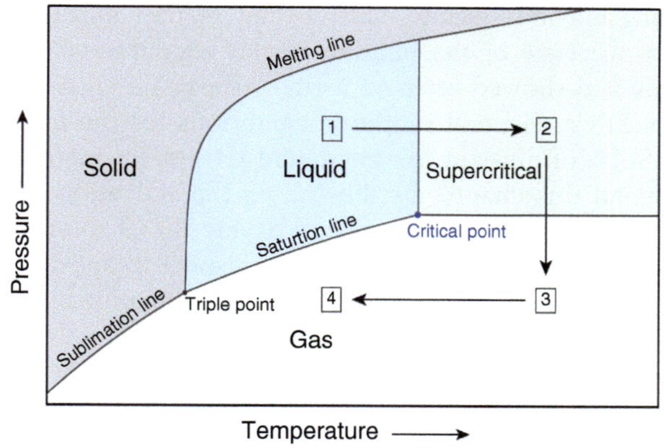

Figure 3.21 Schematic phase diagram of CO_2 illustrating the path for supercritical activation. By passing through the supercritical state the phase transition between liquid and gaseous phase can be avoided, decreasing the mechanical stress on the framework structure since no surface tension resulting in capillary forces is acting on the framework. This method is especially promising when dealing with mesoporous materials.

Subsequent evacuation yields the activated material. For MOFs, whose structures contain large pores (typically >2 nm, mesopores), evacuation often resulted in a (partial) loss of porosity and surface areas lower than the expected values. To understand this finding, it is instructive to consider the processes occurring during the removal of solvent as a function of the pore diameter. For small pores, the forces acting on the framework upon evacuation are comparatively small. Molecules will simply evaporate, and no capillary forces will act on the pores. In contrast, in pores larger than 2 nm (mesopores), capillary forces begin to play a significant role. To avoid "stress" on the framework originating from capillary forces, one has to prevent the evaporation and the concomitant formation of a liquid meniscus within the pores, which can be achieved by supercritical CO_2 exchange [19]. This method circumvents the forces acting on the material upon evaporation of the solvent caused by the phase change from the liquid to the gaseous phase. To avoid passing through the liquid–gas phase boundary the solvent is exchanged to liquid CO_2 at low temperatures and elevated pressure (Figure 3.21). By heating the sample, the supercritical region of the phase diagram is reached. Here, the densities of the liquid phase and vapor phase are equal and upon release of the pressure the gaseous state is reached without passing through a phase boundary. Cooling to room temperature yields the fully activated material [19].

3.6 Summary

In this chapter we outlined the synthesis and chemistry of building units commonly used to synthesize carboxylate-based MOFs. We introduced frequently used routes for the synthesis of the organic linker and gave examples of common linkers with topicity 2–8 alongside strategies to control their metrics and the local symmetry. We gave an overview of the binding modes encountered in metal-carboxylate compounds and showed selected metal-carboxylate clusters encountered in MOFs as the SBUs. General synthesis conditions for the formation of MOFs built from SBUs of divalent, trivalent, and tetravalent metals were outlined and we concluded this chapter by illustrating the activation of MOFs using traditional methods as well as super critical drying. In Chapter 4 we will discuss the structures of MOF in more detail with respect to specific requirements in terms of the chemistry and structure of their building units.

References

1 Wells, A. (1954). The geometrical basis of crystal chemistry. Part 1. *Acta Crystallographica* 7 (8–9): 535–544.
2 Yaghi, O.M., O'Keeffe, M., Ockwig, N.W. et al. (2003). Reticular synthesis and the design of new materials. *Nature* 423 (6941): 705–714.
3 Miyaura, N. and Suzuki, A. (1995). Palladium-catalyzed cross-coupling reactions. *Chemical Reviews* 95 (1): 2457–2483.
4 Sonogashira, K. (2002). Development of Pd–Cu catalyzed cross-coupling of terminal acetylenes with sp^2-carbon halides. *Journal of Organometallic Chemistry* 653 (1): 46–49.
5 Heck, R.F. and Nolley, J.P. (1972). Palladium-catalyzed vinylic hydrogen substitution reactions with aryl, benzyl, and styryl halides. *The Journal of Organic Chemistry* 37 (14): 2320–2322.
6 Schnobrich, J.K., Lebel, O., Cychosz, K.A. et al. (2010). Linker-directed vertex desymmetrization for the production of coordination polymers with high porosity. *Journal of the American Chemical Society* 132 (39): 13941–13948.
7 (a) Chae, H.K., Eddaoudi, M., Kim, J. et al. (2001). Tertiary building units: synthesis, structure, and porosity of a metal-organic dendrimer framework (MODF-1). *Journal of the American Chemical Society* 123 (46): 11482–11483. (b) Zou, Y., Park, M., Hong, S., and Lah, M.S. (2008). A designed metal-organic framework based on a metal-organic polyhedron. *Chemical Communications* (20): 2340–2342.
8 (a) Tranchemontagne, D.J., Mendoza-Cortes, J.L., O'Keeffe, M., and Yaghi, O.M. (2009). Secondary building units, nets and bonding in the chemistry of metal-organic frameworks. *Chemical Society Reviews* 38 (5): 1257–1283. (b) Tasiopoulos, A.J., Vinslava, A., Wernsdorfer, W. et al. (2004). Giant single-molecule magnets: a $\{Mn_{84}\}$ torus and its supramolecular nanotubes. *Angewandte Chemie International Edition* 116 (16): 2169–2173.
9 Lu, W., Wei, Z., Gu, Z.-Y. et al. (2014). Tuning the structure and function of metal-organic frameworks via linker design. *Chemical Society Reviews* 43 (16): 5561–5593.
10 Karagiaridi, O., Bury, W., Mondloch, J.E. et al. (2014). Solvent-assisted linker exchange: an alternative to the *de novo* synthesis of unattainable metal-organic frameworks. *Angewandte Chemie International Edition* 53 (18): 4530–4540.

11 Baes, C.F. and Mesmer, R.E. (1976). *Hydrolysis of Cations*. Wiley.

12 Smart, S.E., Vaughn, J., Pappas, I., and Pan, L. (2013). Controlled step-wise isomerization of the Keggin-type Al_{13} and determination of the γ-Al_{13} structure. *Chemical Communications* 49 (97): 11352–11354.

13 (a) Harris, W.R. and Martell, A.E. (1976). Aqueous complexes of gallium(III). *Inorganic Chemistry* 15 (3): 713–720. (b) Bradley, S.M., Kydd, R.A., and Yamdagni, R. (1990). Detection of a new polymeric species formed through the hydrolysis of gallium(III) salt solutions. *Journal of the Chemical Society, Dalton Transactions* (2): 413–417. (c) Mensinger, Z.L., Gatlin, J.T., Meyers, S.T. et al. (2008). Synthesis of heterometallic group 13 nanoclusters and inks for oxide thin-film transistors. *Angewandte Chemie International Edition* 47 (49): 9484–9486. (d) Radnai, T., Bálint, S., Bakó, I. et al. (2014). The structure of hyperalkaline aqueous solutions containing high concentrations of gallium – a solution X-ray diffraction and computational study. *Physical Chemistry Chemical Physics* 16 (9): 4023–4032. (e) Petrosyants, S. and Ilyukhin, A. (2011). Indium(III) coordination compounds. *Russian Journal of Inorganic Chemistry* 56 (13): 2047–2069. (f) Aldridge, S. and Downs, A.J. (2011). *The Group 13 Metals Aluminium, Gallium, Indium, and Thallium: Chemical Patterns and Peculiarities*. Wiley.

14 (a) Richens, D.T. (2005). Ligand substitution reactions at inorganic centers. *Chemical Reviews* 105 (6): 1961–2002. (b) Taube, H. (1952). Rates and mechanisms of substitution in inorganic complexes in solution. *Chemical Reviews* 50 (1): 69–126.

15 (a) Stock, N. (2011). Metal-organic frameworks: aluminium-based frameworks. In: *Encyclopedia of Inorganic and Bioinorganic Chemistry*. Wiley. (b) Czaja, A.U., Trukhan, N., and Müller, U. (2009). Industrial applications of metal-organic frameworks. *Chemical Society Reviews* 38 (5): 1284–1293.

16 Kickelbick, G. and Schubert, U. (1997). Oxozirconium methacrylate clusters: $Zr_6(OH)_4O_4(OMc)_{12}$ and $Zr_4O_2(OMc)_{12}$ (OMc = methacrylate). *European Journal of Inorganic Chemistry* 130 (4): 473–478.

17 Bai, Y., Dou, Y., Xie, L.-H. et al. (2016). Zr-based metal-organic frameworks: design, synthesis, structure, and applications. *Chemical Society Reviews* 45 (8): 2327–2367.

18 Nguyen, H.L., Gándara, F., Furukawa, H. et al. (2016). A titanium-organic framework as an exemplar of combining the chemistry of metal- and covalent-organic frameworks. *Journal of the American Chemical Society* 138 (13): 4330–4333.

19 Nelson, A.P., Farha, O.K., Mulfort, K.L., and Hupp, J.T. (2008). Supercritical processing as a route to high internal surface areas and permanent microporosity in metal-organic framework materials. *Journal of the American Chemical Society* 131 (2): 458–460.

4

Binary Metal-Organic Frameworks

4.1 Introduction

In Chapter 3 we have explored the wide range of linker shapes accessible through organic synthesis and gave an insight into the chemistry of the secondary building units (SBUs), the elements they are typically composed of, and their variety with respect to geometry and connectivity. Combining these two components allows for the formation of a myriad of extended framework structures with different topologies. In this chapter, we will analyze important framework structures categorized by the connectivity of their SBUs starting from low connectivity and work our way up to highly connected SBUs.

4.2 MOFs Built from 3-, 4-, and 6-Connected SBUs

4.2.1 3-Connected (3-c) SBUs

Most structures based on 3-connected (3-c) nets are found in coordination networks built from single metal nodes and neutral donor linkers. In contrast, the 3-c species in metal-organic framework (MOF) structures is most often the organic linker. However, there are examples of MOF structures involving 3-c SBUs, such as the dinuclear paddle wheel $M_2(-COO)_3$ (e.g. M = Cu^{2+}, Zn^{2+}), an SBU that is closely related to the more common 4-c paddle wheel $M_2(-COO)_4$ we encountered earlier. As discussed in Chapter 3, subtle changes in the reaction conditions can have a dramatic impact on which SBU forms and thus also on the structure of the resulting MOF. An example of this is PNMOF-3 ($Zn_4(NH_2\text{-BDC})_3(NO_3)_2(H_2O)_2$). Even though PNMOF-3 is synthesized from the very same starting materials ($Zn(NO_3)_2$ and $(NH_2)\text{-}H_2BDC$) as IRMOF-3 ($Zn_4O(NH_2\text{-BDC})_3$) its structure is fundamentally different [1]. PNMOF-3 comprises 3-c $Zn_2(-COO)_3$ paddle wheel SBUs rather than the 6-c octahedral $Zn_4O(-COO)_6$ SBUs found in IRMOF-3 and the formation of these 3-c paddle wheel SBUs is ascribed to the presence of a cross-linked copolymer in the reaction mixture, hence the name PNMOF-3. The hexagonal crystal structure of PNMOF-3 consists of stacked 2D layers of **hcb** (honeycomb) topology (Figure 4.1) that are stacked in an eclipsed arrangement resulting in hexagonal channels of 14.9 Å diameter running along the crystallographic *c*-axis. The

Introduction to Reticular Chemistry: Metal-Organic Frameworks and Covalent Organic Frameworks,
First Edition. Omar M. Yaghi, Markus J. Kalmutzki, and Christian S. Diercks.

Companion website: www.wiley.com/go/yaghi/reticular

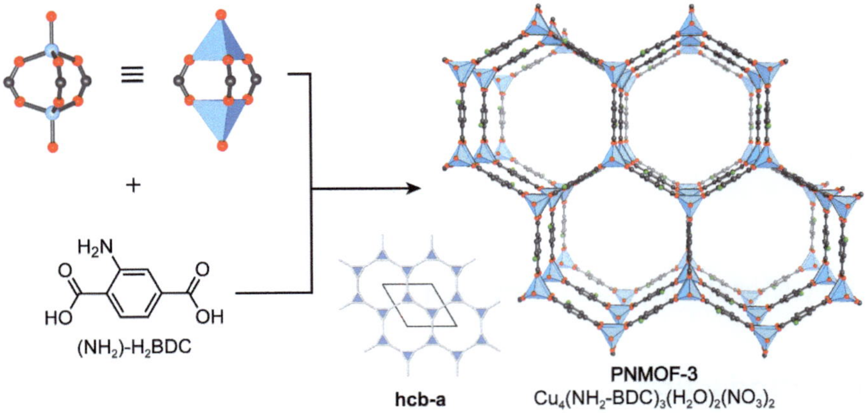

Figure 4.1 Crystal structure of PNMOF-3 formed by reticulation of Zn^{2+} and (NH_2)-H_2BDC in the presence of a cross-linked copolymer. The reticulation of a 3-c SBU and ditopic (NH_2)-H_2BDC linkers affords a framework of **hcb** topology. The eclipsed arrangement of the hexagonal layers leads to the formation of hexagonal channels of 14.9 Å. All hydrogen atoms are omitted for clarity. Color code: Cu, blue; C, gray; N, green; O, red.

charge of the cationic framework is compensated by one terminal nitrate ligand per $Zn_2(-COO)_3$ paddle wheel SBU. Even though the 3-c dinuclear paddle wheel SBU is less common than its 4-c analog ($Zn_2(-COO)_4$), several highly stable porous frameworks based on this SBU are known [2].

4.2.2 4-Connected (4-c) SBUs

The most common 4-connected (4-c) SBU in MOF chemistry is the dinuclear paddle wheel $M_2(-COO)_4$ (e.g. M = Cu^{2+}, Zn^{2+}, Co^{2+}), in which each metal ion is coordinated in a square-pyramidal fashion by four oxygen of the bridging carboxylates and one terminal neutral ligand, most often water. For clarity, the capping ligands are often omitted, and the paddle wheel unit is represented as two squares connected by four carboxylates. In terms of net topology, the paddle wheel is represented by a square 4-c vertex figure (Figure 4.2).

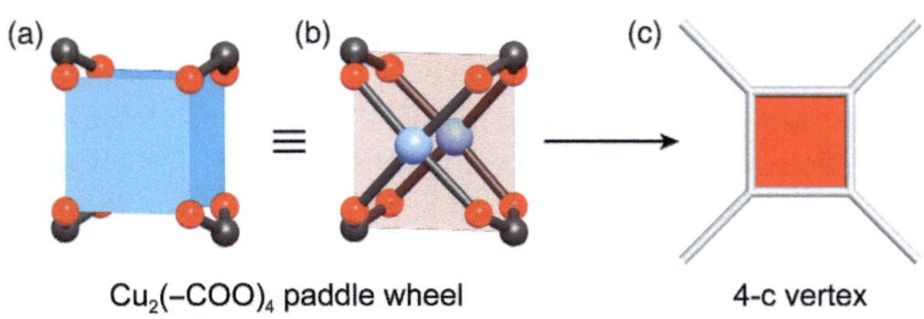

Figure 4.2 Representations of the dinuclear $M_2(-COO)_4$ paddle wheel SBU. (a) Two metal centers form a dinuclear paddle wheel complex. (b) Connecting the carboxylate carbons reduces the paddle wheel to a square building unit with four points of extension. (c) Topological representation of the vertex figure as a 4-c square building unit. Color code: Cu, blue; C, gray; O, red.

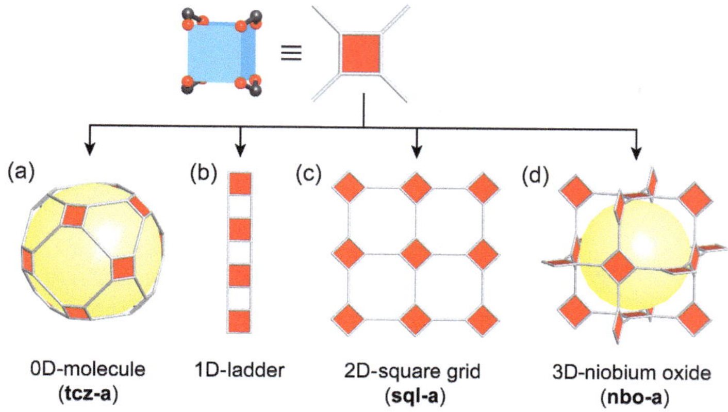

Figure 4.3 Topologies of structures formed by linking 4-c paddle wheel SBUs (represented by red squares) and ditopic linkers (gray lines connecting the squares). The precise geometry of the linker directs the reticular synthesis toward (a) discrete, 0D metal-organic polyhedra (MOPs, here a truncated cuboctahedron [**tcz-a**] is shown), (b) 1D chain/ladder, (c) 2D square layers (**sql-a** or **fes**), and (d) 3D networks (here a niobium oxide net [**nbo-a**] is shown). In the case of the **nbo** net, the yellow ball is placed in the frame of the structure as a visual aid. Note that all ditopic linkers are represented by gray lines regardless of their exact molecular structure.

Linking such 4-c square building units with ditopic linkers can afford a variety of different structures depending on the exact geometry of the linker. Here, we will discuss the geometrical requirement for the formation of (i) discrete 0D metal-organic polyhedra (MOPs)[1], (ii) 1D chains, (iii) 2D layers, and (iv) 3D networks (Figure 4.3).

While the geometry of the paddle wheel SBU is fixed, in the molecular structure of ditopic linkers, three different angles can be deliberately modified: a bending angle θ between the coplanar binding groups can be introduced by altering the positions of the binding groups on the central unit, a dihedral angle φ between the binding groups is created by twisting them about the axis of the linker, and an angle ψ between the planes of the binding groups results from bending them toward each other (Figure 4.4).

Reticulating Cu^{2+} ions and ditopic m-H_2BDC linkers with a bending angle θ of approximately 120° between the coplanar carboxylic binding groups yields a 0D MOP of **tcz** topology, called MOP-1 ($Cu_2(m\text{-}BDC)_2$, Figure 4.5) [3]. MOPs of the same topology can generally be synthesized by reticulating ditopic linkers that have an angle θ between 90° and 120° with metal ions favoring the formation of square 4-c paddle wheel SBUs. The synthesis and design of MOPs is part of reticular chemistry and is further discussed in Chapter 19. Examples illustrating the use of MOPs as tertiary building units (TBUs) in the synthesis of extended structures are discussed in Chapter 5.

Introduction of an angle ψ between the two carboxylate binding groups of the linker, as in 4,4′-DMEDBA, and reticulation of this linker with Cu^{2+}

1 MOPs will be discussed in more detail in Chapter 19.

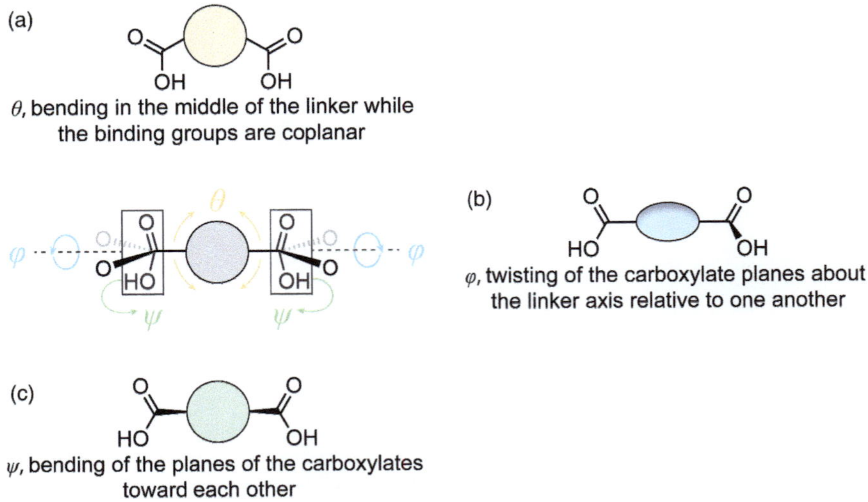

Figure 4.4 Different angles on a ditopic linker that can be altered by precise design of the molecular structure. (a) The bending angle θ refers to a bent linker with coplanar binding groups, (b) the dihedral angle φ refers to non-coplanar binding groups, and (c) the angle ψ refers to binding groups bent toward each other.

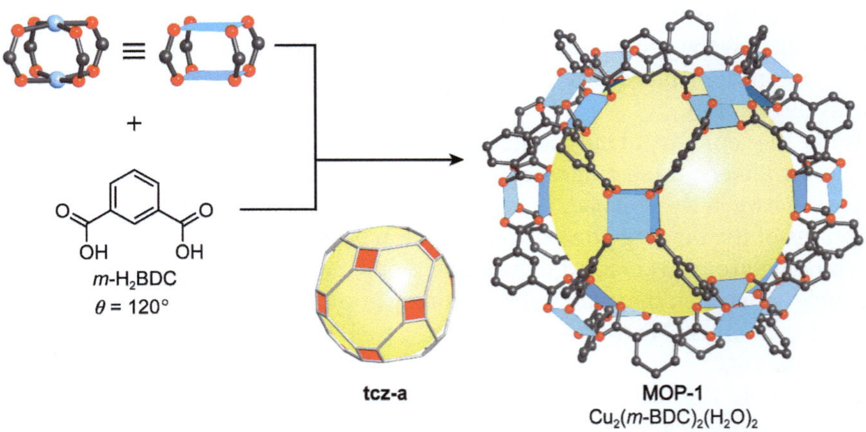

Figure 4.5 Reticulation of m-H_2BDC and Cu^{2+} ions facilitates the formation of a molecular polyhedron of **tcz**, termed MOP-1. Here, the precise adjustment of θ, the bending angle between the coplanar binding groups, dictates the product of the synthesis. The terminal water ligands on the copper paddle wheel as well as all hydrogen atoms are omitted for clarity. The yellow sphere represents the empty space within the **tcz** polyhedron. Color code: Cu, blue; C, gray; O, red.

gives MOF-222 (Cu_2(4,4′-DMEDBA)$_2$) with a structure built from 1D chains (Figure 4.6) [4].

A 2D structure is formed when 4-c copper paddle wheel SBUs are connected by linear ditopic linkers with coplanar carboxylic acid binding groups ($\varphi = 180°$). In this scenario, the binding groups are either coplanar to the backbone of the

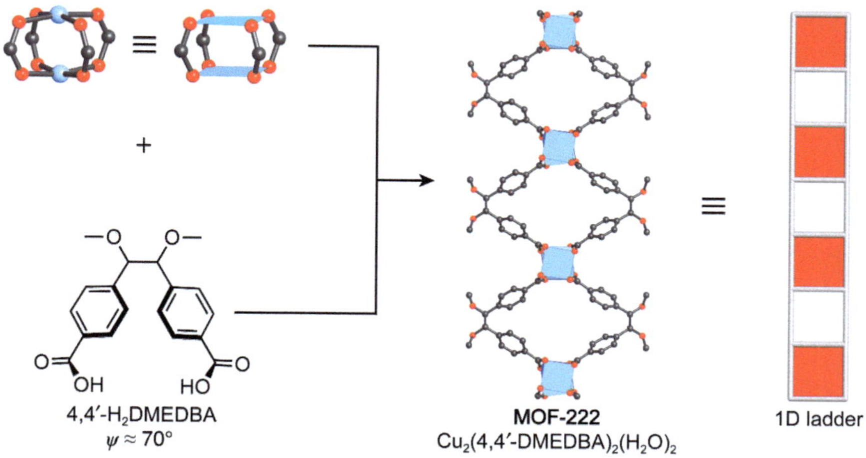

Figure 4.6 Crystal structure of MOF-222. The formation of 1D chains by linking 4-c paddle wheels and ditopic linkers is achieved by introducing an angle ψ (approximately 70°) between the two carboxylate binding groups. The terminal water ligands on the copper paddle wheel SBUs as well as all hydrogen atoms are omitted for clarity. Only a section of one chain is shown. Color code: Cu, blue; C, gray; O, red.

linker, as in the case of the H_2BDC (MOF-2, see Figure 1.14), or they are rotated out of plane with respect to the central aryl unit, as in the case of (Cl_2)-H_2BDC (Figure 4.7). The steric hindrance of the chlorine substituents increases the rotational barrier, thereby fixing the binding groups in a coplanar orientation perpendicular to the aryl core unit. Joining 4-c paddle wheel SBUs through BDC or (Cl_2)-BDC affords 2D **sql** structures termed MOF-2 and MOF-102 $(Cu_2(Cl_2\text{-BDC})_2)$, respectively (Figure 4.7) [5].

A combination of ditopic linkers and 4-c paddle wheel units can further be used to construct extended 3D frameworks. This is achieved when a dihedral angle ($\varphi \neq 0°$) between the two binding groups of the ditopic linker is introduced, resulting in MOFs with **nbo** (niobium oxide) topology as exemplified by MOF-101 ($Cu_2(o\text{-Br-BDC})_2$) illustrated in Figure 4.8 [6]. In this example, the sterically demanding bromide substituent on the linker increases the rotational barrier of one of the carboxylic acid groups, fixing the binding groups in a perpendicular orientation with respect to each other. While the synthesis at low temperatures affords MOF-101, at elevated temperatures, where the rotational barrier can be overcome by thermal energy, a framework of **sql** topology is formed.

The above shown examples illustrate the importance of the precise geometrical design of the molecular structure of linker molecules to direct the synthesis into either 0D, 1D, 2D, or 3D structures. The example of MOF-101 also highlights that it is important to adjust the reaction conditions to ensure an unaltered linker geometry throughout the reticulation process. These considerations emphasize that careful analysis of the geometry of the building units as well as their physical and chemical properties is required to enable the *a priori* synthesis of MOFs [7].[2]

2 Further examples on linking up squares are reported in a 2002 review article [5].

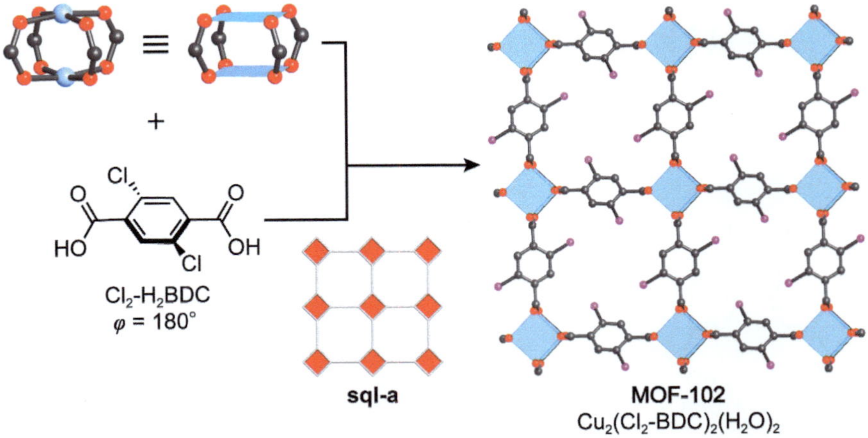

Figure 4.7 Crystal structure of MOF-102. The combination of 4-c square SBUs and linear ditopic linkers with coplanar binding groups ($\varphi = 180°$) leads to the formation of a 2D framework with **sql** topology. The **sql** layers are stacked along the *c*-direction, here, only one layer is shown for clarity. The terminal water ligands on the copper paddle wheel SBUs as well as all hydrogen atoms are omitted for clarity. Color code: Cu, blue; C, gray; O, red; Cl, pink.

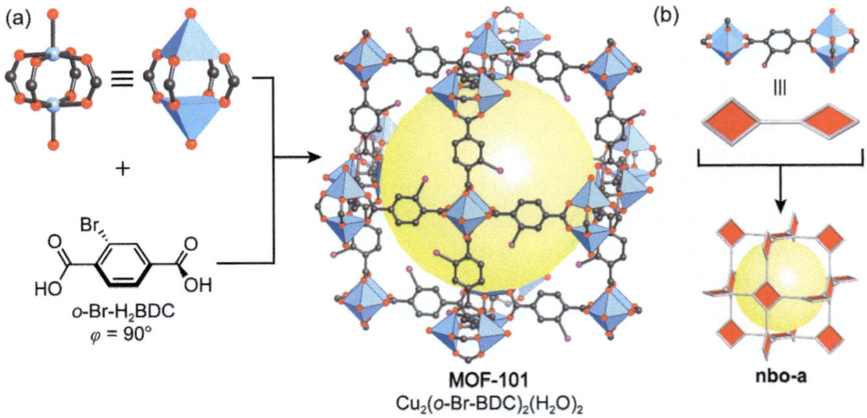

Figure 4.8 (a) The cubic crystal structure of MOF-101 formed by reticulation of 4-c copper paddle wheel SBUs and linear ditopic *o*-Br-H_2BDC linkers. (b) The structure has an underlying **nbo** topology. The formation of this net is achieved by introducing a dihedral angle ($\varphi = 90°$) between the two binding groups. All hydrogen atoms are omitted for clarity. Color code: Cu, blue; C, gray; O, red; Br, pink.

Another way to form 3D frameworks using 4-c paddle wheel SBUs is to use organic linkers with more than two points of extension. An early example of this approach is HKUST-1 (HKUST = Hong Kong University of Science and Technology) ($Cu_3(BTC)_2$, BTC = benzenetricarboxylate), a MOF based on a 3,4-connected **tbo** (twisted boracite) net (Figure 4.9) [8]. The structure comprises 4-c copper paddle wheel SBUs and tritopic BTC linkers, and encompasses an

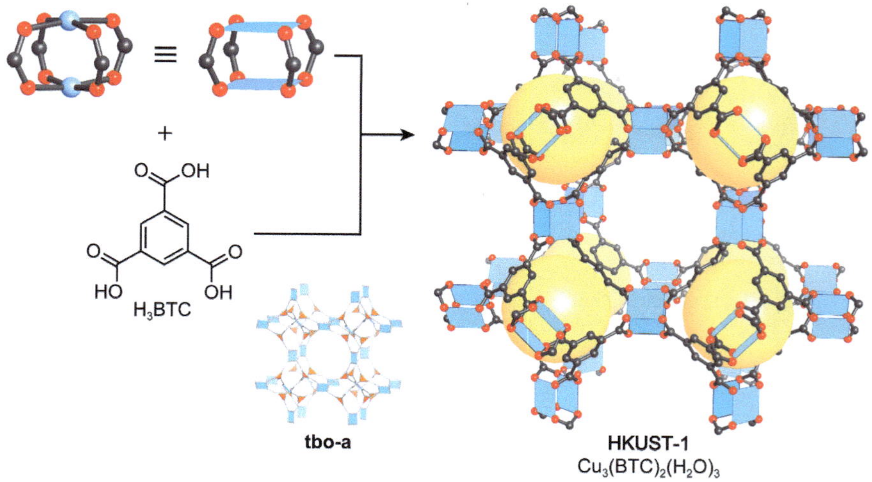

Figure 4.9 Crystal structure of HKUST-1. The **tbo** net is formed by linking 4-c $Cu_2(-COO)_4$ SBUs with trigonal tritopic BTC linkers (D_{3h} symmetry). Each large cubic pore is surrounded by four smaller octahedral pores. The large pores are connected through square windows and form a 3D intersecting pore system. The inset shows the augmented **tbo-a** net. The terminal water ligands on the copper paddle wheel SBUs and all hydrogen atoms are omitted for clarity. Color code: Cu, blue; C, gray; O, red.

intersecting 3D system with square pore openings of 9 × 9 Å. The overall void space is 40.7%, a value comparable to that of many zeolites.

The terminal water ligands bound axially to the paddle wheel SBUs in HKUST-1 can be removed by heating in dynamic vacuum, which results in open metal sites and a color change of the material from blue to purple. Initial gas adsorption measurements yielded relatively low surface areas since the MOF was not properly activated, more recent measurements gave surface areas of up to 2200 $m^2\ g^{-1}$. The high density of open metal sites in fully activated HKUST-1 makes it one of the best performing materials with respect to natural gas and hydrogen storage [9]. HKUST-1 structures containing metals other than copper are architecturally unstable and collapse upon activation as evidenced by the observed negligible nitrogen uptake [10]. The modular structure of HKUST-1 initiated the development of many isoreticular expanded versions such as PCN-6 (PCN = Porous Coordination Network) or MOF-399 [11]. The latter still holds the world record for the lowest density MOF (0.126 $g\ cm^{-3}$).

The H_3BTC linker employed in the synthesis of HKUST-1 has D_{3h} symmetry, with all carboxylic acid groups being coplanar to the central aryl unit. As discussed earlier, the precise geometry of the linker dictates whether a specific topology is favored over another. Hence, when an expanded version of H_3BTC is designed it is required that the overall planar conformation is retained. Symmetric elongation of H_3BTC by one benzene ring gives H_3BTB. Here, the steric repulsion between the aromatic protons of the central aryl unit and those of the terminal benzoic acid groups enforces a dihedral angle between them causing

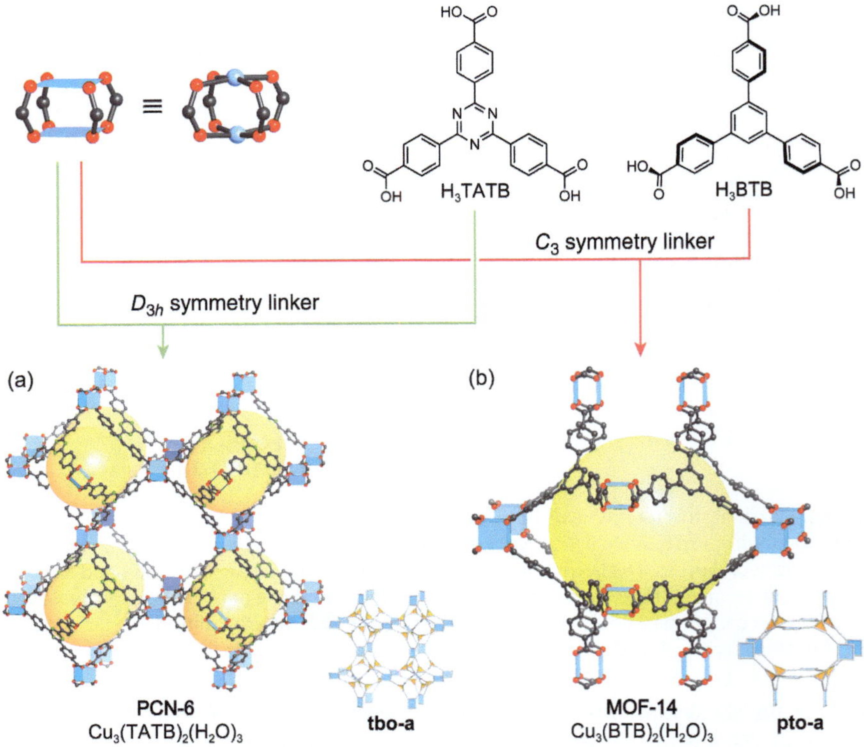

Figure 4.10 Comparison of the crystal structures of PCN-6 and MOF-14. Both structures represent 3,4-connected nets. (a) The D_{3h} symmetry of the H_3TATB linker affords the formation PCN-6 (**tbo**), (b) whereas the lower C_3 symmetry of H_3BTB directs the synthesis toward the formation of MOF-14 (**pto**). Terminal water ligands on the copper paddle wheel SBUs, all hydrogen atoms, and all interpenetrating frameworks in the structure of MOF-14 are omitted for clarity. Color code: Cu, blue; C, gray; N, green; O, red.

H_3BTB to have C_3 (propeller shape) instead of D_{3h} symmetry. Therefore, the reticulation of H_3BTB and Cu^{2+} cannot afford a framework of **tbo** topology and consequently the formation of an interpenetrated framework of **pto** (platinum oxide) topology, termed MOF-14 ($Cu_3(BTB)_2$), is favored (Figure 4.10b) [12]. Despite interpenetration, MOF-14 has large cavities of 16.4 Å diameter and a pore aperture of 7.66 × 14 Å. To avoid the nonplanar conformation of the linker due to steric repulsion between the aromatic protons a triazine core unit can be used. In this way, the perfectly planar expanded equilateral tritopic linker H_3TATB (D_{3h} symmetry) can be synthesized, which allows for the formation of a framework with an underlying **tbo** topology, termed PCN-6 ($Cu_3(TATB)_2$) (Figure 4.10) [13].

4.2.3 6-Connected (6-c) SBUs

6-c SBUs can be built from a wide range of metals and can have different geometries ranging from highly symmetric octahedral, trigonal prismatic, and hexagonal planar, to lower more distorted symmetries. Here we will discuss the most frequently encountered 6-c SBUs in MOF chemistry and important structures built therefrom.

The octahedral $Zn_4O(-COO)_6$ SBU is one of the most common building units in MOF chemistry. In Chapter 2 we have discussed the structures of MOF-177 and its isoreticular expanded analogs having the **qom** topology (hetero-dual, highest symmetry embedding $P\bar{3}1c$, No. 159).[3] Because the **qom** topology is not the default topology for the combination of trigonal and an octahedral building units, frameworks built from a combination of these building units are expected to crystallize in the higher symmetry 3,6-connected **pyr** topology (self-dual, highest symmetry embedding $Pa\bar{3}$, No. 205). While the **qom** net is a hetero-dual net, and structures of this topology are thus not expected to interpenetrate upon isoreticular expansion of **pyr** frameworks (self-dual net) results in interpenetrated frameworks. Here, we want to compare two related structures, MOF-150 ($Zn_4O(TCA)_2$) and MOF-177 ($Zn_4O(BTB)_2$) with underlying **pyr** and **qom** topology, respectively, and present possible factors leading to the formation of one topology or the other (Figure 4.11). Since the synthesis conditions for both MOFs are similar, no structure directing agent is employed, and the SBUs in both compounds are identical, the formation of the two different topologies must originate from the linkers. Both MOFs are built from equilateral trigonal tritopic linkers with terminal benzoic acid groups that only differ in their respective core unit. The H_3TCA linker employed to prepare MOF-150 is constructed around a monoatomic nitrogen core that facilitates the free and independent rotation of the three terminal benzoate units. In contrast, due to steric repulsion between aromatic hydrogen atoms within the H_3BTB linker used in the synthesis of MOF-177, more constraints are imparted with respect to the dihedral angle between the aryl core and the terminal benzoic acids, and their orientation relative toward each other. Linking basic zinc carboxylate SBUs and BTB gives a framework of **qom** topology, whereas the reticulation of the same SBU and TCA linkers directs the framework formation toward the higher symmetry **pyr** topology. Here, the edge-transitive **pyr** net is the thermodynamic and the **qom** net is the kinetic product.

With more than 400 reported MOF structures, trigonal prismatic SBUs of the general formula $M_3OL_3(-COO)_6$ ($M = Al^{3+}$, In^{3+}, Cr^{3+}, V^{3+}) are among the most common inorganic building units of MOFs [14]. A structurally similar SBU of general formula $M_3(-COO)_6$ can be formed from divalent metals such as copper. In the following text, we will discuss some examples of MOFs constructed from the $M_3OL_3(-COO)_6$ SBU and linear ditopic (MIL-101), trigonal tritopic (MIL-100), and rectangular tetratopic (**soc**-MOF) linkers [14c, d, f, g].

Both, MIL-100 ($[M_3OL_3](BTC)_2$) and MIL-101 ($[M_3OL_3](BDC)_3$) (MIL = Materials Institute Lavoisier), are built from trigonal prismatic $M_3OL_3(-COO)_6$ SBUs and share the same topology even though the linker used in their synthesis are of different topocity; tritopic H_3BTC is used to prepare MIL-100 and ditopic H_2BDC is employed in the synthesis of MIL-101 [14b–d]. This may seem confusing at first but can be explained in terms of choosing larger

3 For more information on the terminology used in the topological description of crystal structures see Chapter 18.

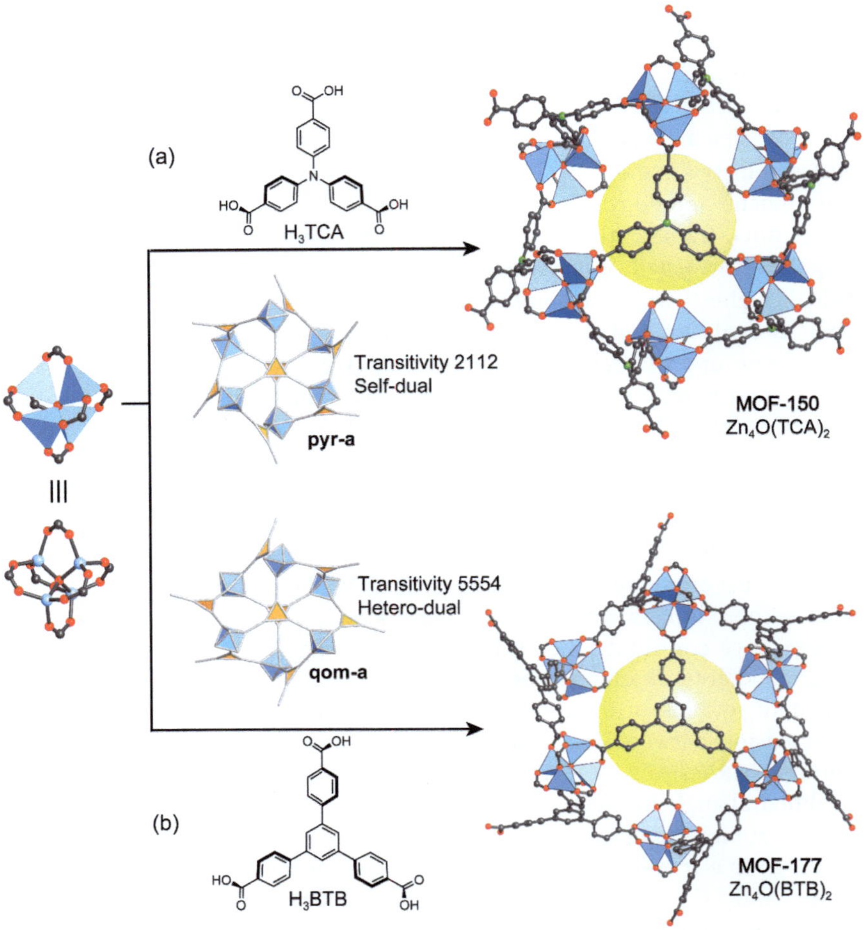

Figure 4.11 Comparison of the crystal structures and topologies of MOF-150 and MOF-177. (a) A framework with the default self-dual **pyr** net (transitivity 2112) is formed when H_3TCA is used, (b) while employing H_3BTB affords the formation of a framework based on the **qom** net (transitivity 5554), a hetero-dual net that is unlikely to support interpenetration. The corresponding augmented nets are shown next to the respective crystal structure. All hydrogen atoms are omitted for clarity. Color code: Cu, blue; C, gray; N, green; O, red.

"tertiary building units" rather than secondary building units to simplify the overall structure. In the case of MIL-100 and MIL-101, these TBUs are tetrahedral units formed by linking four 6-c SBUs by either four BTC or six BDC units (Figure 4.12, top). In MIL-100 the tritopic BTC linker is positioned on the faces of these tetrahedra, whereas in MIL-101 the ditopic linkers constitute their edges (Figure 4.12). By further linking these tetrahedral TBUs in a vertex-sharing manner two differently sized cages are formed that are connected through 5-membered rings to give a framework with an overall **mtn** topology, a tetrahedral topology commonly found in zeolites (Figure 4.12, bottom) [15].

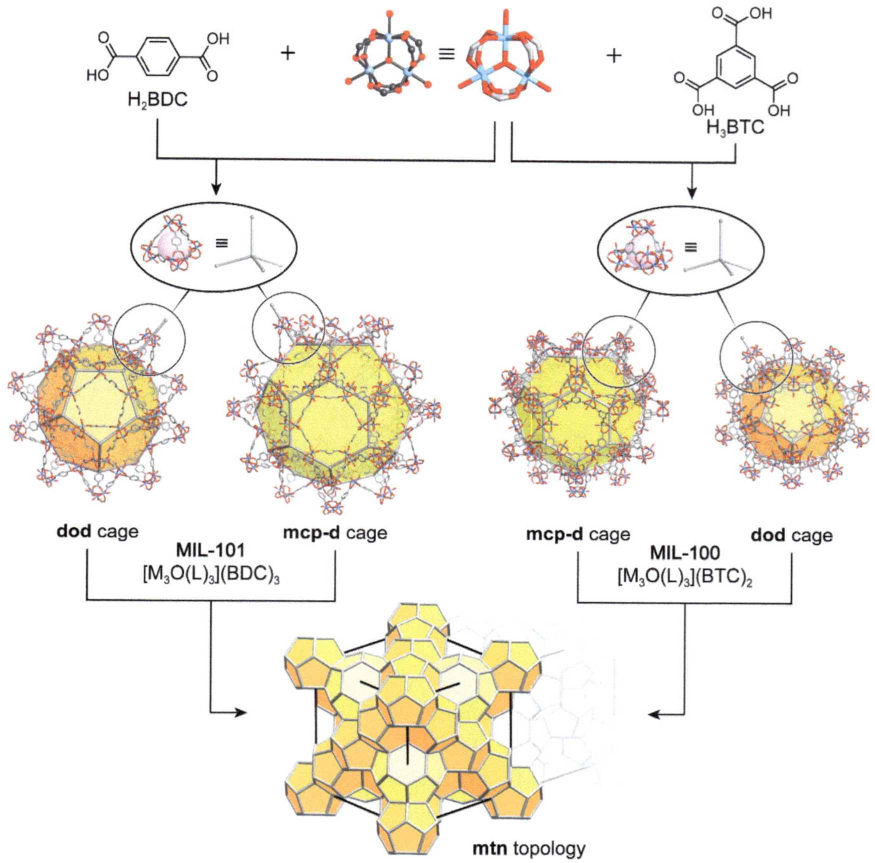

Figure 4.12 Topological analysis of the structures of MIL-100 and MIL-101. In both structures, tetrahedral TBUs are formed by joining trinuclear $M_3OL_3(\text{–COO})_6$ SBUs with trigonal tritopic BTC or linear ditopic BDC linkers, respectively. These TBUs are further linked to form two differently sized cages (a small **dod** and a large **mcp-d** cage) that are fused to give frameworks of **mtn** topology. All hydrogen atoms are omitted for clarity. Color code: Metal, blue; C, gray, O, red; **dod** cages are represented as orange, **mcp-d** cages as yellow polyhedra.

In MIL-100 these cages have diameters of 25 and 29 Å with pore openings of 5 and 9 Å, respectively [14c]. With diameters of 29 and 34 Å and pore openings of 12 and 16 Å, the cages in MIL-101 are slightly larger [14d]. Both materials displayed unprecedented chemical stability at the time of their discovery, which in combination with their zeolite-like structure (small pore apertures, large cages) made them interesting candidates for their application in catalytic processes. Accordingly, the incorporation of functional groups, discrete molecules, and metal nanoparticles into the pores of MIL-101 was pursued. The resulting materials typically show high catalytic activities for a variety of organic transformations [14e, 16]. The expansion of the linkers used in the synthesis of MIL-100 and MIL-101 yields an isoreticular series of frameworks with pore sizes up to 68 Å (see Figure 2.13) [17].

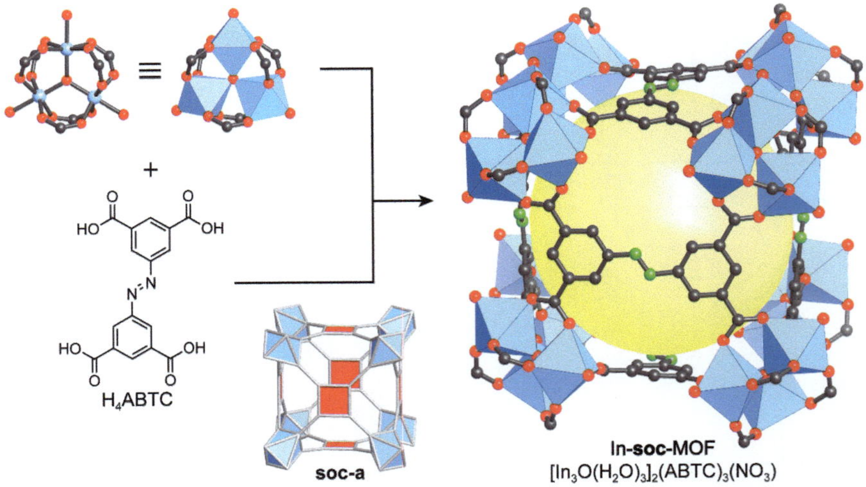

Figure 4.13 Crystal structure of In-**soc**-MOF and its topology representation as a **soc** net. Removal of the terminal water ligands affords the creation of open metal sites pointing into the narrow pores. The augmented **soc** net is depicted next to the crystal structure of In-**soc**-MOF. The NO_3 counter ions residing in the open pores of the framework as well as all hydrogen atoms are omitted for clarity. Color code: In, blue; C, gray; N, green; O, red.

MOFs with an underlying 4,6-connected **soc**[4] topology are another important example of framework structures involving $M_3OL_3(-COO)_6$ SBUs. The structure of In-**soc**-MOF ($[In_3O(H_2O)_3]_2(ABDC)_3NO_3)$) is built from rectangular offset tetratopic ABTC linkers and trigonal prismatic $In_3O(H_2O)_3(-COO)_6$ SBUs (Figure 4.13). The chemical formula of the SBU suggests a cationic framework where the charge is neutralized by $(NO_3)^-$ anions residing in the pores of the open framework structure [14f]. The crystal structure of In-**soc**-MOF and the underlying **soc** topology are given in Figure 4.13.

The terminal water ligands on the $In_3O(H_2O)_3(-COO)_6$ SBUs can be removed by activation under dynamic vacuum at elevated temperatures, leaving behind open metal sites pointing into the narrow pores of the MOF. This results in a high localized charge density within the pores that is advantageous for gas adsorption, in particular for hydrogen storage. This is indeed the case for In-**soc**-MOF, which has a comparatively high hydrogen storage capacity of 2.61% at 78 K and 1.2 atm. As expected, H_2 is strongly bound to the open metal sites (primary adsorption site) and at higher loadings, other adsorption sites with lower binding energy are occupied as evidenced by inelastic neutron scattering experiments [14f]. MOFs with the same underlying topology, but built from a lighter metal, hold great promise with respect to gas storage applications, since an increase in gravimetric gas uptake is expected while the high volumetric uptake is retained. This principle is used in the construction of an isoreticular

4 The denotation **soc**, or "square-octahedron" refers to the reticulation of square and octahedral building units.

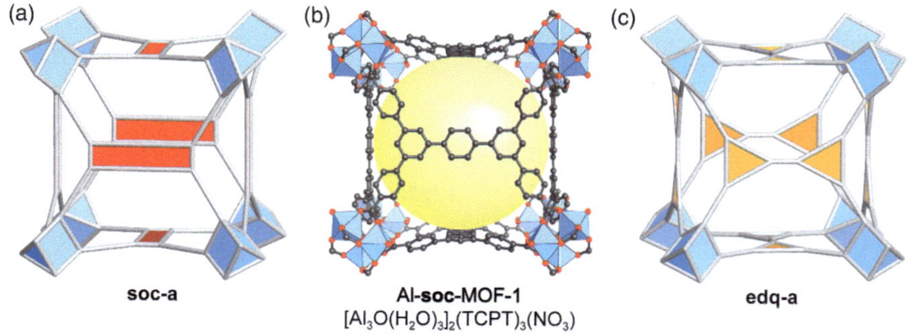

Figure 4.14 Comparison of the structure of Al-**soc**-MOF and the **soc** and **edq** topology. (b) Unit cell of Al-**soc**-MOF-1. The framework is built from trigonal prismatic $Al_3OL_3(-COO)_6$ SBUs that are linked by TCPT moieties. (a, c) The structure can be deconstructed into the **soc** or the **edq** topology. In the **soc** net the linker is represented as one tetratopic building unit (left), and as two linked trigonal tritopic building units in the **edq** net (right). Color code: Al, blue; C, gray; O, red.

series of highly porous aluminum-based **soc**-MOFs. The Al-**soc**-MOF platform meets the challenging Department of Energy (DOE) dual target for methane storage of $0.5\,g\,g^{-1}$ (gravimetric) and $264\,cm^3\,cm^{-3}$ (volumetric) [14g]. The structure of Al-**soc**-MOF-1 ($[In_3O(H_2O)_3]_2(TCPT)_3(NO_3)$) is illustrated in Figure 4.14. From a topological point of view the tetratopic rectangular linker applied in the synthesis of all **soc**-MOFs can also be described as two triangular vertex figures that are connected by an edge and thus the Al-**soc**-MOF can also be described by the 3,6-connected **edq** net. Both descriptions are acceptable, and it is the chemical backbone of the organic linker that makes one or the other the more plausible choice (Figure 4.14).

The chemistry of zirconium carboxylates, especially that of $Zr_6(\mu_3\text{-}O)_4(\mu_3\text{-}OH)_4(RCOO)_{12}$ clusters, is well established in molecular inorganic chemistry and SBUs with an analogous structure are known in MOF chemistry. Such SBUs are based on the same Zr_6O_8-core and, in contrast to their molecular analogs, they can have a connectivity ranging between 6 and 12 [18]. This is because not all carboxylates have to originate from linkers connecting the SBUs, but monofunctional carboxylic acids or $-OH$ and $-OH_2$ moieties can act as terminal ligands.

For a 6-c SBU based on the Zr_6O_8-core, two distinct geometries can be envisioned: trigonal antiprismatic and hexagonal planar (Figure 4.15). Both geometries have been found in frameworks involving trigonal tritopic linkers. Which of these two geometries is adopted in a given framework strongly depends on the exact geometry of the linker as well as the reaction conditions, in particular the amount and nature of modulator employed in the synthesis. Modulators are additives that influence the rate of formation of the SBU and the framework, and therefore the crystallinity of the resulting MOF. Among the most common modulators in carboxylate MOF chemistry are organic monocarboxylic acids (e.g. formic, acetic, benzoic acid) or inorganic acids (e.g. HCl, HNO_3).

Reticulating Zr^{4+} ions and H_3BTC linkers (D_{3h} symmetry) leads to the formation of trigonal antiprismatic SBUs that are linked by BTC units to form

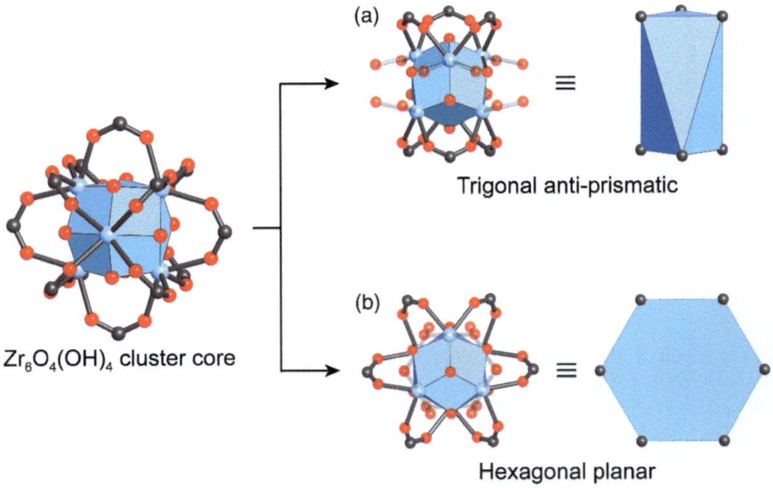

Figure 4.15 SBUs based on the Zr_6O_8-core can have up to 12 points of extension. For a 6-c version, two geometries are possible: (a) trigonal antiprismatic and (b) hexagonal planar. Which of these SBUs forms depends on the precise geometry of the linker. Color code: Zr, blue; C, gray; O, red.

a crystalline framework termed MOF-808 ($Zr_6O_4(OH)_4(HCOO)_6(BTC)_2$) with an overall **spn** topology, the default topology for the combination of trigonal and trigonal antiprismatic building units [19]. The trigonal antiprismatic $[Zr_6(\mu_3\text{-}O)_4(\mu_3\text{-}OH)_4]^{12+}$ core of the SBUs is connected to six BTC linkers and the remaining uncoordinated metal sites are occupied by formate ligands. Each BTC linker is connected to three neighboring trigonal anti-prismatic SBUs to form tetrahedral subunits with an internal pore diameter of 4.8 Å. These tetrahedral units are further linked in a vertex-sharing manner to give adamantane-like cages with a pore diameter of 18.4 Å. The structure can be described by either the **spn** topology (6-c trigonal anti-prismatic and 3-c trigonal building units) or, alternatively a **dia** topology, where larger tetrahedral TBUs are defined (Figure 4.16).

As discussed earlier in this chapter expanded analogs of the H_3BTC linker employed in the synthesis of MOF-808 can be prepared. Replacing the three terminal carboxylic acid binding groups by benzoic acid groups is accompanied by a decrease in symmetry (H_3BTB has C_3 rather than D_{3h} symmetry). This distortion can be circumvented by using a triazine core unit as in the case of H_3TATB (D_{3h} symmetry). Consequently, using the H_3TATB linker it is possible to prepare an isoreticular expanded version of MOF-808 termed PCN-777 ($Zr_6O_4(OH)_4(HCOO)_6(TATB)_2$), whereas a MOF with a different topology is formed when the nonplanar H_3BTB linker is used. The fact that the carboxylic acid binding groups in H_3BTB are not coplanar with respect to each other prevents the formation of an **spn** topology framework [20]. Hence, the resulting MOF (UMCM-309a, UMCM = University of Michigan Crystalline Material, $Zr_6O_4(OH)_4(BTB)_6(OH)_6(H_2O)_6$) has a 2D layered structure with a **kgd** (kagome dual) instead of the default **spn** topology [21]. In the structure of

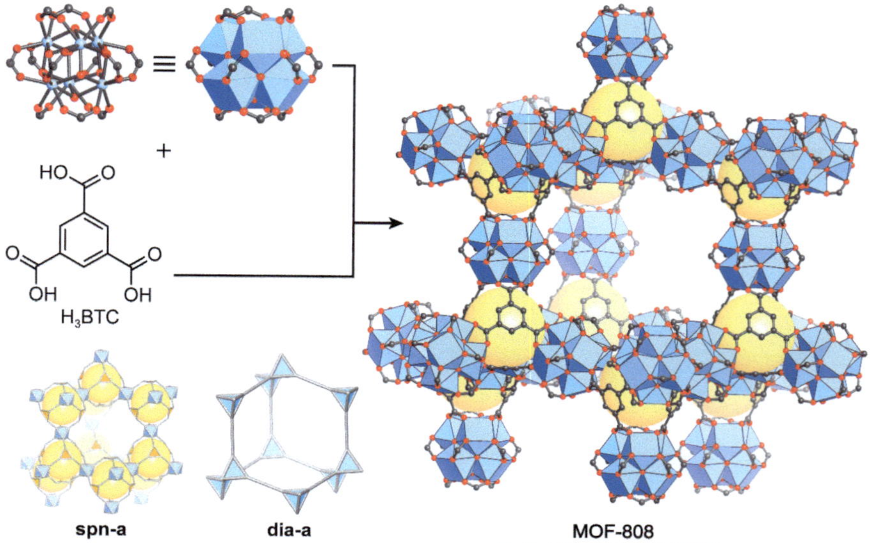

Figure 4.16 Crystal structure of MOF-808. Trigonal antiprismatic zirconium SBUs are linked by trigonal tritopic BTC linkers to form tetrahedral units. These are further connected in a vertex-sharing manner resulting in the formation of adamantine-shaped cages. Thus, the crystal structure of MOF-808 is described by either the **spn** or **dia** net depending on which subunits are chosen as the vertices. The augmented **spn** and **dia** net are depicted in the inset. All hydrogen atoms are omitted for clarity. Color code: Zr, blue; C, gray; O, red.

UMCM-309a the $[Zr_6(\mu_3\text{-}O)_4(\mu_3\text{-}OH)_4]^{12+}$ clusters are connected to six BTB linkers in a hexagonal planar fashion and the remaining uncoordinated metal sites are occupied by six –OH and six $-OH_2$ ligands, resulting in a 6-connected layered structure. It should be noted that these layers can form a 3D interpenetrated structure, which is avoided by using sterically demanding modulators in the synthesis [22]. A comparison of both structures and their underlying topologies is given in Figure 4.17.

4.3 MOFs Built from 7-, 8-, 10-, and 12-Connected SBUs

4.3.1 7-Connected (7-c) SBUs

Building units with a low point group symmetry are rarely used in MOF chemistry. This is due to the fact that the formation of highly symmetric structures is more favorable and such structures are generally easier to crystallize. In Chapter 3, we saw that low symmetry linkers can be used to construct MOFs and, in a similar way, SBUs with low symmetry can yield crystalline frameworks. An example is a MOF constructed from a 7-c zinc-based SBUs – an SBU that is closely related to the well-studied 6-c $Zn_4O(-COO)_6$ SBU – that are linked by tetratopic H4BPTC linkers. Consequently, the framework crystallizes in a previously unreported 4,7-connected net [23].

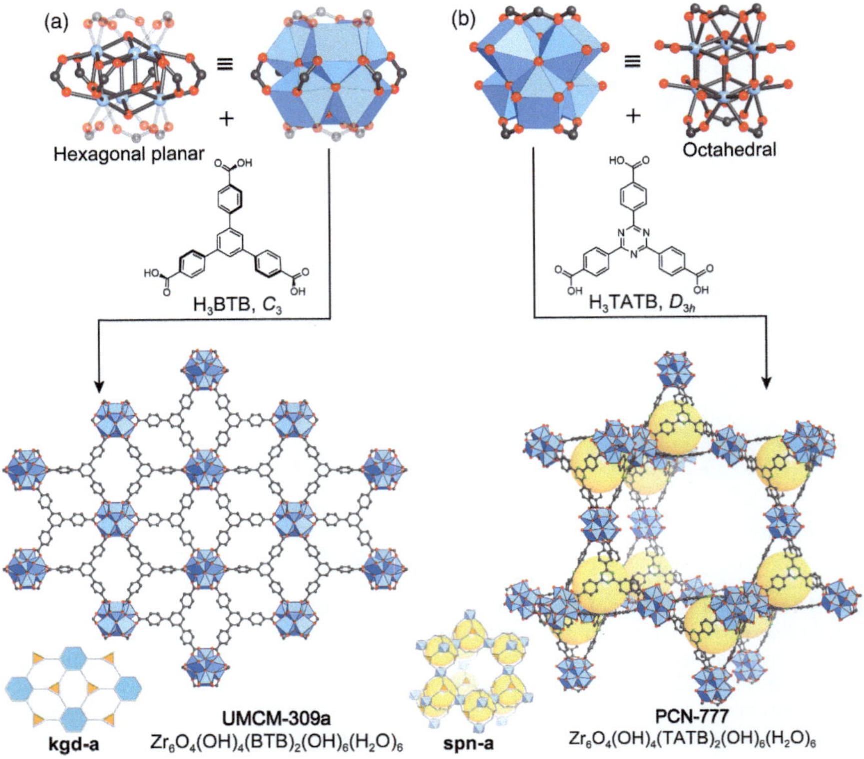

Figure 4.17 Comparison of the crystal structures of UMCM-309a and PCN-777, both built from 6-c zirconium SBUs and trigonal tritopic linkers. The lower symmetry of H_3BTB (C_3) used in the synthesis of UMCM-309a compared to H_3TATB (D_{3h}) directs the reticulation to a (a) **kgd** net rather than (b) an **spn** net. Topology representation of the augmented versions of **kgd** and **spn** are given next to the corresponding structure. All hydrogen atoms are omitted for clarity. Color code: Zr, blue; C, gray; N, green; O, red.

4.3.2 8-Connected (8-c) SBUs

As we will see later in this chapter, most SBUs with a connectivity higher than six are typically based on highly charged metal ions such as Al^{3+}, Cr^{3+}, Ti^{4+}, Zr^{4+}, or Hf^{4+}. There are however rare examples of 8-c SBUs built from divalent metal ions as exemplified by the cubic $Cu_4Cl(-COO)_8$ SBU. Reticulation of these 8-c $Cu_4Cl(-COO)_8$ SBUs and trigonal tritopic H_3BTC linkers yields an anionic **the** framework with chemical formula $(Cu_4Cl_3)(BTC)_8(R_4N)_3$ (R = methyl, ethyl, propyl) (Figure 4.18). Even though the **the** net describes the structure of this MOF accurately, an alternative description based on tetrahedral units can be used to further simplify the structure, in a manner akin to what was described earlier in this chapter for MIL-100 and MIL-101. When each $Cu_4Cl(-COO)_8$ is represented as a 4-membered ring of linked tetrahedra, a **sod** (sodalite) net results. Despite their low surface areas of about 800 m^2 g^{-1},

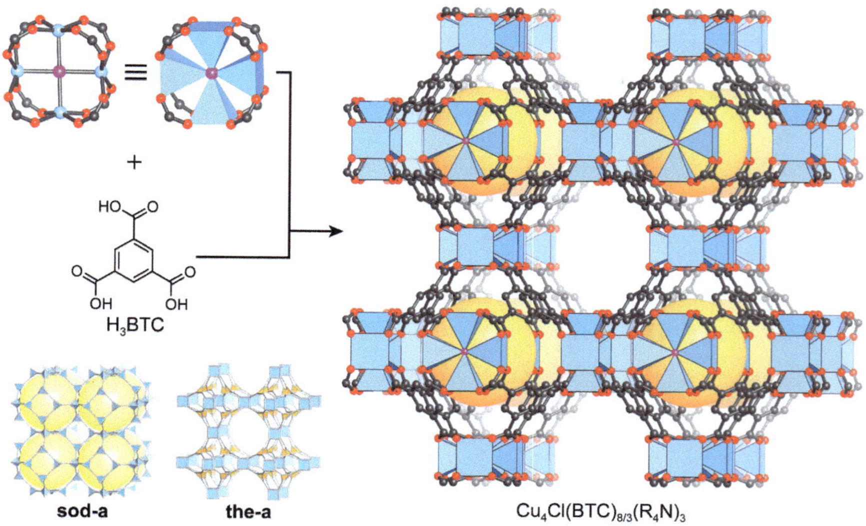

Figure 4.18 Crystal structure of $Cu_4Cl(BTC)_8(R_4N)_3$. The cubic $Cu_4Cl(-COO)_8$ SBUs are joined by tritopic BTC linkers to form cages of **tro** (truncated octahedron, sodalite) topology. The resulting structure is therefore best described by the **sod** rather than the **the** net. The augmented **sod** and **the** net are depicted in the inset. All hydrogen atoms and counter ions residing in the pores are omitted for clarity. Color code: Cu, blue; C, gray; O, red; Cl, pink.

$(Cu_4Cl_3)(BTC)_8(R_4N)_3$ materials have a high sorption capacity for CO_2, which is attributed to the pore partitioning effect of the anionic framework [24].

There are many MOF structures involving 8-c zirconium or hafnium SBUs. Here, we will discuss only selected examples, illustrating the linker-directed synthesis of such frameworks. Linking cubic 8-c SBUs by ditopic linkers can lead to the formation of frameworks with two different topologies: **bcu** (body centered cubic) and **reo** (rhenium oxide). Both can be targeted for MOFs based on Zr_6O_8-core SBUs by adjusting the geometry of the linker. The default topology for linking cubes through ditopic linkers is the **bcu** net (highest symmetry embedding $I\,m\bar{3}m$, No. 229, transitivity 1111), which is expected to form when linear ditopic linkers are reticulated with cubic 8-c SBUs. In contrast, bent linkers are expected to direct the framework formation toward the **reo** topology (highest symmetry embedding $P\,m\bar{3}m$, No. 223, transitivity 1122). Examples for these two cases are PCN-700 ($Zr_6O_4(OH)_4(Me_2\text{-BPDC})_4(OH)_4(H_2O)_4$) and DUT-67 (DUT = Dresden University of Technology, $Zr_6O_6(OH)_2(TDC)_4(CH_3COO)_2$, TDC = thiophene-2,5-dicarboxylic acid), respectively. Both MOFs are synthesized from a Zr^{4+} source that forms cubic 8-c SBUs, and ditopic carboxylate linkers of linear and bent geometry, respectively [25]. The formation of a **bcu** net by linking cubic 8-c Zr_6O_8-core SBUs requires a linear ditopic linker with perpendicular terminal carboxylic acid binding groups (dihedral angle of 90°). Introduction of methyl substituents in the 2- and 2′-position of the planar H_2BPDC linker introduces a twist (rotation around φ) due to steric hindrance (see Figure 4.4). Therefore, the reticulation of $(Me)_2$-BPDC with

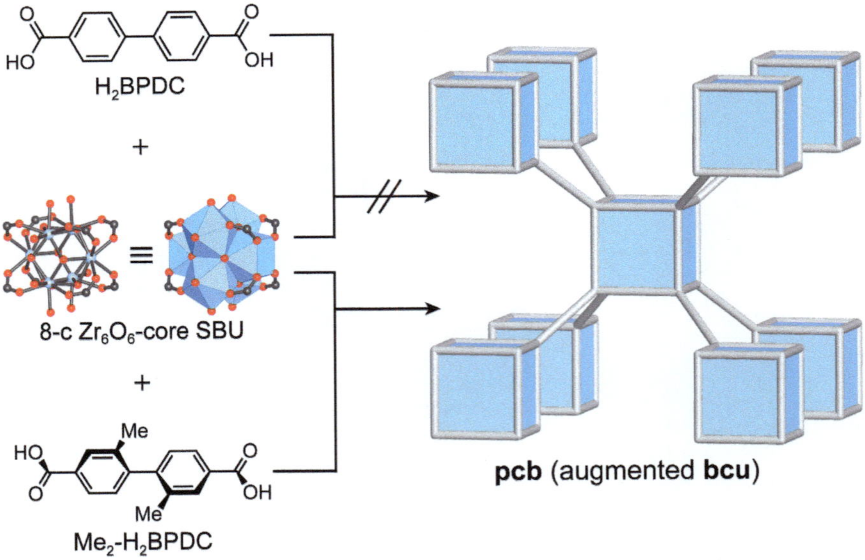

Figure 4.19 Appending methyl substituents in the 2- and 2′-position of the H_2BPDC linker engenders a dihedral angle of 90° between the two binding groups. This geometry is a prerequisite for the formation of a **bcu** net from cubic 8-c zirconium SBUs. The augmented **bcu** net is referred to as the **pcb** (polycubane) net. In contrast, employing unsubstituted H_2BPDC under otherwise identical conditions does not afford a **bcu** framework (see 12-c SBUs). Color code: Zr, blue; C, gray, O, red.

cubic 8-c Zr_6O_8-core SBUs as illustrated in Figure 4.19 yields PCN-700 with the anticipated **bcu** topology.[5]

When ditopic linkers with an angle θ (see Figure 4.4) between coplanar carboxylic acid groups are used, the formation of the default **bcu** topology is avoided, and the formation of **reo** frameworks is favored. This approach is used for the synthesis of DUT-67, which is built from cubic 8-c Zr_6O_8-core SBUs that are joined by bent ditopic TDC linkers featuring an angle of $\theta = 147.9°$ between the coplanar binding groups (Figure 4.20). The synthesis of an isoreticular expanded form of DUT-67 is however challenging because it requires the angle between the two carboxylic acid groups of the linker to remain the same. In H_2DTTDC, fusing three thiophene units results in an angle $\theta = 148.6°$ ($\theta_{(TDC)} = 147.9°$) between the coplanar carboxylic acid groups. At the same time the distance between the two carboxylates is increased by almost 4 Å compared to the parent H_2TDC (Figure 4.20). Using this linker, DUT-51, the isoreticular expanded analog of the **reo** MOF DUT-67, can be prepared.

Figure 4.21 shows a comparison of the structures of PCN-700 and DUT-67. The **bcu** topology is formed with linear ditopic linkers featuring perpendicular binding groups, whereas a framework of **reo** topology results when bent linkers with a precisely adjusted angle θ are used.

5 An isostructural framework with a mixed Zr_6Ni_4 SBU can be prepared by post-synthetic metalation [26].

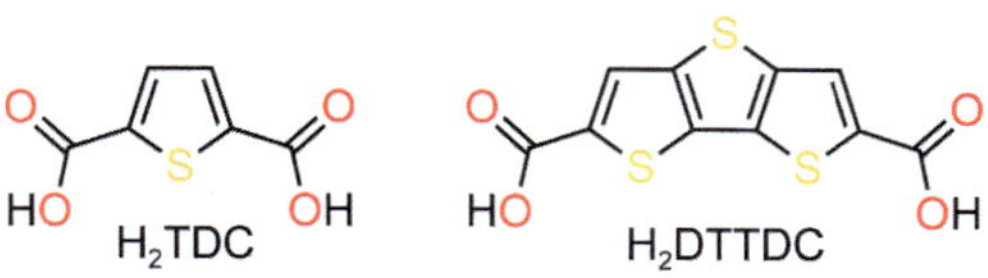

Figure 4.20 Bent ditopic linkers employed in the synthesis of zirconium-based frameworks with **reo** topology. An angle $\theta = 147.9°$ in H_2TDC is required for the formation of **reo** frameworks. Isoreticular expansion requires the angle θ to remain the same, which is achieved by fusing three thiophene units together to give H_2DTTDC ($\theta_{(DTTDC)} = 148.6°$).

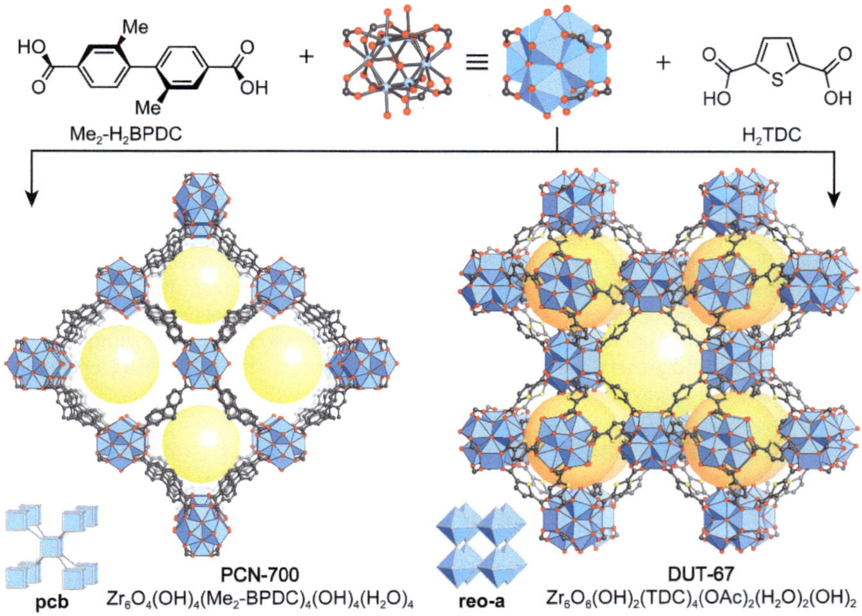

Figure 4.21 Comparison of the crystal structures and topologies of PCN-700 and DUT-67. Both frameworks are built from 8-c zirconium SBUs and ditopic linkers. Employing a linear linker with a dihedral angle $\varphi = 90°$ favors the formation of frameworks of **bcu** topology, whereas the use of a bent linker with an angle θ of around 148° between the coplanar binding groups directs the synthesis toward **reo** topology frameworks. Topology representations of **pcb** and **reo-a** are given next to the corresponding crystal structure. All hydrogen atoms are omitted for clarity. Color code: Zr, blue; C, gray; O, red; S, yellow.

Connecting 8-c Zr_6O_8-core SBUs by tetratopic linkers is expected to give 4,8-connected frameworks with varying topologies. The default net for the combination of cubic 8-c SBUs and tetratopic tetrahedral linkers is the "fluorite" (**flu**) net (e.g. MOF-841), whereas for the combination of cubic 8-c SBUs and tetratopic square linkers three different nets can result: **csq** (e.g. MOF-545), **scu** (e.g. NU-902, NU = Northwestern University), and **sqc** (e.g. PCN-225) [27].[6]

6 The letter "s" and "c" in all three topologies (**csq**, **sqc**, and **scu**) refer to "cube" and "square" and all three nets are referred to as "SC-nets."

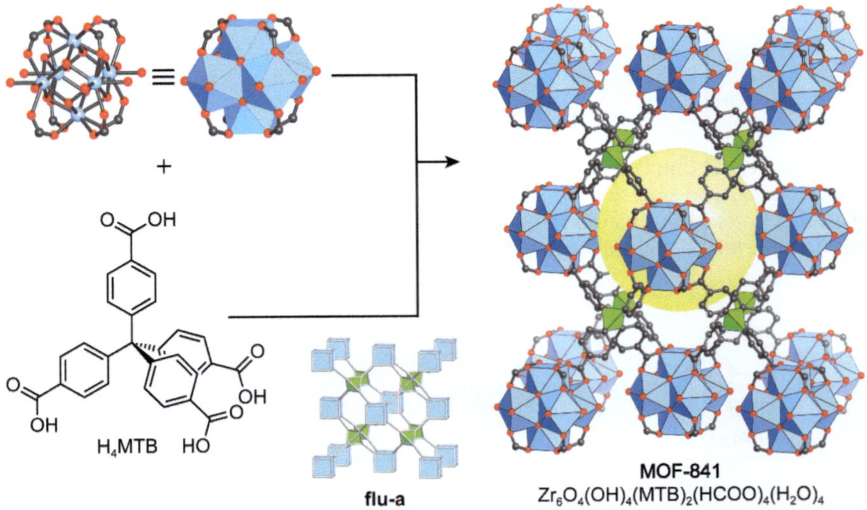

Figure 4.22 Crystal structure of MOF-841. The combination of a tetrahedral linker (H_4MTB) and a cubic 8-c zirconium SBU results in a framework of **flu** topology. Here, the zirconium clusters form a face centered packing while the MTB linkers occupy the tetrahedral holes. The large pore in the center of the unit cell corresponds to the octahedral hole in the structure of CaF_2. The inset shows a topology representation of the augmented **flu** net. All hydrogen atoms are omitted and green tetrahedra are placed at the center of the MTB linkers for clarity. Color code: Zr, blue; C, gray; O, red.

The structure of MOF-841 ($Zr_6O_4(OH)_4(MTB)_2(HCOO)_4(H_2O)_4$) is built from tetrahedral MTB linkers connecting four cubic $Zr_6(\mu_3\text{-}O)_4(\mu_3\text{-}OH)_4(-COO)_8$ SBUs (Figure 4.22). The combination of cubic and tetrahedral building units gives rise to the highly symmetric default topology (**flu**, edge transitive), which is derived from the structure of the mineral CaF_2 (Fluorite). Consequently, the arrangement of the zirconium SBUs is described as cubic face centered and all tetrahedral holes are filled by MTB linkers. The structure of MOF-841 encompasses one type of pore with a diameter of 11.6 Å that corresponds to the octahedral holes in the fluorite structure [19].

The combination of 8-c cubic SBUs and tetrahedral tetratopic linkers can only result in the formation of the **flu** net. In contrast, when 8-c Zr_6O_8-core SBUs and a tetratopic porphyrin-based linker ($H_4TCPP\text{-}H_2$) are reticulated, frameworks with three different topologies are formed by adjusting the pH, the concentration of the modulator, and the reaction temperature. When considering topological isomers (nets built from identical building units but having different topologies), as a rule of thumb, the isomer having the topology with the most symmetric "highest symmetry embedding" is the thermodynamically favored phase, whereas those with less symmetric highest symmetry embeddings are the kinetic products. This allows us to sort the three possible topologies for the combination of squares and cubes as follows: the **csq** net is the thermodynamic product (highest symmetry embedding $P\,6/mmm$, No. 191, transitivity 2155), whereas the topologies with a less symmetric highest symmetry embedding, namely **sqc** (highest symmetry embedding $P\,4_1/amd$, No. 141, transitivity 2132) and **scu** (highest symmetry embedding $P4/mmm$, No. 123, transitivity 2133) correspond to kinetic products. High reaction temperatures combined with a strongly binding modulator in high concentrations are thus expected to drive the reaction toward the thermodynamically favored frameworks of **csq** topology. In contrast, gradually lowering the strength of the modulator as well as the reaction temperatures will favor the kinetically controlled formation of frameworks with **sqc** and eventually with **scu** topology.

We illustrate this principles with three topological isomers: MOF-545 ($Zr_6O_4(OH)_4(TCPP\text{-}H_2)_2(H_2O)_8$) (**csq**), PCN-225 ($Zr_6O_4(OH)_4(TCPP\text{-}H_2)_2(H_2O)_4(OH)_4$) (**sqc**), and NU-902 ($Zr_6O_4(OH)_4(TCPP\text{-}H_2)_2(H_2O)_4(OH)_4$) (**scu**) [27]. All three MOFs are prepared by reacting a Zr^{4+} salt and $H_4TCPP\text{-}H_2$ in *N*,*N*-dimethylformamide (DMF) or *N*,*N*-diethylformamide (DEF) using the following modulators and temperatures: the addition of formic acid ($pK_a = 3.75$) and a high reaction temperature of 130 °C in DMF gives MOF 545 (**csq**), the addition of benzoic acid ($pK_a = 4.19$) and heating to 120 °C in DEF yields PCN-225 (**sqc**), and the use of acetic acid ($pK_a = 4.75$) combined with a low reaction temperature of 90 °C in DMF affords NU-902 (**scu**). A comparison of the structures of MOF-545, PCN-225, and NU-902 is given in Figure 4.23. It is worth noting that not only the modulator and the temperature influence the formation of a particular phase but also the solvent, especially considering that it is the source of the base released upon decomposition. DMF releases dimethylamine (DMA, $pK_b = 10.7$), whereas DEF releases diethylamine (DEA, $pK_b = 11.1$). DEA is a slightly stronger base than DMA and thus the formation of the kinetic product is more favorable from DEF solutions.

4.3.3 10-Connected (10-c) SBUs

In the discussion of 8-c SBUs we saw that the reticulation of ditopic bent linkers and Zr^{4+} ions results in frameworks of **reo** topology as exemplified by DUT-67. Using the same linker (H_2TDC) employed in the synthesis of DUT-67 and reacting it with a zirconium salt while decreasing the concentration of the modulator (acetic acid) compared to the synthesis conditions of DUT-67 directs the reaction toward the formation of a 10-c zirconium SBU, and consequently a framework of **btc** (body-centered tetragonal) topology, termed DUT-69 ($Zr_6O_4(OH)_4(TDC)_5(CH_3COO)_2$). To rationalize this finding, the role of the modulator in the crystallization of MOFs must be considered in more detail. Monofunctional carboxylic acids compete with the carboxylate-based linkers and thus prevent adjacent SBUs from aggregating into larger fragments by modulating the rate of framework formation. Hence, in MOF synthesis such compounds are referred to as modulators. A higher connectivity of the SBU is observed for lower concentrations or less strongly binding modulators, or higher concentrations of the linker. Such reaction conditions lead to a higher likelihood for open binding sites of the SBU to be occupied by the linker or the more facile replacement of modulator molecules by the linker, respectively. This is indeed the case for DUT-69 where a net with higher connectivity (10-c **bct** net) is formed using starting materials identical to those used to prepare

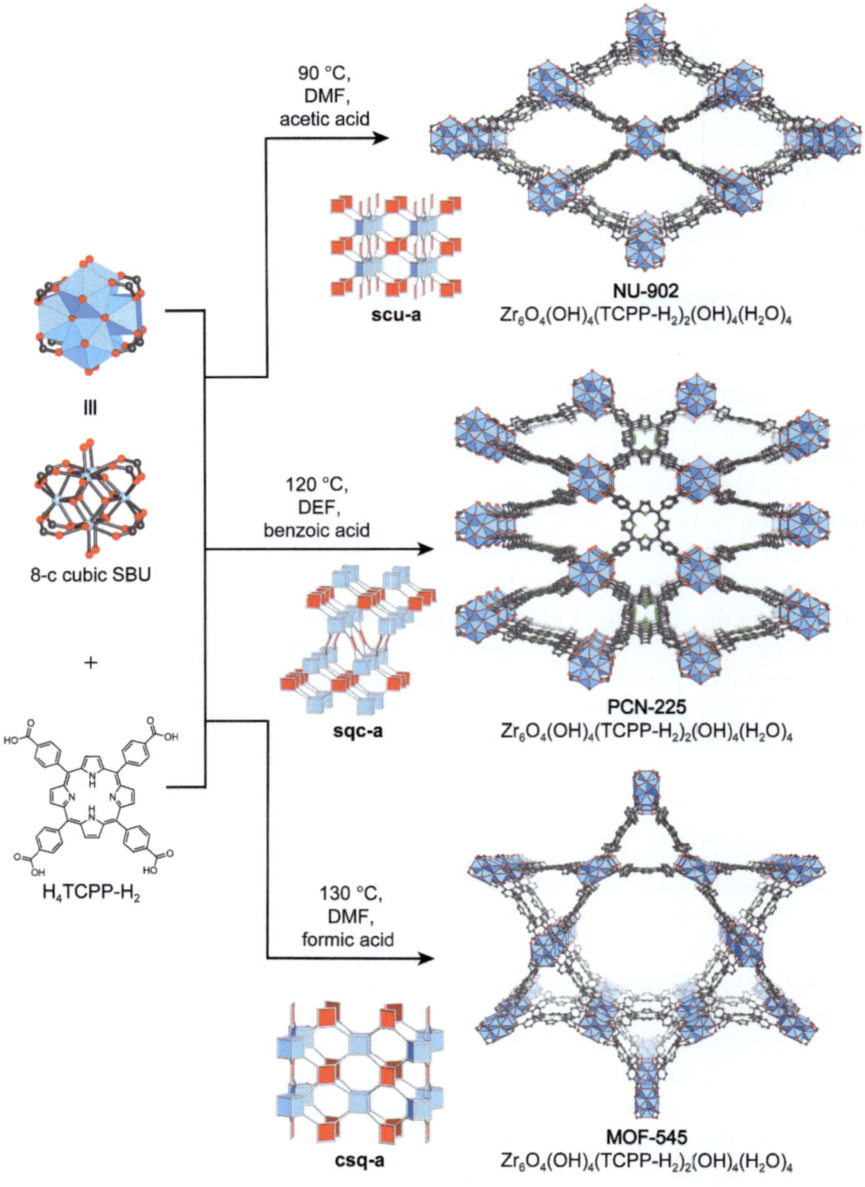

Figure 4.23 Comparison of the crystal structures and topologies of NU-902 (**scu**), PCN-225 (**sqc**), and MOF-545 (**csq**) (top to bottom). The topology with the most symmetric highest symmetry embedding is formed as the thermodynamic product (**csq**, MOF-545). Milder reaction conditions result in the formation of **sqc** (PCN-225) and **scu** (NU-902) nets. Topology representations of **scu**, **sqc**, and **csq** are given next to the corresponding crystal structures. All hydrogen atoms are omitted for clarity. Color code: Zr, blue; C, gray; N, green; O, red.

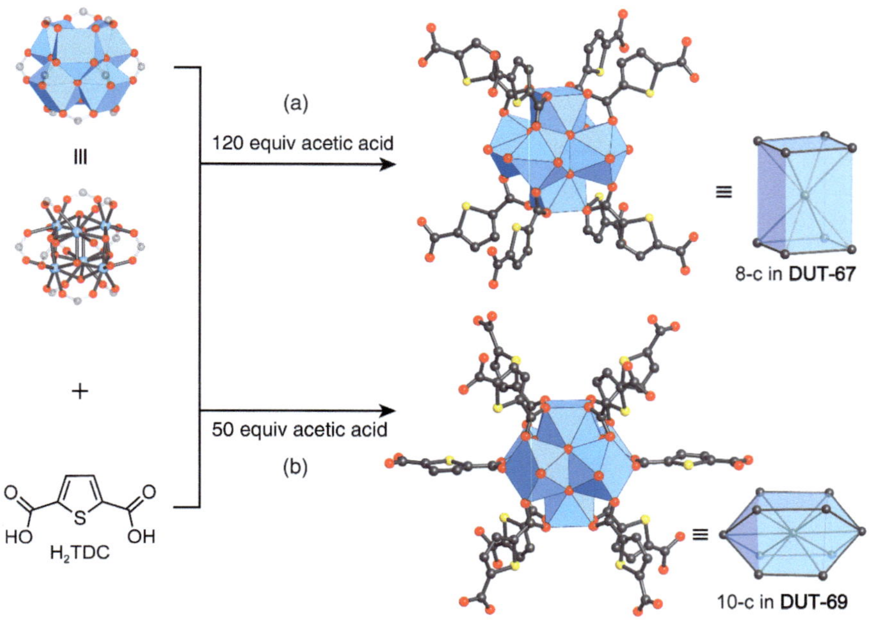

Figure 4.24 Comparison of the SBUs in the structures of (a) DUT-67 (8-c, **reo** net) and (b) DUT-69 (10-c, **bct** net) and their respective vertex figures. A higher connectivity is achieved by lowering the concentration of the modulator under otherwise similar conditions. All hydrogen atoms are omitted for clarity. Color code: Zr, blue; C, gray; O, red; S, yellow.

DUT-67 (8-c **reo** net) but in the presence of a lower concentration of the acetic acid modulator (Figure 4.24).

In the crystal structure of DUT-69 each 10-c Zr_6O_8-core SBU is linked to 10 TDC linkers giving rise to octahedral cages of 5 Å in diameter. Rectangular channels with a pore aperture of 9.15 × 2.66 Å propagate along the crystallographic *c*-axis. The crystal structure of DUT-69 alongside the corresponding **bct-a** net is shown in Figure 4.25.

4.3.4 12-Connected (12-c) SBUs

Similar to the example described above, the connectivity of the Zr_6O_8-core SBU can be further increased by adjusting the ratio of the metal source and the linker employed in the synthesis. The structures of MOF-841 and MOF-812 ($Zr_6O_4(OH)_4(MTB)_3(H_2O)_2$) are both built from tetrahedral MTB linkers and Zr_6O_8-core SBUs [19]. While MOF-841 crystallizes in the 4,8-connected **flu** net (Figure 4.22), in the case of MOF-812 a 12-connected SBU and consequently a structure with an underlying **ith** (icosahedron-tetrahedron) topology forms (Figure 4.26). This is realized by changing the ratio of starting materials employed in the synthesis. MOF-841 is prepared from a mixture of H_4MTB and zirconyl chloride ($ZrOCl_2$) in a ratio of 1 : 4 whereas the higher connected structure of MOF-812 forms when the relative concentration of the linker is increased (linker to metal ratio is decreased to 1 : 2) under otherwise almost identical conditions [19].

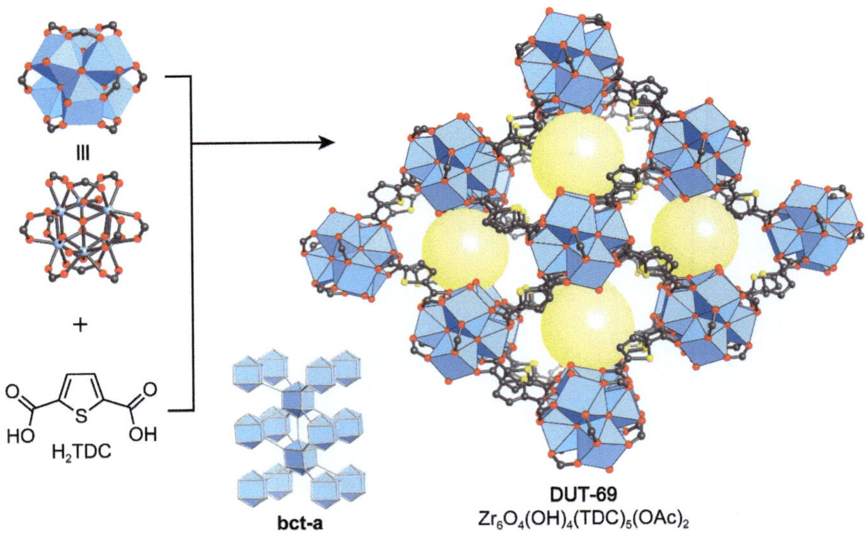

Figure 4.25 Crystal structure of DUT-69 projected along the *c*-direction. 10-c zirconium SBUs are linked by bent TDC linkers to form a 3D framework of **bct** topology with channels of 9.15 × 2.66 Å running along the crystallographic *c*-axis. The topology representation shows the augmented **bct** net viewed along the *a*-direction. All hydrogen atoms are omitted for clarity. Color code: Zr, blue; C, gray; O, red; S, yellow.

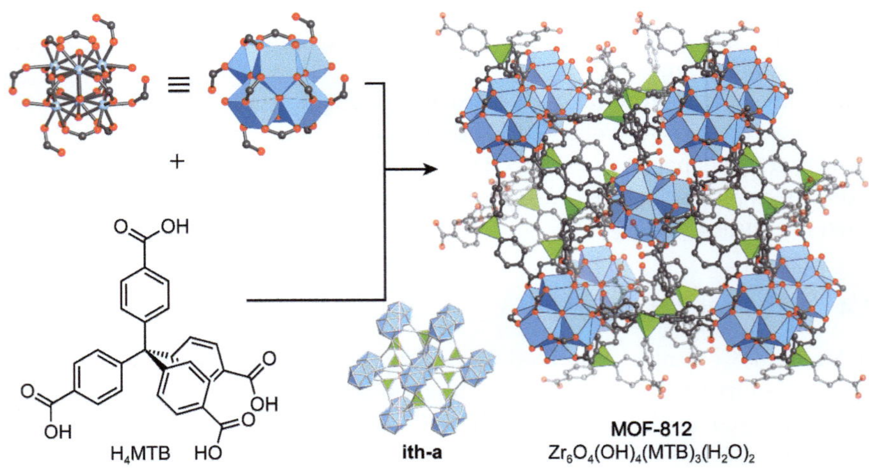

Figure 4.26 The **ith** net of MOF-812 is formed by reticulation of Zr^{4+} and tetrahedral tetratopic H_4MTB linkers. Four MTB linkers are bound to the SBU in a monodentate fashion and eight in a bridging fashion, resulting in an overall "8 + 4-connectivity" of the SBU. The structure has one type of pore with a diameter of 5.6 Å. The inset shows the augmented **ith** net. All hydrogen atoms are omitted and green tetrahedra are placed at the center of the MTB linkers for clarity. Color code: Zr, blue; C, gray; O, red.

The last two examples illustrate the influence of the concentration of the modulator and the ratio of the starting materials on the connectivity of the resulting SBUs and thus the structure of the resulting MOF. When considering nets of identical connectivity that arise from the same set of starting materials (not building units), adjusting the reaction conditions helps to select which of these nets is preferred. To illustrate this, we take a closer look at two 4,12-connected nets that are formed by linking square planar linkers ($H_4TCPP\text{-}H_2$) and 12-c Zr_6O_8-core SBUs. Two geometries exist for a 12-c zirconium SBU: cuboctahedral and hexagonal prismatic. Linking such SBUs with square planar building units leads to the formation of frameworks of **ftw** (e.g. MOF-525, $Zr_6O_4(OH)_4(TCPP\text{-}H_2)_3$) or **shp** (e.g. PCN-223, $Zr_6O_4(OH)_4(TCPP\text{-}H_2)_3$) topology, respectively (Figure 4.27). Both are the default topologies for the combination of their respective vertex geometries [27c]. While PCN-223 is formed in the presence of high concentrations of the modulator and at high reaction temperatures (120 °C), MOF-525 forms under milder conditions, meaning a lower concentration of modulator and relatively low reaction temperatures (65 °C). The pores in the cubic **ftw** structure of MOF-525 are surrounded by eight cuboctahedral $Zr_6(\mu_3\text{-}O)_4(\mu_3\text{-}OH)_4(-COO)_{12}$ SBUs and six $TCPP\text{-}H_2$ linkers. In contrast, PCN-223 has trigonal 1D pores and a lower surface area than MOF-525 as a consequence of the closer packing of hexagonal prismatic $Zr_6(\mu_3\text{-}O)_4(\mu_3\text{-}OH)_4(-COO)_{12}$ SBUs in the **shp** net.

The first reported, and still most prominent, 12-c MOF based on zirconium SBUs is UiO-66 (UiO = University of Oslo, $Zr_6O_4(OH)_4(BDC)_{12}$) [28]. UiO-66 crystallizes in in a **fcu** (face-centered cubic) net constructed from cuboctahedral 12-c Zr_6O_8-core SBUs and linear ditopic BDC linkers. The open 3D framework structure encompasses tetrahedral and octahedral pores where each octahedral pore is surrounded by eight tetrahedral ones in a face-sharing manner (Figure 4.28). The highly connected and therefore relatively dense structure is prone to defects in the form of missing linkers or missing SBUs [29]. Hence, UiO-66 was subject to many studies on defects in MOFs and properties resulting from them. All MOFs of the isoreticular series based on UiO-66 feature high architectural and chemical stability.

Zirconium-based MOFs comprise by far the largest fraction of MOFs containing 12-c SBUs. However, it should be noted, that even though they are rare, 12-c SBUs of di- and trivalent metals such as beryllium, aluminum, and chromium, as well as some zinc-based examples exist.

A synthetic strategy of matching the geometry between tetrahedrally coordinated silicon and zinc ions is employed in the synthesis of a framework based on 12-c $Zn_8(SiO_4)(-COO)_{12}$ SBUs. These SBUs are composed of a central tetrahedral SiO_4 unit that is connected to four tetrahedrally coordinated Zn^{2+} ions in a vertex-sharing manner [30]. Connecting the SBUs by 12 BDC linkers that propagate in doubles along six directions results in a double cross-linked interpenetrated 6-c framework ($Zn_8(SiO)_4(BDC)_6$) with an overall **pcu** topology (Figure 4.29). This MOF exhibits outstanding thermal (up to 520 °C) and chemical stability, arguably due to the double cross-linking of the SBUs in combination with framework interpenetration. The 12-c $Zn_8(SiO_4)(-COO)_{12}$ SBU can be used to prepare further frameworks with linkers such as *m*-BDC [31].

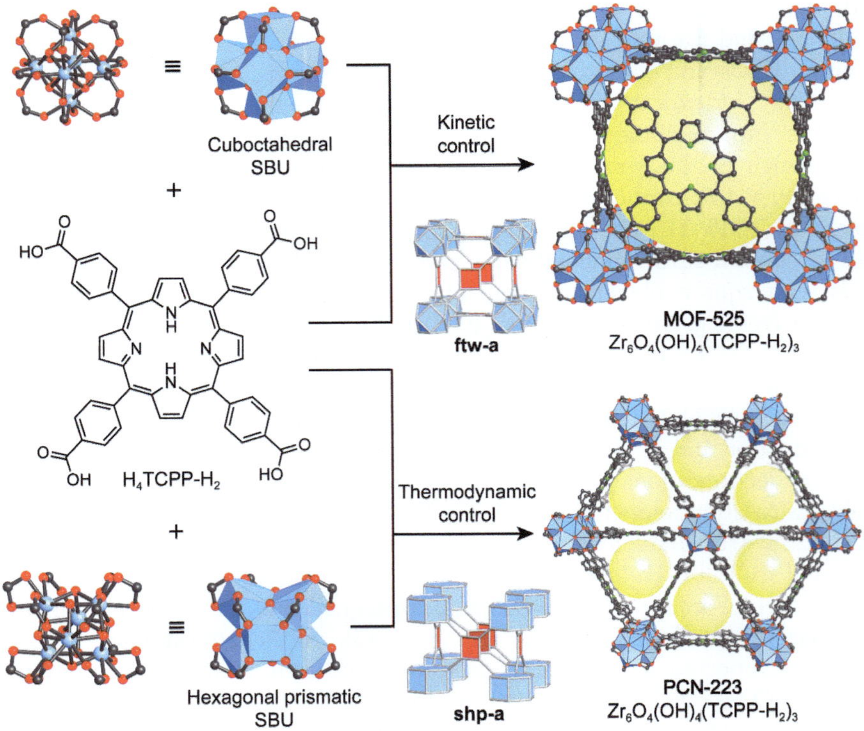

Figure 4.27 Comparison of the crystal structures and underlying topologies of MOF-525 (top) and PCN-223 (bottom). Both structures are based on 4,12-connected nets. The **shp** framework PCN-223 is formed under comparatively harsh conditions, whereas the **ftw** MOF-525 forms under mild reaction conditions. While MOF-525 has discrete cubic pores, PCN-223 has a 1D pore system. The closer packing of hexagonal prismatic SBUs in PCN-223 results in smaller pores and a lower surface area compared to MOF-525. Topology representations of both, **ftw** and **shp**, are given next to the corresponding crystal structures. All hydrogen atoms are omitted for clarity. Color code: Zr, blue; C, gray; N, green; O, red.

A different strategy for making MOFs with unusual 12-c zinc SBUs is used to prepare bio-MOF-100 ($Zn_4O_2(AD)_4(BPDC)_6(Me_2NH_2)_4$) [32]. Its 12-c SBU of chemical formula $Zn_8O_2(AD)_4(-COO)_{12}$ (ZABU) is built from eight tetrahedrally coordinated Zn^{2+} ions and four adeninates (AD). It is terminated by 12 carboxylates of 12 BPDC linkers that are bound to the SBU in a monodentate fashion. The ZABU SBU can be described as a truncated tetrahedron and the framework formed by linking these SBUs with linear ditopic BPDC linkers is a triple cross-linked 4-connected net with overall **lcs** topology (Figure 4.30). The cubic structure of bio-MOF-100 contains only mesopores and activation by supercritical drying yields a highly porous material with an internal surface area of 4300 $m^2\ g^{-1}$ and a free pore volume of about 85%. bio-MOF-100 is an ideal platform for post-synthetic linker exchange reactions, a post-synthetic modification that is discussed in more detail in Chapter 6 [33].

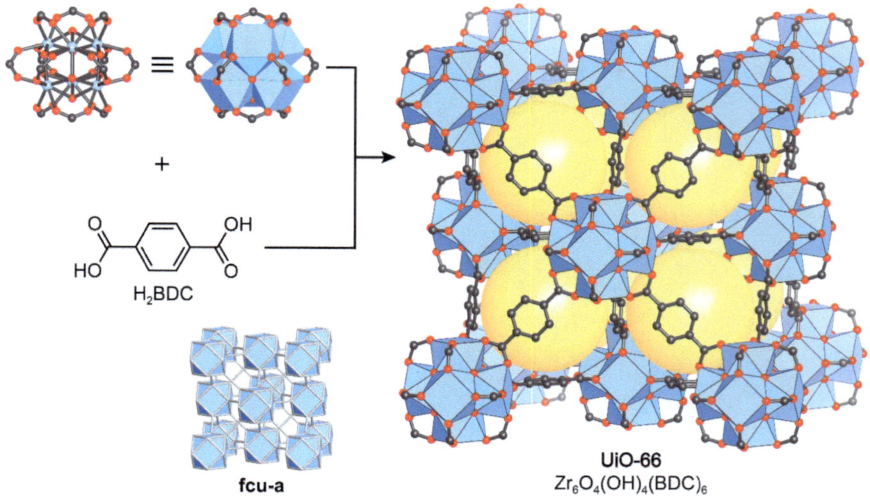

Figure 4.28 Crystal structure and **fcu** topology of UiO-66. 12-c zirconium SBUs are connected by linear ditopic BDC linkers resulting in a face centered arrangement of the SBUs (**fcu** net). The structure has two different pores that are equivalent to the tetrahedral (orange spheres) and octahedral holes (not shown for clarity) in a face-centered cubic arrangement. The inset shows the augmented **fcu** net. All hydrogen atoms are omitted for clarity. Color code: Zr, blue; C, gray; O, red.

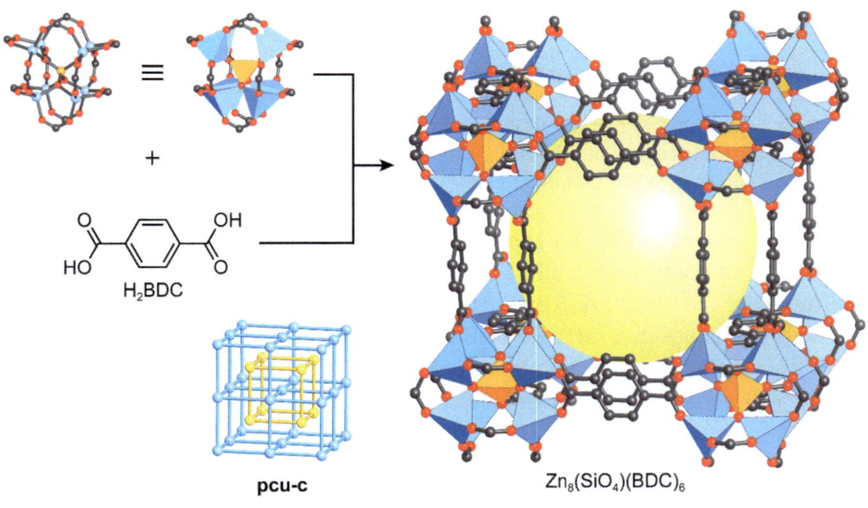

Figure 4.29 Crystal structure of $Zn_8(SiO_4)(BDC)_6$. The 12-c SBUs are connected in six directions in a double cross-linked manner, resulting in a framework of **pcu** topology. The open space encompassed by this framework is partially filled by an interpenetrating framework (catenated, **pcu-c**). The double cross-linking as well as the interpenetration endow $Zn_8(SiO_4)(BDC)_6$ with high thermal stability. All hydrogen atoms are omitted for clarity. Color code: Zn, blue; Si, orange; C, gray; O, red.

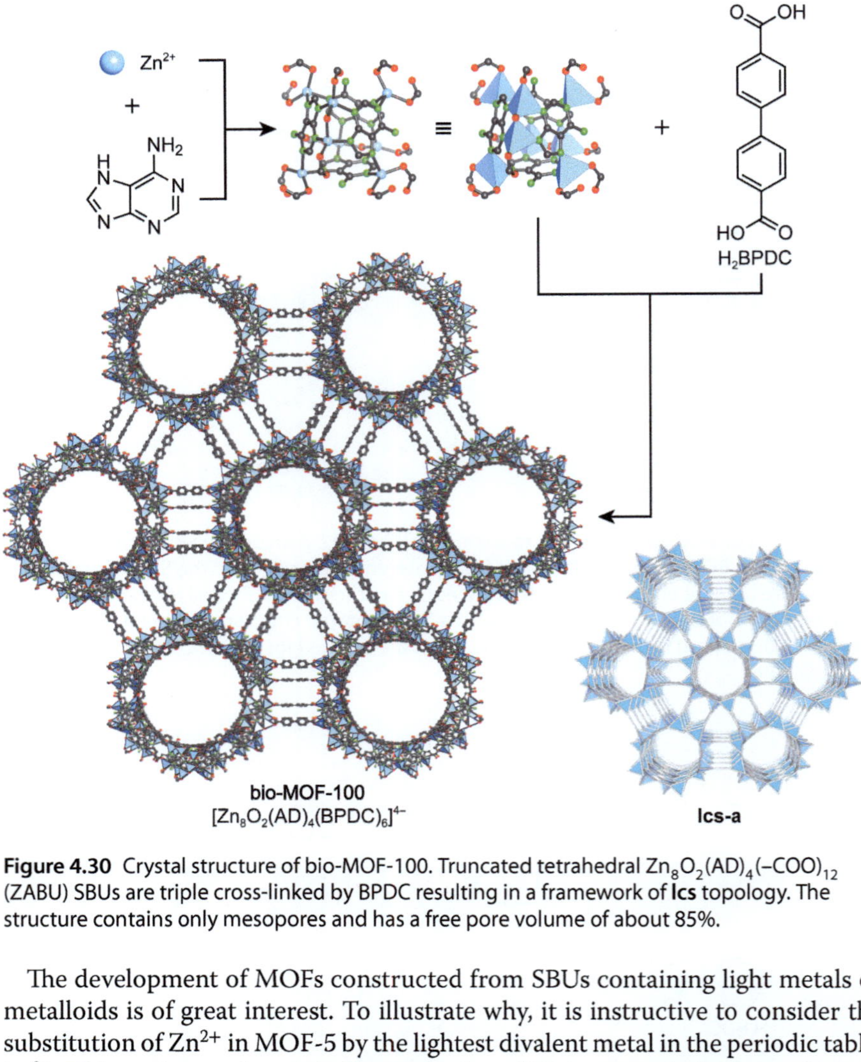

Figure 4.30 Crystal structure of bio-MOF-100. Truncated tetrahedral $Zn_8O_2(AD)_4(-COO)_{12}$ (ZABU) SBUs are triple cross-linked by BPDC resulting in a framework of **lcs** topology. The structure contains only mesopores and has a free pore volume of about 85%.

The development of MOFs constructed from SBUs containing light metals or metalloids is of great interest. To illustrate why, it is instructive to consider the substitution of Zn^{2+} in MOF-5 by the lightest divalent metal in the periodic table, Be^{2+}. This exchange theoretically results in a 40% increase in both gravimetric surface area ($m^2\ g^{-1}$) and gravimetric hydrogen storage capacity ($cm^3\ g^{-1}$) while the high volumetric capacity of MOF-5 is retained [34]. Inspired by this idea, a MOF of formula $Be_{12}(OH)_{12}(BTB)_4$ was synthesized, marking the first structurally characterized beryllium-based MOF [34d]. The highly porous framework structure is built from saddle-shaped $[Be_{12}(OH)_{12}]^{12+}$ SBUs composed of 12 tetrahedrally coordinated Be^{2+} ions. The inside edge of these ring-SBUs is bridged by –OH groups and the periphery is linked by carboxylates of the trigonal BTB linkers connecting the SBUs (Figure 4.31a). The 12-membered ring SBU constitutes a novelty as it was previously unknown in the realm of molecular metal carboxylates. Owing to its low weight, $Be_{12}(OH)_{12}(BTB)_4$ shows a high gravimetric hydrogen storage capacity of 9.2 wt% at 100 bar and 77 K.

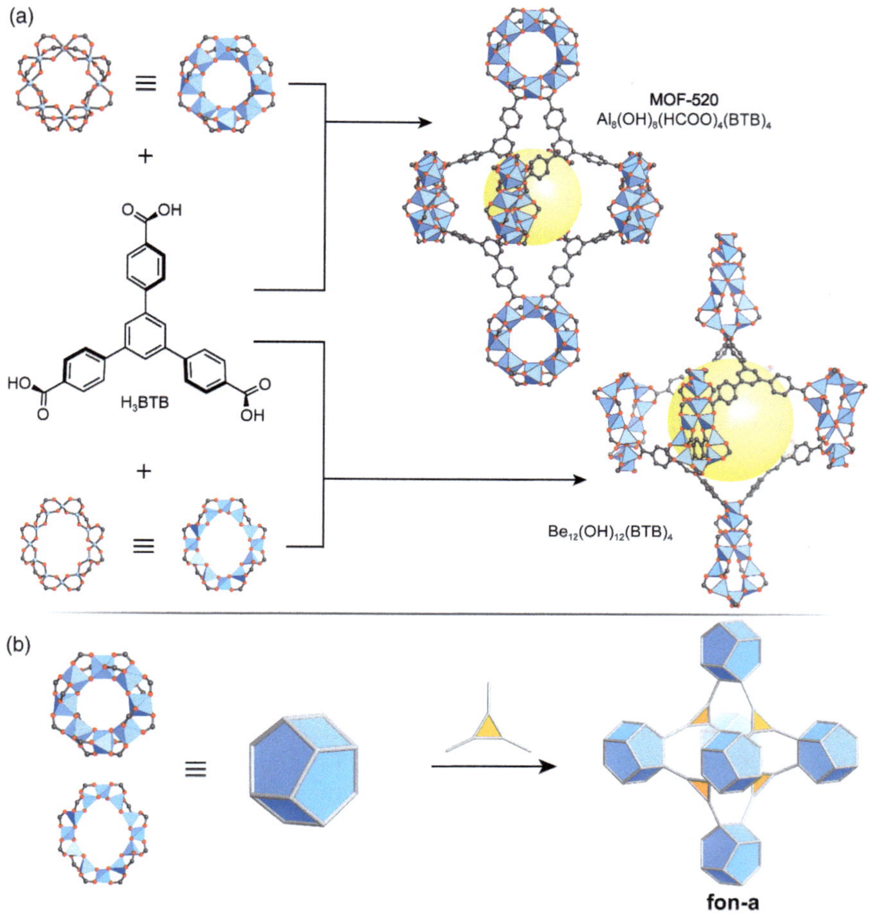

Figure 4.31 (a) Comparison of the crystal structure of MOF-520 (top) and $Be_{12}(OH)_{12}(BTB)_4$ (bottom). Both frameworks have a 3,12-connected **fon** topology. $Be_{12}(OH)_{12}(BTB)_4$ is built from 12-membered –OH bridged saddle-shaped SBUs, whereas the ring-shaped SBUs in MOF-520 consist of eight vertex-sharing hydroxyl bridged aluminum octahedra. (b) Topology representation of MOF-520 and $Be_{12}(OH)_{12}(BTB)_4$. The 12-c SBUs are shown in blue, the tritopic BTB linkers are represented by orange triangles. All hydrogen atoms are omitted for clarity. Color code: Zr, blue; C, gray; O, red.

A similar ring-shaped aluminum-based SBU is found in the structure of MOF-520 ($Al_8(OH)_8(HCOO)_4(BTB)_4$), a framework synthesized by the reticulation of Al^{3+} ions with trigonal tritopic H_3BTB linkers. MOF-520 shares the same topology with the aforementioned $Be_{12}(OH)_{12}(BTB)_4$ [35]. The 3,12-connected **fon** net of MOF-520 is built from octanuclear $Al_8(OH)_8(HCOO)_4(-COO)_{12}$ SBUs consisting of eight octahedrally coordinated vertex-sharing aluminum atoms that are connected by doubly bridging hydroxyl ions (Figure 4.31). Each SBU is connected to 12 BTB linkers and 4 terminal formates, which allows for large pores of 16.2 × 9.9 Å.

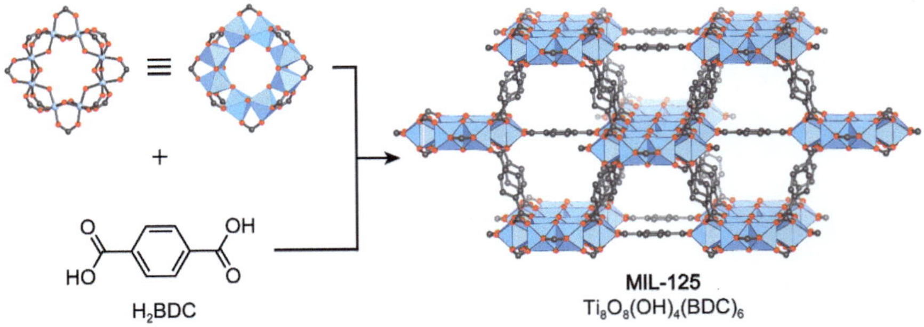

Figure 4.32 Crystal structure of MIL-125. The structure can be described as a body-centered arrangement of ring-shaped $Ti_8O_8(OH)_4(-COO)_{12}$ SBUs connected by linear ditopic BDC linkers. The resulting augmented **bcu** net has two distinct pores corresponding to the tetrahedral and octahedral holes of a body-centered cubic packing. All hydrogen atoms are omitted for clarity. Color code: Ti, blue; C, gray; O, red.

Among all metals employed in the construction of SBUs, titanium is one of the most attractive candidates, largely due to its low toxicity, redox activity, photocatalytic properties, and the high stability of carboxylate-based titanium MOFs. MIL-125 ($Ti_8O_8(OH)_4(BDC)_6$) is the first reported titanium MOF (Figure 4.32) [36]. Its structure is composed of cyclic octamers of corner-sharing $TiO_5(OH)$ octahedra connected by linear ditopic BDC linkers to form a porous structure with an underlying **bcu** topology. The structure can be described as a body-centered cubic arrangement of ring-shaped $Ti_8O_8(OH)_4(-COO)_{12}$ SBUs. Each SBU is connected to 12 neighboring SBUs by 12 BDC linkers, 4 of which are in the plane of the ring-shaped SBU, 4 are above, and 4 below. This arrangement results in two distinct pores with an accessible pore diameter of 6.13 and 12.55 Å that can be thought of as the octahedral and tetrahedral holes in a body-centered cubic packing, respectively [36].

4.4 MOFs Built from Infinite Rod SBUs

So far, we have only discussed discrete 0D SBUs with a defined connectivity; however, 1D SBUs with infinite rod-like structures are also known in MOF chemistry. It is worthwhile noting that such SBUs are not known as isolated molecular entities but are exclusive to MOF chemistry. Even though most MOFs containing rod SBUs are built from rare earth metals, we will focus on materials built from transition metals as we did for the discrete 0D SBUs. For a comprehensive compilation as well as an in-depth discussion of MOFs with rod SBUs including such involving rare earth metals, the reader is referred elsewhere [37].

We already encountered MOFs with rod SBUs in Chapter 2 when we discussed the structure of MOF-74 and its isoreticular expanded analogs. Another important example of a series of MOFs based on rod SBUs is the MIL-53 series (M(BDC)(OH), where M = Cr^{3+}, Fe^{3+}, Al^{3+}, V^{3+}) [38]. The structure of MIL-53 is built from chains of octahedrally coordinated M^{3+} ions connected by BDC

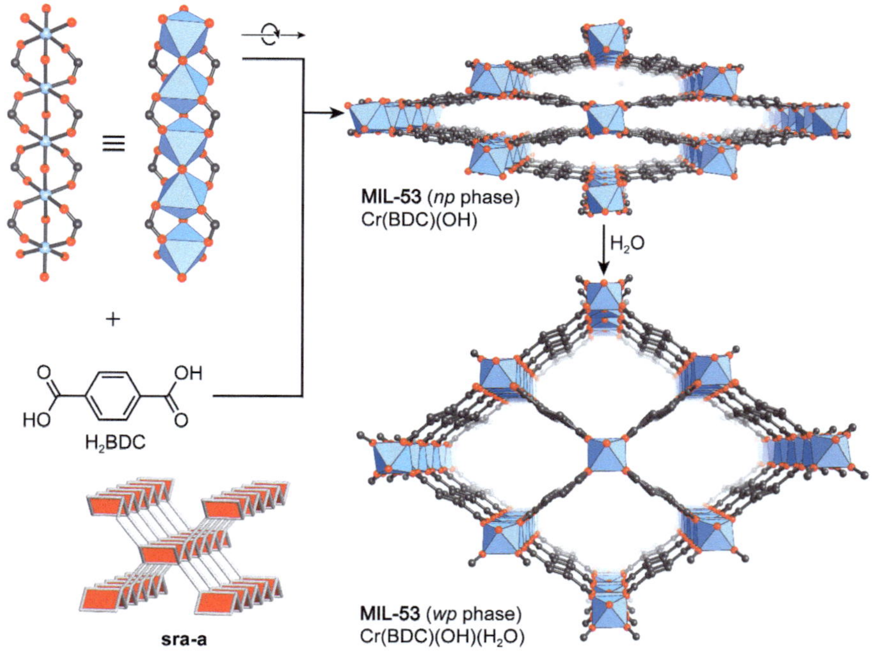

Figure 4.33 Crystal structure of MIL-53, a MOF built from $[Cr(-COO)_{4/4}(OH)_2]_\infty$ rod SBUs linked by linear ditopic BDC linkers to form a framework of **sra** topology. Interaction of the framework with molecular guests possessing large quadrupole or dipole moments results in a breathing motion of the framework. The insert illustrates the augmented **sra** net. All hydrogen atoms are omitted for clarity. Color code: Cr, blue; C, gray; O, red.

linkers to form a 3D microporous structure of **sra** topology with an array of 1D channels running along the crystallographic *c*-axis (Figure 4.33). The rod SBUs are built from $[Cr(-COO)_{4/4}(OH)_{2/2}]$ octahedra where the carboxylates of the two linkers connect two metal centers in a bridging mode. MIL-53 is particularly interesting as an unusual structural distortion upon CO_2 adsorption is observed. A "breathing" motion of the whole framework caused by the quadrupole moment of CO_2 affords a structural transition from its pristine narrow pore (*np*) geometry into an open wide pore (*wp*) phase [39]. This phenomenon was also studied in the presence of other gases with large dipole or quadrupole moments such as water where a similar deformation of the structure is observed.

Rod SBUs can also be of a helical shape. An illustrative example for this is the helical rod SBU in the structure of a chiral MOF of formula $Ni_{2.5}(OH)(\text{L-Asp})_2$ [40]. The structure of this framework is composed of helical $[M_4(OH)_2(-COO)_2]_\infty$ rod SBUs and $[Ni(\text{L-Asp})_2]^{2-}$ metallo-linkers. The helical SBUs in this chiral MOF of **srs** topology are built from edge- and corner-sharing Ni octahedra and are all of the same hand. Linking these helices leads to the formation of chiral helical channels of 5 × 8 Å running along the crystallographic *c*-axis (Figure 4.34).

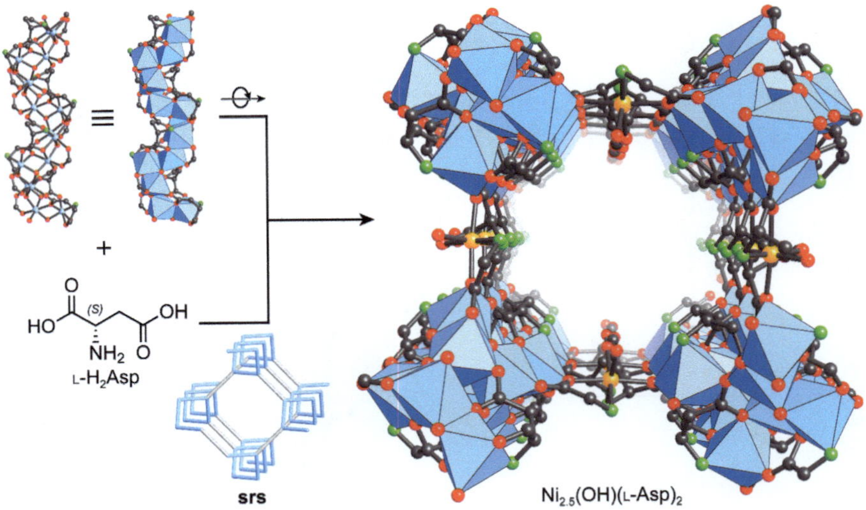

Figure 4.34 Crystal structure of $Ni_{2.5}(OH)(L\text{-}Asp)_2$, a framework built from 1D helical SBUs linked into an **srs** topology by linear $[Ni(L\text{-}Asp)_2]^{2-}$ metallo-linkers. The augmented **srs** net is depicted in the insert. All hydrogen atoms are omitted for clarity. Color code: Ni within the SBU, blue; Ni within the metallo-linker, orange; C, gray; N, green; O, red.

4.5 Summary

In this chapter we discussed the structures of MOFs built from discrete SBUs with a connectivity ranging from 3 to 12 as well as those of frameworks built from rod-like SBUs. We showed that in many cases the precise local geometry of the linker and the SBU, their chemical nature, and the reaction conditions strongly influence the structure of the resulting framework. A classification of structures built from building units of identical local symmetry into thermodynamic and kinetic products was introduced and we showed that this distinction can be made based on the analysis of the corresponding network topology. We provided a general understanding of why specific structures form and how they can be targeted by careful analysis of the requirements to their building units and adjustment of the reaction conditions. All structures discussed in this chapter are built from one type of SBU and one type of linker. It is however possible to synthesize MOFs that combine three or more distinct building units within one structure, and these will be the subject of Chapter 5.

References

1 Grzesiak, A.L., Uribe, F.J., Ockwig, N.W. et al. (2006). Polymer-induced heteronucleation for the discovery of new extended solids. *Angewandte Chemie International Edition* 45 (16): 2553–2556.

2 (a) Han, L., Xu, L.-P., and Zhao, W.-N. (2011). A novel 2D (3,5)-connected coordination framework with $Zn_2(COO)_3$ SBU. *Journal of Molecular Structure*

1000 (1–3): 58–61. (b) Wu, M., Jiang, F., Wei, W. et al. (2009). A porous polyhedral metal-organic framework based on $Zn_2(COO)_3$ and $Zn_2(COO)_4$ SBUs. *Crystal Growth & Design* 9 (6): 2559–2561. (c) Zhang, X., Zhang, Y.-Z., Zhang, D.-S. et al. (2015). A hydrothermally stable Zn(II)-based metal-organic framework: structural modulation and gas adsorption. *Dalton Transactions* 44 (35): 15697–15702. (d) Yang, J., Wang, X., Dai, F. et al. (2014). Improving the porosity and catalytic capacity of a zinc paddlewheel metal-organic framework (MOF) through metal-ion metathesis in a single-crystal-to-single-crystal fashion. *Inorganic Chemistry* 53 (19): 10649–10653.

3 Eddaoudi, M., Kim, J., Wachter, J. et al. (2001). Porous metal-organic polyhedra: 25 Å cuboctahedron constructed from 12 $Cu_2(CO_2)_4$ paddle-wheel building blocks. *Journal of the American Chemical Society* 123 (18): 4368–4369.

4 Furukawa, H., Kim, J., Ockwig, N.W. et al. (2008). Control of vertex geometry, structure dimensionality, functionality, and pore metrics in the reticular synthesis of crystalline metal-organic frameworks and polyhedra. *Journal of the American Chemical Society* 130 (35): 11650–11661.

5 Eddaoudi, M., Kim, J., Vodak, D. et al. (2002). Geometric requirements and examples of important structures in the assembly of square building blocks. *Proceedings of the National Academy of Sciences* 99 (8): 4900–4904.

6 Eddaoudi, M., Kim, J., O'Keeffe, M., and Yaghi, O.M. (2002). $Cu_2[o\text{-Br-}C_6H_3(CO_2)_2]_2(H_2O)_2(DMF)_8(H_2O)_2$: a framework deliberately designed to have the NbO structure type. *Journal of the American Chemical Society* 124 (3): 376–377.

7 Yaghi, O.M., O'Keeffe, M., Ockwig, N.W. et al. (2003). Reticular synthesis and the design of new materials. *Nature* 423 (6941): 705–714.

8 Chui, S.S.-Y., Lo, S.M.-F., Charmant, J.P.H. et al. (1999). A chemically functionalizable nanoporous material $[Cu_3(TMA)_2(H_2O)_3]_n$. *Science* 283 (5405): 1148–1150.

9 (a) Millward, A.R. and Yaghi, O.M. (2005). Metal-organic frameworks with exceptionally high capacity for storage of carbon dioxide at room temperature. *Journal of the American Chemical Society* 127 (51): 17998–17999. (b) Mason, J.A., Veenstra, M., and Long, J.R. (2014). Evaluating metal-organic frameworks for natural gas storage. *Chemical Science* 5 (1): 32–51.

10 Feldblyum, J.I., Liu, M., Gidley, D.W., and Matzger, A.J. (2011). Reconciling the discrepancies between crystallographic porosity and guest access as exemplified by Zn-HKUST-1. *Journal of the American Chemical Society* 133 (45): 18257–18263.

11 (a) Spanopoulos, I., Tsangarakis, C., Klontzas, E. et al. (2016). Reticular synthesis of HKUST-like tbo-MOFs with enhanced CH_4 storage. *Journal of the American Chemical Society* 138 (5): 1568–1574. (b) Ma, S., Sun, D., Ambrogio, M. et al. (2007). Framework-catenation isomerism in metal-organic frameworks and its impact on hydrogen uptake. *Journal of the American Chemical Society* 129 (7): 1858–1859. (c) Furukawa, H., Go, Y.B., Ko, N. et al. (2011). Isoreticular expansion of metal-organic frameworks with triangular and square building units and the lowest calculated density for porous crystals. *Inorganic Chemistry* 50 (18): 9147–9152.

12 Chen, B., Eddaoudi, M., Hyde, S. et al. (2001). Interwoven metal-organic framework on a periodic minimal surface with extra-large pores. *Science* 291 (5506): 1021–1023.

13 Sun, D., Ma, S., Ke, Y. et al. (2006). An interweaving MOF with high hydrogen uptake. *Journal of the American Chemical Society* 128 (12): 3896–3897.

14 (a) Sudik, A.C., Côté, A.P., and Yaghi, O.M. (2005). Metal-organic frameworks based on trigonal prismatic building blocks and the new "acs" topology. *Inorganic Chemistry* 44 (9): 2998–3000. (b) Férey, G., Serre, C., Mellot-Draznieks, C. et al. (2004). A hybrid solid with giant pores prepared by a combination of targeted chemistry, simulation, and powder diffraction. *Angewandte Chemie* 116 (46): 6456–6461. (c) Férey, G., Mellot-Draznieks, C., Serre, C., and Millange, F. (2005). Crystallized frameworks with giant pores: are there limits to the possible? *Accounts of Chemical Research* 38 (4): 217–225. (d) Férey, G., Mellot-Draznieks, C., Serre, C. et al. (2005). A chromium terephthalate-based solid with unusually large pore volumes and surface area. *Science* 309 (5743): 2040–2042. (e) Hong, D.-Y., Hwang, Y.K., Serre, C. et al. (2009). Porous chromium terephthalate MIL-101 with coordinatively unsaturated sites: surface functionalization, encapsulation, sorption and catalysis. *Advanced Functional Materials* 19 (10): 1537–1552. (f) Liu, Y., Eubank, J.F., Cairns, A.J. et al. (2007). Assembly of metal-organic frameworks (MOFs) based on indium-trimer building blocks: a porous MOF with soc topology and high hydrogen storage. *Angewandte Chemie International Edition* 46 (18): 3278–3283. (g) Alezi, D., Belmabkhout, Y., Suyetin, M. et al. (2015). MOF crystal chemistry paving the way to gas storage needs: aluminum-based soc-MOF for CH_4, O_2, and CO_2 storage. *Journal of the American Chemical Society* 137 (41): 13308–13318.

15 Eddaoudi, M., Sava, D.F., Eubank, J.F. et al. (2015). Zeolite-like metal-organic frameworks (ZMOFs): design, synthesis, and properties. *Chemical Society Reviews* 44 (1): 228–249.

16 Bromberg, L., Diao, Y., Wu, H. et al. (2012). Chromium(III) terephthalate metal organic framework (MIL-101): HF-free synthesis, structure, polyoxometalate composites, and catalytic properties. *Chemistry of Materials* 24 (9): 1664–1675.

17 (a) Horcajada, P., Chevreau, H., Heurtaux, D. et al. (2014). Extended and functionalized porous iron(III) tri-or dicarboxylates with MIL-100/101 topologies. *Chemical Communications* 50 (52): 6872–6874. (b) Feng, D., Liu, T.-F., Su, J. et al. (2015). Stable metal-organic frameworks containing single-molecule traps for enzyme encapsulation. *Nature Communications* 6: 5979.

18 Kickelbick, G. and Schubert, U. (1997). Oxozirconium methacrylate clusters: $Zr_6(OH)_4O_4(OMc)_{12}$ and $Zr_4O_2(OMc)_{12}$ (OMc = methacrylate). *European Journal of Inorganic Chemistry* 130 (4): 473–478.

19 Furukawa, H., Gándara, F., Zhang, Y.-B. et al. (2014). Water adsorption in porous metal-organic frameworks and related materials. *Journal of the American Chemical Society* 136 (11): 4369–4381.

20 Feng, D., Wang, K., Su, J. et al. (2015). A highly stable zeotype mesoporous zirconium metal-organic framework with ultralarge pores. *Angewandte Chemie International Edition* 54 (1): 149–154.

21 Ma, J., Wong-Foy, A.G., and Matzger, A.J. (2015). The role of modulators in controlling layer spacings in a tritopic linker based zirconium 2D microporous coordination polymer. *Inorganic Chemistry* 54 (10): 4591–4593.

22 Wang, R., Wang, Z., Xu, Y. et al. (2014). Porous zirconium metal-organic framework constructed from 2D → 3D interpenetration based on a 3,6-connected kgd net. *Inorganic Chemistry* 53 (14): 7086–7088.

23 Bai, S., Zhang, W., Ling, Y. et al. (2015). Predicting and creating 7-connected Zn_4O vertices for the construction of an exceptional metal-organic framework with nanoscale cages. *CrystEngComm* 17 (9): 1923–1926.

24 Tan, Y.-X., He, Y.-P., and Zhang, J. (2011). Pore partition effect on gas sorption properties of an anionic metal-organic framework with exposed Cu^{2+} coordination sites. *Chemical Communications* 47 (38): 10647–10649.

25 (a) Yuan, S., Lu, W., Chen, Y.-P. et al. (2015). Sequential linker installation: precise placement of functional groups in multivariate metal-organic frameworks. *Journal of the American Chemical Society* 137 (9): 3177–3180. (b) Bon, V., Senkovskyy, V., Senkovska, I., and Kaskel, S. (2012). Zr(IV) and Hf(IV) based metal-organic frameworks with reo-topology. *Chemical Communications* 48 (67): 8407–8409.

26 Yuan, S., Chen, Y.P., Qin, J. et al. (2015). Cooperative cluster metalation and ligand migration in zirconium metal-organic frameworks. *Angewandte Chemie International Edition* 54 (49): 14696–14700.

27 (a) Jiang, H.-L., Feng, D., Wang, K. et al. (2013). An exceptionally stable, porphyrinic Zr metal-organic framework exhibiting pH-dependent fluorescence. *Journal of the American Chemical Society* 135 (37): 13934–13938. (b) Deria, P., Gómez-Gualdrón, D.A., Hod, I. et al. (2016). Framework-topology-dependent catalytic activity of zirconium-based (porphinato) zinc(II) MOFs. *Journal of the American Chemical Society* 138 (43): 14449–14457. (c) Morris, W., Volosskiy, B., Demir, S. et al. (2012). Synthesis, structure, and metalation of two new highly porous zirconium metal-organic frameworks. *Inorganic Chemistry* 51 (12): 6443–6445.

28 (a) Cavka, J.H., Jakobsen, S., Olsbye, U. et al. (2008). A new zirconium inorganic building brick forming metal organic frameworks with exceptional stability. *Journal of the American Chemical Society* 130 (42): 13850–13851. (b) Feng, D., Gu, Z.-Y., Chen, Y.-P. et al. (2014). A highly stable porphyrinic zirconium metal-organic framework with shp-a topology. *Journal of the American Chemical Society* 136 (51): 17714–17717.

29 Trickett, C.A., Gagnon, K.J., Lee, S. et al. (2015). Definitive molecular level characterization of defects in UiO-66 crystals. *Angewandte Chemie International Edition* 54 (38): 11162–11167.

30 Yang, S., Long, L., Jiang, Y. et al. (2002). An exceptionally stable metal-organic framework constructed from the $Zn_8(SiO_4)$ core. *Chemistry of Materials* 14 (8): 3229–3231.

31 Yang, S., Long, L., Huang, R., and Zheng, L. (2002). $[Zn_8(SiO_4)(C_8H_4O_4)_6]_n$: the firstborn of a metallosilicate-organic hybrid material family ($C_8H_4O_4$ = isophthalate). *Chemical Communications* (5): 472–473.

32 An, J., Farha, O.K., Hupp, J.T. et al. (2012). Metal-adeninate vertices for the construction of an exceptionally porous metal-organic framework. *Nature Communications* 3: 604.

33 Li, T., Kozlowski, M.T., Doud, E.A. et al. (2013). Stepwise ligand exchange for the preparation of a family of mesoporous MOFs. *Journal of the American Chemical Society* 135 (32): 11688–11691.

34 (a) Bragg, W. (1923). Crystal structure of basic beryllium acetate. *Nature* 111: 532. (b) Pauling, L. and Sherman, J. (1934). The structure of the carboxyl group II. The crystal structure of basic beryllium acetate. *Proceedings of the National Academy of Sciences* 20 (6): 340–345. (c) Han, S.S., Deng, W.-Q., and Goddard, W.A. (2007). Improved designs of metal-organic frameworks for hydrogen storage. *Angewandte Chemie International Edition* 46 (33): 6289–6292. (d) Sumida, K., Hill, M.R., Horike, S. et al. (2009). Synthesis and hydrogen storage properties of $Be_{12}(OH)_{12}$(1,3,5-benzenetribenzoate)$_4$. *Journal of the American Chemical Society* 131 (42): 15120–15121.

35 Gándara, F., Furukawa, H., Lee, S., and Yaghi, O.M. (2014). High methane storage capacity in aluminum metal-organic frameworks. *Journal of the American Chemical Society* 136 (14): 5271–5274.

36 Dan-Hardi, M., Serre, C., Frot, T. et al. (2009). A new photoactive crystalline highly porous titanium(IV) dicarboxylate. *Journal of the American Chemical Society* 131 (31): 10857–10859.

37 (a) Schoedel, A., Li, M., Li, D. et al. (2016). Structures of metal-organic frameworks with rod secondary building units. *Chemical Reviews* 116 (19): 12466–12535. (b) Rosi, N.L., Kim, J., Eddaoudi, M. et al. (2005). Rod packings and metal-organic frameworks constructed from rod-shaped secondary building units. *Journal of the American Chemical Society* 127 (5): 1504–1518. (c) Cheng, P. and Bosch, M. (2015). *Lanthanide Metal-Organic Frameworks.* Springer.

38 (a) Millange, F., Serre, C., and Ferey, G. (2002). Synthesis, structure determination and properties of MIL-53as and MIL-53ht: the first Cr^{III} hybrid inorganic–organic microporous solids: $Cr^{III}(OH)\cdot\{O_2C{-}C_6H_4{-}CO_2\}\cdot\{HO_2C{-}C_6H_4{-}CO_2H\}_x$. *Chemical Communications* (8): 822–823. (b) Serre, C., Millange, F., Thouvenot, C. et al. (2002). Very large breathing effect in the first nanoporous chromium(III)-based solids: MIL-53 or $Cr^{III}(OH)\cdot\{O_2C{-}C_6H_4{-}CO_2\}\cdot\{HO_2C{-}C_6H_4{-}CO_2H\}_x\cdot H_2O_y$. *Journal of the American Chemical Society* 124 (45): 13519–13526. (c) Loiseau, T., Serre, C., Huguenard, C. et al. (2004). A rationale for the large breathing of the porous aluminum terephthalate (MIL-53) upon hydration. *Chemistry – A European Journal* 10 (6): 1373–1382.

39 Boutin, A., Coudert, F.-X., Springuel-Huet, M.-A. et al. (2010). The behavior of flexible MIL-53(Al) upon CH_4 and CO_2 adsorption. *The Journal of Physical Chemistry C* 114 (50): 22237–22244.

40 Anokhina, E.V., Go, Y.B., Lee, Y. et al. (2006). Chiral three-dimensional microporous nickel aspartate with extended Ni–O–Ni bonding. *Journal of the American Chemical Society* 128 (30): 9957–9962.

5

Complexity and Heterogeneity in MOFs

5.1 Introduction

Thus far, we have described metal-organic framework (MOF) structures that are built from two different building units, where one kind of a linker and one kind of a secondary building unit (SBU) are linked into a periodic framework structure. It is however possible to combine more than two kinds of building unit together into a single framework. For this scenario, we distinguish two cases: "complexity" and "heterogeneity." This will be the focus of this chapter and we begin by illustrating the difference between heterogeneity and complexity based on examples of MOF structures, all involving $Zn_4O(-COO)_6$ SBUs, but different organic linkers.

When $Zn_4O(-COO)_6$ SBUs are reticulated with linear ditopic BDC or trigonal tritopic BTB linkers, MOF-5 (**pcu**) or MOF-177 (**qom**) are formed, respectively [1]. In contrast, when both linkers are employed in the same synthesis, a different and more complex MOF structure (**muo**, UMCM-1, $Zn_4O(BDC)(BTB)_{4/3}$) results [2]. Notice that in this multinary structure the position of each of the individual building units is crystallographically defined (Figure 5.1) and we refer to this as "complexity within frameworks." Such complexity can be achieved by introducing (i) multiple different linkers, (ii) multiple different SBUs, (iii) multiple different SBUs and linkers, or (iv) ordered vacancies into a single MOF structure.

A different scenario arises when $Zn_4O(-COO)_6$ SBUs are reticulated with linear ditopic BDC or Br-BDC linkers. Here, two isostructural and metrically identical but chemically distinct MOFs, MOF-5 (**pcu**) and IRMOF-2 (**pcu**), respectively, are formed [3]. Since the backbone of both MOFs is identical, it is possible to combine both linkers within one framework while maintaining the parent structure type (**pcu**). Notice that in the resulting structure (MTV-MOF-5) the spatial arrangement of the BDC and Br-BDC linkers is unknown, thus introducing heterogeneity onto the otherwise ordered MOF-5 backbone (Figure 5.2), and we refer to this as "heterogeneity in frameworks" [4]. Heterogeneity can be achieved by introducing (i) multiple interchangeable linkers that are identical in terms of their binding groups and metrics but different with respect to their chemical composition, (ii) multiple metal ions that form the same SBU, and (iii) aperiodic vacancies onto or into the backbone of a single MOF structure. Such MOFs are referred to as multivariate (MTV) MOFs.

Introduction to Reticular Chemistry: Metal-Organic Frameworks and Covalent Organic Frameworks,
First Edition. Omar M. Yaghi, Markus J. Kalmutzki, and Christian S. Diercks.

Companion website: www.wiley.com/go/yaghi/reticular

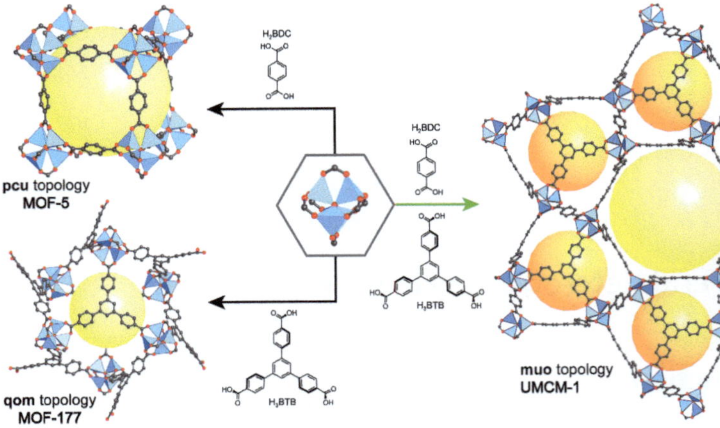

Figure 5.1 Complexity in frameworks is observed in multinary MOFs, where three or more building units of different size and geometry are combined into a single MOF structure. In this example, we consider the reticulation of $Zn_4O(-COO)_6$ SBUs with H_2BDC and H_3BTB linkers. In the binary systems, the products of these reticulations are MOF-5 (**pcu**) and MOF-177 (**qom**), respectively. In contrast, reticulating both linkers with $Zn_4O(-COO)_6$ SBUs into a single MOF structure yields UMCM-1 (**muo**), a MOF that is structurally not related to any of the two binary structures. Since this MOF has a more complex backbone, we refer to this as "complexity in frameworks." All hydrogen atoms are omitted for clarity. Color code: Zn, blue; C, gray; O, red.

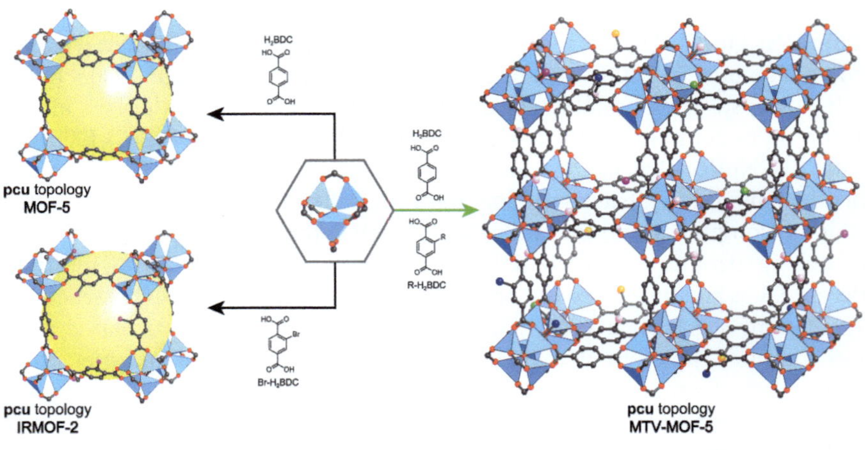

Figure 5.2 Heterogeneity in frameworks is observed, when three or more building units, of which at least two are structurally interchangeable, are reticulated into a single framework. In this example, we consider the reticulation of $Zn_4O(-COO)_6$ SBUs with H_2BDC and Br-H_2BDC linkers. In the binary systems, the products of these reticulations are MOFs of the same **pcu** topology, MOF-5 and IRMOF-2, respectively. MOF-5 and IRMOF-2 are isoreticular (isostructural) but differ in terms of their chemical composition. Thus, reticulating both linkers with $Zn_4O(-COO)_6$ SBUs into a single structure again yields a MOF (MTV-MOF-5) of the same structure type (**pcu**). The spatial arrangement of the BDC and Br-BDC linker in MTV-MOF-5 imposes heterogeneity onto its crystalline backbone, which is thus referred to as "heterogeneity in frameworks." All hydrogen atoms are omitted for clarity. Color code: Zn, blue; C, gray; O, red; Br, pink; Undefined substituents of the R-BDC linkers in MTV-MOF-5 have been given various colors.

In this chapter, we explore MOF structures that go beyond binary systems and illustrate how complexity and heterogeneity within frameworks can afford MOFs with properties that go beyond those of the corresponding binary counterparts.

5.2 Complexity in Frameworks

5.2.1 Mixed-Metal MOFs

Introducing multiple different SBUs into a MOF structure expands the scope of accessible net topologies. In the case of MOF structures containing two geometrically and/or metrically distinct SBUs, these are (i) either all constructed from the same metal or (ii) different metals that from distinct SBUs within the same material. Forming multiple distinct types of SBU in a one-pot reaction poses a challenge because, as discussed in Chapters 3 and 4, meticulously adjusted reaction conditions are required to target the formation of a specific SBU. Thus, in many cases, MOF structures with more than one unique SBU are formed by chance rather than by design. In the following text we will discuss two general approaches for the synthesis of complex MOF structures built from more than one type of SBU.

5.2.1.1 Linker De-symmetrization

The approach of "vertex-directed linker de-symmetrization" is used in the design and synthesis of UMCM-150 ($Cu_3(BHTC)_2(H_2O)_3$), a MOF built from two different SBUs, the 4-c $Cu_2(-COO)_4$ paddle wheel and the comparatively rare 6-c $Cu_3(-COO)_6$ paddle wheel (Figure 5.3) [5]. The de-symmetrized trigonal H_3BHT linker (C_{2h}) is made by asymmetric expansion of the trigonal H_3BTC linker (D_{3h}). In the structure of UMCM-150, each BHTC linker is connected to one 6-c and two 4-c paddle wheel SBUs and the resulting framework has an overall **agw** topology. As depicted in Figure 5.3, the 3,4,6-connected **agw** net consists of triangles (BHTC), squares ($Cu_2(-COO)_4$), and trigonal prisms ($Cu_3(-COO)_6$). Dissecting the BHTC linker into a benzoate and isophthalate subunit reveals that the structure of UMCM-150 can alternatively be described as hexagonal kagome (**kgm**, the augmented net of **kgm** is called **fxt**) layers built from 4-c paddle wheel SBUs connected through ditopic isophthalate units that are sixfold cross-linked through the 6-c trigonal prismatic paddle wheel units to give a 3D framework.

5.2.1.2 Linkers with Chemically Distinct Binding Groups

To gain more control over the reticulation process when targeting complex MOF structures, linkers bearing multiple different binding groups favoring the formation of different types of SBU can be used. This approach is employed in the synthesis of MOF-325 where reticulating Cu^{2+} ions and H_2PyC (4-pyrazolecarboxylic acid) yields a framework of the chemical formula $Cu_3(H_2O)_3[(Cu_3O)(PyC)_3(NO_3)L_2]_2$ (Figure 5.4) [6]. This framework contains two different types of SBU. Trigonal planar $Cu_3OL_3(PyC)_3$ SBUs (L = NO_3 and/or solvent, PyC = 4-pyrazolecarboxylate) are formed by the pyrazolate portion of PyC, and the carboxylates connect two Cu^{2+} centers to form

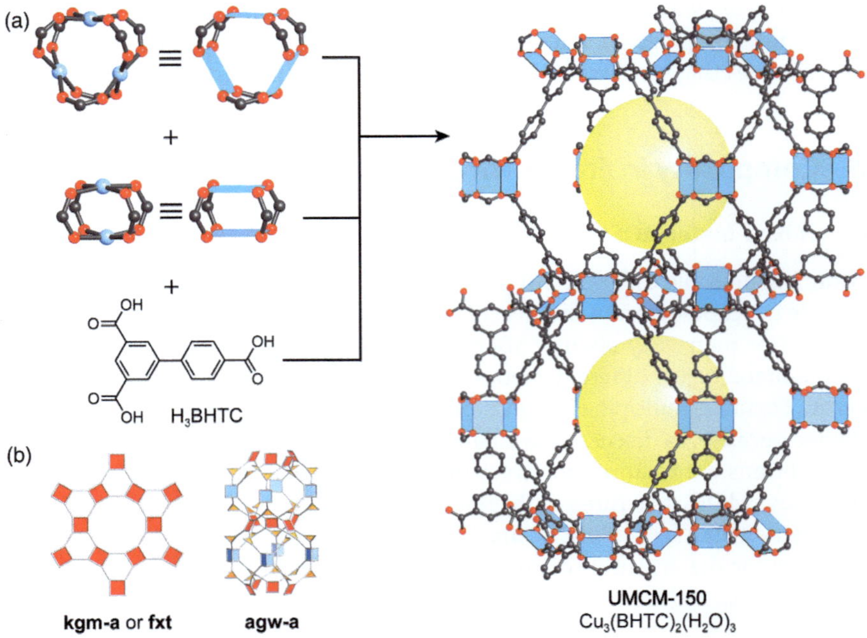

Figure 5.3 (a) Crystal structure of UMCM-150. In this example, vertex-directed linker de-symmetrization directs the synthesis toward the formation of two distinct SBUs; a 4-c and a 6-c copper paddle wheel. The resulting structure can be described as kagome layers (**kgm**) formed from 4-c paddle wheel SBUs and the isophthalate portion of the BHTC linker, that are triply cross-linked to form a 3D structure of **agw** topology. (b) Topology representations of the augmented **kgm** (**fxt**) and **agw** net, the 4-c paddle wheel SBUs are represented by red squares, the 6-c paddle wheel SBUs by blue trigonal prisms, and the BHTC linkers by orange triangles. All hydrogen atoms are omitted for clarity. Color code: Cu, blue; C, gray; O, red.

$Cu_2(-COO)_4$ paddle wheel SBUs [6]. Consequently, these trigonal planar and square planar SBUs form a framework of **tbo** topology. MOF-325 is isostructural to HKUST-1, where the trigonal planar BTC (benzene-tricarboxylate) linkers (D_{3h}) are replaced by trigonal planar $Cu_3OL_3(PyC)_3$ SBUs (D_{3h}). Since these trinuclear $Cu_3OL_3(PyC)_3$ SBUs are larger than the BTC linker in HKUST-1, the pores of MOF-325 (19.6 Å) are also larger than those in the structure of HKUST-1 (18 Å) [7].

To prepare even more complex framework structures, the linker used in the synthesis of MOF-325 (i.e. H_2PyC) is reticulated with two different metals to enforce the formation of two distinct types of SBU. Reticulation of H_2PyC with Cu^{2+} and Zn^{2+} ions facilitates the formation of a highly complex MOF structure (FDM-3, where FDM = Fudan Materials) of chemical formula $[(Zn_4O)_5(Cu_3OH)_6(PyC)_{22.5}(OH)_{18}(H_2O)_6][Zn(OH)(H_2O)_3]_3$ [8]. The structure of FDM-3 contains multiple different types of SBU; one copper-based $Cu_3OH(-PyC)_3$ SBU and six different zinc-based SBUs that are structurally related to the $Zn_4O(-COO)_6$ SBU. Four of these zinc-based SBUs are octahedral ($ZnO_4(-COO)_3R_3$ where R = –COO or –NN) and two are square pyramidal ($ZnO_4(-COO)_4R$). The highly complex structure of FDM-3 crystallizes in the

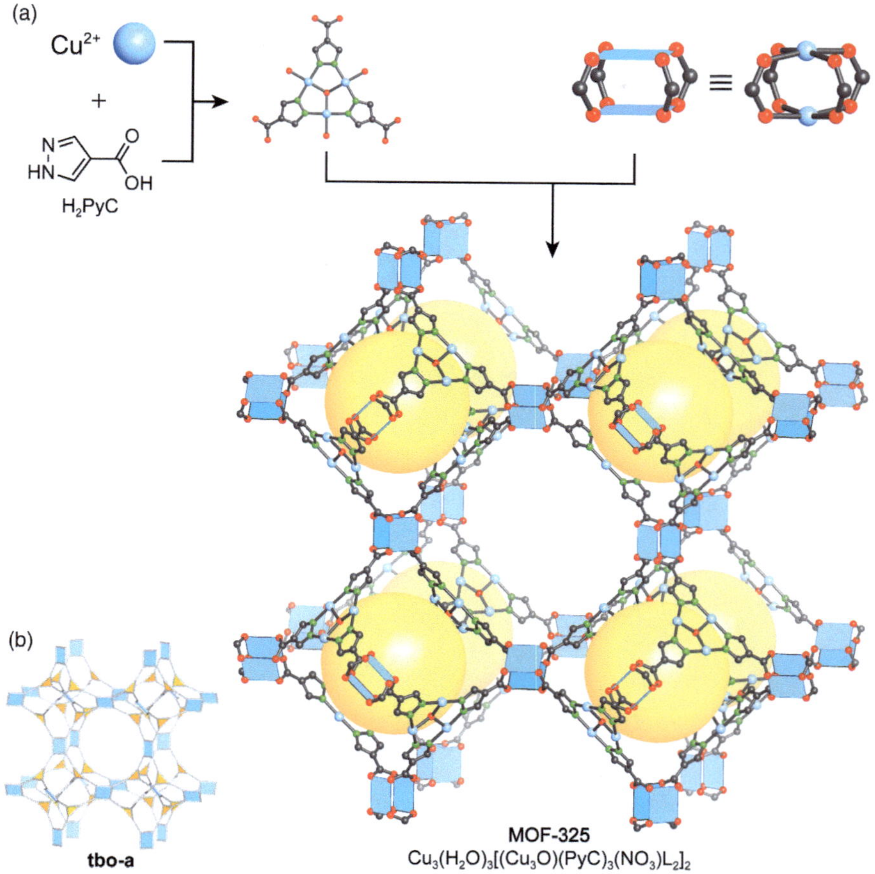

Figure 5.4 (a) Two distinct copper SBUs sustain the structure of MOF-325. The trigonal 3-c $Cu_3OL_3(PyC)_3$ SBU has D_{3h} symmetry and connects 4-c paddle wheel SBUs to form a framework of **tbo** topology. The large pores (19.6 Å) are connected in all three directions giving rise to a 3D intersecting pore system. (b) Representation of the augmented **tbo** net. Copper paddle wheel SBUs are represented by blue squares and $Cu_3OL_3(PyC)_3$ units are represented by orange triangles. The terminal water ligands on the copper paddle wheel, terminal ligands on the $Cu_3OL_3(PyC)_3$ SBUs, and all hydrogen atoms are omitted for clarity. Color code: Cu, blue; N, green; C, gray; O, red.

3,5,6-connected **ott** net (Figure 5.5) and the pores of the anionic framework are partially filled with charge balancing $[Zn(OH)(H_2O)_3]^+$ ions.

The overall structure of FDM-3 is composed of four kinds of cage, two of which are microporous and the other two are mesoporous (Figure 5.6). The cubic cage (cage I) has a diameter of 7.6 Å and is positioned at the center and on all edges of the cubic unit cell similar to a cubic close packed arrangement. The second cage (cage II) can be described by a gyrobifastigium (two face-regular trigonal prisms rotated toward each other by 90° and joined along corresponding square faces) and has a size of approximately 8.0 × 8.0 Å. The mesoporous cage III is topologically identical to the largest cage in MIL-101 and has a pore diameter of 23.4 Å. Cage IV, is the largest cage in FDM-3. It is of pseudo-octahedral shape

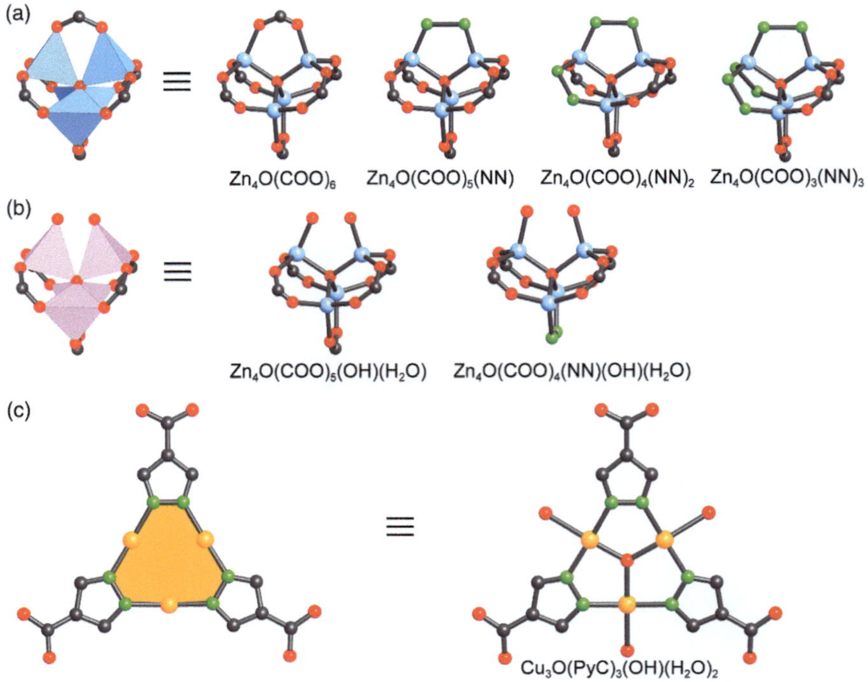

Figure 5.5 The structure of FDM-3 contains seven different types of SBU that are built from two different metals (zinc and copper) and have a connectivity ranging from 3 to 6. (a) 6-c octahedral zinc-based SBUs (blue) of chemical formula $Zn_4O(COO)_6$, $Zn_4O(COO)_5(NN)$, $Zn_4O(COO)_4(NN)_2$, and $Zn_4O(COO)_4(NN)_3$. (b) 5-c square pyramidal zinc-based SBUs (pink) of chemical formula $Zn_4O(COO)_5(OH)(H_2O)$ and $Zn_4O(COO)_4(NN)(OH)(H_2O)$. (c) 3-c trigonal copper-based SBU (orange) of chemical formula $Cu_3(OH)(H_2O)_2(PyC)_3$. All hydrogen atoms are omitted for clarity. Color code: Zn, blue; Cu, orange; C, gray; N, green; O, red.

and has a pore diameter of 28.8 Å. One single unit cell of FDM-3 contains a total of 28 microporous cages (4 × cage I, 24 × cage II) and 11 mesoporous cages (8 × cage III, 3 × cage IV) and the fully activated materials has a surface area of $2585\ m^2\ g^{-1}$.

5.2.2 Mixed-Linker MOFs

Similar to the synthesis of MOFs built from more than one kind of SBU, employing more than one kind of linker in a MOF synthesis can result in the formation of a complex multinary MOF structure. It is however difficult to synthesize such MOFs in a designed way, because the presence of multiple building units in the reaction mixture introduces a degree of unpredictability. To illustrate this, we look at the example used in the beginning of this chapter: MOF structures that are formed by the reticulation of Zn^{2+} ions with linear ditopic H_2BDC and/or trigonal tritopic H_3BTB (see Figure 5.1). The reticulation of Zn^{2+} ions with H_3BTB leads to the formation of MOF-177 (**qom**), and under almost identical conditions, reticulation of Zn^{2+} with H_2BDC yields MOF-5 (**pcu**).

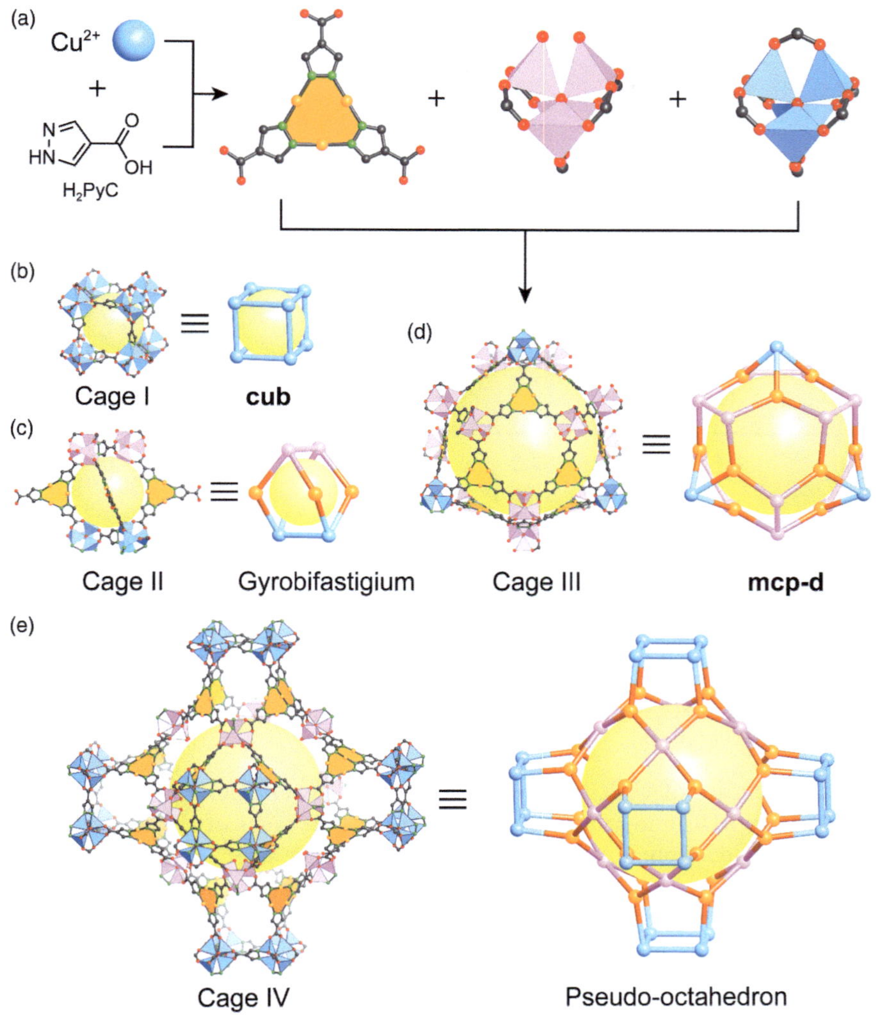

Figure 5.6 (a) The combination of the seven different types of SBUs shown in Figure 5.5 within one MOF structure (FDM-3) gives a highly complex framework structure with **ott** topology. The structure contains four differently sized cages; two micro- and two mesoporous. (b) Cage I (**cub**) is built from only one type of SBU, has a diameter of 7.8 Å, and resembles the pores in the structure of MOF-5. (c) Cage II has the shape of a gyrobifastigium and is about 8.0 × 8.0 Å large. (d) Cage III is of **mcp-d** topology and measures 23.4 Å in diameter. (e) The largest cage (cage IV) is of pseudo-octahedral shape, is built from 60 SBUs, and has a diameter of 28.8 Å. A single unit cell of FDM-3 contains a total of 39 cages (4 × cage I, 24 × cage II, 8 × cage III, and 3 × cage IV).

In contrast, for a synthesis involving both linkers, it is not yet possible to predict whether one or the other binary MOF, or even a new complex mixed-linker MOF will form. Studies of the ternary system $Zn^{2+}/H_2BDC/H_3BTB$ reveal that by altering the ratio of linkers, MOF-5, MOF-177, and a mixed-linker MOF (UMCM-1, $Zn_4O(BDC)(BTB)_{4/3}$, **muo**), that contains both linkers can be isolated (Figure 5.7) [9].

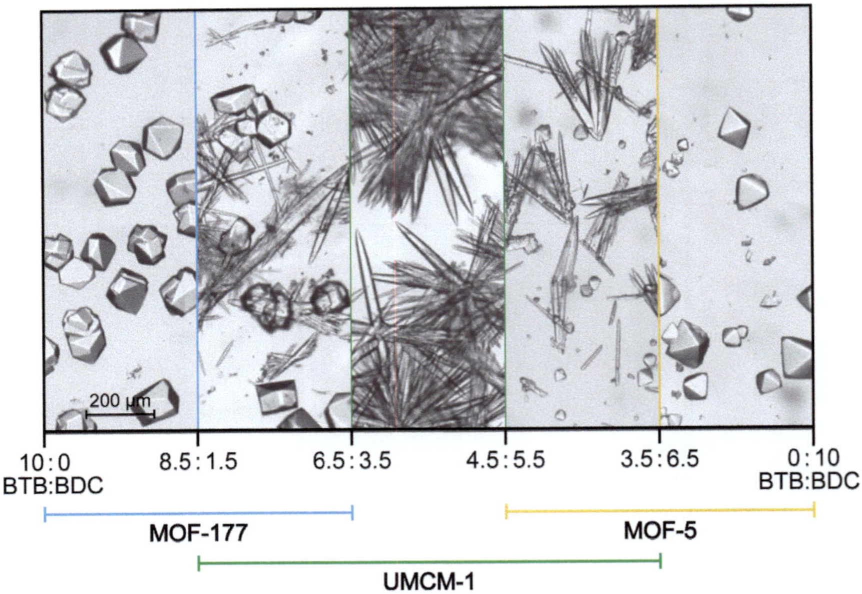

Figure 5.7 Formation of different MOF phases depending on the ratio of the linkers (H_2BDC:H_3BTB) employed in the synthesis. MOF-5 (**pcu**) and MOF-177 (**qom**) are formed from the reaction of Zn^{2+} and only one (or an excess of one) linker, while for the reaction with linker mixtures with a ratio of H_2BDC:H_3BTB between 6 : 4 and 4 : 6 the formation of phase pure UMCM-1 (**muo**) is observed.

In the structure of UMCM-1, each octahedral $Zn_4O(-COO)_6$ SBU is connected to two BDC and four BTB linkers resulting in cages of about 14 × 17 Å, each constructed from nine SBUs connected through six BDC and five BTB linkers (Figure 5.8). These cages are fused in an edge-sharing manner to form an arrangement with 1D hexagonal channels of 24 × 29 Å in diameter. UMCM-1 combines both, the high surface area ($5730\,m^2\,g^{-1}$) common to most MOFs and the large pore apertures common to mesoporous silicates/aluminosilicates. Efforts to prepare an expanded version of UMCM-1 (**muo**) by replacing H_2BDC with H_2T^2DC (thieno[3,2-*b*]thiophene-2,5-dicarboxylate) have resulted in the formation of a framework with a different structure, termed UMCM-2 ($Zn_4O(T^2DC)(BTB)_{4/3}$, **umt** topology) [10]. When both, the geometry and the metrics (length ratios) of all constituents is maintained, the isoreticular expansion of multinary MOFs is feasible, as exemplified by the expansion of UMCM-2 to DUT-32 ($Zn_4O(BPDC)(BTCTB)_{4/3}$) [11].

In a similar fashion, another mixed-linker MOF can be prepared by reticulation of Zn^{2+} ions with trigonal tritopic H_3BTE and linear ditopic H_2BPDC linkers. The use of Zn^{2+} ions and H_2BPDC or H_3BTE separately yields the binary MOFs IRMOF-10 (**pcu**) and MOF-180 (**qom**), respectively [1, 3]. Combining both of these linkers within one structure results in a multinary framework of chemical formula ($(Zn_4O)_3(BPDC)_4(BTE)_3$) named MOF-210 (Figure 5.9), that crystallizes in a **toz** topology [1]. This is especially interesting, considering that UMCM-1 is built from building units that are represented by the same general vertex figures (octahedron, triangle, and linear edge) but are differ with respect to their metrics. The structure of MOF-210 contains two distinct cages, where the larger

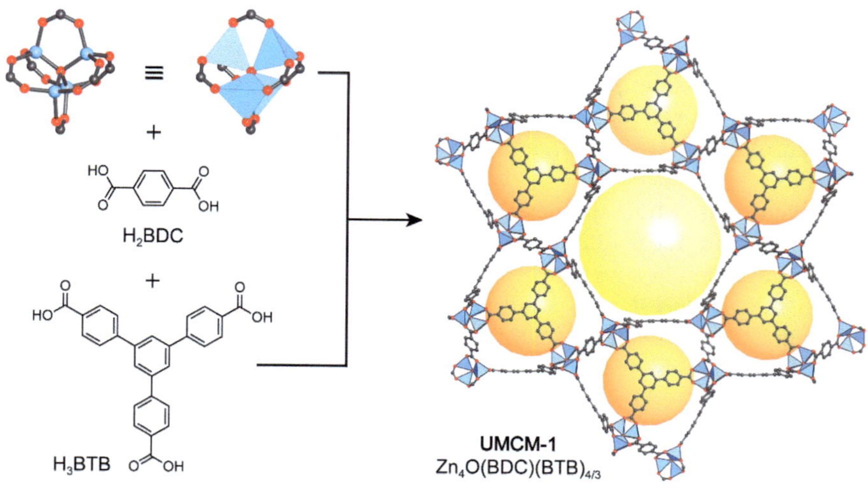

Figure 5.8 Crystal structure of UMCM-1 with underlying **muo** topology.Small cages of about 14×17 Å (orange spheres) are formed by linking nine SBUs through six BDC and five BTB linkers. These small cages are arranged in an edge-sharing manner to create large 1D hexagonal channels of 24×29 Å (yellow sphere). All hydrogen atoms are omitted for clarity. Color code: Zn, blue; C; gray; O, red.

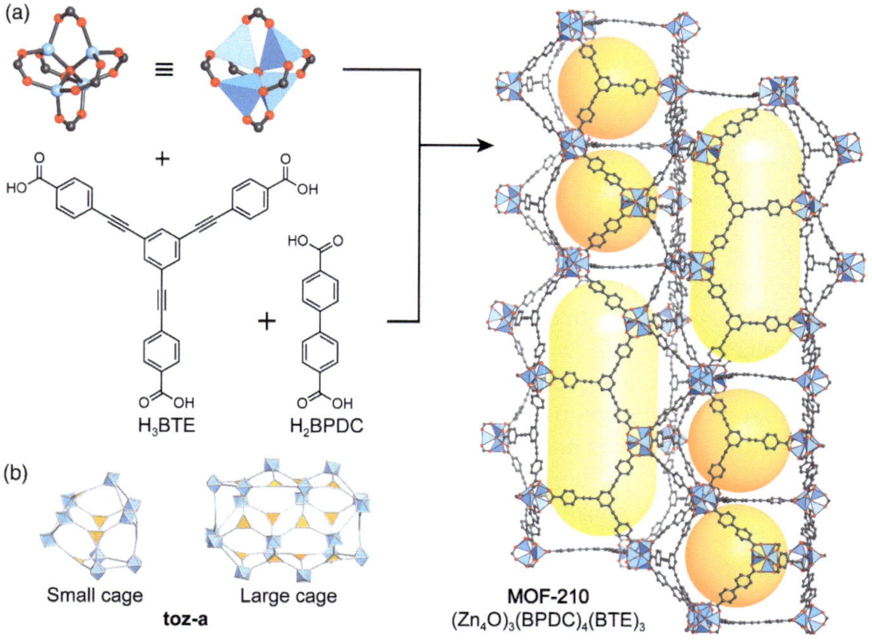

Figure 5.9 (a) Crystal structure of MOF-210. Reticulation of Zn^{2+} ions with H_3BTE and H_2BPDC leads to the formation of a complex framework structure with **toz** topology. The structure comprises two differently sized pores. The smaller pores (orange spheres) have a diameter of approximately 11 Å, and the larger elliptical pores (yellow ellipsoids) measure 27×48 Å. This complex architecture gives rise to an ultrahigh surface area of $6240\ m^2\ g^{-1}$. (b) Topology representation of the two cages in the **toz** net of MOF-210: BTE linkers are represented by orange triangles and zinc SBUs are shown as blue octahedra. All hydrogen atoms are omitted for clarity. Color code: Zn, blue, C, gray; O, red.

cage comprises 18 $Zn_4O(-COO)_6$ SBUs connected by 14 BTE and 6 BPDC linkers, encompassing an elliptical pore of about 27 × 48 Å. The smaller cage has a diameter of approximately 11 Å and is framed by 9 $Zn_4O(-COO)_6$ SBUs, 5 BTE, and 6 BPDC linkers. The complex framework architecture of MOF-210 gives rise to an ultrahigh surface area of 6240 $m^2\ g^{-1}$ and outstanding gas sorption properties.

All mixed-linker MOFs discussed thus far are built from a combination of trigonal tritopic and linear ditopic linkers, and $Zn_4O(-COO)_6$ SBUs, but they all crystallize in different structure types. This highlights the fact that for multinary systems, the *a priori* design is not trivial. An exact ratio of the linker lengths must be maintained to direct the reticular synthesis toward a specific topology as illustrated by the subtle variations in the linker length ration between MOF-210 (0.76), UMCM-1 (0.786), and UMCM-2 (0.79). This makes the isoreticular expansion of such structures challenging and sometimes even impossible as in the case of MOF-205 ($Zn_4O(BTB)_{4/3}(NDC)$) [1].

Multinary MOFs can also be built from more than three distinct building units, and several MOFs combining three different kinds of linker within a single structure are known. Among them are MOFs with pillared-layered structures. In MOFs of that type most often only the 2D layers are formed by joining the SBUs through carboxylate-based linkers and the pillars are typically O- or N-donor linkers (e.g. 4,4′-bipyridine, DABCO), and similar structures containing only carboxylate-based linkers are rare. An example of such frameworks is a series of isoreticular MOFs based on UMCM-4, a quaternary pillared-layered MOF where all building units are stitched together solely by strong bonds. In all structures (UMCM-4, 10, 11, 12), $Zn_4O(-COO)_6$ SBUs are connected by tritopic TPA and ditopic BDC linkers to form layers with a **cru** topology. These layers are in turn pillared by BDC (UMCM-4), Me_4-BPDC (UMCM-10, Me4-BPDC = 2,2′,6,6′-tetramethylbiphenyl-4,4′-dicarboxylic acid), EDDB (UMCM-11, EDDB = 4,40-(ethyne-1,2-diyl)dibenzoic acid), or TMTPDC (UMCM-12, TMTPDC = 2′,3′, 5′,6′-tertramethylterphenyl-4,4″-dicarboxylic acid) to give rise to 3D extended framework structures [12]. In contrast to other mixed-linker systems that do not possess a pillared-layered structure, here, an anisotropic isoreticular expansion is possible by design. Introduction of a longer ditopic linker into the 2D layers leads to a disturbance of their symmetry, which is why selective implementation of the expanded ditopic linkers is only observed between the layers. The tunability of the pore window by expansion of the linker is a powerful tool for tweaking the selectivity in molecular sieving.

Another example for a complex, yet well-ordered, quaternary 3D framework is MUF-7a ($(Zn_4O)_3\ (BTB)_{4/3}(BDC)_{1/2}(BPDC)_{1/2}$, MUF = Massey University Metal-Organic Frameworks) with an underlying **ith-d** topology [13]. MUF-7a is formed by reticulation of Zn^{2+} ions with trigonal tritopic H_3BTB, linear ditopic H_2BDC, and linear ditopic H_2BPDC linkers. Its structure has two different cages that measure 10 and 20 Å in diameter (Figure 5.10). The precise design of the linker and matching of its symmetry to that of the sites it occupies in the crystal structure helps to avoid randomness and disorder. This approach allows for the synthesis of MOFs with "programmed pores," featuring multiple functional

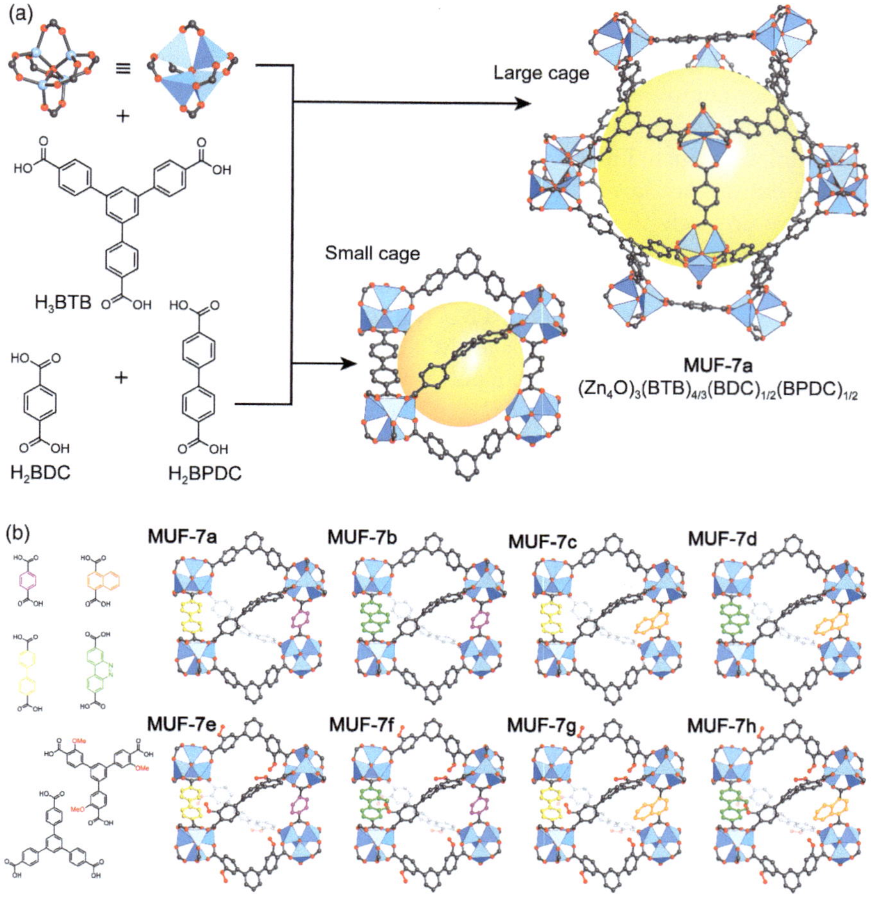

Figure 5.10 (a) Fragments of the crystal structure of the quaternary MOF MUF-7a showing the small (orange sphere) and large pore (yellow sphere) that assemble to form a framework with an overall **ith-d** topology (not shown). (b) Employing derivatized linkers that are symmetry matched to their respective positions within the crystal structure allows for the formation of eight different isoreticular materials (MUF-7a-h) with programmed pores. Only the small tetrahedral pores are shown. BTB and $(OMe)_3$-BTB linkers are only shown partially and all other derivatized linkers are highlighted in different colors. All hydrogen atoms are omitted for clarity. Color code if not noted differently: Zn, blue; C, gray; O, red.

groups arranged in a predetermined pattern within the periodic lattice. MUF-7 analogs with programmed pores show an enhancement of the CO_2 adsorption capacity by almost 100% [14].

Sequential linker installation (SLI) is another approach to the synthesis of MOFs with highly complex structures, albeit less generally applicable [15]. SLI allows for the systematic increase of the connectivity of SBUs making it possible to prepare frameworks that cannot be synthesized directly. This approach is discussed in more detail in Chapter 6 in the context of the post-synthetic modification of MOF structures.

5.2.3 The TBU Approach

Not only SBUs, but also larger molecular building units, can be used to design and synthesize complex MOF structures. If such building units are formed during the reticulation process, they are referred to as tertiary building units (TBUs) and examples for such TBUs are metal-organic polyhedra (MOPs, see Chapter 20). Such entities are formed *in situ* by linking of multiple individual building units (more precisely the segments of the organic linkers bearing the binding groups and the inorganic SBUs) into discrete 0D polyhedra. Here, the organic linker does not only connect adjacent SBUs, but also encodes the information for the *in situ* formation of the targeted TBU, thus enabling the design of structures built from "predefined" molecular cages. When analyzing the structure of MOP-1 (Cu_2(*m*-BDC)$_2$(H_2O)$_2$), a molecular polyhedron built from 4-c copper paddle wheel SBUs connected by *m*-BDC linkers, it becomes clear that multiple MOP-1 units can potentially be linked into a geometrically well-defined arrangement by functionalization of the isophthalate linkers to provide additional binding groups. This approach is illustrated in Figure 5.11 [16].

To allow for more predictability and thus the deliberate design of extended structures based on TBUs, nets of minimum transitivity that are exclusive for a certain combination of building units (vertex geometries) must be targeted. For the example shown in Figure 5.11 this means that in a first step the connectivity of the modified MOP-1 molecule needs to be determined. Each MOP-1 entity is built from 24 *m*-BDC linkers, and since the 5-positions of these linkers act as the points of extension, the truncated cuboctahedral MOP-1 TBU represent a 24-c rhombicuboctahedral (**rco**) vertex figure (Figure 5.11b). Such 24-c TBUs can form frameworks of **rht** topology (transitivity 2123) when linked by

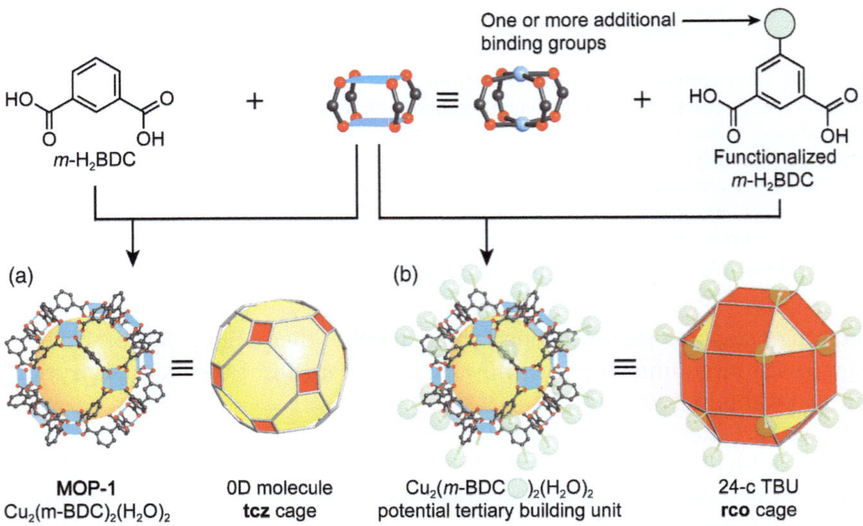

Figure 5.11 (a) Reticulation of *m*-H_2BDC linkers and 4-c Cu_2(—COO)$_4$ paddle wheel SBUs affords the formation of a discrete **tcz** cage. (b) Functionalization of the *m*-BDC linker in the 5-position yields a molecular building unit with 24 points of extension.

trigonal linkers. In the following text we will illustrate two strategies for linking MOP-1-based TBUs into 3D extended structures either through additional SBUs or deliberate linker design. When planning the synthesis of such compounds, it is important to keep in mind that the chemical nature of the linker must not interfere with the formation of the MOP-1 TBUs.

5.2.3.1 Linking TBUs Through Additional SBUs

A MOF constructed from rhombicuboctahedral TBUs linked through additional SBUs is **rht**-MOF-1 (Figure 5.12) [17]. Here, H_3TZI (5-tetrazolylisophthalic acid) is reticulated with Cu^{2+} ions. The isophthalate unit of the TZI linkers connects

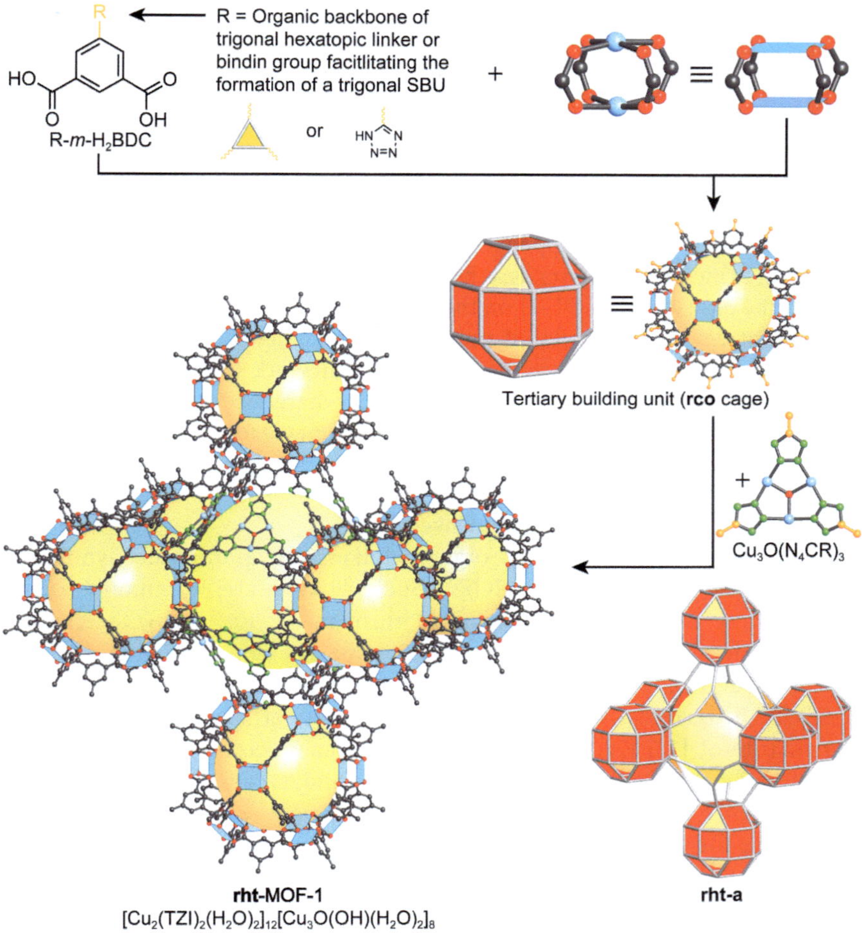

Figure 5.12 Crystal structure of **rht**-MOF-1. Copper paddle wheel SBUs are linked into MOP-1 TBUs (**tcz**) through the isophthalate segment of the H_2TZI linker. The tetrazole moieties installed at the 5-position of the isophthalate (C5 and C6 of the linker are highlighted in orange) are linked into trigonal $Cu_3OL_3(TZI)_3$ SBUs (D_{3h}) that connect the TBUs to yield an extended 3D structure of **rht** topology. The topology representation of the **rco** cage as well as the overall **rht** topology are shown. All hydrogen atoms are omitted for clarity. Color code: Cu, blue; C, gray; N, green; O, red.

two copper centers to form $Cu_2(-COO)_4$ paddle wheel SBUs thus yielding 24-c cuboctahedral TBUs bearing tetrazole group in the 5-position of each bridging isophthalate unit. The additional tetrazole binding groups are located precisely on the vertices of the rhombicuboctahedral TBUs and consequently each TBU is connected to 12 neighboring TBUs through 3-c $Cu_3OL_3(TZI)_3$ SBUs formed by linking three tetrazole groups of three adjacent TBUs [17a]. This affords a 3,24-connected structure with an overall **rht** topology. The structure encompasses three differently sized pores of **tcz**, **tcu**, and **rdo-a** topology. By identifying the two different types of SBU and the tritopic TZI linker as the building units of this MOF, its structure can also be described by the 3,3,4-connected **ntt** net; however, the resulting structure description is more complicated (see Figure 2.17).

For the deconstruction of the crystal structures of **rht**-MOFs in terms of an **rht** net, selecting the carbonyl carbons as the points of extension is not helpful and it is more meaningful to define new points of extension that help in simplifying the structure. In the example above, the 5-position of the isophthalate terminus provides the connection to neighboring TBUs; hence, from a topological point of view, a description of the isophthalate as having two points of extension is not meaningful and the description of isophthalate as one point of extension, located at its 5-position, is the more appropriate choice.

A general principle that describes the influence of linker terminus (e.g. isophthalate in the case of **rht**-MOF-1) on the hierarchical pore system, facilitates the design of **rht**-MOFs with tailored pore sizes. Changing the distance between the two carboxylate binding groups affords a framework with expanded TBUs (**tcz** cage, see linker of **rht**-MOF-2 in Figure 5.13) whereas changing the distance between the tetrazolate and the isophthalate segments leads to the expansion of the **rdo-a** and **tcu** cages (see linker of **rht**-MOF-3 in Figure 5.13).

5.2.3.2 Linking TBUs Through Organic Linkers

In an analogous fashion, purely organic linkers of appropriate symmetry can be used to prepare **rht**-MOFs. Such linkers are commonly composed of three isophthalate units joined through a core unit with trigonal symmetry. Here we will discuss a MOF with **rht** topology that is built from 24-c cuboctahedral TBUs and a trigonal hexatopic linker, H_6TTATP (5,5′,5″-(1,3,5-triazine-2,4,6-triyl)tris(azanediyl)triisophthalic acid). The molecular structure of H_6TTATP is composed of a triazine core unit connected to three isophthalate binding groups through secondary amines (see linker of **rht**-MOF-7 in Figure 5.13). Reticulation of Cu^{2+} ions with this linker leads to the *in situ* formation of MOP-1 TBUs that are consequently linked into an **rht** framework. In contrast to **rht**-MOP-1, a second SBU is not needed for the formation of **rht**-MOF-7. The metrics of all **rht**-MOFs can be modified following the same general design principles outlined earlier for tetrazole-based **rht**-MOFs. Different organic units of suitable geometry can be employed as the core of the hexatopic linker, which allows for the synthesis of isoreticular **rht**-MOFs with a wide variety of pore sizes (Figure 5.13) [17b].

The assembly of discrete metal-organic TBUs facilitates the rational design and synthesis of complex MOF structures, and MOPs appear to be ideal building units in this approach. Their high connectivity reduces the number of

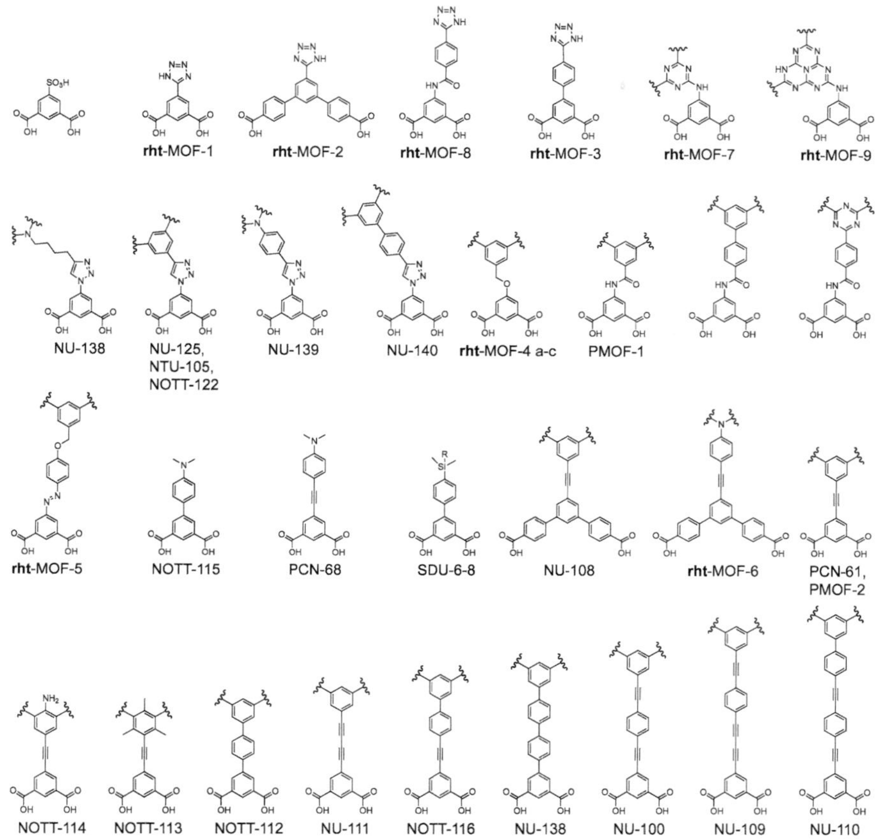

Figure 5.13 Compilation of linkers used in the preparation of MOFs with **rht** topology. Sulfonic acid and tetrazole derivatives rely on the *in situ* formation of trigonal SBUs to link up the individual TBUs. A variety of extending units have been used, and the largest linker affording a framework with **rht** topology (NU-110) is shown on the far bottom right corner.

accessible topologies when linked up into 3D frameworks and thus increases the predictability with respect to the targeted structure. Additionally, the high connectivity of such TBUs allows for the formation of framework structures that cannot be accessed otherwise. In analogy to the examples discussed in this chapter, a variety of different MOPs can be used as TBUs to yield highly complex MOFs [18, 19]. For a more detailed review on such MOFs the reader is referred elsewhere [17b, 19].

5.3 Heterogeneity in Frameworks

In contrast to framework complexity, heterogeneity is introduced through the presence of (i) multiple interchangeable linkers, with similar chemical behavior and metrics, but different chemical composition, (ii) multiple metal ions forming the same type of SBU, or (iii) aperiodic vacancies within the same structure.

Such MTV systems possess the same overall topology as the parent MOF. Since these structures feature at least two building units that have no topological preference as to what position in the crystal structure they occupy, these "elements of heterogeneity" result in an undefined spatial arrangement. This means that the possible emerging sequences are a result of the chemical nature of the individual constituents and the reaction conditions. Interestingly, phase separation is typically not observed and pure phase MTV-MOFs can be prepared from an appropriate mixture of starting materials. In the following text we will elaborate three approaches to introduce heterogeneity into frameworks: (i) mixed-linker MTV-MOFs, (ii) mixed-metal MTV-MOFs, and (iii) disordered vacancies.

5.3.1 Multi-Linker MTV-MOFs

MTV-MOF-5 marks the first report of a multivariate MOF. Here, a variety of X-H_2BDC derivatives (X = H, NH_2, Br, $(Cl)_2$, NO_2, $(CH_3)_2$, C_4H_4, $(OC_3H_5)_2$, and $(OC_7H_7)_2$) were implemented into the backbone of MOF-5 (Figure 5.14) [4]. To prepare such materials, a mixture of up to eight differently substituted X-H_2BDC linkers is employed in the synthesis of a single MOF. The resulting materials are isostructural to MOF-5 and possess both, the crystallinity and microporosity inherent to the parent MOF-5. Because the atomic coordinates of the BDC backbone of all linkers within the structure of MOF-5 are related by symmetry (MTV-MOF-5 crystallizes in the space group $Fm\bar{3}m$), the distribution of differently substituted linkers within the framework structure cannot be fully elucidated by diffraction techniques. Different scenarios may be envisioned for the arrangement of linkers within MTV-MOF-5: (i) a random distribution, (ii) a well-ordered alternating pattern, or (iii) clustering of functionalities. A combination of solid-state MAS NMR studies and molecular dynamics simulations can help to elucidate the determination of the apportionment differently substituted linkers within this structure. While some functionalities favor the formation of small clusters (e.g. $-NH_2$ and $-CH_3$), other functionalities prefer a random arrangement (e.g. $-NO_2$, $-(OC_3H_5)_2$, and $-(OC_7H_7)_2$), which suggests that the distribution of functionalities is highly dependent on their chemical nature. Interestingly, MTV-MOF-5 materials show gas adsorption properties superior to those of the parent MOF-5, an observation that is explained by the larger number of adsorption sites within the pores of the MTV system. A similar nonlinear enhancement of properties has been reported for other MTV systems [20].

5.3.2 Multi-Metal MTV-MOFs

Another approach to the synthesis of MTV-MOFs is the introduction of different metals on symmetry-equivalent sites within the SBUs, thus introducing heterogeneity into the backbone of the MOF. This is achieved either by a one-pot reaction or by post-synthetic metal exchange reactions (see Chapter 6). Co-doped mixed-metal (Zn)MOF-5:Co^{2+} containing 8 and 21% of cobalt is prepared in a one-pot reaction and while the incorporation of Co^{2+} into the zinc SBUs of the framework can be confirmed, the precise spatial distribution of Co^{2+} cannot be

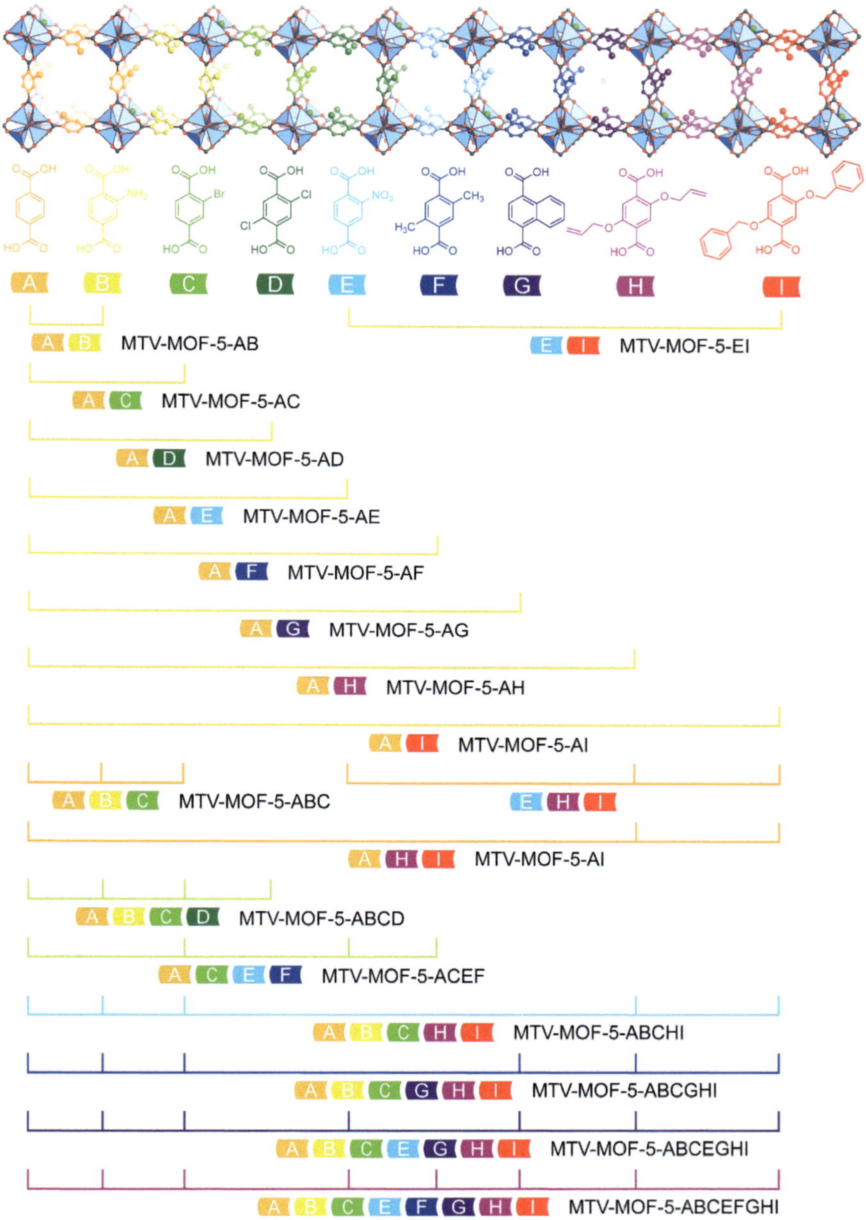

Figure 5.14 Different linker combinations used to prepare MTV-MOF-5. Up to eight chemically distinct BDC derivatives are employed in the syntheses yielding MTV-MOF-5 materials that encompass heterogeneous pore environments.

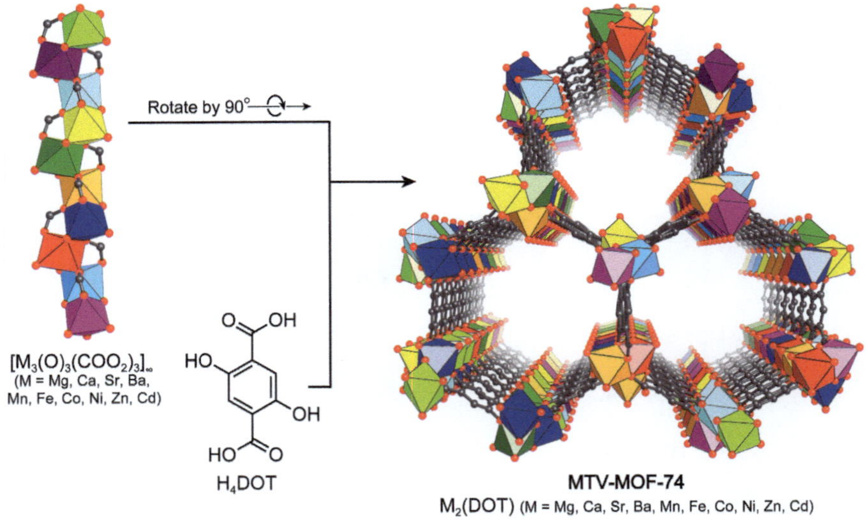

Figure 5.15 Crystal structure representation of MTV-MOF-74 with view along the crystallographic *c*-axis. The **etb** framework is formed by linking 1D $[M_3(O)_3(COO)_3]_\infty$ rod SBUs through DOT linkers. The resulting framework has hexagonal 1D channels running along the crystallographic *c*-axis. The SBUs of MTV-MOF-74 contain up to 10 different metals (Mg, Ca, Sr, Ba, Mn, Fe, Ni, Co, Zn, and Cd), out of which there are metals not present in binary MOF-74 structures. The metals are shown as their respective coordination polyhedra, each color represents a different metal. All hydrogen atoms are omitted for clarity. Color code: C, gray; O, red.

determined [21]. Mixed-metal analogs of MOF-5 containing Ti^{3+}, $V^{2+/3+}$, $Cr^{2+/3+}$, Mn^{2+}, and Fe^{2+} can be prepared by post-synthetic metal exchange reactions [22]. To incorporate trivalent metals, these are introduced in a reduced oxidation state and subsequently oxidized without effecting the original structure, crystallinity, and porosity of MOF-5.

Up to 10 different metals (Mg, Ca, Sr, Ba, Mn, Fe, Co, Ni, Zn, and Cd) can be incorporated into the SBUs of MOF-74 in a one-pot reaction yielding MTV-MOF-74 (Figure 5.15) [23]. This method allows to incorporate metal ions that cannot be employed in the synthesis of the pure phase MOF (i.e. Ca, Sr, Ba, and Cd). The determination of the spatial distribution of metal ions within such heterogeneous structures is a difficult task. Energy dispersive X-ray spectroscopy can help to elucidate the distribution of metal centers within such mixed-metal MOFs. Using this method, the distribution of different metals in the SBUs of MTV-MOF-74 was shown to be nonuniform.

Another example of mixed-metal MTV-MOFs are compounds of the general formula $[M_3OL_3]_2(TCPP\text{-}M)_3$, built from trinuclear $M_3OL_3(-COO)_6$ SBUs connected through tetratopic porphyrin-based linkers [24]. Using five different metals to construct the SBUs of the MOF and six different linkers (unmetalated $TCPP\text{-}H_2$ and five metalated derivatives TCPP-M) 36 isostructural MOFs of the general formula $[M_3OL_3]_2(TCPP\text{-}M)_3$ can be prepared (Figure 5.16). For the resulting structures, two scenarios arise: (i) formation of domains, i.e. different SBUs, each of which is composed of one sort of metal, form domains within the

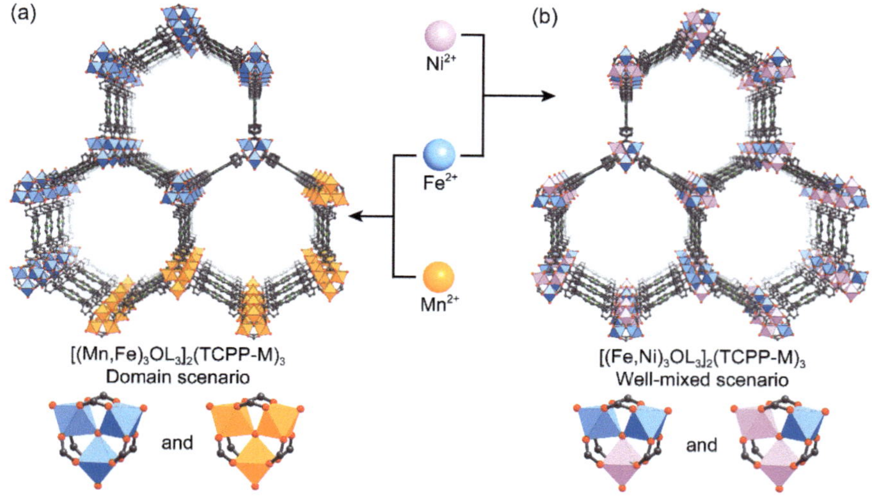

Figure 5.16 Crystal structure of $[M_3OL_3]_2(TCPP\text{-}M)_3$ with an underlying **stp** topology. (a) Metals with significantly different radii and electronegativity (Mn^{2+} and Fe^2) tend to form SBUs containing only one kind of metal, which results in a domain arrangement (b) Metals of similar radius and electronegativity (Fe^{2+} and Ni^{2+}) form mixed-metal SBUs, resulting in a well-mixed arrangement.

MOF structure and (ii) the well-mixed case, where different metals are present within the same SBU and the MOF structure is built from only mixed-metal SBUs. For all materials prepared in the way described earlier, no phase separation is observed as evidenced by energy dispersive X-ray spectroscopy, which is indicative of the formation of the targeted mixed-metal MTV-MOFs. As a general principle, metals of similar radius and electronegativity form mixed-metal SBUs (well-mixed arrangement), whereas metals with significantly different radii and electronegativity form SBUs built from one kind of metal (domain arrangement). This is well in accordance with the principles established for intermetallic or ionic solid-state materials and can experimentally be assessed by X-ray photoelectron spectroscopy.

5.3.3 Disordered Vacancies

Similar to the formation of MTV systems constructed from linkers bearing multiple different functionalities or mixed-metal centers within the SBU, the introduction of defects into a MOF structure can also yield an MTV system [25]. Furthermore, the presence of defects can enhance the properties of MOFs with respect to certain applications.

To illustrate this approach, we take a closer look at defects in the structure of UiO-66. Here, defects can be introduced by adding a strong modulator, such as trifluoroacetic acid, to the reaction mixture. During framework formation, the modulator competes with the BDC linker for binding sites on the SBU, resulting in missing linker defects whose presence can be verified by X-ray diffraction techniques [26]. Interestingly, the resulting defect rich material shows an

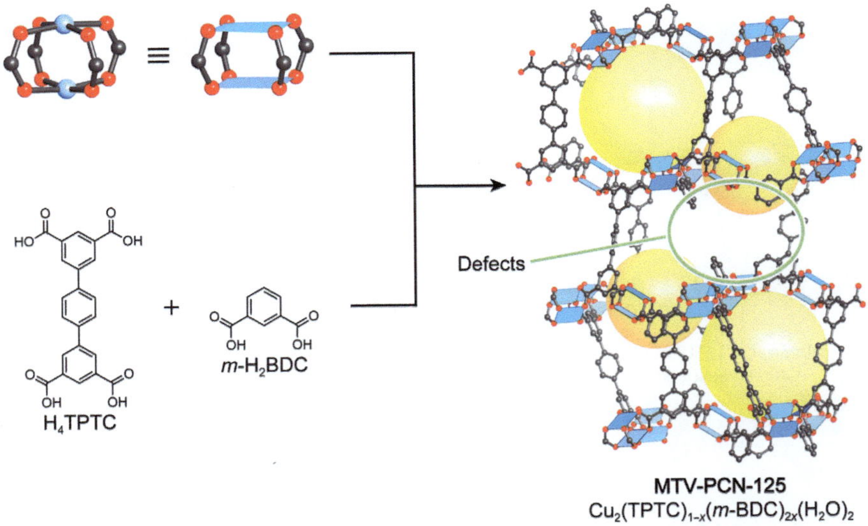

Figure 5.17 Disordered defects in the structure of PCN-125. The defects are introduced by adding m-H_2BDC to the reaction mixture. Even though defects are present in the structure, the material is crystalline. Disordered vacancies impart the pores with heterogeneity.

increased catalytic activity in the cyclization of citonellal to isopulegol compared to pristine UiO-66 and a similar enhancement of the properties of other defect rich MOFs is known [27]. Modulators used to introduce defect sites are often structurally related to the linker molecules used in the synthesis [25, 28]. An example making use of this strategy is the formation of defect rich PCN-125 ($[Cu_2(H_2O)_2](TPDC)$) [28a]. Here, m-H_2BDC is added to the reaction mixture of Cu^{2+} ions and H_4TPTC to introduce missing-linker defects (Figure 5.17). Even though the resulting material displays a powder X-ray diffraction pattern identical to that of pristine PCN-125, the presence of m-BDC in the defect structure can be corroborated by digestion NMR. A schematic drawing of PCN-125 showing the nature of these defects is given in Figure 5.17. The presence of defects in PCN-125 increases the CO_2 uptake capacity which is attributed to the increased pore size of defect rich PCN-125. In a manner akin, defects can be introduced into other MOF structures [25, 28b].

Another approach to introducing defects is the rapid formation of MOF by fast precipitation [29]. Typical solvothermal reactions that yield single crystalline materials require reaction times of more than 12 hours, and fast crystal formation (<1 minute) is therefore expected to result in materials containing many defect sites [30].

A unique method for the preparation of MTV-MOFs with homogeneous pore sizes is micro-mesoporous MOF-5. By addition of DBA (4-(dodecycloxy)benzoic acid) to the reaction mixture, pomegranate-like crystals of pmg-MOF-5 consisting of a mesoporous core and a microporous shell are formed. Further increasing the amount of DBA in the reaction mixture even results in meso, macroporous sponge-like crystals [31].

Thus far, we only discussed the direct synthesis of MOFs providing us with the concepts of design in terms of the topology and metrics of MOF structures. However, another great advantage of MOFs over other porous materials is that their molecular building units can be addressed like molecules even though they are part of an extended solid. In Chapter 6, we will discuss ways to modify MOFs in terms of their structure and functionality to tailor the properties of a given material toward specific applications.

5.4 Summary

In this chapter we showed that MOFs built from more than two distinct building units can be classified as "complex" or "heterogeneous" and we made this distinction based on the crystallographic order of the backbone of the structure; the different positions the building units occupy within the structure. We showed MOF structures illustrating both concepts and derived general ways for constructing such structures. The expansion of the scope of structure accessible using the mixed-linker, mixed-metal, or MTV approach was highlighted and we saw that heterogeneity within frameworks can result in enhanced materials properties compared to the simple binary counterparts. Thus far, we discussed the direct synthesis of MOF by linking individual building units. The porous nature of MOFs further allows for the modification of their structures post-synthetically as well as for the formation of well-defined composite materials, and such modification and functionalization reactions will be the focus of Chapter 6.

References

1 (a) Li, H., Eddaoudi, M., O'Keeffe, M., and Yaghi, O.M. (1999). Design and synthesis of an exceptionally stable and highly porous metal-organic framework. *Nature* 402 (6759): 276–279. (b) Furukawa, H., Ko, N., Go, Y.B. et al. (2010). Ultrahigh porosity in metal-organic frameworks. *Science* 329 (5990): 424–428.

2 Koh, K., Wong-Foy, A.G., and Matzger, A.J. (2008). A crystalline mesoporous coordination copolymer with high microporosity. *Angewandte Chemie International Edition* 47 (4): 677–680.

3 Eddaoudi, M., Kim, J., Rosi, N. et al. (2002). Systematic design of pore size and functionality in isoreticular MOFs and their application in methane storage. *Science* 295 (5554): 469–472.

4 Deng, H., Doonan, C.J., Furukawa, H. et al. (2010). Multiple functional groups of varying ratios in metal-organic frameworks. *Science* 327 (5967): 846–850.

5 Schnobrich, J.K., Lebel, O., Cychosz, K.A. et al. (2010). Linker-directed vertex desymmetrization for the production of coordination polymers with high porosity. *Journal of the American Chemical Society* 132 (39): 13941–13948.

6 Tranchemontagne, D.J., Park, K.S., Furukawa, H. et al. (2012). Hydrogen storage in new metal-organic frameworks. *The Journal of Physical Chemistry C* 116 (24): 13143–13151.

7 Chui, S.S.-Y., Lo, S.M.-F., Charmant, J.P.H. et al. (1999). A chemically functionalizable nanoporous material $[Cu_3(TMA)_2(H_2O)_3]_n$. *Science* 283 (5405): 1148–1150.

8 Tu, B., Pang, Q., Ning, E. et al. (2015). Heterogeneity within a mesoporous metal-organic framework with three distinct metal-containing building units. *Journal of the American Chemical Society* 137 (42): 13456–13459.

9 Koh, K., Wong-Foy, A.G., and Matzger, A.J. (2010). Coordination copolymerization mediated by $Zn_4O(CO_2R)_6$ metal clusters: a balancing act between statistics and geometry. *Journal of the American Chemical Society* 132 (42): 15005–15010.

10 Koh, K., Wong-Foy, A.G., and Matzger, A.J. (2009). A porous coordination copolymer with over 5000 m^2/g BET surface area. *Journal of the American Chemical Society* 131 (12): 4184–4185.

11 (a) Lee, S.J., Doussot, C., and Telfer, S.G. (2017). Architectural diversity in multicomponent metal-organic frameworks constructed from similar building blocks. *Crystal Growth & Design* 17 (6): 3185–3191. (b) Grunker, R., Bon, V., Muller, P. et al. (2014). A new metal-organic framework with ultra-high surface area. *Chemical Communications* 50 (26): 3450–3452.

12 Dutta, A., Wong-Foy, A.G., and Matzger, A.J. (2014). Coordination copolymerization of three carboxylate linkers into a pillared layer framework. *Chemical Science* 5 (10): 3729–3734.

13 Liu, L., Konstas, K., Hill, M.R., and Telfer, S.G. (2013). Programmed pore architectures in modular quaternary metal-organic frameworks. *Journal of the American Chemical Society* 135 (47): 17731–17734.

14 Liu, L. and Telfer, S.G. (2015). Systematic ligand modulation enhances the moisture stability and gas sorption characteristics of quaternary metal-organic frameworks. *Journal of the American Chemical Society* 137 (11): 3901–3909.

15 Yuan, S., Lu, W., Chen, Y.-P. et al. (2015). Sequential linker installation: precise placement of functional groups in multivariate metal-organic frameworks. *Journal of the American Chemical Society* 137 (9): 3177–3180.

16 Eddaoudi, M., Kim, J., Wachter, J.B. et al. (2001). Porous metal-organic polyhedra: 25 Å cuboctahedron constructed from 12 $Cu_2(CO_2)_4$ paddle-qheel building blocks. *Journal of the American Chemical Society* 123 (18): 4368–4369.

17 (a) Nouar, F., Eubank, J.F., Bousquet, T. et al. (2008). Supermolecular building blocks (SBBs) for the design and synthesis of highly porous metal-organic frameworks. *Journal of the American Chemical Society* 130 (6): 1833–1835. (b) Eubank, J.F., Nouar, F., Luebke, R. et al. (2012). On demand: the singular rht net, an ideal blueprint for the construction of a metal-organic framework (MOF) platform. *Angewandte Chemie International Edition* 51 (40): 10099–10103. (c) Luebke, R., Eubank, J.F., Cairns, A.J. et al. (2012). The unique rht-MOF platform, ideal for pinpointing the functionalization and CO_2 adsorption relationship. *Chemical Communications* 48 (10): 1455–1457. (d) Pham, T., Forrest, K.A., Hogan, A. et al. (2014). Simulations of hydrogen sorption in rht-MOF-1: identifying the binding sites through explicit polarization and quantum rotation calculations. *Journal of Materials Chemistry A* 2 (7): 2088–2100.

18 (a) Zheng, B., Bai, J., Duan, J. et al. (2010). Enhanced CO_2 binding affinity of a high-uptake rht-type metal-organic framework decorated with acylamide groups. *Journal of the American Chemical Society* 133 (4): 748–751. (b) Yuan, D., Zhao, D., Sun, D., and Zhou, H.C. (2010). An isoreticular series of metal-organic frameworks with dendritic hexacarboxylate ligands and exceptionally high gas-uptake capacity. *Angewandte Chemie International Edition* 49 (31): 5357–5361. (c) Farha, O.K., Yazaydın, A.Ö., Eryazici, I. et al. (2010). De novo synthesis of a metal-organic framework material featuring ultrahigh surface area and gas storage capacities. *Nature Chemistry* 2 (11): 944–948.

19 Guillerm, V., Kim, D., Eubank, J.F. et al. (2014). A supermolecular building approach for the design and construction of metal-organic frameworks. *Chemical Society Reviews* 43 (16): 6141–6172.

20 (a) Park, T.-H., Koh, K., Wong-Foy, A.G., and Matzger, A.J. (2011). Nonlinear properties in coordination copolymers derived from randomly mixed ligands. *Crystal Growth & Design* 11 (6): 2059–2063. (b) Zhang, Y.-B., Furukawa, H., Ko, N. et al. (2015). Introduction of functionality, selection of topology, and enhancement of gas adsorption in multivariate metal-organic framework-177. *Journal of the American Chemical Society* 137 (7): 2641–2650.

21 Botas, J.A., Calleja, G., Sánchez-Sánchez, M., and Orcajo, M.G. (2010). Cobalt doping of the MOF-5 framework and its effect on gas-adsorption properties. *Langmuir* 26 (8): 5300–5303.

22 Brozek, C.K. and Dincă, M. (2013). Ti^{3+}-, $V^{2+/3+}$-, $Cr^{2+/3+}$-, Mn^{2+}-, and Fe^{2+}-aubstituted MOF-5 and redox reactivity in Cr- and Fe-MOF-5. *Journal of the American Chemical Society* 135 (34): 12886–12891.

23 Wang, L.J., Deng, H., Furukawa, H. et al. (2014). Synthesis and characterization of metal-organic framework-74 containing 2, 4, 6, 8, and 10 different metals. *Inorganic Chemistry* 53 (12): 5881–5883.

24 Liu, Q., Cong, H., and Deng, H. (2016). Deciphering the spatial arrangement of metals and correlation to reactivity in multivariate metal-organic frameworks. *Journal of the American Chemical Society* 138 (42): 13822–13825.

25 Fang, Z., Dürholt, J.P., Kauer, M. et al. (2014). Structural complexity in metal-organic frameworks: simultaneous modification of open metal sites and hierarchical porosity by systematic doping with defective linkers. *Journal of the American Chemical Society* 136 (27): 9627–9636.

26 (a) Wu, H., Chua, Y.S., Krungleviciute, V. et al. (2013). Unusual and highly tunable missing-linker defects in zirconium metal-organic framework UiO-66 and their important effects on gas adsorption. *Journal of the American Chemical Society* 135 (28): 10525–10532. (b) Cliffe, M.J., Wan, W., Zou, X. et al. (2014). Correlated defect nano-regions in a metal-organic framework. *Nature Communications* 5: 4176.

27 Vermoortele, F., Bueken, B., Le Bars, G. et al. (2013). Synthesis modulation as a tool to increase the catalytic activity of metal-organic frameworks: the unique case of UiO-66(Zr). *Journal of the American Chemical Society* 135 (31): 11465–11468.

28 (a) Park, J., Wang, Z.U., Sun, L.-B. et al. (2012). Introduction of functionalized mesopores to metal-organic frameworks via metal–ligand–fragment coassembly. *Journal of the American Chemical Society* 134 (49): 20110–20116. (b) Barin, G., Krungleviciute, V., Gutov, O. et al. (2014). Defect creation by linker fragmentation in metal-organic frameworks and its effects on gas uptake properties. *Inorganic Chemistry* 53 (13): 6914–6919.

29 Ravon, U., Savonnet, M., Aguado, S. et al. (2010). Engineering of coordination polymers for shape selective alkylation of large aromatics and the role of defects. *Microporous and Mesoporous Materials* 129 (3): 319–329.

30 Huang, L., Wang, H., Chen, J. et al. (2003). Synthesis, morphology control, and properties of porous metal-organic coordination polymers. *Microporous and Mesoporous Materials* 58 (2): 105–114.

31 Choi, K.M., Jeon, H.J., Kang, J.K., and Yaghi, O.M. (2011). Heterogeneity within order in crystals of a porous metal-organic framework. *Journal of the American Chemical Society* 133 (31): 11920–11923.

6

Functionalization of MOFs

6.1 Introduction

As illustrated in the previous chapters, metal-organic frameworks (MOFs) can be prepared in an almost infinite variety of structures and compositions, allowing for the targeted synthesis of porous materials with tailored pore sizes and metrics. It is therefore, that, over the past decade, MOFs have become the most studied class of porous solids. However, this does not represent their full synthetic potential.

Following the isoreticular principle, MOFs can be functionalized prior to their synthesis by adding substituents to the backbone of the organic linker, and the composition of their SBUs can be influenced by employing a variety of different metals in the synthesis. However, these methods are limited since the formation of a given MOF is susceptible to subtle changes in reaction parameters, the geometry, sterics, and chemical nature of the linker, as well as the electronic configuration of both its constituents, namely, the metal ion and the linker. These limitations for the pre-synthetic functionalization of MOFs made it necessary to develop a tool box to chemically modify framework structures after they had been synthesized. This method termed "post-synthetic modification" (PSM) allows for the deliberate functionalization of as-synthesized MOFs while retaining their basic structure, crystallinity, and porosity. The general principle of PSM is not a new one. It is well established for carbon nanotubes, zeolites and mesoporous silica, organosilicates, and biopolymers; however, it should be noted that the highly ordered crystalline structure and the metal-organic nature of MOFs provide a more suitable platform, as both the position and the degree of functionalization can be controlled. Trivial PSMs have been known early on, as exemplified by ion exchange or solvent removal, but over the last decade a variety of more elaborate methods both, organic and inorganic, have been developed. These methods facilitate the synthesis of finely tuned materials that are optimized through combinatorial synthesis. In this chapter, we will discuss the methods and limitations of *in situ* and pre-synthetic functionalization, and demonstrate how post-synthetic methods can be used to prepare a wide range of isoreticular MOFs with tailored functionalities that are not accessible by direct synthesis. In the context of this chapter we classify the methods of PSMs according to the strength of the interactions between the framework and the added unit: weak, strong, and covalent interactions (Figure 6.1).

Introduction to Reticular Chemistry: Metal-Organic Frameworks and Covalent Organic Frameworks,
First Edition. Omar M. Yaghi, Markus J. Kalmutzki, and Christian S. Diercks.

Companion website: www.wiley.com/go/yaghi/reticular

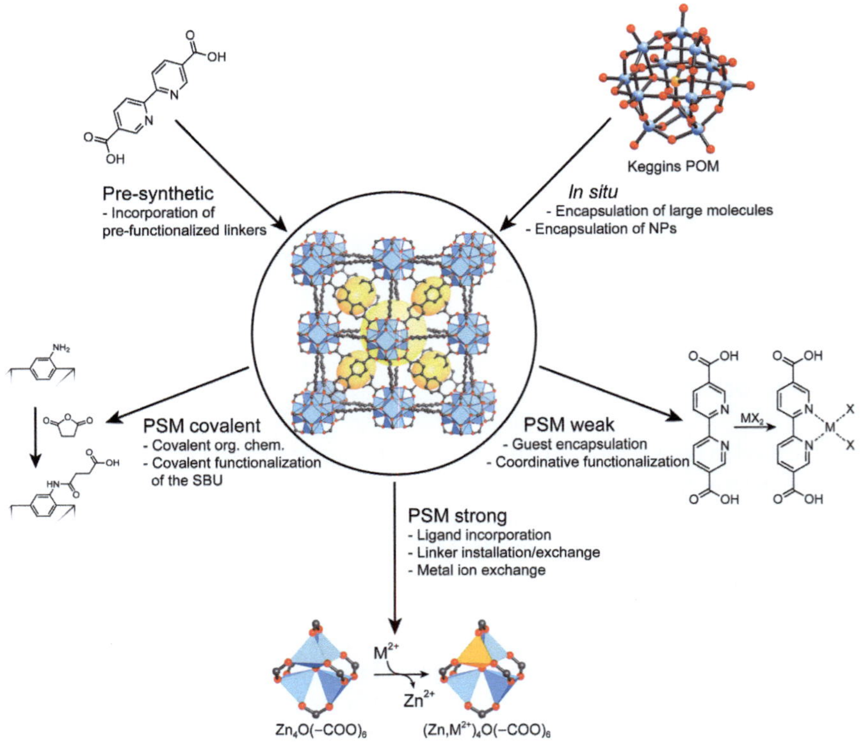

Figure 6.1 Possibilities of chemical functionalization of MOFs. Using pre-synthetic and *in situ* functionalization, functionalized linkers and other functional entities can be incorporated into MOFs during the synthesis, while post-synthetic modifications allow for chemical alteration of the as-synthesized MOF. Post-synthetic modifications are grouped according to the type of interaction arising from them. In contrast to pre-synthetic and *in situ* functionalization, post-synthetic modifications are independent of the reticulation process and thus allow for a wider range of chemical modifications.

6.2 *In situ* Functionalization

Functionality can be imparted by guests that either reside within the pores of a MOF or are embedded into a MOF matrix. When guests are introduced during the synthesis, we refer to this as *in situ* functionalization. Interesting guests are typically large complex organic molecules, inorganic clusters, or metal nanoparticles. The incorporation of guests can also be achieved by PSM, which will be discussed later in this chapter.

6.2.1 Trapping of Molecules

To trap a given molecule in the pores of a MOF the framework has to fulfill one of the following prerequisites: (i) the pore aperture is significantly smaller than the kinetic diameter of the guest molecule or (ii) the framework has to possess binding sites within the pores that can interact with the targeted molecule, thus fixing it in the pores. Molecular species of interest are either large inorganic molecules with catalytic (e.g. POMs = polyoxometallates), optical (e.g. quantum dots), or magnetic (e.g. Fe_3O_4 nanoparticles) properties, or large organic moieties such as enzymes and drug molecules [1].

This approach can be used to immobilize a variety of discrete molecules in the pores of MIL-101 by trapping them during framework formation. Composite materials formed by encapsulation of Keggin-type POMs in the pores of MIL-101 are Brønsted acidic catalysts with high activities for a variety of organic transformations. Such materials are commonly referred to as POM@MOF. In the structure of POM@MIL-101, the POMs randomly occupy the large cages in the **mtn** structure. By choosing a MOF with pore sizes adjusted to the kinetic diameter of the targeted POM, it can be crystallized within the pores of the MOF (e.g. **rht**-MOF-1) and its atomic positions and occupancies can be determined crystallographically [2].

6.2.2 Embedding of Nanoparticles in MOF Matrices

Embedding metal nanoparticles into a MOF matrix can prevent the Ostwald ripening of these particles, and the resulting materials are therefore interesting candidates for a variety of applications, mainly in the realm of catalysis. Here, the catalytic activity of the nanoparticle itself can be exploited or an increased activity of the MOF–catalyst composite can be harnessed due to synergistic effects [3]. Hybrid materials consisting of nanoparticles encapsulated into single crystals or crystalline powders of MOF can be prepared either by adding pre-synthesized nanoparticles of the desired size and shape to the reaction mixture or by coating a nanoparticle with an oriented thin film of MOF, the latter procedure being synthetically more demanding. There are different synthetic approaches to the synthesis of such mesoscopic constructs. Here, we will illuminate the synthesis of MOF-coated silver nanocrystals, denoted as O_h-nano-Ag⊂Al-PMOF-1 (Figure 6.2) [4].[1] In the first step, the silver nanocrystals are coated with a layer of Al_2O_3 by atomic layer deposition. Subsequent reaction of that Al_2O_3 layer with $H_4TCPP\text{-}H_2$ linkers in solution affords the formation of a layer of $Al_2(OH)_2(TCPP\text{-}H_2)$ around the silver nanocrystals. By adjusting the thickness of the alumina layer that is deposited on the nanocrystals, the thickness of the MOF layer can be controlled.

Encapsulating copper nanoparticles into single crystals of UiO-66 (denoted as Cu⊂UiO-66) improves both the activity and the selectivity of the copper catalyst with respect to CO_2 hydrogenation [3a]. This is attributed to the strong interaction between the nanoparticles and the $Zr_6(\mu_3\text{-}O)_4(\mu_3\text{-}OH)_4(-COO)_{12}$ secondary building units (SBUs) of the MOF. Constructs of 18 nm copper nanoparticles encapsulated by a single crystal of UiO-66 outperform many benchmark catalysts (e.g. Cu, ZnO, Al_2O_3), giving a steady eightfold enhanced yield as well as a selectivity of 100% for methanol. An illustration of the crystal structure of UiO-66, SEM (scanning electron microscope) images of UiO-66 single crystals, and TEM (transmission electron microscope) images of Cu⊂UiO-66 are given in Figure 6.3.

1 O_h-nano-Ag⊂MOF refers to octahedral silver nanocrystals coated by MOF.

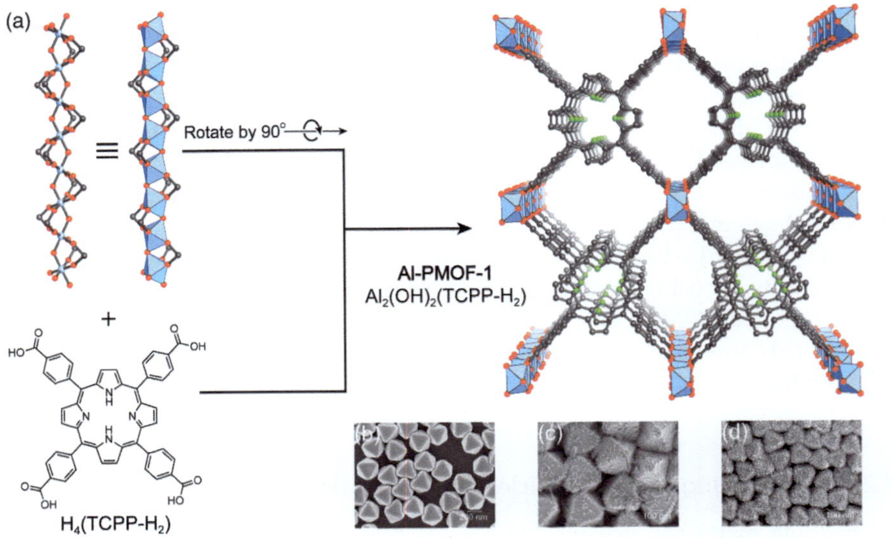

Figure 6.2 (a) Crystal structure of Al-PMOF-1 composed of 1D $[Al(OH)(-COO)]_{\infty}$ SBUs linked through tetratopic TCPP-H_2 linkers forming a framework structure with underlying **frz** topology. (b, c) Al-PMOF-1 is grown onto silver nanocrystals whose surface has been modified by atomic layer deposition of Al_2O_3 by reaction of alumina with H_4TCPP-H_2. (b) SEM images of the pristine silver nanocrystals, (c) silver nanocrystals coated with a thin layer, and (d) a thick layer of $Al_2(OH)_2$(TCPP-H_2). All hydrogen atoms are omitted for clarity. Color code: Al, blue; C, gray; N, green; O, red.

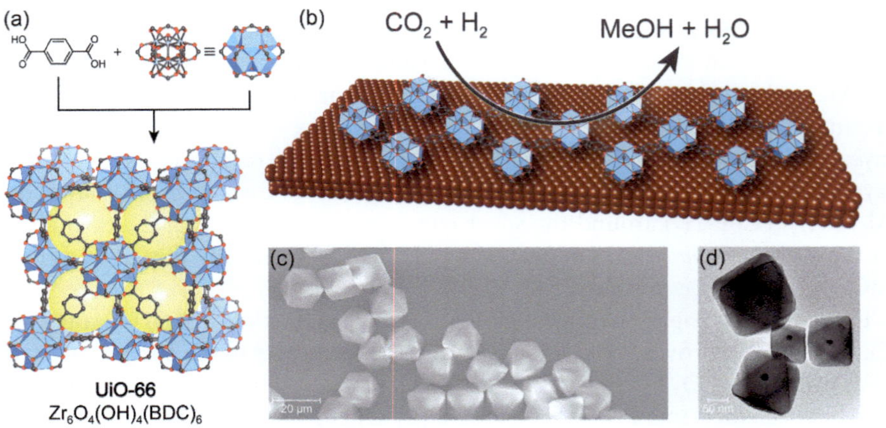

Figure 6.3 Trapping of copper nanoparticles in single crystals of UiO-66. (a) The **fcu** net of UiO-66 is constructed from linear ditopic BDC linkers and 12-c $Zr_6O_4(OH)_4(-COO)_{12}$ SBUs. (b) Synergistic effects at the interface of the copper nanoparticles and the zirconium SBUs in UiO-66 facilitate the hydrogenation of CO_2 with a high selectivity for methanol. (c) SEM image of as-synthesized octahedral UiO-66 single crystals. (d) TEM image of copper nanoparticles trapped inside UiO-66 single crystals. All hydrogen atoms are omitted for clarity. Color code: Zr, blue; C, gray; O, red.

6.3 Pre-Synthetic Functionalization

In contrast to the encapsulation of functional guests during the synthesis, functionality can be introduced by derivatization of the organic linker prior to the MOF synthesis in an approach called "pre-synthetic functionalization." The substituents appended to the linker have to be chosen carefully so that they do not interfere with the formation of the targeted MOF, which severely limits the number of adequate substituents. Amino-substituted or bipyridine-based linkers are often used because they provide appropriate sites for subsequent covalent or coordinative functionalization of the as-synthesized MOF by PSM [5]. The use of protected functional groups is a useful strategy for the pre-synthetic incorporation of functional groups that are otherwise not compatible with the synthesis conditions. However, to take advantage of these functionalities, post-synthetic deprotection is required.

An illustrative example of the effect that pre-synthetic functionalization of the linker can have on the properties of the MOF is the comparison of MIL-125 and its $-NH_2$ functionalized analog. Pristine MIL-125, a **bcu** topology MOF composed of cyclic octamers of corner-sharing $TiO_5(OH)$ octahedra connected by linear ditopic BDC, is prepared from the reaction of titanium tetraisopropoxide and H_2BDC (see Figure 4.32) [6]. By replacing the H_2BDC linker used in the synthesis of MIL-125 with NH_2–H_2BDC, an isoreticular framework termed MIL-125(NH_2) is formed [7]. While pristine MIL-125 only exhibits strong absorption of light in the UV region of the electromagnetic spectrum, the absorption spectrum of MIL-125(NH_2) shows a second absorption band in the visible region (centered around 400 nm) due to a ligand-to-metal charge transfer (LMCT). This is attributed to the electron donating nature of the $-NH_2$ substituent. Upon irradiation with visible light an LMCT state is achieved and consequently reduced Ti^{3+} centers are created, which can photocatalytically reduce CO_2 to formate ($HCOO^-$). The regeneration of the catalyst is facilitated by a sacrificial donor such as TEOA (triethanolamine).

6.4 Post-Synthetic Modification

The preceding sections highlighted that *in situ* and pre-synthetic functionalization methods suffer from limitations that render it difficult to make them generally applicable to all MOF structures. This is because many functional groups can interfere with the formation of the targeted MOF structure or because their chemical and physical properties are not compatible with the synthesis conditions. The more complex the functionalities appended to the linker, the more difficult it is to prepare MOFs bearing such functionalities following the isoreticular principle. Therefore, PSMs present an appealing way to functionalize frameworks while avoiding the aforementioned restrictions by simply carrying out the modification after the framework is formed.

PSMs dealing with weak interactions include ion exchange reactions, solvent exchange reactions, and the incorporation of guests into the pores. Such modification may appear trivial at first, but the removal of guests from the pores of as-synthesized MOFs forms the basis of their application as porous materials and is thus of great importance in MOF chemistry. Not only the removal but also the exchange of guests can endow MOFs with interesting properties originating from the guests themselves or from their interaction with the framework. There are many ways to use coordination chemistry to modify MOFs such as attaching organic molecules to open metal sites or the formation of metal complexes involving pre-synthetically functionalized linkers (e.g. bipyridine). Another way to introduce functionality is to modify components of the framework held together by strong bonds. This approach includes the substitution of metal ions within the SBU and the substitution or addition of ligands and linkers. The functionalization of the linker by means of covalent organic chemistry represents the most extensively studied type of PSM. In the following text we will discuss strategies for the PSM of MOFs organized by the nature and strength of the interactions involved.

6.4.1 Functionalization Involving Weak Interactions

One route to functionalize MOFs is to take advantage of weak coordinative interactions of the framework with functional molecules. This includes the encapsulation of guest molecules, the removal of neutral terminal ligands from SBUs to create open metal sites and subsequent reaction with donor ligands, and metalation of pre-synthetically installed linkers with coordination sites using a variety of different metals. In the following text we show selected examples of MOFs where these modifications have been carried out and highlight the properties arising from therefrom.

6.4.1.1 Encapsulation of Guests

The possibility to exchange guest molecules residing in the pores of a MOF while retaining the crystallinity of the framework not only forms the basis of all gas adsorption-based applications, but also allows for the adsorption of functional guest molecules. There are many examples where the host–guest chemistry of MOFs is used to modify or enhance the properties of MOFs by loading their pores with functional molecules or metal nanoparticles. This method is only limited by the pore opening that puts a limit to the size of the guest molecules and often renders it impossible to encapsulate pre-synthesized metal nanoparticles or large molecules directly. This issue can be addressed by diffusing organometallic precursors (e.g. metal salts) into the pores of the MOF by chemical vapor infiltration or wetness impregnation, followed by reduction under mild conditions [8]. This process is illustrated in Figure 6.4 for the synthesis of palladium nanoparticles in MIL-101. Materials prepared in this way often show outstanding catalytic properties and therefore attract growing interest.

In a similar way, small functional molecules can be absorbed into the pores of MOFs. This approach has some drawbacks compared to trapping of large molecules by *in situ* functionalization, since the molecules are only bound to the framework by weak interactions and cannot be trapped in the pores by spatial confinement [9].

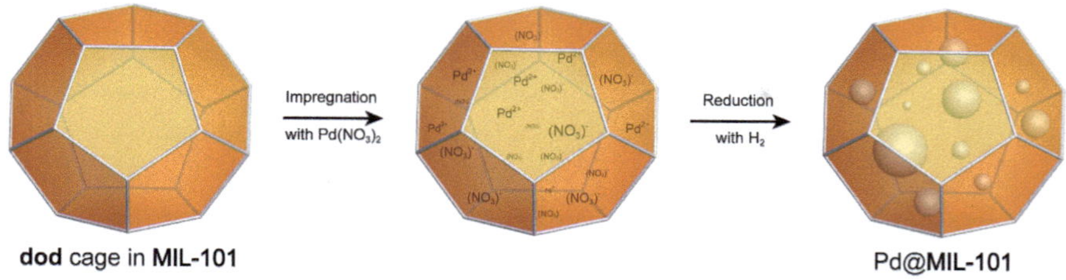

Figure 6.4 A schematic illustration of the trapping of palladium nanoparticles within the cages in the structure of MIL-101. Here, the smaller pore (**dod** topology) is shown. In the first step, the MOF is impregnated with a $Pd(NO_3)_2$ solution. Subsequent reduction of Pd^{2+} affords Pd nanoparticles trapped within the pores of MIL-101.

6.4.1.2 Coordinative Functionalization of Open Metal Site

The removal of solvent from the pores of MOFs forms the foundation of all applications related to gas adsorption. In some cases, the evacuation of a given MOF at elevated temperatures results in the dissociation of terminal neutral ligands from the SBUs, leaving behind Lewis acidic open metals sites, while the overall structure and crystallinity of the MOF is retained. Such open metal sites impart the MOF with new properties and open yet another way for further coordinative functionalization.[2]

Both, MIL-100 and MIL-101, are built from the same trigonal prismatic trinuclear $Cr_3O(H_2O)_2F(-COO)_6$ SBUs. Removal of the terminal water ligands from these SBUs provides open metal sites on the Cr^{3+} centers [1a, 10]. Experimentally, this is achieved by heating the MOF under dynamic vacuum. Subsequent refluxing of fully activated MIL-101 in a solution of an organic amine yields amine-grafted MIL-101, which exhibits remarkably high activities in base-catalyzed reactions while also acting as a size-selective molecular sieve (Figure 6.5) [11]. When multifunctional amines are used, the pores of MIL-101 can be further functionalized by coordination of noble metals such as Pd, Pt, and Au to the free amines dangling in the pores. Subsequent reduction of these metals with $NaBH_4$ yields nanoparticles trapped within the pores. These composites (i.e. Pd@amine-grafted MIL-101) can catalyze organic reactions such as Heck coupling reactions [11].

Open metals sites also increase the sorption capacity of MOFs, especially that of nonpolar gases such as hydrogen, because their polar Lewis acidic nature renders them strong adsorption sites (see Chapters 2 and 15) [12].

6.4.1.3 Coordinative Functionalization of the Linker

Metals can be coordinated to pre-synthetically functionalized linkers such as bipyridine, BINOL, or porphyrins. As discussed earlier, the synthesis of MOFs built from linkers bearing sites for metal coordination is challenging as these sites compete with the binding groups of the linker for the metals present in the reaction mixture. Therefore, coordination sites are typically introduced by PSM

2 This was discussed for In-**soc**-MOF in Chapter A4.

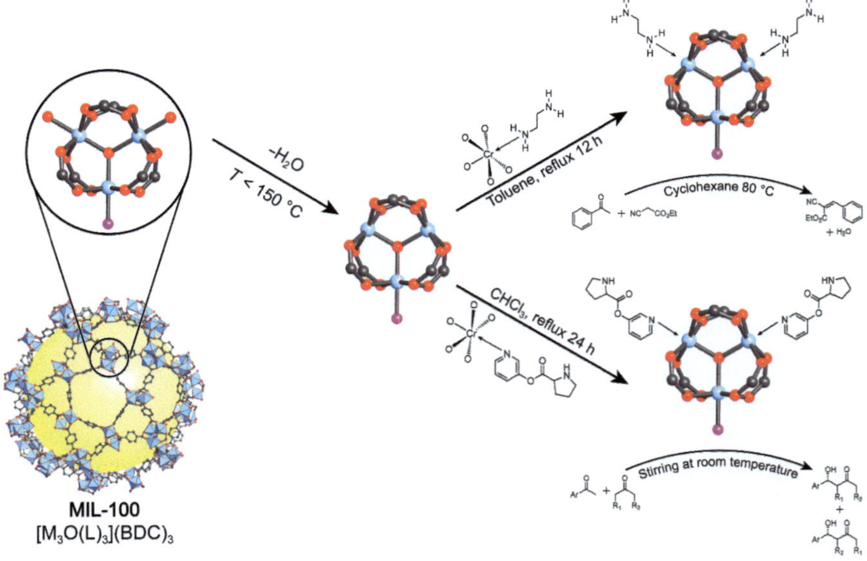

Figure 6.5 Coordinative functionalization of open metal sites in MIL-101. Heating to temperatures above 150 °C under a dynamic vacuum facilitates the removal of terminal water ligands from the $Cr_3O(H_2O)_2L(COO)_6$ SBUs. The resulting open metal sites can subsequently be functionalized using N-donor ligands. Amine-grafted MIL-101 species were shown to have high activities in base-catalyzed reactions. All hydrogen atoms are omitted for clarity. Color code: Cr, blue; C, gray; O, red; L, pink.

involving organic reactions, an approach discussed in Section 6.4.3. Here, we will discuss the incorporation of N-donor sites into the organic backbone of MOF structures.

Incorporation of bipyridinedicarboxylate (BPyDC) into MOFs with SBUs built from highly charged metal ions such as Zr^{4+} is relatively easy due to the poor interaction of the hard acid (zirconium) and the comparatively soft bipyridine base. UiO-67, a MOF built from 12-c $Zr_6O_4(OH)_4(-COO)_{12}$ SBUs and linear ditopic BPDC linkers, can be functionalized by employing a mixture of H_2BPDC and its bipyridine analog (H_2BPyDC = bipyridine-4-4′-dicarboxylic acid) in the synthesis [5b]. The BPyDC linker can be metallated with different metals by PSM, yielding a wide range of highly functional MOFs. A selection of transition metal complexes that can be introduced into pre-synthetically modified UiO-67 analogs is shown in Figure 6.6. These metal complex derivatized MOFs show high catalytic activities for reactions ranging from water oxidation (L_1–L_3), over photochemical CO_2 reduction (L_4), to photocatalytic organic transformations (e.g. aza-Henry reaction, aerobic amine coupling, and aerobic thioanisole oxidation) [13].

The photocatalytic activity of L_4 derivatized UiO-67 is further improved by encapsulation of plasmonic silver nanocubes (denoted as Ag⊂UiO-67(L_4)). The optical nearfield of the silver nanocubes enhances the conversion of CO_2 to CO under visible light by a factor of 7. This example illustrates how different methods of functionalization when applied subsequently evoke materials properties

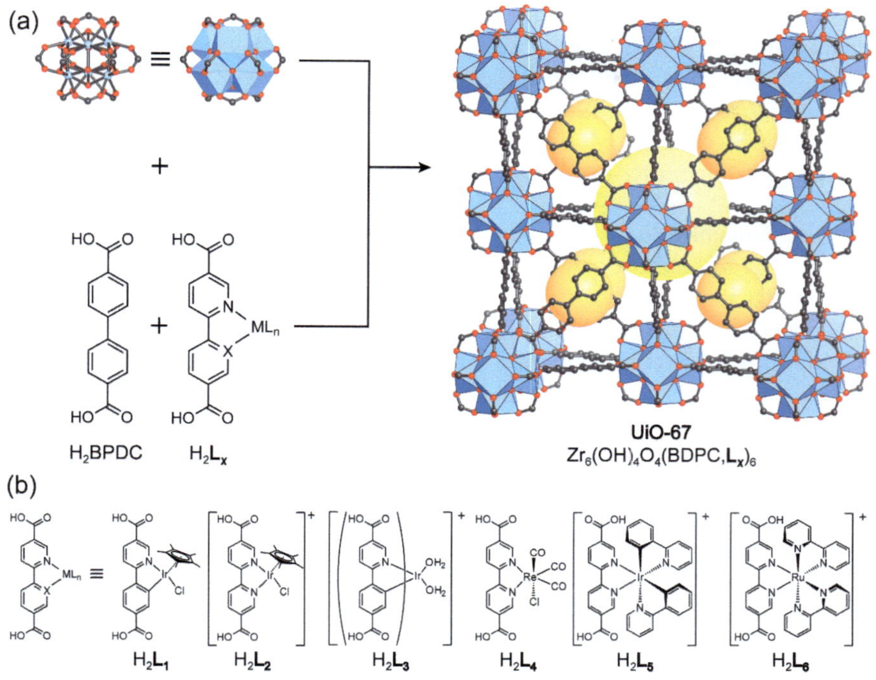

Figure 6.6 Coordinative functionalization of UiO-67 by implementation of functionalized linkers during the synthesis and subsequent coordinative functionalization, yielding a wide range of metal complexes. Note that some of the complexes shown below can also be incorporated during the synthesis. (a) Crystal structure of UiO-67. The **fcu** net is built from 12-c zirconium SBUs and linear ditopic BPDC linkers. (b) Selection of different metal complexes that have been incorporated into the structure of UiO-67. All hydrogen atoms are omitted for clarity. Color code: Zr, blue; C, gray; O, red.

surpassing those of the individual components. This method of functionalization in generally applicable and many other metal complex derivatized MOFs can be prepared in a manner akin to that described for UiO-67 [14].

6.4.2 PSM Involving Strong Interactions

PSMs involving strong interactions include reactions such as the exchange or addition of framework constituents, either the linkers and ligands, or the metal ions in a controlled way. This facilitates the preparation of highly diverse systems that are not accessible by direct synthesis [15]. By replacing charged terminal ligands on the SBU, a process commonly referred to as SALI (solvent assisted ligand incorporation), it is possible to introduce a wide variety of functionalities in the form of functional ligands, which can later on be modified using further covalent organic reactions [16]. In a similar way, organic molecules that cannot be crystallized from solution can be aligned with a MOF backbone to allow for the determination of their crystal structures. Sequential linker installation (SLI) is used to transform frameworks from one topology to another and prepare frameworks not accessible by direct synthesis. Similarly, it is often possible to exchange linkers within a framework structure and this process is given the acronyms PSE (post-synthetic linker exchange) and SALE (solvent assisted linker exchange) [17]. Most of the above mentioned PSMs are applicable to a wide variety of different MOFs and in the following text, we will discuss each PSM using an illustrative example.

6.4.2.1 Coordinative Functionalization of the SBUs by AIM

Terminal ligands on zirconium SBUs are Lewis bases and can therefore act as ligands that coordinate to additional metal ions in a bridging fashion. A good platform for this kind of coordinative functionalization is NU-1000 (NU = Northwestern University, $Zr_6(\mu_3\text{-}OH/O)_8(H_2O,OH)_8(TBAPy)_2$), a MOF of **csq** topology that is built from rectangular TBAPy linkers and 8-c Zr_6O_8-core SBUs. The high thermal stability common to zirconium MOFs, the 1D mesopores of about 30 Å diameter, and the presence of spatially isolated –OH and $–OH_2$ groups on the $[Zr_6(\mu_3\text{-}O)_4(\mu_3\text{-}OH)_4(OH)_4(OH_2)_4]^{8+}$ nodes make NU-1000 an ideal candidate for atomic layer deposition in a MOF, also known by the acronym AIM [18]. In this procedure, NU-1000 is treated with $Al(CH_3)_3$ or $In(CH_3)_3$ to prepare aluminum and indium AIM-NU-1000 analogs. In these MOFs, Al^{3+}, or In^{3+} are bound to the zirconium SBUs by terminal oxygens to form heterometallic clusters [19]. The crystal structure of NU-1000 and the modeled reaction pathway of the AIM process using $In(CH_3)_3$ are given in Figure 6.7.

The zirconium SBUs in the structure of UiO-68 can be functionalized using earth-abundant metals such as cobalt and iron to yield highly active catalysts for a wide range of organic transformations [20]. In contrast to the modification of NU-100 with $In(CH_3)_3$ and $Al(CH_3)_3$, here, the terminal –OH groups on the SBUs are first deprotonated using *n*-BuLi. Subsequent reaction with $CoCl_2$ or $FeBr_2{\cdot}2THF$ in tetrahydrofuran (THF) gives structures where cobalt or iron is bound to two terminal oxygens of the SBU in a bridging manner.

6.4.2.2 Post-Synthetic Ligand Exchange

The terminal –OH and $–OH_2$ ligands on 8-c Zr_6O_8-core SBUs can not only be functionalized by coordination to metal species, but can also be replaced by other Lewis basic charged ligands in a process termed SALI. SALI implies the solvent-assisted ligand exchange by soaking a MOF in a concentrated solution of a ligand molecule at elevated temperatures for a prolonged period of time. The partial replacement of the terminal –OH and $–OH_2$ groups on the SBUs in NU-1000 by perfluoroalkyl carboxylic acids using the SALI method results in enhanced CO_2 capture capacities attributed to the higher polarity of the pores due to the C–F dipole moment (Figure 6.8) [21]. Other carboxylate-based alkyl and aryl moieties can be incorporated in a similar way and further organic reactions can subsequently be carried out on these molecules [22]. Ligand substitution with phenylphosphate (PPA) using the SALI method yields more chemically stable frameworks owing to the chelating and shielding nature (tridentate versus bidentate) of the phosphate group [23].

Similarly to the ligand exchange reactions discussed above, acidic moieties can be introduced into MOFs. To allow for the introduction of acidic or even

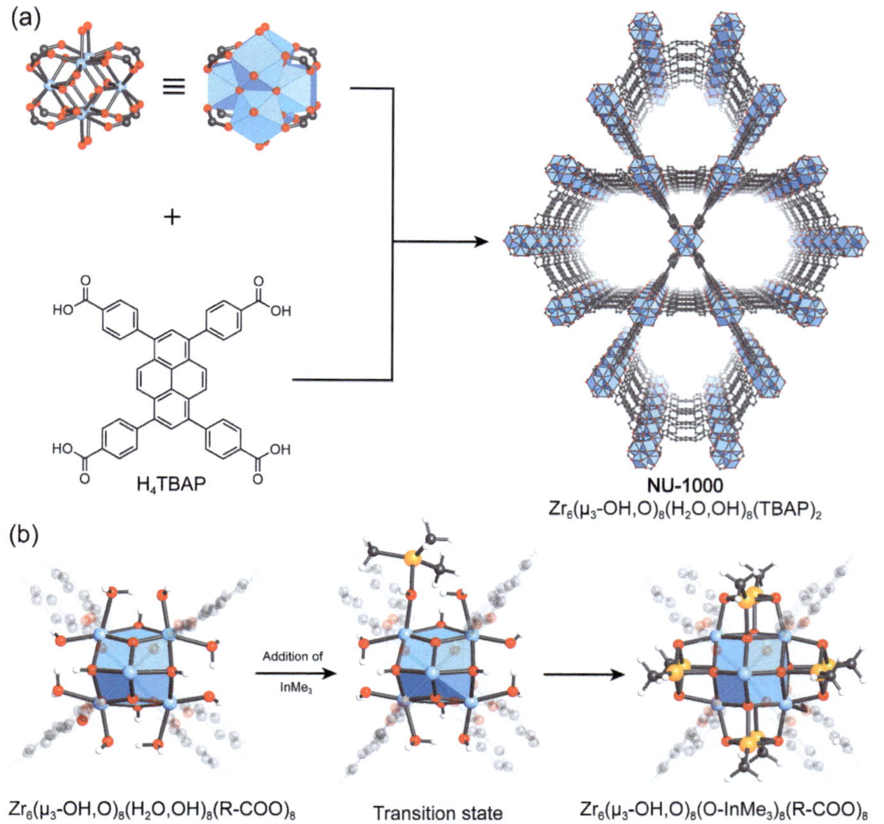

Figure 6.7 (a) Crystal structure of NU-1000. The framework of **csq** topology is formed by reticulation of Zr^{4+} ions (forming 8-c Zr_8O_6-core SBUs) and rectangular tetratopic H_4TBAP linkers. The structure can be viewed as linked stacks of kagome (**kgm**) layers. (b) Reaction pathway of the AIM process using $InMe_3$. Coordination of indium to the terminal –OH and $-OH_2$ ligands on the SBU affords the formation of an SBU of chemical formula $Zr_6(\mu_3\text{-OH-O})_8(\text{O-InMe})_8(-\text{COO})_8$. All hydrogen atoms are omitted for clarity. Color code: Zr, blue; In, orange; C, gray; O, red.

superacidic sites into a MOF structure, the MOF must be chemically very robust. Zirconium frameworks generally show high thermal and chemical stability, and as we discussed earlier, the 8-c zirconium SBUs also allow for the substitution of the terminal ligands, and thus present the prospect of incorporating acidic ligand molecules into such frameworks. An example where this approach is applied to acidify a MOF is S-MOF-808 ($Zr_6O_5(OH)_3(BTC)_2(SO_4)_{2.5}(H_2O)_{2.5}$), which is prepared from MOF-808 ($Zr_6O_5(OH)_3(BTC)_2(HCOO)_5(H_2O)_2$), see Figure 4.16) in a single-crystal to single-crystal transformation by treatment with aqueous sulfuric acid (Figure 6.9) [24]. S-MOF-808 has a Hammett acidity function of $H_0 \leq -14.5$, and is thus considered a super acid. The source of such strong acidity are terminal water ligands that form hydrogen bonds with adjacent chelating sulfate group [25]. In analogy to sulfated zirconia, a commercial catalyst material, S-MOF-808 catalyzes the dimerization of 1-butene but shows a higher selectivity than commercial catalysts.

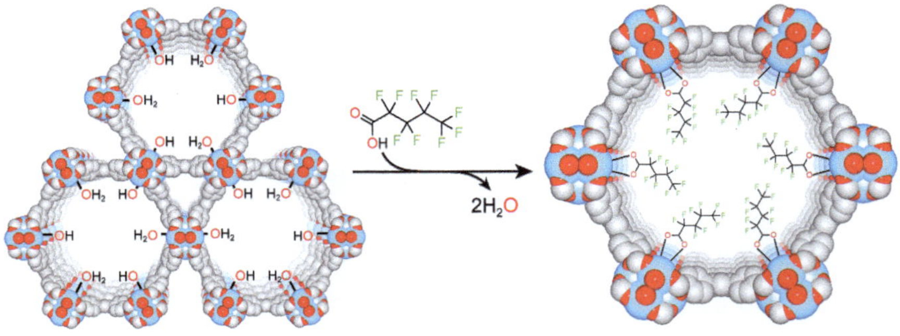

Figure 6.8 Functionalization of the 8-c zirconium SBUs in the structure of NU-1000 using the SALI method. Terminal –OH and –OH$_2$ ligands on the SBU can be replaced using organic Lewis bases with appropriate binding groups such as carboxylates. Here, the exchange with perfluoroalkyl carboxylic acids is shown. The crystal structure is shown as the space filling model. The terminal –OH and –OH$_2$ ligands are only shown partially and all hydrogen atoms are omitted for clarity. Color code: Zr, blue; C, gray; O, red; F, green.

6.4.2.3 Coordinative Alignment

Organic molecules can be aligned to a chiral MOF backbone in a method called the CAL-method (CAL = coordinative alignment) [26]. Here, the MOF acts as a substrate to crystallize organic molecules and helps in the determination of their structures by single crystal X-ray diffraction. A suitable substrate for this method is MOF-520, an aluminum-based MOF built from ring shaped 12-c SBUs connected by BTB linkers, that crystallizes in the non-centrosymmetric space group $P4_32_12$ (see Figure 6.14a). Each octametallic SBU in MOF-520 is connected to 12 BTB linkers and an additional 4 formate ligands complete the octahedral coordination. These capping formate ligands can be replaced by molecules bearing appropriate binding groups such as carboxylic acids, primary alcohols, or 1,2-diols. Analogous to the SALI method, this is achieved by soaking MOF-520 in a concentrated solution of the targeted molecule in DMF at elevated temperatures (Figure 6.10a). The CAL method allows for a precise discrimination of single and double bonds in complex molecules and the non-centrosymmetric MOF backbone serves as a reference in the structure solution, allowing for an unambiguous assignment of the absolute configuration of the bound molecule (Figure 6.10b).

6.4.2.4 Post-Synthetic Linker Exchange

Post-synthetic linker exchange of neutral N-donor linkers can control the catenation of framework structures as exemplified by many examples of MOFs with layered-pillared structures, where 2D layers are pillared by N-donor linkers [16, 27]. Such linker exchange reaction are however synthetically more challenging when strongly bound, charged linker such as carboxylates are involved.

The structure of bio-MOF-100 was discussed in detail in Chapter 4 and we already pointed out that this compound provides an ideal platform for linker exchange reactions [28]. The structure of bio-MOF-100 is built from 12-c $Zn_8O_2(AD)_4(-COO)_{12}$ SBUs that are connected to 12 carboxylates of the BPDC linkers in a monodentate fashion to form a triple cross-linked 4-connected net

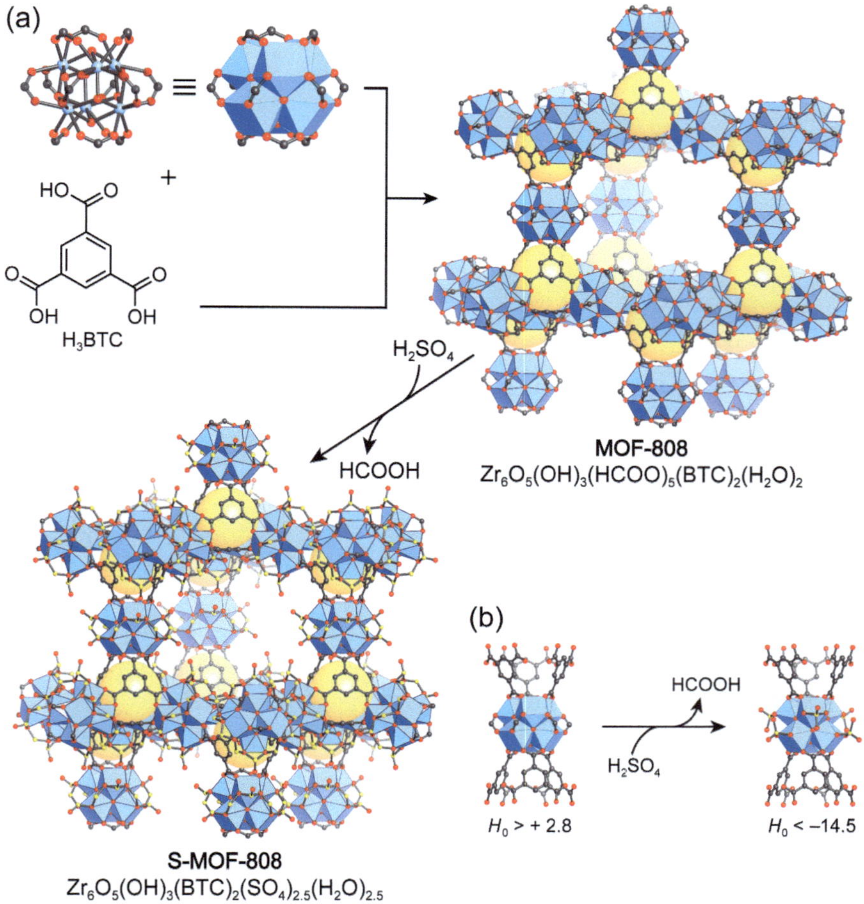

Figure 6.9 (a) Synthesis of MOF-808 by reticulation of Zr^{4+} and BTC. Subsequent treatment with aqueous sulfuric acid yields S-MOF-808. (b) The equatorial formates on the $Zr_6O_5(OH)_3(BTC)_2(HCOO)_5(H_2O)_2$ SBUs are partially replaced by $(SO_4)^{2-}$ moieties, imparting S-MOF-808 with superacidity. All hydrogen atoms and guest molecules are omitted for clarity. Color code: Zr, blue; C, gray; O, red; S, yellow.

with **lcs** topology [28b]. Whereas isoreticular analogs based on a shorter linker (i.e. H_2NDC, bio-MOF-101) can be prepared by direct synthesis, this is not possible for expanded bio-MOF-100 analogs. Direct synthesis is however not the only approach to the synthesis of isoreticular MOFs. Both, the NDC linkers in bio-MOF-101 and the BPDC linkers in bio-MOF-100 are readily replaced by longer linear ditopic linkers by soaking the MOF in a concentrated solution of the expanded linker. Using this approach two expanded frameworks, namely, bio-MOF-102 (ABDC linker) and bio-MOF-103 (NH_2-TPDC linker) are prepared in single-crystal to single-crystal transformations. This procedure allows for the increase of the pore diameter from 2.1 to 2.9 nm (Figure 6.11). The ease of isoreticular expansion of these compounds using linker exchange reactions

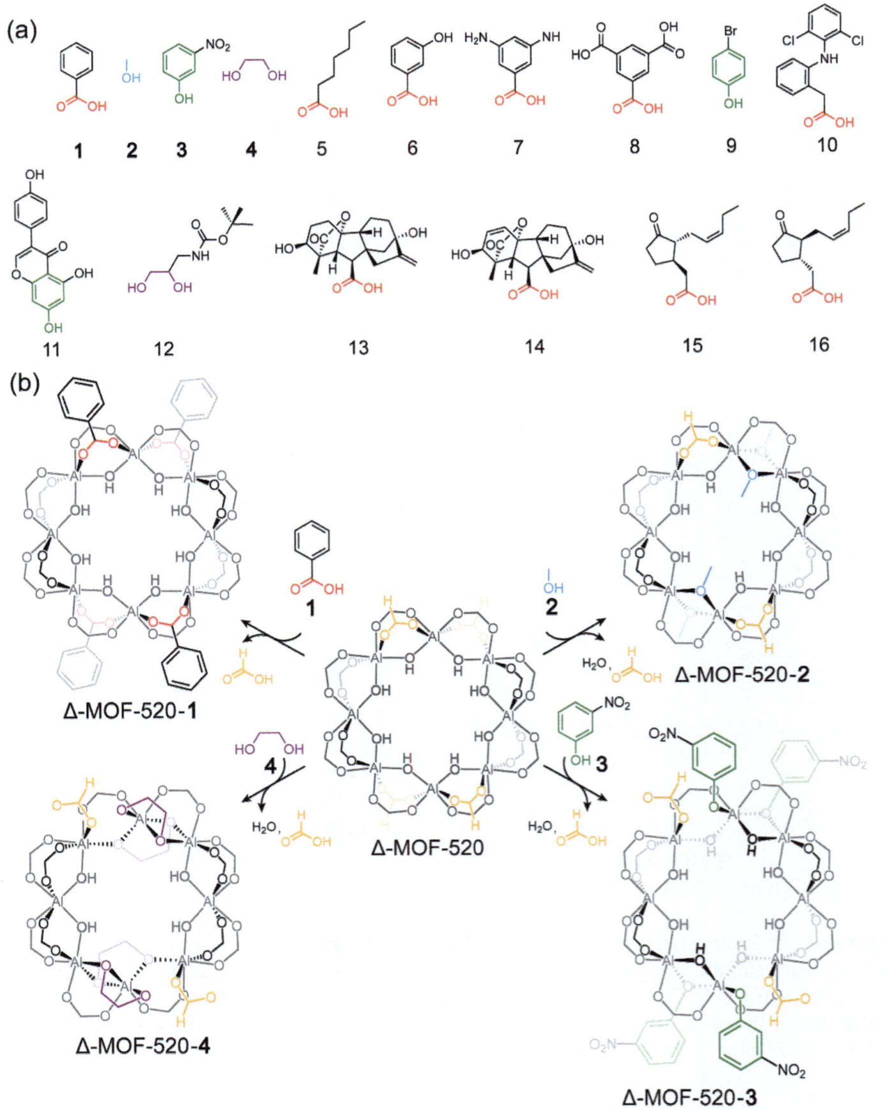

Figure 6.10 The chemical structure of the octametallic aluminum SBU in MOF-520 allows for an exchange of capping formate ligands with organic molecules bearing either carboxylic acids, alcohols, phenols, or 1-2 diols. (a) Molecules that can be incorporated into the framework, and whose structure and absolute configuration can be determined crystallographically using the CAL method. (b) The backbone of the chiral MOF serves as a reference in determining the absolute configuration of the aligned molecules. The alignment of benzoic acid (**1**), methanol (**2**), *m*-nitrophenol (**3**), and ethylene glycol (**4**) in the structure of Δ-MOF-520 are shown.

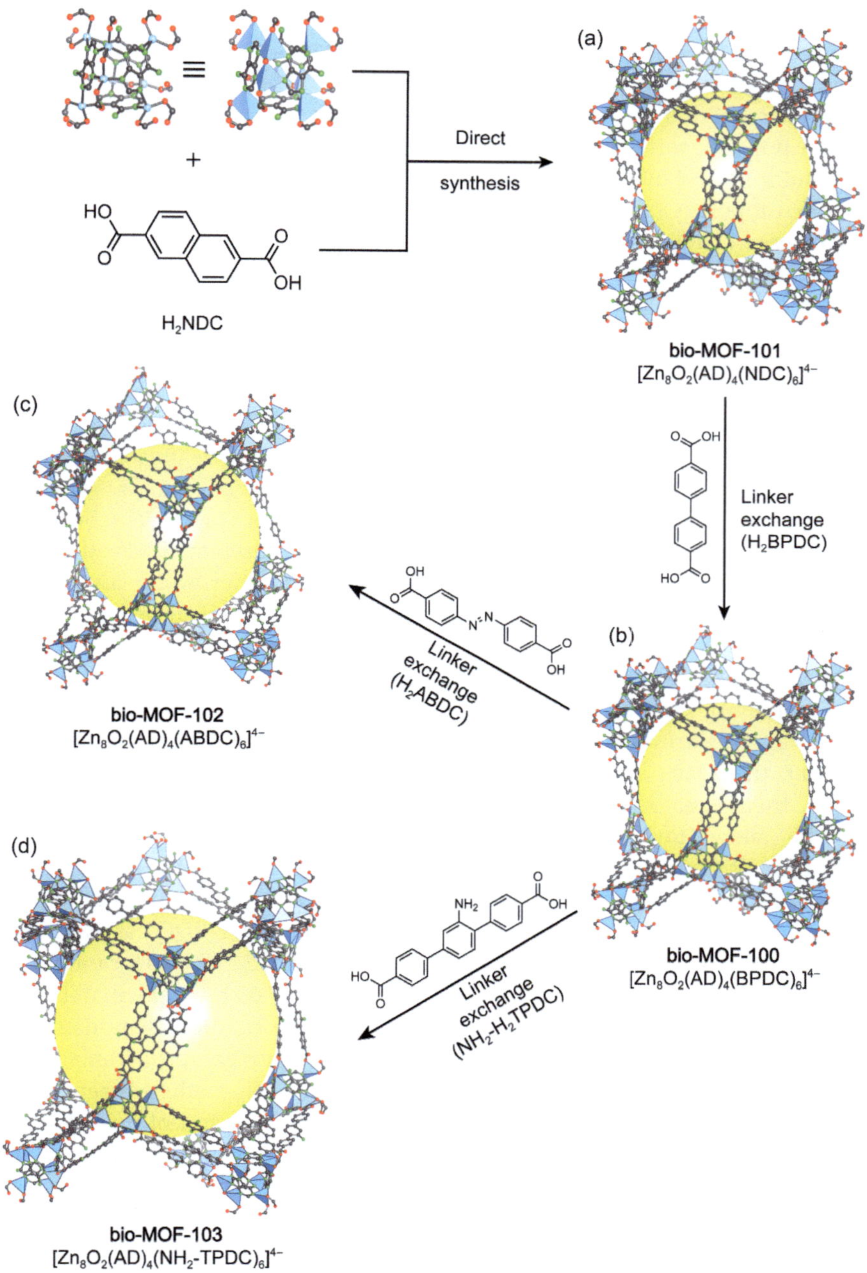

Figure 6.11 Isoreticular expansion by linker exchange reactions in the bio-MOF-10x series. (a) bio-MOF-101 and (b) bio-MOF-100 are the only two members of this series that are accessible by direct synthesis. (c) bio-MOF-102 and (d) bio-MOF-103 are exclusively accessible by linker exchange reactions starting from bio-MOF-100, using H_2ABDC and NH_2–H_2TPDC, respectively. All hydrogen atoms are omitted for clarity. Color code: Zn, blue; C, gray; N, green, O, red.

may be attributed to the monodentate binding mode of the carboxylates, and is thus not generally applicable to other MOF structures.

A linker exchange by functionalized linkers with retention of the metrics of the parent MOF is more common than the isoreticular expansion shown for the bio-MOF-10x series. One such example is the linker exchange in UiO-66, a zirconium MOF built from 12-c Zr_6O_8-core SBUs and BDC linkers with an overall **fcu** topology (see Figure 6.3). Figure 6.12 shows the replacement of the BDC linkers in UiO-66 by functionalized analogs such as NH_2–BDC, Br–BDC, N_3–BDC, OH–BDC, and $(OH)_2$–BDC. The exchange ratio can be controlled by adjusting the reaction conditions. Once incorporated into the framework structure, these functionalized linkers allow for further PSM by carrying out organic transformations [29].

6.4.2.5 Post-Synthetic Linker Installation

As illustrated for NU-1000, the terminal –OH and $-OH_2$ ligands on its 8-c Zr_6O_8-core SBU can be replaced by a manifold of charged ligands, such as organic carboxylates and phosphates [21–23]. In a similar way, linkers bearing appropriate binding groups can be used to link up adjacent SBUs, thereby transforming frameworks from one topology into another while imparting functionality or altering the chemical, physical, and mechanical properties of the MOF.

Sequential linker installation (SLI) is used to construct complex MOFs with precisely positioned functional groups and we will illustrate this for the topological transformation of PCN-700 (Figure 6.13) [30]. In the **bcu** framework PCN-700, the space between two adjacent 8-c $Zr_6O_4(OH)_8(H_2O)_4(-COO)_8$ SBUs in their equatorial plane represents two distinct "pockets" of different size. This allows for the regioselective installation of additional linkers by substitution of terminal $-OH/-OH_2$ ligands. Figure 6.13 shows the synthesis of an 11-connected net that is prepared by soaking PCN-700 in solutions of linear ditopic linkers of different metrics (H_2BDC and Me_2-H_2BTDC) at elevated temperatures. The product of this reaction, a MOF termed PCN-703, cannot be prepared by direct synthesis. The modification of PCN-700 by sequential linker installation in a single-crystal to single-crystal transformation demonstrates that SLI is a powerful tool in the construction of highly complex framework architectures with precisely positioned functionalities. This method is generally applicable and similar topological transformations by insertion of linkers into pockets to connect adjacent SBUs are possible.

The influence of post-synthetically installed linkers on the structural properties as well as the mechanical and architectural stability of MOFs has been investigated for retrofitted MOF-520. MOF-520 is molecularly retrofitted by installation of rigid BPDC linkers that bridge adjacent SBUs. These BPDC "girders" endow the resulting framework (MOF-520-BPDC) with enhanced mechanical stability by modulating the structural deformation under ultrahigh pressure (up to several GPa) [31]. The more rigid architecture of MOF-520-BPDC prevents its expansion upon increasing the hydrostatic pressure by tightly fixing adjacent SBUs. As a result, retrofitted MOF-520 is stable under high hydrostatic compression and subsequent decompression in a diamond-anvil cell at pressures

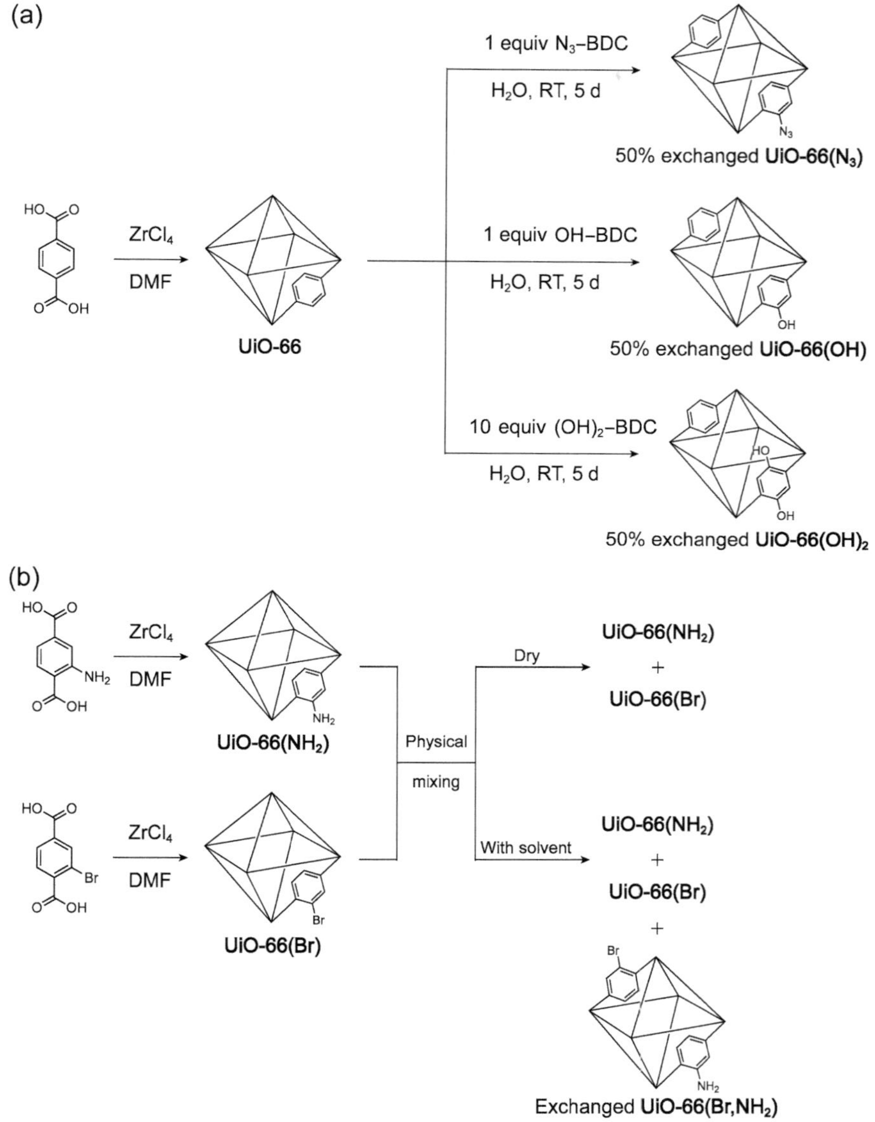

Figure 6.12 Post-synthetic linker exchange in UiO-66. (a) Soaking pristine UiO-66 in solutions containing functionalized BDC derivatives affords the formation of linker exchanged UiO-66 analogs (yield 50%). (b) Physical mixing of UiO-66(NH_2) and UiO-66(Br), synthesized by pre-synthetic modification, in the presence of solvent leads to (partial) formation of a mixed linker (MTV) version of UiO-66, containing both Br–BDC and NH_2–BDC. Since the linker exchange is solvent assisted it is not observed when the physical mixing is carried out in the absence of solvent.

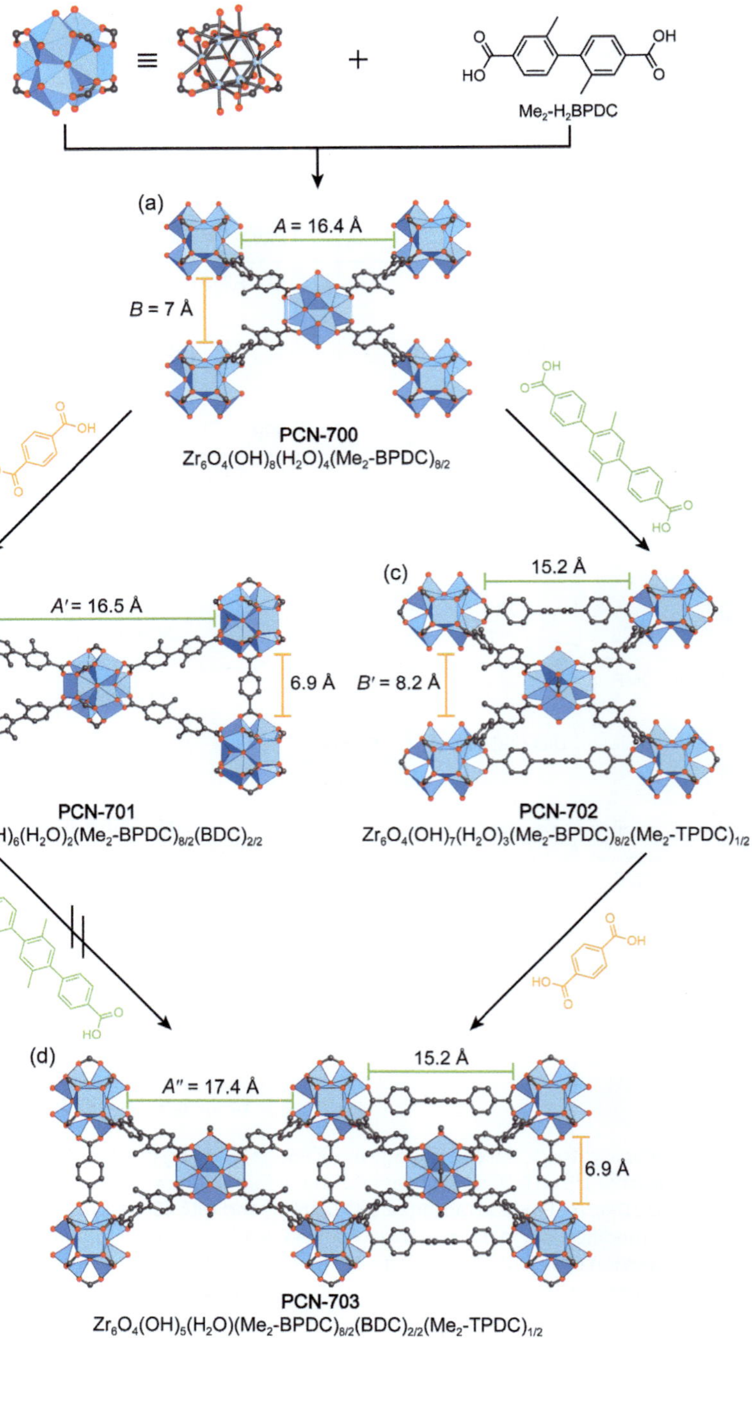
Me2-H2BPDC
(a)
A = 16.4 Å
B = 7 Å
PCN-700
Zr6O4(OH)8(H2O)4(Me2-BPDC)8/2
(b)
A′ = 16.5 Å
6.9 Å
PCN-701
Zr6O4(OH)6(H2O)2(Me2-BPDC)8/2(BDC)2/2
(c)
15.2 Å
B′ = 8.2 Å
PCN-702
Zr6O4(OH)7(H2O)3(Me2-BPDC)8/2(Me2-TPDC)1/2
(d)
A″ = 17.4 Å
15.2 Å
6.9 Å
PCN-703
Zr6O4(OH)5(H2O)(Me2-BPDC)8/2(BDC)2/2(Me2-TPDC)1/2

Figure 6.13 Sequential linker installation in PCN-700. (a) The crystal structure of PCN-700 is built from linear ditopic Me_2-TPDC linkers and 8-c zirconium SBUs. The **bcu** net has two distinct pockets, $A = 7$ Å and $B = 16.4$ Å, allowing to connect adjacent SBUs by replacing the terminal –OH and $–OH_2$ ligands with (b) H_2BDC (PCN-701) and (c) Me_2-H_2TPDC (PCN-702), respectively. (d) PCN-702 can be further modified by installing additional BDC linkers in pocket B' (8.2 Å) resulting in an 11-c SBU (PCN-703), whereas installation of additional Me_2-TPDC linkers in pocket A' of PCN-701 is not possible. All hydrogen atoms are omitted for clarity. Color code: Zr, blue; C, gray; O, red.

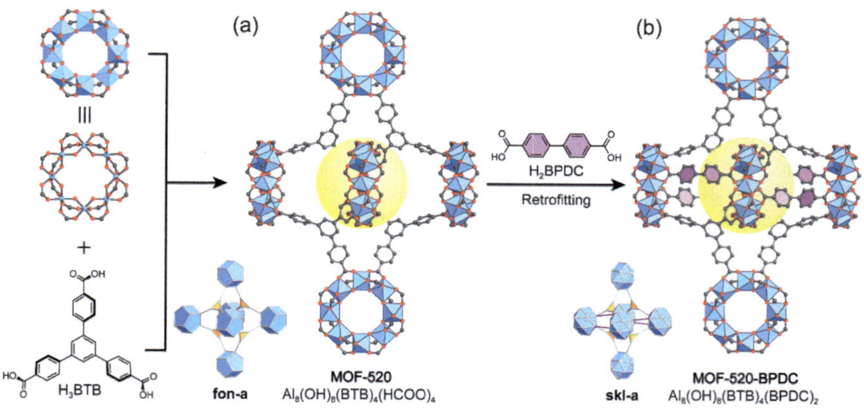

Figure 6.14 (a) Crystal structure of MOF-520. Octametallic ring-shaped 12-c aluminum SBUs are connected by trigonal tritopic BTB linkers to give a framework of **fon** topology. Four coordination sites of each SBU are occupied by capping formate ligands that can be replaced by other molecules bearing carboxylate binding groups (for other appropriate Lewis basic binding groups, see CAL-method). (b) The distance between adjacent SBUs matches the length of BPDC, which can be introduced as a girder to link neighboring SBUs, affording a transition from the **fon** into the **skl** topology. Retrofitting of MOF-520 significantly increases the mechanical stability of the MOF architecture. All hydrogen atoms are omitted for clarity. Color code: Al, blue; C, gray, O, red; the BPDC girder is highlighted in violet.

up to 5.5 GPa. In contrast, pristine MOF-520 shows amorphization at pressures lower than 3 GPa, indicative of the enhanced mechanical stability of MOF-520-BPDC. A comparison of the structures of MOF-520 and MOF-520-BPDC is given in Figure 6.14.

6.4.2.6 Introduction of Ordered Defects

Similar to the examples discussed earlier, where introduction of additional linkers results in a transition from one topology to another, controlled removal of linkers from a MOF structure can have the same result. Upon removal of a quarter of the Zn^{2+} ions and half of the PyC (4-pyrazolecarboxylate) linkers from the structure of $Zn_4O(PyC)_3$ (a MOF-5 analog), the formation of ordered vacancies leads to a transition from a **pcu** into an **srs** topology framework (Figure 6.15) [32]. Interestingly, the original **pcu** net can be restored by re-metalation with Li^+, Co^{2+}, Cd^{2+}, or La^{3+} and the addition of H_2PyC or R–H_2PyC derivatives. This procedure results in the formation of a structure where the spatial arrangement

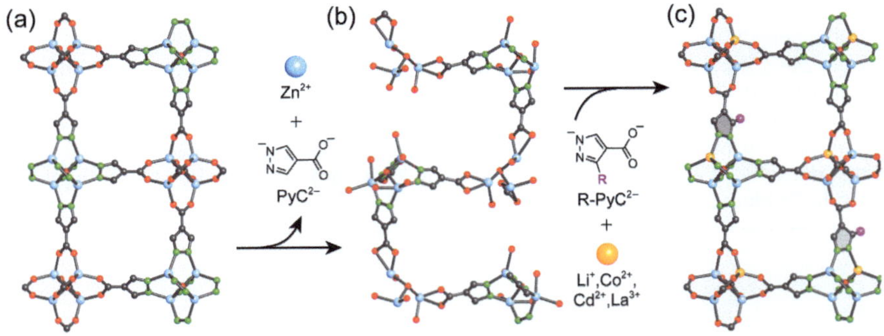

Figure 6.15 (a, b) Removal of a quarter of the PyC linkers and Zn^{2+} ions in the structure of $Zn_4O(PyC)_3$ affords a transition from a **pcu** into an **srs** net, which can be considered a defect variant of the **pcu** net. (c) The framework is subsequently re-metallated by addition of mono-, di-, or trivalent metal ions (Li^+, Co^{2+}, Cd^{2+}, La^{3+}) and linker molecules affording the formation of an MTV version of the original **pcu** topology framework. All hydrogen atoms are omitted for clarity. Color code: Zr, blue; $Li^+/Co^{2+}/Cd^{2+}/La^{3+}$, orange; C, gray; N, green; O, red; substituents on PyC are shown in pink.

of the functionalized linkers and metal constituents is well defined and follows a periodic pattern. The transition from a **pcu** to an **srs** framework and subsequent formation of a multivariate **pcu** framework derivative is illustrated in Figure 6.15.

6.4.2.7 Post-Synthetic Metal Ion Exchange

The installation of additional linkers and the replacement of already existing linkers facilitate the preparation of frameworks not accessible by direct synthesis. Similarly, in many cases, a given structure can only be prepared from one type of metal ion as the targeted SBU only forms from specific metals. This limitation can be overcome by metal ion exchange reactions [33]. This route is also applicable to the preparation of mixed-metal SBUs that are unlikely to form in a direct synthesis. Examples include the preparation of titanium- and hafnium-substituted UiO-66 analogs [17a]. Here, titanium is of special interest because it is not known to form an SBU isostructural to the Zr_6O_8-core SBUs of UiO-66. As illustrated in Figure 6.16, soaking UiO-66 in a DMF solution containing a Ti^{4+} salt affords the partial exchange of Zr^{4+} by Ti^{4+}. The successful metal ion exchange can be confirmed experimentally by positive-ion aerosol time-of-flight mass spectroscopy.

Redox-active di- and trivalent first row transition metals can be incorporated into MOF-5 by post-synthetic metal exchange. This facilitates the synthesis of materials inaccessible by typical synthetic pathways. The method used to achieve this is termed "post-synthetic ion metathesis." Here, the pristine MOF is soaked in concentrated DMF solutions of an appropriate metal salt for an extended period of time. Following this approach, it is possible to prepare MOF-5 with SBUs partially substituted with Ti^{3+}, V^{2+}, V^{3+}, Cr^{2+}, Cr^{3+}, Mn^{2+}, and Fe^{3+}. The metal ions within these SBUs have unique coordination environments, not found in molecular complexes, which makes them interesting candidates for application in catalysis. For example, (Cr)MOF-5 and (Fe)MOF-5 show

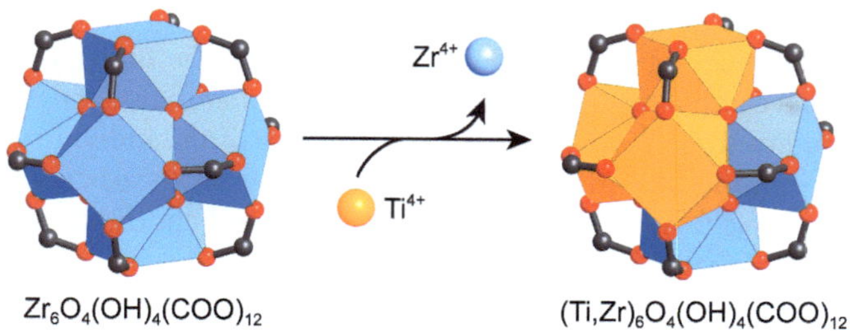

Figure 6.16 Metal exchange in the SBUs of UiO-66. A Ti^{4+} based SBU with a structure analogous to the $Zr_6O_4(OH)_4(COO)_{12}$ SBUs of UiO-66 cannot be prepared by direct synthesis. Metal ion exchange can afford a mixed-metal SBU containing both, Ti^{4+} and Zr^{4+}. Color code: Zr, blue; Ti, orange; C, gray; O, red.

more efficient outer-sphere electron transfer in the activation of NO than any other MOF [34]. To achieve the complete exchange of metal ions in the SBU, rather than partial substitution as in the previous examples, a method called "post-synthetic ion metathesis and oxidation" can be employed. Here, a MOF with labile M—O bonds is selected as a template to allow for a facile metal ion exchange [35]. The metal ion in the SBUs of this MOF template is exchanged with kinetically labile low-oxidation state metal ions. Subsequent oxidation drives the metal exchange to completion. This is typically done in a single-crystal to single-crystal transformation, enabling the preparation of single crystals of MOFs that can otherwise only be obtained as powders (e.g. Cr^{3+} MOFs).

6.4.3 PSM Involving Covalent Interactions

Methods that facilitate the making and breaking of covalent bonds are powerful tools in chemical synthesis and have been perfected in the realm of organic chemistry. Considering the presence of bonds of different strength in MOF structures, it appears challenging to perform chemical reactions involving covalent bonds without deterioration of the overall crystalline structure of the framework. The periodic nature of MOF structures, however, imparts them with remarkable thermal and chemical stability, and their open framework architecture enables the facile diffusion of liquid and gaseous reactants in and out of the framework. This allows to address the individual constituents in a facile way, giving rise to "framework chemistry." Various types of reactions performed on organic linkers including amide coupling, imine condensation, N-alkylation, bromination, reduction, and click reactions have been explored. These modifications often rely on the presence of simple functional groups appended to the linker, which, as discussed earlier, can be introduced by pre-synthetic functionalization or in a post-synthetic linker exchange reaction. The covalent derivatization of the bridging –OH groups of the SBU is independent of preinstalled functionalities, and is therefore a highly interesting way of imparting functionality into MOFs. In the following text we will give examples of prominent PSMs involving covalent interactions. This compilation is not conclusive, but other reactions PSM involving the formation of covalent bonds rely on similar principles.

6.4.3.1 Covalent PSM of Amino-Functionalized MOFs

Simple organic reactions involving amino groups are well studied in molecular organic chemistry. However, applying these methods to solid materials such as MOFs is challenging due to the altered reactivity and stability of the individual building units within the framework in addition to possible stability issues with respect to the framework itself. Additionally, the purification and characterization of solids pose problems that need to be overcome.

Amide Coupling To illustrate the principles of covalent PSM of MOFs using organic reactions, we will discuss the functionalization of MOFs by amide coupling using a simple amino-functionalized MOF and its reaction with a series of anhydrides. As the MOF substrate we select IRMOF-3, the amino-functionalized isoreticular analog of MOF-5, because it features a simple cubic structure with large pores and pore apertures, high crystallinity, and $-NH_2$ groups covalently appended to the linker [36]. In a molecular system, the acetylation of NH_2-H_2BDC with acetic anhydride leads to the formation of acetamido$-H_2BDC$. The same reaction is expected to occur on the linkers in the structure of IRMOF-3 when treated with acetic anhydride. Indeed, treatment of IRMOF-3 with 2 equiv of acetic anhydride results in the desired amide-functionalized IRMOF-3(AM1) (Figure 6.17). Powder X-ray diffraction measurements suggest that the crystallinity of IRMOF-3 is retained, and single-crystal X-ray analysis provides further evidence of the presence of the acetamido-functionalized linker. In molecular chemistry, the yield of this type of reaction is independent of the length of the alkyl chain on the anhydride employed in the reaction. This is however different when such reactions are carried out in a MOF due to the introduction of spatial constraints. When investigating the degree of amidation in IRMOF-3 using a series of 10 symmetric straight chain alkyl anhydrides as a function of the length of the alkyl chain it was found, that amides with shorter chains ($n \leq 5$) are incorporated almost quantitatively whereas amides with longer chains ($n \leq 18$) show low degrees of incorporation of less than 10% [37]. This is attributed to the steric control of this reaction in heterogeneous systems, indicating that the relative size of the reactant to the pore diameter is a critical parameter. Functionalization

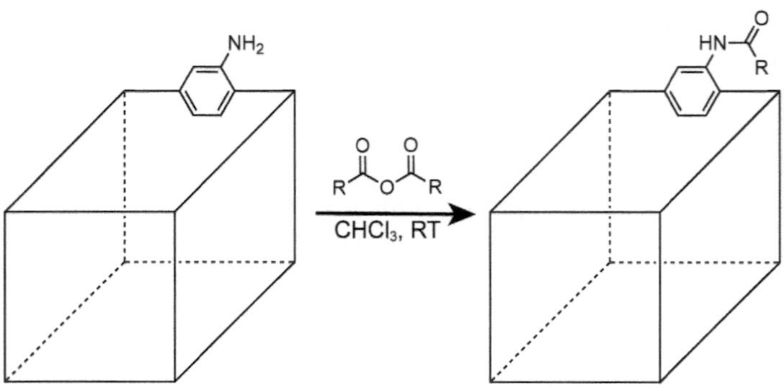

Figure 6.17 Post-synthetic amide coupling of an amino-functionalized linker and an acid anhydride affording the corresponding amide-functionalized MOF.

by amidation has been reported for many MOFs such as DMOF-1(NH_2) (see Figure 17.5), UMCM-1(NH_2) (see Figure 5.8), and MIL-125(NH_2) (see Figure 4.32) underpinning the general applicability of this method. Similarly, chiral anhydrides can be used to convert achiral into chiral MOFs, thus further expanding the scope of possible products.

By judicious choice of appropriate substrates, new coordinative sites for the complexation of metals can be introduced into MOF structures by covalent PSM. Such reactions often yield catalytically active materials that are superior to their molecular analogs in terms of activity and stability. The reaction of UMCM-1(NH_2) with cyclic anhydrides under mild conditions yields the corresponding amides with full retention of the structural integrity of the framework (35–50% yield). Addition of Cu^{2+} and Fe^{3+} salts gives the corresponding copper and iron complexes with yields as high as 50%. The resulting MOFs are termed UMCM-1(AM Cupz) and UMCM-1(AM Fesal), respectively (Figure 6.18). UMCM-1(Am Fesal) catalyzes Mukaiyama aldol reactions at room temperature, and even though the activity of this catalyst is moderate, the robust framework retains its full catalytic activity and its crystallinity over multiple cycles.

Other PSMs Involving Amino-Functionalized Linkers Following a similar reaction pathway to that described above, amino-functionalized MOFs can undergo reactions with a variety of other substrates. Figure 6.19 provides a compilation of such PSMs.

All reactions shown in Figure 6.19 are well studied in homogeneous molecular organic chemistry but when carried out within the pores of a MOF, the same general limitations as for the amide coupling apply. To introduce more complex

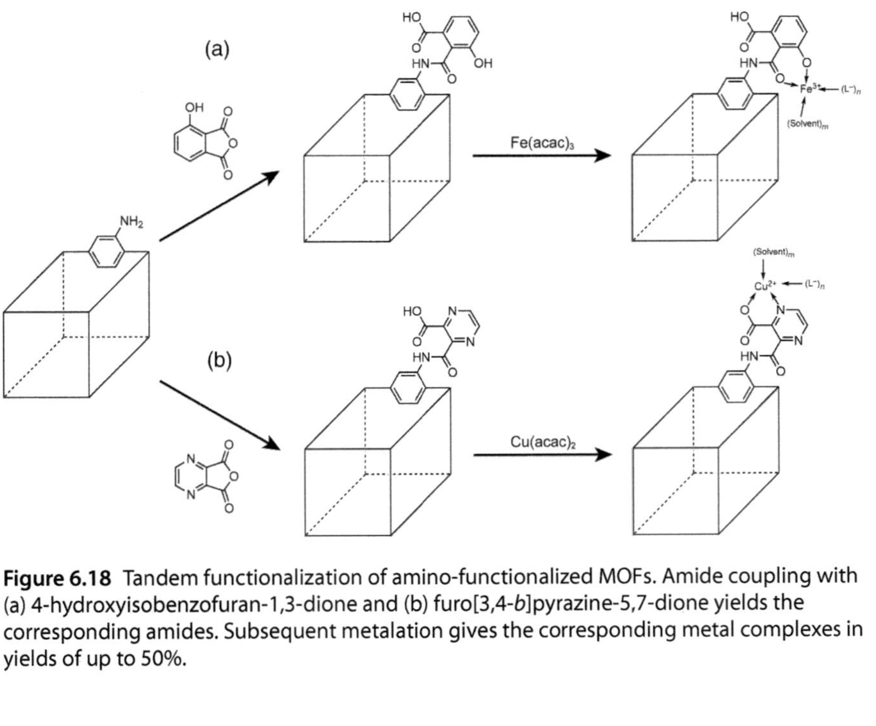

Figure 6.18 Tandem functionalization of amino-functionalized MOFs. Amide coupling with (a) 4-hydroxyisobenzofuran-1,3-dione and (b) furo[3,4-*b*]pyrazine-5,7-dione yields the corresponding amides. Subsequent metalation gives the corresponding metal complexes in yields of up to 50%.

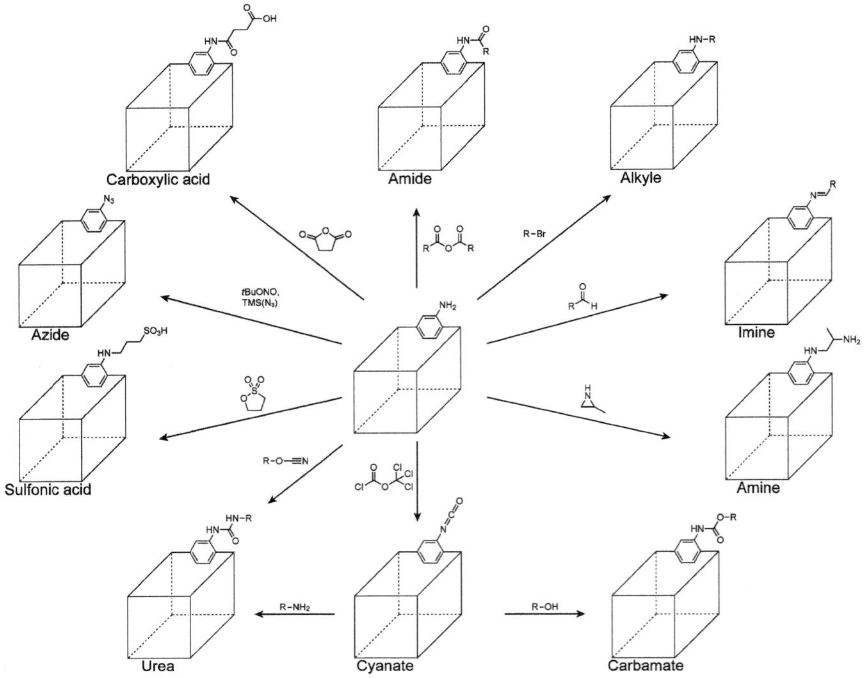

Figure 6.19 Amino-functionalized linkers can be covalently modified by reactions involving a wide range of substrates. This facilitates the introduction of functionalities such as carboxylic acids, amines, carbamates, cyanates, urea, and azides onto the organic backbone of the MOF by amide or imine linkages.

organic functional groups or additional coordination sites for the complexation of metal ions requires multistep reactions. The lack of methods for the purification of the reaction products due to the solid nature of the substrate, combined with the fact that most conversions do not proceed quantitatively, results in materials containing a mixture of all possible products of the individual functionalization steps. Carrying out multiple PSMs in a sequence is termed "tandem functionalization," which includes the sequential installation of different functionalities and the subsequent interconversion of functional groups, both of which are illustrated in Figure 6.20. Other types of tandem functionalization are known and they all underlie the same general limitations with respect to the steric demand of the reactants and the purification of the reaction products discussed earlier.

Employing multiple PSM reactions in sequence can lead to MOFs featuring enzyme-like complexity. Figure 6.21 illustrates the installation of a peptide sequence into the pores of IRMOF-74-III by carrying out a series of seven peptide bond formation and deprotection reactions.

6.4.3.2 Click Chemistry and Other Cycloadditions

The copper(I) catalyzed Huisgen cycloaddition of an alkyne with an azido functionality to form a triazole species is commonly referred to as "click chemistry" [38]. This kind of reaction generally proceeds with high efficiency

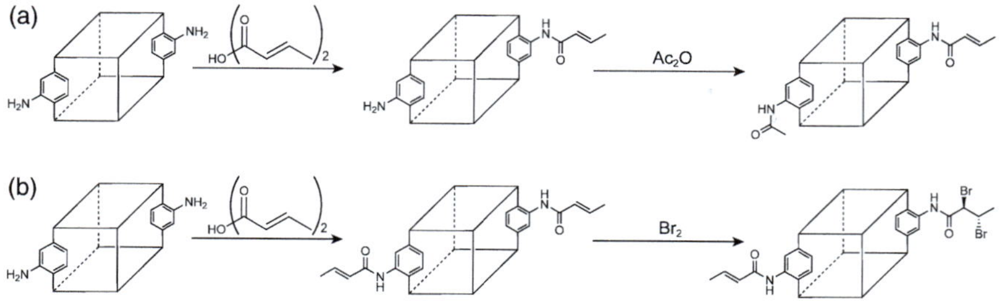

Figure 6.20 Owing to the lack of purification methods, a sequence of consecutive covalent PSMs results in materials bearing multiple functionalities. (a) Sequence of two consecutive amide couplings using different acid anhydrides. (b) Amid coupling using an unsaturated acid anhydride and subsequent bromination of the double bond. In both cases an MTV–MOF is formed since, generally, PSMs do not perform quantitatively.

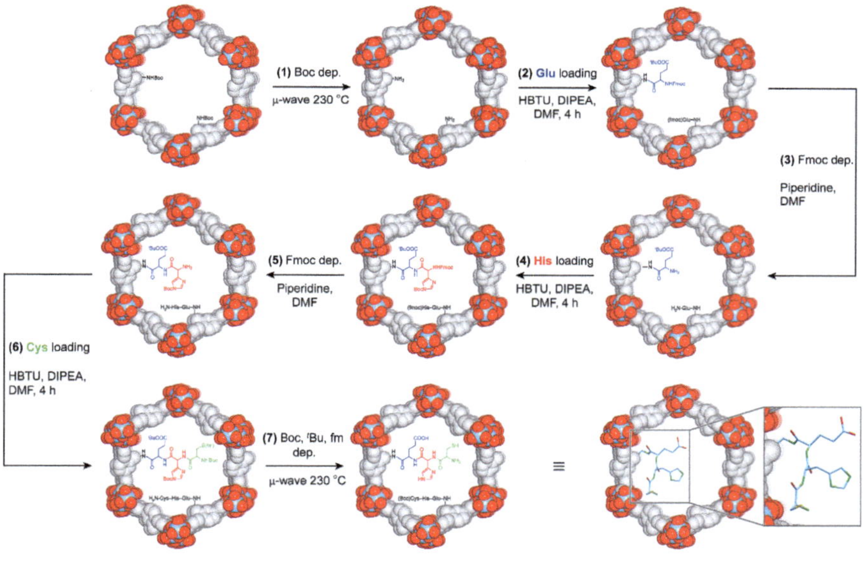

Figure 6.21 Seven consecutive covalent PSMs starting from IRMOF-74-III(NHBoc) that lead to a pore environment with enzyme-like complexity. The amino acid sequence Glu-His-Cys is introduced by a series of amide couplings and deprotection steps. One pore of IRMOF-74-III is shown in the space filling representation and the amino acid sequence Glu-His-Cys is shown in the wire frame representation. All hydrogen atoms are omitted for clarity. Color code: Mg, blue; C, gray; O, red.

at low concentrations of reactants and displays a high tolerance toward other functional groups; features that make it an ideal reaction for the covalent PSM of MOFs.

In order to carry out click reactions on the organic backbone of a MOF, the linkers have to bear either a terminal azide or an alkyne moiety that can undergo cycloaddition reactions with acetylenes or azides, respectively [39]. Figure 6.22 shows two strategies to prepare MOFs having these functionalities and their subsequent transformation by cycloaddition reactions. The yield of such reactions displays a strong dependence on the MOF used, ranging from less than 1% to

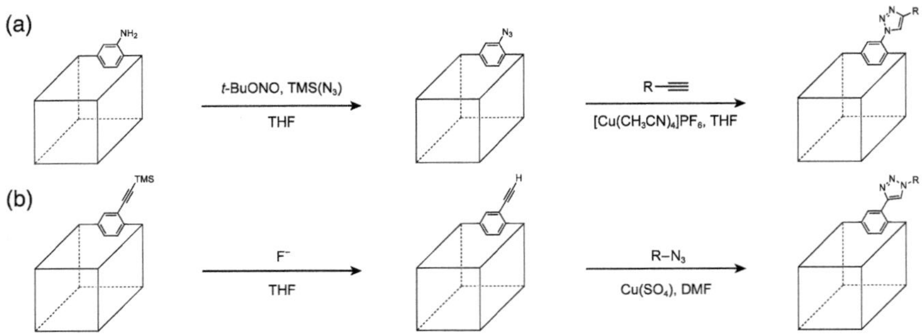

Figure 6.22 Click chemistry performed on a functionalized MOF linker. (a) Transformation of the amine into the corresponding azide allows for a cycloaddition with alkynes and the formation of 4-substituted triazoles. (b) Linkers bearing a TMS (TMS = trimethylsilyl) protected alkyne group can be incorporated into a MOF by direct synthesis. After deprotection under mild conditions, the resulting alkyne is reacted with an azide derivative, yielding the corresponding 1-substituted triazole.

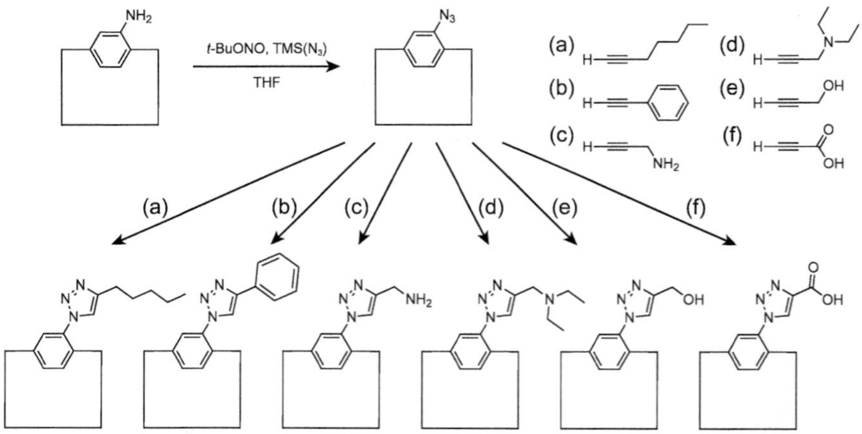

Figure 6.23 Overview of functional groups that can be introduced into MOFs using click chemistry ranging from (a) alkanes, (b) aryls, (c) primary amines, (d) tertiary amines, and (e) alcohols, to (f) carboxylic acids. Both, the alkyl reagents (a–f) and the corresponding products (bottom) are shown.

almost quantitative conversion [39]. As the transformation of amines into azides by their corresponding diazonium salts is typically carried out under acidic conditions and the diazonium salts are highly explosive intermediates, milder conditions have to be applied to achieve analogous transformation in MOFs.

This type of PSM is applicable to a wide range of MOFs and can be performed with good yields, even in the presence of bulky groups, due to the low steric demand of this reaction. Figure 6.23 shows a range of functional groups that can be installed following this approach.

Another important cycloaddition reaction used to post-synthetically modify MOFs is the Diels–Alder reaction. MOFs with furan moieties appended to their linkers can undergo Diels–Alder cycloadditions with a variety of substrates. Figure 6.24 shows the reaction of a furan functionalized MOF with maleimide.

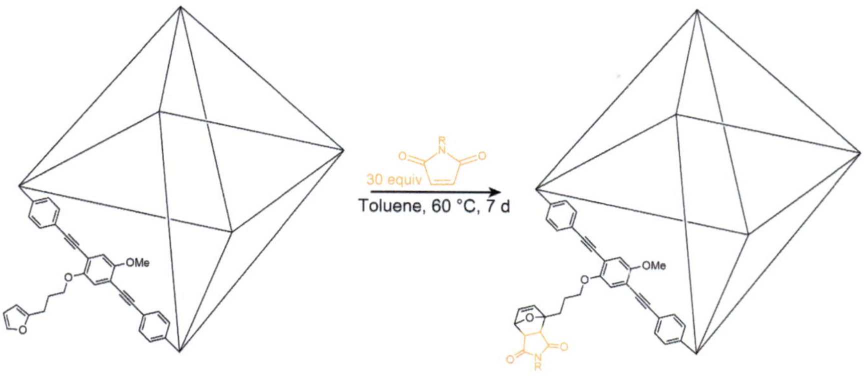

Figure 6.24 Post-synthetic modification of a furan-functionalized MOF by Diels–Alder cycloaddition. Here, the cycloadditon reaction with maleimide is shown, but other substrates can be used as well in analogy to similar reaction under homogeneous reaction conditions. Diels–Alder cycloadditions typically perform with high yields.

This reaction performs with yields of up to 98% and a modest exo/endo selectivity of 24% [40].

6.4.4 Covalent PSM on Bridging Hydroxyl Groups

In analogy to zeolites whose surface hydroxyl groups can be used to functionalize the interior of their pores, bridging –OH within the SBUs of MOFs show a similar behavior. This type of PSM is less developed than those based on organic reactions but since it provides another approach to the covalent functionalization of MOFs it is still worth mentioning. As discussed before, the terminal –OH and $-OH_2$ ligands of Zr_6O_8-core SBUs can coordinate to metal ions resulting in metal-substituted SBUs (see AIM). Bridging hydroxyl groups show quite a different behavior. The bridging –OH in the rod SBU of MIL-53 (see Figure 4.33) can undergo a silylation reaction when treated with highly reactive 1,1′-ferrocenediyldimethylsilane under solvent-free gas-phase conditions. The silylation is performed with a yield of 25% and the resulting MOF is catalytically active in the oxidation of benzene in liquid phase [41]. The modified MOF suffers from stability issues; nonetheless, this example demonstrates that the covalent functionalization of MOFs is not limited to the organic backbone.

6.5 Analytical Methods

A wide range of analytical methods is employed to identify and quantify functional groups or functionalities incorporated by PSM, and multiple methods must be used to validate the proposed outcome. We will discuss a selection of the most important tools commonly used to fulfill this task.

Powder X-ray diffraction (PXRD) is a routine method for the analysis of crystalline products. It gives insight into structural changes, the formation of crystalline side phases, and the structural degradation of crystalline materials. Modern refinement methods even allow for the determination of crystal structures from powder diffraction data. PXRD can only be used to analyze crystalline phases that may not constitute the entire sample. It also has a weak response to subtle structural changes of crystalline materials such as those evoked by the introduction of organic functional groups. If the functionalization can be achieved in a single-crystal to single-crystal transformation, single crystal X-ray diffraction provides the most powerful tool as it allows for the characterization, structural determination, and quantification of the incorporated functionalities. Adsorption measurements and consequent calculations of the surface area can indicate whether bulky groups reducing the accessible pore volume have been incorporated. Elemental analysis of an activated material can help to determine the chemical formula of a functionalized MOF, but one must keep in mind that improper activation will result in an unrepresentative chemical formula due to the presence of guest molecules in the pore. The incorporation of new elements can often be assessed by means of X-ray photoelectron spectroscopy (XPS), energy-dispersive X-ray spectroscopy (EDX), X-ray absorption spectroscopy (XAS), or several inductively coupled plasma (ICP) based methods, but it is sometimes hard to determine the yield of functionalization, and whether the material is only functionalized on the outer or on the inner surface as well. Thermal analysis (thermogravimetric analysis (TGA), differential scanning calorimetry (DSC)) can give information on whether a functionalization was successful and may allow for its quantification. The formation of a new kind of bond, not present in the parent material, can be confirmed by FT–IR spectroscopy, and the incorporation of metal complexes can be traced by UV–vis spectroscopy, both showing characteristic bands for the installed functionalities. Covalent modifications can be analyzed by NMR spectroscopy. For this purpose, MOFs are digested in acidic or basic conditions and subsequently analyzed by NMR, which requires the functional group in question to be stable under digestion conditions. Acid- or base- catalyzed reactions in the NMR tube can complicate the correct determination of yields due to bond cleavage or other side reactions under the harsh acidic or basic conditions required for the digestion of the MOF. This section illustrates that a critical assessment of the chemistry underlying a certain modification is vital for the appropriate selection of analytical methods for the qualitative and quantitative characterization of the resulting materials.

6.6 Summary

In this chapter we introduced the concept of functionalization and modification of MOFs. We distinguished between pre-synthetic, *in situ*, and post-synthetic methods, and categorized these methods into weak, strong, and covalent modifications by the strength of interaction they are based on. We illustrated the synthesis of well-defined mesoscopic constructs, introduced the concepts behind trapping molecules and nanoparticles in the pores of MOFs, and outlined the resulting properties with respect to catalysis. Rational approaches to tailored materials by PSM of as-synthesized MOFs, also referred to as “framework chemistry," were introduced. We discussed both organic and inorganic reactions that allow for the precise adjustment of structural, electronic, and chemical properties of MOFs and give rise to unprecedented properties, and thus the potential for their utilization in many different applications.

References

1 (a) Hong, D.-Y., Hwang, Y.K., Serre, C. et al. (2009). Porous chromium terephthalate MIL-101 with coordinatively unsaturated sites: surface functionalization, encapsulation, sorption and catalysis. *Advanced Functional Materials* 19 (10): 1537–1552. (b) Bromberg, L., Diao, Y., Wu, H. et al. (2012). Chromium(III) terephthalate metal organic framework (MIL-101): HF-free synthesis, structure, polyoxometalate composites, and catalytic properties. *Chemistry of Materials* 24 (9): 1664–1675. (c) Buso, D., Jasieniak, J., Lay, M.D.H. et al. (2012). Highly luminescent metal-organic frameworks through quantum dot doping. *Small* 8 (1): 80–88. (d) Hu, L., Wu, N., Zheng, J. et al. (2014). Preparation of a magnetic metal organic framework composite and its application for the detection of methyl parathion. *Analytical Sciences* 30 (6): 663–668. (e) Lian, X., Fang, Y., Joseph, E. et al. (2017). Enzyme-MOF (metal-organic framework) composites. *Chemical Society Reviews* 46 (11): 3386–3401. (f) Horcajada, P., Chalati, T., Serre, C. et al. (2010). Porous metal-organic-framework nanoscale carriers as a potential platform for drug delivery and imaging. *Nature Materials* 9 (2): 172.

2 Sun, J.-W., Yan, P.-F., An, G.-H. et al. (2016). Immobilization of polyoxometalate in the metal-organic framework rht-MOF-1: towards a highly effective heterogeneous catalyst and dye scavenger. *Scientific Reports* 6: 25595.

3 (a) Rungtaweevoranit, B., Baek, J., Araujo, J.R. et al. (2016). Copper nanocrystals encapsulated in Zr-based metal-organic frameworks for highly selective CO_2 hydrogenation to methanol. *Nano Letters* 16 (12): 7645–7649. (b) Choi, K.M., Kim, D., Rungtaweevoranit, B. et al. (2016). Plasmon-enhanced photocatalytic CO_2 conversion within metal-organic frameworks under visible light. *Journal of the American Chemical Society* 139 (1): 356–362.

4 Zhao, Y., Kornienko, N., Liu, Z. et al. (2015). Mesoscopic constructs of ordered and oriented metal-organic frameworks on plasmonic silver nanocrystals. *Journal of the American Chemical Society* 137 (6): 2199–2202.

5 (a) Eddaoudi, M., Kim, J., Rosi, N. et al. (2002). Systematic design of pore size and functionality in isoreticular MOFs and their application in methane storage. *Science* 295 (5554): 469–472. (b) Fei, H. and Cohen, S.M. (2014). A robust, catalytic metal-organic framework with open 2,2′-bipyridine sites. *Chemical Communications* 50 (37): 4810–4812.

6 Dan-Hardi, M., Serre, C., Frot, T. et al. (2009). A new photoactive crystalline highly porous titanium(IV) dicarboxylate. *Journal of the American Chemical Society* 131 (31): 10857–10859.

7 Fu, Y., Sun, D., Chen, Y. et al. (2012). An amine-functionalized titanium metal-organic framework photocatalyst with visible-light-induced activity for CO_2 reduction. *Angewandte Chemie International Edition* 51 (14): 3364–3367.

8 (a) Hermes, S., Schröder, F., Amirjalayer, S. et al. (2006). Loading of porous metal-organic open frameworks with organometallic CVD precursors: inclusion compounds of the type $[L_nM]_a$@MOF-5. *Journal of Materials Chemistry* 16 (25): 2464–2472. (b) Schröder, F., Esken, D., Cokoja, M. et al. (2008). Ruthenium nanoparticles inside porous $[Zn_4O(bdc)_3]$ by hydrogenolysis of adsorbed [Ru(cod)(cot)]: a solid-state reference system for surfactant-stabilized ruthenium colloids. *Journal of the American Chemical Society* 130 (19): 6119–6130. (c) Müller, M., Lebedev, O.I., and Fischer, R.A. (2008). Gas-phase loading of $[Zn_4O(btb)_2]$ (MOF-177) with organometallic CVD-precursors: inclusion compounds of the type $[L_nM]_a$@MOF-177 and the formation of Cu and Pd nanoparticles inside MOF-177. *Journal of Materials Chemistry* 18 (43): 5274–5281. (d) Sabo, M., Henschel, A., Fröde, H. et al. (2007). Solution infiltration of palladium into MOF-5: synthesis, physisorption and catalytic properties. *Journal of Materials Chemistry* 17 (36): 3827–3832.

9 Aulakh, D., Pyser, J.B., Zhang, X. et al. (2015). Metal-organic frameworks as platforms for the controlled nanostructuring of single-molecule magnets. *Journal of the American Chemical Society* 137 (29): 9254–9257.

10 (a) Férey, G., Mellot-Draznieks, C., Serre, C., and Millange, F. (2005). Crystallized frameworks with giant pores: are there limits to the possible? *Accounts of Chemical Research* 38 (4): 217–225. (b) Férey, G., Mellot-Draznieks, C., Serre, C. et al. (2005). A chromium terephthalate-based solid with unusually large pore volumes and surface area. *Science* 309 (5743): 2040–2042.

11 Hwang, Y.K., Hong, D.Y., Chang, J.S. et al. (2008). Amine grafting on coordinatively unsaturated metal centers of MOFs: consequences for catalysis and metal encapsulation. *Angewandte Chemie International Edition* 47 (22): 4144–4148.

12 Murray, L.J., Dincă, M., and Long, J.R. (2009). Hydrogen storage in metal-organic frameworks. *Chemical Society Reviews* 38 (5): 1294–1314.

13 Wang, C., Xie, Z., deKrafft, K.E., and Lin, W. (2011). Doping metal-organic frameworks for water oxidation, carbon dioxide reduction, and organic photocatalysis. *Journal of the American Chemical Society* 133 (34): 13445–13454.

14 (a) Chen, R., Zhang, J., Chelora, J. et al. (2017). Ruthenium(II) complex incorporated UiO-67 metal-organic framework nanoparticles for enhanced two-photon fluorescence imaging and photodynamic cancer therapy. *ACS Applied Materials & Interfaces* 9 (7): 5699–5708. (b) Tang, Y., He, W., Lu, Y. et al. (2014). Assembly of ruthenium-based complex into metal-organic framework with tunable area-selected luminescence and enhanced photon-to-electron conversion efficiency. *The Journal of Physical Chemistry C* 118 (44): 25365–25373. (c) Braglia, L., Borfecchia, E., Lomachenko, K.A. et al. (2017). Tuning Pt and Cu sites population inside functionalized UiO-67 MOF by controlling activation conditions. *Faraday Discussions* 201 (0): 265–286.

15 Deria, P., Mondloch, J.E., Karagiaridi, O. et al. (2014). Beyond post-synthesis modification: evolution of metal-organic frameworks via building block replacement. *Chemical Society Reviews* 43 (16): 5896–5912.

16 Bury, W., Fairen-Jimenez, D., Lalonde, M.B. et al. (2013). Control over catenation in pillared paddlewheel metal-organic framework materials via solvent-assisted linker exchange. *Chemistry of Materials* 25 (5): 739–744.

17 (a) Kim, M., Cahill, J.F., Fei, H. et al. (2012). Postsynthetic ligand and cation exchange in robust metal-organic frameworks. *Journal of the American Chemical Society* 134 (43): 18082–18088. (b) Karagiaridi, O., Bury, W., Mondloch, J.E. et al. (2014). Solvent-assisted linker exchange: an alternative to the *de novo* synthesis of unattainable metal-organic frameworks. *Angewandte Chemie International Edition* 53 (18): 4530–4540.

18 Planas, N., Mondloch, J.E., Tussupbayev, S. et al. (2014). Defining the proton topology of the Zr_6-based metal-organic framework NU-1000. *The Journal of Physical Chemistry Letters* 5 (21): 3716–3723.

19 Kim, I.S., Borycz, J., Platero-Prats, A.E. et al. (2015). Targeted single-site MOF node modification: trivalent metal loading via atomic layer deposition. *Chemistry of Materials* 27 (13): 4772–4778.

20 Manna, K., Ji, P., Lin, Z. et al. (2016). Chemoselective single-site Earth-abundant metal catalysts at metal-organic framework nodes. *Nature Communications* 7: 12610.

21 Deria, P., Mondloch, J.E., Tylianakis, E. et al. (2013). Perfluoroalkane functionalization of NU-1000 via solvent-assisted ligand incorporation: synthesis and CO_2 adsorption studies. *Journal of the American Chemical Society* 135 (45): 16801–16804.

22 Deria, P., Bury, W., Hupp, J.T., and Farha, O.K. (2014). Versatile functionalization of the NU-1000 platform by solvent-assisted ligand incorporation. *Chemical Communications* 50 (16): 1965–1968.

23 Deria, P., Bury, W., Hod, I. et al. (2015). MOF functionalization via solvent-assisted ligand incorporation: phosphonates vs carboxylates. *Inorganic Chemistry* 54 (5): 2185–2192.

24 Jiang, J., Gándara, F., Zhang, Y.-B. et al. (2014). Superacidity in sulfated metal-organic framework-808. *Journal of the American Chemical Society* 136 (37): 12844–12847.

25 Trickett, C.A., Osborn Popp, T.M., Su, J., Yan, C., Weisberg, J., Huq, A., Urban, P., Jiang, J., Kalmutzki, M.J., Liu, Q., Baek, J., Head-Gordon, M.P., Somorjai, G.A., Reimer, J.A., Yaghi, O. (2018) Identification of the strong Brønsted Acid Site in a metal-organic framework solid acid catalyst. *Nature Chemistry*, 11: 170–176.

26 Lee, S., Kapustin, E.A., and Yaghi, O.M. (2016). Coordinative alignment of molecules in chiral metal-organic frameworks. *Science* 353 (6301): 808–811.

27 Burnett, B.J., Barron, P.M., Hu, C., and Choe, W. (2011). Stepwise synthesis of metal-organic frameworks: replacement of structural organic linkers. *Journal of the American Chemical Society* 133 (26): 9984–9987.

28 (a) Li, T., Kozlowski, M.T., Doud, E.A. et al. (2013). Stepwise ligand exchange for the preparation of a family of mesoporous MOFs. *Journal of the American Chemical Society* 135 (32): 11688–11691. (b) An, J., Farha, O.K., Hupp, J.T. et al. (2012). Metal-adeninate vertices for the construction of an exceptionally porous metal-organic framework. *Nature Communications* 3: 604.

29 Kim, M., Cahill, J.F., Su, Y. et al. (2012). Postsynthetic ligand exchange as a route to functionalization of “inert” metal-organic frameworks. *Chemical Science* 3 (1): 126–130.

30 Yuan, S., Lu, W., Chen, Y.-P. et al. (2015). Sequential linker installation: precise placement of functional groups in multivariate metal-organic frameworks. *Journal of the American Chemical Society* 137 (9): 3177–3180.
31 Kapustin, E.A., Lee, S., Alshammari, A.S., and Yaghi, O.M. (2017). Molecular retrofitting adapts a metal-organic framework to extreme pressure. *ACS Central Science* 3 (6): 662–667.
32 Tu, B., Pang, Q., Wu, D. et al. (2014). Ordered vacancies and their chemistry in metal-organic frameworks. *Journal of the American Chemical Society* 136 (41): 14465–14471.
33 Brozek, C. and Dincă, M. (2014). Cation exchange at the secondary building units of metal-organic frameworks. *Chemical Society Reviews* 43 (16): 5456–5467.
34 Brozek, C.K. and Dincă, M. (2013). Ti^{3+}-, $V^{2+/3+}$-, $Cr^{2+/3+}$-, Mn^{2+}-, and Fe^{2+}-substituted MOF-5 and redox reactivity in Cr-and Fe-MOF-5. *Journal of the American Chemical Society* 135 (34): 12886–12891.
35 Liu, T.-F., Zou, L., Feng, D. et al. (2014). Stepwise synthesis of robust metal-organic frameworks via postsynthetic metathesis and oxidation of metal nodes in a single-crystal to single-crystal transformation. *Journal of the American Chemical Society* 136 (22): 7813–7816.
36 Wang, Z. and Cohen, S.M. (2007). Postsynthetic covalent modification of a neutral metal-organic framework. *Journal of the American Chemical Society* 129 (41): 12368–12369.
37 Tanabe, K.K., Wang, Z., and Cohen, S.M. (2008). Systematic functionalization of a metal-organic framework via a postsynthetic modification approach. *Journal of the American Chemical Society* 130 (26): 8508–8517.
38 (a) Hansen, T.V., Wu, P., Sharpless, W.D., and Lindberg, J.G. (2005). Just click it: undergraduate procedures for the copper(I)-catalyzed formation of 1,2,3-triazoles from azides and terminal acetylenes. *Journal of Chemical Education* 82 (12): 1833. (b) Kolb, H.C., Finn, M.G., and Sharpless, K.B. (2001). Click chemistry: diverse chemical function from a few good reactions. *Angewandte Chemie International Edition* 40 (11): 2004–2021.
39 (a) Gadzikwa, T., Lu, G., Stern, C.L. et al. (2008). Covalent surface modification of a metal-organic framework: selective surface engineering via Cu^{I}-catalyzed Huisgen cycloaddition. *Chemical Communications* (43): 5493–5495. (b) Goto, Y., Sato, H., Shinkai, S., and Sada, K. (2008). "Clickable" metal-organic framework. *Journal of the American Chemical Society* 130 (44): 14354–14355.
40 Roy, P., Schaate, A., Behrens, P., and Godt, A. (2012). Post-synthetic modification of Zr-metal-organic frameworks through cycloaddition reactions. *Chemistry A European Journal* 18 (22): 6979–6985.
41 Meilikhov, M., Yusenko, K., and Fischer, R.A. (2009). Turning MIL-53(Al) redox-active by functionalization of the bridging OH-group with 1,1′-ferrocenediyl-dimethylsilane. *Journal of the American Chemical Society* 131 (28): 9644–9645.

Part II

Covalent Organic Frameworks

7

Historical Perspective on the Discovery of Covalent Organic Frameworks

7.1 Introduction

Reticular chemistry is the study of linking discrete chemical entities through strong bonds into extended frameworks and large discrete molecular structures. In Chapters 1–6, we covered the principles of reticular chemistry for metal-organic frameworks (MOFs) where polynuclear metal clusters are joined by organic linkers. The formation of crystalline extended solids can also be realized by linking only organic building units through covalent bonds between light elements (e.g. H, B, C, N, O, and Si) into covalent organic frameworks (COFs) [1]. The synthetic challenges encountered in the synthesis of COFs are different from those pertaining to MOFs. The great challenge in synthesizing MOFs is controlling the metrics of the building units to open up their structures in a rational way to make crystalline and porous extended solids. This was achieved for the first time by joining polynuclear metal clusters with organic linkers affording architecturally stable and permanently porous materials. As such the challenge in synthesizing MOFs was to make them as crystalline materials and establish their porosity [2]. In organic chemistry, no crystalline extended 2D and 3D structures were known and as such the challenge in realizing COFs was first and foremost to obtain them in crystalline form. To understand the significance of the synthetic realization of COFs and the potential of this class of materials, we provide a historical perspective on how this field has emerged and what were the discoveries of synthetic organic chemistry that influenced its development.

In 1916, Gilbert N. Lewis published his seminal paper introducing the concept of the covalent bond to describe the nature of bonding within molecules [3]. In this work, titled "the atom and the molecule," he addressed the fundamental question of how atoms can be linked to form molecules and how these molecules can be described in the context of structure and reactivity. Since then, drawing from this concept, organic chemists have mastered the craft of synthesizing increasingly elaborate molecules leading to the art and science of total synthesis. Despite the fact that the toolset of organic chemists has been steadily extended over the last century, this methodology was never applied to the synthesis of organic 2D and 3D extended structures, a point highlighted by Roald Hoffmann in 1993: "*Organic chemists are masterful at exercising control in zero dimensions. One subculture of organic chemists has learned to exercise*

Introduction to Reticular Chemistry: Metal-Organic Frameworks and Covalent Organic Frameworks, First Edition. Omar M. Yaghi, Markus J. Kalmutzki, and Christian S. Diercks.

Companion website: www.wiley.com/go/yaghi/reticular

control in one dimension. These are polymer chemists, the chain builders … but in two or three dimensions, it's a synthetic wasteland" [4]. This begs the question as to why this chemistry remained undeveloped for such a long time. One of the reasons for the lack of crystalline organic extended structures is that they would be linked by inert covalent bonds. Thus, reticulating organic building units into crystalline extended structures requires that the process is performed under thermodynamic control to allow for microscopic reversibility. This challenge was addressed with the development of COFs, a class of extended 2D and 3D crystalline extended organic solids [5]. In the context of Lewis' original work, COFs have become the natural extension of the molecule into 2D and 3D [6]. Akin to how molecules pin down atoms in specific arrangements, COFs are geometric arrangements of molecules linked into periodic structures through covalent bonds. The development of COF chemistry has been strongly correlated to the progress made in synthetic organic chemistry. In the following sections, we will highlight important milestones in synthetic organic chemistry, starting with Lewis' conceptual introduction of the covalent bond, and illustrate the impact these developments have had on the field of COFs.

7.2 Lewis' Concepts and the Covalent Bond

Today, the covalent bond represents an integral part of our understanding of chemical bonding. It is defined by IUPAC as "a region of relatively high electron density between nuclei which arises at least partly from sharing of electrons and gives rise to an attractive force and characteristic inter-nuclear distance" [7]. The term "covalent" expresses the notion that electrons are shared between atoms. The origin of this concept can be traced back to 1902 when Gilbert N. Lewis, in an attempt to explain the concept of valence to his students, conveyed his first ideas on chemical bonding on paper. He visualized the outer shell of atoms as a cube with electrons at each vertex. This "cubic atom" explained the concept that chemical bonds are formed by mutual sharing of electrons to give each atom a complete set of eight valence electrons (an "octet"). This was in stark contrast to the widely accepted theory of bonding between atoms at the time. Helmholtz's electron theory of valence assumed that an electron either does or does not completely pass from one atom to another [8]. While this theory had proved useful for the description of polar molecules and ionic compounds, it did not accurately represent the empirical observations for nonpolar molecules.

The cubic atom model based on covalent bonds between atoms explained the behavior observed in the bonding of molecular compounds that could not be accurately described by previous theories. One such example is the behavior of the diatomic halogen molecules as exemplified for the case of iodine in Figure 7.1. I_2 does not form charged ion pairs in solution indicating that the description of a full electron transfer from one iodine atom to the other to form a hypothetical I^+ I^- ion pair is inaccurate. According to Lewis' theory, iodine could now be described by the edge sharing of two cubic iodine atoms resulting in a shared electron pair between them and thus the formation of one

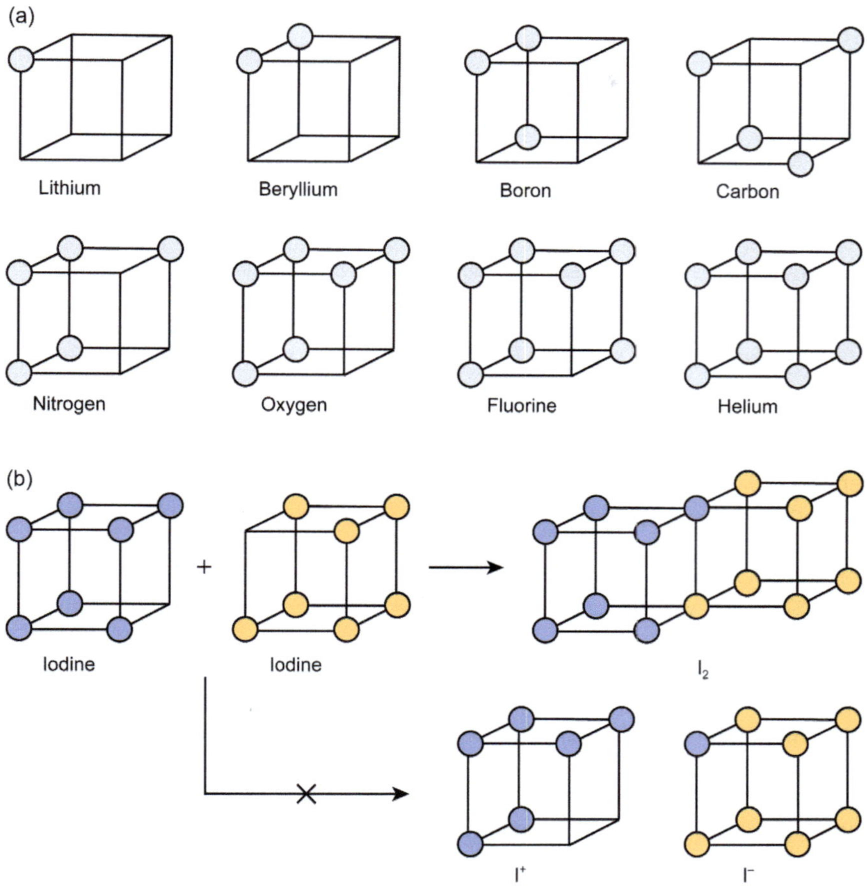

Figure 7.1 Illustration of the concept of Lewis' cubic atom. (a) Valence electrons of lithium, beryllium, boron, carbon, nitrogen, oxygen, fluorine, and helium; highlighting the increase in the number of valence electrons from a single electron in lithium to a closed "shell" of electrons for helium. (b) Atoms can achieve full valence by sharing of electrons as exemplified for the case of iodine. Lewis' theory assumes the sharing of electrons between two iodine atoms to form dimers, which are covalent in character. Here, both iodine atoms achieve full valence. In contrast, Helmholtz' theory is based on the transfer of electrons from one atom to another, where a positively and a negatively charged iodine ion would be formed, only one of which achieves the desired full valence. Color code: (a) valence electrons, gray; (b) valence electrons, light blue and orange.

charge-neutral as opposed to two charged species. This was in good agreement with experimental observations.

Lewis' theory of chemical bonding continued to evolve and, in 1916, he published his seminal article "the atom and the molecule," defining the covalent bond as a pair of electrons shared by two atoms [3]. In this article he explained chemical bonding based on his previously developed model of the cubical atom but moved on to lay out the concept of "Lewis dot structures." This model explained in a comprehensive manner some important empirical observations such as the clear

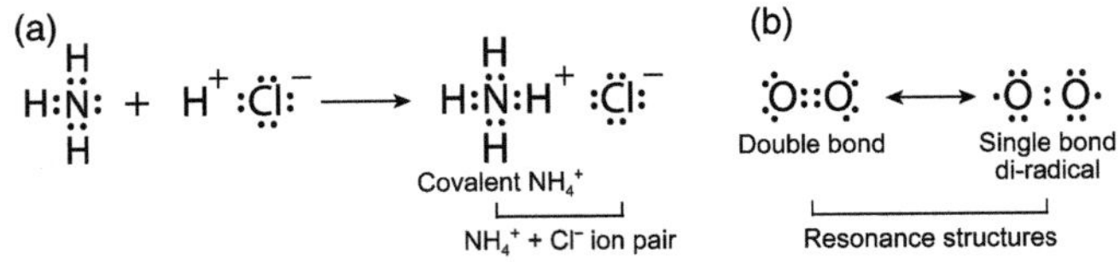

Figure 7.2 Structures of molecules as depicted in Lewis dot structure models. (a) Reaction of NH_3 with H^+Cl^- and the formation of $NH_4^+Cl^-$. A covalent bond is formed between NH_3 and H^+, whereas the resulting NH_4^+ forms an ion pair with the Cl^-. (b) Resonance structures of O_2. The di-radical form is in good agreement with the observed reactivity of oxygen at low temperatures.

distinction between the sharing of an electron pair between ammonia and a proton to form the ammonium cation as opposed to its bonding to chloride ions as a loose ion pair (Figure 7.2a). The case of ammonium ions had long troubled the chemical community as different enantiomers of alkyl-substituted analogs had been isolated indicating a four connected nitrogen atom in spite of the proposed trivalence of nitrogen. Evidently, this observation could not have been rationalized without the concept of the covalent bond and the implied sharing of electrons between atoms. Another problem that previous theories of valence had failed to explain was the structure of oxygen. It was well known at the time that at low temperatures reactions with oxygen often resulted in the formation of peroxo species; however, this contradicted prevailing bonding theories that generally assumed that oxygen as a divalent element forms a double bond in the oxygen molecule. In the Lewis dot structure it becomes apparent that there are two possible isomers for oxygen, one di-radical and one with a double bond, but only the di-radical explained the reactivity of oxygen sufficiently (Figure 7.2b). The implications of the concept of the covalent bond provided the necessary theoretical foundation for the evolving field of organic chemistry, which has since progressed from being purely empirical to a rational and systematic field of science.

7.3 Development of Synthetic Organic Chemistry

In the following decades the discovery of covalent macromolecular structures (1D polymers) by Hermann Staudinger instigated the field of organic performance materials [9]. While the term polymer was coined as early as 1833, the concept of polymers as covalently bonded macromolecular structures proposed by Staudinger in 1920 was a direct consequence of Lewis' work on the theory of the covalent bond. Previously it was assumed that the interactions between the constituents of polymers could be described by mere agglomeration rather than actual chemical bonding. Staudinger realized that macromolecular compounds include many important natural products/compounds such as proteins, enzymes, and nucleic acids as well as a large number of fully synthetic plastics and artificial fibers [9]. The importance of his discovery was recognized by awarding him the Nobel Prize in chemistry in 1953.

Another major milestone was the total synthesis of complex organic natural products. In the years following the conceptual elaboration of the covalent bond, synthesis routes for natural products of increasing complexity were established (Figure 7.3). The first reported synthesis of a naturally occurring molecule, urea, from ammonium cyanate was reported by Friedrich Wöhler in 1828 [10]. At the time it was still believed that organic matter is alive because of a special vital force, a theory commonly referred to as vitalism, and that organic molecules can thus not be synthesized artificially. Wöhler refuted this belief but, nonetheless, this first discovery was a product of serendipity and not the outcome of a devised synthetic strategy. With the elaboration of the covalent bond and the subsequent development of synthetic methodologies organic molecules could now be targeted in a rational manner. In the following decades increasingly complex natural products were synthesized. Initially, small molecules such as tropinone and camphor were reported [11]. The lessons learned from their syntheses were subsequently employed to target more elaborate molecules as exemplified by the landmark report of Robert. B. Woodward's synthesis of strychnine [12]. The most famous examples of how far the synthetic control over organic matter has progressed during the course of the twentieth century are the total synthesis of Vitamin B_{12}, achieved in a joint effort of the groups of Albert Eschenmoser and R. B. Woodward in 1972 [13], and that of Paclitaxel (taxol) by Kyriacos C. Nicolaou and coworkers in 1994 [14]. The latter molecule highlights the impact this development has had. Taxol is used in the chemotherapy of various types of cancer and as such illustrates the importance of the development of synthetic organic chemistry on the pharmaceutical industry.

The synthesis of organic molecules with increasingly complex structures as depicted in Figure 7.3 is a powerful illustration of how synthetic organic chemistry had matured throughout the twentieth century from a science based on empirical observations toward the multistep retrosynthesis of highly complex natural products [16]. The ability to control matter on the atomic level is what sets organic synthesis apart from other fields of chemistry. It allows for *de novo* synthesis of macromolecules and their precise functionalization with high regio- and stereoselectivity.

7.4 Supramolecular Chemistry

With the aforementioned improvements in synthetic methodologies being established, in the 1960s the focus of research shifted from the making of elaborate molecules toward their assembly into more sophisticated architectures. While synthetic chemists were able to target specific molecules, the question of how to deliberately align those molecules with respect to each other in solid materials remained unanswered. The development of macromolecular chemistry has shown that linking of molecular building units resulted in 1D polymers with well-defined sequences. The question then turned to how to expand the synthetic control of organic chemistry from discrete molecules and 1D polymers into 2D and 3D. An early attempt on addressing this problem was the development of supramolecular chemistry, which is concerned with non-covalent interactions between molecules and the resulting phenomena such as molecular recognition, self-assembly, and template synthesis [17].

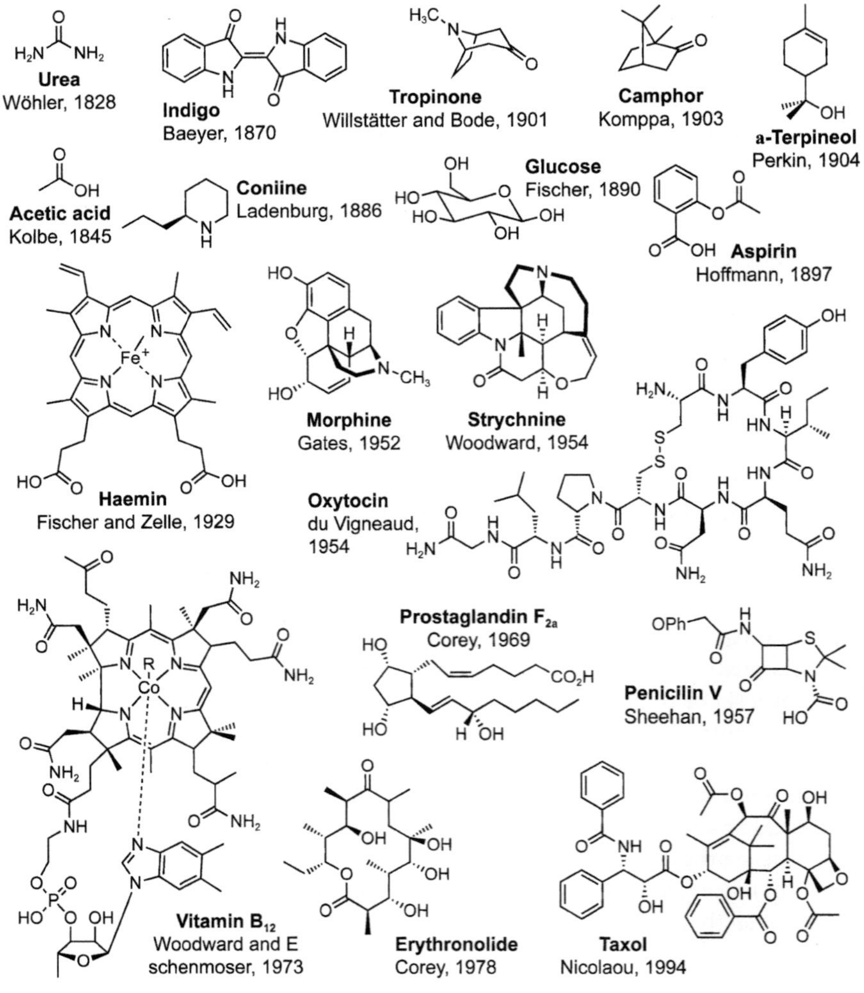

Figure 7.3 Progression of synthetic molecules starting from urea, the first reported synthesis of an organic molecule in 1828, toward increasingly elaborate molecules [10, 15]. The degree to which this field has matured can best be illustrated by the syntheses of the natural product Vitamin B_{12}, which was first synthesized in 1972. The progress in synthetic methodologies had a profound impact on the pharmaceutical industry as illustrated by the designed synthesis of the complex chemotherapy drug taxol in 1994 [13, 14].

Such non-covalent interactions were first observed in the selective molecular recognition of different metal ions by macrocyclic organic molecules. In 1967, Charles J. Pedersen attempted to prepare a complexing agent for divalent cations and accidentally discovered a simple method for the synthesis of crown ethers. In an attempt to link two catecholate moieties through one hydroxyl group on each molecule by a Williamson ether synthesis, he observed the formation

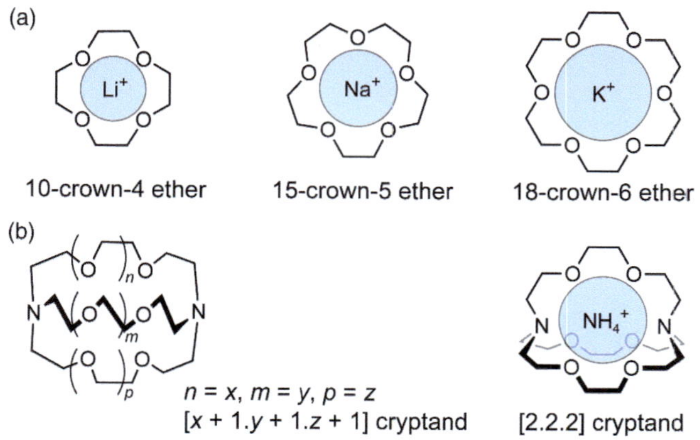

Figure 7.4 (a) Schematics of the 12-crown-4 complex of Li^+, the 15-crown-5 complex of Na^+, and the 18-crown-6 complex of K^+ illustrating the effect of the ring size on the selectivity for different alkali metal cations. (b) Illustration of the structure and nomenclature of a cryptand and the complex of a {2.2.2} cryptand with NH_4^+.

of a cyclic dibenzo-16-crown-4 polyether as a by-product, which complexed the potassium ion from the reaction mixture. Pedersen realized that this high affinity was due to non-covalent interactions between the oxygen atoms of the crown ether and the positively charged potassium ion. Accordingly, he studied the effect of the ring size on the selectivity for alkaline ions of different sizes [18]. Soon after, Jean-Marie Lehn showed that the selectivity of binding can be increased by an order of magnitude by transitioning from planar macrocyclic crown ethers to organic cages, the so-called cryptands (Figure 7.4) [19].

These examples highlight the importance of exercising control over the metrics and dimensionality of a given system for the rational optimization of host–guest interactions. The study of selective non-covalent interactions inspired researchers to go beyond molecular recognition and guide the organization of molecular building units into large supramolecular architectures through complementary non-covalent interactions in a process termed self-organization [20]. An illustrative example for compounds emanating from this development is the formation of the so-called circular helicate reported by Lehn and coworkers, which forms spontaneously from five linear tris-bipyridine TBPy struts (5,5′-bis(2-(5′-methyl-[2,2′-bipyridin]-5-yl)ethyl)-2,2′-bipyridine) guided by coordination of the bipyridine units to 5 equiv of $FeCl_2$ upon heating in ethylene glycol at 170 °C (Figure 7.5). The center of the resulting pentameric circular helicate has the ideal size to host one of the chloride counter anions. It was shown later that by replacing the $FeCl_2$ starting material with $Fe(SO_4)$, the slightly bigger sulfate ions direct the product away from the pentameric and toward a hexameric circular helicate where the center of the assembly is occupied by a sulfate, rather than a chloride ion (Figure 7.5) [21]. This shows that the assembly process through non-covalent interactions is microscopically reversible and can be carried out

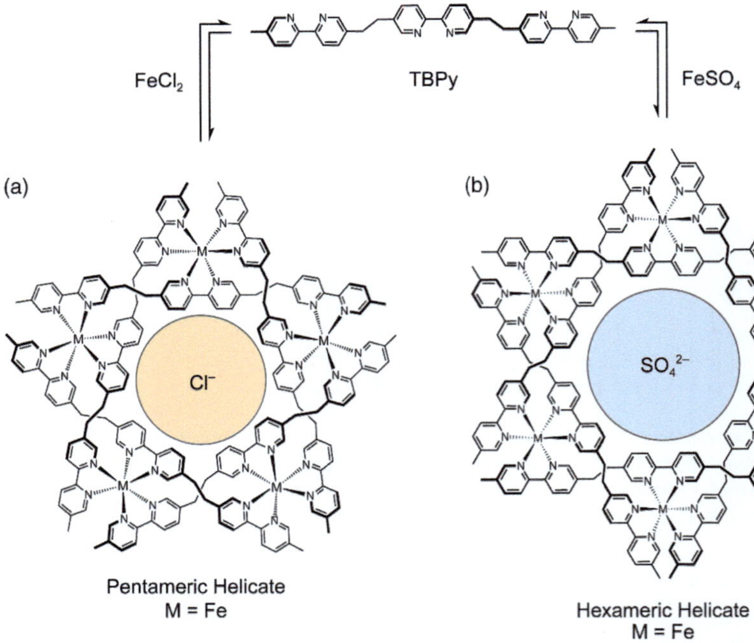

Figure 7.5 Synthesis of pentameric and hexameric circular helicates assisted by metal templation. Reacting linear TBPy with $FeCl_2$ yields the ninefold negatively charged pentameric circular helicate. The central pocket in the assembly is of the right size to accommodate a chloride ion, which templates the formation of the structure. In contrast, reaction of the linker with $Fe(SO_4)$ results in the formation of the 12-fold negatively charged hexameric circular helicate. Here, the sulfate ion resides in the central pocket. The only difference in the reaction conditions between the two systems is the counter anion, which controls the formation of one structure over the other. This fact highlights the dynamic nature of the assembly process.

under thermodynamic control. This is very powerful as it allows to generate the resulting products with 100% yield and without the need for purification.

The idea of supramolecular chemistry, the chemistry beyond the molecule, has initiated the field of crystal engineering where crystal structures of molecular assemblies are designed using complementary non-covalent interactions between the individual components. Here again researchers made use of the fact that dynamic error correction in such assemblies allows for their crystallization in one step. As such, self-organization, for the first time, enabled organic chemists to align molecular building units into solids with predetermined structures [22]. The early examples of how non-covalent interactions between organic molecules and metal ions can lead to the formation of coordination networks has been covered in Chapter 1 in the context of the historical development of MOFs and we refer the reader to this section for a more detailed description. Supramolecular assemblies are difficult to be modified without losing their structural integrity, which can be attributed to the following: functionalization of the building units alters the interactions between the constituents, which in turn often leads to different assemblies and, therefore, performing chemistry on or within assemblies held together by weak interactions is difficult and often results in structural rearrangement. Nonetheless, supramolecular chemistry in general and crystal engineering in particular have had a profound impact on the expansion of organic chemistry beyond the molecule and toward crystalline extended organic solids [22a].

7.5 Dynamic Covalent Chemistry

When comparing supramolecular assemblies to the way in which nature employs self-organization to create complex architectures such as those of proteins or DNA, it becomes apparent that biological macromolecules comprise a primary structure of atoms held together by strong covalent bonds. Only the secondary and tertiary structure that govern the spatial arrangement of the individual building units are due to non-covalent interactions. Inspired by the well-defined covalent backbone prevalent in nature, the formation of covalent macromolecular architectures in synthetic systems and thus strategies for the alignment of molecules not by weak interactions, but by covalent bonds became the focus of attention. At this point it is instructive to take a step back and think about why the formation of large synthetic architectures from molecular building units had thus far exclusively been realized in systems held together by non-covalent bonds but not in entirely covalent materials. Supramolecular assemblies are formed under mild reaction conditions and (largely) thermodynamic control due to the inherent microscopic reversibility of their underlying weak interactions. In contrast, the formation of covalent bonds is in general less reversible because of their inertness. Such kinetically controlled reactions do not exclusively yield the desired product but also side-products. This is not an issue in the realm of molecular chemistry where reaction products can be purified but it inevitably poses a problem for extended crystalline solids where purification is impossible and the formation of pure, crystalline products has to proceed in a single step. It is therefore desirable to carry out covalent organic reactions under conditions allowing for microscopic reversibility and thus for error correction. The reversibility of several organic transformations had been reported but the notion of strategically carrying out organic chemistry under thermodynamic control was not conceptualized until 1999 [23]. The development of what was termed "dynamic covalent chemistry" was in large carried by the field of mechanically interlocking molecules [24]. The simplest example of such species are catenanes (from Latin "catena" for "chain") where two or more molecular rings are interlocking, thus being held together by mechanical rather than chemical bonding. The first synthetic realization of a molecular catenane was reported in 1983 by Jean-Pierre Sauvage and coworkers. In the synthesis, a CBP (Cu(I)bis-4,4′-(1,10-phenanthroline-2,9-diyl)diphenol) complex is used as a template, which assumes a tetrahedral geometry with the two embracing phenanthroline ligands serving as points of registry for the following ring-closing step. By reacting this template with 2 equiv of DIT (1,14-di-iodo-3,6,9,12-tetraoxy-tetradecane) in a Williamson ether synthesis and subsequent demetalation of the resulting structure it is converted into two discrete interlocking macrocycles (Figure 7.6a) [25]. The template helps to bring the constituents together thus improving the overall yield of the reaction;

however, the formation of the catenane is still only achieved with 72% yield. This is not a big drawback when only two of these rings are entangled but becomes detrimental when more rings are involved. It is therefore not surprising that the field readily absorbed the development of dynamic covalent chemistry as the microscopic reversibility of these reactions allows for significantly higher yields in the final ring-closing step. The library of accessible dynamic covalent reactions has been steadily expanded and in particular reversible Schiff-base chemistry was employed in the synthesis of catenanes, such as the molecular Borromean rings by Sir James F. Stoddart in 2004 (Figure 7.6b). Here, six DFP (2,6-pyridinedicarboxaldehyde) and six bipyridine based diamine moieties (DAB, (([2,2′-bipyridine]-5,5′-diylbis(oxy))bis(4,1-phenylene))dimethanamine) are reacted with 6 equiv of $ZnCl_2$ to form an 18 component molecular architecture of a Borromean ring topology (Figure 7.6b). In this structure, each of the six Zn^{2+} ions is coordinated by one of the six *exo*-bidentate bipyridyl groups formed *in situ* and one of the six *endo*-diiminopyridyl ligands. Upon demetalation, the three molecular macrocycles, while not interlocking with each other, are mechanically entangled in such a way that if only one of the rings is removed the other two can part company. The realization of this structure is possible only due to the microscopic reversibility of the non-covalent coordination of the ligands to the labile Zn^{2+} ions, and more importantly the formation of the

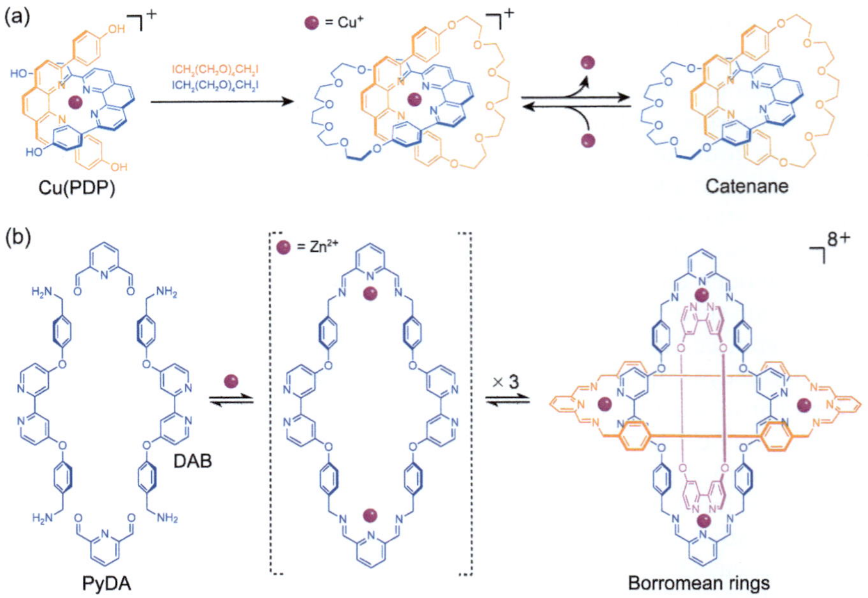

Figure 7.6 Synthetic approach toward (a) a molecular catenane and (b) molecular Borromean rings. (a) Reaction of the CBP complex, serving as a point of registry, with DIT, followed by demetalation of the structure with potassium cyanide results in the formation of a molecular catenane. (b) Reversible imine bond formation between DFP and DAB in the presence of trifluoro acetic acid as a catalyst ensures the formation of the macrocycles with 100% yield where the Zn^{2+} ions serve as templates to bring the individual rings together in the exact manner necessary for the Borromean ring topology.

covalent imine bonds between the organic building units under thermodynamic control in the presence of trifluoro acetic acid as a catalyst.

The strategy of forming molecular catenanes under thermodynamic control has since been applied to the synthesis of a large variety of interlocking molecules. For this purpose, many different dynamic covalent reactions such as imine bond formation, disulfide bond formation, and alkene metathesis have been employed [26, 27]. The development of the field of mechanically interlocking molecules has played a pivotal role in the progression of organic chemistry and both Stoddart and Sauvage were awarded the Nobel Prize in chemistry in 2016.

7.6 Covalent Organic Frameworks

The synthesis of crystalline extended 2D and 3D organic solids, where molecular building units are stitched together through covalent bonds remained an undeveloped area of research throughout the twentieth century as it was widely believed that the product of such reticulation would inevitably be amorphous. The core values at the heart of organic chemistry, the fact that compounds can be made in a controlled fashion, with atomic precision and in pure phase, can, in the case of extended 2D and 3D solids, only be guaranteed if they possess crystallinity [28]. Prior to the development of dynamic covalent chemistry, it was widely accepted that the microscopic reversibility required for the crystallization of organic molecules into covalently linked extended framework structures is difficult, if not impossible, to achieve and this challenge was commonly referred to as the "crystallization problem" [29].

In 2005, Omar M. Yaghi and coworkers reported the first crystalline extended organic frameworks [5a]. The synthesis of the two reported frameworks, termed COF-1 ($[BDBA]_{boroxine}$, where BDBA = 1,4-phenylenediboronic acid) and COF-5 ($[(HHTP)_2(BDBA)_3]_{boronate\ ester}$, where HHTP = 2,3,6,7,10,11-hexahydroxyterphenylene), was achieved by making use of the reversibility of boroxine and boronate ester formation, respectively. The challenge in the crystallization of COFs formed by such condensation reactions is twofold: the reaction needs to be slowed down and water has to remain in the reaction mixture to allow for full reversibility of the bond formation and thus for error correction. COF-1 is formed from the self-condensation of BDBA at 120 °C. To slow down the reaction, it is carried out in a 1 : 1 v/v solvent mixture of dioxane/mesitylene in which the starting material is not fully soluble. To ensure the reversibility of framework formation, the reaction is carried out in a sealed Pyrex tube preventing the evaporation of water liberated in the condensation. X-ray powder diffraction of the material unambiguously confirmed that COF-1 crystallizes in a layered honeycomb (**hcb**) topology. The layers of COF-1 have openings of 15.1 Å and stack in a staggered conformation along the crystallographic *c*-axis with an interlayer distance of 3.3 Å (Figure 7.7a). This arrangement leads to the formation of zig–zag channels of 7 Å in diameter propagating along the crystallographic *c*-axis thus rendering COF-1 porous. COF-5 is made by the cross-condensation of BDBA with HHTP resulting in a boronate ester

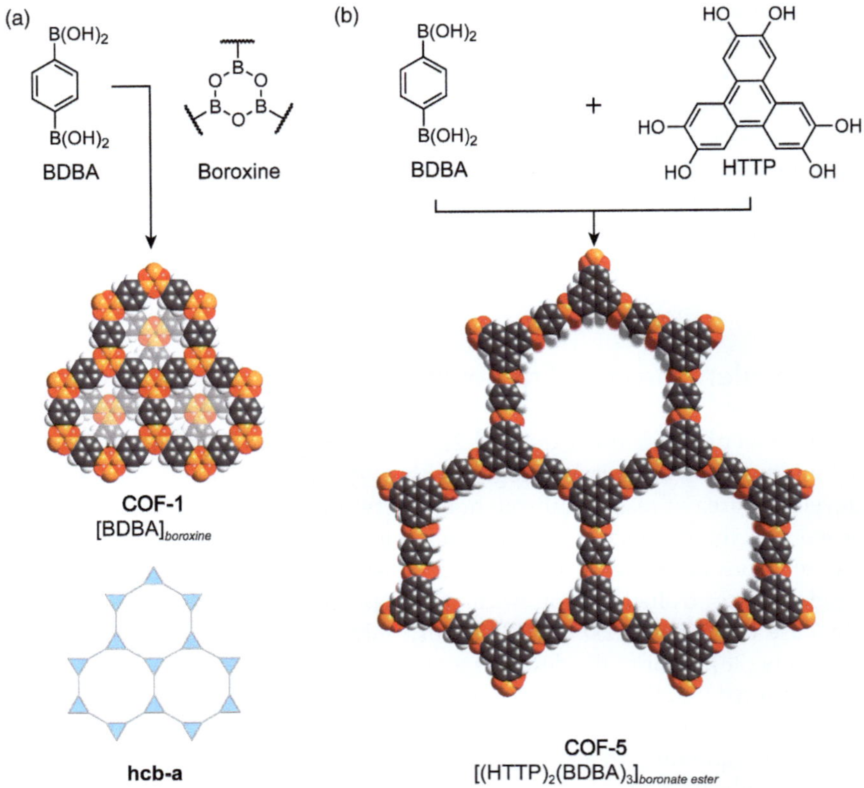

Figure 7.7 Synthetic scheme for the formation of the first examples of covalent organic frameworks, COF-1 and COF-5. (a) Self-condensation of BDBA results in boroxine linkages to yield COF-1 (**hcb** topology). (b) Reticulation of BDBA with HHTP leads to COF-5 through the formation of boronate ester linkages between the constituents. Color code: H, white; B, orange; C, gray; O, red.

linked framework. The condensation reaction requires analogous conditions to those described for COF-1. The resulting framework is also of **hcb** topology; however, here the 2D layers are stacked in an eclipsed fashion resulting in large mesoporous channels of 27 Å in width along the *c*-direction (Figure 7.7b).

The covalent nature of COF-1 and COF-5 endows these materials with excellent thermal stability up to 600 °C. Heating under dynamic vacuum results in the removal of residual solvent molecules from the pores and the activated materials of COF-1 and COF-5 sustain permanent porosity with BET surface areas of 711 and 1590 $m^2\ g^{-1}$, respectively (Figure 7.8).

This first realization of crystalline extended structures composed solely of light atoms and prepared by rational synthesis highlights the added value of reticulating organic building units through covalent bonds as opposed to weak, non-covalent interactions. The strong, directional bonds endow COFs with high thermal and architectural stability and thereby allow for the removal of solvent molecules from their pores. In stark contrast to supramolecular assemblies,

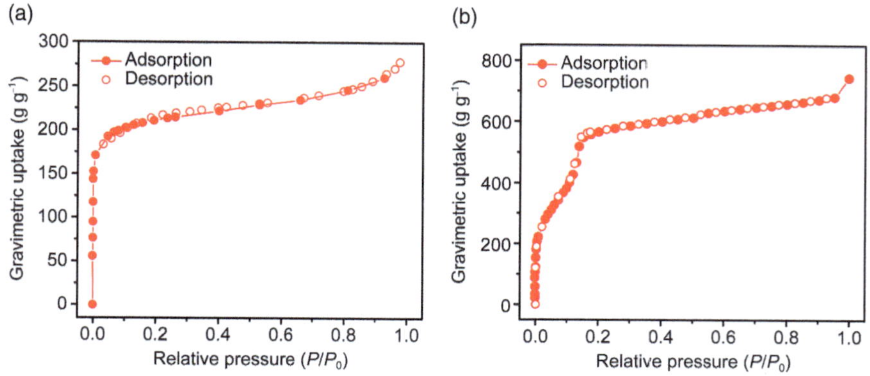

Figure 7.8 N_2 adsorption isotherms of COF-1 and COF-5 at 77 K. (a) COF-1 is microporous and features a Type I isotherm with an uptake of 278 $cm^3\,g^{-1}$ and a surface area of 711 $m^2\,g^{-1}$. (b) COF-5 is mesoporous, which is manifested in a Type IV isotherm with an uptake of 744 $cm^3\,g^{-1}$ and a surface area of 1590 $m^2\,g^{-1}$. The fact that the desorption branches perfectly trace the adsorption branches in the isotherms corroborates the architectural stability of COF-1 and COF-5 and gives further evidence for their permanent porosity. The adsorption and desorption branches are illustrated as filled and empty circles, respectively.

COFs exhibit permanent porosity which opens up the field of organic chemistry to applications in areas such as gas storage, gas separation, and catalysis to name a few [30].

After the synthesis of extended 2D COFs, the next obvious progression was to expand this chemistry into 3D [5b]. Toward this end the cross-condensation of HHTP with tetrahedral tetratopic TBPM (tetra(4-dihydroxyborylphenyl)methane) gave the crystalline 3D framework COF-108 ($[(TBPM)_3(HHTP)_4]_{boronate\ ester}$), cross-condensation of HHTP with TBPS (tetra(4-dihydroxyborylphenyl)silane) yielded COF-105 ($[(TBPS)_3(HHTP)_4]_{boronate\ ester}$), and self-condensation of TBPM or TBPS was found to afford COF-102 ($[TBPM]_{boroxine}$) and COF-103 ($[TBPS]_{boroxine}$), respectively. The synthesis conditions for these COFs follow the basic considerations laid out for their 2D counterparts, the main difference being that the reactions are carried out at a lower temperature of 85 °C as opposed to 120 °C. The reticulation of tetrahedral and trigonal building units leads to 3,4-c nets, and there are two possible topologies for this scenario. COF-102, COF-103, and COF-105 crystallize in a **ctn** topology, whereas COF-108 crystallizes in a **bor** net. Here, we will focus on COF-105 and COF-108. COF-105 has large pores that measure 18.3 Å in diameter. COF-108 has two differently sized pores of 15.2 and 29.2 Å, the latter marking the first mesopore in a 3D COF (Figure 7.9). The large 3D pores in combination with the fact that these materials are entirely constructed from light elements (C, Si, B, O, H) endows these two frameworks with exceptionally low densities of 0.18 and 0.17 $g\,cm^{-3}$ for COF-105 and COF-108, respectively. These values are markedly lower than those of highly porous MOFs such as MOF-5 (0.59 $g\,cm^{-3}$) or MOF-177 (0.42 $g\,cm^{-3}$) and COF-105 and COF-108 constituted the lowest density crystals known at the time.

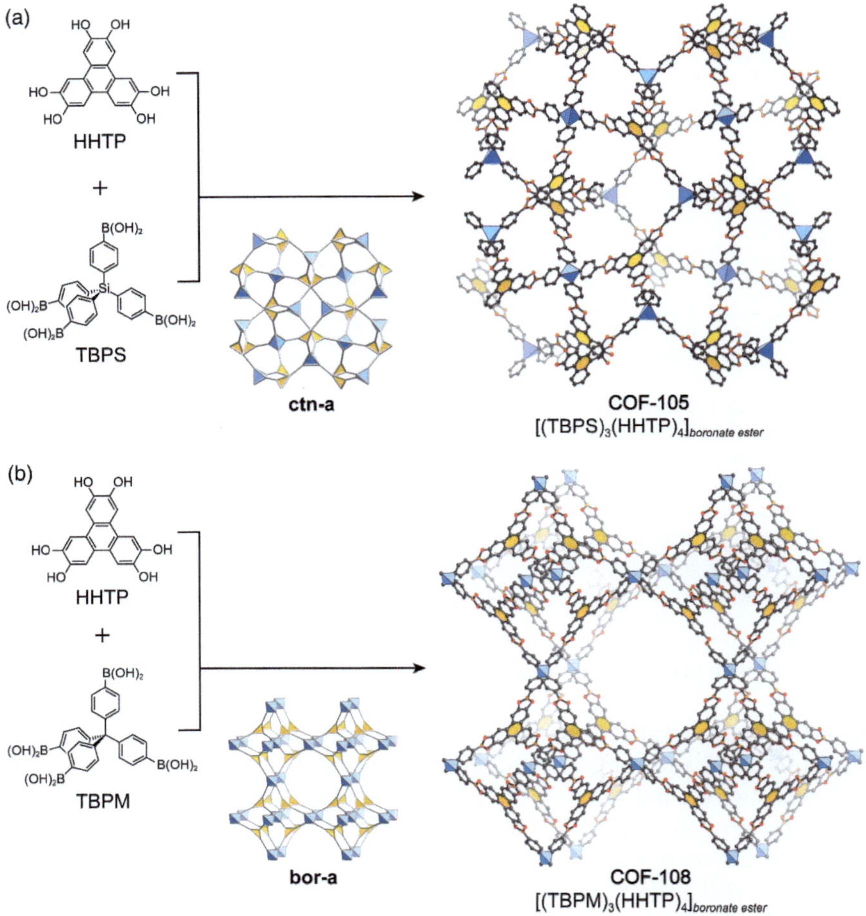

Figure 7.9 Synthetic schemes for COF-108 and COF-105. (a) Reticulation of TBPM with HHTP through boronate ester bonds affords COF-108 (**bor**). (b) Reticulation of TBPS with HHTP affords COF-105 (**ctn**). All hydrogen atoms are omitted for clarity. Tetrahedral carbon and silicon, blue; core of the tritopic linker, orange polygons. Color Code: B, orange; C, gray; O, red.

7.7 Summary

In this chapter we covered the historical development of COFs starting from the conceptual elaboration of the covalent bond by G. N. Lewis. The concept of covalent bonding in organic molecules laid the conceptual foundation that transformed organic chemistry from an observation driven to a systematic and rational field of research. The twentieth century has seen a rapid expansion of the synthetic toolbox of organic chemists as evidenced by the retrosynthesis of complex natural products such as Taxol and Vitamin B_{12}. The large majority of organic transformations employed in such syntheses are kinetically controlled and microscopically irreversible, which hinders their application in the formation of crystalline extended 2D and 3D frameworks. With the advent of supramolecular chemistry, concerned with molecular recognition and self-assembly through microscopically reversible weak intermolecular interactions, an interest in the formation of large macromolecular organic assemblies emerged. The development of dynamic covalent chemistry served as precedent that assembly processes using strong covalent bonds are feasible. The synthetic realization of COF-1 and COF-5, linked by boroxine and boronate ester linkages, respectively, was the first example of reticulation of organic building units into crystalline porous 2D COFs. Subsequently, the generality of this approach toward the synthesis of organic extended structures was demonstrated by the designed synthesis of 3D COFs: COF-102, COF-103, COF-105, and COF-108. In Chapter 8 we will discuss the different linkage chemistries that have since been developed according to the fundamental principles derived in this chapter.

References

1 Waller, P.J., Gándara, F., and Yaghi, O.M. (2015). Chemistry of covalent organic frameworks. *Accounts of Chemical Research* 48 (12): 3053–3063.

2 Schoedel, A. and Yaghi, O.M. (2016). Porosity in metal-organic compounds. In: *Macrocyclic and Supramolecular Chemistry: How Izatt-Christensen Award Winners Shaped the Field*, vol. 2, 200–219. Weinheim: Wiley-VCH.

3 Lewis, G.N. (1916). The atom and the molecule. *Journal of the American Chemical Society* 38 (4): 762–785.

4 Hoffmann, R. (1993). How should chemists think? *Scientific American* 268 (2): 66–73.

5 (a) Côté, A.P., Benin, A.I., Ockwig, N.W. et al. (2005). Porous, crystalline, covalent organic frameworks. *Science* 310 (5751): 1166–1170. (b) El-Kaderi, H.M., Hunt, J.R., Mendoza-Cortés, J.L. et al. (2007). Designed synthesis of 3D covalent organic frameworks. *Science* 316 (5822): 268–272.

6 Diercks, C.S. and Yaghi, O.M. (2017). The atom, the molecule, and the covalent organic framework. *Science* 355 (6328): eaal1585.

7 Muller, P. (1994). Glossary of terms used in physical organic chemistry (IUPAC recommendations 1994). *Pure and Applied Chemistry* 66 (5): 1077–1184.

8 Helmholtz, P. (1881). On the modern development of Faraday's conception of electricity. *Science* 2 (43): 182–185.

9 Staudinger, H. (1920). Über polymerisation. *European Journal of Inorganic Chemistry* 53 (6): 1073–1085.

10 Wöhler, F. (1828). Ueber künstliche bildung des harnstoffs. *Annalen der Physik* 88 (2): 253–256.

11 (a) Willstätter, R. and Bode, A. (1901). Ueberführung von tropinon in r cocain. *European Journal of Inorganic Chemistry* 34 (2): 1457–1461. (b) Komppa, G. (1903). Die vollständige synthese der camphersäure und dehydrocamphersäure. *European Journal of Inorganic Chemistry* 36 (4): 4332–4335.

12 Woodward, R.B., Cava, M.P., Ollis, W.D. et al. (1954). The total synthesis of strychnine. *Journal of the American Chemical Society* 76 (18): 4749–4751.

13 Woodward, R.B. (1973). The total synthesis of vitamin B_{12}. *Pure and Applied Chemistry* 33 (1): 145–178.
14 Nicolaou, K.C., Yang, Z., Liu, J.J. et al. (1994). Total synthesis of taxol. *Nature* 367 (6464): 630–634.
15 (a) Baeyer, A.v. and Emmerling, A. (1870). Reduction des isatins zu indigoblau. *European Journal of Inorganic Chemistry* 3 (1): 514–517. (b) Kolbe, H. (1845). Beiträge zur kenntniss der gepaarten verbindungen. *European Journal of Organic Chemistry* 54 (2): 145–188. (c) Ladenburg, A. (1886). Synthese der activen coniine. *European Journal of Inorganic Chemistry* 19 (2): 2578–2583. (d) Fischer, E. (1890). Synthesen in der zuckergruppe. *European Journal of Inorganic Chemistry* 23 (2): 2114–2141. (e) Fischer, H. and Zeile, K. (1929). Synthese des haematoporphyrins, protoporphyrins und haemins. *European Journal of Organic Chemistry* 468 (1): 98–116. (f) Gates, M. and Tschudi, G. (1952). The synthesis of morphine. *Journal of the American Chemical Society* 78 (7): 1380–1393. (g) du Vigneaud, V., Ressler, C., Swan, J.M. et al. (1954). The synthesis of oxytocin 1. *Journal of the American Chemical Society* 76 (12): 3115–3121. (h) Corey, E.J., Weinshenker, N.M., Schaaf, T.K., and Huber, W. (1969). Stereo-controlled synthesis of dl-prostaglandins F2.alpha. and E2. *Journal of the American Chemical Society* 91 (20): 5675–5677. (i) Sheehan, J.C. and Henery-Logan, K.R. (1957). The total synthesis of penicillin V. *Journal of the American Chemical Society* 81 (12): 3089–3094. (j) Corey, E.J., Kim, S., Yoo, S.-E. et al. (1978). Total synthesis of erythromycins. 4. Total synthesis of erythronolide B. *Journal of the American Chemical Society* 100 (14): 4620–4622. (k) Perkin, W.H. (1904). LXVI.—Experiments on the synthesis of the terpenes. Part I. Synthesis of terpin, inactive terpineol, and dipentene. *Journal of the Chemical Society, Transactions* 85: 654–671. (l) Hoffmann, F. (1898). Acetyl salicylic acid. US 644077.
16 Corey, E.J. (1989). *The Logic of Chemical Synthesis.* Рипол Классик.
17 (a) Lehn, J.-M. (1995). *Supramolecular Chemistry.* Weinheim: Wiley-VCH. (b) Lehn, J.M. (1990). Perspectives in supramolecular chemistry – from molecular recognition towards molecular information processing and self-organization. *Angewandte Chemie International Edition* 29 (11): 1304–1319. (c) Reinhoudt, D. and Crego-Calama, M. (2002). Synthesis beyond the molecule. *Science* 295 (5564): 2403–2407.
18 (a) Pedersen, C.J. (1967). Cyclic polyethers and their complexes with metal salts. *Journal of the American Chemical Society* 89 (26): 7017–7036. (b) Pedersen, C.J. (1988). The discovery of crown ethers (noble lecture). *Angewandte Chemie International Edition* 27 (8): 1021–1027.
19 (a) Dietrich, B., Lehn, J.-M., Sauvage, J.P., and Blanzat, J. (1973). Cryptates – X: syntheses et proprietes physiques de systemes diaza-polyoxa-macrobicycliques. *Tetrahedron* 29 (11): 1629–1645. (b) Lehn, J.-M. (1988). Supramolecular chemistry – scope and perspectives molecules, supermolecules, and molecular devices (nobel lecture). *Angewandte Chemie International Edition* 27 (1): 89–112.

20 (a) Lehn, J.-M. (2002). Toward complex matter: supramolecular chemistry and self-organization. *Proceedings of the National Academy of Sciences* 99 (8): 4763–4768. (b) Cram, D.J. and Cram, J.M. (1997). *Container Molecules and Their Guests*. Cambridge, U.K: Royal Society of Chemistry.

21 (a) Hasenknopf, B., Lehn, J.-M., Kneisel, B.O. et al. (1996). Self-assembly of a circular double helicate. *Angewandte Chemie International Edition in English* 35 (16): 1838–1840. (b) Hasenknopf, B., Lehn, J.-M., Boumediene, N. et. al. (1997). Self-assembly of tetra- and hexanuclear circular helicates. *Journal of the American Chemical Society* 119 (45): 10956–10962.

22 (a) Kinoshita, Y., Matsubara, I., Higuchi, T., and Saito, Y. (1959). The crystal structure of bis(adiponitrilo)copper(I) nitrate. *Bulletin of the Chemical Society of Japan* 32 (11): 1221–1226. (b) Desiraju, G.R. (2001). Chemistry beyond the molecule. *Nature* 412 (6845): 397–400.

23 Lehn, J.-M. (1999). Dynamic combinatorial chemistry and virtual combinatorial libraries. *Chemistry – A European Journal* 5 (9): 2455–2463.

24 Rowan, S.J., Cantrill, S.J., Cousins, G.R. et al. (2002). Dynamic covalent chemistry. *Angewandte Chemie International Edition* 41 (6): 898–952.

25 Dietrich-Buchecker, C.O., Sauvage, J.P., and Kintzinger, J.P. (1983). Une nouvelle famille de molecules: les metallo-catenanes. *Tetrahedron Letters* 24 (46): 5095–5098.

26 Chichak, K.S., Cantrill, S.J., Pease, A.R. et al. (2004). Molecular borromean rings. *Science* 304 (5675): 1308–1312.

27 (a) Ponnuswamy, N., Cougnon, F.B., Clough, J.M. et al. (2012). Discovery of an organic trefoil knot. *Science* 338 (6108): 783–785. (b) Leigh, D.A., Pritchard, R.G., and Stephens, A.J. (2014). A star of david catenane. *Nature Chemistry* 6 (11): 978–982.

28 Diercks, C.S., Kalmutzki, M.J., and Yaghi, O.M. (2017). Covalent organic frameworks – organic chemistry beyond the molecule. *Molecules* 22 (9): 1575.

29 O, Keeffe, M. (2009). Design of MOFs and intellectual content in reticular chemistry: a personal view. *Chemical Society Reviews* 38 (5): 1215–1217.

30 (a) Ding, S.-Y. and Wang, W. (2013). Covalent organic frameworks (COFs): from design to applications. *Chemical Society Reviews* 42 (2): 548–568. (b) Feng, X., Ding, X., and Jiang, D. (2012). Covalent organic frameworks. *Chemical Society Reviews* 41 (18): 6010–6022. (c) Furukawa, H. and Yaghi, O.M. (2009). Storage of hydrogen, methane, and carbon dioxide in highly porous covalent organic frameworks for clean energy applications. *Journal of the American Chemical Society* 131 (25): 8875–8883. (d) Lin, C.Y., Zhang, D., Zhao, Z., and Xia, Z. (2018). Covalent organic framework electrocatalysts for clean energy conversion. *Advanced Materials* (5): 30. (e) Biswal, B.P., Chaudhari, H.D., Banerjee, R., and Kharul, U.K. (2016). Chemically stable covalent organic framework (COF)-polybenzimidazole hybrid membranes: enhanced gas separation through pore modulation. *Chemistry – A European Journal* 22 (14): 4695–4699.

8

Linkages in Covalent Organic Frameworks

8.1 Introduction

The decade following the realization of crystalline 2D and 3D covalent organic frameworks (COFs) in 2005 and 2007, respectively, has seen an exponential growth of this field of research. In contrast to metal-organic frameworks (MOFs), which are mostly constructed from carboxylate–metal linkages, a lot of research in the field of COFs has been devoted toward expanding the scope of accessible linkage chemistries [1]. For every new linkage, the "crystallization problem" has to be overcome and conditions for the formation of crystalline framework materials need to be devised. To afford extended crystalline solids, the formation of the linkage between the individual building units needs to be reversible under the given reaction conditions and the reaction rates have to be fast enough to enable sufficient self-correction of defects [2]. Using covalent bonds as linkages presents a special challenge for crystallization due to their tendency to form irreversibly under mild conditions. COFs are constructed from organic molecular building units which need to maintain their structural integrity during the reticulation process. Thus, conditions must be identified where reversible bond formation is possible without resorting to extreme temperatures or pressures so as to avoid destroying the organic building blocks. In the following text, we will cover the linkages that have been employed in the crystallization of COFs and show the challenges that had to be overcome for each new type of linkage [1c].

8.2 B–O Bond Forming Reactions

8.2.1 Mechanism of Boroxine, Boronate Ester, and Spiroborate Formation

Historically, the problems encountered in the crystallization of COFs were first addressed for boroxine and boronate ester linked frameworks using condensation reactions where stoichiometric quantities of water are generated. As described in Chapter 7, crystallinity in COFs based on these linkages is achieved by controlling the amount of water in the system to modulate the equilibrium of the COF forming reaction [1a,b]. In practice, this is accomplished by using a solvent

Introduction to Reticular Chemistry: Metal-Organic Frameworks and Covalent Organic Frameworks,
First Edition. Omar M. Yaghi, Markus J. Kalmutzki, and Christian S. Diercks.

Companion website: www.wiley.com/go/yaghi/reticular

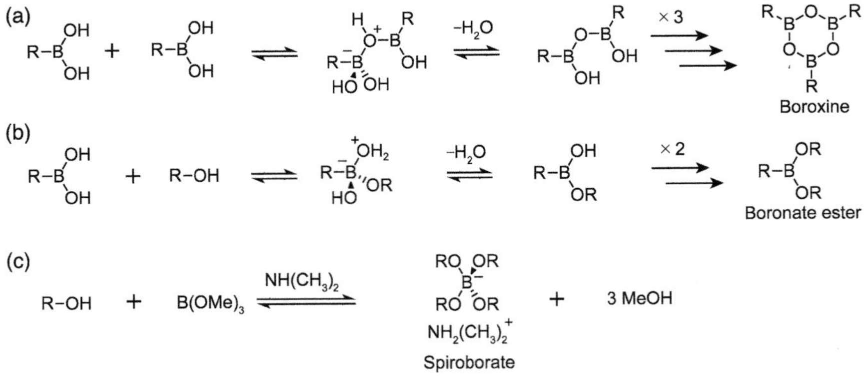

Figure 8.1 Reversible condensation reactions of boronic acids, boronates, and spiroborates. (a) Condensation of three boronic acids yields boroxine molecules. (b) Cross-condensation of a boronic acid with two alcohols or one diol yields boronate esters. (c) Reaction of four alcohols (or two diols) with 1 equiv of trimethoxy borate in the presence of a weak base results in transesterification and the formation of anionic spiroborates. All of the reactions are fully reversible and as such promising candidates for COF forming reactions [3].

system (1 : 1 v/v mesitylene/dioxane), which limits the solubility of the starting materials thus slowing down the reaction, and by carrying out the reactions in a closed system (flame-sealed pyrex tubes) to keep water in the reaction equilibrium (head space of the tube and the solution mixture). The fact that boronic acids can condense reversibly with vicinal diols is well known from research in sugar chemistry [1a,b]. The mechanism of boroxine bond formation starts with the attack of the Lewis acidic boron by a hydroxyl group of another boronic acid molecule. Intermolecular proton transfer and subsequent elimination of water result in the formation of a new B—O bond. Two analogous condensation reactions involving a third boronic acid moiety yield the boroxine unit (Figure 8.1a). All steps in the mechanism of boroxine bond formation are fully reversible.

Boronate ester bond formation also relies on reversible condensation reactions of boronic acids but here the nucleophilic hydroxyl group that attacks the boron center is an alcohol, and reversible condensation of two alcohols (or one diol) with the boronic acid yields boronate esters (Figure 8.1b).

Finally, alcohols can condense with trimethyl borate in the presence of a weak base to yield anionic spiroborates and methanol as a by-product (Figure 8.1c) [3]. In the following text we will show how these microscopically reversible reactions are used to expand the scope of accessible COFs beyond boroxine and boronate ester linkages [3].

8.2.2 Borosilicate COFs

In Chapter 7, we discussed the formation of the boronate ester linked 3D frameworks COF-105 and COF-108. Boronate ester formation is not restricted to the cross-condensation of boronic acids with alcohols and can also involve silanols to

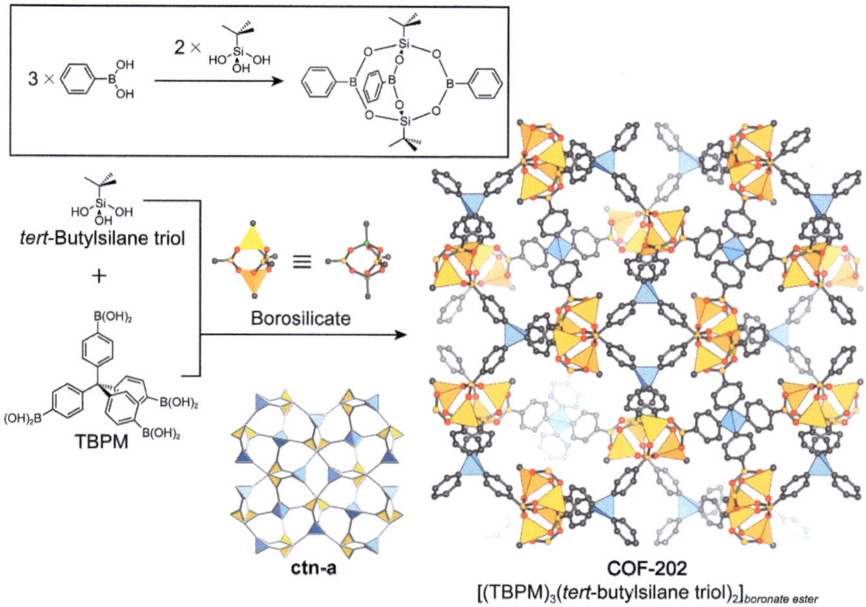

Figure 8.2 Synthetic scheme for the formation of the borosilicate linked framework COF-202. Reticulation of *tert*-butylsilane triol with TBPM by borosilicate bond formation yields the **ctn** topology framework COF-202. The bond forming borosilicate reaction is depicted in the inset. All hydrogen atoms are ommited for clarity. The triangles in **ctn-a** correspond to the three boron atoms in the borosilicate unit. Tetrahedral C, blue tetrahedra; tetrahedral Si, orange tetrahedra. Color code: B, orange; C, gray; O, red; Si, green (in the ball and stick representation of the borosilicate unit).

yield borosilicates. This is utilized in the synthesis of the **ctn** topology structure COF-202 ([(TBPM)$_3$(*tert*-butylsilane triol)$_4$]$_{borosilicate}$), a framework constructed from TBPM and *tert*-butylsilane triol (Figure 8.2) [4]. The crystallization conditions for COF-202 are similar to those of the related boronate ester COFs. To achieve microscopic reversibility, the reaction is carried out at 120 °C in a 1 : 2 v/v mixture of dioxane/toluene in sealed pyrex tubes. Again, the elevated temperature is necessary to promote dynamic error correction during bond formation, the sealed pyrex tubes are needed to keep water in the reaction equilibrium, and the solvent mixture is chosen such that the starting materials are not fully soluble to slow down the crystallization process. The formation of the B–O–Si linkage in COF-202 is confirmed by FT-IR spectroscopy where a characteristic B–O–Si stretch at 1310 cm^{-1}, as well as the expected attenuation in the hydroxyl stretching band of the *tert*-butylsilane triol and the TBPM starting materials are observed. The completeness of the reaction is further corroborated by ^{11}B multiple-quantum magic-angle spinning NMR data of COF-202, which is in good agreement with the spectrum obtained for a discrete molecular model compound. COF-202 crystallizes in the cubic space group $I\bar{4}_3d$ and has a high BET surface area of 2690 m^2 g^{-1}, as determined by argon sorption at 87 K.

8.2.3 Spiroborate COFs

The formation of the **sql** topology spiroborate framework ICOF-1 ([(OHM)(TMB)$_2$]$_{spiroborate}$) is achieved by the transesterification reaction of the octahydroxy functionalized macrocycle OHM with TMB (trismethoxy borate) (Figure 8.3). As shown in Figure 8.1c, the bond formation of spiroborates from diols and borates is fully reversible in the presence of a weak base. To determine the conditions for reversible bond formation, a model reaction between the monotopic OHM analog (3,6-di(prop-1-yn-1-yl)-9*H*-fluorene-9,9-diyl) dimethanol) and TMB is evaluated. The reaction proceeds to completion in 60 minutes at room temperature in the presence of dimethyl amine as a base, whereas in the absence of the base a more sluggish reaction is observed. To crystallize ICOF-1, the reaction is thus carried out in DMF (*N*,*N*′-dimethylformamide) as a solvent at 120 °C for seven days. DMF slowly decomposes at this temperature, thus gradually releasing dimethylamine into the reaction mixture. The *in situ* generation of base ensures the slow controlled nucleation and thereby promotes crystallization. In the ^{13}C cross-polarization magic angle spinning (CP-MAS) NMR spectrum of ICOF-1, singlets at 54.5, 70.2, and 90.1 ppm are assigned to the quaternary carbon of the fluorene units, the methylene carbon adjacent to the hydroxyl groups, and the ethynylene carbons, respectively. A singlet at 38.6 ppm is assigned to the methyl carbon of $[Me_2NH_2]^+$ thus corroborating the presence of the counter ion in the pores of the framework. ICOF-1 shows permanent porosity albeit with a BET surface area of only 210 m^2 g^{-1}, attributed to the large cations residing in the pores of the framework. Consequently, the exchange of the bulky $[Me_2NH_2]^+$ counter ions

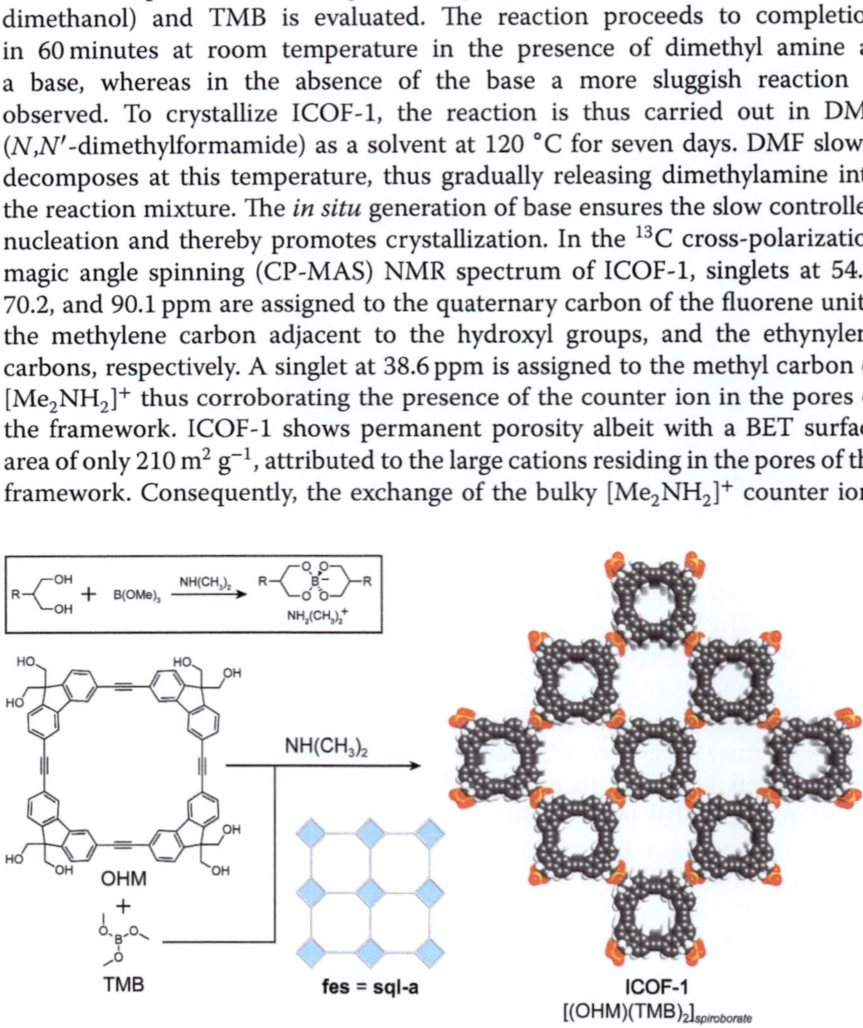

Figure 8.3 Synthetic Scheme for the formation of the anionic spiroborate linked framework, ICOF-1, constructed from macrocyclic tetratopic OHM building units that are reticulated with TMB to yield a framework of **sql** topology (The space group and structural parameters of ICOF-1 were not reported. For visualization, the structure was modeled in the tetragonal space group *P*422.). The spiroborate formation reaction is depicted in the inset. Color code: H, white; B, orange; C, gray, O, red.

with smaller Li^+ ions increases the surface area to $1022\,m^2\,g^{-1}$. Investigation of the chemical stability of the spiroborate COF toward water and base reveals that ICOF-1 is substantially more stable under those conditions as compared to boroxine-, boronate ester-, or borosilicate- linked COFs. No significant changes in the BET surface area and the powder X-ray diffraction (PXRD) pattern of the material are observed after immersion of the sample in water or 1 M aqueous LiOH solution for two days. This can be rationalized based on the additional Lewis base coordination to boron and the chelating nature of the spirocyclic ester rendering hydrolysis pathways less favorable [5].

8.3 Linkages Based on Schiff-Base Reactions

Schiff-base reactions are among the most extensively studied reactions in dynamic covalent chemistry. The advantage of these reactions is that reversibility in the bond formation is achieved by employing acid as a catalyst. Consequently, the formed bonds are relatively stable at neutral pH, even in the presence of water. The mechanism of Schiff-base reactions is generally initiated by protonation of a carbonyl group (i.e. aldehyde, ketone) to render the carbonyl carbon more electrophilic. Subsequent nucleophilic attack by an amine group as well as intramolecular proton transfer yield a hemiaminal intermediate that forms the Schiff-base upon elimination of water [6]. Depending on the substituent on the amine moiety this chemistry can yield imines, hydrazones, azines, β-ketoenamies, phenazines, or benzoxazoles and we will discuss COFs formed through these linkages in this section (see Figure 8.4).

8.3.1 Imine Linkage

8.3.1.1 2D Imine COFs

To enhance the chemical stability of COFs and render them chemically more inert, reversible imine condensation reactions can be employed as linkages [7].

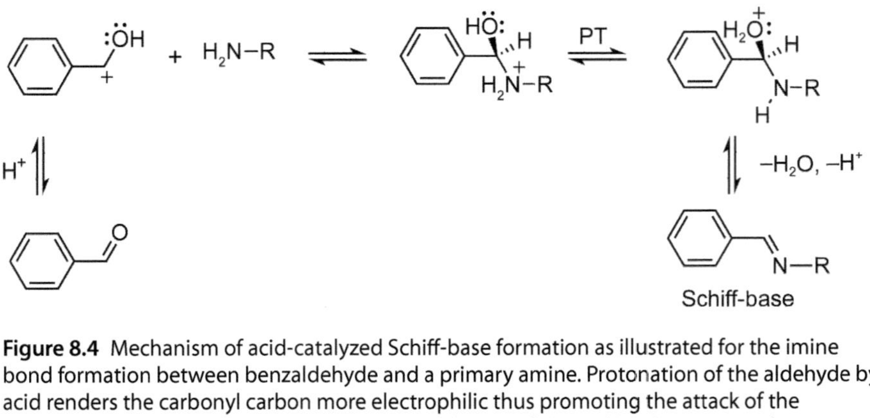

Figure 8.4 Mechanism of acid-catalyzed Schiff-base formation as illustrated for the imine bond formation between benzaldehyde and a primary amine. Protonation of the aldehyde by acid renders the carbonyl carbon more electrophilic thus promoting the attack of the nucleophilic amine. Intramolecular proton transfer (PT), followed by elimination of water yields the targeted Schiff-base.

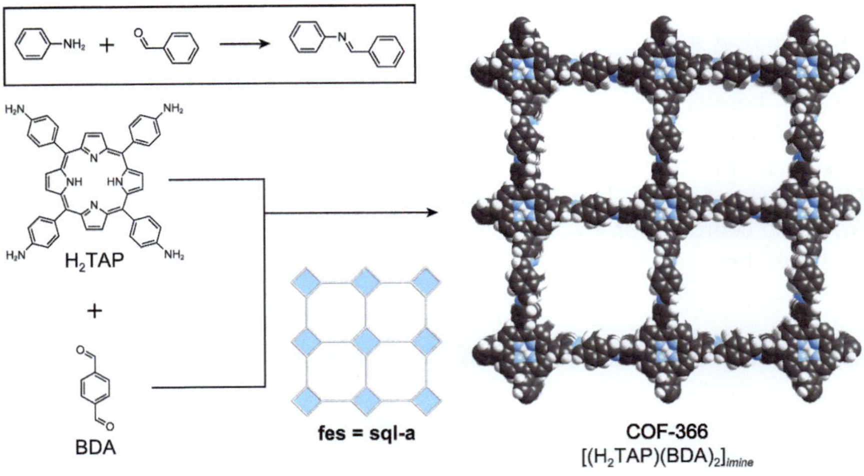

Figure 8.5 Synthetic Scheme for the formation of COF-366, the first 2D imine COF. Reticulation of H_2TAP with BDA affords COF-366 with an underlying **sql** net. The imine bond forming reaction is depicted in the inset. Color code: H, white; C, gray, N, blue.

The imine linked framework COF-366 ([(H_2TAP)(BDA)$_2$]$_{imine}$) is constructed from square planar H_2TAP (5,10,15,20-tetrakis(4-amino-phenyl)porphyrin) and linear BDA (terephthaldehyde) units, which are reacted in a 1 : 1 v/v mixture of mesitylene and dioxane at 120 °C to form a framework of **sql** topology (Figure 8.5). To enhance the reversibility of the reaction, 6 M aqueous acetic acid is added as a catalyst. To ensure that the equilibrium is not shifted by the loss of water from the reaction mixture, the reaction is carried out in sealed pyrex tubes as discussed in the context of the condensation reactions of boroxine and boronate ester linked COFs. Upon activation, COF-366 features 2-nm-wide open tetragonal channels along the crystallographic *c*-axis and the open, permanently porous structure translates into a BET surface area of 735 m^2 g^{-1}. The completion of the reaction of imine-based COFs is confirmed by FT-IR, where the carbonyl stretch of BDA at 1590 cm^{-1} and the amine stretch of H_2TAP at 3174 cm^{-1} disappear upon reticulation and a new imine stretch attributed to the formed imine bond appears at 1620 cm^{-1}. 2D imine linked COFs attract a lot of attention because they tend to feature extended π-conjugated systems, which hold promise for applications in the field of organic electronics and semiconductors. COF-366 is a hole-conductor with a charge carrier mobility of 8.1 cm^2 V^{-1} s^{-1}, exceeding the performance of silicon, the state-of-the-art material in the semiconductor industry [8].

The combination of high thermal stability, charge-carrier mobility, electrical conductivity, and pore accessibility holds great promise for the application of 2D imine linked COFs in organic electronics and optoelectronic systems. In this context, it is preferable to have well-defined structures with long-range order to minimize the number of grain boundaries within the material, which is why the low crystallinity of the first generation of 2D imine COFs posed a substantial problem. The lower crystallinity of 2D COFs as compared to their 3D counterparts can be rationalized by the fact that in the structure the layers are held together only by weak, non-directional interactions and, consequently, layered COFs suffer from inherent stacking disorder rendering the crystallization of these materials more challenging. Consequently, periodicity in layered COF structures was initially only achieved on smaller length scales of typically a few tens to hundreds of nanometers.

To address the apparent shortcomings of the first generation of 2D imine COFs with regard to crystallinity, a number of strategies have been devised: (i) The use of monodentate aromatic amines (aniline) as additives, analogous to modulators in the synthesis of MOFs, helps to improve the crystallinity of 2D COFs. (ii) Addition of Lewis acidic scandium triflate catalyst speeds up the formation of imine bonds in molecular systems by orders of magnitude (Chapter 11) [9]. When adapted to COF formation by imine condensation, the enhanced microscopic reversibility not only results in the formation of highly crystalline 2D COFs but further allows for the synthesis of COFs at ambient conditions (room temperature) as opposed to high reaction temperatures required when organic acids were used as catalysts. (iii) The use of propeller-shaped building units serving as molecular docking sites minimize the occurrence of stacking faults, thus guiding the attachment of building units between layers to improve crystallinity [10].

8.3.1.2 3D Imine COFs

An example of a 3D imine linked framework is COF-300 ($[(TAM)(BDA)_2]_{imine}$) [11]. Here, the tetrahedral tetratopic building unit TAM (tetra-(4-aminophenyl) methane) is reticulated with BDA to form a fivefold interpenetrated framework of **dia** topology (Figure 8.6). The crystallinity of the material is confirmed by PXRD

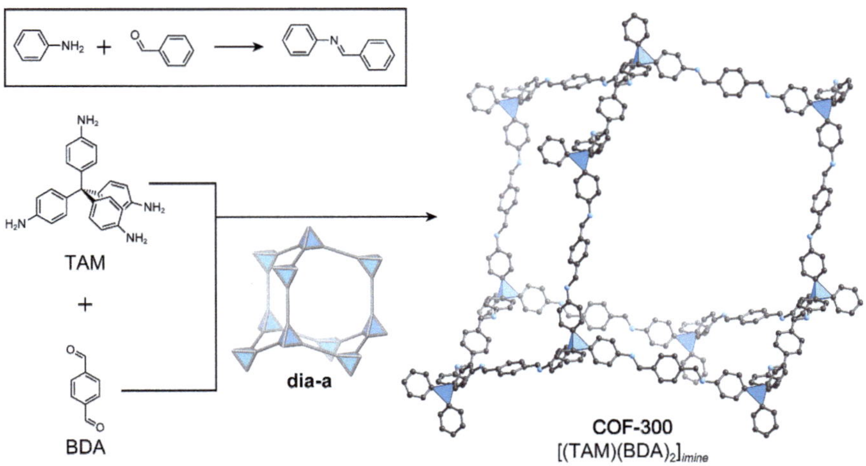

Figure 8.6 Synthetic scheme for COF-300. The **dia** topology of the fivefold interpenetrated framework is formed by reticulation of tetratopic TAM with ditopic BDA building units. The bond forming reaction is depicted in the inset. All hydrogen atoms and interpenetrating frameworks are omitted for clarity. Tetrahedral C atoms, blue. Color code: C, gray, N, blue.

and COF-300 crystallizes in the cubic space group $I4_1/a$. The crystal structure of COF-300 features 7.8 Å large channels propagating along the crystallographic *c*-axis and after activation under dynamic vacuum the material shows permanent porosity with a BET surface area of 1360 m^2 g^{-1}. Since the realization of this first imine COF, imines have become the most frequently employed linkage. The reaction conditions established in this first report are still used today in the synthesis of imine linked COFs, and only minor modifications have been implemented.

The formation and cleavage of covalent bonds involves multiple reaction partners and proceeds through several intermediates. Thus, these processes will inherently be more complex and typically less reversible than the mechanistically simple Lewis acid base reactions leading to the formation of MOFs. In the synthesis of imine COFs, the formation of an extended amorphous solid is observed at first, and only over time the material is transformed into a pure, crystalline phase by error correction mechanisms attributed to the microscopic reversibility of the bond formation under the selected reaction conditions [12]. As a consequence, the COF crystallites prepared by such reactions are generally only nanosized and structure determination is therefore commonly achieved by analysis of PXRD data, substantiated by structural models devised based on general geometry considerations. Without analyzing single-crystal samples of COFs, some uncertainty still remains regarding their precise atomic structure. This issue can be addressed by solving and refining the single-crystal structure of a COF from data collected by RED (3D rotation electron diffraction), a technique allowing for the unambiguous structure determination of micro- and nanosized single crystals [13]. COF-320 ($[(TAM)(BPDA)_2]_{imine}$, where BPDA = 4,4′-biphenyldialdehyde), an isoreticular expanded analog of COF-300, is prepared by cross-condensation of TAM with BPDA to yield an imine linked 3D COF with underlying **dia** topology. Using RED at 298 K, the crystal structure of COF-320 can unambiguously be determined. The dimensions of the adamantane cages in the framework are 28 × 31 × 71 Å and as a consequence of the large open cages the structure is isolated as a ninefold interpenetrated framework. The crystal structure shows square-shaped 1D channels with an aperture size of 11.5 × 11.5 Å running along the *c*-direction. Comparison of the PXRD pattern of an activated sample of COF-320 with that calculated from single crystal data shows good agreement, indicating the formation of a pure phase. The importance of solving the single crystal structure of COF-320 is highlighted by the fact that due to the high degree of interpenetration and the pronounced framework distortion, PXRD data and computer modeling alone cannot provide an unambiguous structure solution.

Imine bond formation can not only be used in the synthesis of COFs but is also applicable to MOFs. The synthesis of MOFs requires the knowledge of the reaction conditions that are suitable for the formation of the targeted secondary building unit (SBU), as well as for the reversible reticulation of these SBUs with the linker. In contrast, for COFs, once reaction conditions for the reversible formation of a particular linkage are found, they can be broadly applied to a large variety of structure types built from many chemically distinct building units.

The inorganic SBUs in MOFs are desirable building units in the construction of frameworks as they endow the material with both structural diversity and functionality. MOF-901 ($Ti_6O_6(OCH_3)_6(AB)_6$), where AB = 4-aminobenzoate) combines the best of both, MOFs and COFs [14]. It is synthesized by linking amine decorated titanium oxo-clusters ($Ti_6O_6(OCH_3)_6(AB)_6$, that are formed *in situ* and are subsequently linked into a 2D extended structure of **hxl** topology by imine condensation with BDA. This strategy shows that MOF and COF chemistry should not be regarded as separate disciplines; in fact, there are no boundaries between these fields of research and, when combined, there still remains a lot to be explored under the umbrella of reticular chemistry.

8.3.1.3 Stabilization of Imine COFs Through Hydrogen Bonding

From the very beginning of COF chemistry, research was focused on enhancing the chemical stability of COFs. One approach is the stabilization of imines through hydrogen bonding [15]. Reticulation of square planar tetratopic H_2TAP with linear ditopic BDA-$(OH)_2$ (2,5-dihydroxy-1, 4-benzenedialdehyde) building units affords $[(H_2TAP)(BDA\text{-}(OH)_2)_2]_{imine}$ of **sql** topology (Figure 8.7). The reaction is carried out under conditions similar to those for the parent COF-366 with minor modification to the solvent system. Instead of the 1 : 1 v/v mixture of dioxane and mesitylene, a 1 : 1 v/v mixture of ethanol and *o*-dichlorobenzene is employed in the synthesis. The more polar solvent mixture likely improves the solubility of the starting materials thereby

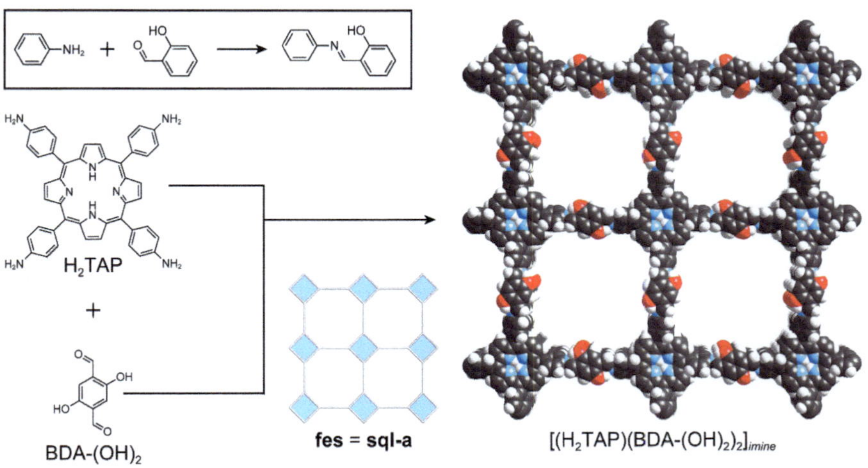

Figure 8.7 Imine linked $[(H_2TAP)(BDA\text{-}(OH)_2)_2]_{imine}$ stabilized by hydrogen bonding of hydroxyl groups on the linker with the imine bond. In the synthesis, square planar tetratopic H_2TAP building units are stitched together with linear ditopic BDA-$(OH)_2$ linkers bearing hydroxyl groups in the *o*-position of the aldehyde functionality. Upon reticulation, the imines are hydrogen bonded to the pre-synthetically installed hydroxyl groups to stabilize the framework. The hydrogen bonding motif is depicted in the inset. All hydrogen atoms are omitted for clarity. Color code: H, white; C, gray; N, blue; O, red.

enhancing the reversibility of the imine bond formation and thus improving the crystallinity of the resulting material.

$[(H_2TAP)(BDA\text{-}(OH)_2)_2]_{imine}$ has the same pore metrics and crystallizes in the same space group as COF-366. However, the stability toward acids is increased substantially. Boiling $[(H_2TAP)(BDA\text{-}(OH)_2)_2]_{imine}$ in water for three days and even exposure to 3N hydrochloric acid for one week has no apparent effect on the crystallinity of the sample and only a 5% weight loss is recorded for the treated material. The stability of the COF under these conditions is furthermore substantiated by nitrogen adsorption measurements carried out before and after treatment with water and acid. In the case of treatment with water, the BET surface area of 1305 $m^2\ g^{-1}$ is retained. Treatment with acid leads to a loss in surface area of the COF and the resulting material only retains a surface area of 570 $m^2\ g^{-1}$. While the loss in surface area shows the limitations of this hydrogen-bonding approach, the increase in stability of the material is still remarkable when compared to the parent unfunctionalized COF-366 where complete structural degradation is observed after treatment under identical conditions.

8.3.1.4 Resonance Stabilization of Imine COFs

Hydrogen bonding is a viable approach toward enhancing the chemical stability of the imine linkage itself within the layers of 2D COFs. Decomposition can however also occur due to delamination of the COF layers [16]. It is believed that in layered COFs with eclipsed conformations the polarized imine bonds in adjacent layers repel each other, rendering the interactions between adjacent layers less strong. An approach to reduce this electronic repulsion is based on resonance stabilization. The **hcb** topology framework $[(TAPB)_2(BDA\text{-}(OMe)_2)_3]_{imine}$ is constructed from trigonal tritopic TAPB (1,3,5-tris(4-aminophenyl)benzene) and linear ditopic BDA-$(OMe)_2$ (2,5-dimethoxy-1,4-terephthaldehyde) building units (Figure 8.8) [17]. The methoxy groups on the aromatic ring in close proximity to the imine bond donate electrons into the π system through resonance effects. This directly translates into an electron donating effect on the carbon of the imine bond carrying a partial negative charge, as well as an electron withdrawing effect on the otherwise partially negatively charged nitrogen atom.

Studies exemplify that the combination of these effects decreases the repulsion between adjacent layers and the electrophilicity of the imine bond carbon, rendering it more chemically inert with respect to destructive nucleophilic substitution pathways. Indeed, the resulting material shows high stability to common organic solvents, boiling water, 12 M aqueous HCl, and even 14 M aqueous NaOH solutions. Despite minor weight losses observed after treatment of $[(TAPB)_2(BDA\text{-}(OMe)_2)_3]_{imine}$ with boiling water and 12 M HCl, both, crystallinity and surface area are fully retained. The fact that this material is so chemically robust makes it an interesting platform for various applications: (i) functionalization of $[(TAPB)_2(BDA\text{-}(OMe)_2)_3]_{imine}$ with chiral proline moieties yields an efficient organocatalytic asymmetric C—C bond forming catalyst [17], and (ii) loading the pores of the framework with *N*-heterocyclic proton makes $[(TAPB)_2(BDA\text{-}(OMe)_2)_3]_{imine}$ a heterogeneous solid-state proton

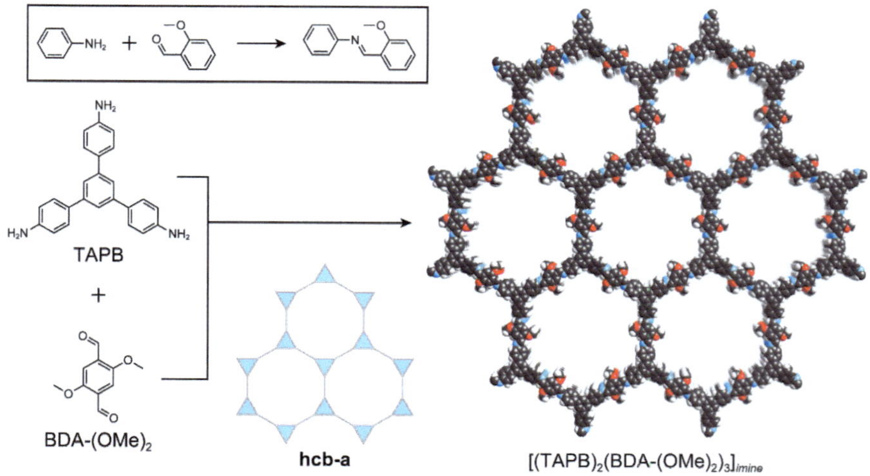

Figure 8.8 Stabilization of imine COFs by resonance effects. The **hcb** topology framework $[(TAPB)_2(BDA\text{-}(OMe)_2)_3]_{imine}$ is formed by reticulation of TAPB with BDA-$(OMe)_2$ through reversible imine bond formation. Upon reaction, the methoxy groups in the *o*-position of the imine bond donate electron density to the partially positively charged imine carbons thus decreasing their nucleophilicity and making them more chemically inert. The decreased positive partial charge on the imine carbon furthermore renders the interlayer interactions more favorable thus enhancing the stability of the framework with respect to delamination. The bond forming reaction is depicted in the inset. Color code: H, white; C, gray; N, blue; O, red.

conductor [18]. However, the stabilizing effect observed for this particular framework is not universally applicable [15].

8.3.2 Hydrazone COFs

Reversible Schiff-base chemistry is not limited to imines but can be further applied to the formation of COFs based on hydrazone linkages by reaction of organic hydrazides with aldehydes. Just as in the case of imine COFs, this reaction is well investigated in the context of dynamic covalent chemistry and reversibility can be achieved under reaction conditions similar to those established for imine COFs. Hydrazone linked COFs are formed under reaction conditions similar to their imine linked counterparts [19]. The reaction of trigonal tritopic TFB (1,3,5-triformyl-benzene) and linear ditopic BDH-$(OEt)_2$ (2,5-diethoxyterephtalohydrazide) building units yields the 2D hydrazone linked COF-42 ($[(TFB)_2(BDH\text{-}(OEt)_2)_2]_{hydrazone}$) with **hcb** topology (Figure 8.9). In analogy to imine-based COFs, the hyrdazone bond formation yielding COF-42 is carried out at 120 °C in a 1 : 1 v/v mixture of mesitylene/dioxane in the presence of catalytic amounts of aqueous acetic acid. The use of mixed solvent systems for the formation of layered COF structures with large aromatic linkers is a common approach in COF chemistry. It is speculated that, for COFs based on Schiff-base chemistry, a polar solvent is required to enhance the reversibility

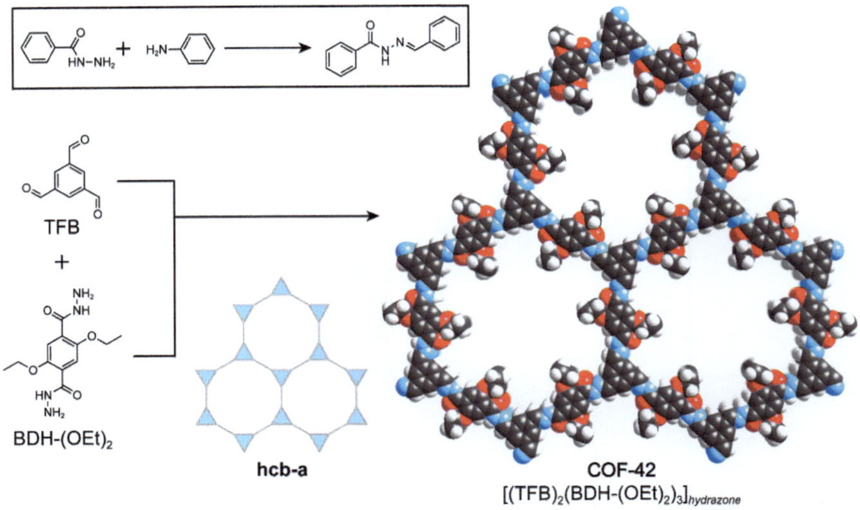

Figure 8.9 Synthetic scheme for the formation of COF-42 constructed by hydrazone bond formation. TFB and BDH-$(OEt)_2$ are reticulated to yield the **hcb** topology framework. The bond forming reaction is depicted in the inset. Color code: H, white; C, gray; N, blue; O, red.

of the hydrazone bond formation whereas an aromatic solvent is used to solubilize the large aromatic building units by disrupting their aggregation thus aiding in the reversible formation of the π–π stacking between adjacent layers. Consequently, since aromatic hydrazides are typically poorly soluble in common organic solvents, solubilizing substituents are appended to the building units. For COF-42, appending two ethoxy groups to the linker increases the solubility enough to allow for the formation of a crystalline product. This approach further enables the activation of COF-42 as it facilitates the removal of unreacted starting materials from the large 28 Å wide channels of the as-synthesized material. After activation under dynamic vacuum, COF-42 shows permanent porosity and a BET surface area of 710 $m^2\ g^{-1}$. An isoreticular expanded version of COF-42, termed COF-43 ($[(TFPB)_2(BDH\text{-}(OEt)_2)_2]_{hydrazone}$), is formed by reticulation of the trigonal tritopic TFPB (1,3,5-tris-(4-formylphenyl)-benzene) with linear ditopic (BDH)-$(OEt)_2$ and feature large channels of 35 Å in diameter. Since hydrazone COFs are more hydrolytically stable than imine COFs, they are used as a platform for visible light driven photocatalytic proton reduction. Continuous hydrogen production from water under visible light is observed for a hydrazone COF loaded with platinum nanoparticles. This hybrid material does not show any signs of degradation over the course of 24 hours of irradiation. The performance of the platinum loaded framework with respect to photocatalytic proton reduction is comparable to the best nonmetal photocatalysts and represents a lightweight, well-ordered model system that can, in principle, be readily tuned using the toolbox of reticular chemistry.

8.3.3 Squaraine COFs

Another approach to expanding the chemical versatility of Schiff-base chemistry in COFs is the development of squaraine linkages. Squaraines are formed from the reaction between SQ (squaric acid) with amines. This chemistry is analogous to the previously reported imine or hydrazone formation, but nonetheless, it is an interesting addition because the resulting linkage has a prominent zwitterionic resonance structure that renders squaraines interesting dyes for applications in areas such as imaging, nonlinear optics, and photovoltaics [20].

The reticulation of square planar tetratopic Cu(TAP) [(5,10,15,20-tetrakis(4-aminophenyl)porphinato]copper) with linear ditopic SQ affords a framework of **sql** topology, termed $[(\text{Cu(TAP)})(\text{SQ})_2]_{squaraine}$ (Figure 8.10) [21]. The material is permanently porous, albeit with a BET surface area of only 539 m^2 g^{-1}, a value that is significantly lower than the calculated surface area of 2289 m^2 g^{-1} indicating incomplete activation of the material. To corroborate the structure model obtained from PXRD data and molecular modeling, the pore size distribution of the material is derived from the experimental sorption data by fitting the isotherm using a nonlocal density functional theory (NLDFT) model. The derived pore-size distribution is in good agreement with the predicted 2.1 nm channels expected from the model of the structure. Deriving the pore-size distribution for COFs is good practice as it gives an additional confirmation of the structure which is otherwise only supported by the obtained PXRD pattern. This becomes especially important when the structure of the material cannot unambiguously be confirmed due to low crystallinity, as is often the

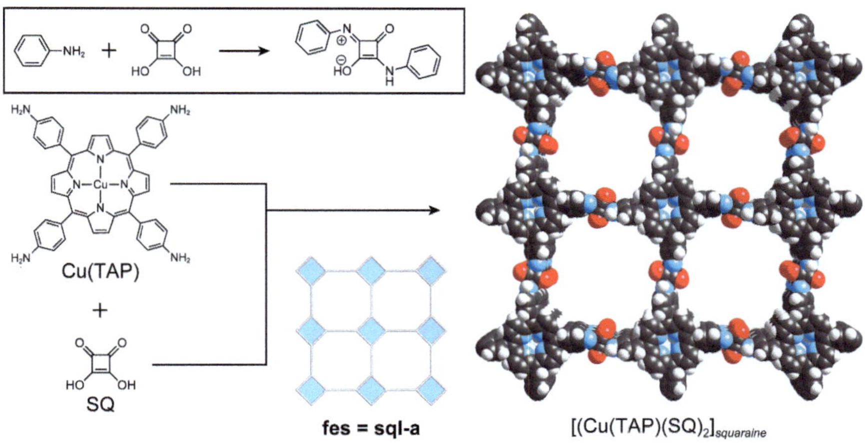

Figure 8.10 Synthesis of $[(\text{Cu(TAP)(SQ)}_2]_{squaraine}$ by reticulation of the square planar tetratopic Cu(TAP) with linear ditopic SQ affording an **sql** topology framework. The framework has a pronounced zwitterionic resonance structure which manifests itself in a decreased band gap and improved light harvesting capacity. The bond forming reaction is depicted in the inset. Color code: H, white; C, gray; N, blue; O, red; Cu, pink.

case for layered materials where stacking between the COF sheets is ill-defined. The pore-size distribution can help distinguish between the staggered and the eclipsed conformation since different pore sizes result from these two scenarios. $[(Cu(TAP)(SQ)_2]_{squaraine}$ is unique in that the squaraine building unit endows it with light-harvesting capacity and a decreased bandgap compared to conventional imine COFs.

8.3.4 β-Ketoenamine COFs

While the previously mentioned strategies are based on improving the stability of imine bonds themselves, a different approach is taken for the synthesis of β-ketoenamine linked COFs [22]. Here, the idea is to form a COF based on the well-established reversible imine chemistry which, upon completion, tautomerizes into an irreversible and thus chemically more inert β-ketoenamine.

To realize this in the formation of COFs, trigonal tritopic TFP (triformylphloroglucinol) is reacted with linear ditopic PDA (1,4-phenylenediamine) under conditions identical to those used in the formation of the first imine COFs: dioxane as the solvent, 3 M aqueous acetic acid as the catalyst, and elevated reaction temperatures of 120 °C in a sealed pyrex tube (Figure 8.11). Initially, an imine bond forms by reaction of the amines of PDA with the aldehydes on the TFP linker. Upon formation of all three imine bonds, an irreversible keto-enol tautomerization takes place to transform the imine bonds into chemically more inert β-ketoenamine linkages. The formation of the linkage is confirmed by FT-IR where upon completion of the COF formation the imine stretch disappears and instead a prominent C=C stretch at 1578 cm^{-1} and a C–N stretch at 1255 cm^{-1}

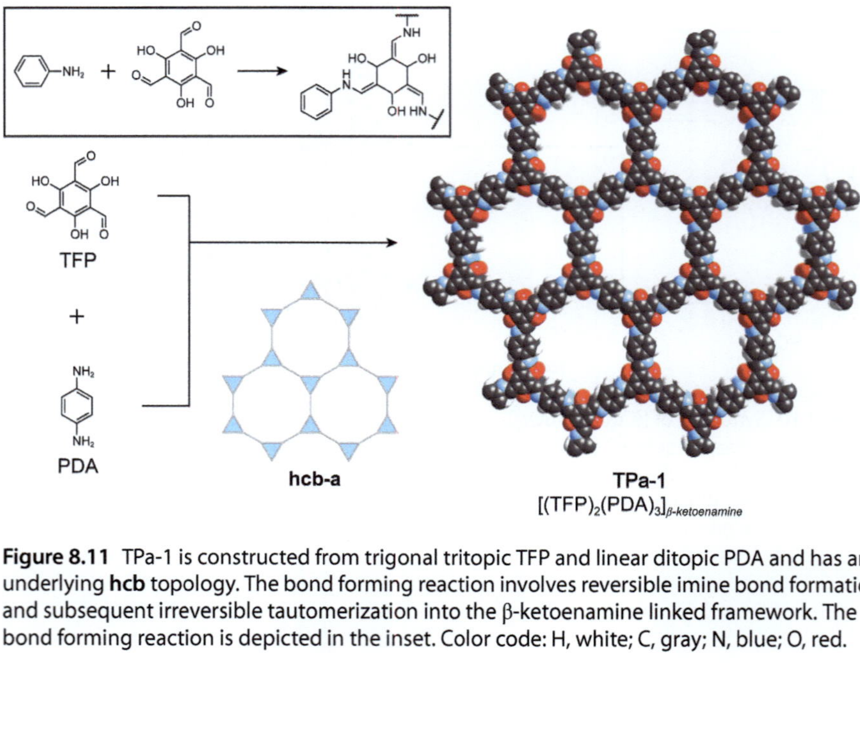

Figure 8.11 TPa-1 is constructed from trigonal tritopic TFP and linear ditopic PDA and has an underlying **hcb** topology. The bond forming reaction involves reversible imine bond formation and subsequent irreversible tautomerization into the β-ketoenamine linked framework. The bond forming reaction is depicted in the inset. Color code: H, white; C, gray; N, blue; O, red.

are observed. The tautomerization affords a COF with **hcb** topology, termed TPa-1 ($[(TFP)_2(PDA)_3]_{\beta\text{-}ketoenamine}$). The material is stable in both 9 M aqueous HCl and 9 M aqueous NaOH. The combination of a reversible followed by an irreversible step is a viable approach for the crystallization of chemically robust COFs as it circumvents the inherent disadvantage of microscopic reversibility in the context of framework stability. TPa-1 has a moderate BET surface area of 530 m^2 g^{-1} and its structure encompasses large 1.8 nm wide hexagonal channels propagating along the crystallographic *c*-axis. This, in combination with the unprecedented chemical stability toward both acid and bases makes hydrazone linked COFs interesting candidates for incorporation into devices. Isoreticular analogs of TPa-1 were employed as capacitive energy storage materials [23].

8.3.5 Phenazine COFs

Different linkages based on reversible imine condensations can be accessed not only by tautomerization, but also by the reaction of vicinal di-ketones with 1,2 diamines, which results in the formation of an aromatic pyrazine moiety by two consecutive imine condensations. Here, the second condensation step of the reaction is irreversible due to stabilization of the product by aromaticity.

The reaction of trigonal tritopic HATP (2,3,6,7,10,11-hexaaminoterphenylene) with linear ditopic PT (*tert*-butyl pyrene tetraone) leads to the formation of CS-COF ($[(HATP)_2(PT)_3]_{phenazine}$), which is built from 2D **hcb** layers stacked in an eclipsed fashion (Figure 8.12) [24]. The reaction conditions of CS-COF differ significantly from those of other imine COFs. 3 M aqueous acetic acid is used as the catalyst and the reaction is carried out in ethylene glycol as the solvent to allow for higher reaction temperatures of 160 °C and to avoid

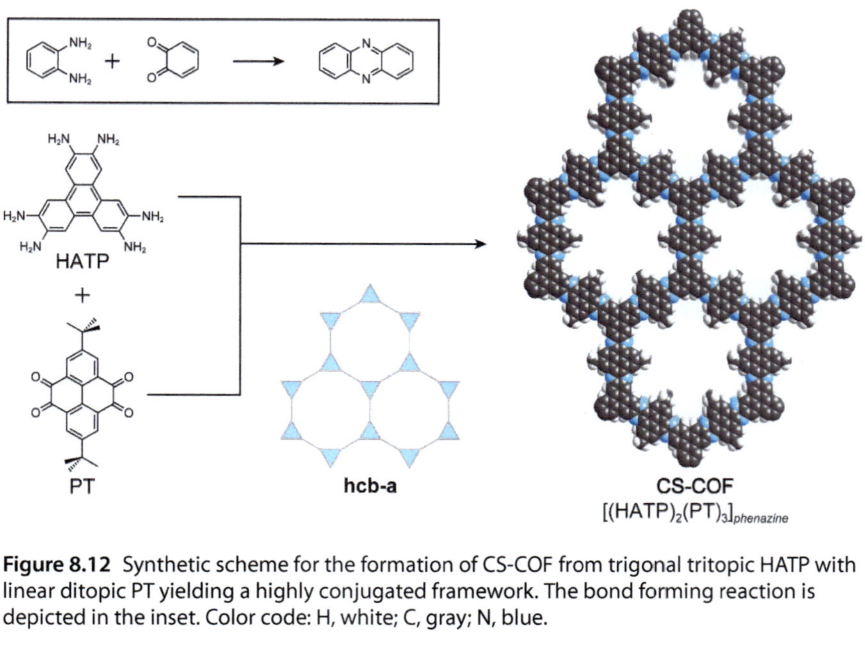

Figure 8.12 Synthetic scheme for the formation of CS-COF from trigonal tritopic HATP with linear ditopic PT yielding a highly conjugated framework. The bond forming reaction is depicted in the inset. Color code: H, white; C, gray; N, blue.

pressure-induced breaking of the flame-sealed pyrex tubes that are used as the reaction vessel. The higher reaction temperatures used are necessary to enhance the reversibility of the reaction. It should be noted in this context that the flame-sealed pyrex tubes not only serve the purpose of keeping water in the reaction equilibrium, but also help to avoid oxidation of the molecular building units. This is especially important when molecules bearing multiple amino functionalities, such as HATP, are used (Figure 8.12). Another point to note in the context of this COF is the use of PT: a pyrene tetraone derivative with two *tert*-butyl groups was chosen to enhance the solubility of this otherwise highly insoluble polyaromatic hydrocarbon, similar to what was previously discussed in the synthesis of hydrazone COFs. The highly conjugated structure of CS-COF enhances its chemical stability to common organic solvents, 1 M aqueous HCl, and 1 M aqueous NaOH, and the high degree of conjugation furthermore renders this framework interesting with regard to its electronic properties. CS-COF is a high-rate hole conducting framework with a charge-carrier mobility of $4.2\,cm^2\,V^{-1}\,s^{-1}$.

8.3.6 Benzoxazole COFs

Stabilization of COFs by aromatization is also used in the synthesis of benzoxazole linked COFs [25]. Benzoxazoles are formed by the reaction of 2-hydroxy anilines with aldehydes. The first step of this reaction is an imine condensation, followed by a nucleophilic attack of the hydroxyl group on the imine carbon and subsequent oxidation to form a 5-membered heteroaromatic system.

Two benzoxazole linked frameworks, termed BBO-COF-1 ($[(TFB)_2(PDA\text{-}(OH)_2)_3]_{benzoxazole}$) and BBO-COF-2 ($[(TFPB)_2(PDA\text{-}(OH)_2)_3]_{benzoxazole}$), are formed from reticulation of trigonal tritopic TFB or TFPB with linear ditopic PDA-$(OH)_2$ (2,5-diamino-1,4-benzenediol), respectively (Figure 8.13). The reaction conditions employed in the synthesis strongly differ from the prototypical synthesis of imine COFs. The starting materials are dissolved in DMF for 3 hours at −15 °C to ensure slow formation of the phenolic imine linked intermediate and to avoid rapid precipitation. Afterward, the mixture is slowly heated to room temperature overnight before 1 equiv of a methanolic solution of sodium cyanide is added to catalyze the formation of the benzobisoxazole species, which is carried out over the course of four days at 130 °C. While the crystallinity of the resulting samples is not very high, a structure model could be derived by comparing results from adsorption measurements and subsequent calculation of the pore-size distribution with those gained from the modeled crystal structure. BBO-COF-1 and BBO-COF-2 both form frameworks of **hcb** topology and crystallize in the hexagonal space group $P6$. Both compounds exhibit permanent porosity with surface areas of $891\,m^2\,g^{-1}$ for BBO-COF-1 and $1106\,m^2\,g^{-1}$ for BBO-COF-2. CO_2 adsorption data collected for these two frameworks reveals a high uptake capacity and high binding affinities as exemplified by the respective Q_{st} values of 30.2 and $27.8\,kJ\,mol^{-1}$ for BBO-COF-1 and BBO-COF-2.

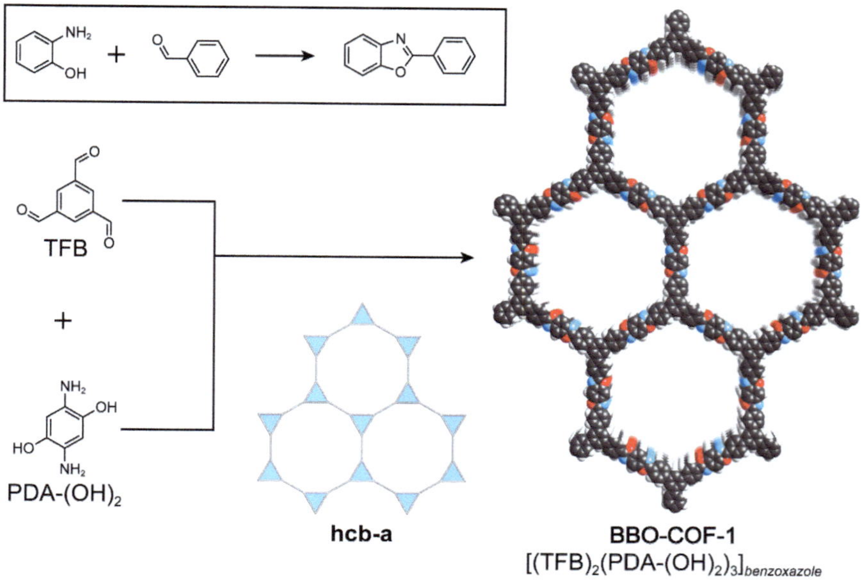

Figure 8.13 Synthesis scheme for the **hcb** topology BBO-COF-1 formed by benzoxazole linkages. Trigonal tritopic TFB and linear ditopic PDA-$(OH)_2$ are reticulated in a one-pot synthesis to first yield imine linked frameworks that undergo nucleophilic attack by the hydroxyl group on the linker and subsequent oxidation to yield the benzoxazole linked BBO-COF-1. The bond forming reaction is illustrated in the inset. Color code: H, white, C, gray; N, blue; O, red.

8.4 Imide Linkage

In contrast to reversible Schiff-base chemistry, imide bond formation has not been studied in dynamic covalent chemistry. Depending on the reaction conditions formation of the imide bond can be carried out under reversible conditions. This is illustrated in Figure 8.14 for the example of phthalimide formation from phthalic anhydride by reaction with an amine. Reaction of formed phthalimide with a second equivalent of amine yields phthalamide that can convert back at high temperatures. Alternatively, phthalamide can react with water to yield 2-carbamoylbenzoic acid. 2-Carbamoylbenzoic acid can convert back in the presence of an amine. At high temperature, 2-carbamoylbenzoic acid dehydrates into phthalimide and this process can be reverted by hydration. Alternatively, 2-carbamoylbenzoic acid can be transformed into phthalic acid by reaction with water. This reaction can again be reversed in the presence of amine. 2-Carbamoylbenzoic acid can also hydrolyze in the presence of protons to yield the phthalic anhydride and can be reformed by reaction with an amine. Last but not least, phthalic anhydride can be hydrolyzed to form phthalic acid and this transformation can be reversed at elevated temperatures. Taking stock of these reactions we can summarize that in the presence of base and at

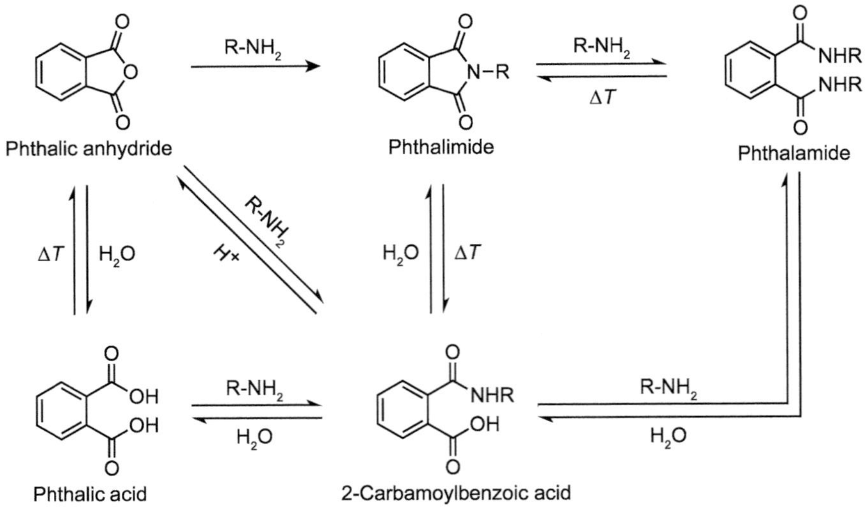

Figure 8.14 Reaction of phthalic anhydride in the presence of an amine. A complex equilibrium between phthalic anhydride, phthalic acid, phthalimide, 2-carbamoylbenzoic acid, and phthalamide is expected at high reaction temperatures in the presence of base. Elevated reaction temperatures and the absence of water favor the formation of the phthalimide over other reaction products [26].

high temperatures an equilibrium between several different products can be established with phthalimide being the thermodynamically favored product at high temperatures (Figure 8.14) [26].

8.4.1 2D Imide COFs

COFs constructed from imide linkages are formed from acid anhydride and amine building units [27]. Unlike the previously covered carbonyl chemistry, the formation of the imide bond is not reversible under acidic conditions but reversibility can be achieved under basic conditions.

The imidization reaction has not been studied in the realm of dynamic covalent chemistry but a lot of research has been devoted to polyimide polymers, which are known for their high thermal and chemical stability, and outstanding mechanical properties. Mesoporous 2D imide linked COFs can be formed from PMDA (pyromellitic dianhydride) as the linear ditopic linker. Reticulation of PMDA with the trigonal tritopic building units, TAPA (tris(4-aminophenyl)amine), TAPB, or TABPB (1,3,5-tris[4-amino (1,1biphenyl-4-yl)]benzene) affords three isoreticular frameworks of **hcb** topology termed PI-COF-1 ($[(TAPA)_2(PMDA)_3]_{imide}$), PI-COF-2 ($[(TAPB)_2(PMDA)_3]_{imide}$), and PI-COF-3 ($[(TABPB)_2(PMDA)_3]_{imide}$), respectively (Figure 8.15). The synthesis conditions for imide-based COFs differ significantly from those based on Schiff-base chemistry. The PI-COF series is synthesized in a solution of NMP (*N*-methyl-2-pyrrolidone) and mesitylene in the presence of a basic isoquinoline catalyst at high reaction temperatures between 200 and 250 °C. Polar aprotic solvents such as NMP have high boiling points and readily dissolve organic molecules. The choice of a high boiling point solvent for the reaction is rooted in the need for high reaction temperatures to assure sufficient

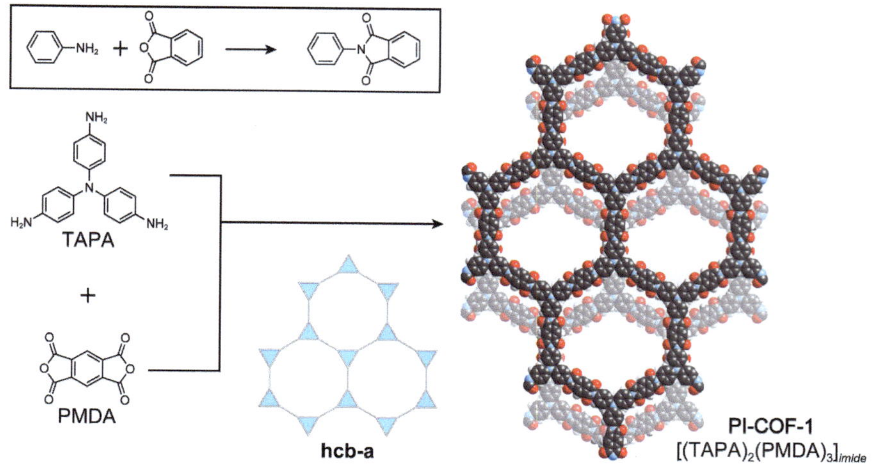

Figure 8.15 Synthetic scheme for the formation of PI-COF-1. Reticulation of trigonal tritopic TAPA with linear ditopic PMDA yields an **hcb** topology framework. PI-COF-1 forms under basic conditions and at high reaction temperatures. The bond forming reaction is depicted in the inset. Color code: H, white; C, gray; N, blue; O, red.

reversibility for the formation of crystalline products. The choice of NMP over more commonly used DMF or DEF (*N,N′*-diethylformamide) is because unlike these solvents NMP does not decompose at elevated temperatures. Since the imidization reaction is catalyzed by base, the fast liberation of additional base from solvent decomposition of DMF or DEF at such high reaction temperatures would complicate the controlled crystallization process. Mesitylene is added to minimize stacking disorder and afford crystalline materials. PXRD data suggests that in all three COFs, PI-COF-1, PI-COF-2, and PI-COF-3, the layers stack in a slightly offset fashion. Large open channels (21, 20, and 36 Å, respectively) propagate along the crystallographic *c*-axes of the materials and endow them with permanent porosity and high BET surface areas of 1027, 1297, and 1760 $m^2\ g^{-1}$, exceeding the values of all known amorphous polyimides.

8.4.2 3D Imide COFs

Two **dia** topology 3D polyimide COFs with high thermal stability and surface area, termed PI-COF-4 ($[(TAA)(PMDA)_2]_{imide}$, where TAA = 1,3,5,7-tetraaminoadamantane) and PI-COF-5 ($[(TAM)(PMDA)_2]_{imide}$) are synthesized by linking tetrahedral tetratopic and linear ditopic building units through imide bonds [28]. Analogous to 2D imide COFs PMDA serves as the linear ditopic linker which is reacted with tetrahedral tetratopic TAA or TAM to yield PI-COF-4 and PI-COF-5, respectively. The reaction conditions employed in the formation of 3D imide COFs are similar to those of their 2D counterparts, with only minor modifications in reaction temperature and solvent ratios. **dia** nets are self-dual and frameworks of this topology are prone to interpenetration. PI-COF-4 is the first example of a non-interpenetrated **dia** framework in COF chemistry, likely due to the short and sterically demanding building units used in the synthesis. In contrast, PI-COF-5, which is formed using the expanded TAM, has a threefold interpenetrated **dia** net. To corroborate that the imidization reaction is in fact reversible under the applied reaction conditions FT-IR spectroscopy and PXRD of the materials are performed at different time intervals. The FT-IR spectra after one, three, and five days are similar and show the characteristic absorption bands of the 5-membered imide ring indicating completeness of the reaction. Absorption bands around 1775 and 1720 cm^{-1} are attributed to the asymmetric and symmetric stretching vibrations of C=O groups of the 5-membered imide rings, whereas peaks at 1350 cm^{-1} are attributed to the stretching vibration of the C–N–C moieties. In contrast, PXRD patterns show no crystalline phase for the sample after one day and show only weak reflections corresponding to the crystalline phase of the COF after three days. After five days the frameworks show good crystallinity as confirmed by their PXRD patterns. These results suggest that the formation of crystalline imide is slow and that over the course of the reaction defects in the structure are corrected through reversible bond formation. The observed transition from an amorphous, ill-defined phase to a well-defined crystalline phase is analogous to what is observed in COFs made from imine linkages.

8.5 Triazine Linkage

All examples of COF linkages discussed so far are carried out under solvothermal conditions at relatively moderate temperatures, which is possible because of their dynamic and reversible nature under benign reaction conditions. Although, studies suggest that more thermodynamically and kinetically stable COFs can be made if linkages that form reversibly only at very high temperatures are employed, only few organic molecular building units can withstand such harsh reaction conditions and therefore this approach is not frequently employed in the synthesis of COFs. An exception is the formation of porous, covalent triazine linked frameworks which are prepared under ionothermal synthesis conditions [29].

Triazine-based COFs are accessed by a formal [2+2+2] cycloaddition of organic nitriles. The reaction is carried out at 400 °C in molten $ZnCl_2$ to increase the reversibility of bond formation. Such harsh conditions are rare in organic chemistry due to limited thermal stability and thus far only one building unit, DCyB (1,4-dicyanobenzene), has been reticulated into a crystalline COF structure according to this approach (Figure 8.16). The synthesis of CTF-1 ($[DCyB]_{triazine}$) is performed in sealed quartz glass tubes because conventional pyrex tubes (borosilicate) typically employed in the synthesis of COFs cannot withstand such harsh conditions. The closed system serves the purpose of avoiding oxidation of the starting materials upon contact of the reaction mixture with air, especially at high temperature. To confirm the completeness of the reaction, FT-IR spectra of the starting materials and the product are recorded. The aromatic nitrile has a prominent C≡N stretch at 2228 cm^{-1}, which decreases in intensity over the course of the reaction while at the same time new bands corresponding to the triazine core located at 1507 and 1352 cm^{-1} appear. The structure of CTF-1 is composed of eclipsed **hcb** layers and crystallizes in the

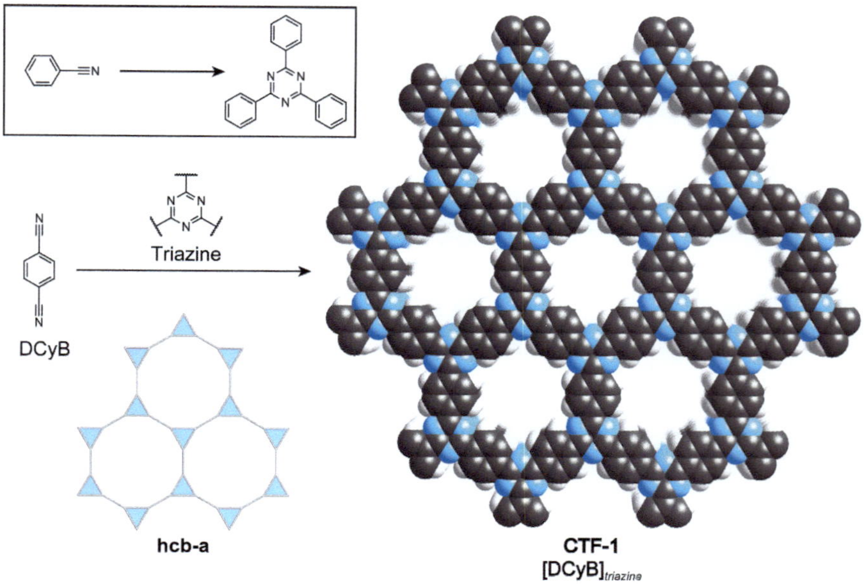

Figure 8.16 Synthesis scheme for CTF-1 which is formed by trimerization of the nitrile groups of DCyB linkers. Reticulation under ionothermal conditions in molten $ZnCl_2$ at 400 °C yields a triazine linked framework of **hcb** topology. The bond forming reaction is depicted in the inset. Color code: H, white; C, gray; N, blue.

hexagonal space group *P* 6/*mmm*. The material is architecturally stable and has a BET surface area of 791 m^2 g^{-1} which is comparable to that of other, noncrystalline, porous triazine frameworks. This value is obscured by the fact that 5 wt% of residual $ZnCl_2$ remains in the sample as evidenced by TGA (thermogravimetric analysis). The COF is initially isolated as a black monolithic material. To facilitate the removal of the salts, the samples are crushed, ground into a powder, and subsequently extensively washed with a diluted HCl solution. Nonetheless, no complete removal of inorganic impurities is achieved.

8.6 Borazine Linkage

The thermal decomposition of arylamine–borane or borontrihalide adducts in aprotic solvents leads to borazine ring formation, which is isostructural to the boroxine and triazine building units found in COFs [30].

Borazine has been used for the fabrication of BN-based ceramics or in organic optoelectronics [31]. To date, the use of borazine as a building unit for the preparation of polymers remains scarce [32]. Polymers linked by B—N bonds are highly sought after, but their preparation remains challenging. Crystalline architectures based on B—N bond formation, such as boron nitride (BN), are usually accessible only through high temperature and high pressure solid-state syntheses.

The synthesis of the borazine linked **hcb** topology COF $[BLP]_{borazine}$ is achieved by thermal decomposition of BLP (1,3,5-(*p*-aminophenyl)-benzene–borane) in a

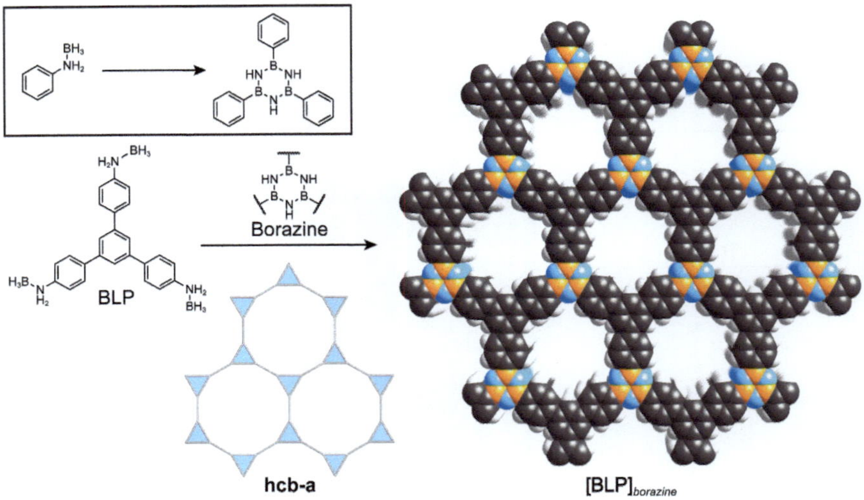

Figure 8.17 $[BLP]_{borazine}$ is reticulated from trigonal tritopic BLP through trigonal tritopic borazine linkages. Trimerization by dehydrogenation of the BLP building unit affords an **hcb** topology framework. The bond forming reaction is depicted in the inset. Color code: H, white; B, orange; C, gray; N, blue.

solvent mixture of 4 : 1 v/v mesitylene/toluene at 120 °C in a flame-sealed pyrex tube (Figure 8.17). These conditions are noticeably milder than those used in the synthesis of boron nitride. The pyrex tubes are necessary because hydrogen, the by-product of the reaction, must remain in the reaction equilibrium to allow for microscopic reversibility. The formation of the borazine ring is confirmed by FT-IR spectroscopy as evidenced by a significant attenuation of the amine bands around 3420 cm^{-1} and the evolution of characteristic bands consistent with the formation of borazine: 2560 cm^{-1} (B–H stretch), 1400 cm^{-1} (B–N stretch). $[BLP]_{borazine}$ crystallizes in the hexagonal space group $P\bar{6}m2$ and is built from **hcb** layers stacked in an eclipsed fashion. The material is thermally stable up to 420 °C and shows architectural stability upon evacuation of solvent from its pores yielding a permanently porous structure with a BET surface area of 1178 $m^2\ g^{-1}$.

8.7 Acrylonitrile Linkage

It has been a major research effort to realize COFs based on C=C bond linkages. This is a difficult feat because the inertness of such bonds generally renders their synthesis irreversible and conditions to crystallize extended structures based on this linkage are therefore difficult to establish. One approach is to render the C=C bond of the linkage more polar by introducing strong electron withdrawing groups into its vicinity. According to this strategy, acrylonitrile linkages can form reversibly. In Figure 8.18, a model reaction of this bond formation is illustrated for the case of the reaction of 2-phenylacetonitrile with benzaldehyde. In the first step, 2-phenylacetonitrile is deprotonated by base. The resulting nucleophilic carbanion intermediate attacks the electrophilic carbonyl carbon. An intermolecular

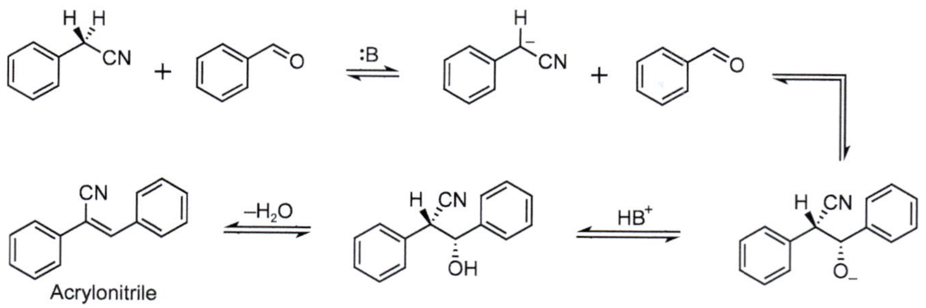

Figure 8.18 Reaction mechanism for the formation of acrylonitrile linkages. In the presence of base as a catalyst 2-phenylacetonitrile is deprotonated and the carbanionic intermediate attacks the electrophilic carbonyl carbon of benzaldehyde. An intramolecular proton transfer followed by dehydration yields the acrylonitrile linkage [33].

proton transfer followed by a dehydration step yields the acrylonitrile product. In the presence of base the whole process can be carried out in a reversible manner.

The formation of C=C coupled COFs comprising an entirely sp^2 hybridized backbone is synthetically challenging [33]. sp^2C-COF ($[(TFPPy)(PDAN)_2]_{acrylonitrile}$) is formed by reacting TFPPy (4,4′,4″,4‴-(pyrene-1,3,6,8-tetrayl)tetrabenzaldehyde) with PDAN (2,2′-(1,4-phenylene)diacetonitrile) by acrylonitrile bond formation (Figure 8.19) [34]. In the synthesis, the starting materials are dissolved in a 1 : 5 v/v solution of mesitylene/dioxane. 4 M aqueous NaOH solution is added to the reaction mixture as a catalyst and the reaction is carried out at 90 °C over the course of three days. Solid-state ^{13}C NMR confirms the completeness of the reaction. The methylene carbon peak for PDAN at

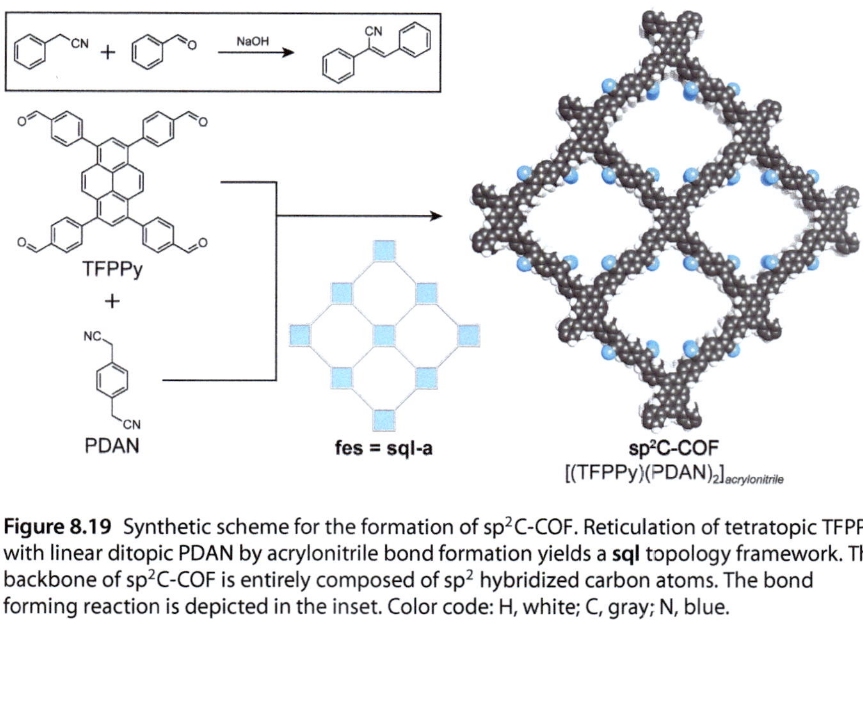

Figure 8.19 Synthetic scheme for the formation of sp^2C-COF. Reticulation of tetratopic TFPPy with linear ditopic PDAN by acrylonitrile bond formation yields a **sql** topology framework. The backbone of sp^2C-COF is entirely composed of sp^2 hybridized carbon atoms. The bond forming reaction is depicted in the inset. Color code: H, white; C, gray; N, blue.

24.2 ppm disappears upon reticulation, and the peak assigned to the cyano carbon at 120.4 ppm is shifted to 107.7 ppm Similar shifts are observed for a molecular model compound thus corroborating the formation of the expected structure. sp^2C-COF crystallizes in the orthorhombic space group $C\,2/m$ with 1.8 nm large channels running along the crystallographic c-axis. Upon activation the material sustains permanent porosity with a BET surface area of 692 $m^2\ g^{-1}$. The backbone of the resulting **sql** topology framework is entirely composed of sp^2 hybridized carbons and endows the material with interesting electronic properties. sp^2C-COF is a semiconductor with a discrete band gap of 1.9 eV. Nonetheless, in electrical conductivity measurements sp^2C-COF is found to be an insulator with an ohmic conductivity value of only $6.1 \times 10^{-14}\ S\ m^{-2}$. Chemical oxidation of the framework with iodine can enhance the materials conductivity by up to 12 orders of magnitude giving an ohmic conductivity of $7.1 \times 10^{-2}\ S\ m^{-2}$. This is due to the generation of radicals on the pyrene moieties yielding a paramagnetic material with a high number of charge carriers.

8.8 Summary

In this chapter, we covered the linkage chemistries of COFs. The mechanism of bond formation of the different bond types was discussed and conditions under which the reaction can be carried out with microscopic reversibility were deduced. For B–O linkages (boronate ester, boroxine, borosilicate, and spiroborate) no catalyst is needed and the reaction is reversible if the by-product of the condensation reaction, water, remains in equilibrium. For linkages based on Schiff-base chemistry (imines, hydrazones, squaraines, β-ketoenamines, benzoxazoles, and phenazines) acid is needed as a catalyst to provide for the necessary microscopic reversibility and, consequently, the chemical stability of such frameworks in the absence of catalyst is increased. Strategies to enhance the chemical stability in these systems even more, such as hydrogen bonding or resonance effects, were discussed. COFs based on imide linkages are formed at elevated temperatures and in the presence of base. These reaction conditions result in a dynamic equilibrium of multiple different species with the thermodynamic product being the imide linked frameworks. Trimerization of nitriles can yield COFs with triazine linkages. Reversibility in these systems is achieved only at high temperatures under ionothermal reaction conditions and only few molecular building units can withstand these harsh conditions. Borazine linkages are formed from trimerization of arylamine-borane adducts with hydrogen as a by-product. To render this reaction reversible, it is crucial to keep the hydrogen molecules in the reaction equilibrium. Finally, acrylonitrile linkages form reversibly under basic conditions and yield frameworks with an entirely sp^2 hybridized backbone. For all linkages illustrative examples for resulting COF structures are given and the reaction conditions yielding crystalline materials rationalized. In Chapter 9 we will discuss how COFs of different topologies are formed and elaborate on the design considerations that are relevant to their targeted synthesis.

References

1 (a) Côté, A.P., Benin, A.I., Ockwig, N.W. et al. (2005). Porous, crystalline, covalent organic frameworks. *Science* 310 (5751): 1166–1170. (b) El-Kaderi, H.M., Hunt, J.R., Mendoza-Cortés, J.L. et al. (2007). Designed synthesis of 3D covalent organic frameworks. *Science* 316 (5822): 268–272. (c) DeBlase, C.R. and Dichtel, W.R. (2016). Moving beyond boron: the emergence of new linkage chemistries in covalent organic frameworks. *Macromolecules* 49 (15): 5297–5305.

2 O'Keeffe, M. (2009). Design of MOFs and intellectual content in reticular chemistry: a personal view. *Chemical Society Reviews* 38 (5): 1215–1217.

3 Clayden, J., Greeves, N., Warren, S., and Wothers, P. (2012). *Organic Chemistry*. Oxford: Oxford University Press.

4 Hunt, J.R., Doonan, C.J., LeVangie, J.D. et al. (2008). Reticular synthesis of covalent organic borosilicate frameworks. *Journal of the American Chemical Society* 130 (36): 11872–11873.

5 Du, Y., Yang, H., Whiteley, J.M. et al. (2016). Ionic covalent organic frameworks with spiroborate linkage. *Angewandte Chemie International Edition* 55 (5): 1737–1741.

6 Vollhardt, K.P.C. and Schore, N.E. (2014). *Organic Chemistry: Structure and Function*, Palgrave version. Basingstoke: Palgrave Macmillan.

7 Wan, S., Gándara, F., Asano, A. et al. (2011). Covalent organic frameworks with high charge carrier mobility. *Chemistry of Materials* 23 (18): 4094–4097.

8 Schiff, E.A. (2006). Hole mobilities and the physics of amorphous silicon solar cells. *Journal of Non-Crystalline Solids* 352 (9–20): 1087–1092.

9 Matsumoto, M., Dasari, R.R., Ji, W. et al. (2017). Rapid, low temperature formation of imine-linked COFs catalyzed by metal triflates. *Journal of the American Chemical Society* 139 (14): 4999–5002.

10 Ascherl, L., Sick, T., Margraf, J.T. et al. (2016). Molecular docking sites designed for the generation of highly crystalline covalent organic frameworks. *Nature Chemistry* 8 (4): 310–316.

11 Uribe-Romo, F.J., Hunt, J.R., Furukawa, H. et al. (2009). A crystalline imine-linked 3-D porous covalent organic framework. *Journal of the American Chemical Society* 131 (13): 4570–4571.

12 Smith, B.J., Overholts, A.C., Hwang, N., and Dichtel, W.R. (2016). Insight into the crystallization of amorphous imine-linked polymer networks to 2D covalent organic frameworks. *Chemical Communications* 52 (18): 3690–3693.

13 Zhang, Y.-B., Su, J., Furukawa, H. et al. (2013). Single-crystal structure of a covalent organic framework. *Journal of the American Chemical Society* 135 (44): 16336–16339.

14 Nguyen, H.L., Gándara, F., Furukawa, H. et al. (2016). A titanium–organic framework as an exemplar of combining the chemistry of metal– and covalent–organic frameworks. *Journal of the American Chemical Society* 138 (13): 4330–4333.

15 Kandambeth, S., Shinde, D.B., Panda, M.K. et al. (2013). Enhancement of chemical stability and crystallinity in porphyrin-containing covalent

organic frameworks by intramolecular hydrogen bonds. *Angewandte Chemie International Edition* 125 (49): 13290–13294.
16 Bunck, D.N. and Dichtel, W.R. (2013). Bulk synthesis of exfoliated two-dimensional polymers using hydrazone-linked covalent organic frameworks. *Journal of the American Chemical Society* 135 (40): 14952–14955.
17 Xu, H., Gao, J., and Jiang, D. (2015). Stable, crystalline, porous, covalent organic frameworks as a platform for chiral organocatalysts. *Nature Chemistry* 7 (11): 905–912.
18 Xu, H., Tao, S., and Jiang, D. (2016). Proton conduction in crystalline and porous covalent organic frameworks. *Nature Materials* 15 (7): 722–726.
19 Uribe-Romo, F.J., Doonan, C.J., Furukawa, H. et al. (2011). Crystalline covalent organic frameworks with hydrazone linkages. *Journal of the American Chemical Society* 133 (30): 11478–11481.
20 (a) Chiba, Y., Islam, A., Watanabe, Y. et al. (2006). Dye-sensitized solar cells with conversion efficiency of 11.1%. *Japanese Journal of Applied Physics* 45 (7L): L638–L640. (b) Colin, H. (1996). 2, 4-Bis [4-(*N*,*N*-dibutylamino) phenyl] squaraine: X-ray crystal structure of a centrosymmetric dye and the second-order non-linear optical properties of its non-centrosymmetric Langmuir–Blodgett films. *Journal of Materials Chemistry* 6 (1): 23–26. (c) Luo, S., Zhang, E., Su, Y. et al. (2011). A review of NIR dyes in cancer targeting and imaging. *Biomaterials* 32 (29): 7127–7138.
21 Nagai, A., Chen, X., Feng, X. et al. (2013). A squaraine-linked mesoporous covalent organic framework. *Angewandte Chemie International Edition* 52 (13): 3770–3774.
22 Kandambeth, S., Mallick, A., Lukose, B. et al. (2012). Construction of crystalline 2D covalent organic frameworks with remarkable chemical (acid/base) stability via a combined reversible and irreversible route. *Journal of the American Chemical Society* 134 (48): 19524–19527.
23 (a) DeBlase, C.R., Silberstein, K.E., Truong, T.-T. et al. (2013). β-Ketoenamine-linked covalent organic frameworks capable of pseudocapacitive energy storage. *Journal of the American Chemical Society* 135 (45): 16821–16824. (b) Mulzer, C.R., Shen, L., Bisbey, R.P. et al. (2016). Superior charge storage and power density of a conducting polymer-modified covalent organic framework. *ACS Central Science* 2 (9): 667–673.
24 Guo, J., Xu, Y., Jin, S. et al. (2013). Conjugated organic framework with three-dimensionally ordered stable structure and delocalized π clouds. *Nature Communications* 4: 2736.
25 Pyles, D.A., Crowe, J.W., Baldwin, L.A., and McGrier, P.L. (2016). Synthesis of benzobisoxazole-linked two-dimensional covalent organic frameworks and their carbon dioxide capture properties. *ACS Macro Letters* 5 (9): 1055–1058.
26 Kurti, L. and Czakó, B. (2005). *Strategic Applications of Named Reactions in Organic Synthesis*. Burlington, MA: Elsevier.
27 Fang, Q., Zhuang, Z., Gu, S. et al. (2014). Designed synthesis of large-pore crystalline polyimide covalent organic frameworks. *Nature Communications* 5: 4503.

28 Fang, Q., Wang, J., Gu, S. et al. (2015). 3D porous crystalline polyimide covalent organic frameworks for drug delivery. *Journal of the American Chemical Society* 137 (26): 8352–8355.
29 Kuhn, P., Antonietti, M., and Thomas, A. (2008). Porous, covalent triazine-based frameworks prepared by ionothermal synthesis. *Angewandte Chemie International Edition* 47 (18): 3450–3453.
30 Jackson, K.T., Reich, T.E., and El-Kaderi, H.M. (2012). Targeted synthesis of a porous borazine-linked covalent organic framework. *Chemical Communications* 48 (70): 8823–8825.
31 (a) Gervais, C., Maquet, J., Babonneau, F. et al. (2001). Chemically derived BN ceramics: extensive 11B and 15N solid-state NMR study of a preceramic polyborazilene. *Chemistry of Materials* 13 (5): 1700–1707. (b) Wang, Q.H., Kalantar-Zadeh, K., Kis, A. et al. (2012). Electronics and optoelectronics of two-dimensional transition metal dichalcogenides. *Nature Nanotechnology* 7 (11): 699.
32 Sánchez-Sánchez, C., Brüller, S., Sachdev, H. et al. (2015). On-surface synthesis of BN-substituted heteroaromatic networks. *ACS Nano* 9 (9): 9228–9235.
33 Zhuang, X., Zhao, W., Zhang, F. et al. (2016). A two-dimensional conjugated polymer framework with fully sp^2-bonded carbon skeleton. *Polymer Chemistry* 7 (25): 4176–4181.
34 Jin, E., Asada, M., Xu, Q. et al. (2017). Two-dimensional sp^2 carbon–conjugated covalent organic frameworks. *Science* 357 (6352): 673–676.

9

Reticular Design of Covalent Organic Frameworks

9.1 Introduction

In Chapter 8 the linkages used to stitch together building units in the synthesis of covalent organic frameworks (COFs) were discussed. Here, the different structure types that can form upon reticulation of various molecular geometric units are presented. The designed synthesis of extended structures of a specific topology is an aspect of organic synthesis that is exclusive to COFs as they are the only class of extended organic materials with periodicity and long-range order. This is an important synthetic advance leading to the expansion of organic chemistry beyond the molecule while preserving definitiveness of structure and a rational synthetic approach [1]. In this chapter, we cover the reticular design of COFs of specific structure types and point out what aspects of geometry and connectivity of the molecular building units play a role in selecting a specific topology [2]. For a more detailed discussion on the topic of topology, the reader is referred to Chapter 18.

The general approach for the reticular synthesis of COFs is carried out in five steps (Figure 9.1). In step 1, a target framework topology is identified. The highest symmetry embedding of this topology is dissected into its vertices by breaking up the edges connecting them.

In step 2, the vertices are evaluated based on their number of points of extension, and the geometry (e.g. tetrahedral versus square planar for 4-c vertices) and idealized angles with which they are predisposed. Knowing the precise geometry of the vertices is crucial because based on connectivity alone a large number of possible topologies exist for each set of such units.

In step 3, molecular equivalents of the vertices are identified. Figure 9.1 illustrates a number of linker backbones alongside the vertices they can represent. Often, large poly-aromatic molecules are chosen as building units because their rigidity helps to predispose the linking functionality in a precise geometry.

In step 4, the COF is formed by stitching together the molecular building units through covalent bonds. We have covered linkages that can be employed in this process and the general thought process that goes into this step was the subject of Chapter 8. Synthetic conditions must be identified for crystalline products to form. This is achieved by balancing the microscopic reversibility of the linkages

Introduction to Reticular Chemistry: Metal-Organic Frameworks and Covalent Organic Frameworks, First Edition. Omar M. Yaghi, Markus J. Kalmutzki, and Christian S. Diercks.

Companion website: www.wiley.com/go/yaghi/reticular

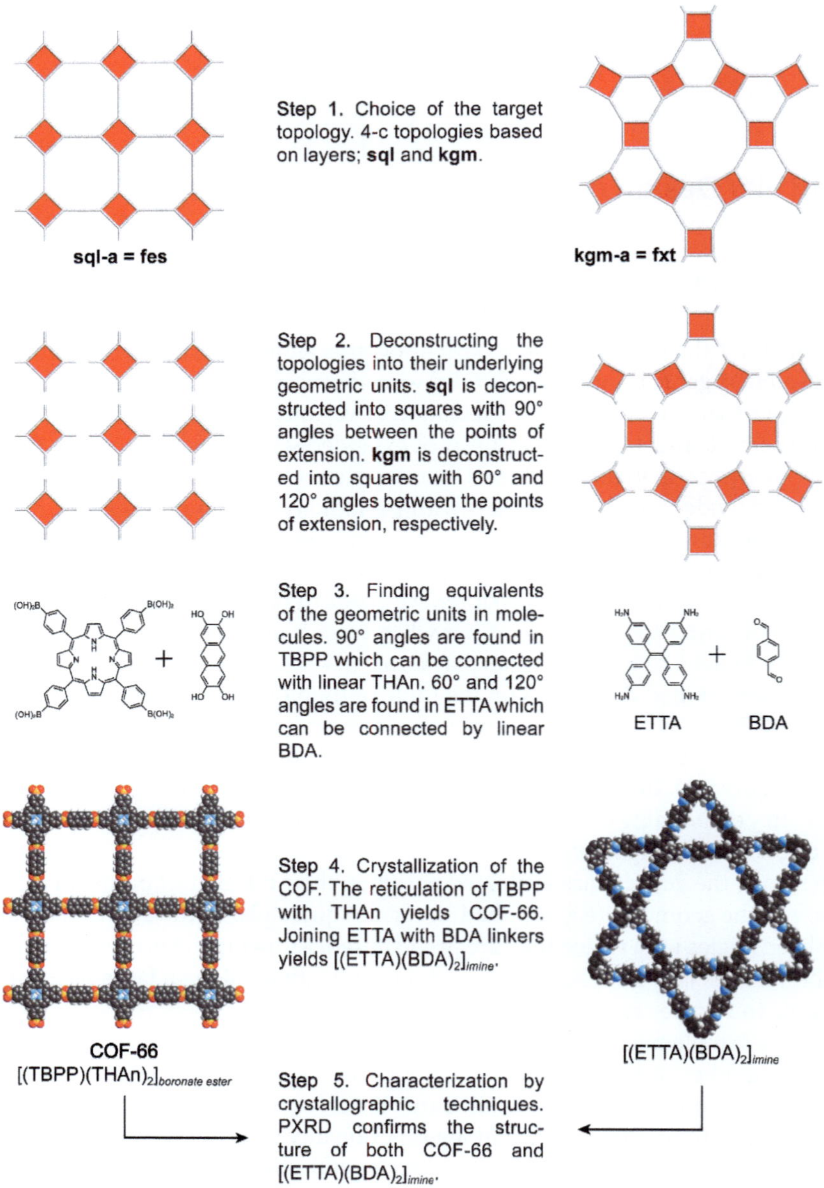

Figure 9.1 General Approach for the reticular synthesis of COFs. Color code: H, white; B, orange; C, gray; N, blue; O, red.

between the building units (thermodynamics), and the rate of the COF-forming reaction (kinetics).

Finally, in step 5, the product of reticulation needs to be structurally characterized to confirm the formation of the expected framework structure. In discrete compounds, analysis by ^{1}H-NMR generally suffices to provide accurate information of the atomic connectivity within a molecule. In contrast, NMR techniques generally cannot unambiguously confirm the exact structure of extended solids. Frameworks of different topologies can be of identical composition and local connectivity and therefore cannot be differentiated spectroscopically. To elucidate its structure the material needs to be characterized by diffraction techniques. When crystals suitable for single crystal X-ray diffraction or rotating electron diffraction cannot be obtained, methods combining structure modeling and fitting to the experimental powder diffraction patterns are used to evaluate whether the targeted structure has formed.

To date, COFs with nine different underlying topologies have been reported: five 2D layered and four 3D nets. In the following text, we will give an overview of the different topologies and illustrate the design considerations necessary to afford frameworks of these specific structure types.

9.2 Linkers in COFs

Linkers employed in COF chemistry are generally based on (poly)aromatic core and extending units similar to the linkers used in MOF chemistry (see Chapter 3) with the functional groups needed for the formation of the linkage attached to them. Typically, they have two, three, four, or six points of extension. We have already covered a number of different linkers in Chapters 7 and 8. Since many linkers of similar shape that differ only in the functional group appended to them can be utilized in the construction of COFs it is more instructive to consider what general shapes of building units are utilized. Linear linkers in COFs are linear, bent, or they can feature an offset angle. Tritopic linkers can be symmetrical with 120° angles between the respective points of extension. They can be de-symmetrized by manipulating the angles or by expanding the linker in one direction. Tetratopic linkers can be square planar with 90° angles, rectangular with 60° and 120° angles, or tetrahedral with 109.5° angles between their points of extension (Figure 9.2).

9.3 2D COFs

The structures of COFs are commonly built from aromatic building units that endow the frameworks with mechanical and architectural stability. Many linkages also comprise conjugated sp^2-hybridized atoms, which extend this conjugation between the building units upon framework formation. As a consequence of strong π–π stacking interactions between the resulting large aromatic π-systems, it is often favorable for COFs to crystallize as layered structures.

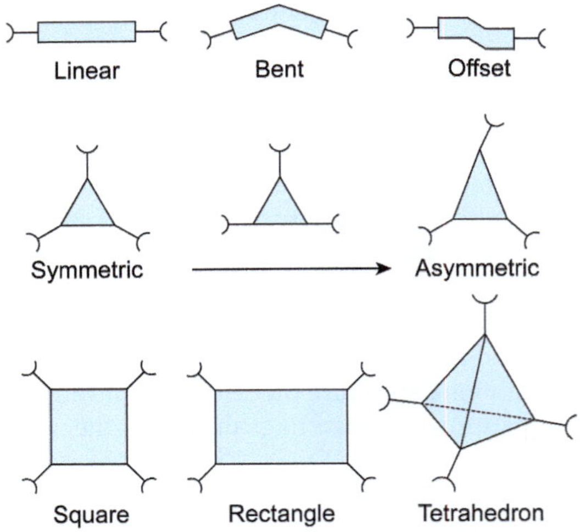

Figure 9.2 Linker geometries found in COFs. Ditopic linkers can be linear, bent, or have an offset between the appended functional groups utilized in the formation of the linkage. Tritopic linkers can be symmetric with 120° angles between the points of extension. The symmetry can be decreased by altering the respective length in different directions. Tetratopic linkers can be square shaped with 90° angles between the points of extension. Their symmetry can be decreased to rectangular linkers with 120° and 60° angles between the points of extension, respectively. Finally, tetratopic linkers can also assume tetrahedral geometries.

The products of reticular synthesis are generally edge-transitive nets and there are exactly five possible nets constructed by linking building units with just one kind of edge into periodic 2D extended structures: **hcb** (111), **sql** (111), **kgm** (121), **hxl** (111), and **kgd** (211), where (*pqr*) represents the transitivity of the respective net [3]. It must be noted here that one kind of vertex and one kind of link merely refers to the topological description and that in the synthesis of COFs multiple different building units with the same connectivity can be employed to form frameworks of these topologies. COFs of these structure types are the most frequently observed products of reticular synthesis and **hcb**, **sql**, **hxl**, and **kgd** represent the default topologies for the combination of a given set of vertices with well-defined geometries. These nets are constructed from one kind of vertex ($p = 1$; uninodal) and one kind of link ($q = 1$; edge transitive) with the exception of **kgd** (kagome dual), which is a binodal net built from two vertices ($p = 2$) linked by one kind of edge.

In the context of layered frameworks, linking of building units with three points of extension with ditopic linkers always affords frameworks of **hcb** topology. Similarly, linking of hexagonal hexatopic building units with ditopic linkers can only result in the formation of an **hxl** net. Combinations of building units with six and three points of extension yield **kgd** layers. In contrast, linking tetratopic building units through ditopic linkers into 2D extended structures can lead to the formation of frameworks of two different nets: the default **sql** or the **kgm** net. For the purpose of selectively targeting frameworks of **kgm** topology, geometric constraints need to be introduced into the building units to rule out the formation of COFs based on the **sql** default net and we will cover these strategies in more detail in Section 9.3.3 [4]. The five possible edge-transitive 2D topologies are illustrated

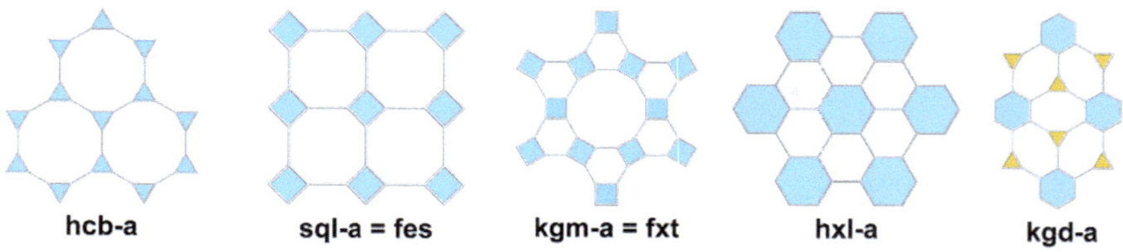

Figure 9.3 List of the five edge-transitive 2D topologies represented as their respective augmented nets. Linking triangles with one kind of edge yields an **hcb** net of transitivity 111. Linking of squares can yield the default **sql** topology (transitivity 111) or a **kgm** layer (transitivity 112). **hxl** nets are formed when hexagons are linked (transitivity 111), and the combination of triangles and hexagons yields a **kgd** topology (transitivity 211). Color code: polygons representing the building units, blue and orange.

in Figure 9.3. In the following text, we will give examples of designed syntheses of frameworks having these topologies.

9.3.1 hcb Topology COFs

We have described the two **hcb** topology frameworks, COF-1 and COF-5, in Chapter 7 [3a]. Both structures have an underlying **hcb** topology but are fundamentally different with respect to the stacking modes between adjacent layers (Figure 9.4). In the structures of COF-1 and COF-5 a staggered confirmation and an eclipsed conformation are observed, respectively. This finding illustrates that in the realm of layered COFs, the design principles of reticular chemistry only allow for the topological design of the individual sheets held together by strong bonds. The stacking between the layers is dominated by non-covalent interactions and is difficult to predict. COF-1 is formed from the self-condensation of linear ditopic BDBA (1,4-benzenediboronic acid) through tritopic boroxin linkages to form **hcb** layers with 15.1 Å wide pore openings. The framework crystallizes in the hexagonal space group *P*63/*mmc* and the structure can be described in analogy to the crystal structure of graphite where the sp^2 carbon atoms are replaced by boroxin linkages and the C—C bonds are replaced by the phenylene units.

The structure of COF-5 is formed through reticulation of linear BDBA with trigonal tritopic HHTP (2,3,6,7,10,11-hexahydroxyterphenylene) by boronate ester linkages leading to the formation of eclipsed **hcb** layers with an interlayer distance of 3.3 Å. COF-5 crystallizes in the hexagonal space group *P*6/*mmm* and its structure can be derived from that of boron nitride, where HHTP occupies the positions of boron and nitrogen, and BDBA adopts the role of the B—N bond. The size of the pore apertures of 27 Å corresponds to the size of the hexagonal channels running along the *c*-direction of the structure. This example illustrates the impact of the stacking modes of the layers on the pore metrics of the framework.

Since, with regard to layered structures, the reticulation of tritopic building units and ditopic linkers can only afford the formation of extended structures with **hcb** topology, the tritopic building unit can be de-symmetrized to achieve more complex structures. One such example is HP-COF-1, where, to achieve the vertex de-symmetrization, a tritopic building unit of C_{2v} symmetry FPI

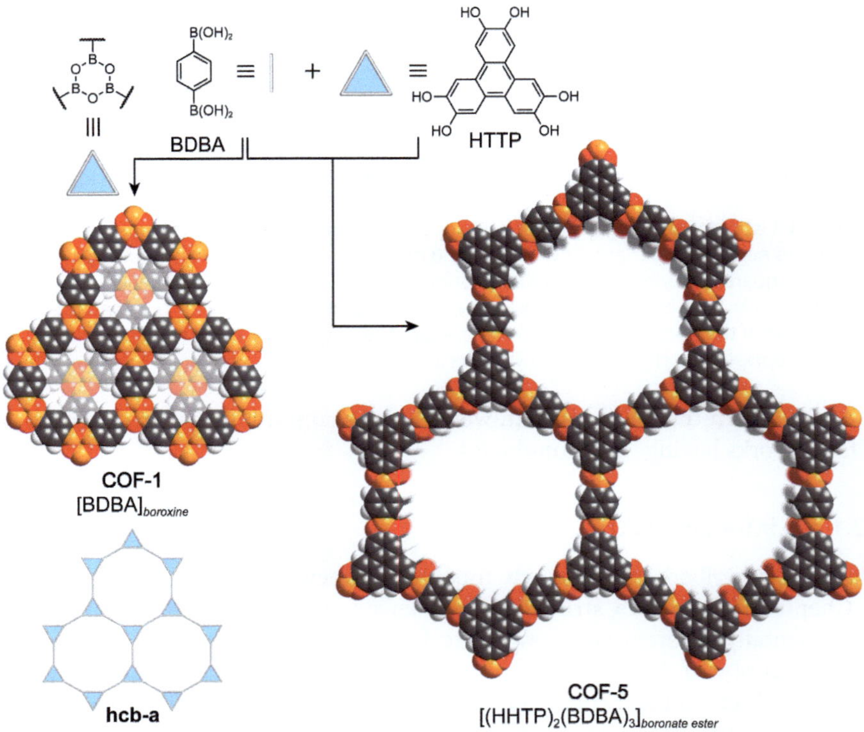

Figure 9.4 Synthetic scheme for the formation of COF-1 and COF-5, both of **hcb** topology. Self-condensation of the linear ditopic BDBA linker through trigonal tritopic boroxin linkages affords COF-1 with layers stacked in a staggered conformation. Cross-condensation of trigonal tritopic HHTP with linear ditopic BDBA yields COF-5 having an eclipsed stacking motif. Color code: H, white; B, orange; C, gray; O, red.

(5-(4-formylphenyl)-isophtaldehyde) was selected instead of building units of the higher D_{3h} symmetry commonly used in the synthesis of **hcb** COFs. Upon reaction with hydrazine, HP-COF-1 ($[(FPI)_2(hydrazine)_3]_{imine}$), a framework composed of eclipsed **hcb** layers is formed. Owing to the de-symmetrized tritopic building unit, the structure of HP-COF-1 has a hierarchical pore system with two distinct kinds of pore (Figure 9.5). The fully activated material sustains permanent porosity with a BET surface area of 1197 m^2 g^{-1}. The pore-size distribution calculated from experimental nitrogen sorption data using an NLDFT (nonlinear density functional theory) model confirms the presence of two differently sized pores of 1.06 and 1.96 nm, respectively. These values are in good agreement with those predicted from the calculated structural model. This example illustrates the usefulness of topological considerations for the rational design of hierarchical structures with complex pore systems [5].

In the aforementioned example, the geometry of the building unit is modified by alteration of the length of the edges but the angles between the points of extension remain unaltered. Modification of these angles is another way to design **hcb** frameworks with modified pore geometries. This is illustrated by

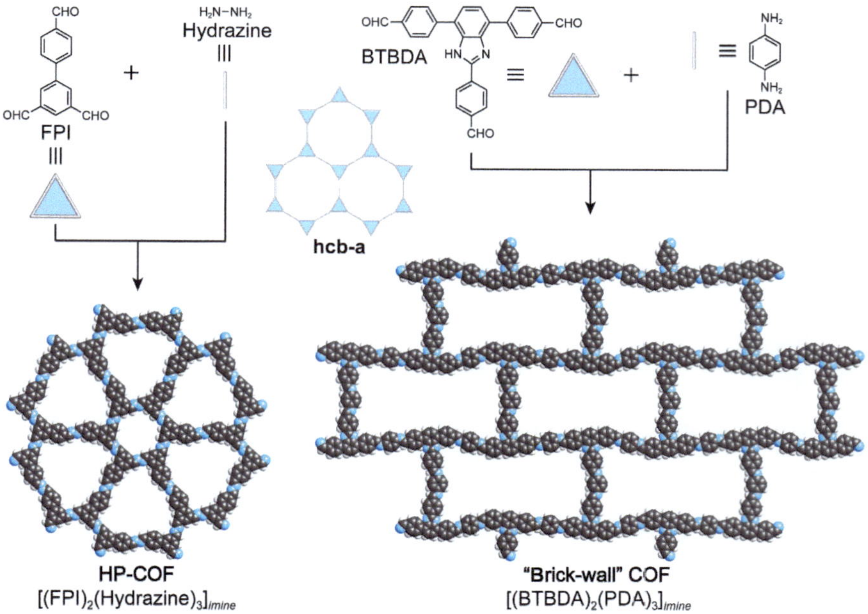

Figure 9.5 **hcb** topology COFs constructed from tritopic linkers with decreased symmetry. Reticulation of C_{2v} symmetric tritopic FPI with hydrazine yields the azine-linked HP-COF. Owing to the low symmetry of the linker the framework features a hierachichal pore system with two distinct pores of 1.06 and 1.96 nm in diameter. Reticulation of the T-shaped BTBDA with linear ditopic PDA yields the "brick-wall COF" with tetragonal pores. Color code: H, white, C, gray, N, blue.

the structure of the so-called "brick-wall-COF" ($[(BTBDA)_2(PDA)_3]_{imine}$, where BTBDA = 4,4′,4″-(1H-benzo[*d*]imidazole-2,4,7-triyl)tribenzaldehyde and PDA = 1,4-phenylenediamine) (Figure 9.5). In this framework, T-shaped tritopic building units with angles of 90°, 90°, and 180° between the points of extension are linked by linear ditopic PDA. The resulting framework has large, rectangular pores, which at first glance do not bear resemblance to the expected **hcb** structure. From a topological point of view, the structure is still a 3-connected honeycomb lattice since topology does not take into account the angles or chemical information but is only concerned with the connectivity of the individual constituents (Chapter 18). The layers of this framework are packed in a staggered conformation resulting in 2-nm-wide rectangular channels that run along the crystallographic *c*-axis. The permanently porous framework has a BET surface area of 401 $m^2\ g^{-1}$ the above shown examples highlight that differences in stacking modes, vertex de-symmetrization, and modulation of angles within the building units are ways to impart structural diversity even within the same topology.

9.3.2 sql Topology COFs

The boronate ester condensation of square-shaped tetratopic $(OH)_8PcNi$ (2,3,9,10,16,17,23,24-octahydroxyphthalocyaninato)nickel(II)) building units

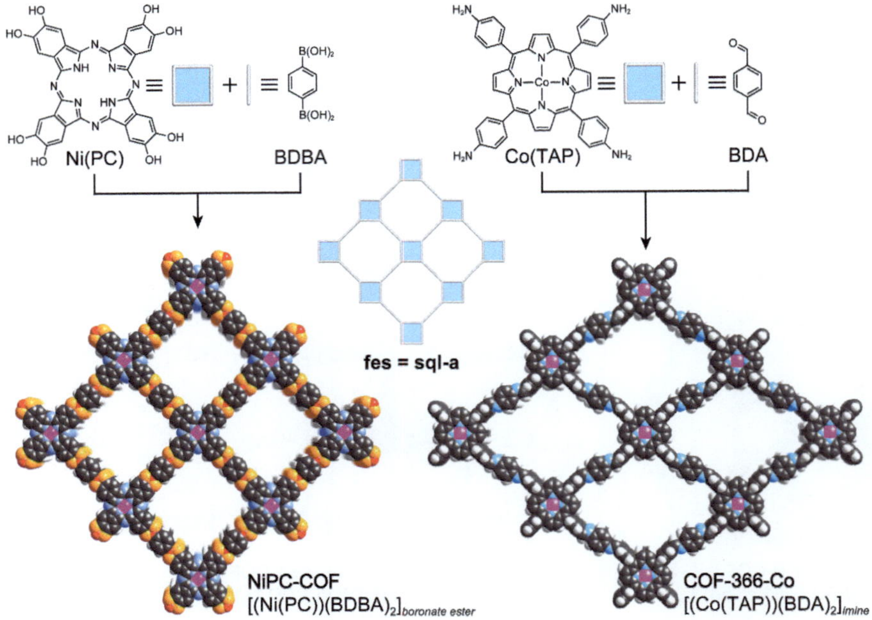

Figure 9.6 Synthetic scheme for NiPC-COF and COF-366-Co, both of **sql** topology. NiPC-COF is constructed from D_{4h} symmetry square-shaped tetratopic $(OH)_8PcNi$ building units that are connected by linear ditopic BDBA struts. COF-366-Co is formed by reticulation of the D_{4h} square-shaped tetratopic (Co)TAP with linear ditopic BDA. Color code: H, white; B, orange; C, gray; N, blue; M (Ni, Co), pink.

and linear ditopic BDBA linkers yields the boronate ester linked NiPC-COF ($[((OH)_8PcNi)(BDBA)_2]_{boronate\ ester}$) (Figure 9.6) [3b]. The $(OH)_8PcNi$ building unit has a point group of D_{4h} and perfect 90° angles between their points of extension, which are the exact angles found in the highest symmetry embedding of the **sql** topology. In the case of 4-connected layers, the precise adjustment of the angles between the building units is critical since two different topologies, **sql** and **kgm**, can be obtained by virtue of adjusting the angles between the constituents. In NiPC-COF the layers stack in an eclipsed fashion and the framework crystallizes in the tetragonal space group *P4/mmm*. The crystal structure encompasses 1.9-nm-large square-shaped channels propagating along the crystallographic *c*-direction. The eclipsed stacking mode of the layers provides a unique way to align the large π-systems of the metallo-phthalocyanine macrocycles, which is very interesting in the context of imparting the structure with high charge carrier mobility and for triggering photochemical events. Such molecular alignment is not attainable in conventional polymers and is difficult to realize in supramolecular architectures. Owing to the ordered stacking of the phthalocyanine units, NiPc-COF shows enhanced light-harvesting capability in the visible and near-infrared region, has a small bandgap (semiconductor), and facilitates charge carrier transport. NiPC-COF is highly photoconductive and exhibits a panchromatic light response and exceptional sensitivity to visible and near-infrared photons [6].

COF-366-Co is composed of the square-shaped tetratopic (Co)TAP ([5,10,15, 20-tetrakis(4-aminophenyl)porphinato]cobalt) and linear ditopic BDA building units (Figure 9.6). The point symmetries of the building units of this COF are identical to those in NiPC-COF; however, the symmetry of the resulting framework is decreased significantly from the tetragonal space group *P*4/*mmm* to the orthorhombic space group *Cmmm*. The main difference between the two frameworks is the kind of linkage used to join the individual building units. The formation of NiPC-COF is achieved through boronate ester bond formation whereas COF-366-Co is formed by imine condensation. Boronate esters are more directional than imine bonds in that the linkage can only adopt one conformation with respect to the other linkages. In imine-linked COFs, the imine bonds can either all point in the same direction (homodromous orientation) or alternate between pointing to opposing directions (heterodromous orientation). In COF-366-Co, the imine bonds around the porphyrin core adopt a heterodromous orientation resulting in a decrease in symmetry of the overall structure. The symmetry of the tetratopic building unit is decreased from a fourfold to a twofold point symmetry forcing a translationengleiche (t_2, see International Tables of Crystallography Section A, https://it.iucr.org/A/) transition into the orthorhombic subgroup *Cmmm*. While the pores of NiPC-COF represent perfect squares, the pores of COF-366-Co are of a trapezoidal shape. COF-366-Co has superior in-plane charger carrier mobility because reticulation with sp^2-hybridized imine bonds yields a fully conjugated π-system making it an interesting candidate for electrocatalysis. Molecular cobalt-porphyrins are well-studied CO_2 reduction electrocatalysts and COF-366-Co and its derivatives selectively reduce CO_2 to CO under mild conditions with high activity, selectivity, and efficiency. The catalyst can operate without degradation for up to 140 hours and outperforms state-of-the-art molecular and heterogeneous catalyst systems [7].

9.3.3 kgm Topology COFs

The kagome (**kgm**) net is not the default topology for linking tetratopic building units into 2D layers, and the more symmetric **sql** topology is generally more likely to form. To afford the formation of COFs built from **kgm** layers by design, the angles between the points of extension on the molecular building units have to be adjusted in such a way that they are closely matched to those in the highest symmetry embedding *P*6/*mmm* of the targeted **kgm** topology. The chemistry of COFs is ideal in this respect since the molecular structure and geometry of the building units remain unchanged throughout the synthesis and are joined by directional covalent bonds. The formation of a **kgm** net in its highest symmetry embedding requires that the angles between the points of extension of the tetratopic building units be 120° and 60°. These angles are found in ETTA (4,4′,4″,4‴-(ethene-1,1,2,2-tetrayl)-tetraaniline). Reticulating tetratopic ETTA with linear ditopic BDA units leads to the formation of $[(ETTA)(BDA)_2]_{imine}$, which crystallizes in the hexagonal space group *P*6 and indeed has an underlying **kgm** topology (Figure 9.7) [3c, 8]. To confirm the formation of the targeted product, a comparison of the experimental powder X-ray diffraction (PXRD)

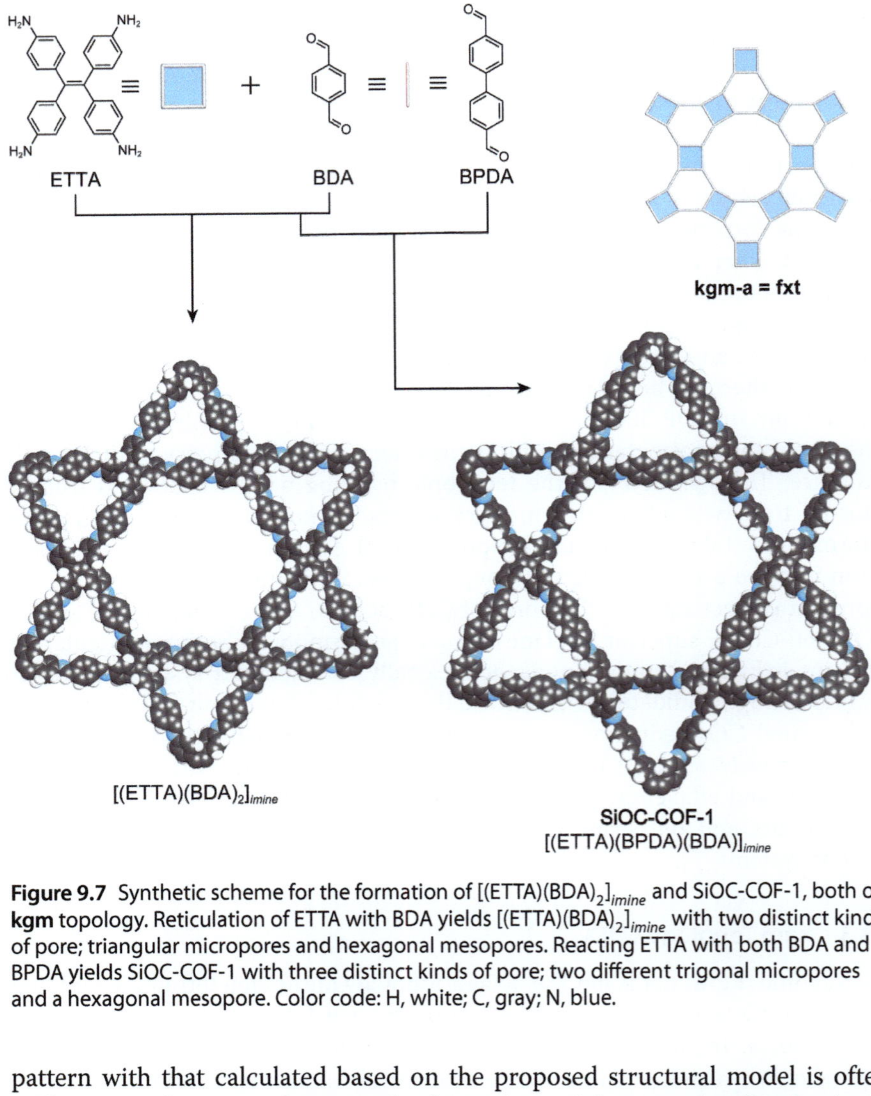

Figure 9.7 Synthetic scheme for the formation of [(ETTA)(BDA)$_2$]$_{imine}$ and SiOC-COF-1, both of **kgm** topology. Reticulation of ETTA with BDA yields [(ETTA)(BDA)$_2$]$_{imine}$ with two distinct kinds of pore; triangular micropores and hexagonal mesopores. Reacting ETTA with both BDA and BPDA yields SiOC-COF-1 with three distinct kinds of pore; two different trigonal micropores and a hexagonal mesopore. Color code: H, white; C, gray; N, blue.

pattern with that calculated based on the proposed structural model is often sufficient. In this particular example, the analysis of the pore-size distribution is another tool to corroborate the structural model created based on topological considerations. In an **sql** topology framework, the structure has just one kind of pore, while **kgm** topology frameworks have two distinct pores, a hexagonal one and a triangular one. The pore-size distribution for [(ETTA)(BDA)$_2$]$_{imine}$ calculated from nitrogen adsorption data shows two distinct maxima, corresponding to the presence of two different pores of 7.3 and 25.2 Å in diameter, which is in good agreement with the proposed triangular and hexagonal channels in the structural model (7.1 and 26.2 Å, respectively). As such, **kgm** layers are a way to target COFs with hierarchical pore systems [9].

The complexity of **kgm** topology frameworks can be further increased and materials with three distinct pores are obtained by employing two different linear struts of varying length in the framework formation. This is illustrated by the reticulation of ETTA, BDA, and BPDA (4,4′-biphenyldicarboxaldehyde) which yields SiOC-COF-1 ($[(ETTA)(BPDA)(BDA)]_{imine}$ (Figure 9.7). The resulting COF features two distinct triangular micropores and one hexagonal mesopore. The smaller triangular pore is bordered by three BDA linkers and is identical in size to the small pore in the structure of $[(ETTA)(BDA)_2]_{imine}$ as indicated by a maximum in the pore-size distribution at 7.3 Å. The larger triangular pore is bordered by three BPDA linkers and measures 11.8 Å in diameter, and the large hexagonal pore, measuring 30.6 Å in diameter, is bordered by three BPDA and three BDA linkers. The low surface area of 478.41 m^2 g^{-1} indicates incomplete activation. Nonetheless, the fact that one can rationally target a structure with three distinct kinds of pore highlights the power of reticular chemistry in the design of complex hierarchical frameworks [10].

9.3.4 Formation of hxl Topology COFs

Imine condensation between hexagonal hexatopic HPB (1,2,3,4,5,6-hexa(4-aminophenyl)benzene) or HBC (1,2,3,4,5,6-(4-aminophenyl)hexabenzocorronene) with linear ditopic BDA yields $[(HPB)(BDA)_3]_{imine}$ and $[(HBC)(BDA)_3]_{imine}$, respectively (Figure 9.8). The PXRD patterns calculated from structural models based on the expected **hxl** frameworks are in good agreement with the experimental PXRD data.[1] COFs of **hxl** topology feature exclusively triangular pores, whereas 2D COFs of other topologies are either entirely or at least partially based on square or hexagonal pore systems. This is an important consideration with respect to the design of COFs with small micropores because, in general, the use of large polyaromatic building units in COFs yields large pore structures. The diameter of the inscribed circle of a pore is close to the real pore diameter and can be approximated by 0.28a for a trigonal pore, 0.5a for tetragonal pores, and 0.86a for hexagonal pores, where a is the length of the linker (Chapter 2). This illustrates that with the same length a of the linker the smallest pores are obtained for a triangular pore system. The 1.2-nm-large pores of $[(HPB)(BDA)_3]_{imine}$ and the 1.9-nm-large pores of $[(HBC)(BDA)_3]_{imine}$ that propagate along the crystallographic c-axis are among the smallest pores reported in 2D COFs to date, despite the use of the comparably large molecular building units. There is a second attribute related to the column density that makes COFs with triangular pore systems relevant for use organic electronics.[2] The smaller pores in triangular pore systems compared to those of square and hexagonal ones directly translate into a higher density of π-columns in the corresponding COF structure (assuming the use of polyaromatic building units). $[(HPB)(BDA)_3]_{imine}$ has a π-column density of 0.25 nm^{-2},

1 Since both the space group and unit cell parameters for the two materials were not reported, we have constructed models according to the structures described in the original report. Both were modeled in the space group $P6$ and we found that the PXRD patterns calculated for these models are in good agreement with the reported experimental data.

2 The values given are calculated according to the following equations:

$A_{\text{triangle}} = \frac{\sqrt{3}}{4}a^2, A_{\text{square}} = a^2, A_{\text{hexagon}} = \frac{3}{2}a^2\sqrt{3}.$

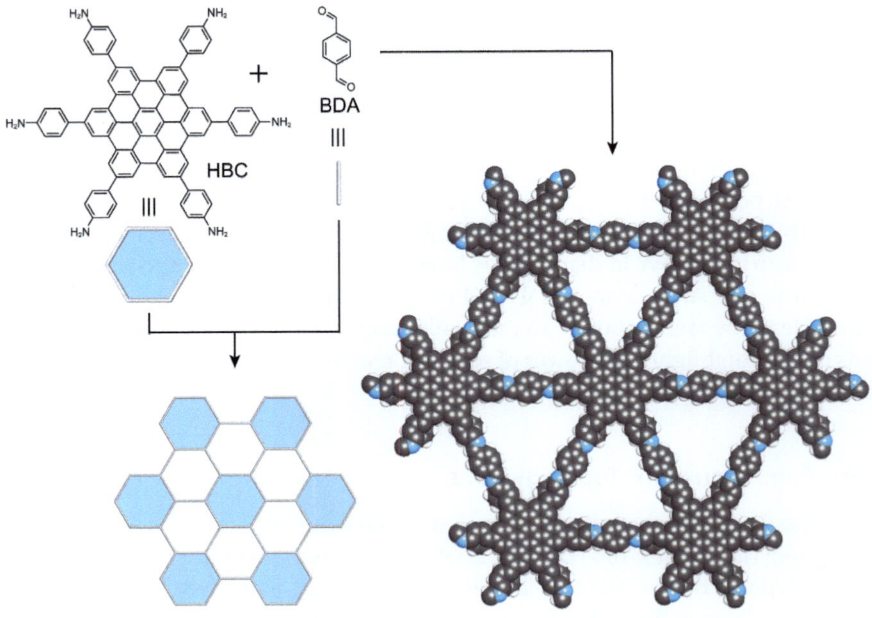

Figure 9.8 Synthetic scheme for $[(HBC)(BDA)_3]_{imine}$ of **hxl** topology. $[(HBC)(BDA)_3]_{imine}$ is formed by reticulation of hexagonal hexatopic HBC with linear ditopic BDA to form a structure featuring triangular-shaped pores of 1.2 nm in diameter. This is among the smallest values of pore sizes in layered COFs despite the large size of the HBC linker. Reticulation of hexagonal hexatopic HBP with BDA yields $[(HPB)(BDA)_3]_{imine}$ with 1.9-nm-wide trigonal pores. Color code: H, white; C, gray; N, blue.

the highest value obtained in COFs. This is important because in organic semiconductors and optoelectronic devices ordered and dense π-units are desirable for improved performance. The hole mobility of $[(HBC)(BDA)_3]_{imine}$ was measured and evaluated to be as high as $0.7\,cm^2\ V^{-1}\ s^{-1}$ which is 40–70 times higher than those of triphenylene-based COFs of **hcb** topology ($0.01\,cm^2\ V^{-1}\ s^{-1}$) [11].

9.3.5 kgd Topology COFs

The structures of all previously mentioned COFs are based on uninodal edge-transitive nets meaning that they are all constructed from one kind of vertex joined by one kind of edge. The dual of the kagome net, termed **kgd** is the only binodal 2D edge-transitive net. In **kgd** layers, hexatopic and tritopic vertices are linked by one kind of edge. Reticulation of HAT (2,3,6,7,10,11-hexaaminoterphenylene) with NTBA (4,4′,4′′-nitrilotribenzaldehyde) or NTBCA (4′,4′′′,4′′′′′-nitrilotris([1,1′-biphenyl]-4-carbaldehyde)) results in the formation of two COFs with **kgd** topology, termed $[(HAT)(NTBA)_2]_{imine}$ and $[(HAT)(NTBCA)_2]_{imine}$, respectively (Figure 9.9). In the highest symmetry embedding of the **kgd** net, the angles between the points of extension

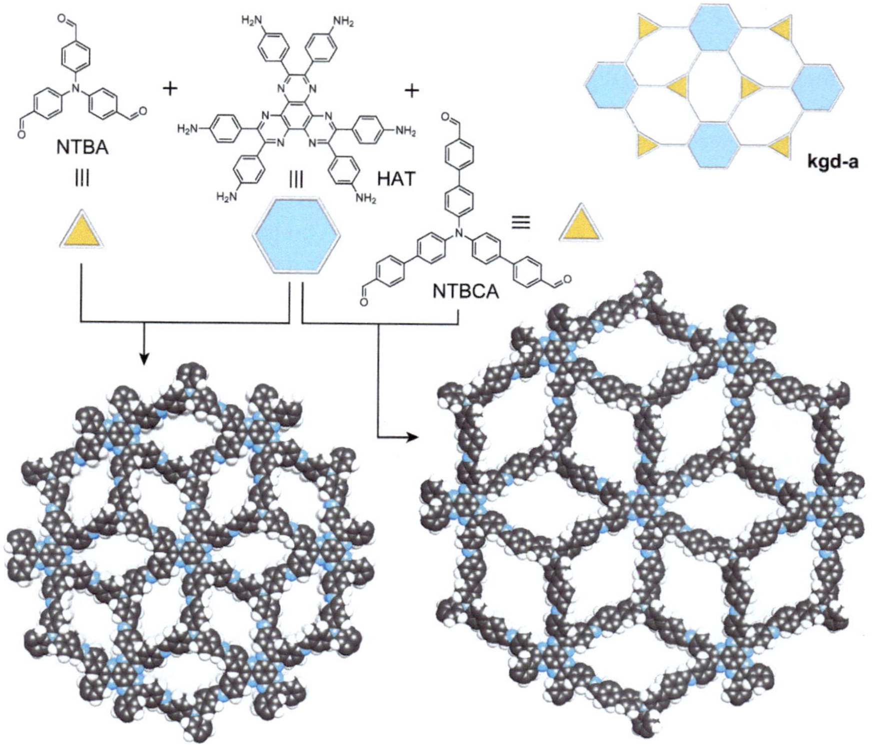

Figure 9.9 Synthetic scheme for the formation of $[(HAT)(NTBA)_2]_{imine}$ and $[(HAT)(NTBCA)_2]_{imine}$, both of **kgd** topology. Reticulation of hexatopic HAT with trigonal tritopic NTBA or NTBCA yields $[(HAT)(NTBA)_2]_{imine}$ and $[(HAT)(NTBCA)_2]_{imine}$, respectively. The structures feature trapezoidal channels of 9.5 and 12.7 Å, respectively, which propagate along the crystallographic along the *c*–axis of the respective framework. Color code: H, white; C, gray; N, blue.

in the 3-c vertex are 120° and those in the 6-c vertex are 60°. These values perfectly match those of the angles in HAT and NTBA/NTBCA, respectively. The crystallinity of both materials was confirmed by means of PXRD.[3] Both structures feature one kind of trapezoidal channel that propagates along the crystallographic *c*-axis. The pore-size distribution of the two COFs calculated from nitrogen adsorption yields values that are in good agreement with those determined from the structural models (10.2 and 13.5 Å for $[(HAT)(NTBA)_2]_{imine}$ and $[(HAT)(NTBCA)_2]_{imine}$, respectively). Both frameworks sustain permanent porosity and feature BET surface areas of 628.0 and 439.9 $m^2\ g^{-1}$ for $[(HAT)(NTBA)_2]_{imine}$ and $[(HAT)(NTBA)_2]_{imine}$, respectively [3e].

3 Both structures were modeled in space group *P*1. Considering the highly symmetric building units used to construct the framework, as well as the fact that the highest symmetry embedding of the **kgd** net is *P*6/*mmm*, it is unlikely that this COF crystallizes in a low symmetry like *P*1. We have remodeled the structures in space group *P*6 and found the calculated PXRD pattern to be in good agreement with the experimental data.

9.4 3D COFs

The majority of reported COFs today are 2D layered structures. Despite the fact that there are far more 3D than 2D topologies, up until now, only four of these have been reported in COFs. As mentioned earlier, the structures of COFs are generally based on rigid aromatic building units and these usually favor the formation of stacked layers. To make 3D frameworks, polyhedral rather than polygonal linkers must be employed or geometric constraints must be placed on the building units. Thus far, all reported 3D COFs have been made according to the first approach, relying on the use of polyhedral building units. Specifically, the combination of tetrahedral and linear building units typically results in the formation of frameworks of **dia** topology, tetrahedral and trigonal tritopic building units tend to afford structures of **bor** and **ctn** nets, and the combination of tetrahedral and square planar building units generally yields frameworks of **pts** topology (Figure 9.10) [12].

9.4.1 dia Topology COFs

The reticulation of tetrahedral TAM and linear BDA affords a COF of **dia** topology [12a]. The fivefold interpenetrating framework termed COF-300 crystallizes in the orthorhombic space group $I4_1/a$. The crystal structure features channels of 7.8 Å in diameter running along the crystallographic *c*-axis, and upon activation under dynamic vacuum COF-300 shows permanent porosity with a BET surface area of 1360 m^2 g^{-1}. **dia** nets are self-dual and thus prone to the formation of interpenetrating frameworks. To a certain extent, the degree of interpenetration can be controlled by varying the metrics of the building units, and in general it is found that the larger the pores, the higher the degree of interpenetration. This is illustrated by COF-320 ($[(TAM)(BPDA)_2]_{imine}$), the isoreticular expanded version of COF-300, obtained by the condensation reaction of TAM and BPDA [13]. The expansion of the adamantane cages in the framework translates into a larger accessible pore volume when compared to the parent COF-300. The large open space encompassed by COF-320 results in a ninefold interpenetrated framework as opposed to the fivefold interpenetrated structure of COF-300.

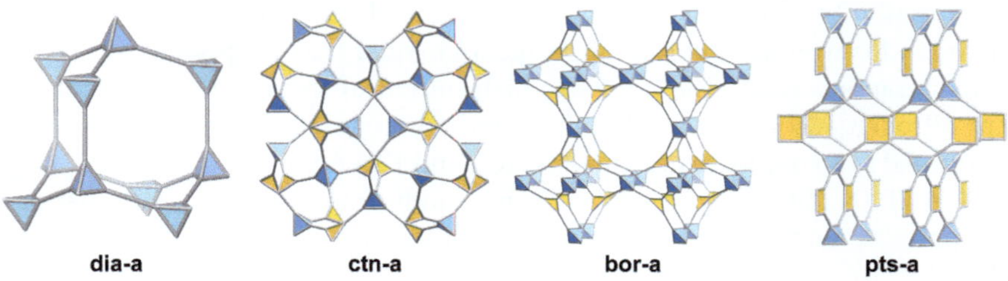

Figure 9.10 3D topologies accessed in COFs represented as augmented nets. All 3D topologies realized in COF chemistry rely on the use of tetrahedral building units. The default topology for linking of tetrahedra is **dia.** Combination of tetrahedral and trigonal nodes affords **ctn** and **bor** nets. The use of tetrahedral and square planar building units yields a **pts** net. Color code: polyhedra representing the building units, blue; polygons representing the building units, orange.

While most examples of **dia**-based COFs interpenetrate, there is one example of a non-interpenetrated **dia** COF, termed PI-COF-4 ($[(TAA)(PMDA)_2]_{imide}$) [14]. PI-COF-4 is synthesized through imidization of TAA (1,3,5,7-tetraaminoadamantane) and PMDA (pyromellitic dianhydride). The non-interpenetrated structure can be explained by the steric hindrance around the TAA building unit after reaction with PMDA. In the structure of PI-COF-4, each adamantane-like cage has a size of 17 Å, and the 3D non-interpenetrated open framework structure possesses channels with a pore aperture of about 15 Å in diameter running along the crystallographic *a*- and *b*-axes.

9.4.2 ctn and bor Topology COFs

For each of the previously discussed combinations of building units of different topicity, there was exactly one possible default net. In contrast, the combination of tetrahedral and trigonal building units can result in the formation of two equally favorable 3,4-connected nets, **ctn** and **bor**, whose highest symmetry embeddings have almost identical space groups $I\bar{4}3d$ and $P\bar{4}3m$, respectively. For both of these topologies the transitivity is 2122 meaning that they comprise two kinds of vertex, connected by one kind of edge, with two different kinds of tile each having two distinct faces. Since these two topologies are equally likely to form from a topological point of view, the question arises whether it is possible to target one of the two. Among the only structures having 3,4-connected nets are the first 3D COFs ever reported, and we have introduced them in Chapter 7. Self-condensation of tetrahedral tetra(4-dihydroxyborylphenyl)methane (TBPM), its silane analog (tetra(4-dihydroxyborylphenyl)silane, TBPS), and co-condensation of TBPS with triangular HHTP affords crystalline 3D COFs termed COF-102, COF-103, and COF-105, respectively, all of which have the **ctn** topology.[4] Cross-condensation of TBPM and HHTP on the other hand yields a framework of **bor** topology, termed COF-108 [12b]. There is no difference in geometry between the linker combinations that yield COF-105 and COF-108, and thus an explanation as to why empirically, frameworks of **ctn** topology seem to form more readily cannot be given in terms of linker geometry (Figure 9.11). Instead, it is likely that the main reason is the difference in topological density. The topological density is defined as "the number of vertices per unit volume for the given embedding." Nature favors the formation of dense structures. Thus, when identical linkers are used for COFs of two different topologies (e.g. **ctn** and **bor**), the COF carrying the net of lower topological density will be of lower (actual) density. The respective topological densities of **ctn** and **bor** are 0.5513 and 0.4763 [15]. The formation of a **bor** net in the case of COF-108, is an exception to the rule. As a result, at the time that COF-108 was first reported, it represented the material with the lowest density at room temperature ($0.17\ \mathrm{g\,cm^{-3}}$) and it is still one of the least dense materials known today.

---

4 Note that in the case of the self-condensation reaction, the trigonal unit needed to form a framework of **ctn** topology is the linkage itself.

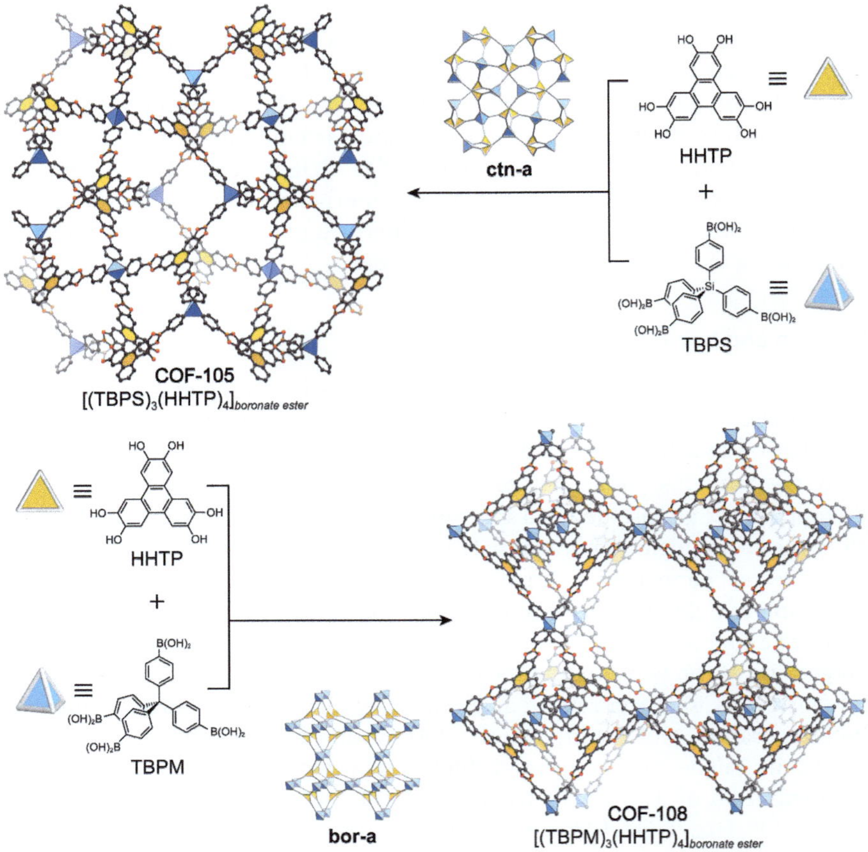

Figure 9.11 Synthetic scheme for the formation of COF-105 and COF-108 of **ctn** and **bor** topology, respectively. Reticulation of TBPS with HHTP yields COF-105 and the reaction of TBPM with HHTP yields COF-108. The **bor** net has a lower topological density than the **ctn** net and because the building units of the two structures are almost identical, COF-108 has a lower actual density than COF-105. All hydrogen atoms are omitted for clarity. Tetrahedral carbon and silicon, blue; core of the tritopic linker, orange polygons. Color Code: B, orange; C, gray; O, red.

9.4.3 COFs with pts Topology

The first synthesis of a 4,4-connected 3D COF was achieved through imine condensation between TAPM (tetra(*p*-aminophenyl)methane) and TFPP (1,3,6,8-tetrakis(4-formylphenyl)pyrene) to give a framework termed 3D-Py-COF (Figure 9.12). The default net of the combination of tetratopic tetrahedral and square building units is the **pts** net, and this is the underlying topology of 3D-Py-COF ($[(\text{TAPM})(\text{TFPP})]_{imine}$). Analysis of the PXRD pattern aided by molecular modeling reveals that the material crystallizes in the orthorhombic space group *Cmmm* and has a twofold interpenetrated structure. After activation, the material displays permanent porosity with a BET surface area of $1290\,m^2\,g^{-1}$. The pore-size distribution of 3D-Py-COF shows just one peak centered at 0.6 nm, which is in good agreement with the values estimated from

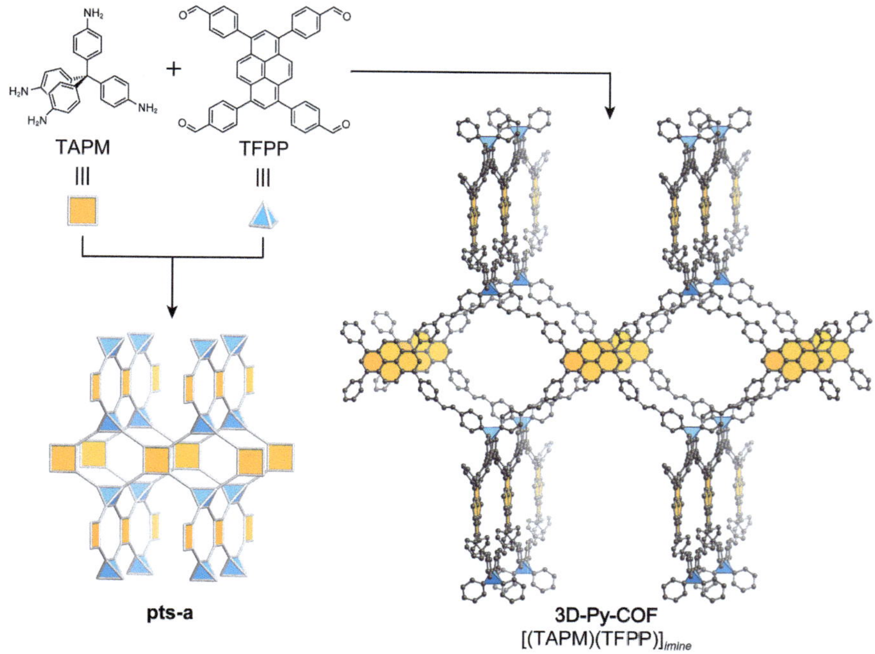

Figure 9.12 Synthetic scheme for 3D-Py-COF of **pts** topology. The twofold interpenetrated 3D-Py-COF is formed by reticulation of TAPM with TFPP. Owing to its 3D open structure, and the concomitant lack of π–π interactions between the linkers, 3D-Py-COF shows an intense yellow/green fluorescence upon excitation at 408 nm. All hydrogen atoms and the interpenetrating net are omitted for clarity. Tetrahedral carbon, blue; core of the tetratopic linker, orange polygons. Color code: C, gray; N, blue.

the structural model. 3D-Py-COF shows an intense yellow/green luminescence and constitutes the first fluorescent 3D COF. Upon excitation at 408 nm, a suspension of 3D-Py-COF in DMF emits at 484 nm. This is interesting, especially when taking into consideration that all reported pyrene-based 2D imine-COFs are non-fluorescent [16]. The fluorescence of 3D-Py-COF originates from the isolated imine-functionalized pyrene units in the 3D network preventing aggregation-induced quenching. In many instances, for applications in the realm of luminescence it is necessary to prevent π–π interactions between the constituents. This highlights the importance of the development of new 3D COFs.

9.5 Summary

In this chapter we covered the underlying design principles governing the reticular synthesis of COFs. A general approach toward the design of COFs of a given topology in five steps was discussed. The connectivity and angles of the building units need to be chosen judiciously to target frameworks of a specific topology, especially when this topology is not the default net for the combination of a given set of building units. General considerations about the linker shape were elaborated and their influence was discussed in the context of exemplary COF structures. The most common topologies encountered in COFs are the five edge-transitive 2D layered nets: **hcb**, **sql**, **kgm**, **hxl**, and **kgd**. Illustrative examples for frameworks of these structure types were given and design considerations for each example highlighted. The most common 3D topologies in COFs are the **dia**, **ctn**, **bor**, and **pts** nets, all of which make use of tetrahedral building units to disfavor the often thermodynamically favorable formation of 2D layered nets. Examples for COFs of these topologies were discussed and, when applicable, design principles regarding the selective formation of COFs of these respective nets explained. In Chapter 10 we will discuss how functionality can be introduced into COFs and introduce different protocols for their pre- and post-synthetic modification.

References

1 (a) Diercks, C.S. and Yaghi, O.M. (2017). The atom, the molecule, and the covalent organic framework. *Science* 355 (6328): eaal1585. (b) Diercks, C.S., Kalmutzki, M.J., and Yaghi, O.M. (2017). Covalent organic frameworks – organic chemistry beyond the molecule. *Molecules* 22 (9): 1575.

2 (a) Delgado-Friedrichs, O., O'Keeffe, M., and Yaghi, O.M. (2007). Taxonomy of periodic nets and the design of materials. *Physical Chemistry Chemical Physics* 9 (9): 1035–1043. (b) Ockwig, N.W., Delgado-Friedrichs, O., O'Keeffe, M., and Yaghi, O.M. (2005). Reticular chemistry: occurrence and taxonomy of nets and grammar for the design of frameworks. *Accounts of Chemical Research* 38 (3): 176–182. (c) Yaghi, O.M., O'Keeffe, M., Ockwig, N.W. et al. (2003). Reticular synthesis and the design of new materials. *Nature* 423 (6941): 705–714.

3 (a) Cote, A.P., Benin, A.I., Ockwig, N.W. et al. (2005). Porous, crystalline, covalent organic frameworks. *Science* 310 (5751): 1166–1170. (b) Ding, X., Guo, J., Feng, X. et al. (2011). Synthesis of metallophthalocyanine covalent organic frameworks that exhibit high carrier mobility and photoconductivity. *Angewandte Chemie International Edition* 50 (6): 1289–1293. (c) Zhou, T.-Y., Xu, S.-Q., Wen, Q. et al. (2014). One-step construction of two different kinds of pores in a 2D covalent organic framework. *Journal of the American Chemical Society* 136 (45): 15885–15888. (d) Dalapati, S., Addicoat, M., Jin, S. et al. (2015). Rational design of crystalline supermicroporous covalent organic frameworks with triangular topologies. *Nature Communications* 6: 7786. (e) Xu, S.-Q., Liang, R.-R., Zhan, T.-G. et al. (2017). Construction of 2D covalent organic frameworks by taking advantage of the variable orientation of imine bonds. *Chemical Communications* 53 (16): 2431–2434.

4 Eddaoudi, M., Kim, J., Vodak, D. et al. (2002). Geometric requirements and examples of important structures in the assembly of square building blocks. *Proceedings of the National Academy of Sciences* 99 (8): 4900–4904.

5 Zhu, Y., Wan, S., Jin, Y., and Zhang, W. (2015). Desymmetrized vertex design for the synthesis of covalent organic frameworks with periodically heterogeneous pore structures. *Journal of the American Chemical Society* 137 (43): 13772–13775.

6 (a) Ding, X., Feng, X., Saeki, A. et al. (2012). Conducting metallophthalocyanine 2D covalent organic frameworks: the role of central metals in controlling π-electronic functions. *Chemical Communications* 48 (71): 8952–8954. (b) Dogru, M. and Bein, T. (2014). On the road towards electroactive covalent organic frameworks. *Chemical Communications* 50 (42): 5531–5546.

7 (a) Lin, S., Diercks, C.S., Zhang, Y.-B. et al. (2015). Covalent organic frameworks comprising cobalt porphyrins for catalytic CO_2 reduction in water. *Science* 349 (6253): 1208–1213. (b) Diercks, C.S., Lin, S., Kornienko, N. et al. (2017). Reticular electronic tuning of porphyrin active sites in covalent organic frameworks for electrocatalytic carbon dioxide reduction. *Journal of the American Chemical Society* 140 (3): 1116–1122.

8 Ascherl, L., Sick, T., Margraf, J.T. et al. (2016). Molecular docking sites designed for the generation of highly crystalline covalent organic frameworks. *Nature Chemistry* 8 (4): 310–316.

9 Tian, Y., Xu, S.-Q., Liang, R.-R. et al. (2017). Construction of two heteropore covalent organic frameworks with Kagome lattices. *CrystEngComm* 19 (33): 4877–4881.

10 Pang, Z.-F., Xu, S.-Q., Zhou, T.-Y. et al. (2016). Construction of covalent organic frameworks bearing three different kinds of pores through the heterostructural mixed linker strategy. *Journal of the American Chemical Society* 138 (14): 4710–4713.

11 Feng, X., Chen, L., Honsho, Y. et al. (2012). An ambipolar conducting covalent organic framework with self-sorted and periodic electron donor–acceptor ordering. *Advanced Materials* 24 (22): 3026–3031.

12 (a) Uribe-Romo, F.J., Hunt, J.R., Furukawa, H. et al. (2009). A crystalline imine-linked 3-D porous covalent organic framework. *Journal of the American Chemical Society* 131 (13): 4570–4571. (b) El-Kaderi, H.M., Hunt, J.R., Mendoza-Cortés, J.L. et al. (2007). Designed synthesis of 3D covalent organic frameworks. *Science* 316 (5822): 268–272. (c) Lin, G., Ding, H., Yuan, D. et al. (2016). A pyrene-based, fluorescent three-dimensional covalent organic framework. *Journal of the American Chemical Society* 138 (10): 3302–3305.

13 Zhang, Y.-B., Su, J., Furukawa, H. et al. (2013). Single-crystal structure of a covalent organic framework. *Journal of the American Chemical Society* 135 (44): 16336–16339.

14 Fang, Q., Wang, J., Gu, S. et al. (2015). 3D porous crystalline polyimide covalent organic frameworks for drug delivery. *Journal of the American Chemical Society* 137 (26): 8352–8355.

15 O'Keeffe, M., Peskov, M.A., Ramsden, S.J., and Yaghi, O.M. (2008). The reticular chemistry structure resource (RCSR) database of, and symbols for, crystal nets. *Accounts of Chemical Research* 41 (12): 1782–1789.

16 (a) Dalapati, S., Jin, S., Gao, J. et al. (2013). An azine-linked covalent organic framework. *Journal of the American Chemical Society* 135 (46): 17310–17313. (b) Chen, X., Huang, N., Gao, J. et al. (2014). Towards covalent organic frameworks with predesignable and aligned open docking sites. *Chemical Communications* 50 (46): 6161–6163.

10

Functionalization of COFs

10.1 Introduction

In Chapters 7 through 9 we have outlined how the structure type, metrics, and type of linkage in covalent organic frameworks (COFs) can be tailored. These parameters dictate the physical properties of COFs such as their chemical and thermal stability, as well as crystallinity and porosity. To devise materials with specific function or to fine-tune the properties of a given COF for a targeted application, functionality needs to be introduced. This is achieved by the introduction of guest molecules in the pores of the COF, by covalent modification of the frameworks' backbone, and by metalation of specific binding sites within the structure. Historically, much of what has been reported in the functionalization of COFs was inspired by work on discrete molecules and the vast toolbox of synthetic organic and organometallic chemistry. The organic backbone of COFs can be functionalized both pre- and post-synthetically to introduce functional groups into specific positions within the framework. Similarly, pre- and post-synthetic metalation of frameworks at predesigned metal coordination sites can be achieved. Here, it is important to consider that applying this know-how to the solid state strictly requires a porous material. In the case of pre-synthetic modifications, the porous nature ensures the necessary space for the introduction of the functionality into the framework without affecting changes to the overall structure metrics or structure type. With respect to post-synthetic modification, porosity is required to guarantee the addressability of the framework constituents to the reagents. The functionalization of COFs exceeds what can be done in molecular chemistry because their porosity further enables additional modes of functionalization such as trapping of functional guests in the pores and embedding of molecular entities within the extended crystals of COFs.

10.2 *In situ* Modification

A simple way to introduce functionality into COFs is to embed large functional guests within their pores. This mode of functionalization is affected *in situ* during the COF synthesis because the introduced guests exceed the

Introduction to Reticular Chemistry: Metal-Organic Frameworks and Covalent Organic Frameworks,
First Edition. Omar M. Yaghi, Markus J. Kalmutzki, and Christian S. Diercks.

Companion website: www.wiley.com/go/yaghi/reticular

size of the pores (ship-in-a-bottle), which precludes post-synthetic introduction of these species into the framework. Theoretically, a large number of potential guests such as biomolecules, polyoxometallates, or metal nanoparticles can be introduced with this strategy but thus far the feasibility of this approach has only been demonstrated for the incorporation of metal-oxide nanoparticles.

10.2.1 Embedding Nanoparticles in COFs

The formation of nanoparticles embedded in COFs is difficult because the controlled precipitation of these materials from a homogeneous solution poses a significant challenge, an issue that will be covered in more detail in Chapter 11. One strategy to circumvent this is to coat the nanoparticles with an amorphous imine polymer of controllable thickness in a first step and to subsequently transform it to a crystalline COF in a second step (Figure 10.1a). $[(TFP)_2(BZ)_3]_{\beta\text{-}ketoenamine}$ is constructed from trigonal tritopic TFP and linear ditopic BZ (benzidine) building units (Figure 10.1b). Under conventional COF forming conditions and in the presence of an acid catalyst, $[(TFP)_2(BZ)_3]_{\beta\text{-}ketoenamine}$ precipitates from solution in an uncontrolled fashion [1]. To grow the COF around Fe_3O_4 nanoparticles they are therefore first coated with an amorphous imine polymer of the same composition to achieve control over the nucleation process. This is achieved in the absence of acid, and the imine polymer grows exclusively around the nanoparticles added to the solution. After isolation of the nanoparticles coated with a layer of amorphous imine polymer (Fe_3O_4⊂Polyimine) of controllable thickness the material is subjected to a solution containing 10% of the organic base pyrrolidine resulting in reversible error correction of the material and the formation of a crystalline $[(TFP)_2(BZ)_3]_{\beta\text{-}ketoenamine}$ shell around the nanoparticles ($Fe_3O_4\subset[(TFP)_2(BZ)_3]_{\beta\text{-}ketoenamine}$) (Figure 10.1b). This amorphous to crystalline transformation is not only corroborated by powder X-ray diffraction but also by a substantial improvement in the surface area of the respective materials. $Fe_3O_4\subset[(TFP)_2(BZ)_3]_{\beta\text{-}ketoenamine}$ has a surface area of 1346 $m^2\ g^{-1}$, whereas the amorphous Fe_3O_4⊂Polyimine has a surface area of 255 $m^2\ g^{-1}$, and the isolated nanoparticles have a surface area of just 123 $m^2\ g^{-1}$ [2].

The generality of the amorphous to crystalline transformation approach for the formation of NP⊂COF (NP = nanoparticle) core-shell structures is further supported by its applicability to imine-linked COFs such as $[(TAPB)(TFB)]_{imine}$, a material composed of trigonal tritopic TFB and trigonal tritopic TAPB building units (Figure 10.1c). Nanoparticles of various chemical nature and size can be incorporated: Fe_3O_4 (9.8 nm), Au (9.0 nm), and Pd (3.3 nm). Analogous to the case of the ketoenamine linked framework, the strategy for coating nanoparticles with $[(TAPB)(TFB)]_{imine}$ relies on the formation of an amorphous imine polymer to control the nucleation. In contrast, the subsequent crystallization step to yield the crystalline imine-linked COF is carried out in the presence of aqueous acetic acid [3].

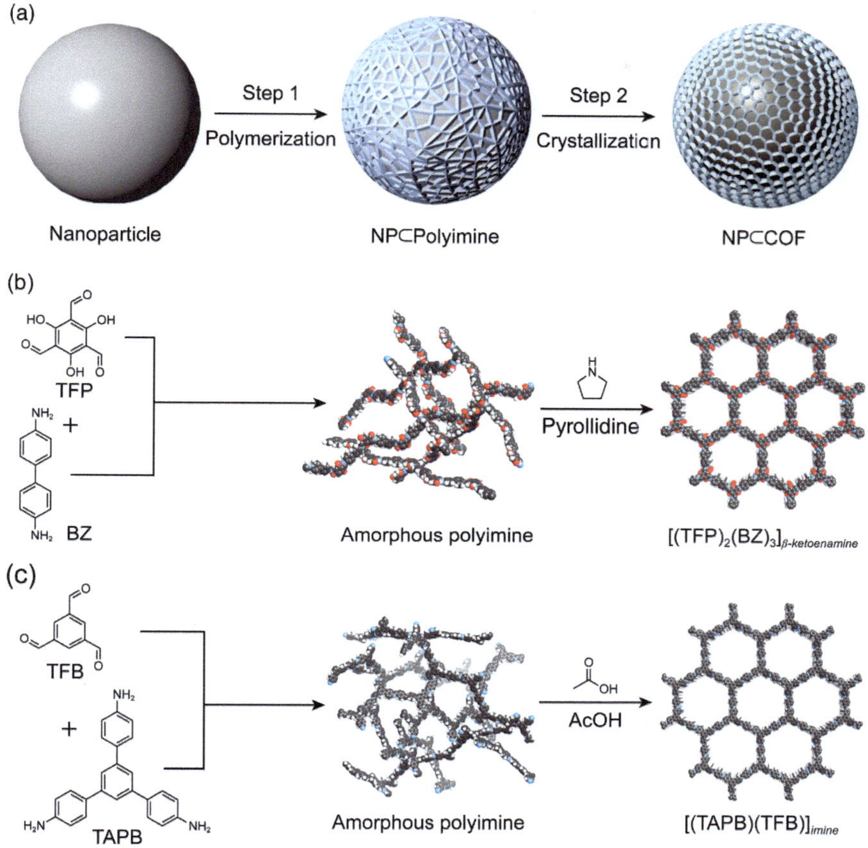

Figure 10.1 Formation of COF-coated nanoparticles in a two-step process (a). In the first step the nanoparticles (gray spheres) are coated by an amorphous imine polymer in the absence of a catalyst under homogeneous conditions, thus allowing for controlled nucleation. In a second step the polymer is converted into a crystalline COF through dynamic error correction in the presence of a catalyst. (b) Reaction of trigonal tritopic TFP with linear ditopic BZ in the absence of catalyst yields an amorphous imine polymer that can subsequently be converted into crystalline $[(TFP)_2(BZ)_3]_{\beta\text{-}ketoenamine}$ by exposing the material to a 10% solution of pyrollidine as a catalyst at elevated temperatures. (c) Reaction of TFB and TAPB yields an amorphous imine polymer. Exposing the polymer to a solution of acetic acid at elevated temperatures yields crystalline $[(TAPB)(TFB)]_{imine}$. Color code in (b) and (c): H, white; C, gray; N, blue; O, red.

10.3 Pre-Synthetic Modification

We have shown previously, that by choosing linkers of the same connectivity and geometry but with different metrics, isoreticular expansion of COFs is performed to adjust the pore metrics of a given framework. The isoreticular principle is not limited to adjusting the pore metrics but can furthermore be applied to the functionalization of frameworks, where metal ions are coordinated to, or functional groups appended onto specific sites of the linker without interfering with the overall structure metrics of the resulting COF. If the appended functionality does not interfere with the COF forming reaction these modifications can be affected pre-synthetically, which enables the formation of the functionalized framework in one step. Pre-synthetic modification of building units is the most common approach for the functionalization of COFs, and many different of functional groups can be incorporated according to this strategy. A large number of examples of pre-synthetic modification will be highlighted throughout this chapter as this type of functionalization also forms the basis for subsequent post-synthetic modifications. Consequently, we restrict this section to highlighting the versatility of this concept using illustrative examples that convey the general underlying concepts of and motivation for pre-synthetic modification.

10.3.1 Pre-Synthetic Metalation

The imine-linked COF-366 shows promise with respect to applications in organic electronics owing to its inherent high charge carrier mobility of 8.1 $cm^2 V^{-1} s^{-1}$ [4]. COF-366 is constructed from square planar tetratopic H_2TAP and linear ditopic BDA (terephthaldehyde) building units by imine bond formation, yielding a framework of **sql** topology. To introduce functionality into the structure the porphyrin core can serve as a metal coordination site. An isoreticular functionalized analog of the framework, termed COF-366-Co $[(Co(TAP))(BDA)_2]_{imine}$, is constructed from the metalated Co(TAP) (tetra(4-aminophenyl)porphinato cobalt) and BDA (Figure 10.2). The Co(TAP) building unit is stable to the COF forming reaction, allowing for its introduction during the synthesis. Cobalt

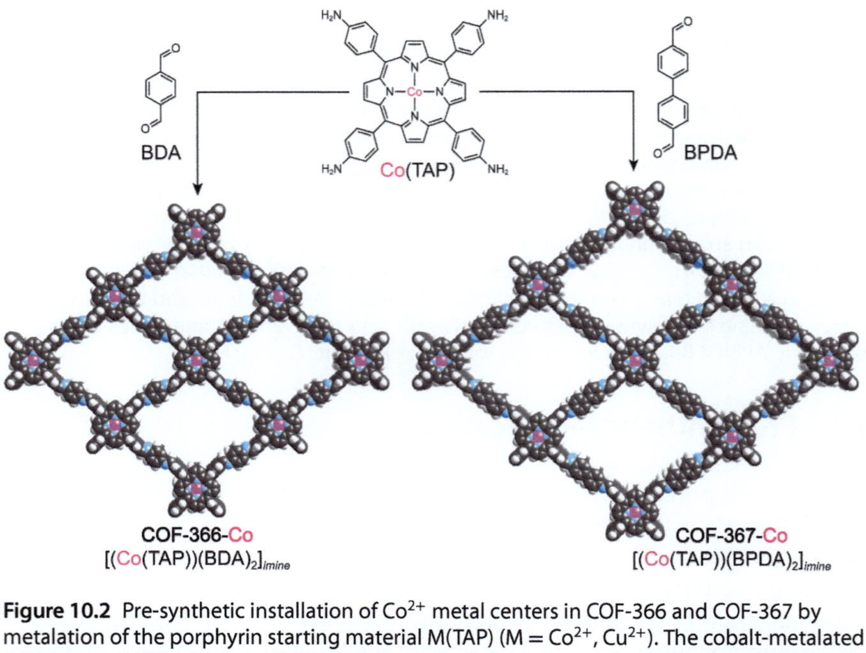

Figure 10.2 Pre-synthetic installation of Co^{2+} metal centers in COF-366 and COF-367 by metalation of the porphyrin starting material M(TAP) (M = Co^{2+}, Cu^{2+}). The cobalt-metalated COFs termed COF-366-Co and COF-367-Co are built from pre-synthetically metalated H_2TAP building units and Co^{2+} (Co(TAP)). The metal centers in these frameworks serve as active sites for the electrocatalytic conversion of CO_2 to CO. Color code: Co, pink; H, white; C, gray; N, blue.

porphyrins are active electrochemical CO_2 reduction catalysts [5]. Similarly to discrete porphyrin catalysts, COF-366-Co can electrocatalytically convert CO_2 to CO but the permanent interface of the framework with the electrode significantly enhances the performance of the material as compared to the molecular analog. An even higher performance is observed in the isoreticular expanded version, termed COF-367-Co $[(Co(TAP))(BPDA)_2]_{imine}$, constructed by reticulating the Co(TAP) with the linear ditopic BPDA (Figure 10.2).

In general, the metalation of COFs is not restricted to one single kind of metal and a series of isoreticular frameworks is accessible when the Co(TAP) building units in the synthesis are entirely or partially replaced with Cu(TAP) [6].

10.3.2 Pre-Synthetic Covalent Functionalization

COF-366-Co can be further modified covalently by making use of the framework backbone. Analogous to the optimization of molecular transition metal catalysts

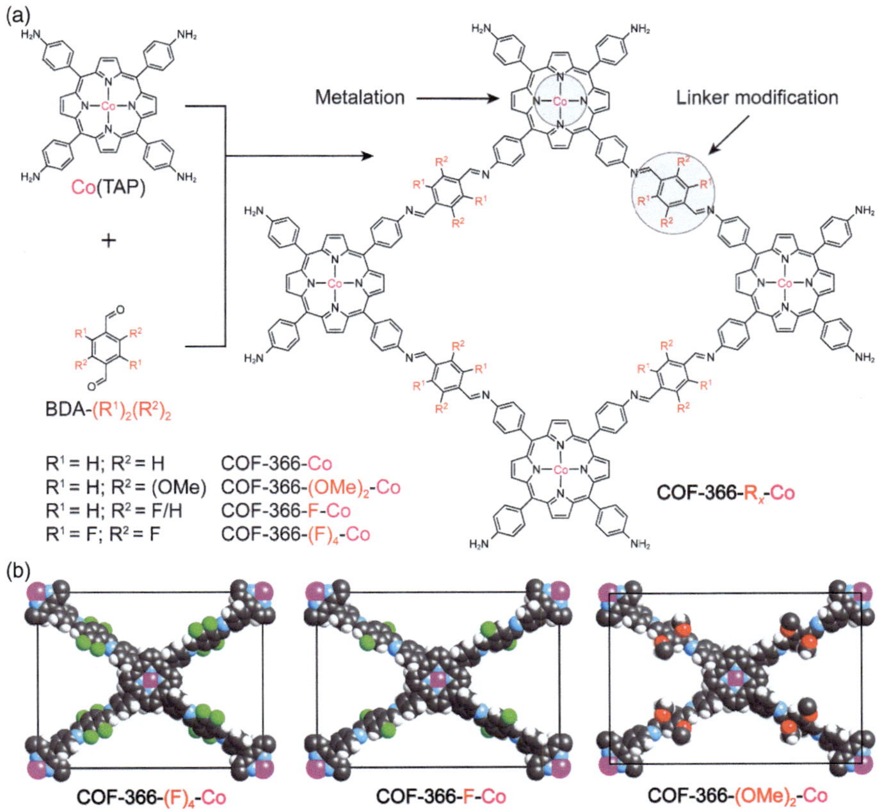

Figure 10.3 Reticular tuning of the pre-synthetically incorporated Co^{2+} active sites in a series of COF-366-Co analogs by modification of the parent structure through covalent pre-synthetic modification of the linker. (a) Substitution of the parent BDA linker with BDA-$(OMe)_2$, BDA-(F), and BDA-$(F)_4$ results in isostructural frameworks termed COF-366-Co, COF-366-$(OMe)_2$-Co, COF-366-(F)-Co, and COF-366-$(F)_4$-Co. (b) The unit cells of these COFs highlight that the underlying structure and the metrics of the substituted frameworks remain unaltered. Color code: Co, pink; H, white; C, gray; N, blue; F, green.

by modification of their ligands, COF catalysts can be optimized by covalent modification of their backbone. Inductive effects on the metal center in the cobalt porphyrins affect their catalytic performance [7]. Substitution of the BDA linker in COF-366-Co by BDA-(F)$_4$ (2,3,5,6-tetrafluoroterephthaldehyde), BDA-(F) (2-fluoroterephthaldehyde), and BDA-(OMe)$_2$ (2,5-dimethoxyterephthaldehyde) yields three functionalized COFs termed COF-366-(F)$_4$-Co, COF-366-(F)-Co, and COF-366-(OMe)$_2$-Co, respectively (Figure 10.3). The frameworks show substantial differences in reactivity as a consequence of changes in the electronic structure imparted by the pre-synthetically modified linkers, corroborating the importance of covalent functionalization of the organic linkers for the reticular tuning of COFs for specific applications [8].

10.4 Post-Synthetic Modification

In many cases functional groups interfere with the COF synthesis conditions, and consequently these functionalities need to be introduced post-synthetically. There are different means to achieve this: (i) the interior of COFs can serve as a host for the trapping of functional molecules, biomacromolecules, or nanoparticles, (ii) organic transformations can be performed on the organic backbone of the framework to append functional groups in specific positions, (iii) metal ions can be incorporated into predesigned metal coordination sites within the structure, (iv) linkers can be exchanged post-synthetically with full retention of crystallinity and definitiveness of structure, and (v) the linkage in the parent COF can be modified to alter the inherent physical and chemical properties of the framework. In the following sections we will highlight examples for these different types of post-synthetic modification.

10.4.1 Post-Synthetic Trapping of Guests

A versatile strategy to impart functionality into COFs is to make use of their large accessible pores to trap guest molecules. This is a potent mode of modification as it allows for a large variety of functional organic and inorganic molecules, biomacromolecules, or metal nanoparticles to be incorporated. In contrast to embedding of species in the pores, the incorporated guests introduced according to this principle are not limited by the need for compatibility with the COF forming reaction conditions.

10.4.1.1 Trapping of Functional Small Molecules

$[(TAPB)_2(BDA\text{-}(OMe)_2)_3]_{imine}$, a mesoporous framework constructed from trigonal tritopic TAPB and linear ditopic BDA-(OMe)$_2$, serves as a platform for solid-state proton conductors [9]. The framework crystallizes in the hexagonal space group *P*6 and features mesoporous hexagonal channels of 3.3 nm in width running along the crystallographic *c*-axis. The framework is capable of trapping large amounts of *N*-heterocyclic proton carriers such as triazole (180 wt%) and imidazole (164 wt%) inside of its mesopores (Figure 10.4). While the parent COF shows a negligible proton conductivity of 10^{-12} S cm^{-1}, the

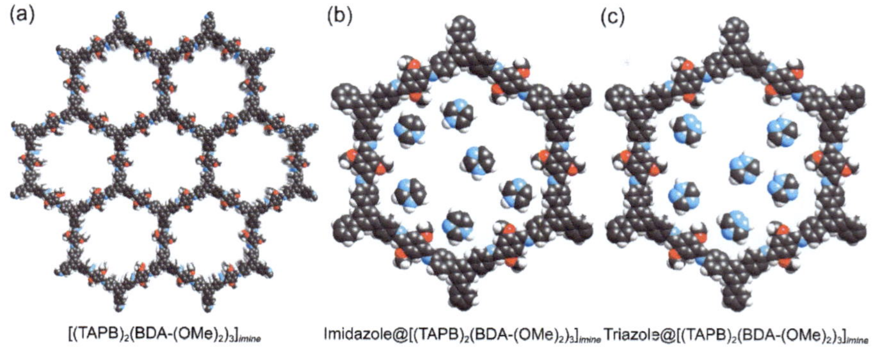

Figure 10.4 Trapping of *N*-heterocycles in the 3.3-nm-wide mesopores of the (a) imine-linked $[(TAPB)_2(BDA\text{-}(OMe)_2)_3]_{imine}$. (b) Impregnation of the material with imidazole (164 wt%) and (c) triazole (180 wt%) yields the solid-state proton conductors imidazole@$[(TAPB)_2(BDA\text{-}(OMe)_2)_3]_{imine}$ and triazole@$[(TAPB)_2(BDA\text{-}(OMe)_2)_3]_{imine}$ with conductivities of 4.37×10^{-3} S cm^{-1} and 1.1×10^{-3} S cm^{-1}, respectively. Color code: H, white; C, gray; N, blue; O, red.

triazole- and imidazole-loaded frameworks feature an increased proton conductivity (at 130 °C) of 1.1×10^{-3} S cm^{-1} and 4.37×10^{-3} S cm^{-1}, respectively. The materials retain 99.3% of the *N*-heterocyclic proton carriers under 100% relative humidity for more than 15 days [10].

10.4.1.2 Post-Synthetic Trapping of Biomacromolecules and Drug Molecules

Trypsin, a globular protein with hydrodynamic radius of 3.8 nm, can be trapped inside a COF constructed from trigonal tritopic TAPB and linear ditopic BDA-$(OH)_2$ (2,5-dihydroxyterephthalaldehyde), termed $[(TAPB)_2(BDA\text{-}(OH)_2)_3]_{imine}$ [9b]. The framework crystallizes in the hexagonal space group *P*6 and has an **hcb** topology. It features large hexagonal channels with a pore-size distribution derived from the nitrogen isotherm centered around 3.7 nm and a high surface area of 1500 m^2 g^{-1}. The hydrogen bonding in this COF endows the framework with chemical stability making it a suitable platform for protein adsorption from water. While the average pore size of the COF is marginally smaller than the hydrodynamic radius of Trypsin, this does not affect the adsorption of this biomacromolecule because soft molecules such as enzymes can adjust their conformation to fit inside the pores. The maximum storage capacity of Trypsin for this COF is 15.5 mmol g^{-1} and the loaded samples retain about 60% of the activity of the free enzyme. $[(TAPB)_2(BDA\text{-}(OMe)_2)_3]_{imine}$ can furthermore be employed for loading and release of the anticancer drug doxorubicin (DOX). The DOX loading capacity of $[(TAPB)_2(BDA\text{-}(OMe)_2)_3]_{imine}$ is 0.35 mg g^{-1} and the drug shows a slow release profile with a 42% decrease over the course of seven days in a pH 5 phosphate buffer [11].

10.4.1.3 Post-Synthetic Trapping of Metal Nanoparticles

A 3D framework formed by self-condensation of tetrahedral tetratopic TBPM termed COF-102 ($[TBPM]_{boroxine}$) is constructed from trigonal

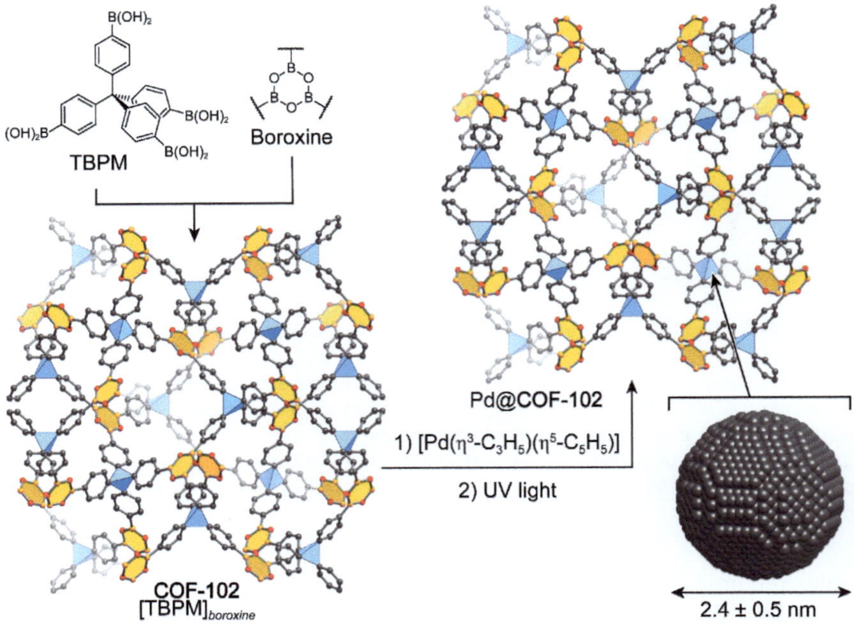

Figure 10.5 Trapping of palladium nanoparticles in the pores of the 3D boronate ester COF-102 by a precursor chemical infiltration technique. COF-102 is constructed from the self-condensation of tetrahedral tetratopic TBPM through the formation of trigonal tritopic boroxine linkages. By infiltration of the pores with the $Pd(\eta^3\text{-}C_3H_3)(\eta^5\text{-}C_5H_5)$ precursor and subsequent exposure to UV light, palladium nanoparticles are formed in the interconnected pores of the **ctn** topology framework, termed Pd@COF-102. All hydrogen atoms are omitted for clarity. Central tetrahedral carbon, blue tetrahedra; boroxine linkage, orange polygons. Color code: C, gray; N, blue; B, orange; O, red.

tritopic boroxine linkages to give a framework of **ctn** topology (Figure 10.5). Palladium nanoparticles can be trapped in the interconnected pores of COF-102 by a precursor chemical infiltration technique. The framework crystallizes in the cubic space group $I\bar{4}3d$ and features 0.9-nm-wide pores that are connected by face-sharing 1-nm-wide pore openings. Diffusion of the volatile and light-sensitive $Pd(\eta^3\text{-}C_3H_5)(\eta^5\text{-}C_5H_5)$ complex in the dark followed by irradiation with UV light results in the formation of Pd@COF-102 (Figure 10.5). TEM shows the formation of the metal-nanoparticle-loaded COF with particles of narrow size distribution centered around 2.5 nm, even at high loadings of up to 30 wt%. Notably, this size is larger than the cavity size of the pores in COF-102 of 0.9 nm, which is rationalized by the fact that the pores are interconnected thus allowing for the formation of larger, interconnected particles. Samples loaded with 9.5 wt% of palladium nanoparticles show a two- to threefold enhancement of the H_2 storage capacity of COF-102 at room temperature and 20 bar. The hydrogen uptake is reversible but cannot exclusively be rationalized by the formation of palladium hydride [12].

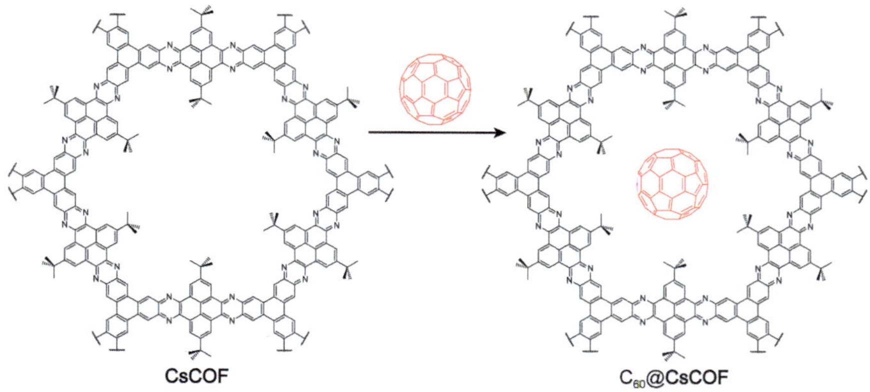

Figure 10.6 Trapping of fullerene C_{60} in the highly conjugated phenazine-linked CsCOF. The pore size of the framework allows for exactly one molecule of C_{60} per cross-section of each pore. The C_{60}@CsCOF construct serves as an active layer for efficient photoenergy conversion and shows a power conversion efficiency of 0.9% and an open-circuit voltage of 0.98 V.

10.4.1.4 Post-Synthetic Trapping of Fullerenes

CS-COF is a framework constructed from trigonal tritopic HATP (2,3,6,7,10,11-hexaminotriphenylene) and linear ditopic PTO (2,7-di-*tert*-butyl-pyrene-4,5,9,10-tetraone) through phenazine bond formation (Figure 10.6) [13]. The highly conjugated framework is a high rate hole-conducting framework with an exceptional mobility of 4.2 $cm^2\ V^{-1}\ s^{-1}$. By impregnation with fullerene molecules with a loading of 25 wt% an ordered bi-continuous donor–acceptor system termed C_{60}@CS-COF is obtained. Owing to the offset stacking of the COF layers as well as the presence of bulky *tert*-butyl groups on the linker pointing into the channels, the pore size of CS-COF is reduced to about 1.6 nm resulting in the presence of exactly one C_{60} molecule in the cross-section of each pore (Figure 10.6). C_{60}@CS-COF can serve as the active layer for photoenergy conversion and shows a power conversion efficiency of 0.9% with a large open-circuit voltage of 0.98 V upon irradiation.

10.4.2 Post-Synthetic Metalation

Post-synthetic metalation of COFs is utilized to incorporate transition-metal complexes into COFs that cannot be incorporated pre-synthetically. The metal coordination site can either be the linkage itself which forms *in situ* and thus cannot be metalated pre-synthetically, or it can be a binding site on the linker that does not withstand the COF forming reactions. Here, we provide examples for both scenarios.

10.4.2.1 Post-Synthetic Metalation of the Linkage

Imine type ligands are versatile binding motifs in coordination chemistry. This is exploited in COF chemistry for the metalation of imine-linked frameworks as exemplified by the **hcb** topology COF LZU-1 ($[(TFB)_2(PDA)_3]_{imine}$) (Figure 10.7a) [14]. LZU-1 is built from trigonal tritopic TFB and linear ditopic

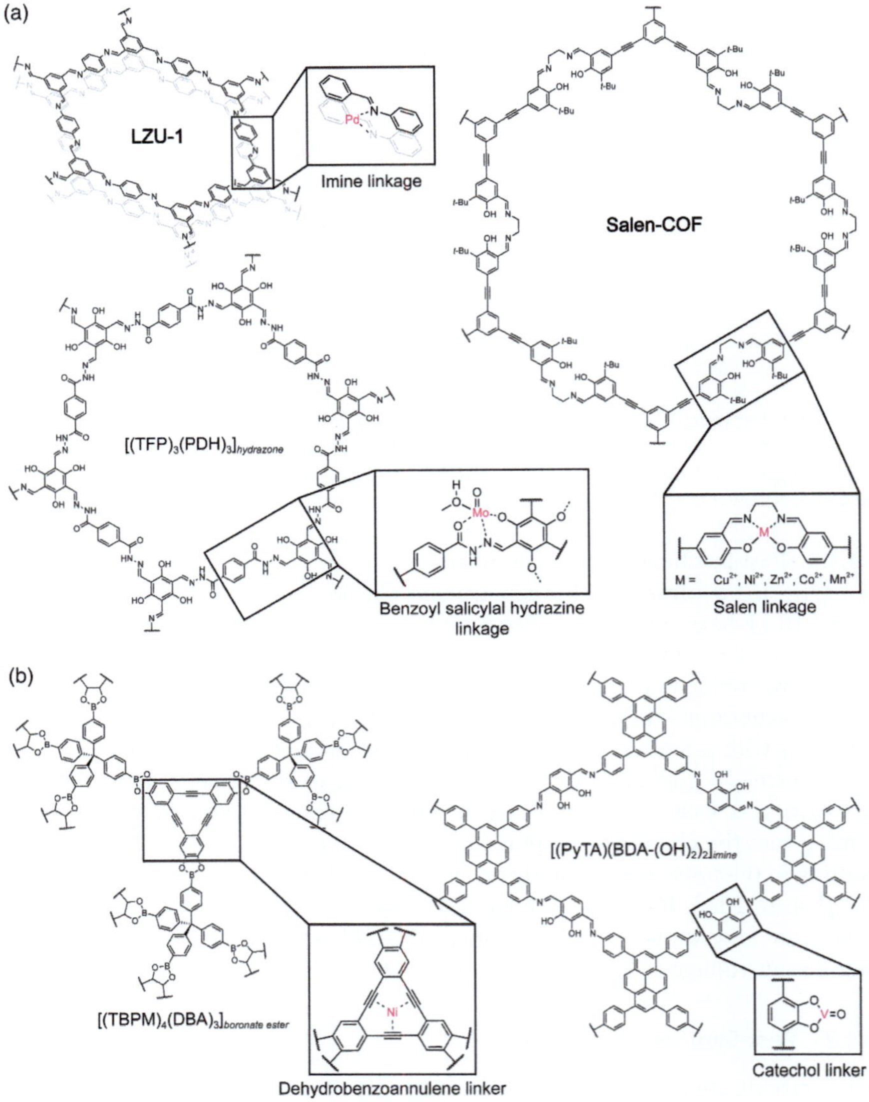

Figure 10.7 Post-synthetic metalation is carried out to introduce metal ions into COFs when they cannot be incorporated pre-synthetically due to limitations with respect to complex stability or because the binding sites are generated during the reticulation process. (a) Linkages of COFs can serve as metal coordination sites as in the case of Salen linkages in Salen-COF, benzoyl salicylal hydrazine linkages in $[(TFP)_2(PDH)_3]_{hydrazone}$, or the imine linkages in LZU-1. (b) Metalation sites can also be introduced into the frameworks as linkers. Examples for such coordination sites are dehydrobenzannulene units in $[(TBPM)_3(DBA)_4]_{boronate\ ester}$ or catechol moieties in $[(PyTA)(BDA\text{-}(OH)_2)_2]_{imine}$.

PDA. The framework crystallizes in the hexagonal space group *P6/m* and features 1.8-nm-wide channels running along the crystallographic *c*-axis. The distance between the imine groups of adjacent layers is 3.7 nm, which constitutes the ideal distance for binding of a metal ion to two imine bonds of neighboring COF layers. Treatment of LZU-1 with a solution of a $Pd(OAc)_2$ results in the formation of the corresponding metalated COF termed LZU-1-Pd. Changes in the binding energy of Pd^{2+} by 0.7 eV from 338.4 eV for the free $Pd(OAc)_2$ to 337.7 eV for LZU-1-Pd are observed by X-ray photoelectron spectroscopy, thus confirming the binding of the metal species to the framework. Molecular palladium complexes are known catalysts for cross-coupling reactions but their homogeneous nature complicates catalyst recycling [15]. LZU-1 catalyzes Suzuki–Miyaura cross-coupling reactions with a broad substrate scope and excellent yields of conversion (96–98%). The high stability and recyclability of the COF catalyst renders this approach promising for the heterogenization of molecular catalysts (Figure 10.7a).

The molybdenum-functionalized framework $[(TFP)_2(PDH)_3]_{hydrazone}$ is synthesized by reticulation of trigonal tritopic TFP with linear ditopic PDH (1,4-dicarbonyl-phenyl-dihydrazide). The linkages of this hydrazone-linked COF are benzoyl salicylal hydrazine ligands, which can coordinate to $MoO_2(acac)_2$ (acac = acetylacetonate). The introduction of the molybdenum species into the framework is achieved by immersing the COF in a methanol solution of $MoO_2(acac)_2$ resulting in an efficient organomolybdenum catalyst with a high active site density of 2.0 mmol g^{-1}. The binding of the COF to molybdenum is confirmed by XPS studies, which indicate the strong coordination of Mo^{2+} to the benzoyl salicylic hydrazine groups of $[(TFP)_2(PDH)_3]_{hydrazone}$. The metalated COF shows high catalytic activity for the epoxidation of cyclohexene with conversion above 99%. Recovery of the catalyst is achieved by filtration and the COF can be reused for more than four cycles with retention of the catalytic activity (Figure 10.7a) [16].

10.4.2.2 Post-Synthetic Metalation of the Linker

A metal-binding motif that is introduced as a linker is found in a framework constructed from tetratopic PyTA (4,4′,4′′,4′′′(pyrene-1,3,6,8-tetrayl)tetraaniline) and linear ditopic BDA-$(OH)_2$ building units, termed $[(PyTA)(BDA\text{-}(OH)_2)_2]_{imine}$ (Figure 10.7b) [17]. In this structure, the catechol moieties of the BDA-$(OH)_2$ linkers point into the 2.4-nm-wide trapezoidal channels running along the crystallographic *c*-axis of this **sql** topology framework. Treatment of these catechol groups with vanadium(IV)-oxy acetylacetonate results in the formation of V=O moieties bound to the catechol binding sites with near-quantitative conversion (0.96 V=O moieties per catechol unit).

Dehydrobenzoannulene macrocycles are known to coordinate to a number of different metal species [18]. The binding of metal ions in such organometallic complexes is weak and thus this binding motif can only be realized in COFs through post-synthetic metalation pathways. The **ctn** topology framework $[(TBPM)_3(DBA)_4]_{boronate\ ester}$ (where DBA = hexahydroxy-dehydrobenzoannulene) is constructed from tetrahedral tetratopic TBPM and trigonal tritopic DBA through boronate ester bond formation (Figure 10.7b) [19]. The high porosity of the framework is highlighted by its large BET surface area of 5083 m^2 g^{-1}. Metalation of the dehydrobenzoannulene core of the DBA linkers with $Ni(COD)_2$ (COD = 1,5-cyclooctadiene) yields the metalated $([(TBPM)_3(DBA)_4]_{boronate\ ester})$-Ni. Only a small decrease in the gravimetric surface area is observed upon metalation (4763 m^2 g^{-1}), attributable to the increase in framework density. Comparison of the UV–vis diffuse-reflectance spectra of $[(TBPM)_3(DBA)_4]_{boronate\ ester}$ and $([(TBPM)_3(DBA)_4]_{boronate\ ester})$-Ni show that the parent material exhibits a broad absorbance between 300 and 420 nm whereas in the metalated analog a new absorption band centered around 575 nm is observed, analogous to what is observed in the discrete DBA–Ni^0 complex. The introduced Ni^0 centers impart $([(TBPM)_3(DBA)_4]_{boronate\ ester})$-Ni with luminescent properties with λ_{max}, the wavelength of highest intensity, located at 510 nm.

10.4.3 Post-Synthetic Covalent Functionalization

Covalent post-synthetic modification of the linker is performed to introduce functional groups into the framework that would interfere with the COF forming reaction. This interference can either be due to the functional group reacting with the building units that form the framework (e.g. amines in the synthesis of imine linked COFs) or because the functionality manipulates the COF forming equilibrium (e.g. carboxylic acids in the formation of imine COFs). To introduce these functionalities post-synthetically, reactions need to be chosen that occur with high yields and under conditions that ensure that the integrity of the framework is retained. In COF chemistry, covalent modification is often achieved by copper(I)-catalyzed click reactions, and either alkynes or azides can be anchored to the COF backbone pre-synthetically to serve as sites of modification (Figure 10.8a). A large variety of functional groups such as alkyls, hydroxyls, esters, anhydrides, or amines can be incorporated into COFs according to this approach. Other reactions that are commonly employed in COF chemistry are succinic anhydride ring-opening reactions to form carboxylic acids, nitro reductions to introduce amines, or aminolysis to yield amides (Figure 10.8b,c).

10.4.3.1 Post-Synthetic Click Reactions

COFs bearing highly functionalized backbones are difficult to obtain by means of pre-synthetic functionalization of their constituents since for each additional appended functionality synthetic conditions need to be identified to crystallize the COF. Additionally, the functional group tolerance of the solvothermal COF synthesis conditions restricts what type of functionalities can be incorporated in this way. One strategy to introduce a wide array of functional groups post-synthetically is based on copper(I)-catalyzed click reactions, which occur under mild reaction conditions and are orthogonal to a lot of chemical functionalities. COFs incorporating alkynyl-functionalized building units can serve as platforms for covalent post-synthetic functionalization using this protocol. Reactions of BDA-(H_2C—C≡CH) (2,5-bis(2-propynyloxy)terephthalaldehyde) and BDA-$(OMe)_2$ at different molar ratios with $H_2(TAP)$ yield an **sql** topology framework termed COF-366-(X%[H_2C—C≡CH]) (X = 0–100) with different amounts of ethynyl groups pointing into the 1.8-nm-wide square-shaped channels of

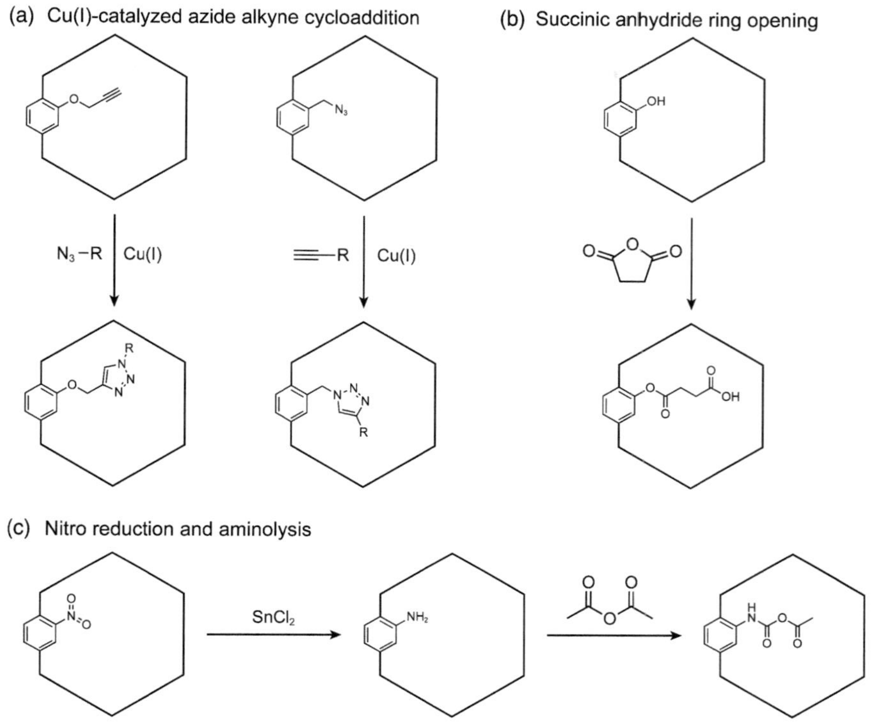

Figure 10.8 Covalent post-synthetic modification reactions employed in COFs. (a) Reaction of alkynes on the COF backbone with organic azides are performed by copper(I)-catalyzed click reactions. Similarly, the COF backbone can be functionalized with organic azides that are reacted with alkynes to yield triazole derivatives. (b) COFs bearing hydroxyl groups are reacted with succinic anhydride to yield frameworks containing carboxylic acid functionalities. (c) Nitro groups in the framework are reduced by $SnCl_2$ to yield COFs with amino groups. In a second step, aminolysis of amino functionalities yields amide-grafted COFs.

the framework (Figure 10.9). Post-synthetic click reactions are employed to introduce the stable organic radical TEMPO (4-azido-2,2,6,6-tetramethyl-1-piperidinyloxy) to yield the corresponding functionalized framework (Figure 10.9). Owing to the large number of accessible radicals of the TEMPO substituents in the functionalized framework, the COF can undergo rapid and reversible redox reactions, leading to capacitive energy storage with high capacitance, high-rate kinetics, and robust cycle stability (Figure 10.9) [20].

Similarly, quantitative click reactions between the ethynyl units on the COF-366-(X%[$H_2C—C{\equiv}CH$]) backbone and azide compounds can be performed to anchor ethyl, acetate, hydroxyl, carboxylic acids, and amino groups. Investigation of the CO_2 uptake capacities of such COFs shows that the functionalization has a profound impact on the sorption behavior. The material functionalized with 50% of amino substituents shows the highest uptake capacity of 157 $mg\,g^{-1}$, thus illustrating the utility of post-synthetic covalent modifications for performance screening of a large number of framework structures (Figure 10.9) [21].

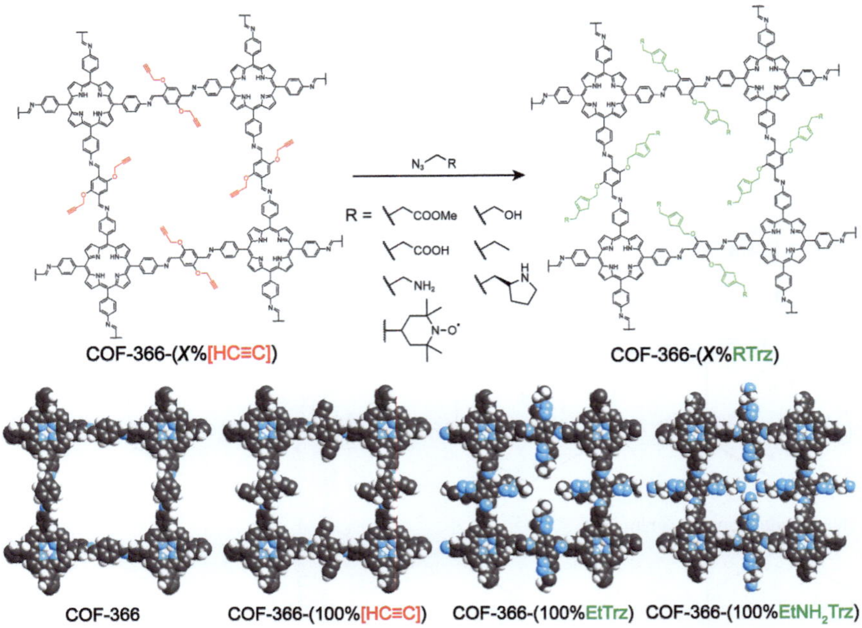

Figure 10.9 Covalent modification of the imine-linked COF-366 by copper(I)-catalyzed click chemistry. A variety of functional groups including TEMPO, ethyl, ester, hydroxyl, acetic anhydride, and primary amine moieties are introduced into the channels of the pre-synthetically functionalized framework COF-366-(X%[H_2C—C≡CH]) bearing various amounts of alkyne functionalities. The parent COF-366, the pre-synthetically functionalized COF-366-(100%[H_2C—C≡CH]), and post-synthetically functionalized COF-366-(100%EtTrz) (Trz = triazine) and COF-366-(100%EtNH_2Trz) are illustrated as examples. Color code: H, white; C, black; N, blue.

COFs with chiral substituents can be created by introducing (*S*)-pyrrolidine into the backbone of COF-366-(*X*%[H_2C—C≡CH]) (*X* = 25%, 50%, 75%, and 100%). Subjecting these frameworks to post-synthetic modification yields COFs functionalized with different amounts of (*S*)-pyrrolidine (Figure 10.9). The frameworks show catalytic activity for enantioselective Michael-addition reactions. Benchmarking against the molecular (*S*)-4-(phenoxymethyl)-1-(pyrrolidin-2-ylmethyl)-1*H*-1*H*-1,2,3-triazole catalyst shows that the molecular catalyst requires 3.3 hours for completion and catalyzes the reaction with an enantiomeric excess of 49%. In contrast, the COF catalysts achieve similar enantiomeric excess (44–51%) but the framework functionalized at 25% completes the reaction in just one hour, thus outperforming not only the molecular model catalyst but also the other COF catalysts with higher catalyst loading. This highlights the necessity for the optimization of the interplay between pore aperture and the amount of exposed functional groups on the rate of the catalytic transformations [22].

Owing to the benign reaction conditions, click chemistry as a tool for covalent post-synthetic modification is not limited to imine COFs but can also be applied to the functionalization of boronate ester COFs. Many substituents interfere in boronate ester COF formation due to their inherent low hydrolytic

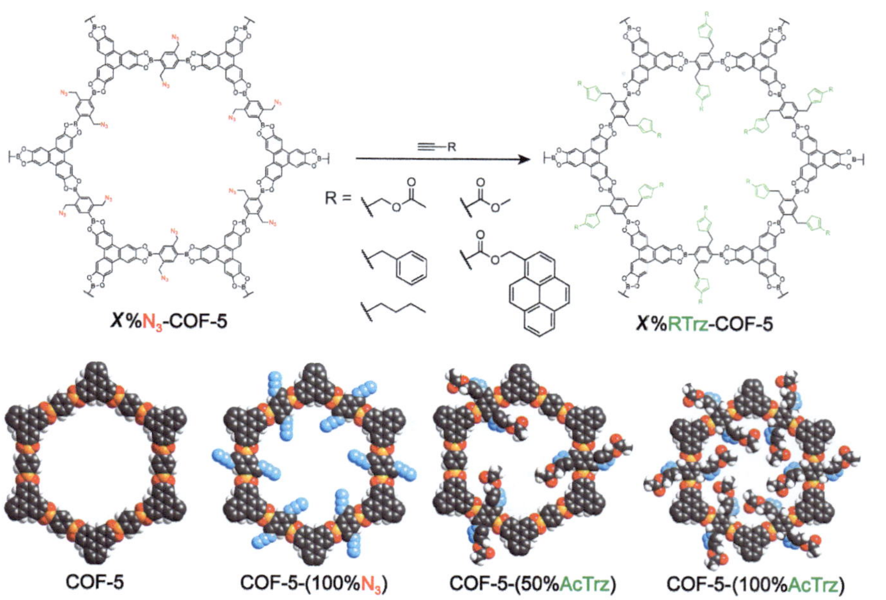

Figure 10.10 Covalent modification of the boronate ester linked COF-5 by copper(I)-catalyzed click chemistry. A variety of functional groups including acetate, butyl, phenyl, ester, and pyrene moieties can be introduced into the channels of pre-synthesized COF-5-X%N_3 bearing different amounts of azide functionalities by means of copper-catalyzed click reactions. This covalent functionalization of the framework occurs without compromising the fidelity and crystallinity of the underlying structure. The parent COF-5 structure, COF-5-(100%N_3), COF-5-(50%AcTrz), and COF-5-(100%AcTrz) are illustrated as examples. Color code: H, white; C, gray; N, blue; B, orange; O, red.

stability. As such, post-synthetic covalent modification is utilized to impart these frameworks with functionality. In the case of the pre-synthetic functionalization of COF-5, the framework is reticulated from HHTP with varying ratios of BDBA (1,4-phenylene diboronic acid) and its azide functionalized analog BDBA-($H_2C{-}N_3$) (Figure 10.10). Here, the post-synthetic functionalization of the framework is achieved by reacting the predisposed azide functionalities with functionalized alkynes, again by copper(I)-catalyzed click chemistry. Various functional groups such as acetyl, butyl, phenyl, methyl esters, or pyrene moieties can be appended with full retention of the crystallinity of the parent structure [23].

10.4.3.2 Post-Synthetic Succinic Anhydride Ring Opening

A different strategy is applied for the introduction of carboxylic acid groups. Carboxylic acids interfere with the formation of Schiff-base COFs since they act as catalysts for imine formation and as such disturb the reaction equilibrium. To introduce carboxylic acids post-synthetically the use of ring-opening reactions of phenolic units on the linker with succinic anhydride are employed. The feasibility of this approach is once again demonstrated for COF-366, in this case formed from H_2TAP and varying ratios of BDA and BDA-$(OH)_2$ (Figure 10.11).

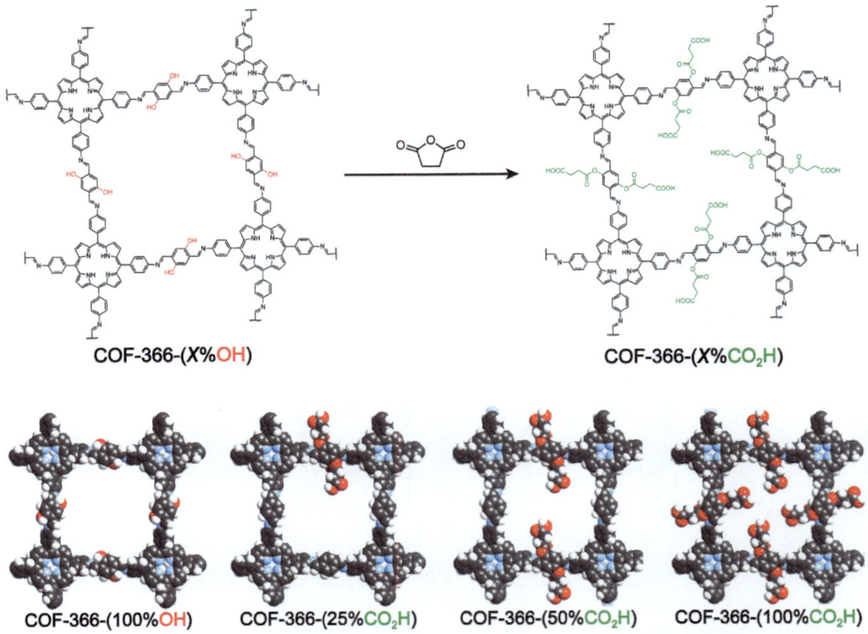

Figure 10.11 Post-synthetic succinic anhydride ring-opening reaction on COF-366-(X%OH). Hydroxyl groups are introduced into the parent COF-366 structure by substituting (partially or completely) BDA with BDA-(OH) during the synthesis. Pre-synthetically functionalized COF-366-(100%OH) and post-synthetically modified COF-366-(25%CO_2H), COF-366-(50%CO_2H), and COF-366-(100%CO_2H) are shown as examples for this approach. Color code: H, white; C, gray; N, blue; O, red.

Frameworks with varying amounts of hydroxyl groups and their carboxyl group functionalized derivatives differ significantly with respect to their CO_2 sorption behavior. The binding affinity of the framework for CO_2 correlates with the nature and amount of the functional groups present in the pores. The framework with 100% carboxylic acid functionalities shows the highest binding affinity with a Q_{st} value of $-43.5\,kJ\,mol^{-1}$ that far exceeds the value for the hydroxyl group functionalized progenitor of $-36.4\,kJ\,mol^{-1}$. Notably, compared to click chemistry this type of reaction requires no metal catalysts and proceeds cleanly, representing a promising strategy for covalent post-synthetic modification [21].

10.4.3.3 Post-Synthetic Nitro Reduction and Aminolysis

The post-synthetic introduction of amides into COFs can be achieved by means of a two-step post-synthetic modification. The **hcb** topology framework $[(TFP)_2(BZ\text{-}(NO_2)_3]_{\beta\text{-}ketoenamine}$ is constructed from trigonal tritopic TFP and linear ditopic BZ-$(NO_2)_2$ (2,2′-dinitrobenzidine) through β-ketoenamine bond formation (Figure 10.12) [24]. Amines on the linker would interfere with the reticulation as they react with the TFP building units, and therefore a post-synthetic approach needs to be employed. In the first step, the nitro groups on the as-synthesized COF are reduced by $SnCl_2$ to afford the amine functionalized framework, termed $[(TFP)_2(BZD\text{-}(NH_2)_3]_{\beta\text{-}ketoenamine}$. The resulting

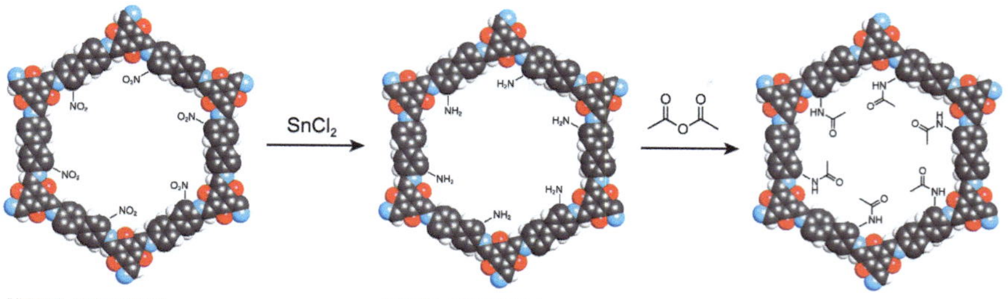

Figure 10.12 Two-step post-synthetic conversion of nitro groups in the mesoporous $[(TFP)_2(BZ\text{-}(NO_2)_3]_{β\text{-}ketoenamine}$: reduction of the preinstalled nitro groups on the linker to amines using $SnCl_2$ followed by aminolysis of acetic anhydride yield the amide functionalized $[(TFP)_2(BZ\text{-}((NHCOCH_3)_2)_3]_{β\text{-}ketoenamine}$. Owing to the high inherent chemical stability of the frameworks' β-ketoenamine linkages and the basic character of the introduced functionalities the resulting materials are utilized for vapor sorption of lactic acid. Color code: H, white; C, gray; N, blue; O, red.

framework can now make use of the amine moieties on the linker and this is illustrated by aminolysis of acetic anhydride as the second step to obtain the amide functionalized COF, termed $[(TFP)_2(BZ\text{-}((NHCOCH_3)_2)_3]_{β\text{-}ketoenamine}$ (Figure 10.12). Owing to the acid stability of these frameworks endowed by the β-ketoenamine linkages and the basic character of the introduced functionalities they are employed as a sorbent for lactic acid. Comparison of $[(TFP)_2(BZ\text{-}(NO_2)_3]_{β\text{-}ketoenamine}$, and $[(TFP)_2(BZ\text{-}((NHCOCH_3)_2)_3]_{β\text{-}ketoenamine}$ reveals that the amine-functionalized material has the highest uptake (6.6 wt%), followed by the amide-functionalized analog (4.0 wt%), whereas the parent framework only shows an uptake of 2.5 wt%.

10.4.3.4 Post-Synthetic Linker Exchange

Construction of an imine-linked COF from tetratopic TCA ((1,1′,3′,1″-terphenyl)-3,3″,5,5″-tetracarbaldehyde) and linear ditopic BZ building units yields $[(TCA)(BZ)_2]_{imine}$. This framework crystallizes in the hexagonal space group *P*6 and in an unusual **fxt** net. The framework has three distinct kinds of pore. One micropore that is not gas accessible and two different mesopores with the maxima in the pore-size distribution centered around 2.56 and 3.91 nm. The BZ linkers in this framework can be exchanged to shorter PDA linkers without losing the crystallinity of the framework. Owing to the increased electron density on the two nitrogen atoms introduced by the inductive effect of the respective electron donating amine functionality in the para-positions, the amines of the shorter PDA linker are more nucleophilic than the ones in BZ. Hence, under reversible conditions the linkers exchange, resulting in the formation of an isoreticular analog of the framework with decreased pore size (Figure 10.13). Exposing $[(TCA)(BZ)_2]_{imine}$ to PDA in the presence of acetic acid as a catalyst yields a new COF termed $[(TCA)(PDA)_2]_{imine}$ over the course of just four hours. The expected change in unit cell dimensions is confirmed by PXRD and the position of the respective maxima in the pore-size distribution shift to smaller

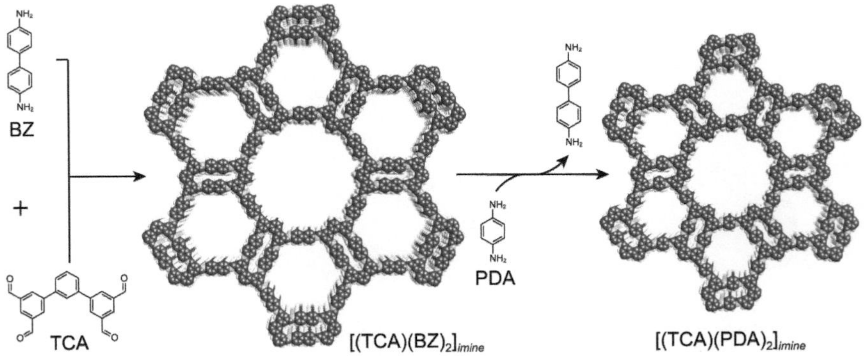

Figure 10.13 An isoreticular analog of the **fxt** topology framework $[(TCA)(BZ)_2]_{imine}$ is obtained by post-synthetic linker exchange of the frameworks' BZ linkers with the shorter PDA. The resulting framework, termed $[(TCA)(PDA)_2]_{imine}$ retains the overall topology and symmetry of its progenitor but features smaller pore sizes. Color code: H, white, C, gray; N, blue.

values of 1.61 and 3.18 nm, in good agreement with the proposed structural model [24].

10.4.3.5 Post-Synthetic Linkage Conversion

Many physico-chemical properties of COFs are determined by the linkage that is used to reticulate the building units into an extended framework structure. Consequently, a lot of research efforts have been devoted to devising new linkage chemistries [25]. For every new linkage a new set of conditions needs to be identified under which the reaction occurs with microscopic reversibility to allow for error correction and thus the formation of the material without defects and in crystalline form. This severely limits the number of suitable linkages in COF chemistry, as many reactions are difficult to run under thermodynamic control and we have covered this challenge in Chapter 8. A strategy to circumvent this very issue is to post-synthetically convert linkages within the crystallized COF. This is illustrated by the conversion of the two COFs, $[(TAPB)_2(BDA)_3]_{imine}$ and $[(ETTA)(BDA)_2]_{imine}$ (where ETTA = 1,1,2,2-tetrakis(4-aminophenyl)ethane), from their imine progenitors to the respective amide-linked counterparts. $[(TAPB)_2(BDA)_3]_{imine}$ is formed from trigonal tritopic TAPB by reticulation with linear ditopic BDA. The resulting framework crystallizes in the hexagonal space group *P*6 and has an underlying **hcb** topology (Figure 10.14). In contrast, $[(ETTA)(BDA)_2]_{imine}$ is constructed from tetratopic ETTA (1,1,2,2-tetrakis(4-aminophenyl)ethane) and BDA building units and has an underlying **kgm** net. Both frameworks feature wide mesoporous hexagonal channels along the crystallographic *c*-direction and therefore allow for facile access of reagents to the sites of conversion throughout the entire material. The conversion of the two frameworks to their amide-linked counterparts is effected by a Pinnick-style oxidation in a mixture of dioxane, acetic acid, and $NaOCl_2$ as an oxidant in the presence of 2-methyl-2-butene as a proton scavenger. The oxidation occurs at room temperature over the course of two days. The resulting amidated frameworks termed $[(TAPB)_2(BDA)_3]_{amide}$

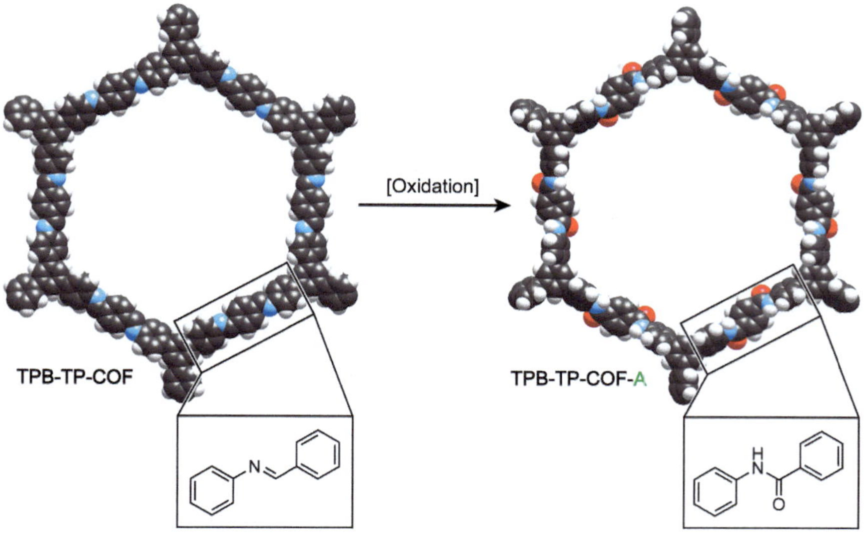

Figure 10.14 Post-synthetic linkage modification of $[(TAPB)_2(BDA)_3]_{imine}$ by oxidation of the imine linkages to chemically more inert amide bonds. The symmetry and structure metrics of the parent framework are retained in the resulting $[(TAPB)_2(BDA)_3]_{amide}$. The modified framework shows a significantly improved chemical stability toward acid and base. Color code: H, white; C, black, N, blue; O, red.

and $[(ETTA)(BDA)_2]_{amide}$ fully retain the crystallinity of their progenitors, albeit with lower surface areas (Figure 10.14). The conversion of the frameworks is confirmed by FT-IR spectroscopy where the complete disappearance of the C=N stretch of the imine bond at ~1620 cm^{-1} and the appearance of the C=O stretch of the amide carbonyl bond at ~1655 cm^{-1} are observed. The conversion is further substantiated by isotopically enriched ^{13}C CP-MAS (^{13}C cross-polarization magic angle spinning) NMR spectra, which show a clear shift of the imine carbon in the starting material at 157 ppm upon oxidation to the amide carbonyl carbon at 166 ppm. The amide-linked COFs show improved chemical stability to both acid and base. Treatment with 12 M aqueous HCl and 1 M aqueous NaOH for 24 hours results in the amorphization and partial dissolution of the imine-linked $[(TAPB)_2(BDA)_3]_{imine}$; however, the amide-linked $[(TAPB)_2(BDA)_3]_{amide}$ fully retains its crystallinity under these conditions [26].

10.5 Summary

Functionalization of covalent organic frameworks is a powerful tool for imparting functionality onto the ordered backbone of these materials. The organic backbone of COFs can be covalently modified pre- and post-synthetically by making use of the extensive toolbox of synthetic organic chemistry and both, the building units as well as the linkage can be targeted. Similarly both, the linkages, as well as the building units can be used as binding sites for the incorporation of metal centers. The pores of COFs can trap functional molecules, biomacromolecules, or nanoparticles. Finally, nanoparticles can be embedded within the crystallites of the extended framework *in situ*. The fact that COFs are porous is crucial for all of the aforementioned functionalization strategies as this not only provides the space for functional groups to be added to the framework without compromising the fidelity and overall metrics of the structure but also because it enables the access to these active sites. Consequently COFs can be considered a true extension of synthetic organic chemistry to the solid state. This gives rise to the notion of treating crystals as molecules. In Chapter 11 we will cover how the macroscopic and nanoscopic morphology of COFs can be structured.

References

1 Smith, B.J., Overholts, A.C., Hwang, N., and Dichtel, W.R. (2016). Insight into the crystallization of amorphous imine-linked polymer networks to 2D covalent organic frameworks. *Chemical Communications* 52 (18): 3690–3693.

2 Tan, J., Namuangruk, S., Kong, W. et al. (2016). Manipulation of amorphous-to-crystalline transformation: towards the construction of covalent organic framework hybrid microspheres with NIR photothermal conversion ability. *Angewandte Chemie International Edition* 55 (45): 13979–13984.

3 Rodríguez-San-Miguel, D., Yazdi, A., Guillerm, V. et al. (2017). Confining functional nanoparticles into colloidal imine-based COF spheres by a sequential encapsulation-crystallization method. *Chemistry - A European Journal* 23 (36): 8623–8627.

4 Wan, S., Gándara, F., Asano, A. et al. (2011). Covalent organic frameworks with high charge carrier mobility. *Chemistry of Materials* 23 (18): 4094–4097.

5 Behar, D., Dhanasekaran, T., Neta, P. et al. (1998). Cobalt porphyrin catalyzed reduction of CO_2. Radiation chemical, photochemical, and electrochemical studies. *The Journal of Physical Chemistry A* 102 (17): 2870–2877.

6 Lin, S., Diercks, C.S., Zhang, Y.-B. et al. (2015). Covalent organic frameworks comprising cobalt porphyrins for catalytic CO_2 reduction in water. *Science* 349 (6253): 1208–1213.

7 Leung, K., Nielsen, I.M., Sai, N. et al. (2010). Cobalt-porphyrin catalyzed electrochemical reduction of carbon dioxide in water. 2. Mechanism from first principles. *The Journal of Physical Chemistry A* 114 (37): 10174–10184.

8 Diercks, C.S., Lin, S., Kornienko, N. et al. (2017). Reticular electronic tuning of porphyrin active sites in covalent organic frameworks for electrocatalytic carbon dioxide reduction. *Journal of the American Chemical Society* 140 (3): 1116–1122.

9 (a) Xu, H., Gao, J., and Jiang, D. (2015). Stable, crystalline, porous, covalent organic frameworks as a platform for chiral organocatalysts. *Nature Chemistry* 7 (11): 905–912. (b) Rosa, D.P., Pereira, E.V., Vasconcelos, A.V.B. et al. (2017). Determination of structural and thermodynamic parameters of bovine α-trypsin isoform in aqueous-organic media. *International Journal of Biological Macromolecules* 101: 408–416.

10 Xu, H., Tao, S., and Jiang, D. (2016). Proton conduction in crystalline and porous covalent organic frameworks. *Nature Materials* 15: 722–726.

11 Kandambeth, S., Venkatesh, V., Shinde, D.B. et al. (2015). Self-templated chemically stable hollow spherical covalent organic framework. *Nature Communications* 6: 6786.

12 (a) Cote, A.P., Benin, A.I., Ockwig, N.W. et al. (2005). Porous, crystalline, covalent organic frameworks. *Science* 310 (5751): 1166–1170. (b) Kalidindi, S.B., Oh, H., Hirscher, M. et al. (2012). Metal@COFs: covalent organic frameworks as templates for Pd nanoparticles and hydrogen storage properties of Pd@ COF-102 hybrid material. *Chemistry - A European Journal* 18 (35): 10848–10856.

13 Guo, J., Xu, Y., Jin, S. et al. (2013). Conjugated organic framework with three-dimensionally ordered stable structure and delocalized π clouds. *Nature Communications* 4: 2736.

14 Ding, S.-Y., Gao, J., Wang, Q. et al. (2011). Construction of covalent organic framework for catalysis: Pd/COF-LZU1 in Suzuki–Miyaura coupling reaction. *Journal of the American Chemical Society* 133 (49): 19816–19822.

15 Clayden, J., Greeves, N., Warren, S., and Wothers, P. (2001). *Organic Chemistry*, 1e. Oxford: Oxford University Press.

16 Zhang, W., Jiang, P., Wang, Y. et al. (2014). Bottom-up approach to engineer a molybdenum-doped covalent-organic framework catalyst for selective oxidation reaction. *RSC Advances* 4 (93): 51544–51547.

17 Chen, X., Huang, N., Gao, J. et al. (2014). Towards covalent organic frameworks with predesignable and aligned open docking sites. *Chemical Communications* 50 (46): 6161–6163.

18 Campbell, K., McDonald, R., Ferguson, M.J., and Tykwinski, R.R. (2003). Functionalized macrocyclic ligands: big building blocks for metal coordination. *Organometallics* 22 (7): 1353–1355.

19 Baldwin, L.A., Crowe, J.W., Pyles, D.A., and McGrier, P.L. (2016). Metalation of a mesoporous three-dimensional covalent organic framework. *Journal of the American Chemical Society* 138 (46): 15134–15137.

20 Xu, F., Xu, H., Chen, X. et al. (2015). Radical covalent organic frameworks: a general strategy to immobilize open-accessible polyradicals for high-performance capacitive energy storage. *Angewandte Chemie International Edition* 54 (23): 6814–6818.

21 Huang, N., Krishna, R., and Jiang, D. (2015). Tailor-made pore surface engineering in covalent organic frameworks: systematic functionalization for performance screening. *Journal of the American Chemical Society* 137 (22): 7079–7082.

22 Xu, H., Chen, X., Gao, J. et al. (2014). Catalytic covalent organic frameworks via pore surface engineering. *Chemical Communications* 50 (11): 1292–1294.

23 Nagai, A., Guo, Z., Feng, X. et al. (2011). Pore surface engineering in covalent organic frameworks. *Nature Communications* 2: 536.

24 Lohse, M.S., Stassin, T., Naudin, G. et al. (2016). Sequential pore wall modification in a covalent organic framework for application in lactic acid adsorption. *Chemistry of Materials* 28 (2): 626–631.

25 (a) DeBlase, C.R. and Dichtel, W.R. (2016). Moving beyond boron: the emergence of new linkage chemistries in covalent organic frameworks. *Macromolecules* 49 (15): 5297–5305. (b) Waller, P.J., Gándara, F., and Yaghi, O.M. (2015). Chemistry of covalent organic frameworks. *Accounts of Chemical Research* 48 (12): 3053–3063. (c) Diercks, C.S. and Yaghi, O.M. (2017). The atom, the molecule, and the covalent organic framework. *Science* 355 (6328): eaal1585.

26 Waller, P.J., Lyle, S.J., Osborn Popp, T.M. et al. (2016). Chemical conversion of linkages in covalent organic frameworks. *Journal of the American Chemical Society* 138 (48): 15519–15522.

11

Nanoscopic and Macroscopic Structuring of Covalent Organic Frameworks

11.1 Introduction

In Chapters 7–10 we have outlined how the existing toolbox of molecular organic chemistry can be applied to the construction and modification of crystalline porous extended organic solids. This chapter covers a challenge that emerges with the transition from solution-based molecular chemistry to the chemistry of insoluble solids such as covalent organic frameworks (COFs). We have highlighted the importance of crystallinity for the elucidation of the structure of COFs on the atomic level. This knowledge is crucial as it enables structure–function correlations and thus rational optimization of the material with respect to its structural properties (topology and structure metrics) and chemical properties (chemical stability of the linkage, appended functionality on the backbone). While crystallinity gives information about the long-range order on the atomic scale it is not strictly related to the nanoscopic or macroscopic morphology of the material. This is an important point to consider with respect to applications of heterogeneous materials where the morphology has a profound impact on the performance. COFs have received a lot of attention in the context of organic electronics because they exhibit high charge carrier mobility, are pseudocapacitive energy storage materials, and can be employed as electrocatalysts [1]. All of these applications benefit from nanostructured forms of COFs. In general, organic materials can be processed (e.g. thin films, crystals) from solution, but COFs, unlike conventional 1D polymers or molecular organic materials, are insoluble in common organic solvents or water and are generally isolated as aggregated microcrystalline powders. With regard to processing powder samples they can be exfoliated into dispersible few- or monolayer sheets by top–down approaches such as sonication, grinding, or chemical exfoliation. Alternatively, a bottom–up approach can be taken and controlled nucleation on substrates, polymerization at the liquid–liquid interface, flow chemistry, or the assembly on surfaces using chemical vapor deposition techniques in ultrahigh vacuum are being explored. In the following text, we will give an overview of synthetic strategies for structuring COFs and illustrate advantages and disadvantages of the respective approaches.

Introduction to Reticular Chemistry: Metal-Organic Frameworks and Covalent Organic Frameworks,
First Edition. Omar M. Yaghi, Markus J. Kalmutzki, and Christian S. Diercks.

Companion website: www.wiley.com/go/yaghi/reticular

11.2 Top–Down Approach

One broadly applicable strategy to access thin films of COFs is to exfoliate crystallites of layered frameworks. Such structures are generally constructed from aromatic building units and an energy input is required to break up the strong π–π interactions between layers. This energy can be applied to the system by different means such as sonication or mechanical grinding. The advantage of these top–down approaches is that they are broadly applicable to a wide range of COFs as they do not have any specific requirements with respect to the structure or chemical composition of the material. Unfortunately, the thicknesses of the films obtained this way often show a broad size distribution. Alternatively, the layers in COF crystallites can be manipulated chemically by reaction with bulky substituents to set the layers apart and break the π–π interactions, an approach referred to as chemical exfoliation. Here, the advantage is that very thin films of uniform thickness can be obtained but the applicability of this approach is limited by the necessity for specific functionalities to carry out the chemical transformation inducing the exfoliation.

11.2.1 Sonication

Exposure of COF-43, a hydrazone-linked COF, to certain organic solvents leads to a substantial decrease in crystallinity (Figure 11.1a) [2]. This is surprising given the fact that based on chemical intuition, a hydrazone-linked framework should be stable under such conditions. It was found that samples of COF-43 exposed to organic solvents such as THF (tetrahydrofuran), chloroform, toluene, and MeOH (methanol) retained their crystallinity, while those exposed to dioxane, water, and DMF became amorphous. FT-IR spectra before and after solvent

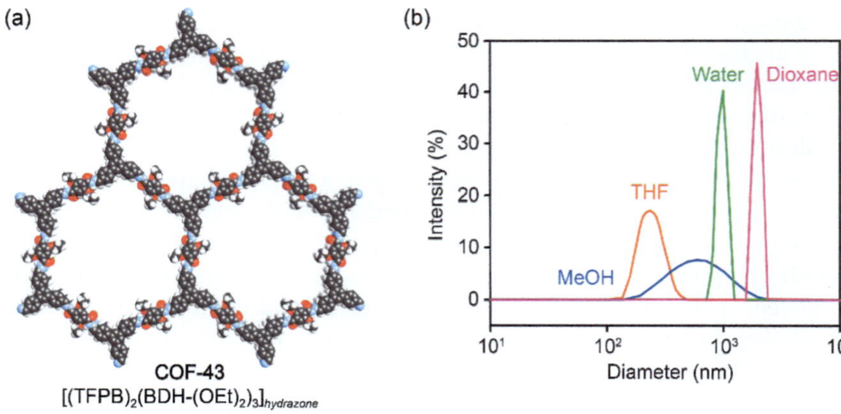

Figure 11.1 (a) Structure of the mesoporous hydrazone-linked COF-43. (b) Average size of COF-43 dispersions derived from dynamic light scattering of solutions of microcrystalline COF-43 powder after brief sonication in THF, MeOH, water, and dioxane, respectively. Larger dispersed fragments are observed in exfoliating solvents (water, dioxane) as compared to non-exfoliating solvents (THF, MeOH). Color code for (a): H, white; C, black; N, blue; O, red.

exposure show that the characteristic C=O and C=N stretching vibrations at 1656 and 1597 cm^{-1} of the hydrazone linkages are fully retained. The size of dispersed COF particles from different solvents, determined by dynamic light scattering (DLS), depends on whether the solvent induces apparent crystallinity loss. Atomic force microscopy (AFM) studies of COF particles from suspensions sonicated in dioxane indicate platelets with lateral widths of 200 nm and average heights of 1.32 ± 0.37 nm, corresponding to the thickness of 3–5 layers. More importantly, structures as thin as 3.3 Å are obtained from suspensions of COF-43 in water, corresponding to the formation of bi- or even single-layer structures. In contrast, COF suspensions in THF yield particles of morphologies similar to the as-synthesized COF powders. The size dependence of the COF particles from different organic solvents is illustrated in Figure 11.1b. TEM (transmission electron microscopy) confirms that the exfoliated layers retain their hexagonal symmetry and long-range order. The simplicity of this approach to the formation of COF thin films holds promise for large-scale fabrication from microcrystalline powders.

11.2.2 Grinding

The β-ketoenamine-linked **hcb** topology framework TPa-1 is prepared by reacting TFP with BDA. The resulting microcrystalline COF powder can be delaminated by grinding in a mortar for 30 minutes in the presence of a few drops of methanol. The powder of the delaminated COF can be suspended in methanol with concentrations of around 0.04 mg mL^{-1} (~8 wt%) [3]. FT-IR spectra of the material show full retention of the characteristic C=O stretch and C–N stretch at 1580 and 1250 cm^{-1}, respectively. Powder X-ray diffraction (PXRD) patterns of the sheets corroborate that the integrity of the structure is fully retained, albeit with decreased intensity and broadening of the observed reflections. TEM images of the material show 100 nm to 1 μm thin graphene-like sheets. The laminar structure of the COFs was further assessed by AFM measurements, which reveal flat nanosheet-like structures with lengths and widths of several micrometers and thicknesses ranging from 3 to 10 nm. This corresponds to 10–30 COF layers, in good agreement with the TEM data. The advantage of mechanical grinding for the exfoliation of COFs is rooted in the notion that expensive organic solvents are not necessary to obtain the films.

11.2.3 Chemical Exfoliation

The anthracene-based $[(TFP)_2(DAA)_3]_{\beta\text{-}ketoenamine}$, is synthesized by the reticulation of linear ditopic DAA (2,6-diaminoanthracene) with trigonal tritopic TFP (2,4,6-triformylphloroglucinol). $[(TFP)_2(DAA)_3]_{\beta\text{-}ketoenamine}$ crystallizes in a **hcb** topology in the hexagonal space group $P6/mmm$ and features large 2.2 nm wide hexagonal channels along the crystallographic c-direction (Figure 11.2). The anthracene building units can post-synthetically be modified by [4+2]-Diels–Alder cycloaddition reactions with N-hexylmaleimide, thereby introducing bulky substituents between the individual COF layers to set them apart and delaminate the material [4]. Following the reaction by FT-IR shows

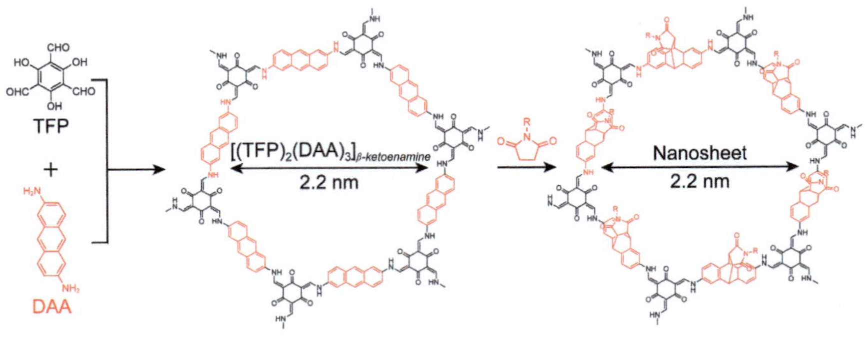

Figure 11.2 Chemical delamination of the β-ketoenamine-linked $[(TFP)_2(DAA)_3]_{\beta\text{-}ketoenamine}$ by Diels–Alder reaction with *N*-hexylmaleimide. Illustrated are the chemical structure of $[(TFP)_2(DAA)_3]_{\beta\text{-}ketoenamine}$ and the reaction with *N*-hexylmaleimide to disrupt the π–π stacking between the layers and form the $[(TFP)_2(DAA)_3]_{\beta\text{-}ketoenamine}$ sheets. Only one pore is shown for clarity.

that the characteristic C=C and C–N stretching bands at 1590 and 1270 cm^{-1} of $[(TFP)_2(DAA)_3]_{\beta\text{-}ketoenamine}$ are retained, while additional stretching bands at 2937 and 2857 cm^{-1} corresponding to C–H vibrations arise from the *N*-hexyl groups of *N*-hexylmaleimide and the characteristic imide C=O bond stretch at 1695 cm^{-1} of the maleimide itself. Analyses of the exfoliated sheets by scanning electron microscopy (SEM) demonstrate that the ribbon-like aggregated morphology of $[(TFP)_2(DAA)_3]_{\beta\text{-}ketoenamine}$ changes to micrometer-sized lateral rippled sheets. This phenomenon is also reflected in the TEM images, where $[(TFP)_2(DAA)_3]_{\beta\text{-}ketoenamine}$ again shows a ribbon-like morphology (100–200 nm in length, 20–40 nm in width). In contrast, the delaminated sheets exhibit a thin-sheet-like morphology (500 nm in width and 200 nm in thickness). The larger lateral feature size of the delaminated COFs is attributed to non-covalent interactions between small functionalized layers. AFM analysis of the delaminated $[(TFP)_2(DAA)_3]_{\beta\text{-}ketoenamine}$ shows an average thickness of about 17 nm. Owing to the presence of long alkyl chains and the resulting dipole interactions, the delaminated $[(TFP)_2(DAA)_3]_{\beta\text{-}ketoenamine}$ assembles into multilayer sheets with increased thickness. When the layers are suspended in dichloromethane and irradiated with a red laser (λ = 650 nm), a characteristic Tyndall effect is observed, indicative of single layer sheets in solution.

A scalable thin film composed of the delaminated $[(TFP)_2(DAA)_3]_{\beta\text{-}ketoenamine}$ can be obtained at the liquid–air interface. Toward this end, water is chosen as the liquid phase since the hydrophobic alkyl chains prevent the diffusion of the COF layers into the solvent. To fabricate a self-standing thin film, a suspension of delaminated $[(TFP)_2(DAA)_3]_{\beta\text{-}ketoenamine}$ in dichloromethane is added dropwise onto the water surface where it evaporates, leading to the formation of a semi-transparent thin film. TEM images show the aggregation of small nanosheets of 60–80 nm in width. The thickness of these films can be tuned from 1.2 to 1.6 nm (indicative of a single layer) up to 1.0–2.5 μm, by adjusting the concentration of the delaminated COF in the suspension. This high degree of control over the film thickness down to a single monolayer renders chemical delamination a powerful approach for the formation of thin films of COFs.

11.3 Bottom–Up Approach

In contrast to top–down approaches, where the COF is first isolated as a microcrystalline powder and subsequently exfoliated to obtain nanostructured materials, bottom–up approaches have two significant advantages: (i) a higher degree of control over the feature size of the material, (ii) the potential for obtaining them in morphologies other than thin films, and (iii) the potential for direct interfacing with specific substrates. The big challenge in the bottom–up formation of nanostructured COFs is due to practical limitations with respect to the synthesis conditions employed. In Chapter 8 we have discussed the mechanism of bond formation for the various COF linkages that have been developed. It is important to consider that the mechanism of how these materials crystallize is much more complex than that. Understanding and manipulating the nucleation and subsequent seed growth during the crystallization of COFs is at the heart of controlling their morphology. Under conventional synthesis conditions COFs are isolated as aggregated microcrystalline powders and to obtain them as nanoparticles of different sizes or as films with uniform and controllable thickness several challenges need to be addressed. Conventional COF growth is generally carried out in suspensions rather than in homogeneous solutions. As a result, there is no control over the initial nucleation step and consequently particles of various sizes and shapes are obtained. To address this and figure out the parameters that need to be tuned to control the frameworks' morphology and feature size, the mechanism of COF crystallization needs to be deciphered. In the following, we will limit the discussion to frameworks constructed from boronate ester formation and reversible Schiff-base chemistry as for those linkages the crystallization mechanism has been thoroughly investigated.

11.3.1 Mechanism of Crystallization of Boronate Ester COFs

The mechanism of boronate ester COF formation has been studied for the case of COF-5. Homogeneous growth conditions were developed as opposed to the conventional reticulation of the starting materials from suspensions. The starting materials, HHTP and BDBA, for the formation of COF-5 are soluble in a 4 : 1 v/v mixture of dioxane/mesitylene with a small amount (15 equiv with respect to HHTP) of methanol in the reaction mixture. These homogeneous growth conditions enable the determination of the rate of precipitation of COF-5 by measuring the solution turbidity as a function of reaction time (Figure 11.3a). Under the homogeneous reaction conditions, a reproducible induction period during which soluble oligomers form and nucleation occurs is observed. After a few minutes at 90 °C, the crystalline COF starts to precipitate, at which point a change in the turbidity of the solution is recorded.

To test whether the reaction occurs reversibly a monofunctional catechol competitor, 4-(*tert*-butyl)benzene-1,2-diol, is added (Figure 11.3c). The addition,

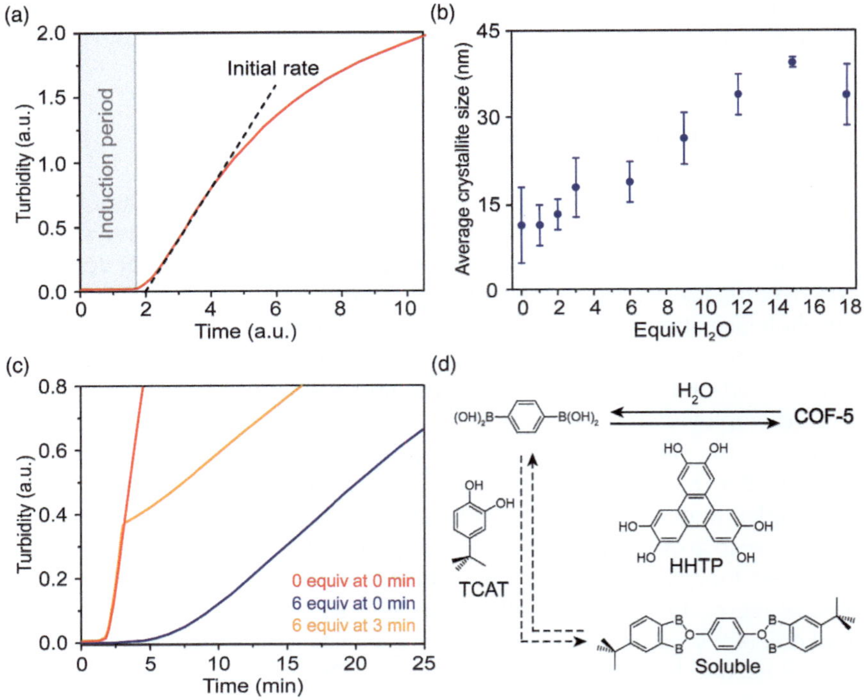

Figure 11.3 Mechanistic studies for the formation of the boronate ester-linked framework COF-5. The system is tested under homogeneous reaction conditions starting from HHTP and BDBA in a 4 : 1 v/v mixture of dioxane/mesitylene. 15 equiv of methanol with respect to HHTP are added to enhance solubility. (a) The reaction occurs after an initial induction period. (b) Addition of water slows down the reaction but also increases the average crystallite size of COF-5. (c) This induction period is prolonged by addition of the monofunctional catechol competitor, *o*-cresol, at the beginning of the reaction. Addition at a later time, where the COF has already partially precipitated, does not result in re-dissolution of the framework but slows down further COF formation. (d) The experiments indicate that in the crystallization mechanism the mono-catechol competitor TCAT slows down the reaction by reversibly competing for the starting materials but does not participate in the reaction equilibrium. Water slows down the reaction by actively modulating the COF-forming equilibrium and this increased reversibility improves the crystallinity of the sample.

even in large excess, slows down but does not inhibit the reaction. This fact, in conjunction with the observation that the modulator is not incorporated into the final product, implies reversible bond formation at the early stages of the reaction. Addition of the modulator after partial precipitation of the COF slows down the formation of additional product but does not redissolve already formed crystallites, suggesting reversibility only in the early stages of seed growth and the existence of an irreversible step further down the crystallization/precipitation pathway. The addition of water (which is known to hydrolyze boronate ester COFs) to the reaction mixture also slows down COF formation, but in contrast to the addition of *o*-cresol, has a profound impact on the crystallite size (Figure 11.3b). By varying the amount of water added to the solution, the average crystallite size is doubled as determined by applying

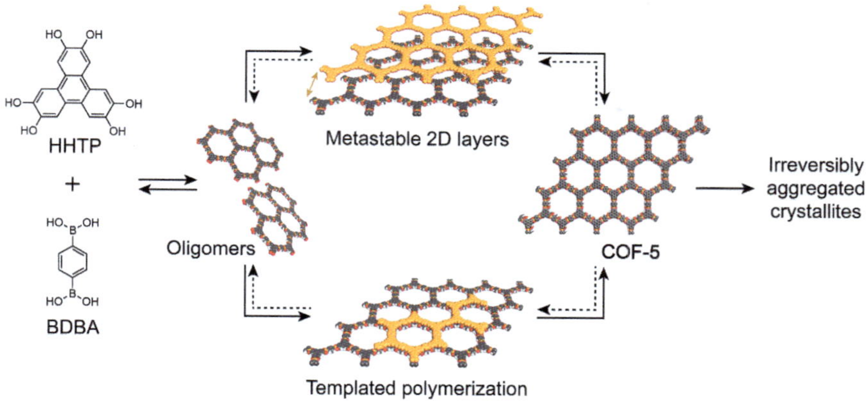

Figure 11.4 Mechanism of the formation of COF-5. In a first reversible step HHTP and BDBA react and form soluble oligomers in solution. COF layers are formed upon further reaction. These metastable layers either stack with other layers or serve as nucleation sites for templated polymerization. The products of this step are colloidal COF-5 crystals. In a last irreversible step, individual COF-5 crystallites aggregate and precipitate from solution. COF-5 layers are partially highlighted in orange for clarity. Color code: H, white; B, orange; C, gray; O, red.

the Scherrer equation to the reflections observed in the powder patterns of the resulting COF. Taking stock of these findings, a mechanism for the formation of COF-5 can be proposed (Figure 11.3d). In the first step, small oligomers form from the starting materials in solution in a reversible manner. These oligomers function as seeds for the subsequent reversible growth of individual COF sheets. After formation of these sheets the material grows into individual COF crystallites in a reversible manner and subsequently aggregates and precipitates, at which point the material is withdrawn from equilibrium. This last irreversible step can be reversed to a certain extent by the addition of water and concomitant partial framework hydrolysis (Figure 11.4) [5].

11.3.1.1 Solution Growth on Substrates

Based on the mechanism of boronate ester COF formation, the first challenge that needs to be addressed with respect to processing is to control the crystallite nucleation. A strategy to preferentially nucleate and thus control the growth of COF thin films makes use of substrates composed of single-layer graphene (SLG) supported on SiO_2 or copper wavers (Figure 11.5). The substrates are added as preferred nucleation sites to favor the formation of COF films on the SLG supported SiO_2 over random nucleation and crystallite growth in solution. For the formation of COF-5 thin films the substrates are directly added to the COF-forming reaction containing BPDA and HHTP. GIWAXS (grazing incidence wide angle X-ray scattering) shows that the layered films grow with preferred orientation of the layers normal to the SLG surface, which can be rationalized based on the maximized π–π interactions in this arrangement. The coverage and thickness of the films on the SLG surface are evaluated using SEM. A top–down micrograph of the COF films grown on SLG/Cu for 30 minutes indicates complete coverage of the graphene surface. A few bulk crystallites

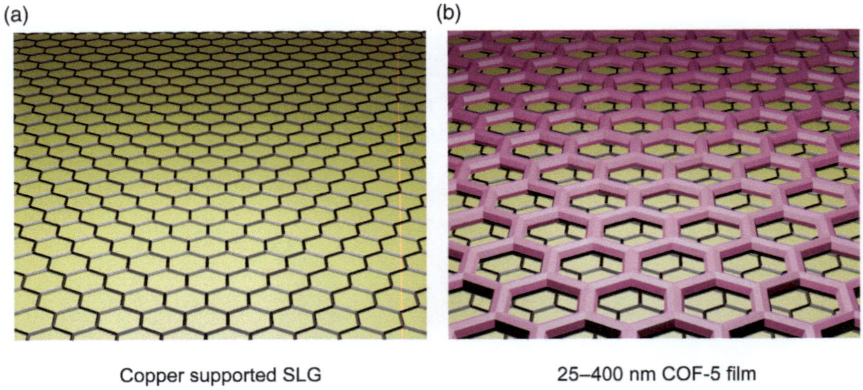

Figure 11.5 Formation of COF-5 on a copper supported single layer graphene substrate. (a) Illustration of the copper supported single layer graphene. (b) Oriented thin films form where the COF layers crystallize parallel to the graphene sheet with a thickness of 195 ± 20 nm. Color code: Cu, gold; graphene, black; COF-5, pink.

are scattered on top of the film but can mostly be removed by sonicating the substrate in dry toluene for 10 seconds, after which the films are uniform over approximately 100 mm^2 areas. The cross section of a film grown for 30 minutes has a uniform thickness of 195 ± 20 nm, which corresponds to approximately 580 layers of COF. The limitations of this approach toward the bottom–up formation of boronate ester-linked COFs is the lack of control over the COF nucleation in solution resulting in the formation of aggregated crystallites that precipitate, some of which end up on the substrate thus resulting in less uniform films [6].

11.3.1.2 Seeded Growth of Colloidal Nanocrystals

The mechanism of boronate ester COF formation implies that the precipitation of the framework from solution is due to irreversible agglomeration of individual crystallites. These crystallites are on the order of 100 nm in diameter and if their aggregation was prevented they could be processed as colloids from solution. Circumventing the precipitation requires the determination of conditions under which polymerization occurs but aggregation is inhibited. The addition of different amounts of nitrile solvents such as acetonitrile or benzonitrile results in stable colloidal suspensions of COF-5 where aggregation does not take place. The average particle size and polydispersity index for the colloids was determined for reaction mixtures of varying monomer and acetonitrile concentrations after reaction at 90 °C for 20 hours. For the tested concentrations of acetonitrile between 15 and 95 vol% the colloids exhibit Gaussian size distributions and low polydispersity indices. The average particle size is invariant at 45–60 nm for acetonitrile concentrations higher than 55%, which suggests stabilization of discrete crystallites at high acetonitrile concentrations. Larger colloids of dimensions between 100 and 240 nm are observed at 15 and 35 vol% acetonitrile, respectively, likely due to the aggregation of smaller crystallites. Once formed, the COF-5 colloids are stable for more than a month. Notably, the size of the colloids does not change with time and can also not be modified by further addition of acetonitrile, indicating that once formed the particles are kinetically inert without appreciable monomer exchange. Unlike COF-5 in microcrystalline form, the colloidal solution is readily processable by evaporation of the solvent, and free-standing films of 10 μm in thickness can be obtained, which retain both the crystallinity and porosity of the parent material [7].

Making use of conditions where aggregation and precipitation are circumvented, stable colloids can function as seeds for further growth of larger single-crystalline COF particles. This is achieved when 80% of acetonitrile is included as a cosolvent in the COF-5-forming reaction to yield a COF-5 dispersion with colloids of 30 nm in size. Fast addition of monomers to this colloidal solution results in limited growth of the seeds and instead in new nucleation sites. This is supported by DLS measurements where the average crystallite size of the colloid decreases. In contrast, slow addition of the monomers favors the growth of existing colloids and only limited new nucleation is observed (Figure 11.6).

When the monomers are added slowly at a rate of 0.10 equiv h^{-1} seeded growth dominates and the average particle size increases. The particle size can be increased from 400 nm to 1 μm upon addition of 4.0 equiv of starting materials. Notably, the size distribution of crystallites as determined by DLS remains monodisperse. Addition of the monomers at a higher rate of 1.0 equiv h^{-1} results in an apparent decrease in the average particle size and a bimodal size distribution

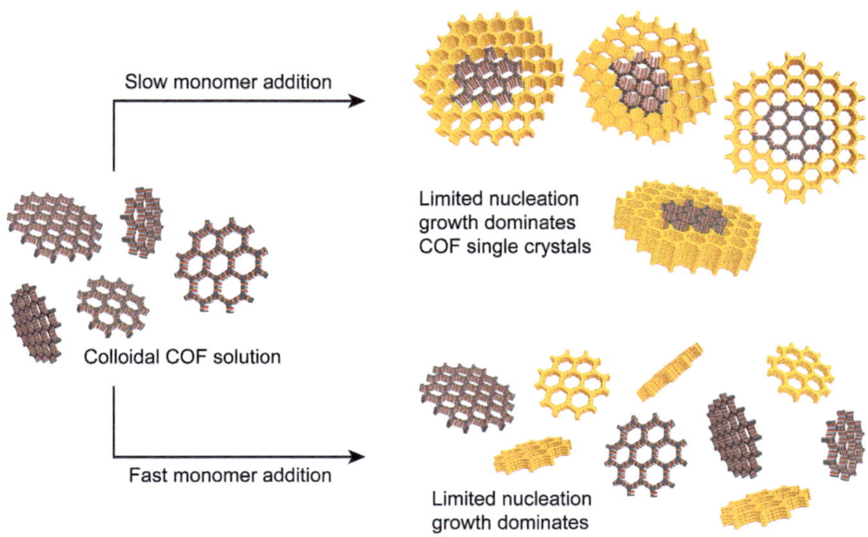

Figure 11.6 Mechanism of seeded growth of COF-5 single crystals. A stable colloidal solution of COF-5 is obtained in reaction mixtures containing organic nitriles such as acetonitrile as solvents. Fast addition of monomers to this colloidal solution favors growth of the existing particles over new nucleation and consequently the particle size of COF-5 crystals is increased up to 1.5 μm. In contrast, when the monomers are added at a higher rate, new nucleation in solution dominates and limits the growth of existing particles. COF-5 formed upon further addition of monomers is highlighted in orange. Color code: H, white; B, orange; C, gray; O, red.

of particles, indicative of newly formed seeds upon addition of starting materials. The particles grown by this seeded growth approach are single crystalline as characterized by TEM but no unambiguous structure solution was reported.

11.3.1.3 Thin Film Growth in Flow

One of the biggest challenges in the formation of COF thin films is the precise control over the film thickness. A protocol that allows incremental adjustment of film thickness is to grow them in flow and monitor the deposition using a quartz crystal microbalance (QCM). The formation of COF-5 and isoreticular analogs was studied by reacting HHTP with 1,4-phenyldiboronic acid (PDA) and expanded versions thereof. The reaction is run under homogeneous growth conditions, which is a necessary requirement for implementation into a flow cell setup. Under these conditions the induction period before precipitation of the framework is two minutes (Figure 11.7b). Accordingly, the length of the tubing needs to be adjusted in such a way that the material reaches the flow cell after two minutes to avoid clogging. By flowing the solution of reactants over the substrate a continuous deposition of mass with time is observed by applying the Sauebrey equation to the frequency response of the QCM (Figure 11.7a). This continuous increase in mass deposition goes along with an increase in resistance, further corroborating the continuous deposition of COF on the substrate. The deposited films are crystalline as evidenced by grazing incidence wide angle X-ray (GIWAX) scattering measurements, and the observed diffraction patterns are in good agreement with the proposed structural models. SEM images show continuous defect-free films without adhering crystallites. Films between 15 and 110 nm are made in a highly controllable manner as confirmed by AFM imaging [8].

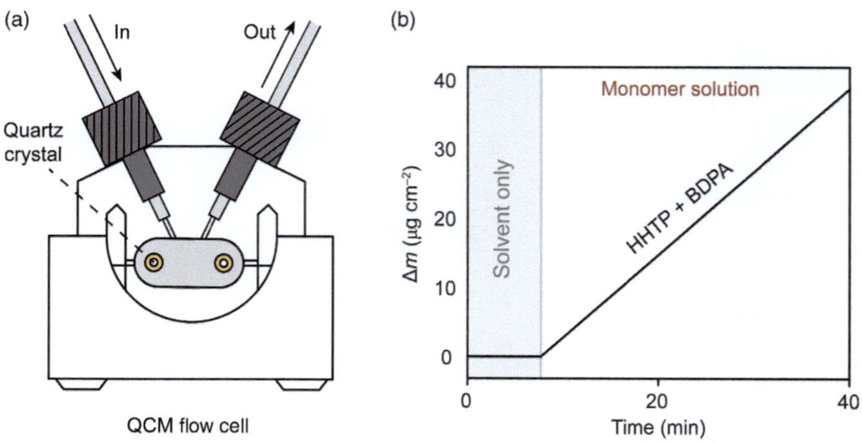

Figure 11.7 Growing films of COF-5 using flow chemistry. (a) Illustration of a flow cell equipped with a quartz crystal microbalance (QCM) used in the experiment to follow the mass deposition of COF (left). (b) Change in mass of COF-5 formed from a solution of HHTP and BDPA in the heated flow cell as a function of the deposition time as determined by QCM measurements.

11.3.1.4 Thin Film Formation by Vapor-Assisted Conversion

Another approach to growing thin films of COFs with a high level of control over the thickness is based on the application of steam-assisted conversion. Here, the idea is that the starting materials are processed into the desired morphology and only then be converted into the COF. Inherently, this approach cannot be carried out in solution. Alternatively, the starting materials are exposed only to vapor of solvents to prevent dissolution and loss of their preprocessed morphology. In practice, HHTP and BDA are dissolved in a mixture of acetone and ethanol and drop cast on a clean glass substrate. This glass substrate is subsequently placed into a desiccator along with a small vessel containing mesitylene and dioxane in a 1 : 1 v/v ratio. To achieve full conversion, the desiccator is stored for 72 hours at room temperature to yield evenly covered films of COF on the glass slides. SEM shows that the films are composed of small intergrown particles forming a continuous coverage on the substrate. Cross-sectional SEM reveals film thicknesses of 300 nm to 7.5 μm depending on the concentration of the precursor solution. PXRD of the films confirms the formation of COF-5 in crystalline form. To verify whether the films are porous, krypton sorption at 77.3 K was carried out after activation of the films at 150 °C under dynamic vacuum yielding a surface area of 990 $m^2 g^{-1}$. TEM micrographs of the polycrystalline COF films display the expected hexagonal honeycomb structure. This method holds promise for the formation of thin films due to the synthetic ease with which the films can be formed [9].

11.3.2 Mechanism of Imine COF Formation

For imine-linked COFs, a vastly different growth mechanism is observed, and we will illustrate this for the mesoporous **hcb** COF$[(TAPB)_2(BDA)_3]_{imine}$ (Figure 11.9). Homogeneous conditions are achieved in a mixture of 4 : 1 v/v dioxane/mesitylene. In contrast to the boronate ester-linked COF-5, which requires heating for the reaction to occur and shows an induction period of several minutes after which the material precipitates, no reaction occurs for $[(TAPB)_2(BDA)_3]_{imine}$ even when heating. When glacial acetic acid is added as a catalyst the reaction proceeds within seconds at room temperature but the resulting material is amorphous. Since the imine bond formation is a condensation reaction the influence of water on the crystallinity was tested and it was found that when both glacial acetic acid and water are added the initially amorphous precipitate transforms into a crystalline material over the course of two days (Figure 11.8a).

Based on these observations a reaction mechanism was proposed in which in the first step, an amorphous polyimine network forms, which over time slowly transforms into the targeted crystalline COF by reversible bond formation and concomitant error correction (Figure 11.8b). The different reaction mechanisms for boronate ester COFs and imine COFs imply that precipitated polyimine can still undergo reversible error correction, while boronate ester COFs precipitate irreversibly and directly in crystalline form (Figure 11.9). This is corroborated by the fact that for imine COFs linker exchange reactions are well established

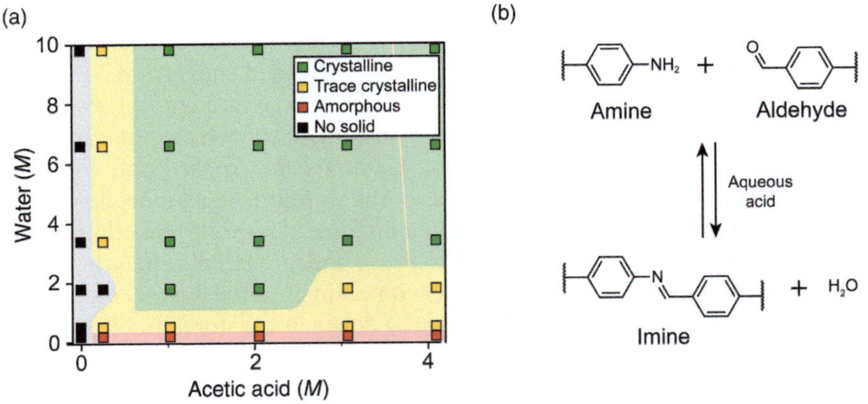

Figure 11.8 Mechanistic studies for the formation of the imine-linked framework TPB-TP-COF. In a 4 : 1 v/v mixture of dioxane/mesitylene at 70 °C the reaction occurs homogeneously. (a) The influence of the addition of water and acetic acid shows that without addition of acetic acid no solid is formed. In contrast, no addition of water results in the formation of amorphous solid. By tuning the amount of both acid and water, crystalline material can be obtained. (b) With respect to the mechanism of COF formation it is concluded that both acid and water have to be added to the reaction mixture. While the acid serves as a catalyst to speed up the reaction, the water modulates the equilibrium between the starting materials and the product.

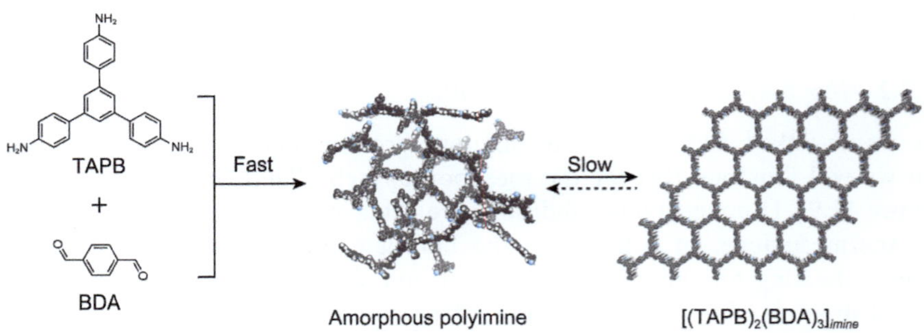

Figure 11.9 Mechanism of $[(TAPB)_2(BDA)_3]_{imine}$ formation. In a first step, TAPB and BDA rapidly react to form an amorphous polyimine network. In a second slow step, the amorphous polymer is transformed into the crystalline COF by reversible bond formation and concomitant error correction. Color code: H, white; C, gray; N, blue.

whereas this has not been reported for boronate ester COFs [10]. The fundamental difference between the two mechanisms is that boronate ester COFs precipitate in crystalline form; their crystallization and precipitation occur simultaneously. For imine COFs the challenge in nanostructuring is that the crystallization and precipitation step occur separately. As a consequence, the strategies used for controlling the morphology of imine COFs are fundamentally different from the ones applied to boronate ester COFs.

11.3.2.1 Nanoparticles of Imine COFs

As discussed earlier, the crystallization of imine COFs is a two-step process where at first an amorphous polyimine forms, which is subsequently converted into a crystalline COF. In practice, both of these steps are usually achieved in a one-pot reaction. However, under the conditions that are required for the COF to crystallize, the polyimine formation is too rapid to effectively control nucleation. One approach to circumvent this issue is to divide the COF crystallization into two separate steps. For the formation of polyimines homogeneous growth conditions have been identified where the nucleation is prolonged and can thus be controlled more effectively through the addition of specific nucleation seeds into the reaction mixture. We have described this approach in Chapter 10 in the context of growing COFs around metal-oxide nanoparticles. When TAPB is reacted with TFB under homogeneous conditions in the absence of water the growth of the polymer proceeds slowly and exclusively on the surface of these added nucleation sites. 9.8 nm Fe_3O_4 nanoparticles, 9.0 nm Au nanoparticles, or 3.3 nm Pd nanoparticles were shown to serve as nucleation sites in this process. The second crystallization step is carried out under regular COF forming conditions and in the presence of acetic acid to yield the respective nanoparticles enclosed in 100 nm spheres of $[(TAPB)(TFB)]_{imine}$. This approach is useful because it represents a facile strategy for interfacing COFs with various other substrates. Furthermore, due to the controlled nucleation process it is in theory possible to control the thickness of the COF layer by adjusting the reaction time or the concentration of starting materials. However, using this approach, the nucleation of COFs only works in the presence of a substrate, which decreases the generality of the approach [11].

In order to address this shortcoming a different approach has to be taken. Instead of running the reaction in two consecutive steps, the initial nucleation period must be slowed down compared to the rate of error correction. In order to control the nucleation process the reaction intermediates need to be readily soluble. Toward this goal, partially Boc-protected (Boc, *tert*-butyloxycarbonyl) amine linkers are employed in COF synthesis. These linkers are slowly deprotected *in situ* thereby slowing down the polymerization step, whereas the rate of error correction remains unaltered. The formation of LZU-1 is carried out under homogeneous conditions when NBPDA(4-(*tert*-butoxycarbonylamino)-aniline) is used (instead of 1,4-benzene diamine) and reacted with TFP (1,3,5-trisformylphenyl benzene) in ethanol as the solvent. TFA (trifluoro-acetic acid) is used as a catalyst and PVP (polyvinylpyrolidone) as a capping agent. In the homogenized imine COF synthesis, nuclei form from clear solution and grow into crystalline frameworks before precipitation. In a typical synthesis, NBPDA and TFB are dissolved in ethanol in the presence of PVP and TFA, and subsequently heated at 120 °C for 30 minutes to yield a red suspension of protonated LZU-1 nanocrystals. This suspension immediately turns yellow upon deprotonation with an ethanol/triethylamine solution. The fact that the COF is protonated during the synthesis is crucial since this renders the nanocrystals polar and allows for PVP to bind and passivate their surface thus regulating the growth in alcoholic solution. The size and morphology of the crystallites were characterized by SEM revealing an average size of 245 ± 25 nm. The size of the LZU-1 nanocrystals can conveniently be tuned by varying the concentration of PVP. Addition of 5 mg mL^{-1} of PVP yields COF particles of 500 ± 52 nm in size and 40 mg mL^{-1} of PVP gives crystals of 112 ± 11 nm. The

COF particles form colloidal solutions in ethanol and remain stable for weeks. The importance of the solvent in the crystallization process is supported by the fact that the morphology of COF nanocrystals strongly correlates with solvent composition. In particular, the addition of toluene to the ethanol solution results in crystallites of distinctly hexagonal shape. This morphology strongly supports the proposed homogeneous nature of the crystal growth process and highlights the high crystallinity of the resulting materials [12].

11.3.2.2 Thin Films of Imine COFs at the Liquid–Liquid Interface

The nucleation of imine COFs takes place only in the presence of acid catalyst. This is exploited to grow COFs at the liquid–liquid interface by interfacing a solution of the COF starting materials in organic media with an aqueous layer containing the acid catalyst. Imine COF formation generally relies on high temperatures to ensure reversibility. This is detrimental in the case of interfacial growth where high reaction temperatures disturb the interface resulting in inhomogeneous films. Furthermore, acetic acid, commonly used as the catalyst in imine COF formation, is readily soluble in organic solvents rendering the formation of a defined interface difficult. Lewis acidic metal triflates can catalyze the imine bond formation, even at room temperature. In the formation of $[(TAPB)_2(BDA)_3]_{imine}$ from TAPB and BDA a series of metal triflates were investigated as potential catalysts. The addition of $In(OTf)_3$ (OTf, triflate) and $Sc(OTf)_3$ was found to lead to crystalline COFs in as little time as one minute at room temperature (Table 11.1) [13].

The advantage of Lewis acidic metal triflate catalysts is that they are readily soluble in water but not soluble in a lot of organic solvents. This, in combination with the fact that they operate at room temperature, allows for the formation of imine COFs at the liquid/liquid interface. When $Sc(OTf)_2$ (0.001 equiv) is applied as a catalyst in the aqueous layer and TATB and BDA are dissolved in a 1 : 4 v/v mixture of mesitylene/dioxane, the COF-forming reaction occurs exclusively at the interface between both solvents since all three reagents are

Table 11.1 Enumeration of catalysts for imine bond formation for the case of $[(TAPB)_2(BDA)_3]_{imine}$.

Entry	Catalyst	T/°C	RT/min	Isolated yield (%)
1	$In(OTf)_3$	20	1	95
2	$Sc(OTf)_3$	20	5	98
3	$Yb(OTf)_3$	20	10	98
4	$Y(OTf)_3$	20	30	96
5	$Eu(OTf)_3$	20	60	98
6	$Zn(OTf)_3$	20	150	95
7	CH_3COOH	20	N/A	0

The reaction rates and isolated yields for metal triflates and acetic acid are compared at 20 °C; RT = reaction time, N/A = not avaliable.

required for imine condensations to occur at room temperature. The reaction proceeds over the course of 72 hours and control over the thickness of the films (20 nm to 100 μm) is achieved by controlling the monomer concentration in the organic layer, as confirmed by AFM. PXRD patterns of the films are consistent with the proposed structural model of the COF. For very thin films the diffraction intensity is too low for PXRD and TEM and the crystallinity cannot be confirmed unambiguously [14].

11.4 Monolayer Formation of Boroxine and Imine COFs Under Ultrahigh Vacuum

A different approach is taken for the formation of COF monolayers under ultrahigh vacuum. Early reports of a monolayer of COF grown this way were achieved by chemical vapor deposition of BDA under ultrahigh vacuum and at 370 K onto a Ag(111) surface. Annealing of the sample results in the formation of the expected structure, COF-1, which can be followed using STM (scanning tunneling microscopy) at room temperature [15]. The small feature size of the resulting material is attributed to the lack of reversibility of the reaction due to water leaving the equilibrium under the ultrahigh vacuum conditions. To address this issue, a small amount of $[Cu(SO_4)]\cdot 5H_2O$ is added to the reaction, which is carried out on a highly ordered pyrolytic graphite (HOPG) surface and in a sealed system. The $[Cu(SO_4)]\cdot 5H_2O$ acts as a water reservoir in order to regulate the chemical equilibrium of the dehydration reaction, resulting in highly ordered networks. The water molecules released from $[Cu(SO_4)]\cdot 5H_2O$ during the heating process shift the equilibrium back to promote error correction under reversible reaction conditions, a necessary requirement for the formation of high-quality films. Upon cooling back to room temperature, the water released from $[Cu(SO_4)]\cdot 5H_2O$ is reabsorbed by the anhydrous $[Cu(SO_4)]$ formed, thus preventing the potential decomposition of the COF film due to dehydration of the boroxine linkage. In addition to the formation of COF-1, this strategy was successfully applied to construct two COFs with different pore sizes by self-condensation of 9,9′-dihexylfluorene-2,7-diboronic acid and 4,4′-biphenyl-diboronic acid. All three frameworks show long-range periodicity and negligible amounts of defects. It was shown that this strategy can further be applied to the formation of imine-linked COFs. Both the cross-condensation of TAPB with BDA and the reaction of TFP (1,3,5-trisformylbenzene) with PDA (1,4-benzenediamine) result in a monolayer of $[(TAPB)_2(BDA)_3]_{imine}$ and LZU-1, respectively. Both materials feature long-range periodicity as evidenced by STM images of the material [16].

11.5 Summary

In this chapter, we covered the top–down and bottom–up approaches for the nanoscopic and macroscopic structuring of COFs. Layered COF crystallites can be exfoliated into single/few layer sheets or thin films of the material. This requires an energy input that can be derived from mechanical forces (grinding, sonication) or can be the result of chemical alteration of the structure (chemical exfoliation). Bottom–up approaches for structuring of COFs require a detailed understanding of the crystallization pathway, which in turn is dependent on the linkage chemistry used to construct the framework. We have covered, in detail, the crystallization mechanisms for imine and boronate ester-linked COFs. By making use of the rate-determining steps of the crystallization or by manipulation of these steps COFs can be isolated as stable nano-sized colloids, micron-sized single crystals or as thin films. In the context of thin film formation of COFs we have introduced different techniques: film formation by controlled nucleation on substrates, nucleation at the liquid–liquid interface, growth in a flow cell, vapor-assisted conversion, and the formation of single COF layers under ultrahigh vacuum. Finally, we have highlighted which COF linkages are amenable to which structuring methodologies based on their underlying crystallization mechanisms.

References

1 (a) Wan, S., Gándara, F., Asano, A. et al. (2011). Covalent organic frameworks with high charge carrier mobility. *Chemistry of Materials* 23 (18): 4094–4097. (b) DeBlase, C.R., Silberstein, K.E., Truong, T.-T. et al. (2013). β-Ketoenamine-linked covalent organic frameworks capable of pseudocapacitive energy storage. *Journal of the American Chemical Society* 135 (45): 16821–16824. (c) Lin, S., Diercks, C.S., Zhang, Y.-B. et al. (2015). Covalent organic frameworks comprising cobalt porphyrins for catalytic CO_2 reduction in water. *Science* 349 (6253): 1208–1213.

2 (a) Uribe-Romo, F.J., Doonan, C.J., Furukawa, H. et al. (2011). Crystalline covalent organic frameworks with hydrazone linkages. *Journal of the American Chemical Society* 133 (30): 11478–11481. (b) Bunck, D.N. and Dichtel, W.R. (2013). Bulk synthesis of exfoliated two-dimensional polymers using hydrazone-linked covalent organic frameworks. *Journal of the American Chemical Society* 135 (40): 14952–14955.

3 Chandra, S., Kandambeth, S., Biswal, B.P. et al. (2013). Chemically stable multilayered covalent organic nanosheets from covalent organic frameworks via mechanical delamination. *Journal of the American Chemical Society* 135 (47): 17853–17861.

4 Khayum, M.A., Kandambeth, S., Mitra, S. et al. (2016). Chemically delaminated free-standing ultrathin covalent organic nanosheets. *Angewandte Chemie International Edition* 55 (50): 15604–15608.

5 (a) Smith, B.J. and Dichtel, W.R. (2014). Mechanistic studies of two-dimensional covalent organic frameworks rapidly polymerized from initially homogenous conditions. *Journal of the American Chemical Society* 136 (24): 8783–8789. (b) Li, H., Chavez, A.D., Li, H. et al. (2017). Nucleation and growth of covalent organic frameworks from solution: the example of COF-5. *Journal of the American Chemical Society* 139 (45): 16310–16318. (c) Koo, B., Heden, R., and Clancy, P. (2017). Nucleation and growth of 2D covalent

organic frameworks: polymerization and crystallization of COF monomers. *Physical Chemistry Chemical Physics* 19 (15): 9745–9754.

6 (a) Colson, J.W., Woll, A.R., Mukherjee, A. et al. (2011). Oriented 2D covalent organic framework thin films on single-layer graphene. *Science* 332 (6026): 228–231. (b) Cote, A.P., Benin, A.I., Ockwig, N.W. et al. (2005). Porous, crystalline, covalent organic frameworks. *Science* 310 (5751): 1166–1170.

7 Smith, B.J., Parent, L.R., Overholts, A.C. et al. (2017). Colloidal covalent organic frameworks. *ACS Central Science* 3 (1): 58–65.

8 Bisbey, R.P., DeBlase, C.R., Smith, B.J., and Dichtel, W.R. (2016). Two-dimensional covalent organic framework thin films grown in flow. *Journal of the American Chemical Society* 138 (36): 11433–11436.

9 Medina, D.D., Rotter, J.M., Hu, Y. et al. (2015). Room temperature synthesis of covalent–organic framework films through vapor-assisted conversion. *Journal of the American Chemical Society* 137 (3): 1016–1019.

10 (a) Smith, B.J., Overholts, A.C., Hwang, N., and Dichtel, W.R. (2016). Insight into the crystallization of amorphous imine-linked polymer networks to 2D covalent organic frameworks. *Chemical Communications* 52 (18): 3690–3693. (b) Qian, C., Qi, Q.-Y., Jiang, G.-F. et al. (2017). Toward covalent organic frameworks bearing three different kinds of pores: the strategy for construction and COF-to-COF transformation via heterogeneous linker exchange. *Journal of the American Chemical Society* 139 (19): 6736–6743.

11 Rodríguez-San-Miguel, D., Yazdi, A., Guillerm, V. et al. (2017). Confining functional nanoparticles into colloidal imine-based COF spheres by a sequential encapsulation-crystallization method. *Chemistry - A European Journal* 23 (36): 8623–8627.

12 (a) Zhao, Y., Guo, L., Gándara, F. et al. (2017). A synthetic route for crystals of woven structures, uniform nanocrystals, and thin films of imine covalent organic frameworks. *Journal of the American Chemical Society* 139 (37): 13166–13172. (b) Ding, S.-Y., Gao, J., Wang, Q. et al. (2011). Construction of covalent organic framework for catalysis: Pd/COF-LZU1 in Suzuki–Miyaura coupling reaction. *Journal of the American Chemical Society* 133 (49): 19816–19822.

13 Matsumoto, M., Dasari, R.R., Ji, W. et al. (2017). Rapid, low temperature formation of imine-linked covalent organic frameworks catalyzed by metal triflates. *Journal of the American Chemical Society* 139 (14): 4999–5002.

14 Matsumoto, M., Valentino, L., Stiehl, G.M. et al. (2018). Lewis-acid-catalyzed interfacial polymerization of covalent organic framework films. *Chem* 4 (2): 308–317.

15 Zwaneveld, N.A., Pawlak, R., Abel, M. et al. (2008). Organized formation of 2D extended covalent organic frameworks at surfaces. *Journal of the American Chemical Society* 130 (21): 6678–6679.

16 (a) Liu, X.-H., Guan, C.-Z., Ding, S.-Y. et al. (2013). On-surface synthesis of single-layered two-dimensional covalent organic frameworks via solid–vapor interface reactions. *Journal of the American Chemical Society* 135 (28): 10470–10474. (b) Guan, C.-Z., Wang, D., and Wan, L.-J. (2012). Construction and repair of highly ordered 2D covalent networks by chemical equilibrium regulation. *Chemical Communications* 48 (24): 2943–2945.

Part III

Applications of Metal-Organic Frameworks

12

The Applications of Reticular Framework Materials

Metal-organic frameworks (MOFs) and other reticular framework materials prepared following the principles outlined in this book have become the largest class of crystalline materials. Today, a large community of scientists and engineers is working in the field of reticular chemistry, expanding it beyond the synthesis and characterization of new materials by exploring a manifold of applications. This wide range of applications of such materials (MOFs, zeolitic imidazolate framework (ZIFs), and covalent organic frameworks (COFs)) is made possible by their crystalline structures with exceptional surface areas, tailorable pore openings and pore sizes, and their extensive structural diversity [1]. As illustrated throughout this book, reticular materials can be modified on an atomic scale, allowing for precise chemical modifications as well as the formation of structural motifs not accessible in the realm of molecular or classical solid-state chemistry. Knowledge of the local structure facilitates the determination of adsorption sites within the pores, a key factor for the design of MOFs with ultrahigh porosity. Controlling the surrounding of catalytically active sites by crafting the sterics and electronics of such sites gives tailored materials with outstanding properties [2].

The development of reticular chemistry has allowed chemists to synthesize materials by design whose properties surpass those of traditional porous materials. Among such properties are ultrahigh gas storage capacities, high selectivity in gas separations, and the capability to harvest water from air [3]. Employing the principles of reticular chemistry not only allows to tune tune the metrics and composition of the resulting materials but also to create structural arrangements such as precisely designed catalytic centers for highly selective transformations and small molecule activation not accessible any other way [4]. Such tunability is also advantageous for adjusting the electronic properties, which led to the discovery of porous conductive frameworks [5]. The ability to make materials that are both conductive and porous opens new fields of applications such as supercapacitors or electrocatalysis; both have been proved fruitful for conductive MOFs and COFs alike [6].

Gaining control over the particle size and being able to process materials into specific shapes facilitates the preparation of thin films, membranes, and shaped bodies for gas separation and catalytic applications, as well as the synthesis of colloidal materials for biomedical applications such as drug delivery systems [7]. Knowledge of the precise crystal structures plays a key role in the development

Introduction to Reticular Chemistry: Metal-Organic Frameworks and Covalent Organic Frameworks, First Edition. Omar M. Yaghi, Markus J. Kalmutzki, and Christian S. Diercks.

Companion website: www.wiley.com/go/yaghi/reticular

of hybrid materials such as encapsulated metal nanocatalysts [8]. Both modeling and quantum chemical calculations lead to a deeper understanding of many phenomena found in porous framework materials and will, without a doubt, become increasingly important for both the discovery of new materials and the understanding of their properties [3j, 9].

Some of the materials covered in this book are being produced on an industrial scale. BASF SE is now scaling up the synthesis of a variety of MOFs and multiple start-up companies produce MOFs for different applications such as gas storage tanks. This illustrates that research in reticular chemistry with respect to possible applications is not limited to research programs at universities worldwide but major chemical and automobile companies are also focusing on the development and commercialization of these porous materials [10]. We will focus on applications of MOFs and ZIFs as these two classes of materials are well developed and many applications have been tested. The study of COFs in various applications is beginning to emerge and will most certainly constitute a major direction in the future. We will also limit our discussion to applications based on the adsorption of gases and vapors since such applications have been studied in detail and a solid understanding of the underlying process has been established. The principles discussed in the following chapters are however also applicable to other reticular materials such as COFs.

References

1 (a) Farha, O.K., Eryazici, I., Jeong, N.C. et al. (2012). Metal-organic framework materials with ultrahigh surface areas: is the sky the limit? *Journal of the American Chemical Society* 134 (36): 15016–15021. (b) Furukawa, H., Ko, N., Go, Y.B. et al. (2010). Ultrahigh porosity in metal-organic frameworks. *Science* 329 (5990): 424–428. (c) Wang, T.C., Bury, W., Gómez-Gualdrón, D.A. et al. (2015). Ultrahigh surface area zirconium MOFs and insights into the applicability of the BET theory. *Journal of the American Chemical Society* 137 (10): 3585–3591. (d) El-Kaderi, H.M., Hunt, J.R., Mendoza-Cortés, J.L. et al. (2007). Designed synthesis of 3D covalent organic frameworks. *Science* 316 (5822): 268–272. (e) Ascherl, L., Sick, T., Margraf, J.T. et al. (2016). Molecular docking sites designed for the generation of highly crystalline covalent organic frameworks. *Natural Chemistry* 8 (4): 310–316. (f) Deng, H., Grunder, S., Cordova, K.E. et al. (2012). Large-pore apertures in a series of metal-organic frameworks. *Science* 336 (6084): 1018–1023. (g) Spitler, E.L., Koo, B.T., Novotney, J.L. et al. (2011). A 2D covalent organic framework with 4.7-nm pores and insight into its interlayer stacking. *Journal of the American Chemical Society* 133 (48): 19416–19421. (h) Ferey, G., Mellot-Draznieks, C., Serre, C. et al. (2005). A chromium terephthalate-based solid with unusually large pore volumes and surface area. *Science* 309 (5743): 2040–2042. (i) Wang, B., Cote, A.P., Furukawa, H. et al. (2008). Colossal cages in zeolitic imidazolate frameworks as selective carbon dioxide reservoirs. *Nature* 453 (7192): 207–211. (j) Cui, X., Chen, K., Xing, H. et al. (2016). Pore chemistry and size control in hybrid porous materials for acetylene capture from ethylene. *Science* 353 (6295): 141–144. (k) Yaghi, O.M., O'Keeffe, M.,

Ockwig, N.W. et al. (2003). Reticular synthesis and the design of new materials. *Nature* 423 (6941): 705–714.

2 (a) Rowsell, J.L.C., Spencer, E.C., Eckert, J. et al. (2005). Gas adsorption sites in a large-pore metal-organic framework. *Science* 309 (5739): 1350–1354. (b) Düren, T., Sarkisov, L., Yaghi, O.M., and Snurr, R.Q. (2004). Design of new materials for methane storage. *Langmuir* 20 (7): 2683–2689.

3 (a) Spanopoulos, I., Tsangarakis, C., Klontzas, E. et al. (2016). Reticular synthesis of HKUST-like tbo-MOFs with enhanced CH_4 storage. *Journal of the American Chemical Society* 138 (5): 1568–1574. (b) Jiang, J., Furukawa, H., Zhang, Y.-B., and Yaghi, O.M. (2016). High methane storage working capacity in metal-organic frameworks with acrylate links. *Journal of the American Chemical Society* 138 (32): 10244–10251. (c) Gándara, F., Furukawa, H., Lee, S., and Yaghi, O.M. (2014). High methane storage capacity in aluminum metal-organic frameworks. *Journal of the American Chemical Society* 136 (14): 5271–5274. (d) Mason, J.A., Oktawiec, J., Taylor, M.K. et al. (2015). Methane storage in flexible metal-organic frameworks with intrinsic thermal management. *Nature* 527 (7578): 357–361. (e) Alezi, D., Belmabkhout, Y., Suyetin, M. et al. (2015). MOF crystal chemistry paving the way to gas storage needs: aluminum-based soc-MOF for CH_4, O_2, and CO_2 storage. *Journal of the American Chemical Society* 137 (41): 13308–13318. (f) Li, B., Wen, H.-M., Wang, H. et al. (2014). A porous metal-organic framework with dynamic pyrimidine groups exhibiting record high methane storage working capacity. *Journal of the American Chemical Society* 136 (17): 6207–6210. (g) Peng, Y., Krungleviciute, V., Eryazici, I. et al. (2013). Methane storage in metal-organic frameworks: current records, surprise findings, and challenges. *Journal of the American Chemical Society* 135 (32): 11887–11894. (h) Mason, J.A., Veenstra, M., and Long, J.R. (2014). Evaluating metal-organic frameworks for natural gas storage. *Chemical Science* 5 (1): 32–51. (i) Fracaroli, A.M., Furukawa, H., Suzuki, M. et al. (2014). Metal-organic frameworks with precisely designed interior for carbon dioxide capture in the presence of water. *Journal of the American Chemical Society* 136 (25): 8863–8866. (j) McDonald, T.M., Mason, J.A., Kong, X. et al. (2015). Cooperative insertion of CO_2 in diamine-appended metal-organic frameworks. *Nature* 519 (7543): 303–308. (k) Nguyen, N.T.T., Furukawa, H., Gándara, F. et al. (2014). Selective capture of carbon dioxide under humid conditions by hydrophobic chabazite-type zeolitic imidazolate frameworks. *Angewandte Chemie International Edition* 53 (40): 10645–10648. (l) Mason, J.A., McDonald, T.M., Bae, T.-H. et al. (2015). Application of a high-throughput analyzer in evaluating solid adsorbents for post-combustion carbon capture via multicomponent adsorption of CO_2, N_2, and H_2O. *Journal of the American Chemical Society* 137 (14): 4787–4803. (m) Nugent, P., Belmabkhout, Y., Burd, S.D. et al. (2013). Porous materials with optimal adsorption thermodynamics and kinetics for CO_2 separation. *Nature* 495 (7439): 80–84. (n) Furukawa, H., Gándara, F., Zhang, Y.-B. et al. (2014). Water adsorption in porous metal-organic frameworks and related materials. *Journal of the American Chemical Society* 136 (11): 4369–4381. (o) Canivet, J., Fateeva, A., Guo, Y. et al. (2014). Water adsorption in MOFs: fundamentals and applications. *Chemical Society Reviews* 43 (16): 5594–5617. (p) Henninger, S.K., Habib,

H.A., and Janiak, C. (2009). MOFs as adsorbents for low temperature heating and cooling applications. *Journal of the American Chemical Society* 131 (8): 2776–2777. (q) Henninger, S.K., Jeremias, F., Kummer, H., and Janiak, C. (2012). MOFs for use in adsorption heat pump processes. *European Journal of Inorganic Chemistry* 2012 (16): 2625–2634. (r) Küsgens, P., Rose, M., Senkovska, I. et al. (2009). Characterization of metal-organic frameworks by water adsorption. *Microporous and Mesoporous Materials* 120 (3): 325–330. (s) Seo, Y.-K., Yoon, J.W., Lee, J.S. et al. (2012). Energy-efficient dehumidification over hierachically porous metal-organic frameworks as advanced water adsorbents. *Advanced Materials* 24 (6): 806–810. (t) Ehrenmann, J., Henninger, S.K., and Janiak, C. (2011). Water adsorption characteristics of MIL-101 for heat-transformation applications of MOFs. *European Journal of Inorganic Chemistry* 2011 (4): 471–474. (u) Akiyama, G., Matsuda, R., and Kitagawa, S. (2010). Highly porous and stable coordination polymers as water sorption materials. *Chemistry Letters* 39 (4): 360–361. (v) Kim, H., Yang, S., Rao, S.R. et al. (2017). Water harvesting from air with metal-organic frameworks powered by natural sunlight. *Science* 356 (6336): 430–434.

4 (a) Manna, K., Ji, P., Lin, Z. et al. (2016). Chemoselective single-site earth-abundant metal catalysts at metal-organic framework nodes. *Nature Communications* 7: 12610. (b) Wang, C., Xie, Z., deKrafft, K.E., and Lin, W. (2011). Doping metal-organic frameworks for water oxidation, carbon dioxide reduction, and organic photocatalysis. *Journal of the American Chemical Society* 133 (34): 13445–13454. (c) Wu, C.-D., Hu, A., Zhang, L., and Lin, W. (2005). A homochiral porous metal-organic framework for highly enantioselective heterogeneous asymmetric catalysis. *Journal of the American Chemical Society* 127 (25): 8940–8941. (d) Metzger, E.D., Brozek, C.K., Comito, R.J., and Dincă, M. (2016). Selective dimerization of ethylene to 1-butene with a porous catalyst. *ACS Central Science* 2 (3): 148–153. (e) Mondloch, J.E., Katz, M.J., Isley, W.C. III, et al. (2015). Destruction of chemical warfare agents using metal-organic frameworks. *Nature Materials* 14 (5): 512–516. (f) Xiao, D.J., Bloch, E.D., Mason, J.A. et al. (2014). Oxidation of ethane to ethanol by N_2O in a metal-organic framework with coordinatively unsaturated iron(II) sites. *Natural Chemistry* 6 (7): 590–595. (g) Feng, D., Gu, Z.-Y., Li, J.-R. et al. (2012). Zirconium-metalloporphyrin PCN-222: mesoporous metal-organic frameworks with ultrahigh stability as biomimetic catalysts. *Angewandte Chemie International Edition* 51 (41): 10307–10310. (h) Dang, D., Wu, P., He, C. et al. (2010). Homochiral metal-organic frameworks for heterogeneous asymmetric catalysis. *Journal of the American Chemical Society* 132 (41): 14321–14323. (i) Manna, K., Zhang, T., Greene, F.X., and Lin, W. (2015). Bipyridine- and phenanthroline-based metal-organic frameworks for highly efficient and tandem catalytic organic transformations via directed C-H activation. *Journal of the American Chemical Society* 137 (7): 2665–2673.

5 (a) Sun, L., Miyakai, T., Seki, S., and Dincă, M. (2013). Mn_2(2,5-disulfhydrylbenzene-1,4-dicarboxylate): a microporous metal-organic framework with infinite $(-Mn-S-)^{\infty}$ chains and high intrinsic charge mobility. *Journal of the American Chemical Society* 135 (22): 8185–8188.

(b) Sun, L., Hendon, C.H., Minier, M.A. et al. (2015). Million-fold electrical conductivity enhancement in Fe_2(DEBDC) versus Mn_2(DEBDC) (E = S, O). *Journal of the American Chemical Society* 137 (19): 6164–6167. (c) Gándara, F., Uribe-Romo, F.J., Britt, D.K. et al. (2012). Porous, conductive metal-triazolates and their structural elucidation by the charge-flipping method. *Chemistry A European Journal* 18 (34): 10595–10601. (d) Sheberla, D., Sun, L., Blood-Forsythe, M.A. et al. (2014). High electrical conductivity in Ni_3(2,3,6,7,10,11-hexaiminotriphenylene)$_2$, a semiconducting metal-organic graphene analogue. *Journal of the American Chemical Society* 136 (25): 8859–8862. (e) Ōkawa, H., Sadakiyo, M., Yamada, T. et al. (2013). Proton-conductive magnetic metal-organic frameworks, $\{NR_3(CH_2COOH)\}[M_a^{II}M_b^{III}(ox)_3]$: effect of carboxyl residue upon proton conduction. *Journal of the American Chemical Society* 135 (6): 2256–2262. (f) Nguyen, N.T.T., Furukawa, H., Gándara, F. et al. (2015). Three-dimensional metal-catecholate frameworks and their ultrahigh proton conductivity. *Journal of the American Chemical Society* 137 (49): 15394–15397. (g) Wan, S., Gándara, F., Asano, A. et al. (2011). Covalent organic frameworks with high charge carrier mobility. *Chemistry of Materials* 23 (18): 4094–4097.

6 (a) Sheberla, D., Bachman, J.C., Elias, J.S. et al. (2017). Conductive MOF electrodes for stable supercapacitors with high areal capacitance. *Nature Materials* 16 (2): 220–224. (b) Mulzer, C.R., Shen, L., Bisbey, R.P. et al. (2016). Superior charge storage and power density of a conducting polymer-modified covalent organic framework. *ACS Central Science* 2 (9): 667–673. (c) Lin, S., Diercks, C.S., Zhang, Y.-B. et al. (2015). Covalent organic frameworks comprising cobalt porphyrins for catalytic CO_2 reduction in water. *Science* 349 (6253): 1208–1213. (d) Miner, E.M., Fukushima, T., Sheberla, D. et al. (2016). Electrochemical oxygen reduction catalysed by Ni_3(hexaiminotriphenylene)$_2$. *Nature Communications* 7: 10942.

7 (a) Chen, Y., Li, S., Pei, X. et al. (2016). A solvent-free hot-pressing method for preparing metal-organic-framework coatings. *Angewandte Chemie International Edition* 55 (10): 3419–3423. (b) Li, Y.-S., Liang, F.-Y., Bux, H. et al. (2010). Molecular sieve membrane: supported metal-organic framework with high hydrogen selectivity. *Angewandte Chemie International Edition* 49 (3): 548–551. (c) Guo, H., Zhu, G., Hewitt, I.J., and Qiu, S. (2009). "Twin copper source" growth of metal-organic framework membrane: $Cu_3(BTC)_2$ with high permeability and selectivity for recycling H_2. *Journal of the American Chemical Society* 131 (5): 1646–1647. (d) Bux, H., Liang, F., Li, Y. et al. (2009). Zeolitic imidazolate framework membrane with molecular sieving properties by microwave-assisted solvothermal synthesis. *Journal of the American Chemical Society* 131 (44): 16000–16001. (e) Bae, T.-H., Lee, J.S., Qiu, W. et al. (2010). A high-performance gas-separation membrane containing submicrometer-sized metal-organic framework crystals. *Angewandte Chemie International Edition* 49 (51): 9863–9866. (f) Rodenas, T., Luz, I., Prieto, G. et al. (2015). Metal-organic framework nanosheets in polymer composite materials for gas separation. *Nature Materials* 14 (1): 48–55. (g) Zornoza, B., Martinez-Joaristi, A., Serra-Crespo, P. et al. (2011). Functionalized flexible MOFs as fillers in mixed matrix membranes for highly

selective separation of CO_2 from CH_4 at elevated pressures. *Chemical Communications* 47 (33): 9522–9524. (h) Bachman, J.E., Smith, Z.P., Li, T. et al. (2016). Enhanced ethylene separation and plasticization resistance in polymer membranes incorporating metal-organic framework nanocrystals. *Nature Materials* 15 (8): 845–849. (i) Rieter, W.J., Taylor, K.M.L., and Lin, W. (2007). Surface modification and functionalization of nanoscale metal-organic frameworks for controlled release and luminescence sensing. *Journal of the American Chemical Society* 129 (32): 9852–9853. (j) Rieter, W.J., Taylor, K.M.L., An, H. et al. (2006). Nanoscale metal-organic frameworks as potential multimodal contrast enhancing agents. *Journal of the American Chemical Society* 128 (28): 9024–9025. (k) Horcajada, P., Chalati, T., Serre, C. et al. (2010). Porous metal-organic-framework nanoscale carriers as a potential platform for drug delivery and imaging. *Nature Materials* 9 (2): 172–178. (l) Wuttke, S., Braig, S., Preiß, T. et al. (2015). MOF nanoparticles coated by lipid bilayers and their uptake by cancer cells. *Chemical Communications* 51 (87): 15752–15755. (m) Zhuang, J., Kuo, C.-H., Chou, L.-Y. et al. (2014). Optimized metal-organic-framework nanospheres for drug delivery: evaluation of small-molecule encapsulation. *ACS Nano* 8 (3): 2812–2819. (n) Zheng, H., Zhang, Y., Liu, L. et al. (2016). One-pot synthesis of metal-organic frameworks with encapsulated target molecules and their applications for controlled drug delivery. *Journal of the American Chemical Society* 138 (3): 962–968.

8 (a) Choi, K.M., Na, K., Somorjai, G.A., and Yaghi, O.M. (2015). Chemical environment control and enhanced catalytic performance of platinum nanoparticles embedded in nanocrystalline metal-organic frameworks. *Journal of the American Chemical Society* 137 (24): 7810–7816. (b) Rungtaweevoranit, B., Baek, J., Araujo, J.R. et al. (2016). Copper nanocrystals encapsulated in Zr-based metal-organic frameworks for highly selective CO_2 hydrogenation to methanol. *Nano Letters* 16 (12): 7645–7649. (c) Zhao, M., Yuan, K., Wang, Y. et al. (2016). Metal-organic frameworks as selectivity regulators for hydrogenation reactions. *Nature* 539 (7627): 76–80. (d) Lu, G., Li, S., Guo, Z. et al. (2012). Imparting functionality to a metal-organic framework material by controlled nanoparticle encapsulation. *Nature Chemistry* 4 (4): 310–316. (e) Zhao, M., Deng, K., He, L. et al. (2014). Core-shell palladium nanoparticle@metal-organic frameworks as multifunctional catalysts for cascade reactions. *Journal of the American Chemical Society* 136 (5): 1738–1741. (f) Kuo, C.-H., Tang, Y., Chou, L.-Y. et al. (2012). Yolk-shell nanocrystal@ZIF-8 nanostructures for gas-phase heterogeneous catalysis with selectivity control. *Journal of the American Chemical Society* 134 (35): 14345–14348.

9 (a) Boyd, P.G., Moosavi, S.M., Witman, M., and Smit, B. (2017). Force-field prediction of materials properties in metal-organic frameworks. *The Journal of Physical Chemistry Letters* 8 (2): 357–363. (b) Yang, D., Bernales, V., Islamoglu, T. et al. (2016). Tuning the durface chemistry of metal organic framework nodes: proton topology of the metal-oxide-like Zr_6 nodes of UiO-66 and NU-1000. *Journal of the American Chemical Society* 138 (46): 15189–15196. (c) Dzubak, A.L., Lin, L.-C., Kim, J. et al. (2012). *Ab initio* carbon capture in open-site metal-organic frameworks. *Nature Chemistry* 4 (10): 810–816. (d) Tsivion, E., Long, J.R., and Head-Gordon, M. (2014). Hydrogen physisorption on metal-organic framework linkers and metalated linkers: a computational study of the factors that control binding strength. *Journal of the American Chemical Society* 136 (51): 17827–17835. (e) Lin, L.-C., Berger, A.H., Martin, R.L. et al. (2012). *In silico* screening of carbon-capture materials. *Nature Materials* 11 (7): 633–641. (f) Frost, H., Düren, T., and Snurr, R.Q. (2006). Effects of surface area, free volume, and heat of adsorption on hydrogen uptake in metal-organic frameworks. *The Journal of Physical Chemistry B* 110 (19): 9565–9570. (g) Fairen-Jimenez, D., Moggach, S.A., Wharmby, M.T. et al. (2011). Opening the gate: framework flexibility in ZIF-8 explored by experiments and simulations. *Journal of the American Chemical Society* 133 (23): 8900–8902.

10 Furukawa, H., Müller, U., and Yaghi, O.M. (2015). "Heterogeneity within order" in metal-organic frameworks. *Angewandte Chemie International Edition* 54 (11): 3417–3430.

13

The Basics of Gas Sorption and Separation in MOFs

13.1 Gas Adsorption

A certain terminology and specific theoretical models are used to describe and evaluate the properties of porous materials with respect to the storage and separation of gases. Some of these terms and models, including the Langmuir and BET theory, were already introduced in Chapter 2. In contrast to the simple evaluation of porosity when considering the application of porous materials, the following parameters gain in importance: (i) working capacity, (ii) system capacity, (iii) kinetics, (iv) cycling stability, (v) isosteric heat of adsorption (Q_{st}), and (vi) selectivity. There are cases where additional differentiation between the excess and total uptake, and the gravimetric and volumetric uptake is helpful. In this chapter, we present definitions and physical principles underlying these parameters.

13.1.1 Excess and Total Uptake

When dealing with gas storage it is important to differentiate between the excess and the total uptake of a given material. Most gas adsorption experiments are carried out between ambient pressure and a pressure of approximately 1 bar. In this regime, the excess and total adsorption values are almost identical. For the adsorption of gases at high pressures (>5 bar), a pressure regime that is of interest for automotive applications, it is instructive to perform high pressure adsorption measurements to simulate application conditions. At high pressures, the excess and total uptake can differ considerably (Figure 13.1). The difference in gas uptake between two containers, one of them filled with a porous adsorbent and the other one being empty and of the same volume as the total pore volume of the porous framework (adsorbent), is referred to as the excess adsorption, also known as Gibbs excess. This quantity is correlated to the adsorption of gas molecules on the inner pore surface of the porous material. The excess adsorption reaches a maximum at elevated pressures (generally between 20 and 40 bar) and subsequently decreases again. This is attributed to the lower efficiency of packing and compressing of gas molecules within the pores of the adsorbent in this pressure range compared to a free volume (e.g. empty gas cylinder).

Introduction to Reticular Chemistry: Metal-Organic Frameworks and Covalent Organic Frameworks,
First Edition. Omar M. Yaghi, Markus J. Kalmutzki, and Christian S. Diercks.

Companion website: www.wiley.com/go/yaghi/reticular

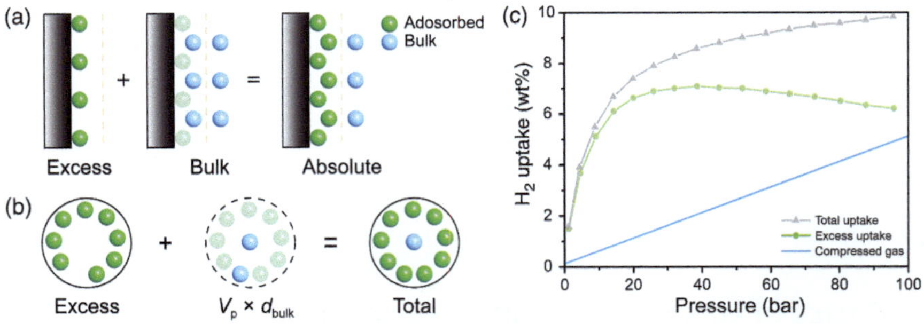

Figure 13.1 (a) Adsorption on a two-dimensional surface: the Gibbs dividing surface (light orange) divides the free volume into adsorbed (green) and bulk (blue) gas molecules [1]. The bulk gas molecules are the molecules that would be present in the pore volume in the absence of adsorption (middle). The absolute adsorption includes all gas molecules in the adsorbed state, meaning the sum of the experimentally measured excess adsorption and the bulk gas molecules (right). (b) The total adsorption in the pores of a porous framework material consists of all gas molecules inside the total pore volume (the sum of the excess adsorption and the bulk gas) [2]. The total adsorption is often used as an approximation for absolute adsorption in microporous materials, since experimental determination of the Gibbs dividing surface of microporous materials is not possible. (c) High-pressure H_2 adsorption measurement on MOF-5 with the total (gray triangles) and excess uptake (green circles) as well as the uptake of a gas cylinder (blue line) representing the bulk gas uptake.

Measurements at pressures beyond the maximum of the excess adsorption are necessary to estimate the total uptake of a material. At high pressures the total amount of gas molecules per volume of adsorbent consists of adsorbed molecules and compressed bulk gas in the pores that does not interact with the surface of the pores. The total uptake includes both the gas molecules adsorbed on the surface and those compressed within the pores of the framework and is described by Eq. (13.1):

$$N_{tot} = N_{ex} + d_{gas} V_{pore} \tag{13.1}$$

where N_{tot} is the total uptake in mg g^{-1}, N_{ex} is the excess uptake in mg g^{-1}, d_{gas} is the density of the gas at a given pressure in mg cm^{-3}, and V_{pore} is the pore volume in cm^3 g^{-1}. The pore volume is calculated from the crystallographic density (d_{cryst} in g cm^{-3}) and the skeletal density ($d_{skeletal}$ in g cm^{-3}, determined by pycnometry) according to Eq. (13.2) or is calculated from gas sorption isotherms.

$$V_{pore} = \frac{1}{d_{cryst}} - \frac{1}{d_{skeletal}} \tag{13.2}$$

For materials with small pores (micropores) the second term in Eq. (13.1) ($d_{gas} V_{pore}$) becomes very small since the free pore volume after (a monolayer of) gas is adsorbed on the surface becomes very small. Consequently, the total uptake for microporous materials is often overestimated. Knowledge of the total uptake facilitates the determination of the volumetric storage density, one of the most important quantities with respect to gas storage in energy applications. As discussed earlier, these fundamental properties do not account for the efficiency of packing in a container. This is however of importance for the overall efficiency of a storage system. Both the excess and total gas uptake are commonly provided in units of weight percentage (wt%) calculated according to Eq. (13.3).

$$\text{wt\%} = \frac{(m_{\text{adsorbate}})}{(m_{\text{adsorbent}} + m_{\text{adsorbate}})} \cdot 100\% \tag{13.3}$$

Since the second term in the denominator is often neglected, leading to overestimated wt% values, it is recommended to use the unit $g\,g^{-1}$ (mass of gas adsorbed per mass of adsorbent) when comparing the uptake of different materials to ensure comparability.

13.1.2 Volumetric Versus Gravimetric Uptake

Owing to their open structures, metal-organic frameworks (MOFs) often have low densities, which is expected to result in high values for the gravimetric gas uptake ($cm^3\,g^{-1}$ or $g\,g^{-1}$). A high gravimetric uptake however acts against achieving a high volumetric uptake ($cm^3\,cm^{-3}$ or $g\,cm^{-3}$). To understand this, we consider a gas that is compressed within a porous material with large open pores. In this scenario, a large fraction of this gas, although floating in the pores, does not interact with the internal surface and therefore behaves like compressed gas in a gas cylinder. In other words, there is no advantage gained by having gas compressed in the pores of such material since gas molecules interacting with the framework are packed more closely than gas molecules that are compressed within the pores at high pressures. Therefore, we expect a high gravimetric uptake but a low volumetric storage capacity for such low-density materials with large pore diameters. This is especially pronounced for covalent organic frameworks (COFs) owing to their ultralow density. In conclusion, this means that a balance must be struck between the size of the pores (gravimetric uptake) and having high volumetric uptake, especially when considering applications such as energy storage in mobile devices where a low volumetric uptake is highly unfavorable. This balance is attained by adjusting the pore diameter and/or the strength of the interaction with the adsorbate, both parameters that are different for different gas species. This point is elaborated further in the respective chapters.

13.1.3 Working Capacity

The maximum gas uptake measured by gas adsorption experiments is a useful quantity for the evaluation of porous materials; however, it does not represent the usable capacity under application conditions. For illustrative purposes, we consider a device powered by a gaseous fuel that requires a minimum delivery pressure to function (typically $P_{\text{min}} = 5\,\text{bar}$). Additionally, a reasonable upper pressure limit must be chosen. In general, the upper limits are $P_{\text{max}} = 35$ or $65\,\text{bar}$, which correspond to the limits of single- and dual-stage compressors, respectively [3]. These prerequisites lower the accessible capacity in comparison to the maximum uptake determined by gas adsorption measurements and the corresponding uptake is therefore commonly referred to as the "working capacity" (Figure 13.2). The working capacity gives the amount of gas that can be delivered to the device when the pressure is lowered from the maximum pressure of the

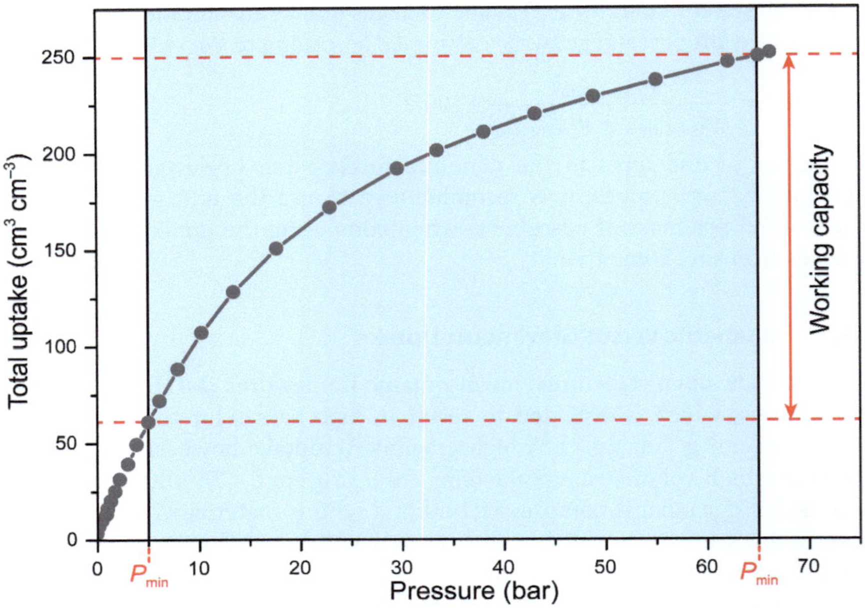

Figure 13.2 An example of a methane isotherm between 0 and 65 bar recorded at room temperature [4]. The working capacity is the deliverable amount of gas between 65 and 5 bar under isothermal conditions and is in this example about 60 $cm^3\ cm^{-3}$ lower than the total uptake. The pressure limits are chosen based on the maximum pressure of a dual-stage compressor and the minimum inlet pressure of a methane-powered combustion engine.

tank to the minimum inlet pressure of the device. With respect to applications, depending on the adsorption and desorption behavior of the material, the total uptake in wt% is not accessible and to estimate the working capacity the uptake at low (<5 bar) and high pressures (>35 or 65 bar) must be excluded.

13.1.4 System-Based Capacity

To estimate the deliverable capacity of gas storage materials on the system level, factors such as the weight and volume of all auxiliary system components including items such as the storage vessel, the working capacity, thermal and pressure management equipment, valves, piping, and sensors need to be considered.

Intrinsic thermal effects in the gas storage system have a dramatic impact on the storage capacity. The heat released upon refueling (Q_{st}) must be dissipated efficiently; otherwise, the adsorbent bed heats up, which results in a lower gas storage capacity. By the same token, if the thermal energy that is consumed during the discharge process is not resupplied, the temperature of the adsorbent bed drops, resulting in the retention of a large amount of gas at low pressure. Therefore, the absolute capacity on a system level of any given porous gas storage material is always lower than its working capacity determined by gas adsorption experiments. The system-based capacity provides an upper limit to the amount of usable gas in a device (Figure 13.3).

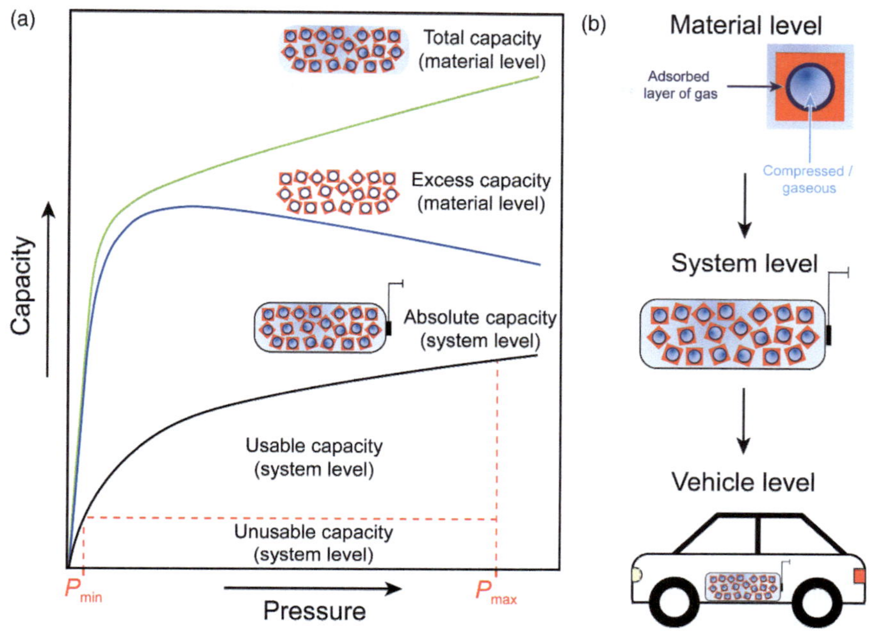

Figure 13.3 (a) Schematic isotherms representing the total (green line), excess (blue line), and absolute (black line) methane storage capacity of a porous adsorbent. Since here the weight of the storage vessel is considered, the gravimetric capacity on the system level is drastically decreased compared to the maximum uptake determined by gas adsorption measurements. The usable capacity is determined by excluding the uptake outside the pressure limits given by the compressor system (upper boundary, P_{max}) and the minimum operating pressure (lower boundary, P_{min}). (b) On the material level the total and excess capacity are distinguished, whereas the system-based capacity is calculated from the total capacity considering the weight of the storage vessel. At the vehicle level, additional factors such as intrinsic thermal effects must be considered.

13.2 Gas Separation

Separation of gases in porous solids follows either a thermodynamic or a kinetic separation mechanism. The separation of gaseous mixtures based on physisorption or chemisorption of the adsorbate to specific adsorption sites within the porous solid is a thermodynamic process and is hence termed "thermodynamic separation." A gaseous mixture can also be separated if the individual components diffuse with different diffusivities or follow different diffusion pathways, a process termed "kinetic separation." Often, both mechanisms are present simultaneously and it is important to understand which of them is rate limiting. The next section will give an insight into the different mechanisms of gas diffusion in porous solids and their importance with respect to gas separation.

13.2.1 Thermodynamic Separation

Thermodynamic or equilibrium separation is found for materials with pore apertures that are large enough to allow for all adsorbate molecules to pass unhindered ($d_{pore} \gg d_{kinetic(gas)}$). For this case, the separation is governed by the differences between the affinities of the individual components of the mixture toward the surface of the adsorbent. Thus, the selectivity can be tuned by introducing specific functionalities or binding sites to modulate the strength of the interaction between a specific component of the mixture and the inner pore surface. The isosteric heat of adsorption (Q_{st}) at zero coverage correlates with the magnitude of the strongest interaction between the adsorbate and the adsorbent, making Q_{st} a significant quantity in the design of porous materials for gas storage and separation. Chemical modifications carried out on a framework material such as covalent modifications, introduction of functional groups, and the creation of open metal sites (see Chapter 6) may result in a change in the Q_{st} value and consequently such modifications are used as a handle for optimizing the thermodynamics of the adsorption process. Q_{st} can be calculated from adsorption isotherms recorded at two different temperatures (typically 77 and 87 K) by fitting the data to either the virial (see Section 13.2.1.1) or the Langmuir–Freundlich equation (see Section 13.2.1.2).

13.2.1.1 Calculation of Q_{st} Using a Virial-Type Equation

The following virial-type equation is used to fit the adsorption data at a given fixed temperature following Eq. (13.4):

$$\ln(N/P) = A_0 + A_1 N + A_2 N^2 + A_3 N^3 + \cdots \tag{13.4}$$

where P is the pressure, N is the amount of gas adsorbed, and A_0, A_1, etc. are the virial coefficients. Q_{st} is calculated as a function of coverage using the Clausius–Clapeyron equation.

$$Q_{st} = R \ln\left(\frac{P_1}{P_2}\right) \frac{T_1 T_2}{T_2 - T_1} \tag{13.5}$$

The virial analysis is mathematically consistent since it can be reduced to Henry's law and allows to estimate the Q_{st} at zero coverage by extrapolation. For higher values of coverage, it shows only minimal deviations for all experimental data points; however, the use of polynomials of too high order can result in an overinterpretation.

13.2.1.2 Calculation of Q_{st} Using the Langmuir–Freundlich Equation

The Langmuir–Freundlich equation can be used to fit adsorption data at a fixed temperature following Eq. (13.6):

$$\frac{N}{N_m} = \frac{BP^{(1/t)}}{1 + BP^{(1/t)}} \tag{13.6}$$

where P is the pressure, N is the amount of gas adsorbed, and B and t are constants. Eq. (13.6) can be rearranged to yield an expression for P (Eq. (13.7)) that allows for the determination of Q_{st} using the Clausius–Clapeyron equation (Eq. (13.5)).

$$P = \left(\frac{N/N_m}{B + BN/N_3}\right) \tag{13.7}$$

13.2.2 Kinetic Separation

Kinetic or nonequilibrium gas separation mechanisms are based on the differences in the diffusion behavior of gases. In general, molecules with high mobility permeate the pore space more quickly than molecules of lower mobility. The diffusivity of a given gas can be influenced by adjustment of the pore metrics and the dimensionality of the pore system. Subtle changes in the pore structure of a porous solid can result in a significant change in the diffusivity of a specific type of molecule by an order of magnitude. In a similar way, structural changes of the diffusing species (e.g. *iso*- and *n*-alkanes) can have a dramatic influence.

13.2.2.1 Diffusion Mechanisms

When considering transport processes of gases within porous solids one must distinguish between convection (transport of gases due to a pressure gradient) and diffusion (transport of gases due to a concentration gradient). In this chapter, we will only consider diffusion mechanisms since diffusion is easily distinguished from convective flow and is the main mechanism in the kinetic separation of gases. Which diffusion mechanism predominates depends on the ratio of the pore diameter and the mean free path of the gas molecules. Within the large free space between particles "normal diffusion" as we are familiar with from gaseous mixtures dominates. Confined pores of a diameter significantly larger than the kinetic diameter of the diffusing gas molecules lead to a transition from normal diffusion to "molecular diffusion." Decreasing the pore size to the regime of the mean free path of the gas molecules leads to "Knudsen diffusion" and for the case of even smaller pore sizes, in the same regime as the kinetic diameter of the gas molecules, "configurational diffusion" is prevalent. Figure 13.4a gives a correlation between the pore size and the mechanism of diffusion. In addition to the mechanisms mentioned above, molecules always undergo adsorption on the surface and move along the surface in a process termed "surface diffusion." We can describe the interplay of the different mechanisms of diffusion analogous to an electric circuit. Molecular and Knudsen diffusion act like resistors in a series circuit, whereas surface diffusion and convective flow are described by resistors wired parallel to molecular and Knudsen diffusion (Figure 13.4b).

Molecular Diffusion Molecular diffusion occurs in porous solids with pores significantly larger than the kinetic diameter of the gas molecules diffusing through the solid. In the absence of a bulk flow (i.e. convection), the steady-state molar flux (j_A) in the z-direction is described by Fick's law:

$$j_A = -D_A \frac{dc_A}{dz} \tag{13.8}$$

where D_A is the diffusivity of the component A through the porous solid and c_A is its concentration. The diffusion through a slab of material of thickness z is determined by integration of Eq. (13.8) based on the assumption of a linear concentration gradient, which leads to Eq. (13.9).

$$j_A = \frac{D_A (c_{A1} - c_{A2})}{z} \tag{13.9}$$

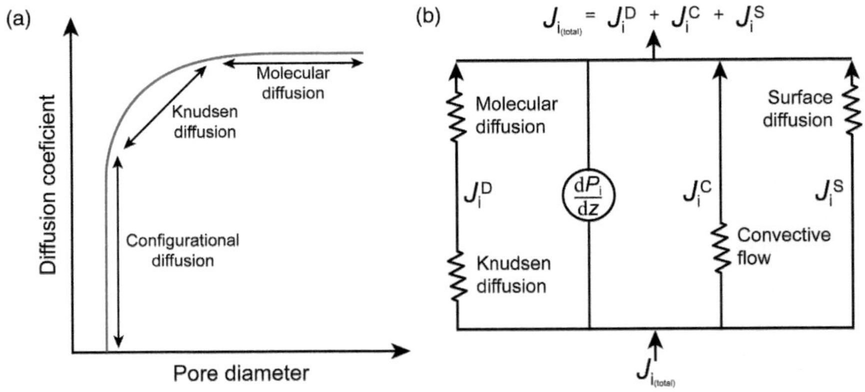

Figure 13.4 (a) Correlation between pore size and diffusion mechanism. For pores that are significantly larger than the kinetic diameter of the diffusing gas molecules, molecular diffusion is observed. From a mechanistic point of view, molecular diffusion is similar to normal diffusion since molecule–molecule collisions are more likely to occur than molecule–pore collisions. For pores in the regime of the mean free path of the gas molecules, Knudsen diffusion is the predominant mechanism and for even smaller pores in the regime of the kinetic diameter of the gas molecules configurational diffusion is observed. (b) Different diffusion mechanisms can be described analogous to an electric circuit where the different types of diffusion are represented by resistors. Molecular and Knudsen diffusion can be described by resistors in a series circuit (flux J_i^D), whereas surface diffusion (flux J_i^S) and convective flow (flux J_i^C) are represented by resistors wired parallel to molecular and Knudsen diffusion.

Here, c_{A1} and c_{A2} are the concentrations of component A at z_1 and z_2, the two opposite sides of the slab of thickness z, respectively. For solids of different shapes, the rate of diffusion (w) is described by Eq. (13.10):

$$w = j_A S_{av} = \frac{D_A S_{av}(c_{A1} - c_{A2})}{z} \tag{13.10}$$

where S_{av} is the average cross section for diffusion through a slab of a porous material of thickness z. This assumption holds true for perfect crystals with uniform pores of a diameter significantly larger than the mean free path of the diffusing gas molecules and of equal length z. Owing to the comparatively large size of the pore compared to the mean free path of the gas molecules the resistance is mainly due to molecule–molecule collisions, and the probability for molecule–pore wall collisions to occur is comparatively low. This, however, changes when the pore diameter is decreased, resulting in diffusion following a different mechanism: Knudsen diffusion.

Knudsen Diffusion Knudsen diffusion is predominant in materials with pore sizes in the regime of the mean free path of the diffusing gas molecules. Owing to the smaller pores, molecule–pore wall collisions dominate over molecule–molecule collisions. Knudsen diffusion is observed for "thin gases" (i.e. low probability of molecule–molecule collisions) in porous solids. The Knudsen diffusivity (D_K in s^{-1} cm^{-2}) of a gas in a cylindrical pore is given by Eq. (13.11):

$$D_{K,\ pore} = \frac{2}{3} r \sqrt{\frac{8}{\pi} \cdot \frac{RT}{M}} \tag{13.11}$$

where r is the pore radius, M is the molar mass of the gas diffusing in the pore, and ⅔ accounts for the cylindrical shape of the pore. Since the presence of exclusively cylindrical pores in a porous solid is relatively rare, an equation allowing for the calculation of the diffusivity in porous materials with more complex pore systems is needed (Eq. (13.12)).

$$D_{K,\ porous\ medium} = D_{K,\ pore} \frac{\varepsilon \cdot \delta}{\tau} \tag{13.12}$$

Here, ε is the porosity (i.e. overall porosity minus pores that are too small for the gas to enter and impasses), δ is the constrictivity (i.e. the deceleration of diffusion due to an increase in viscosity in narrow pores), and τ is the tortuosity (i.e. intricacy), which compensates for the deviation of the pore structure from an array of parallel cylindrical pores.

By decreasing the size of the pores further until a value close to the diameter of the gas molecule is reached, a transition into configurational diffusion occurs. This type of diffusion can only be described with sophisticated models, and for further explanations the reader is referred elsewhere [5].

Surface Diffusion Surface diffusion is an activated process and is strongly correlated to the adsorption–desorption equilibrium on the inner surface of the porous material. This means that the potential on the surface of the adsorbent is not flat and an activation barrier has to be overcome for surface diffusion to occur. Surface diffusion can proceed following two mechanisms:

(i) If the activation energy (E_{diff}) for diffusion is lower than the thermal energy (i.e. kT) molecules diffuse "freely" on the surface, which is typically the case for physisorption.
(ii) If E_{diff} is higher than kT, molecules move following a hopping mechanism typically observed for chemisorbed species. Hopping can occur over long distances since a molecule that has reached the energetically high transition state can migrate along multiple adsorption sites before it has dissipated enough energy to be re-adsorbed on the surface.

13.2.2.2 Influence of the Pore Shape

It can be readily recognized that the metrics of the pores in a porous solid have an impact on the diffusivity since they determine the predominant diffusion mechanisms. However, the shape of the pores is just as important. Differently shaped pores of the same diameter result in significantly different diffusivities. Generally, cylindrical pores allow for the highest diffusivity since both the constrictivity (δ) and tortuosity (τ) can be neglected. Figure 13.5 shows a comparison of the time

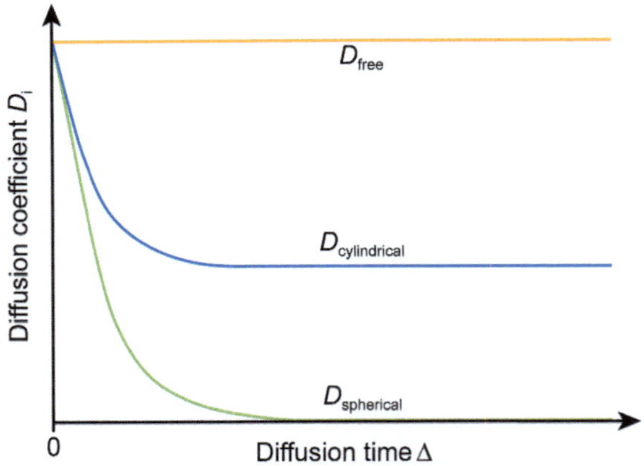

Figure 13.5 Dependence of the diffusion coefficient D_i on the diffusion time Δ. For normal diffusion, the diffusion coefficient (D_{free}) is independent of Δ (orange line). For cylindrical pores, the diffusion is hindered and therefore $D_{cyl.}$ decreases and reaches a lower limit for $\Delta \to \infty$ (blue line). In closed spherical pores $D_{sph.}$ decreases proportional to Δ^2 and reaches zero for $\Delta \to \infty$ (green line). The diffusion coefficient for spherical pores connected through narrow windows shows a dependence on Δ similar to that of cylindrical pores; however, in general a lower value of D is expected for $\Delta \to \infty$. The corresponding graph for 3D pore systems would be situated in the area between that of cylindrical and spherical pores.

dependence of the diffusion coefficient for different pore shapes. At $t = t_0$ the diffusion coefficients of the free gas (D_{free}), and the cylindrical (D_{cyl}) and spherical pores (D_{sph}) are identical. D_{free} is constant whereas D_{cyl} decreases due to an increase in resistance (e.g. increase in viscosity), and D_{sph} eventually approaches zero after clogging of the spherical pores. Diffusion coefficients of 3D pore systems (D_{3D}) containing cages connected through narrow windows show a similar trend to that of spherical pores; however, D_{3D} does not approach zero and the diffusion coefficient reached at $\Delta \to \infty$ depends on the diameter of the pore windows.

13.2.2.3 Separation by Size Exclusion

Separation by size-selective exclusion is found in materials with narrow pores that are smaller than the kinetic diameter of at least one component in a mixture. This means that at least one gas species cannot enter the pore system due to steric hindrance while others with a smaller kinetic diameter can. This type of separation is often undesirable since it is accompanied by a dramatic pressure drop. Therefore, larger pore sizes and a transition from molecular sieving to kinetic separation or partial molecular sieving are more appealing.

13.2.2.4 Separation Based on the Gate-Opening Effect

The gate-opening effect denotes a process associated with a structural transition from a narrow pore (*np*) phase of low porosity to a wide pore (*wp*) porous phase. This effect has been proved useful for many separation processes such as C_1–C_5 separation (see Chapter 16) [6]. The gate-opening effect is often induced by temperature and/or pressure and the uptake and release of individual molecules is controlled by these parameters, for example, by a specific threshold pressure. Owing to the spontaneous release of gas upon gate closing, a large working capacity can be achieved. This type of flexibility is exclusive to reticular materials (MOFs, COFs, and zeolitic imidazolate frameworks (ZIFs)); however, it has thus far not been possible to target such materials by design.

13.2.3 Selectivity

In separation processes the selectivity of the adsorbent toward a specific component in a gas mixture is essential. Such selectivity can be achieved through adjustment of the strength of the adsorption interactions (thermodynamic separation) or by size discrimination and differences in the diffusivities of the individual components in the mixture (kinetic separation). As we saw earlier, thermodynamic separation relies on the fact that different gases have different affinities toward the adsorption sites. Since the gas diffusion in porous solids with sufficiently large pores is unrestricted, the diffusivities of all components in the gas mixture are approximately the same and the equilibrium gas adsorption selectivity is consequently calculated from the ratio of the Henry's constants (K_i, slope of the isotherm at low relative pressures $P/P_0 \rightarrow 0$) of the individual components as in Eq. (13.13).

$$S = \frac{K_1}{K_2} \tag{13.13}$$

Thermodynamic selectivity is based on either a physisorption or chemisorption mechanism and is thus correlated to the difference in physical properties of the various gases in the mixture (e.g. polarizability, quadrupole moment) that influence the isosteric heat of adsorption. To illustrate this, we consider a mixture of CO_2 (15–16%) and N_2 (73–77%), a composition similar to that of post-combustion flue gas. A comparison of the physical parameters relevant for their equilibrium separation reveals that both the polarizability (CO_2, 29.1×10^{-25} cm^{-3}; N_2, 17.4×10^{-25} cm^{-3}) and quadrupole moment (CO_2, 4.3×10^{26} esu^{-1} cm^{-2}; N_2, 1.52×10^{26} esu^{-1} cm^{-2}) of CO_2 are higher than those of N_2. Polar sites located on the pore surface of the capture material will therefore lead to a stronger interaction with and a higher affinity/selectivity for CO_2.

Kinetic separation presupposes a size selectivity caused by appropriately sized pores. A porous material with small pores only permits molecules up to a certain kinetic diameter to diffuse freely, whereas larger molecules will not be able to permeate the capture material due to spatial limitations resulting in nonequilibrium separation [7]. For MOFs with small pores that reach the kinetic diameter of the gas components, the diffusion of these gases becomes increasingly difficult. We saw earlier that in this case the molecular diffusion will gradually evolve into Knudsen or even surface diffusion, thus rendering the diffusivities of the gas components distinguishable. Since surface diffusion is always present, the kinetic separation factor is defined by both the ratio of the Henry's constants (K_i) and the ratio of the diffusivities (D_i) (Eq. (13.14)) [8].

$$S = \frac{K_1}{K_2} \cdot \sqrt{\frac{D_1}{D_2}} \tag{13.14}$$

This kinetic effect may sometimes enhance the equilibrium adsorption selectivity but is more often found to decrease the selectivity, especially if the more strongly adsorbing component diffuses much slower than the less strongly adsorbing component [9]. Kinetic separation is only applicable to mixtures of gases with significantly different kinetic diameters. The previous example of a CO_2/N_2 mixture used to illustrate the thermodynamic selectivity does not fulfill this prerequisite since both molecules have a similar kinetic diameter (3.68 and 3.30 Å, respectively). This would require the pore openings of the capture material to be very small and within a narrow range. Small pore diameters however limit the diffusion of gases throughout the material and additionally the presence of defects (which cannot be avoided) results in larger pores sizes, thus rendering this approach less viable (Figure 13.6b). In contrast, as depicted in Figure 13.6a, a mixture of differently sized gases such as *trans*-butylene and *iso*-butylene fulfills this prerequisite.

Three ways are commonly used to determine the selectivity of a given porous material from experimental data: (i) estimation from single-component isotherms, (ii) calculation following the ideal adsorbed solution theory (IAST), and (iii) from breakthrough experiments.

13.2.3.1 Calculation of the Selectivity from Single-Component Isotherms

To calculate a "selectivity factor" the experimental isotherms for the two gases in question need to be known. It is calculated following Eq. (13.15):

$$S = \frac{q_1/q_2}{P_1/P_2} \tag{13.15}$$

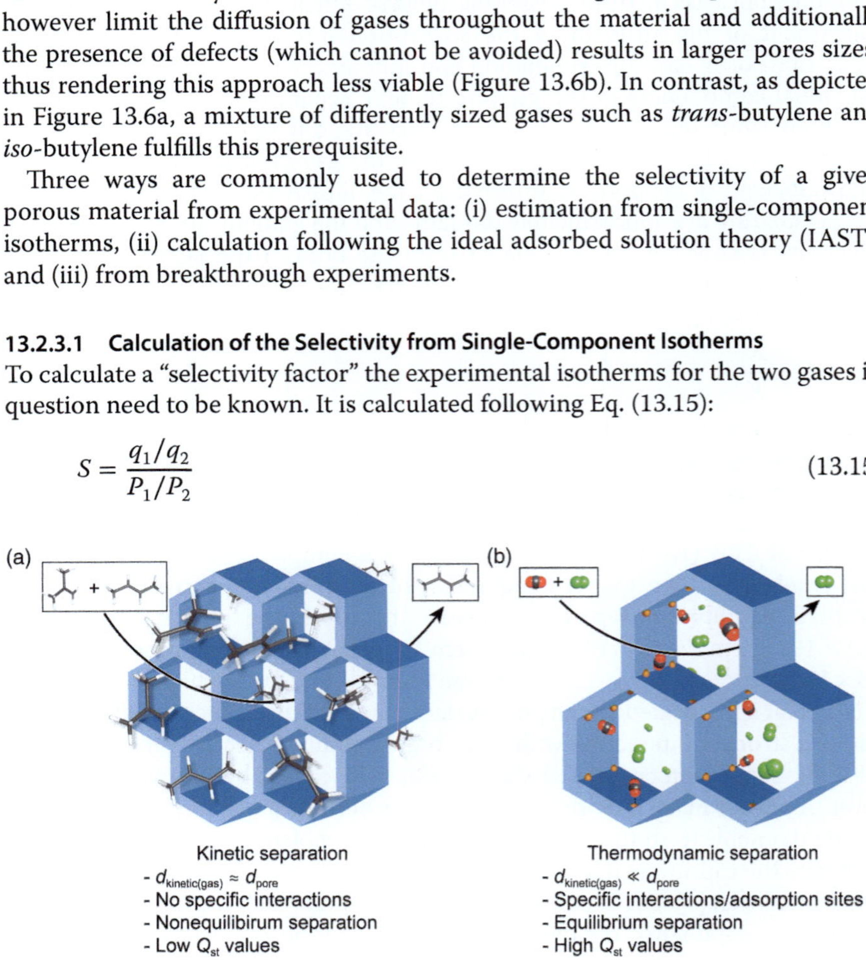

Figure 13.6 Kinetic and thermodynamic separation in porous materials. (a) Kinetic separation relies on different diffusivities of the components of the influent gas mixture. *trans*-Butylene and *iso*-butylene have different kinetic diameters; therefore, only the linear *trans*-butylene can permeate through the pores, whereas *iso*-butylene is held back. (b) Thermodynamic separation relies on specific interactions with one component of the influent gas mixture. CO_2 is bound strongly to specific adsorption sites, whereas N_2 can diffuse through the pores unhindered.

where the selectivity factor S is given as the molar ratio of the adsorbed quantities (q_i) and the partial pressures (P_i) of the gases. Since this estimation originates from single-component adsorption isotherms it does not consider that different types of gas molecules can compete for the same adsorption site. Even though the selectivity factors calculated using this method are not accurate, it allows for the facile qualitative evaluation of different porous adsorbents.

13.2.3.2 Calculation of the Selectivity by Ideal Adsorbed Solution Theory

A more accurate calculation of the selectivity from single-component isotherms is realized when the IAST is used [10]. The IAST is a viable method for gas mixtures where both components mix and behave as ideal gases. To calculate the selectivity following the IAST, isotherms of the single components of the gas mixture must be measured at the same temperature and fitted using mathematical methods. The mole fraction of each species in the adsorbed phase is then calculated by solving the integrals for the single components given by Eq. (13.16):

$$\int_0^{P\cdot y_i/x_i} \frac{\text{Isotherm for component } i(P)}{P} dP = \int_0^{P\cdot y_j/x_j} \frac{\text{Isotherm for component } j(p)}{P} dP \tag{13.16}$$

where P is the total pressure, x_i/x_j are the mole fractions of the adsorbed gas, and y_i/y_j those of the bulk phase of components i and j, respectively. The accuracy of this theory is reduced for high mixture fractions of the less adsorbed component since it requires the integration of the single-component isotherm up to extremely high pressures. The gas adsorption in flexible frameworks is not described accurately by the IAST but other methods applicable to such solids have been developed [11]. The amount of adsorbed gas in the mixture is determined following Eq. (13.17), with n_{tot} being the total number of moles adsorbed in the mixture at a given pressure and n_i^0/n_j^0 the amount of the pure components i and j per gram of adsorbent.

$$\frac{1}{n_{\text{tot}}} = \frac{x_i}{n_i^0} + \frac{x_j}{n_j^0} \tag{13.17}$$

High-quality single-component data are required for accurate simulations using the IAST. It is often advantageous to use isotherms calculated using grand canonical Monte Carlo (GCMC) methods to allow for a better fit, and therefore supporting GCMC simulations are often reported alongside IAST calculations [12]. Both methods have their limitations and it is important to evaluate and scrutinize the selectivity factors calculated with them. The IAST underestimates the CO_2/H_2 selectivity for HKUST-1, which is attributed to the presence of differently shaped and sized pockets within its structure. Such pockets give rise to preferred adsorption of one gas over the other due to their difference in size. This issue is circumvented by substantiating the results gained from the IAST with GCMC simulations. Similarly, selectivity factors calculated for the separation of CO_2 and H_2 in MOF-177 using the IAST do not match the measured data; however, a combination of GCMC and IAST predicts the selectivity in this separation accurately [13].

13.2.3.3 Experimental Methods

Breakthrough experiments give an experimental insight into the selectivity of adsorbents and allow for the experimental evaluation of their performance. In a typical setup, a defined gas mixture that is controlled by a set of mass flow controllers is fed to a measurement cell filled with powder, pellets, or a membrane containing the adsorbent. The eluent gas stream is monitored by mass spectrometry (i.e. MS or GCMS) or a set of specific gas sensors. The result of a typical measurement is depicted in Figure 13.7 where the separation of a 80 : 20 mixture of gases A and B is shown. When the gas mixture is fed to the activated adsorbent, component B is adsorbed selectively due to the lower affinity of the framework toward A, and the eluent gas stream is virtually of the pure component A. As soon as the adsorbent is saturated with B the breakthrough occurs, and the eluent gas is a mixture of A and B of varying composition until it eventually has the same composition as the influent gas mixture (A:B = 80 : 20). The adsorbent bed can subsequently be regenerated to measure multiple cycles. Regeneration

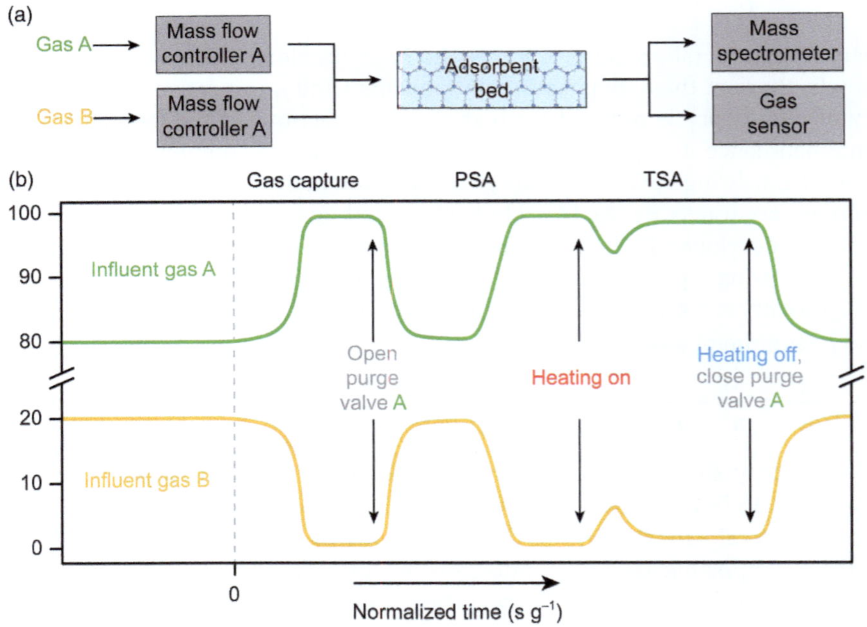

Figure 13.7 (a) Schematic of a typical breakthrough setup. Mass flow controllers are used to prepare a gas mixture with a defined composition. This mixture is fed to the adsorbent bed, which adsorbs selectively one component of the mixture. The eluent gas mixture is analyzed by mass spectrometry or specific gas sensors. Referencing the acquired data and subtraction of the dead time affords a graph similar to that shown in (b). (b) After the gas mixture is fed to the adsorbent one component is adsorbed (here, component B) until the adsorbent is saturated. The adsorbent bed is purged with component A forcing the desorption of B in a pressure swing adsorption cycle (PSA). Heating the adsorbent in a temperature swing adsorption cycle (TSA) liberates the remaining strongly bound component B, thus regenerating the adsorbent bed for another capture cycle.

is commonly realized by purging (in this example with gas A), also called "pressure swing," and subsequent heating ("temperature swing"). More details on temperature and pressure swing adsorption (PSA) are discussed in Chapter 14.

13.3 Stability of Porous Frameworks Under Application Conditions

For porous frameworks to be suitable for applications such as gas storage or separation it is inevitable for them to possess high capacities and fast kinetics. Another important factor that is often neglected is the long-term and cycling stability under application conditions. Depending on the application in question, a different set of testing conditions is required to determine the cycling stability, and these will be discussed in more detail in the chapters focusing on these applications. Even though long-term cycling studies are not necessary at the earlier stages of materials development, it is important to evaluate degradation and the associated loss in storage capacity and/or selectivity of a given material at conditions simulating those of the targeted application. Typically, this is done at a fraction of the targeted number of adsorption–desorption cycles or operating hours. Still, the insight gained from these experiments provides information for the down-selection and optimization of promising materials.

Additionally, thermal and mechanical stress can play an important role. Thermal stress occurs due to adsorption and desorption as these processes release or consume the heat of adsorption (Q_{st}), respectively. We will demonstrate the importance of thermal stability for HKUST-1 when used as a methane storage material in an automotive application. In such applications, a typical storage capacity is 20 kg of CH_4 (approximately 1250 mol or 28 000 l when considering CH_4 to be an ideal gas). The storage capacity of HKUST-1 is $0.15\,g\,g^{-1}$ and therefore a tank filled with about 140 kg of HKUST-1 is required to store 20 kg of methane. The heat of CH_4 adsorption in HKUST-1 is $Q_{st} = 20\,kJ\,mol^{-1}$, equaling $Q = 25\,000\,kJ$ of thermal energy released upon fueling, and consumed upon depletion of the tank. Plugging these numbers and the heat capacity ($c_p = 1.46\,kJ\,kg^{-1}\,K^{-1}$) reported for HKUST-1 into Eq. (13.18), we can estimate an increase in temperature of more than 120 °C upon fueling [14]. This highlights the importance of testing the thermal stability of materials for gas storage applications.

$$\Delta T = \frac{Q}{m \cdot c_p} \tag{13.18}$$

Mechanical stress poses another challenge to the applicability of porous materials. Most applications do not rely on adsorbents in powder form but more commonly shaped bodies are employed. These are typically prepared by pressing or extrusion; hence the material to be processed must be mechanically stable to withstand these shaping processes without collapse of the framework structure [15]. The resulting shaped bodies need to maintain structural integrity for a long time. This gains special importance when considering mobile applications, where the material is exposed to rattling and shaking, which poses an even higher demand on the structural stability of the shaped bodies and the adsorbent alike.

13.4 Summary

In this chapter, we introduced specific terms and theories of gas adsorption and separation used with respect to the application of porous materials in gas storage and gas separation processes. We showed that for specific applications it is important to differentiate between the total and excess, and volumetric and gravimetric uptake, and that the working capacity and system-based capacity gain importance. Different diffusion mechanisms and their influence on the selectivity of separation processes were discussed and we introduced both experimental and mathematical methods for the determination of the selectivity in separation processes. In the following chapters we will use these basics to elucidate the role of MOFs in applications such as CO_2 capture and sequestration (Chapter 14), H_2 and CH_4 storage (Chapter 15), separation processes (Chapter 16), and water adsorption (Chapter 17).

References

1 Gibbs, J. (1928). *The collected Works*, vol. 1. New York: Longmans.

2 (a) Sircar, S. (1999). Gibbsian surface excess for gas adsorption revisited. *Industrial and Engineering Chemistry Research* 38 (10): 3670–3682. (b) Sircar, S. (2001). Measurement of Gibbsian surface excess. *AIChE Journal* 47 (5): 1169–1176.

3 He, Y., Zhou, W., Qian, G., and Chen, B. (2014). Methane storage in metal-organic frameworks. *Chemical Society Reviews* 43 (16): 5657–5678.

4 Li, B., Wen, H.-M., Wang, H. et al. (2015). Porous metal-organic frameworks with Lewis basic nitrogen sites for high-capacity methane storage. *Energy & Environmental Science* 8 (8): 2504–2511.

5 (a) Xiao, J. and Wei, J. (1992). Diffusion mechanism of hydrocarbons in zeolites—I. Theory. *Chemical Engineering Science* 47 (5): 1123–1141. (b) Cui, X., Bustin, R.M., and Dipple, G. (2004). Selective transport of CO_2, CH_4, and N_2 in coals: insights from modeling of experimental gas adsorption data. *Fuel* 83 (3): 293–303.

6 (a) Li, L., Krishna, R., Wang, Y. et al. (2016). Exploiting the gate opening effect in a flexible MOF for selective adsorption of propyne from $C_1/C_2/C_3$ hydrocarbons. *Journal of Materials Chemistry A* 4 (3): 751–755. (b) Gücüyener, C., van den Bergh, J., Gascon, J., and Kapteijn, F. (2010). Ethane/ethene separation turned on its head: selective ethane adsorption on the metal-organic framework ZIF-7 through a gate-opening mechanism. *Journal of the American Chemical Society* 132 (50): 17704–17706.

7 Seoane, B., Castellanos, S., Dikhtiarenko, A. et al. (2016). Multi-scale crystal engineering of metal organic frameworks. *Coordination Chemistry Reviews* 307: 147–187.

8 Do, D.D. (1998). *Adsorption Analysis: Equilibria and Kinetics: (With CD Containing Computer Matlab Programs)*, vol. 2. World Scientific.

9 (a) Nugent, P., Belmabkhout, Y., Burd, S.D. et al. (2013). Porous materials with optimal adsorption thermodynamics and kinetics for CO_2 separation. *Nature* 495 (7439): 80–84. (b) Li, L., Bell, J.G., Tang, S. et al. (2014). Gas storage and diffusion through nanocages and windows in porous metal-organic framework Cu_2(2,3,5,6-tetramethylbenzene-1,4-diisophthalate)$(H_2O)_2$. *Chemistry of Materials* 26 (16): 4679–4695.

10 (a) Myers, A. and Prausnitz, J.M. (1965). Thermodynamics of mixed-gas adsorption. *AIChE Journal* 11 (1): 121–127. (b) Myers, A. and Prausnitz, J. (1965). Prediction of the adsorption isotherm by the principle of corresponding states. *Chemical Engineering Science* 20 (6): 549–556.

11 (a) Coudert, F.-X., Mellot-Draznieks, C., Fuchs, A.H., and Boutin, A. (2009). Prediction of breathing and gate-opening transitions upon binary mixture adsorption in metal-organic frameworks. *Journal of the American Chemical Society* 131 (32): 11329–11331. (b) Coudert, F.-X. (2010). The osmotic framework adsorbed solution theory: predicting mixture coadsorption in flexible nanoporous materials. *Physical Chemistry Chemical Physics* 12 (36): 10904–10913.

12 Richter, E., Wilfried, S., and Myers, A.L. (1989). Effect of adsorption equation on prediction of multicomponent adsorption equilibria by the ideal adsorbed solution theory. *Chemical Engineering Science* 44 (8): 1609–1616.

13 Mason, J.A., Sumida, K., Herm, Z.R. et al. (2011). Evaluating metal-organic frameworks for post-combustion carbon dioxide capture via temperature swing adsorption. *Energy & Environmental Science* 4 (8): 3030–3040.

14 (a) Mu, B. and Walton, K.S. (2011). Thermal analysis and heat capacity study of metal-organic frameworks. *The Journal of Physical Chemistry C* 115 (46): 22748–22754. (b) Koh, H.S., Rana, M.K., Wong-Foy, A.G., and Siegel, D.J. (2015). Predicting methane storage in open-metal-site metal-organic frameworks. *The Journal of Physical Chemistry C* 119 (24): 13451–13458.

15 (a) Czaja, A.U., Trukhan, N., and Muller, U. (2009). Industrial applications of metal-organic frameworks. *Chemical Society Reviews* 38 (5): 1284–1293. (b) Czaja, A., Leung, E., Trukhan, N., and Müller, U. (2011). *Metal-Organic Frameworks*, 337–352. Wiley-VCH.

14

CO_2 Capture and Sequestration

14.1 Introduction

The increasing level of atmospheric carbon dioxide caused by anthropogenic emissions is one of the most pressing problems faced by our planet. About 80% of all CO_2 emissions stem from the combustion of fossil fuels (coal, oil, and natural gas) as evidenced by the steady increase in the burning of fossil fuels and the emission of CO_2 in equal measures over the last century (Figure 14.1) [2]. Further increase in such emissions in the future caused by an increased world population and the economic growth and industrial development of emerging economies is predicted [3]. The resulting need for alternative fuels and the concomitant transition away from carbon-based fossil energy sources has initiated intense research. Realizing the required modifications to the current energy framework can however not be achieved immediately and the large-scale implementation of novel environmentally friendly technologies (e.g. natural gas storage, hydrogen storage, and battery and fuel cell technologies) still requires further development. Thus, there is a demand for efficient CO_2 capture and sequestration (CCS) technologies to lower the CO_2 emission from contemporary energy sources until the energy framework has been adapted to new more environmentally friendly technologies.

The adaptation and implementation of CCS technologies in stationary sources of CO_2 emission such as power plants fired by fossil fuels (e.g. coal and natural gas) appears to be within reach. The generation of electricity from fossil fuels causes about 60% of the worldwide CO_2 emissions [3]. Introducing efficient CCS technologies into the exhaust gas systems of such power plants could therefore lower the worldwide overall CO_2 emissions dramatically. Table 14.1 shows the components of post-combustion flue gas emitted by coal-fired power plants alongside the physical parameters of the emitted gases relevant for carbon capture.

Current technologies for carbon capture from post-combustion flue gas are based mainly on aqueous alkanolamine solutions. A major drawback of this technology with respect to its implementation in power plants is that it carries an energy penalty of approximately 30% of the output of the power plant [4]. This energy loss is associated with the large energy input required to liberate captured CO_2 from the capture medium. As current methods do not allow for

Introduction to Reticular Chemistry: Metal-Organic Frameworks and Covalent Organic Frameworks, First Edition. Omar M. Yaghi, Markus J. Kalmutzki, and Christian S. Diercks.

Companion website: www.wiley.com/go/yaghi/reticular

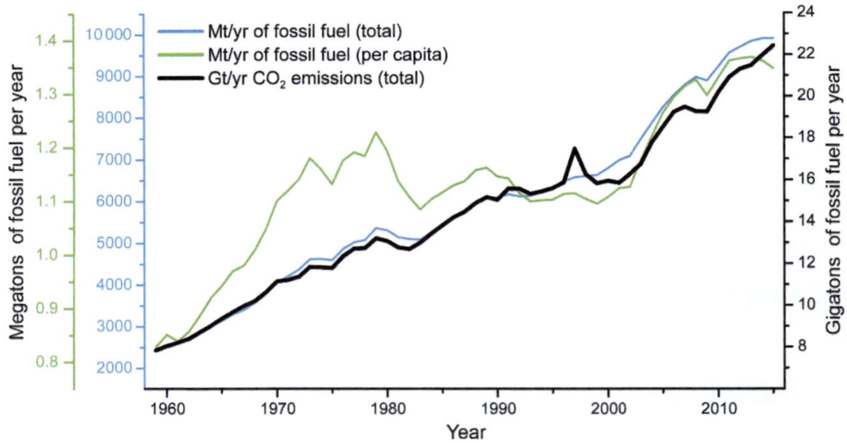

Figure 14.1 Development of the consumption of fossil fuels total (blue line) and per capita (green line) in megatons per year since 1959. The increase in the consumption of fossil fuels correlates with the amount of CO_2 emitted into the atmosphere [1].

Table 14.1 Typical composition of post-combustion flue gas produced by a coal-fired power plant and physical parameters of these gases relevant to the carbon capture process.

Molecule	Flue gas composition (%)	Physical properties			
		Kinetic diameter (Å)	Polarizability (10^{-25} cm^{-3})	Dipole moment (10^{18} esu^{-1} cm^{-1})	Quadrupole moment (10^{26} esu^{-1} cm^{-2})
N_2	73–77%	3.68	17.4	0	1.52
CO_2	15–16%	3.30	29.1	0	4.30
H_2O	5–7%	2.65	14.5	1.85	—
O_2	3–4%	3.46	15.8	0	0.39
SO_2	800 ppm	4.11	37.2–42.8	1.63	—
NO_x [a)]	500 ppm[a)]	—	30.2[a)]	0.316[a)]	—
HCl	100 ppm	3.34	26.3–27.7	1.1086	3.8
CO	20 ppm	3.69	19.5	0.1098	2.5
SO_3	10 ppm	—	—	0	—
Hydrocarbons	10 ppm	—	—	—	—
Hg	1 ppb	—	—	—	—

a) Values for NO_2 and N_2O_4.

the minimization of the input energy required for regeneration of the capture medium it is necessary to develop new materials and methods for CCS with greater energy efficiency [5]. Research in the field of metal-organic frameworks (MOFs) and other reticular framework materials [i.e. zeolitic imidazolate frameworks (ZIFs) and covalent organic frameworks (COFs)] has revealed their potential toward addressing this challenge. MOFs offer the ability to fine-tune the thermodynamics and kinetics of the CO_2 adsorption/desorption, which is crucial to improving the energy efficiency of CO_2 capture systems. Such modular materials are thus an ideal platform for the development of next-generation CO_2 capture materials [6]. The design principles discussed in earlier chapters provide chemists with the necessary tools to design materials suitable for carbon capture, and further optimize such materials by modifying their structures in pre- and post-synthetic approaches. This structural tunability allows for precise adjustment of the affinity toward CO_2, the optimization of a given material to the composition of post-combustion or pre-combustion flue gas, and even for specific site locations. Since CO_2 capture materials are used in a wide variety of scenarios we will limit this chapter to considerations concerning post-combustion (i.e. CO_2 capture from flue gas) and pre-combustion CO_2 capture (i.e. purification of gases such as syngas). Other fields of application include CO_2 capture from transportation emission and the processing of natural gas (CO_2/CH_4 separation). For further information regarding these applications the reader is referred elsewhere [6, 7]. In this chapter, we will mainly focus on CO_2 capture using MOFs and related materials. In general, similar concepts as those discussed below exist for COFs; however, their performance in this application is limited, arguably due to their nonpolar nature. For more information on CO_2 capture in COFs the reader is referred elsewhere [8].

14.2 *In Situ* Characterization

Knowledge of the specific interactions between the MOF and the gas molecules to be captured is crucial in elucidating means of optimizing MOFs with respect to CO_2 capture. A deep understanding of the relationship between the structural and chemical features of the specific MOF and its adsorption characteristics facilitates the development of next-generation framework materials. Aside from gas sorption experiments, three analytical methods are commonly used to gain insight into the adsorption process and they are described in Section 14.2.1.

14.2.1 X-ray and Neutron Diffraction

The most precise way to acquire structural information with respect to the interaction of gases with the inner surface of a MOF is by using X-ray and neutron diffraction techniques. Both methods allow for the precise determination of the position of gas molecules within the framework. Such data are collected by measuring single crystals or powders of activated MOFs enclosed in a measurement cell (often a capillary) that is evacuated and subsequently filled with a defined quantity of high-purity gas. Diffraction experiments followed by structure solution and refinement allow for the quantification and determination of the precise location of adsorbed gas molecules. This provides information that is essential to the evaluation of the interactions involved in the adsorption process. The structure solution and refinement have the same limitations as those in traditional diffraction experiments such as the presence of disorder and the adsorption of X-rays (or neutrons) by the measurement cell. In addition to these factors, the stability of the sample upon evacuation and dosing of the adsorbate must be considered. In Sections 14.2.1.1–14.2.1.3 selected examples demonstrating the power of this method are given.

14.2.1.1 Characterization of Breathing MOFs

Certain MOFs show a reversible structural distortion upon CO_2 adsorption where the entire framework structure interchanges between its pristine geometry and an expanded/contracted form in a "breathing" motion. This structural transition is triggered by temperature, pressure, or the interaction of the framework with the dipole or quadrupole moment of an adsorbate (e.g. CO_2 or H_2O) and can be studied by *in situ* powder X-ray diffraction (PXRD). A breathing effect is only possible for certain net topologies (see Chapter 21). Here, we will illustrate the breathing of MIL-53 in the presence of CO_2. MIL-53 is built from rod secondary building units (SBUs) linked by ditopic BDC linkers to form a framework of **sra** topology with trapezoidal channels running along the crystallographic *c*-axis [9]. This structural arrangement allows for expansion and compression of the channels along one direction, which is referred to as breathing of the framework. Figure 14.2 shows an *in situ* PXRD measurement of (Cr)MIL-53 in the presence of CO_2 at different pressures and the corresponding structures. From the powder pattern, one can discern that at CO_2 pressures at or above 4 bar, the pores of the framework open and the structure transitions from the pristine narrow pore (*np*) into a wide pore (*wp*) phase, resulting in a shift of the reflections toward lower angles. This process is reversible;

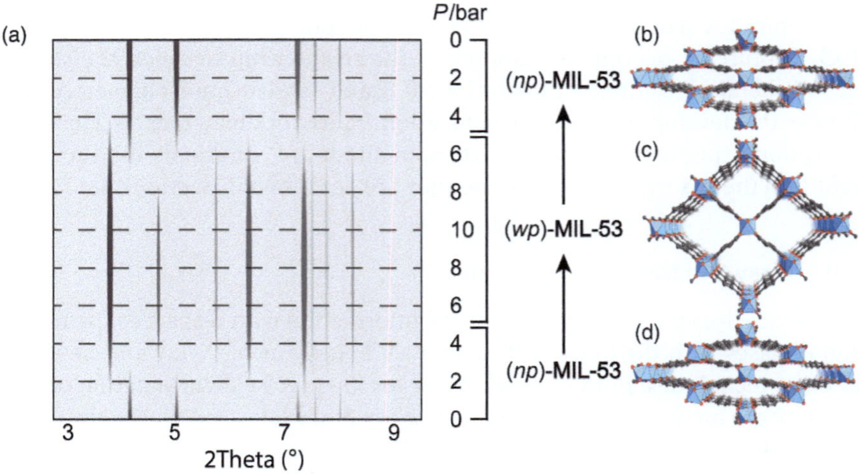

Figure 14.2 Phase transition between the narrow pore (*np*) and wide pore (*wp*) phase of (Cr)MIL-53 upon increasing the CO_2 pressure. (a) The CO_2 pressure-dependent PXRD pattern of (Cr)MIL-53 shows a significant spontaneous shift of the diffraction peaks to lower angles, indicating a phase transition from the *np* to the *wp* phase. (b, d) The *np*-phase is observed up to pressures of 4 bar. (c) At pressures above 4 bar the structure transitions spontaneously into the *wp* phase. All hydrogen atoms are omitted for clarity. Color code: Cr, blue; C, gray; O, red.

after lowering the pressure the framework transitions back to the *np* phase (Figure 14.2). A similar behavior is observed in many other MOFs; however, the synthesis of breathing MOFs in a designed approach has not yet been demonstrated.

14.2.1.2 Characterization of Interactions with Lewis Bases

As mentioned earlier, the central carbon in CO_2 is electrophilic and thus prone to attack by Lewis bases such as amino groups. This interaction can range from weak physisorption to strong covalent interaction (chemisorption) as in the formation of carbamic acid or ammonium carbamate. The former is typically observed for MOFs built from linkers with aromatic amines. An example for such a MOF is $Zn_2(ATZ)_2(OX)$, a framework with a pillared-layered structure comprised of 2D layers of Zn dimers connected by ATZ (5-amino-triazolate) and pillared by OX (oxalate) to form a 3D framework of **pcu** topology (Figure 14.3) [10]. *In situ* single-crystal X-ray diffraction studies of $Zn_2(ATZ)_2(OX)$ in the presence of CO_2 show that the CO_2 molecules interact with the amine moiety in such a way that the CO_2 carbon and the nitrogen of the amine are in close proximity. The C–N distance of about 3.15 Å and the linear molecular geometry of CO_2 indicate that the interaction is best described as physisorption.

14.2.1.3 Characterization of Interactions with Open Metal Sites

The oxygen atoms in CO_2 can interact with Lewis acidic open metal sites. Such open metal sites are found in the structure of (Ni)MOF-74 after the removal of terminal water molecules from its rod SBUs (Figure 14.4; for more detail see Chapter 2). After complete activation, the hexagonal pores of (Ni)MOF-74 are lined with open metal sites. This results in a comparatively high heat of adsorption ($Q_{st} = 42\ kJ\ mol^{-1}$) and a high CO_2 uptake at low pressures (23.9 wt% at 1 bar

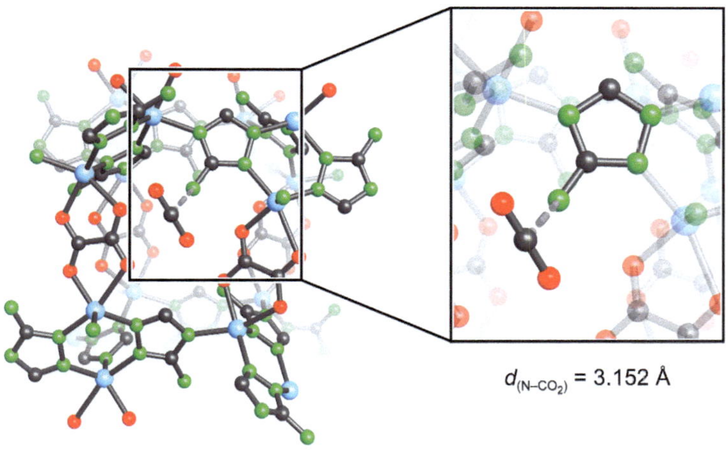

Figure 14.3 Adsorption of CO_2 in $Zn_2(ATZ)_2(OX)_2$. CO_2 is strongly adsorbed on the amines pointing into the pores. The distance between the amine and CO_2 of 3.152 Å indicates strong physisorption. Only one cage and one adsorbed CO_2 molecule are shown. All hydrogen atoms are omitted for clarity. Color code: Zn, blue; C, gray; N, green; O, red.

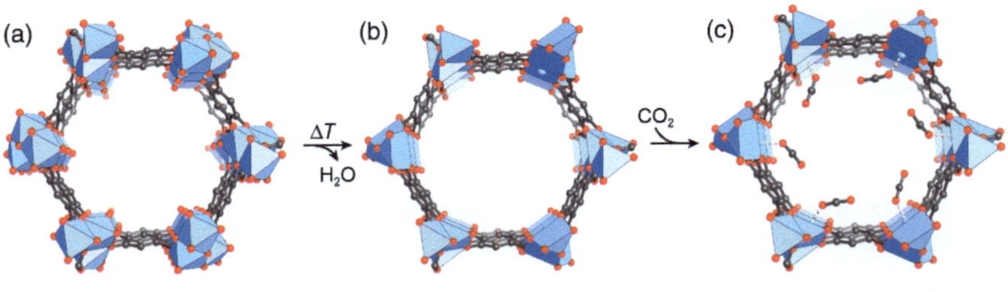

Figure 14.4 (a) Hexagonal pore of (Ni)MOF-74 with view along the crystallographic *c*-axis. (b) Heating (Ni)MOF-74 under dynamic vacuum affords the liberation of water and the creation of open metal sites. (c) Open metal sites are highly polar Lewis acidic structural features that facilitate the physisorption of CO_2 as indicated by short Ni···CO_2 distances (2.29 Å) and a deviation of CO_2 from its linear geometry ($\angle_{(O-C-O)}$ = 162(3)°). Only one pore is shown and all hydrogen atoms are omitted for clarity. Color code: Ni, blue; C, gray; O, red.

and 296 K) [11]. High-resolution powder X-ray diffraction (HR-PXRD) shows that CO_2 is preferentially bound to the open metal sites in an end-on manner [11b]. The relatively short distance between the open metal site and the CO_2 oxygen of 2.29(2) Å and an O–C–O angle of 162(3)° are indicative of strong interactions (Figure 14.4).

14.2.2 Infrared Spectroscopy

The vibrational bands observed in infrared spectra are reliable indicators for the strength of the interaction between CO_2 and the framework. The asymmetric stretching (ν_3 = 2349 cm^{-1}) and bending (ν_2 = 667 cm^{-1}) modes are infrared active, whereas the symmetric stretching mode (ν_1 = 1342 cm^{-1}) is infrared inactive (Figure 14.5). The number of possible molecular vibrational modes equals $3n - 6$ ($3n - 5$ for linear molecules) where n is the number of atoms in the molecule. According to this, CO_2 has four different vibrational modes; however, only three are listed above. This is because the bending vibration ν_2 can occur in two different directions and is therefore degenerated. This degeneracy is lifted when CO_2 interacts strongly with the adsorption site and its linear symmetry is disrupted. Along with the characteristic ν_3 shifts as a diagnostic tool, IR spectroscopy serves as a probe that is sensitive to subtle changes of the local interactions between CO_2 and its chemical surroundings.

The asymmetric stretching vibration (ν_3) of CO_2 adsorbed on the open metal sites of (Ni)MOF-74 displays a redshift of about 8 cm^{-1} compared to gaseous CO_2. The origin of this redshift is the interaction of the oxygen lone pairs with empty d-orbitals of the Ni^{2+} centers (backbonding). Additionally, a splitting of the bending vibration ν_2 into two distinct signals with an energy separation of about 8 cm^{-1} is observed, which supports the slightly bent geometry of CO_2 obtained from HR-PXRD data. The existence of a bond between the Ni^{2+} open metal sites and the adsorbed CO_2 molecules is evidenced by an additional absorption band at 2408 cm^{-1}. This absorption is ascribed to a $\nu_3 + \nu_{M-O}$ combinatorial vibration (ν_{M-O} = 67 cm^{-1}). Collection of temperature-dependent IR spectra allows for the calculation of the isosteric heat of adsorption from a van't Hoff plot, and the values calculated using this method (Q_{st} = −47 kJ mol^{-1}) are

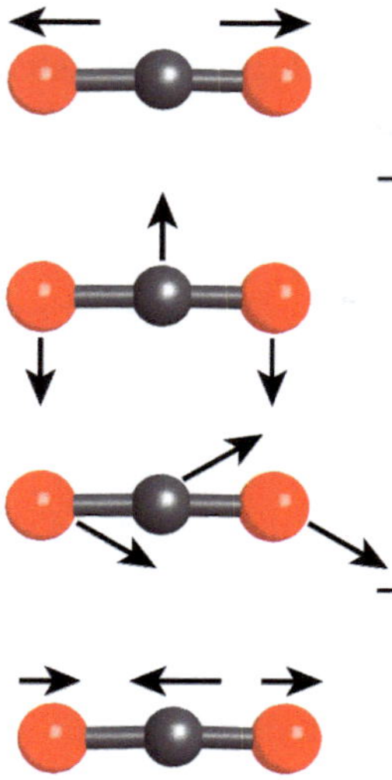

Figure 14.5 (5n−3) vibrational modes of CO_2; only three of them are IR active. The bending vibration ν_2 is degenerate in the gas phase and only two IR bands (IR-active vibrational modes) are observed. Strong interactions of CO_2 with an adsorption site can lift the degeneracy of ν_2 due to deviations from the linear molecule geometry, which results in splitting of this adsorption band. Color code: C, gray; O, red.

in good agreement with those calculated from adsorption data for zero coverage ($Q_{st} = -41\ \mathrm{kJ\ mol^{-1}}$) [12].

Diffuse reflectance infrared Fourier transform spectroscopy (DRIFTS) is a technique used to analyze the interactions of gas molecules with surfaces and can be employed to study the interactions of CO_2 with amine-functionalized MOFs. We illustrate this with amine functionalized CuBTTri (**the** topology). When fully activated, the structure of CuBTTri has open metal sites that can be functionalized by organic amines such as *N,N′*-dimethylethylenediamine (mmen) to give functionalized CuBTTri analogs (mmen-CuBTTri; see Figure 14.6). This functionalized analog exhibits an increased CO_2 uptake compared to pristine CuBTTri due to the chemisorption of CO_2 on the amines in the form of carbamic acid or ammonium carbamate [13].

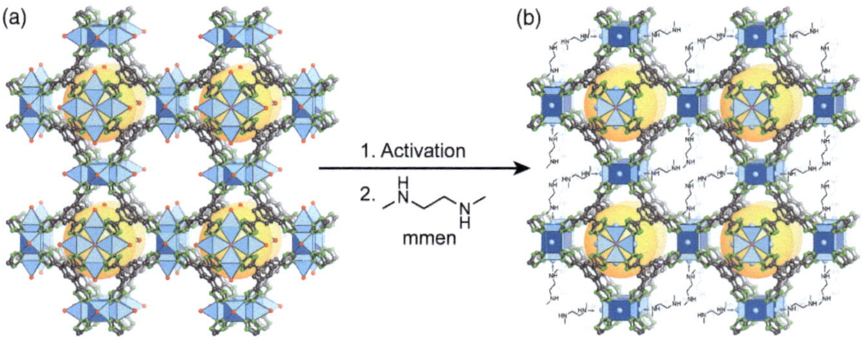

Figure 14.6 (a) Crystal structure of CuBTTri. (b) Activation affords the formation of open metal sites, which can be functionalized using organic diamines such as mmen. The amines dangling into the pores of mmen-CuBTTri can bind CO_2 in the form of carbamic acid resulting in a high selectivity. A drawback of this method is that the amines are only coordinated to the open metals sites and can therefore be exchanged by other potential ligands (e.g. water). The mmen units are drawn as Lewis structures and all hydrogen atoms are omitted for clarity. Color code: Cu, blue; C, gray; N, green; O, red.

Experimentally, the formation of carbamic acid or ammonium carbamate is shown by the disappearance of the N–H stretching vibration at 3282 cm^{-1}. The example of mmen-CuBTTri shows that the modification of framework materials with alkyldiamines can greatly improve the affinity toward CO_2 by strong and selective chemisorption. Interestingly, despite the increased isosteric heat of CO_2 adsorption of mmen-CuBTTri (Q_{st} = −96 kJ mol^{-1}) the saturated material is regenerated by temperature swing adsorption (TSA) under mild conditions.

14.2.3 Solid-State NMR Spectroscopy

The previous example (mmen-BTTri) demonstrates that CO_2 can be chemisorbed by amines dangling into the pores of a framework by formation of carbamic acid or ammonium carbamate. Using infrared spectroscopy these two species cannot be distinguished. In contrast, solid-state (cross polarization-magic angle spinning) CP-MAS NMR allows for discrimination between the two, based on a difference in the chemical shift in both the ^{13}C and ^{15}N spectrum. This is typically accomplished by dosing ^{13}C-isotope labeled $^{13}CO_2$ into a MOF synthesized from or modified with ^{15}N-enriched constituents to increase the concentration of NMR-active isotopes. The ^{13}C carbon of carbamic acid has a characteristic NMR shift of $\delta \approx 160$ ppm (which sometimes overlaps with that of the carbamate species at $\delta \approx 164$ ppm). This difference in chemical shift can be used to verify the formation of carbamic acid in IRMOF-74-III(CH_2NHMe) and IRMOF-74-III(CH_2NH_2) upon CO_2 adsorption [14]. Functionalizing the linker in IRMOF-74-III with two primary amines results in an adsorption behavior under dry conditions similar to that of IRMOF-74-III(CH_2NH_2). However, in the presence of water ammonium carbamate is formed (rather than carbamic acid) as indicated by a shift of the carbonyl peak in the ^{13}C MAS NMR spectrum as well as a shift of the amine peak in the ^{15}N MAS NMR spectrum (Figure 14.7) [15].

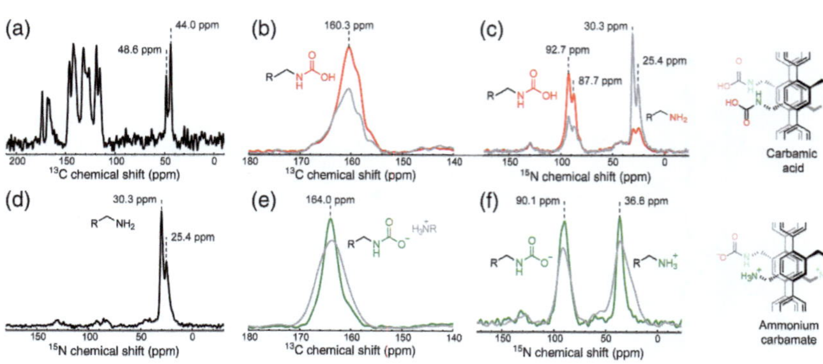

Figure 14.7 (a–c) ^{13}C and (d, c, and f) ^{15}N CP-MAS NMR spectra of isotope-labeled IRMOF-74-III(CH_2NH_2)$_2$ under different measurement conditions (left) and the corresponding chemisorbed species (right). (a, d) Spectra of fully activated IRMOF-74-III(CH_2NH_2)$_2$. (b, c) Spectra after exposure to 675 Torr $^{13}CO_2$ for 24 hours under dry conditions and (e, f) under wet conditions. The gray curves represent the spectra collected after evacuation of the CO_2 treated samples for 24 hours. The corresponding chemisorbed species are shown on the right.

These findings illustrate the power of MAS NMR in distinguishing local chemical surroundings.

14.3 MOFs for Post-combustion CO_2 Capture

In Chapters 2 and 13 we presented an introduction to the general theory of gas adsorption and separation in porous solids. In the following, we will discuss typical structural features that influence the CO_2 capture performance and ways to implement them into MOFs. With respect to CO_2 capture, the design of MOFs and their post-synthetic modification allows the adjustment of the following structural parameters: (i) the presence of open metal sites, (ii) the presence of heteroatoms within the linker and as functionalities appended to the organic backbone or the SBUs, (iii) interactions originating from the SBU, and (iv) the hydrophobicity of the material. For a material to be of interest for CO_2 capture from post-combustion flue gas we must first identify the chemical species present in this type of flue gas (see Table 14.1) to understand the possible interactions they can undergo with the adsorbent. Post-combustion flue gas mainly consists of N_2 (about 77%) and CO_2 (about 16%) and some other minor components such as H_2O, O_2, CO, NO_x, and SO_2. SO_2 is typically removed before the CO_2 capture process. The flue gas is fed to the CO_2 capture material at a pressure of approximately 1 bar and a temperature between 40 and 60 °C. To separate the main components of the gas mixture (i.e. CO_2 and N_2), it is of importance to provide adsorption sites that interact more strongly with CO_2 than with N_2. Since the polarizability (N_2 17.4×10^{-25} cm^{-3}; CO_2 29.1×10^{-25} cm^{-3}) and quadrupole moment (N_2 1.52×10^{26} esu^{-1} cm^{-2}; CO_2 4.30×10^{26} esu^{-1} cm^{-2}) of CO_2 are significantly higher than those of N_2, the separation of these two components following a thermodynamic mechanism is straight forward.

14.3.1 Influence of Open Metal Sites

As illustrated earlier, Lewis acidic open metal sites provide a partial positive charge on the pore surface and are therefore capable of interacting strongly with CO_2. This strong interaction is typically accompanied by high Q_{st} values at low pressures, a high selectivity (in the absence of other molecules with a strong dipole or quadrupole moment), and a high CO_2 uptake. In this context, the most studied group of MOFs with open metal sites is the (M)MOF-74 series (M = Mg^{2+}, Zn^{2+}, Mn^{2+}, Fe^{2+}, and Ni^{2+} among others) [11a, 16]. With 27.5 wt% at 298 K and 1 bar (Mg)MOF-74 has the highest gravimetric uptake of all MOFs in this series [16c]. Every metal center in the rod-like SBUs of MOF-74 (see Chapter 2) has one open metal site and CO_2 binds to these sites in an end-on manner (see Figure 14.4) with an isosteric heat of CO_2 adsorption at zero coverage of -47 kJ mol^{-1}. This is indicative of a strong interaction between the open metal sites and CO_2 and gives rise to the high selectivity for CO_2 in CO_2/N_2 mixtures [12a, 17]. In real world applications, however, flue gas often contains a low percentage of water (about 7%). Water is a very polarizable molecule with a strong dipole moment and thus, it also binds very strongly to open metal sites (see Table 14.1). Consequently, the CO_2 uptake capacity of (M)MOF-74 materials drops dramatically in the presence of water. To mitigate the water problem other polarizable adsorption sites such as Lewis bases that cannot interact with water must be introduced.

14.3.2 Influence of Heteroatoms

14.3.2.1 Organic Diamines Appended to Open Metal Sites

The coordinative functionalization of Lewis acidic open metal sites by amines is known for many MOFs [13, 18]. One such example was discussed earlier in this chapter where the open metal sites in CuBTTri are functionalized using ethylenediamine (en) or mmen to afford en-CuBTTri and mmen-CuBTTri, respectively (see Figure 14.6). Interestingly, both amine functionalized MOFs exhibit an increased CO_2 uptake capacity and higher Q_{st} values when compared to the parent CUBTTri. Similar results are observed for other MOFs functionalized in this way, but most of them suffer from the gradual loss of the coordinated amines over time. In all these materials CO_2 binds to the appended amine dangling in the pore by formation of carbamic acid or ammonium carbamate.

In contrast, alkylethylenediamine-appended variants of Mg_2(DOBPDC) (DOBPDC = 4,4′-dioxidobiphenyl-3,3′-dicarboxylate) adsorb CO_2 with a unique cooperative adsorption mechanism involving the insertion of CO_2 in the N–metal bond by formation of ammonium carbamate chains [19]. The structure of Mg_2(DOBPDC) is shown in Figure 14.8a, the cooperative adsorption mechanism of CO_2 is depicted in Figure 14.8b. The cooperative adsorption leads to a step-shaped isotherm allowing for high working capacities under TSA conditions. Near complete CO_2 desorption of Mg_2(DOBPDC) is realized by heating to 100 °C. This enables a high CO_2 working capacity of 9.1 wt% with a

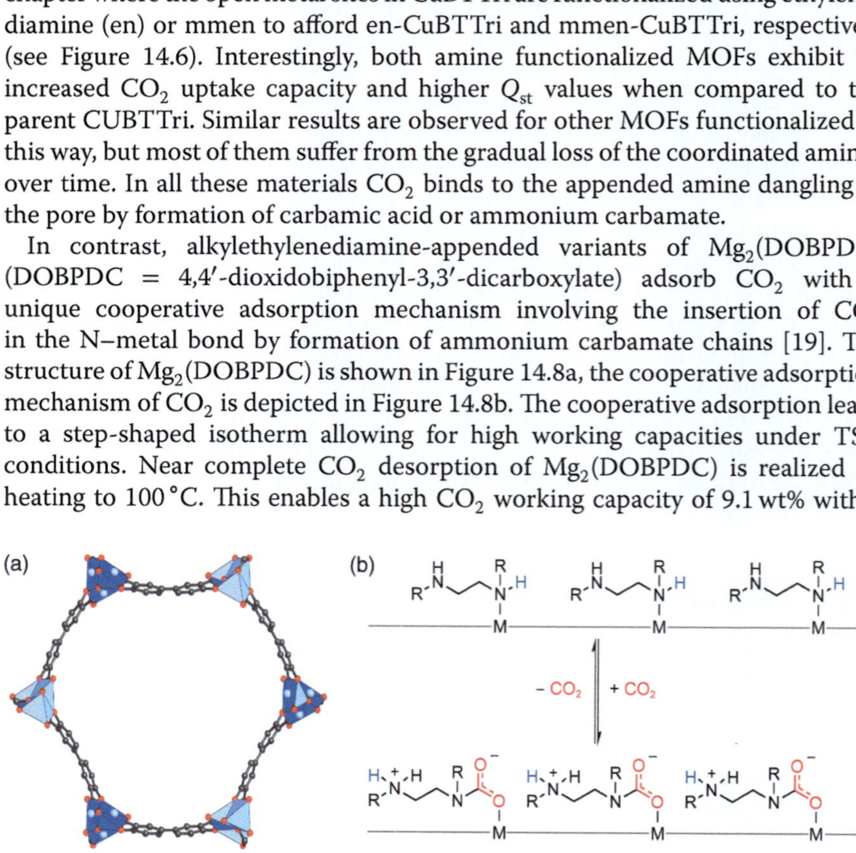

Figure 14.8 Cooperative adsorption mechanism of CO_2 in Mg_2(DOBPDC). (a) The structure of Mg_2(DOBPDC) contains 1D hexagonal channels lined with open metals sites. These open metal sites are functionalized with a series of diaminoalkanes. (b) Schematic adsorption mechanism by insertion of CO_2 into the N–metal bond and formation of ammonium carbamate. Color code: Mg, blue; C, gray; O, red.

modest TSA of only 60 °C. Additionally, the region of steep uptake can be shifted by appending different alkylethylenediamines to the open metal sites.

Even though amine-grafted MOFs show both a higher uptake and high selectivity for CO_2, they generally still suffer from a decrease in adsorption capacity in the presence of water since water can partially replace the diamine grafted onto the open metal site.

14.3.2.2 Covalently Bound Amines

Similar to diamines appended to open metal sites, attaching Lewis basic functional groups with high polarity or even nucleophilic character to the organic backbone of the linker has a stark influence on the CO_2 uptake, as well as on the adsorption mechanisms. An illustrative example for this type of functionalization is amine functionalized (Mg)IRMOF-74-III. (Mg)IRMOF-74-III(CH_2NH_2) has an uptake of 12.5 wt% at 298 K and 1 bar. This might not be as high as the capacity of (Ni)MOF-74; however, in contrast to (Ni)MOF-74, which shows a dramatic decrease of the CO_2 capacity under wet conditions, the performance of (Mg)IRMOF-74-III(CH_2NH_2) is retained [14]. Figure 14.9 shows a breakthrough experiment for (Mg)IRMOF-74-III(CH_2NH_2) that reveals a dynamic adsorption capacity of up to 0.8 mmol g^{-1}, equaling a breakthrough time of 670 ± 10 s g^{-1}.

14.3.3 Interactions Originating from the SBU

The SBUs of MOFs are typically highly polar and strong interactions can arise from these units even in the absence of open metal sites. SIFSIX-2-Cu-i ($Cu(4,4'\text{-dpa})_2(SiF_6)$), a compound built from square $[Cu_2(H_2O)_2](DPA)_{4/2}$ (DPA = 4,4′-dipyridylacetylene) grids that are pillared by SiF_6 units to form a **pcu** net, exhibits the highest CO_2 uptake for any framework material without open metal sites (see Figure 2.12) [20]. The uptake of 19.2 wt% at 298 K and atmospheric pressure is ascribed to interactions of CO_2 with the SiF_6 units. This is supported by the finding that its non-interpenetrated counterpart (SIFSIX-2-Cu) shows a lower CO_2 uptake, and SIFSIX-1-Cu, an isoreticular MOF based on the shorter BIPY (4,4′-bipyridine) linker, shows a comparable CO_2 uptake [21].

Another strategy to increase the interaction between CO_2 and the nonmetallic part of the SBU is to introduce monodentate hydroxyl moieties as capping ligands [22]. An example illustrating the effect of –OH functionalization of the SBUs is the triazolate-based MAF-X25 ($Mn_2^{2+}Cl_2(BBTA)$, where BBTA = 1*H*,5*H*-benzo(1,2-*d*:4,5-*d*′)bistriazole) shown in Figure 14.10a). MAF-X25 is built from rod SBUs that are connected by BBTA (benzo-bistriazole) linkers to form 1D hexagonal channels. The Co^{2+} centers in the SBUs of MAF-X25 can be oxidized using H_2O_2, which leads to the corresponding –OH functionalized analog MAF-X25-ox ($Mn^{2+}Mn^{3+}(OH)Cl_2(BBTA)$; Figure 14.10b). In this example, the introduction of capping hydroxyl groups on the SBU leads to an increase in CO_2 adsorption capacity of up to 50% and a change in the adsorption

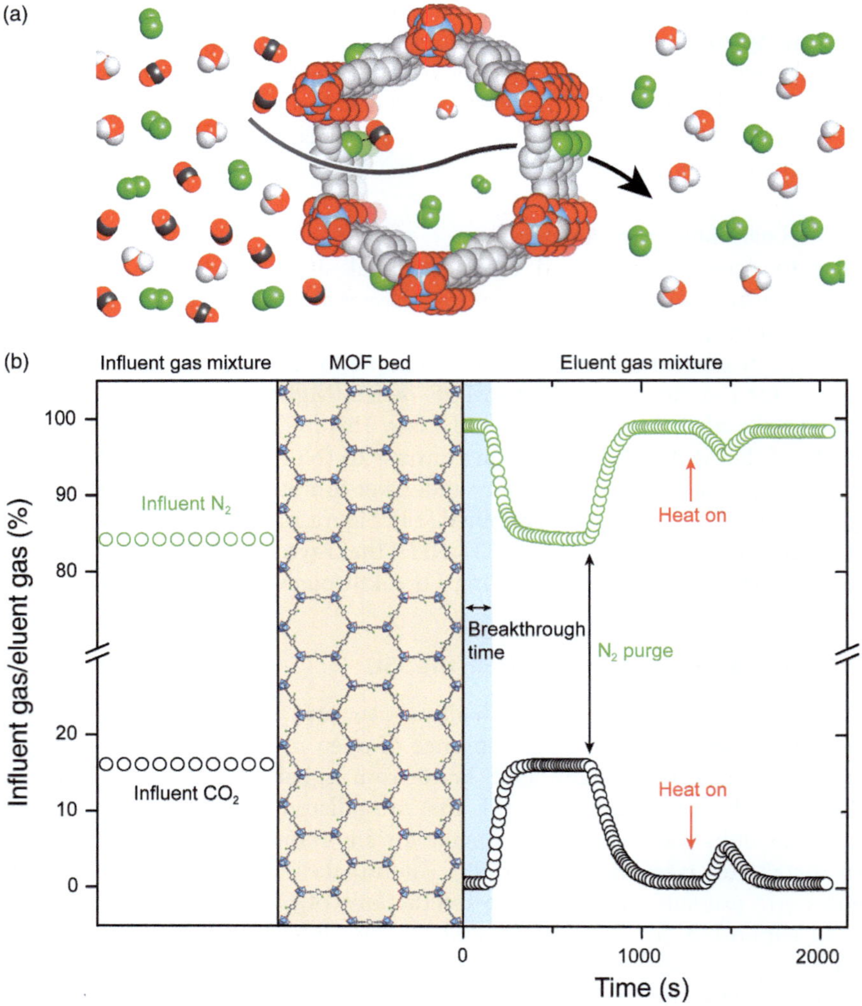

Figure 14.9 (a) Illustration of CO_2 capture by (Mg)IRMOF-74-III(CH_2NH_2). Flue gas is fed to the MOF but only CO_2 is selectively adsorbed. (b) Breakthrough measurement of a N_2/CO_2 (84 : 16) mixture. The MOF adsorbs CO_2 until the working capacity is reached (blue region). Determination of the breakthrough time (s g^{-1}) for a given flow rate allows the calculation of the kinetic gas adsorption capacity (mmol g^{-1}). N_2 purge and subsequent heating regenerate the MOF by liberating all captured CO_2. Color code: Mg, blue; C, gray; O, red.

mechanism. In pristine MAF-X25 CO_2 is adsorbed on the open metal sites while the hydroxyl groups attached to the open metal sites in MAF-X25-ox bind CO_2 covalently as bicarbonate. A similar behavior was observed for the isostructural pair of Mn^{2+} MOFs MAF-X27 and MAF-X27-ox. Interestingly, those materials not only show ultrahigh CO_2 affinity, CO_2 adsorption capacity, and CO_2/N_2 selectivity but also a high cycling stability and good sorption kinetics under dry and wet flue gas conditions.

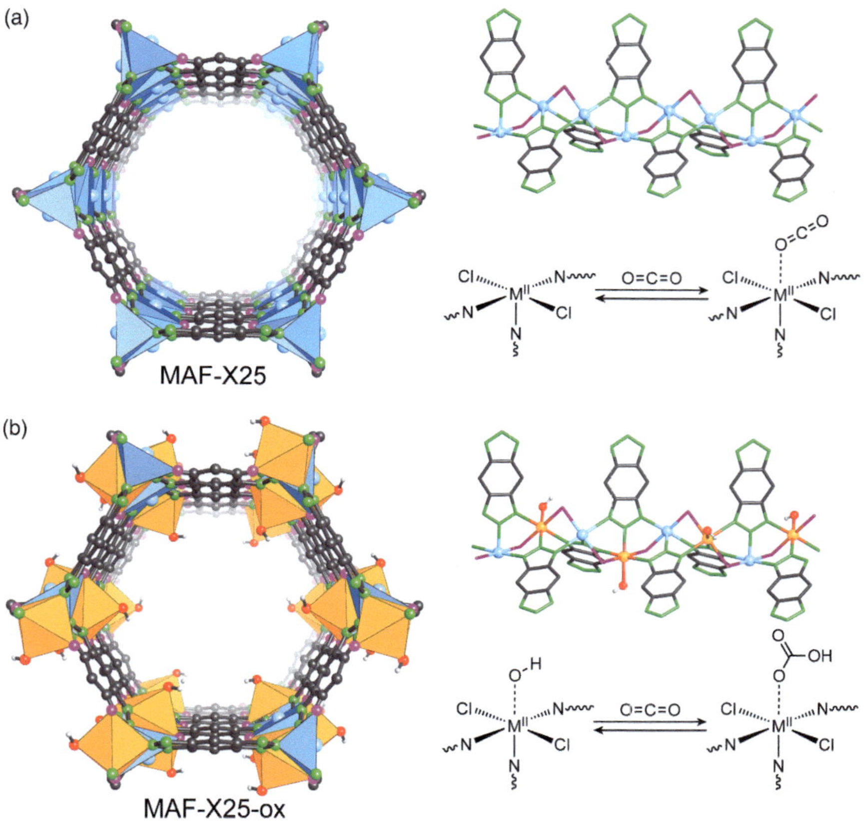

Figure 14.10 Comparison of the adsorption mechanism of CO_2 in MAF-X25 and MAF-X25-ox. (a) In the structure of MAF-X25, CO_2 is adsorbed on the open metal sites generated by activation under dynamic vacuum. (b) The hydroxyl-functionalized open metal sites in MAF-X25-ox bind CO_2 covalently as bicarbonate. Only one pore of the structure is shown and all hydrogen atoms of the linkers are omitted for clarity. Color code: M^{2+}, blue; M^{3+}, orange; C, gray; N, green; O, red; Cl, pink.

14.3.4 Influence of Hydrophobicity

The approaches described above aim at stronger interactions of the framework with CO_2 and therefore higher Q_{st} values to increase the selectivity for CO_2 in the presence of water. Another approach is to prevent water from entering the pores by introducing functionalities that render the pores hydrophobic. To illustrate this approach, we will consider an isoreticular functionalized series of ZIFs: ZIF-300 $Zn(mIM)_{0.86}(bBIM)_{1.14}$ (mIM = 2-methylimidazolate, bBIM = 5-bromo benzoimidazolate), ZIF-301 $Zn(mIM)_{0.94}(cBIM)_{1.06}$ (cBIM = 5-chloro benzoimidazolate), and ZIF-302 $Zn(mIM)_{0.67}(mBIM)_{1.33}$ (mBIM = 5-methyle benzoimidazolate). All three materials are hydrophobic and thus their CO_2 uptake under dry and wet conditions is virtually identical (Figure 14.11) [23]. The drawback of this approach is that materials relying only on hydrophobicity for CO_2/H_2O separation commonly suffer from low CO_2 adsorption capacities due to weak

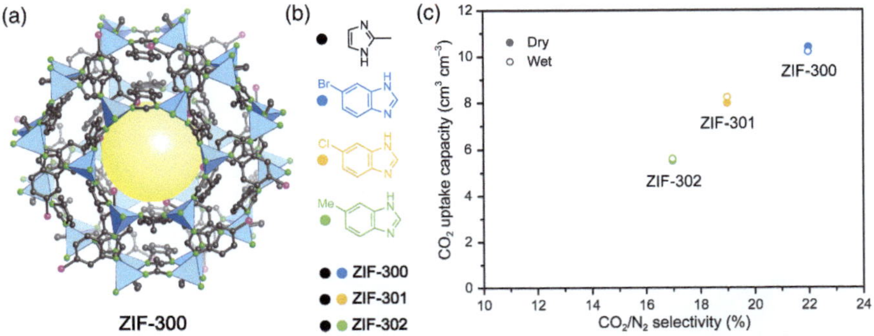

Figure 14.11 (a) Crystal structure of ZIF-300. The molecular structure of the linker molecules and their density in the crystal structure render the resulting materials hydrophobic. (b) Substituted benzimidazolates are used to prepare isostructural ZIFs. (c) Comparison of the CO_2/N_2 selectivity and the CO_2 capacity for ZIF-300, ZIF-301, and ZIF-302 under dry (filled circles) and wet (open circles) conditions. In (a) only one cage is shown and all hydrogen atoms are omitted for clarity. Color code: Zn, blue; C, gray; N, green; Br, pink.

interaction of the framework with CO_2. It would be advantageous to combine all the abovementioned strategies within one material or engineer composites of a hydrophobic shell and a core composed of a material with high selectivity for CO_2 to allow for strong interaction while retaining the initial CO_2 adsorption capacity under wet conditions [24].

14.4 MOFs for Pre-combustion CO_2 Capture

The removal of CO_2 from energy carriers (hydrogen) or fuels (natural gas) is referred to as pre-combustion CO_2 capture. Natural gas, the shifted product of coal gasification, and syngas, produced by steam methane reforming, are both mixtures of various components. The presence of CO_2 in these gas mixtures lowers the energy efficiency when used in combustion processes or fuel cells. Additionally, in the presence of moisture CO_2 is corrosive, which can cause problems with respect to the makeup of containers used. Therefore, it is of interest to purify these gas mixtures by CO_2 separation before the pure gas is either stored or used in power plants, combustion engines, or fuel cells. MOFs suitable for this process, referred to as "sweetening," need to fulfill a set of requirements different from those for post-combustion CO_2 capture. This becomes clear when considering the process conditions and the composition of the gas mixture. Pre-combustion carbon capture is typically carried out at higher pressures (5–40 bar and 40 °C) compared to post-combustion CO_2 capture (about 1 bar) and at a relatively high CO_2 level (e.g. 25–35% for syngas from coal and 15–25% for syngas from methane). This requires MOFs with high CO_2 uptake in the high pressure region [25]. Owing to the larger difference in molecular size in the couples H_2/CO_2 and CH_4/CO_2, kinetic separation (discrimination by pore size) is applicable to pre-combustion CO_2 capture. However, the small pores needed to achieve high selectivity will cause a significant pressure drop, rendering this approach unattractive for large-scale applications. Therefore, thermodynamic separation is favored, and the general principles of structural optimization are similar to those for carbon capture from post-combustion flue gas outlined in Section 14.3. Since the adsorption takes place at elevated pressures, desorption is often achieved by simply lowering the pressure. Optimization regarding both the selectivity and uptake capacity are needed to render MOFs suitable for industrial applications.

14.5 Regeneration and CO_2 Release

An important part of the CO_2 capture cycle is the release of captured CO_2 from the adsorbent. As discussed earlier in this chapter, the release of captured CO_2 from currently employed capture media (alkanolamine solutions) carries a massive energy penalty rendering those technologies economically unattractive [4]. The ease of regeneration of MOFs, often carried out under comparatively mild conditions, makes them an interesting alternative to solution-based CO_2 capture methods. Common technologies for the regeneration of porous CO_2 capture materials include temperature, vacuum, and pressure swing adsorption (TSA, VSA, and PSA, respectively) or combinations thereof (Figure 14.12). In a typical setup, the CO_2 charged capture material is confined in some sort of cell

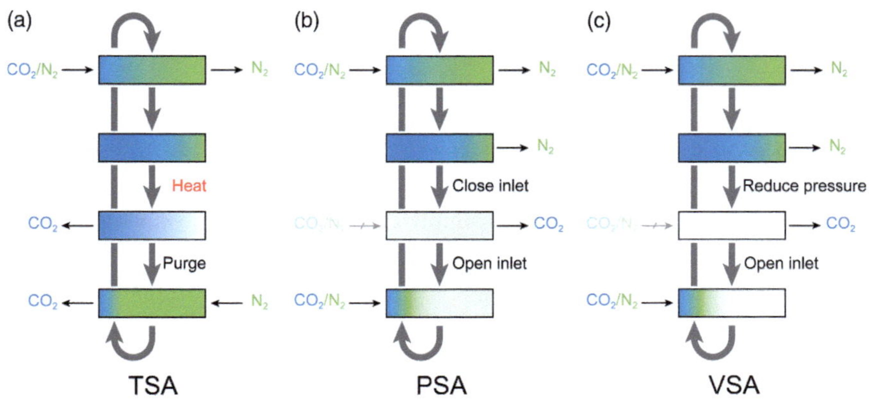

Figure 14.12 Schematic diagrams of temperature swing adsorption (TSA), pressure swing adsorption (PSA), and vacuum swing adsorption (VSA) processes for the regeneration of a fixed bed column used for CO_2 capture from flue gas (N_2/CO_2). (a) The temperature swing adsorption cycle makes use of the different affinities of CO_2 to the adsorbent at different temperatures, facilitating desorption upon increasing the temperature. (b, c) Both pressure and vacuum swing adsorption cycles achieve the regeneration of the adsorbent bed by lowering the partial pressure (closing the inlet valve) or evacuation. This corresponds to moving from the right (high P/P_0) to the left (low P/P_0) on the adsorption isotherm.

or container (e.g. packed bed of powder or pellets) and regenerated by increasing the temperature (TSA) or changing the pressure (PSA and VSA).

14.5.1 Temperature Swing Adsorption

Industrial processes typically produce excess heat that can be used in TSA cycles to regenerate the capture materials that allow for regeneration at low temperatures [26]. By heating the saturated adsorbent at ambient pressure, adsorbed gas molecules desorb from the adsorbent and the increased partial pressure caused by the desorption drives them off the adsorbent bed. An equilibrium state, where no more gas desorbs from the adsorbent, is reached at the optimal desorption temperature of the adsorbent. Desorbed gas molecules remaining in the void spaces of the adsorbent bed are pushed off with a purge, until the desired purity of eluent gas is reached. After cooling the bed to the optimal adsorption/working temperature of the adsorbent the next adsorption cycle is started (Figure 14.12a).

The lower heat capacities of MOFs (about $1\,J\,g^{-1}\,K^{-1}$) compared to conventional monoethanolamine solutions ($3–4\,J\,g^{-1}\,K^{-1}$ depending on the concentration) make them ideal candidates for regeneration by TSA [24, 27]. Regeneration of HKUST-1 by TSA is more successful than regeneration by VSA, which is attributed to the unsuccessful recovery of the open metal sites by VSA [28]. It should, however, be noted that the low thermal conductivity of MOFs may render VSA and PSA more favorable in some cases. This problem is overcome when shaped bodies with appropriate additives, which increase the thermophysical properties of the MOF, are used.

14.5.2 Vacuum and Pressure Swing Adsorption

In PSA and VSA cycles the adsorbent is regenerated by lowering the partial pressure to desorb the captured gas. PSA uses pressures ranging from 8 to 28 bars in the adsorption process whereas the adsorption in VSA is commonly carried out at ambient pressures [25a]. Once the adsorbent is saturated the pressure is lowered. In the case of PSA this is realized by simply closing the inlet valve, which leads to a decrease of the pressure to ambient pressure, whereas in the case of VSA the pressure in the adsorbent bed is lowered to subatmospheric pressures. The adsorbed gas is desorbed by the pressure drop and purged from the adsorbent bed (Figure 14.12b, c). Pressure vacuum swing adsorption (PVSA) can be envisioned as a hybrid of PSA and VSA where an elevated input pressure and a subatmospheric desorption pressure are used. In stationary applications such as power plants post-combustion flue gas is released near ambient pressure, making it necessary to either pressurize the gas for PSA or evacuate the adsorbent bed using VSA. Considering this, TSA represents the most viable process for CO_2 capture from post-combustion flue gas. Pre-combustion carbon capture (e.g. purification of natural gas) typically uses a pressurized gas stream rendering the PSA cycle most appropriate for this application.

Pelletized (Al)MIL-53 that is used in the separation of a mixture of 87% CH_4 and 13% CO_2 at 4 bar and 303 K can be regenerated at 0.1 bar using a PSA cycle yielding methane with 99.4% purity [29]. The regeneration of MOFs by VSA and PVSA is far less studied. In the case of HKUST-1, regeneration using a VSA cycle after exposure to a feed of 13–16% CO_2 in dry N_2 at 2 bar and 308 K is more successful than using a PSA cycle. This is due to the fact that in the VSA cycle all open metal sites are regenerated. The working capacity of HKUST-1 under these conditions is 2.22 mmol g^{-1} with 63% CO_2 recovery [30]. UiO-66 can be regenerated using a PVSA cycle that consists of six consecutive steps: (i) feed pressurization, (ii) adsorption (2 bars and 328 K), (iii) countercurrent blow-down, (iv) concurrent rinse with CO_2, (v) countercurrent evacuation, and (vi) N_2 purge.

14.6 Important MOFs for CO_2 Capture

The performance of an adsorbent with respect to CO_2 capture is characterized by its capacity and selectivity. Since the gas mixtures from which CO_2 is captured contain different gases of similar molecular size, this process is typically based on a thermodynamic separation mechanism and is therefore controlled by the strength of the adsorbate–adsorbent interaction, i.e. the heat of adsorption (Q_{st}). The selectivity and the uptake (in wt%) alongside the primary adsorption sites for MOFs relevant to carbon capture applications are compiled in Tables 14.2 and 14.3, respectively. All values are measured/calculated at low pressure and room temperature conditions similar to those in carbon capture from post-combustion flue gas. From these tables it is clear that MOFs that contain open metal sites or Lewis basic functionalities show higher CO_2 capacities and selectivity. One must keep in mind that open metal sites and diamines appended to them typically lose their selectivity under wet conditions.

The gas uptake determined by adsorption measurements, the Q_{st} values calculation therefrom, and the selectivity calculated using single-component isotherm constitute equilibrium data (see Chapter 13). It is however important to keep in mind that gas separation is a dynamic process and the adsorption/desorption kinetics must be considered in the assessment of a promising adsorbent. For CO_2 capture from flue gas such information is obtained from time-dependent breakthrough experiments. Here, a column filled with activated MOF is exposed to a gas mixture (e.g. CO_2/N_2 simulating flue gas) and the eluent gas breaking through the column is analyzed (see Figure 14.9). Subsequent regeneration of the column can be carried out using any of the abovementioned methods (TSA, PSA, or VSA). Measurements at different concentrations, temperatures, and flow rates provide a more detailed insight into the adsorption/desorption kinetics, information that is useful in the optimization of the parameters relevant for efficient carbon capture (e.g. packing density, structuring of the adsorbent, flow rates, and temperatures, among others).

Table 14.2 Heats of adsorption and selectivity of the best performing MOFs with respect to CO_2 capture.

MOF		CO_2 adsorption selectivity				References
Chemical formula	Common name	Functionality	$-Q_{st}$ (kJ mol^{-1})	Selectivity	*T* (K)	
$H_3[(Cu_4Cl)_3(BTTri)_8(mmen)_{12}]$	mmen- CuBTTri	Amines	96	165	29	[13]
$H_3[(Cu_4Cl)_3(BTTri)_8(en)_{3.75}]$	en-CuBTTri	Amines	90	44	298	[31]
$Cr_3OF(H_2O)_3(BTC)_3$	MIL-101(Cr)	OMS	62	—	—	[32]
$Al(OH)(NH_2\text{-}BDC)$	NH_2-MIL- 53(Al)	Amines	50	—	—	[33]
$Co_2(ade)_2(CO_2CH_3)_2$	bio-MOF-11	Amines	45	65	298	[34]
$Cu_3(TATB)_2$	CuTATB	—	48	24	298	[35]
$Mg_2(DOBDC)$	(Mg)MOF-74	OMS	47	44	303	[12a, 17]
$Ni_2(DOBDC)$	(Ni)MOF-74	OMS	42	30	298	[11, 12b]
$Cu_3(BTC)_2$	HKUST-1	OMS	35	101	293	[36]
$Al(OH)(BDC)$	MIL-53(Al)	—	35	—	—	[37]
$Cu_3(BTC)_2 \cdot 3H_2O$	HKUST-1 (hydrated)	—	30	—	—	[38]
$Zn_4O(NH_2\text{-}BDC)_3$	IRMOF-3	Amine	19	—	—	[39]
$Zn_4O(BDC)_3$	MOF-5	—	17	—	—	[40]
$Zn_4O(BTB)_2$	MOF-177	—	—	4	298	[17]
$Zn_4O(BDC)(BTB)_{4/3}$	UMCM-1	—	12	—	—	[41]

Q_{st} and selectivity for a N_2/CO_2 mixture as well as the primary adsorption sites are given for room temperature and low pressure; conditions similar to those in carbon capture from post-combustion flue gas.

Table 14.3 Gravimetric uptake (wt%) of the best performing MOFs with respect to CO_2 storage.

MOF		CO_2 adsorption capacity				References
Chemical formula	Common name	Primary adsorption site	Pressure (bar)	*T* (K)	Uptake (wt%)	
Mg_2(DOT)	(Mg)MOF-74	OMS	1	298	27.5	[16c]
$Zn_2(TDC)_2$(MA) $Zn_2(TDC)_2$(MA)	$Zn_2(tdc)_2$(MA)	Hybrid	1	298	27.0	[42]
Fe_2(DOT)	(Fe)MOF-74	OMS[a)]	1	298	23.8	[16e]
Mg_2(DOBPDC)	(Mg)IRMOF-74-II	OMS[a)]	1	298	22.0	[19d]
$Al_2(OH)_2$TCPP(Cu)	S_{Cu}	OMS[a)]	1	298	21.7	[43]
$(TEPA)_xMg_2$(DOT)	TEPA-MOF-74	Aliphatic amine	1	298	21.1	[44]
Cu(Me-4py-trz-ia)	Cu(Me-4py-trz-ia)	Hybrid	1	298	21.1	[45]
Cu_3(TDPAT)	Cu-TDPAT	Hybrid	1	298	20.6	[46]
Ni_2(DOT)	(Ni)MOF-74	OMS[a)]	1	298	20.5	[11a]
Cu_3(TDCPAH)	rht-MOF-9	Heteroaromatic amine	1	298	20.2	[47]
$Cu_3(BTC)_2$	HKUST-1	OMS[a)]	1.1	293	19.8	[36a]
$Cu[(PYDC_2)PdCl_2]$	NbO-Pd-1	OMS[a)]	1	298	19.7	[48]
Co_2(DOBDC)	(Co)MOF-74	OMS[a)]	1	298	19.7	[11a]
$[Mg_2(DOT)(N_2H_4)_{1.8}]$	N_2H_4-(Mg)MOF-74	Aliphatic amine	1	298	19.5	[49]
Cu_3(TPBTM)	Cu-TPBTM	OMS[a)]	1	298	19.5	[50]
Cu_2(DBIP)	nbo-Cu_2(DBIP)	OMS[a)]	0.95	298	19.3	[51]
$[Mg_2(DHT)(H_2O)_{0.8}(en)_{1.2}]\cdot 0.2(en)$	CPO-27-Mg-c	Aliphatic amine	1	298	19.2	[52]
Cu(4,4′-dpa)$_2(SiF_6)$	SIFSIX-2-Cu-i	SBU-based interactions	1.1	298	19.2	[20]
$Cu_2[(NH_2)$-PyrDI]	ZJNU-54	Heteroaromatic amine	1	298	19.1	[53]
Cu(4,4′-bpy)$_2(SiF_6)$	SIFSIX-1-Cu	SBU-based interactions	1	298	19.1	[21]
$Cu_6(TADIPA)_3$	JLU-Liu21	Hybrid	1	298	18.8	[54]
Cu_2(iso-QuDI)	ZJNU-44	Heteroaromatic amine	1	296	18.6	[55]
Cu_2(PDAI)	NJU-Bai21, PCN-124	Hybrid	1	298	18.4	[56]
Zn(BTZ)	Zn(BTZ)	Heteroatom	1	298	18.0	[57]
$(PEI)_xCr_3(O)OH(BDC)_3$	PEI-MIL-101	Aliphatic amine	1	298	18.0	[58]

The gravimetric uptake in weight percentage as well as the primary adsorption sites are given for room temperature and low pressure; conditions similar to those in carbon capture from post-combustion flue gas.

a) OMS = open metal site.

14.7 Summary

In this chapter, we discussed different approaches for CO_2 capture from flue gas using MOFs. The different requirements posed to a CCS material by varying compositions of flue gas with respect to the CO_2 capture mechanisms were outlined. We introduced different analytical methods frequently used to characterize the chemical nature of bound CO_2 and thus gain a deeper understanding of possible interactions between CO_2 and MOFs. Using this knowledge, we elucidated the capture mechanisms of MOFs with different structural features, such as open metal sites, –OH functionalized SBUs, and amino functionalities (appended to the SBU or the linker). Different ways for cycled CO_2 adsorption (PSA, TSA, and VSA) were discussed and the chapter concludes with a compilation of MOFs with good CO_2 adsorption properties.

References

1 (a) Le Quéré, C., Andrew, R.M., Canadell, J.G. et al. (2016). Global carbon budget 2016. *Earth System Science Data* 8 (2): 605. (b) Marland, G., Boden, T.A., and Andres, R.J. (2000). Global, Regional, and National Fossil Fuel CO_2 Emissions. Oak Ridge, TN: Carbon Dioxide Information Analysis Center, Oak Ridge National Laboratory, U.S. Department of Energy. (c) Le Quéré, C., Moriarty, R., Andrew, R.M. et al. (2015). Global carbon budget 2014. *Earth System Science Data* 7 (1): 47–85. (d) Houghton, R., Van der Werf, G., DeFries, R. et al. (2012). Chapter G2 Carbon emissions from land use and land-cover change. *Biogeosciences Discussions* 9: 835–878.

2 Edenhofer, O., Pichs-Madruga, R., Sokona, Y. et al. (2015). Intergovernmental panel on climate change. In: *Climate Change 2014: Mitigation of Climate Change*, vol. 3. Cambridge University Press.

3 Pachauri, R.K. and Reisinger, A. (2007). *IPCC Fourth Assessment Report*. Geneva: IPCC.

4 Yu, C.-H., Huang, C.-H., and Tan, C.-S. (2012). A review of CO_2 capture by absorption and adsorption. *Aerosol and Air Quality Research* 12 (5): 745–769.

5 Haszeldine, R.S. (2009). Carbon capture and storage: how green can black be? *Science* 325 (5948): 1647–1652.

6 (a) Li, J.-R., Kuppler, R.J., and Zhou, H.-C. (2009). Selective gas adsorption and separation in metal-organic frameworks. *Chemical Society Reviews* 38 (5): 1477–1504. (b) Morris, R.E. and Wheatley, P.S. (2008). Gas storage in nanoporous materials. *Angewandte Chemie International Edition* 47 (27): 4966–4981.

7 (a) Keskin, S., van Heest, T.M., and Sholl, D.S. (2010). Can metal-organic framework materials play a useful role in large-scale carbon dioxide separations? *ChemSusChem* 3 (8): 879–891. (b) Simmons, J.M., Wu, H., Zhou, W., and Yildirim, T. (2011). Carbon capture in metal-organic frameworks – a comparative study. *Energy & Environmental Science* 4 (6): 2177–2185.

8 Zeng, Y., Zou, R., and Zhao, Y. (2016). Covalent organic frameworks for CO_2 capture. *Advanced Materials* 28 (15): 2855–2873.

9 (a) Millange, F., Serre, C., and Ferey, G. (2002). Synthesis, structure determination and properties of MIL-53as and MIL-53ht: the first Cr^{III} hybrid inorganic–organic microporous solids: $Cr^{III}(OH)\cdot\{O_2C{-}C_6H_4{-}CO_2\}\cdot\{HO_2C{-}C_6H_4{-}CO_2H\}_x$. *Chemical Communications* 8: 822–823. (b) Serre, C., Millange, F., Thouvenot, C. et al. (2002). Very large breathing effect in the first nanoporous chromium(III)-based solids: MIL-53 or $Cr^{III}(OH)\cdot\{O_2C{-}C_6H_4{-}CO_2\}\cdot\{HO_2C{-}C_6H_4{-}CO_2H\}_x\cdot H_2O_y$. *Journal of the American Chemical Society* 124 (45): 13519–13526. (c) Serre, C., Bourrelly, S., Vimont, A. et al. (2007). An explanation for the very large breathing effect of a metal-organic framework during CO_2 adsorption. *Advanced Materials* 19 (17): 2246–2251. (d) Boutin, A., Coudert, F.-X., Springuel-Huet, M.-A. et al. (2010). The behavior of flexible MIL-53(Al) upon CH_4 and CO_2 adsorption. *The Journal of Physical Chemistry C* 114 (50): 22237–22244.

10 Vaidhyanathan, R., Iremonger, S.S., Dawson, K.W., and Shimizu, G.K. (2009). An amine-functionalized metal organic framework for preferential CO_2 adsorption at low pressures. *Chemical Communications* 35: 5230–5232.

11 (a) Özgür Yazaydın, A., Snurr, R.Q., Park, T.-H. et al. (2009). Screening of metal-organic frameworks for carbon dioxide capture from flue gas using a combined experimental and modeling approach. *Journal of the American Chemical Society* 131 (51): 18198–18199. (b) Dietzel, P.D., Johnsen, R.E., Fjellvåg, H. et al. (2008). Adsorption properties and structure of CO_2 adsorbed on open coordination sites of metal-organic framework Ni_2(dhtp) from gas adsorption, IR spectroscopy and X-ray diffraction. *Chemical Communications* 41: 5125–5127.

12 (a) Valenzano, L., Civalleri, B., Chavan, S. et al. (2010). Computational and experimental studies on the adsorption of CO, N_2, and CO_2 on Mg-MOF-74. *The Journal of Physical Chemistry C* 114 (25): 11185–11191. (b) Dietzel, P.D., Besikiotis, V., and Blom, R. (2009). Application of metal-organic frameworks with coordinatively unsaturated metal sites in storage and separation of methane and carbon dioxide. *Journal of Materials Chemistry* 19 (39): 7362–7370.

13 McDonald, T.M., D'Alessandro, D.M., Krishna, R., and Long, J.R. (2011). Enhanced carbon dioxide capture upon incorporation of *N*,*N*′-dimethylethylenediamine in the metal-organic framework CuBTTri. *Chemical Science* 2 (10): 2022–2028.

14 Fracaroli, A.M., Furukawa, H., Suzuki, M. et al. (2014). Metal-organic frameworks with precisely designed interior for carbon dioxide capture in the presence of water. *Journal of the American Chemical Society* 136 (25): 8863–8866.

15 Flaig, R.W., Osborn Popp, T.M., Fracaroli, A.M. et al. (2017). The chemistry of CO_2 capture in an amine-functionalized metal-organic framework under dry and humid conditions. *Journal of the American Chemical Society* 139 (35): 12125–12128.

16 (a) Britt, D., Furukawa, H., Wang, B. et al. (2009). Highly efficient separation of carbon dioxide by a metal-organic framework replete with open

metal sites. *Proceedings of the National Academy of Sciences* 106 (49): 20637–20640. (b) Mason, J.A., McDonald, T.M., Bae, T.-H. et al. (2015). Application of a high-throughput analyzer in evaluating solid adsorbents for post-combustion carbon capture via multicomponent adsorption of CO_2, N_2, and H_2O. *Journal of the American Chemical Society* 137 (14): 4787–4803. (c) Bao, Z., Yu, L., Ren, Q. et al. (2011). Adsorption of CO_2 and CH_4 on a magnesium-based metal organic framework. *Journal of Colloid and Interface Science* 353 (2): 549–556. (d) Caskey, S.R., Wong-Foy, A.G., and Matzger, A.J. (2008). Dramatic tuning of carbon dioxide uptake via metal substitution in a coordination polymer with cylindrical pores. *Journal of the American Chemical Society* 130 (33): 10870–10871. (e) Märcz, M., Johnsen, R.E., Dietzel, P.D., and Fjellvåg, H. (2012). The iron member of the CPO-27 coordination polymer series: synthesis, characterization, and intriguing redox properties. *Microporous and Mesoporous Materials* 157: 62–74. (f) Wang, L.J., Deng, H., Furukawa, H. et al. (2014). Synthesis and characterization of metal-organic framework-74 containing 2, 4, 6, 8, and 10 different metals. *Inorganic Chemistry* 53 (12): 5881–5883. (g) Queen, W.L., Hudson, M.R., Bloch, E.D. et al. (2014). Comprehensive study of carbon dioxide adsorption in the metal-organic frameworks M_2(dobdc)(M = Mg, Mn, Fe, Co, Ni, Cu, Zn). *Chemical Science* 5 (12): 4569–4581.

17 Mason, J.A., Sumida, K., Herm, Z.R. et al. (2011). Evaluating metal-organic frameworks for post-combustion carbon dioxide capture via temperature swing adsorption. *Energy & Environmental Science* 4 (8): 3030–3040.

18 (a) Hwang, Y.K., Hong, D.Y., Chang, J.S. et al. (2008). Amine grafting on coordinatively unsaturated metal centers of MOFs: consequences for catalysis and metal encapsulation. *Angewandte Chemie International Edition* 47 (22): 4144–4148. (b) Montoro, C., Garcia, E., Calero, S. et al. (2012). Functionalisation of MOF open metal sites with pendant amines for CO_2 capture. *Journal of Materials Chemistry* 22 (20): 10155–10158.

19 (a) Milner, P.J., Siegelman, R.L., Forse, A.C. et al. (2017). A diaminopropane-appended metal-organic framework enabling efficient CO_2 capture from coal flue gas via a mixed adsorption mechanism. *Journal of the American Chemical Society* 139 (38): 13541–13553. (b) Siegelman, R.L., McDonald, T.M., Gonzalez, M.I. et al. (2017). Controlling cooperative CO_2 adsorption in diamine-appended Mg_2(dobpdc) metal-organic frameworks. *Journal of the American Chemical Society* 139 (30): 10526–10538. (c) McDonald, T.M., Mason, J.A., Kong, X. et al. (2015). Cooperative insertion of CO_2 in diamine-appended metal-organic frameworks. *Nature* 519 (7543): 303–308. (d) McDonald, T.M., Lee, W.R., Mason, J.A. et al. (2012). Capture of carbon dioxide from air and flue gas in the alkylamine-appended metal-organic framework *mmen*-Mg_2(dobpdc). *Journal of the American Chemical Society* 134 (16): 7056–7065. (e) Jo, H., Lee, W.R., Kim, N.W. et al. (2017). Fine-tuning of the carbon dioxide capture capability of diamine-grafted metal-organic framework adsorbents through amine functionalization. *ChemSusChem* 10 (3): 541–550. (f) Lee, W.R., Jo, H., Yang, L.-M. et al. (2015). Exceptional CO_2 working capacity in a heterodiamine-grafted metal-organic framework. *Chemical Science* 6 (7): 3697–3705. (g) Lee, W.R., Hwang, S.Y.,

Ryu, D.W. et al. (2014). Diamine-functionalized metal-organic framework: exceptionally high CO_2 capacities from ambient air and flue gas, ultrafast CO_2 uptake rate, and adsorption mechanism. *Energy & Environmental Science* 7 (2): 744–751.

20 Nugent, P., Belmabkhout, Y., Burd, S.D. et al. (2013). Porous materials with optimal adsorption thermodynamics and kinetics for CO_2 separation. *Nature* 495 (7439): 80–84.

21 Burd, S.D., Ma, S., Perman, J.A. et al. (2012). Highly selective carbon dioxide uptake by [Cu(bpy-*n*)$_2$(SiF$_6$)](bpy-1 = 4,4′-bipyridine; bpy-2 = 1,2-bis(4-pyridyl)ethene). *Journal of the American Chemical Society* 134 (8): 3663–3666.

22 Liao, P.-Q., Chen, H., Zhou, D.-D. et al. (2015). Monodentate hydroxide as a super strong yet reversible active site for CO_2 capture from high-humidity flue gas. *Energy & Environmental Science* 8 (3): 1011–1016.

23 Nguyen, N.T., Furukawa, H., Gándara, F. et al. (2014). Selective capture of carbon dioxide under humid conditions by hydrophobic chabazite-type zeolitic imidazolate frameworks. *Angewandte Chemie International Edition* 53 (40): 10645–10648.

24 Trickett, C.A., Helal, A., Al-Maythalony, B.A. et al. (2017). The chemistry of metal-organic frameworks for CO_2 capture, regeneration and conversion. *Nature Reviews Materials* 2: 17045.

25 (a) Sircar, S. and Golden, T. (2000). Purification of hydrogen by pressure swing adsorption. *Separation Science and Technology* 35 (5): 667–687. (b) Gupta, R.B. (2008). *Hydrogen Fuel: Production, Transport, and Storage*. CRC Press. (c) Liu, K., Song, C., and Subramani, V. (2009). *Hydrogen and Syngas Production and Purification Technologies*. Wiley.

26 (a) Merel, J., Clausse, M., and Meunier, F. (2008). Experimental investigation on CO_2 post-combustion capture by indirect thermal swing adsorption using 13X and 5A zeolites. *Industrial and Engineering Chemistry Research* 47 (1): 209–215. (b) Berger, A.H. and Bhown, A.S. (2011). Comparing physisorption and chemisorption solid sorbents for use separating CO_2 from flue gas using temperature swing adsorption. *Energy Procedia* 4: 562–567.

27 (a) Kloutse, F.A., Zacharia, R., Cossement, D., and Chahine, R. (2015). Specific heat capacities of MOF-5, Cu-BTC, Fe-BTC, MOF-177 and MIL-53 (Al) over wide temperature ranges: measurements and application of empirical group contribution method. *Microporous and Mesoporous Materials* 217 (Suppl. C): 1–5. (b) Mu, B. and Walton, K.S. (2011). Thermal analysis and heat capacity study of metal-organic frameworks. *Journal of Physical Chemistry C* 115 (46): 22748–22754.

28 Ye, S., Jiang, X., Ruan, L.-W. et al. (2013). Post-combustion CO_2 capture with the HKUST-1 and MIL-101 (Cr) metal-organic frameworks: adsorption, separation and regeneration investigations. *Microporous and Mesoporous Materials* 179: 191–197.

29 (a) Ferreira, A.F., Ribeiro, A.M., Kulaç, S., and Rodrigues, A.E. (2015). Methane purification by adsorptive processes on MIL-53 (Al). *Chemical Engineering Science* 124: 79–95. (b) Serra-Crespo, P., Wezendonk, T.A., Bach-Samario, C. et al. (2015). Preliminary design of a vacuum pressure swing adsorption process for natural gas upgrading based on amino-functionalized MIL-53. *Chemical Engineering and Technology* 38 (7): 1183–1194.

30 Dasgupta, S., Biswas, N., Gode, N.G. et al. (2012). CO_2 recovery from mixtures with nitrogen in a vacuum swing adsorber using metal organic framework adsorbent: a comparative study. *International Journal of Greenhouse Gas Control* 7: 225–229.

31 Demessence, A., D'Alessandro, D.M., Foo, M.L., and Long, J.R. (2009). Strong CO_2 binding in a water-stable, triazolate-bridged metal-organic framework functionalized with ethylenediamine. *Journal of the American Chemical Society* 131 (25): 8784–8786.

32 Llewellyn, P.L., Bourrelly, S., Serre, C. et al. (2008). High uptakes of CO_2 and CH_4 in mesoporous metal-organic frameworks MIL-100 and MIL-101. *Langmuir* 24 (14): 7245–7250.

33 Arstad, B., Fjellvåg, H., Kongshaug, K.O. et al. (2008). Amine functionalised metal organic frameworks (MOFs) as adsorbents for carbon dioxide. *Adsorption* 14 (6): 755–762.

34 An, J., Geib, S.J., and Rosi, N.L. (2009). High and selective CO_2 uptake in a cobalt adeninate metal-organic framework exhibiting pyrimidine-and amino-decorated pores. *Journal of the American Chemical Society* 132 (1): 38–39.

35 Kim, J., Yang, S.-T., Choi, S.B. et al. (2011). Control of catenation in CuTATB-*n* metal-organic frameworks by sonochemical synthesis and its effect on CO_2 adsorption. *Journal of Materials Chemistry* 21 (9): 3070–3076.

36 (a) Aprea, P., Caputo, D., Gargiulo, N. et al. (2010). Modeling carbon dioxide adsorption on microporous substrates: comparison between Cu-BTC metal-organic framework and 13X zeolitic molecular sieve. *Journal of Chemical and Engineering Data* 55 (9): 3655–3661. (b) Wang, Q.M., Shen, D., Bülow, M. et al. (2002). Metallo-organic molecular sieve for gas separation and purification. *Microporous and Mesoporous Materials* 55 (2): 217–230.

37 Bourrelly, S., Llewellyn, P.L., Serre, C. et al. (2005). Different adsorption behaviors of methane and carbon dioxide in the isotypic nanoporous metal terephthalates MIL-53 and MIL-47. *Journal of the American Chemical Society* 127 (39): 13519–13521.

38 (a) Liang, Z., Marshall, M., and Chaffee, A.L. (2009). Comparison of Cu-BTC and zeolite 13X for adsorbent based CO_2 separation. *Energy Procedia* 1 (1): 1265–1271. (b) Liang, Z., Marshall, M., and Chaffee, A.L. (2009). CO_2 adsorption-based separation by metal organic framework (Cu-BTC) versus zeolite (13X). *Energy and Fuels* 23 (5): 2785–2789.

39 Farrusseng, D., Daniel, C., Gaudillere, C. et al. (2009). Heats of adsorption for seven gases in three metal-organic frameworks: systematic comparison of experiment and simulation. *Langmuir* 25 (13): 7383–7388.

40 Choi, J.-S., Son, W.-J., Kim, J., and Ahn, W.-S. (2008). Metal-organic framework MOF-5 prepared by microwave heating: factors to be considered. *Microporous and Mesoporous Materials* 116 (1): 727–731.

41 Mu, B., Schoenecker, P.M., and Walton, K.S. (2010). Gas adsorption study on mesoporous metal-organic framework UMCM-1. *The Journal of Physical Chemistry C* 114 (14): 6464–6471.

42 Lu, Y., Dong, Y., and Qin, J. (2016). Porous pcu-type Zn(II) framework material with high adsorption selectivity for CO_2 over N_2. *Journal of Molecular Structure* 1107: 66–69.

43 Liu, Y., Yang, Y., Sun, Q. et al. (2013). Chemical adsorption enhanced CO_2 capture and photoreduction over a copper porphyrin based metal organic framework. *ACS Applied Materials & Interfaces* 5 (15): 7654–7658.

44 Cao, Y., Song, F., Zhao, Y., and Zhong, Q. (2013). Capture of carbon dioxide from flue gas on TEPA-grafted metal-organic framework Mg_2(dobdc). *Journal of Environmental Sciences* 25 (10): 2081–2087.

45 Forrest, K.A., Pham, T., McLaughlin, K. et al. (2014). Insights into an intriguing gas sorption mechanism in a polar metal-organic framework with open-metal sites and narrow channels. *Chemical Communications* 50 (55): 7283–7286.

46 Li, B., Zhang, Z., Li, Y. et al. (2012). Enhanced binding affinity, remarkable selectivity, and high capacity of CO_2 by dual functionalization of a rht-type metal-organic framework. *Angewandte Chemie International Edition* 51 (6): 1412–1415.

47 Luebke, R., Weseliński, Ł.J., Belmabkhout, Y. et al. (2014). Microporous heptazine functionalized (3,24)-connected rht-metal-organic framework: synthesis, structure, and gas sorption analysis. *Crystal Growth and Design* 14 (2): 414–418.

48 Spanopoulos, I., Bratsos, I., Tampaxis, C. et al. (2016). Exceptional gravimetric and volumetric CO_2 uptake in a palladated NbO-type MOF utilizing cooperative acidic and basic, metal–CO_2 interactions. *Chemical Communications* 52 (69): 10559–10562.

49 Liao, P.-Q., Chen, X.-W., Liu, S.-Y. et al. (2016). Putting an ultrahigh concentration of amine groups into a metal-organic framework for CO_2 capture at low pressures. *Chemical Science* 7 (10): 6528–6533.

50 Zheng, B., Bai, J., Duan, J. et al. (2010). Enhanced CO_2 binding affinity of a high-uptake rht-type metal-organic framework decorated with acylamide groups. *Journal of the American Chemical Society* 133 (4): 748–751.

51 Liang, Z., Du, J., Sun, L. et al. (2013). Design and synthesis of two porous metal-organic frameworks with nbo and agw topologies showing high CO_2 adsorption capacity. *Inorganic Chemistry* 52 (19): 10720–10722.

52 Bernini, M.C., Blanco, A.G., Villarroel-Rocha, J. et al. (2015). Tuning the target composition of amine-grafted CPO-27-Mg for capture of CO_2 under post-combustion and air filtering conditions: a combined experimental and computational study. *Dalton Transactions* 44 (43): 18970–18982.

53 Jiao, J., Dou, L., Liu, H. et al. (2016). An aminopyrimidine-functionalized cage-based metal-organic framework exhibiting highly selective adsorption of C_2H_2 and CO_2 over CH_4. *Dalton Transactions* 45 (34): 13373–13382.

54 Liu, B., Yao, S., Shi, C. et al. (2016). Significant enhancement of gas uptake capacity and selectivity via the judicious increase of open metal sites and Lewis basic sites within two polyhedron-based metal-organic frameworks. *Chemical Communications* 52 (15): 3223–3226.
55 Song, C., Hu, J., Ling, Y. et al. (2015). The accessibility of nitrogen sites makes a difference in selective CO_2 adsorption of a family of isostructural metal-organic frameworks. *Journal of Materials Chemistry A* 3 (38): 19417–19426.
56 Lu, Z., Bai, J., Hang, C. et al. (2016). The utilization of amide groups to expand and functionalize metal-organic frameworks simultaneously. *Chemistry A European Journal* 22 (18): 6277–6285.
57 Cui, P., Ma, Y.-G., Li, H.-H. et al. (2012). Multipoint interactions enhanced CO_2 uptake: a zeolite-like zinc-tetrazole framework with 24-nuclear zinc cages. *Journal of the American Chemical Society* 134 (46): 18892–18895.
58 Lin, Y., Yan, Q., Kong, C., and Chen, L. (2013). Polyethyleneimine incorporated metal-organic frameworks adsorbent for highly selective CO_2 capture. *Scientific Reports* 3: 1859.

15

Hydrogen and Methane Storage in MOFs

15.1 Introduction

Hydrogen is considered an ideal clean energy source because water is the sole product of its combustion and no greenhouse gases or other environmentally harmful compounds are released. Natural gas (NG) is considered an environmentally friendly fossil fuel because its main component, methane (CH_4), has the highest research octane number (RON = 107) among all hydrocarbons and consequently its combustion releases the smallest amount of CO_2. Both hydrogen and natural gas provide the prospect to supersede fossil fuels and thus provide cleaner energy sources.

To commercialize energy systems driven by H_2 or NG, storage materials must be developed. Porous materials have been intensely studied during the past decades owing to their ability to adsorb gases at comparatively low pressures. With respect to energy storage, metal-organic frameworks (MOFs) have attracted a growing interest owing to their modular and highly tunable nature. Their structures combine high surface areas, low densities, and facile functionalization. This allows for the development of storage materials with high gravimetric and volumetric uptake – an outstanding challenge that must be addressed to make gaseous fuels viable alternatives to liquid fossil fuels. For the downselection of materials for this type of application, characteristics such as the working capacity, system capacity, and cycling stability gain in importance. These terms have been defined in Chapter 13.

For hydrogen and natural gas storage in MOFs, upper pressure limits of 35 or 65 bar (maximum pressure achievable with inexpensive single-stage and dual-stage compressors, respectively) and a lower pressure limit of 5 bar (minimum inlet pressure) are typically reported [1]. This implies that in order to maximize the working capacity, not only the storage capacity at high pressures (35–65 bar) needs to be maximized but the amount of gas retained at pressures below 5 bar should also be minimized (see Figure 13.2). In mobile applications such as the installation of adsorbent-based tanks in automobiles, the cycle life requirements are that the storage material should surpass the expected life of the vehicle (more than 250 000 km or 155 343 miles), which equals about 1500 charge–discharge cycles. This prerequisite implies that onboard gas storage

Introduction to Reticular Chemistry: Metal-Organic Frameworks and Covalent Organic Frameworks,
First Edition. Omar M. Yaghi, Markus J. Kalmutzki, and Christian S. Diercks.

Companion website: www.wiley.com/go/yaghi/reticular

systems require materials possessing chemical and architectural stability with respect to cycling, temperature swings of up to 125 °C, and mechanical stress such as vibrations and rattling.

15.2 Hydrogen Storage in MOFs

Hydrogen is an ideal replacement for petroleum especially in automotive applications because it is a zero-emission fuel, abundantly available, and, when liquefied, has approximately triple the energy density of gasoline. While hydrogen is the third most abundant element on the Earth's surface, hydrogen gas (H_2) is very rare (less than 1 ppm per volume) [2]. Gaseous hydrogen is commonly produced by electrolysis of water or steam reforming of natural gas (Figure 15.1). Since the production of hydrogen requires other sources of energy it is considered an energy carrier rather than an energy source.

To use hydrogen in everyday applications the methods of hydrogen production, delivery, and storage, as well as the fuel cell technology must be further developed. Hydrogen storage is considered to play a key role in this development. Current methods of chemical and physical storage present several disadvantages. Metal hydrides, boranes, or imidazolium ionic liquids for chemical storage of hydrogen are often associated with high production costs alongside unfavorable kinetics and a high susceptibility toward contaminations that are typically present in hydrogen gas. Cryogenic storage of liquid hydrogen is accompanied by high operational costs and an increased weight of the cryogenic tanks, rendering it almost impractical. Physical storage of liquid hydrogen has been used by NASA (National Aeronautics and Space Administration) to propel space shuttles into orbit since the 1970s and today some vehicles including buses and trains run on pressurized hydrogen. The high volatility of hydrogen at ambient conditions aggravates the physical storage in pressurized tanks that commonly sustains only low volumetric energy densities, even when expensive multistage compressor systems are used to achieve pressures of up to 300 bar in heavy reinforced tanks. To make hydrogen a fuel that is more germane for everyday applications, low-cost lightweight materials capable of reversible hydrogen uptake (at ambient conditions) and storage densities equal to or even greater than that of liquid hydrogen are needed.

The US Department of Energy (DOE) 2020 targets for hydrogen storage systems are a gravimetric uptake of 0.055 $kg_{(H_2)}\ kg^{-1}$ and a volumetric capacity of 0.00445 $kg_{(H_2)}\ l^{-1}$, while retaining short refueling times and high cycling stability at operating temperatures between 30 and 50 °C. It should be noted that these are requirements for an entire storage system; hence, the performance of the storage material itself must exceed them to compensate for system-based losses (see Chapter 13). Figure 15.2 illustrates how far these requirements can be met by MOF-5.

(a) $CH_4 + H_2O \rightleftharpoons CO + 3H_2$

(b) $2H_2O(l) \underset{\text{Combustion}}{\overset{\text{Electrolysis}}{\rightleftharpoons}} 2H_2 + O_2$

Figure 15.1 Reaction equation for the production of hydrogen gas by (a) steam reforming and (b) electrolysis of water.

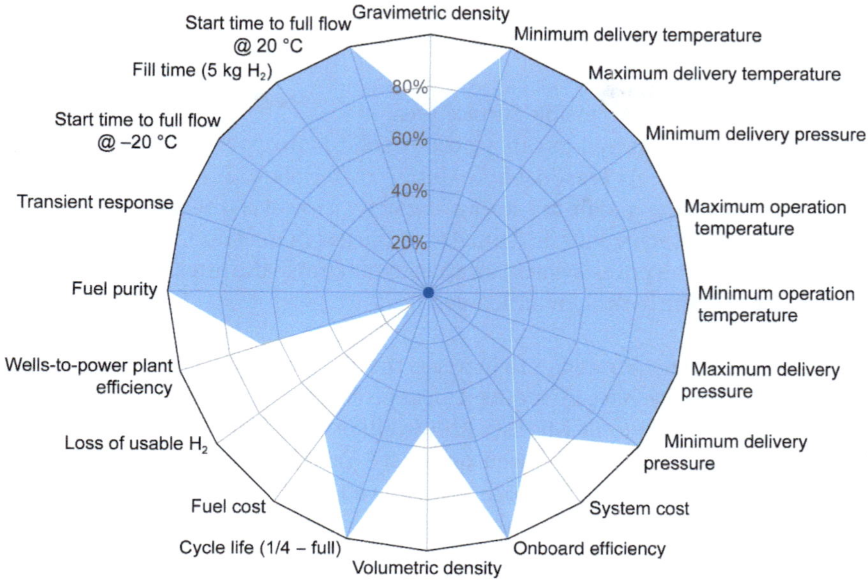

Figure 15.2 Spider graph showing a comparison of the hydrogen storage performance of MOF-5 to the 2020 DOE (Department of Energy) targets. Shown is the data for a projected system (100 bar, 80–160 K, Type I tank, hexcell-loose powder) (https://energy.gov/eere/fuelcells/hydrogen-storage-engineering-center-excellence). Many of the targets set by the DOE have already been achieved; however, further development is needed to render MOFs suitable candidates for large-scale applications.

15.2.1 Design of MOFs for Hydrogen Storage

To design a MOF suitable for hydrogen storage and to optimize it with respect to its storage properties, it is instructive to take a closer look at the interactions between hydrogen and the framework skeleton. Hydrogen can interact with the framework in two ways: physisorption through weak dispersive interactions with the backbone of the framework and strong interactions with specific adsorption sites. The former requires high surface areas to achieve a large hydrogen uptake while the latter requires the introduction of strong adsorption sites into the framework architecture (e.g. open metal sites). Both interactions can be realized in MOFs because they combine high surface areas, low densities, tunable pore sizes, and the possibility for the introduction of specific adsorption sites. General aspects of the design of such structures were discussed in Chapters 1–6. With respect to hydrogen storage, MOFs built from light metals such as Be^{2+}, Al^{3+}, or Mg^{2+} are of interest because they promise to show a high gravimetric uptake in addition to the high volumetric uptake, originating from their highly porous nature. It should be noted that in the temperature regime targeted by the DOE, dispersive forces alone do not lead to substantial hydrogen uptake. Consequently, a polar/polarizable pore environment is required to provide for strong interactions between the pore walls and hydrogen molecules. In the following, we will discuss structural factors that lead to high hydrogen storage capacities, design principles for hydrogen storage materials, and approaches for the optimization of promising candidates.

15.2.1.1 Increasing the Accessible Surface Area

The amount of physisorbed gas in a porous solid is proportional to its surface area. Therefore, it seems reasonable to design materials with high surface areas when targeting a large maximum storage capacity. However, the pore size should be tailored to the specific kinetic diameter of the gas in question, and it is advantageous to design the pores in such a way as to maximize the interaction between their surface and the hydrogen molecules.

Expansion of the Linker Strategies to increase the surface area have been discussed in Chapter 2 where we have introduced an approach for the design of linkers that facilitate the preparation of MOFs with ultrahigh surface areas. We illustrated that trigonal linker molecules based on a 1,3,5-triphenylbenzene core unit are ideal building units for this purpose and showed a series of MOFs (MOF-177, MOF-180, and MOF-200) designed with this idea in mind. Here, the same MOFs (MOF-177 and MOF-200) will help us to determine the correlation between the surface area and the hydrogen storage capacity. MOF-177 and MOF-200 are both built from trigonal tritopic linkers and octahedral $Zn_4O(-COO)_6$ secondary building units (SBUs) and have a **qom** topology (see Figure 2.16) [3]. MOF-177 has a geometric surface area of 4740 $m^2\ g^{-1}$, and with 6400 $m^2\ g^{-1}$ the geometric surface area of the expanded analog MOF-200 is even larger. Consequently, MOF-200 shows a higher hydrogen uptake (163 mg g^{-1}) compared to MOF-177 (75 g mg g^{-1}) [3, 4]. In general, we can say that for isoreticular frameworks the one with the higher surface area also has the higher hydrogen uptake (at low pressures). Aside from the surface area, the structure type (topology) of the framework, along with the consequential pore shapes and metrics, has a big influence on the hydrogen storage capacity. Increasing the surface area by applying the mixed linker approach is therefore often more promising since this frequently leads to high surface areas and complex pore systems that consist of many connected small pores (see Chapter 6). This strategy is used in the design of MOF-205 (**ith-d**) and MOF-210 (**toz**), two tertiary MOFs built from trigonal tritopic linkers, linear ditopic linkers, and octahedral $Zn_4O(-COO)_6$ SBUs [3]. Even though MOF-205 has a lower geometric surface area (4680 $m^2\ g^{-1}$) than MOF-200 (6400 $m^2\ g^{-1}$), its hydrogen uptake (7 wt%) is close to that of MOF-200 (7.4 wt%). MOF-210 has a higher surface area (6240 $m^2\ g^{-1}$) than MOF-205 while maintaining comparatively small pores, which leads to a maximum hydrogen uptake of 8.6 wt%. These findings support the idea that a large surface area leads to a high hydrogen uptake, whereas increasing the pore size above a critical diameter does not help increase the hydrogen uptake. Figure 15.3 shows the correlation between the hydrogen storage capacity of several MOFs and their respective surface areas.

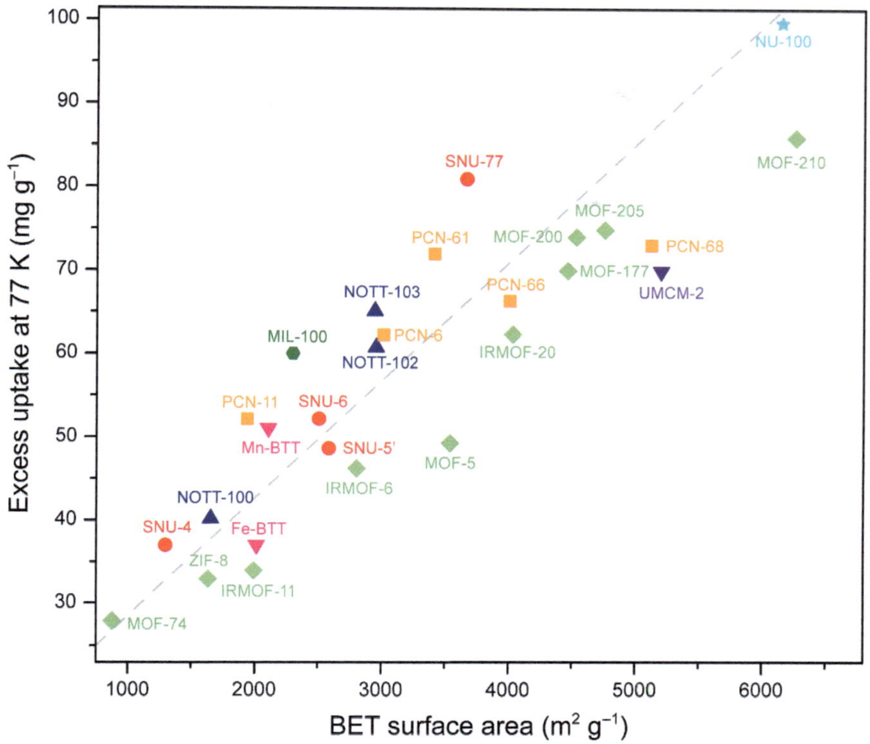

Figure 15.3 Plot of the BET surface area (m^2 g^{-1}) of a compilation of highly porous MOFs versus the hydrogen excess uptake (mg g^{-1}) at 77 K. The gray dotted line represents the trend line that illustrates that the hydrogen adsorption capacity at 77 K correlates strongly with the surfae area.

Interpenetrated Frameworks Hydrogen uptake capacity is not governed by the surface area alone. Different MOFs with similar surface areas exhibit different values for their respective excess hydrogen uptake. This is explained by the difference in their structure or, more specifically, their pore metrics and the fact that some of them are interpenetrated. Interpenetration has a dramatic effect on the pore size but not necessarily on the pore volume. Consequently, interpenetrated structures frequently show a higher excess uptake than their non-interpenetrated counterparts that have similar surface areas but large pores. Smaller pores allow for a stronger interaction between the framework and hydrogen molecules. The optimal pore size for hydrogen adsorption at room temperature is predicted to be approximately 7 Å based on calculations on idealized homogeneous materials. Slit-shaped pores of 7 Å diameter allow for the adsorption of one layer of hydrogen molecules on opposing surfaces, thus minimizing the empty volume of the pore while maximizing the van der Waals interaction of hydrogen with the framework. At cryogenic temperatures the formation of a H_2-sandwich in between two opposing pore surfaces is highly favorable, and the ideal pore size for maximum volumetric hydrogen uptake at 100 bar increases to 10 Å regardless of the pore shape.

15.2.1.2 Increasing the Isosteric Heat of Adsorption

The amount of gas that can be adsorbed on the surface of a framework material is proportional to its surface area. With respect to hydrogen this is only true at cryogenic temperatures and indeed, some MOFs with ultrahigh surface areas meet the DOE targets for hydrogen uptake in terms of weight percent (<5.5 wt%) at cryogenic temperatures. In contrast, a significant drop in the maximum hydrogen uptake is typically observed at elevated temperatures. This is ascribed to the weak van der Waals interactions involved in the physisorption of hydrogen molecules within the framework and results in less than 1/10 of the cryogenic uptake at ambient temperatures. For application in mobile adsorbent-based hydrogen storage tanks, however, storage capacities >5.5% at ambient temperatures are required. This in turn means that stronger adsorption sites are needed to meet this requirement. How strongly a molecule is bound to the surface is defined by the isosteric heat of adsorption (Q_{st}), which is correlated to the strongest interaction between the gas molecules and the framework. Therefore, the affinity of the framework toward hydrogen at ambient temperatures can be optimized by increasing Q_{st}, thus allowing for stronger interactions and thereby higher hydrogen uptake at elevated temperatures. This means that aside from a large free pore volume, the magnitude of Q_{st} is the most important factor in achieving high hydrogen adsorption capacities [5]. The optimum Q_{st} value for high-pressure applications is approximately 20 kJ mol^{-1}.The Q_{st} values for hydrogen adsorption in most MOFs is however in the range of 5–9 kJ mol^{-1} [6]. In this section we will outline way to increase the isosteric heat of adsorption through precise adjustment of structural parameters.

Open Metal Sites and Other Coordinatively Unsaturated Metal Centers The interaction between hydrogen molecules and coordinatively unsaturated metal centers has been studied in molecular chemistry for decades. Figure 15.4 shows the orbitals involved in a side-on coordination of hydrogen to a metal center. The σ-electrons of H_2 interact with the vacant metal d-orbitals of the metal and occupied d-orbitals of the metal donate electrons into the σ*-orbitals of H_2 by backdonation.

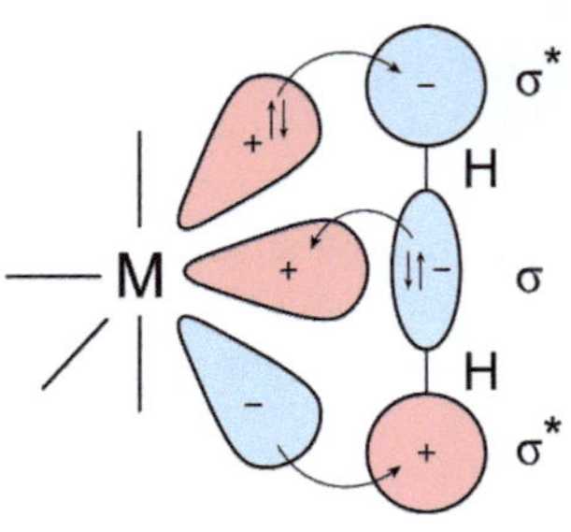

Figure 15.4 Side-on coordination of H_2 to a metal complex. The H_2-σ orbital can bind to a metal-centered σ orbital. Electron density is transferred to the σ* orbital of H_2 by backbonding.

Dissociation energies of 80–90 kJ mol^{-1} have been determined for complexes such as $(C_5H_5)V(CO)_3(H_2)$ and $Mo(CO)_5(H_2)$ [7]. These values are too high for the purpose of gas storage because it takes tremendous amounts of energy (i.e. heat) to liberate the adsorbed hydrogen gas. Desired binding energies for hydrogen storage are within the regime of charge-induced dipole interactions (~20 kJ mol^{-1}); hence, favorable orbital interactions leading to strong metal–H_2 bonds must be avoided. Binding energies in this regime have been observed for interactions between H_2 and Li^+ in the gas phase (~27 kJ mol^{-1}) and consequently the intercalation of Li^+ into porous frameworks (in particular, covalent organic frameworks [COFs]) has been studied. In such materials the charge of the lithium ions is however partially quenched and the heat of adsorption is therefore significantly reduced. Nonetheless, the intercalation of Li^+ into COFs results in a significantly enhanced hydrogen storage capacity. The storage capacity of COF-202 increases from 8.08 g l^{-1} and 1.52 wt% to 25.86 g l^{-1} and 4.39 wt% upon lithium doping, and COF-105 and COF-108 both surpass the DOE target of 6 wt% with 6.73 and 6.84 wt% at 100 bar and 298 K, respectively [8]. Similar effects are achieved by intercalation with other metal ions such as those of Ca, Sc, and Ti, and this approach is also applicable to MOFs [9].

Another way to incorporate coordinatively unsaturated metal centers is the formation of open metal sites in MOFs. Such Lewis acidic sites are considered the most promising way of increasing the isosteric heat of hydrogen adsorption and they are typically formed by removal of terminal ligands from specific SBUs by heating the MOF in dynamic vacuum. Figure 15.5 shows some SBUs with

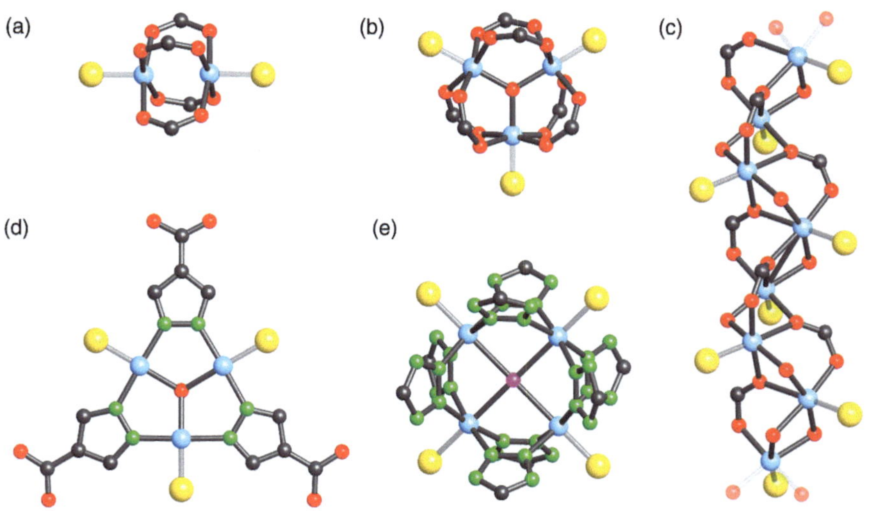

Figure 15.5 SBUs with potential open metal sites. The terminal ligands whose removal leads to the formation of open metal sites are represented by yellow spheres. (a) Square tetratopic $M_2L_2(\text{–COO})_4$ paddle wheel SBU, (b) trigonal prismatic hexatopic $M_3OL_3(\text{–COO})_6$ SBU, (c) rod-like $[M_3O_2L_2(COO)_3]_\infty$ SBU, (d) trigonal tritopic $M_3OL_3(\text{-Py})_3$ SBU, and (e) cubic octatopic $M_4Cl(L_4)(\text{–tetrazolate})_8$ SBU. Color code: metal, blue; N, green; C, gray; O, red; Cl, pink; terminal ligands, yellow.

potential open metal sites highlighting the corresponding adsorption sites. Two prominent examples where the formation of open metal sites leads to a significant increase in the hydrogen storage capacity are HKUST-1 (Figure 15.5a) and $(Mn_4Cl)_3(L)_4(BTT)_8$ (Figure 15.5e) [10]. In both materials, the open metal sites are formed by thermally assisted evacuation of solvent molecules terminally bound to SBUs. To avoid framework collapse these terminal ligands are typically exchanged for more volatile species (e.g. methanol) prior to evacuation. For further information on the formation of open metal sites, the reader is referred elsewhere [11].

Optimization of the Pore Size With respect to the hydrogen sorption capacity small pores are not necessarily a drawback; in fact, very large pores are intrinsically detrimental to achieving a large hydrogen storage capacity. This is because only those hydrogen molecules that are adsorbed on the inner surface of the pore result in a gain in the adsorption capacity (excess uptake), whereas those near the center of the pore do not interact with the surface of the pore and are simply compressed, similar to the situation in a pressurized gas cylinder (bulk uptake, see Figure 13.1). Considering the size of the hydrogen molecule and the fact that in order to increase the heat of adsorption a more efficient packing leading to stronger interactions must be achieved, it becomes clear that a large free pore volume composed of multiple small pores is more desirable than fewer but larger pores. Large pores in a given framework structure can be divided by interpenetration of frameworks, thus creating a more suitable scenario for hydrogen adsorption, and consequently giving materials with an increased hydrogen sorption capacity. However, it is synthetically challenging to control interpenetration, and as discussed earlier not every framework has an interpenetrated counterpart (see Chapter 2). Reaction parameters such as the concentration of the starting materials, the reaction temperature, or the composition of the solvent mixture can influence whether the interpenetrated or the non-interpenetrated structure will form. We will illustrate the effect of interpenetration on the hydrogen storage capacity using PCN-6 ($Cu(TATB)_2$), a framework of **tbo** topology [12]. A comparison of interpenetrated PCN-6 and non-interpenetrated PCN-6 (called PCN-6′) reveals that PCN-6 adsorbs 1.9 wt% of hydrogen whereas PCN-6′ only adsorbs 0.6 wt% (at 1 bar and 77 K) [13]. A pair of isoreticular expanded analogs of PCN-6 and PCN-6′, termed PCN-9 and PCN-9′, is also known; however, these cannot be activated. A comparison of the structures of PCN-6, PCN-9, and PCN-9′ is given in Figure 15.6. The interpenetrating framework fills the large pore avoiding empty pore volume, thereby increasing the interactions with hydrogen molecules.

While interpenetration can endow MOFs with enhanced mechanical, thermal, and chemical stability, interpenetration of frameworks and the concomitant smaller pores should lead to a lower surface area compared to the non-interpenetrated counterpart. This effect is however often less pronounced than expected. The heat of hydrogen adsorption in interpenetrated structures compared to their non-interpenetrated counterparts is only higher at low loadings.

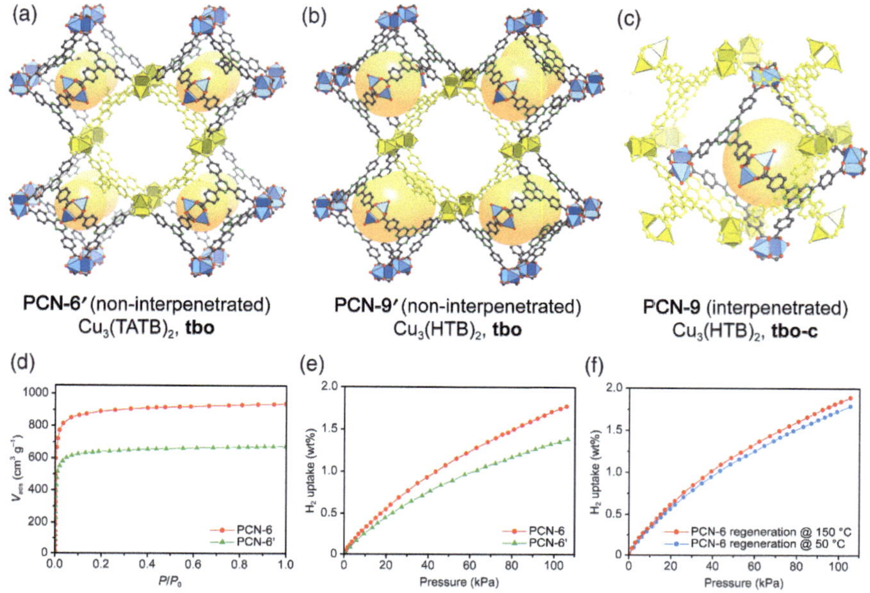

Figure 15.6 (a, b) Comparison of the non-interpenetrated structures of PCN-6′ and PCN-9′ (**tbo**). The small pores are highlighted by orange spheres and the large cubic pores are highlighted as yellow cages. (c) Fragment of the interpenetrated structure of PCN-9. The yellow cage is identical to that shown in the structure of PCN-9′ (b). The interpenetrating framework fills this cage, thereby avoiding large empty pores. Similarly, the structure of PCN-6′ can interpenetrate and the corresponding MOF is termed PCN-6. (d) N_2 adsorption isotherms of PCN-6′ (green triangles) and PCN-6 (red circles) after activation at 50 °C. (e) H_2 adsorption isotherms of PCN-6′ (green triangles) and PCN-6 (red circles) recorded at 77 K after activation at 50 °C. The higher uptake of PCN-6 is ascribed to the smaller pores as a result of framework interpenetration. (f) Comparison of H_2 adsorption isotherms of PCN-6′ activated at 50 °C (blue circles) and 150 °C (red circles) recorded at 77 K. The increased capacity after regeneration at 150 °C is due to the formation of open metal sites. Color code: Cu, blue; C, gray; N, green; O, red.

At intermediate and high loadings, the surface area and the total free pore volume gain importance, respectively. This means that the increase in binding energy due to interpenetration does not outweigh the loss of free pore volume. Consequently, a good material for H_2 sorption should strike a balance between high surface area, a large free pore volume, a high Q_{st} value, and small pore size.

Functionalization The incorporation of functionalized linkers that bear polar or polarizable functional groups has a strong influence on the adsorption behavior of the resulting MOFs. Examples for such a modification are frameworks that contain linkers suitable for metalation by post-synthetic modification (PSM) to create additional strong adsorption sites (Figure 15.7). In some cases, the substitution of one ligand of the metal complexes by hydrogen is facilitated by photolysis as exemplified by $Cr(CO)_3$-functionalized BDC linkers in MOF-5($Cr(CO)_3$) (Figure 15.7c) [11]. Irradiation of this MOF leads to liberation of one CO ligand,

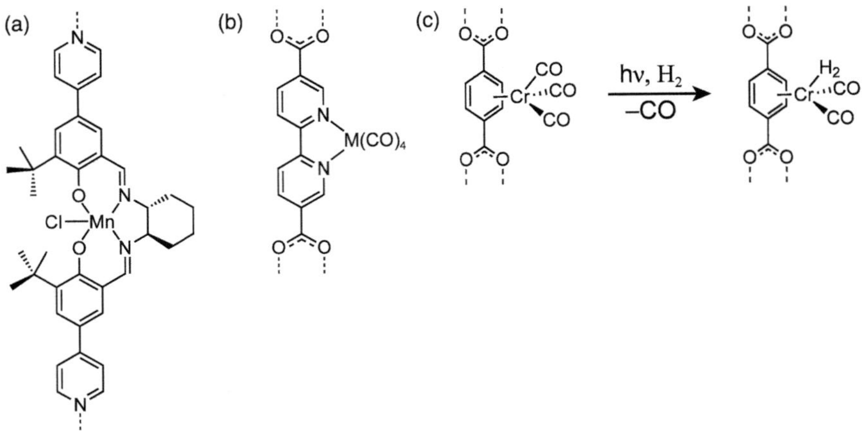

Figure 15.7 (a, b) Introduction of polar adsorption sites by metalation of the organic linker. (c) Functionalization of BDC with $Cr(CO)_3$ and subsequent photochemical decarbonylation creates new binding sites for hydrogen.

and in the presence of hydrogen gas, subsequent H_2 adsorption on the open metal site. The efficiency of this process is however poor and it is not of relevance for applications.

Spillover Hydrogen spillover is the dissociation of H_2 into $H^\cdot$ on a metal surface and the consecutive migration and adsorption of these hydrogen atoms onto other materials. This phenomenon is also observed in mixtures of MOFs and precious metal nanoparticles where hydrogen molecules are split into $H^\cdot$ atoms on the surface of the metal nanoparticle and subsequently migrate into the pores of the MOF. Spontaneous recombination of $H^\cdot$ liberates hydrogen molecules [14]. Drawbacks of such systems are that they typically suffer from low cycling stability and unfavorable desorption kinetics, and in some cases evacuation for more than 12 hours is needed to fully regenerate the material.

15.2.1.3 Use of Lightweight Elements

The formula weight of an adsorbent plays a pivotal role in achieving a high gravimetric uptake. Thus, the use of lightweight elements such as Be^{2+}, Al^{3+}, or Mg^{2+} is highly favorable in designing MOFs with high gravimetric hydrogen uptake. To put this effect into perspective, we consider a comparison of MOF-5 ($Zn_4O(BDC)_3$) and its hypothetical beryllium analog $Be_4O(BDC)_3$. Exchanging the Zn^{2+} ions in MOF-5 by Be^{2+}, the lightest divalent metal, results in an increase in the gravimetric hydrogen storage capacity by 40% [15]. The idea of constructing a compound isostructural to MOF-5 based on $Be_4O(-COO)_6$ SBUs is not too farfetched since molecular analog of this SBU is well-known [16]. The idea of using lightweight elements to construct MOFs with high gravimetric hydrogen capacities is used in the design of $Be_{12}(OH)_{12}(BTB)_4$, the first (and thus far only) beryllium-based MOF. The structure of $Be_{12}(OH)_{12}(BTB)_4$ is built from saddle-shaped 12-c $[Be_{12}(OH)_{12}]^{12+}$ SBUs, a structural element that is unprecedented in molecular beryllium chemistry. These SBUs are joined by

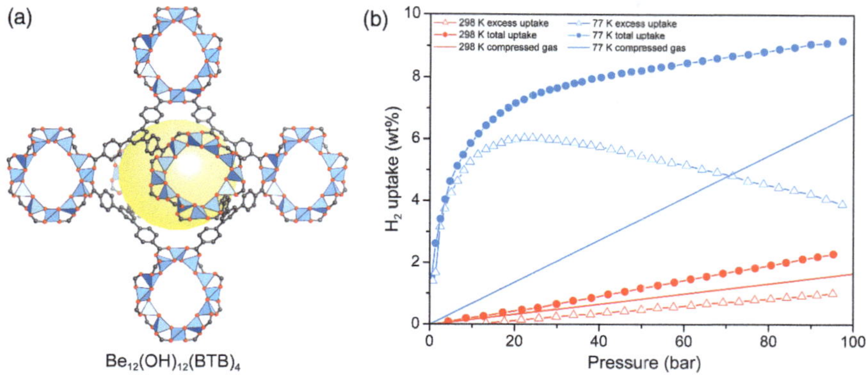

Figure 15.8 (a) Structure of $Be_{12}(OH)_{12}(BTB)_4$. The structure contains a large octahedral pore (yellow ball) surrounded by eight smaller pores (not shown). (b) Hydrogen adsorption isotherms collected at 77 K (blue) and 298 K (red). For both measurements the excess uptake (triangles) and the total uptake (circles) are compared to the density of hydrogen under the respective pressure. Color code: Be, blue; C, gray; O, red.

trigonal BTB linkers to form a framework of **fon** topology [15] (Figure 15.8a). The high gravimetric hydrogen storage capacity of $Be_{12}(OH)_{12}(BTB)_4$ (9.2 wt% at 100 bars and 77 K) is mainly ascribed to its low formula weight since the isosteric heat of hydrogen adsorption is comparatively low ($Q_{st} = -5.5\ \text{kJ mol}^{-1}$) (Figure 15.8b).

15.2.2 Important MOFs for Hydrogen Storage

The concepts discussed in Section 15.2.1 have been employed in the design and synthesis of many MOFs and their hydrogen storage performance has been studied. Table 15.1 gives a compilation of MOFs and a few selected examples of COFs with interesting hydrogen adsorption properties.

15.3 Methane Storage in MOFs

Unlike hydrogen, methane is a direct fuel, not an energy carrier. It occurs naturally as the main component of natural gas (>95%) and has a high gravimetric heat of combustion ($55.7\ \text{MJ kg}^{-1}$), similar to that of gasoline ($46.4\ \text{MJ kg}^{-1}$). Among all hydrocarbons methane has the highest RON (107) and is therefore a comparatively clean fuel, which in combination with its compatibility with modern technologies of combustion-based energy production, renders the development of methane storage materials pivotal in transitioning to cleaner fuels. Two current NG storage technologies are its compression at room temperature (200–300 bar) and cryogenic liquefaction. Liquefied natural gas (LNG) has a volumetric energy density (VED) that is only 64% of that of gasoline ($22.2\ \text{MJ l}^{-1}$ versus $34.2\ \text{MJ l}^{-1}$ at $-161.5\ ^\circ\text{C}$ and ambient conditions, respectively), and requires energy-intensive liquefaction, cooling, and storage in expensive cryogenic vessels that suffer from

Table 15.1 Compilation of the surface area and the pore size as well as the gravimetric (wt%) and volumetric ($g\,l^{-1}$) hydrogen uptake and the Q_{st} values for zero coverage for a selection of promising hydrogen storage materials.

Chemical formula	Common name	Surface area		Pore size (Å)	Pore volume ($cm^3\,g^{-1}$)	OMS	Hydrogen storage capacity					References
		BET ($m^2\,g^{-1}$)	Langmuir ($m^2\,g^{-1}$)				*p* (bar)	*T* (K)	Uptake (wt%)	Uptake ($g\,l^{-1}$)	Q_{st} ($kJ\,mol^{-1}$)	
MOF												
$Be_{12}(OH)_{12}(BTB)_4$		4030	4400			No	1	77	1.6		5.5	[15]
							20 (100)	77	6 (9.2)	(43)		
$Cr_3OF(BDC)_3$	MIL-101		5500	8.6	1.9	Yes	80	77	6.1	1.84	10	[17]
$Cu_3(btc)_2$	HKUST-1		1958			Yes	100	77	3.6		4.5	[18]
$Cu_3(TATB)_2$	PCN-6′		2700		1.045	Yes	1	77	1.35		6	[13]
$Cu_3(TATB)_2$ catenated	PCN-6		3800	5	1.456	Yes	1	77	1.9			[12b]
$Cu_2(tptc)$	NOTT-101	2247		7.3	0.89	Yes	1	77	2.52	43.6	5.5	[19, 20]
$Cu_2(aobtc)$	PCN-10	1047	1779		0.67	Yes	1	77	2.34	18.6	6.8	[21]
$Cu_3(ntei)$	PCN-66	4000	4600		1.63	Yes	1	77	1.79	7.98	6.22	[22]
$Cu_3(ptei)$	PCN-68	5109	6033		2.13	Yes	1	77	1.87	7.2	6.09	[22]
$Cu_3(ttei)$	NU-100	6143			2.82	Yes	1	77	1.82		6.1	[23]
$Cu_3(BHB)$	UTSA-20	1156		3.4, 8.5		Yes	1	77	2.9			[24]
$In_3O(abtc)_{1.5}(NO_3)$	soc-MOF		1417	7.65, 5.95	0.5	Yes	1.2	77	2.61		6.5	[25]
$Mg_2(DOBDC)$	MOF-74	1510				Yes	1	77	2.2		10.3	[26]
$Mn_3[(Mn_4Cl)_3(BTT)_8]_2$	Mn-BTT	2100			0.795	Yes	1.2	77	2.2	43	10.1	[27]

$Mn_3[(Mn_4Cl)_3(tpt\text{-}3tz)_8]_2$		1580	1700			No	80	77	3.7 (4.5)	37		[28]
$Ni_3O(TATB)_2$	PCN-5		225		0.13	No	1	77	0.63			[29]
$Zn(mIM)_2$	ZIF-8	1630	1810		0.64	No	1	77	1.27			[30]
							30	77	3.3			
							55	77	3.01			
$Zn_2(abtc)(DMF)_2$	SNU-4		1460		0.53	No	1	77	2.07			[31]
							50	77	3.7			
$Zn_4O(BBC)_2(H_2O)_3\ 3H_2O$	MOF-200	4530	10 400	3.59		No	80	77	7.4 (16.3)	(36)		[3]
$Zn_4O(BDC)_3$	MOF-5	2296	3840			No	50	77	4.7		3.8	[18]
$Zn_4O(BDC)_3$ catenated	MOF-5		1130	6.7		No	1	77	2.0	23.3	7.6	[32]
$Zn_4O(BTB)_2$	MOF-177	4746	5640		1.59	No	70	77	7.5	32		[4]
$Zn_4O(BTB)_{4/3}(NDC)$	MOF-205	4460	6170		2.16	No	80	77	7.0 (12)	(46)		[3]
$Zn_4O(BTE)_{4/3}(BPDC)$	MOF-210	6240	10 400		3.6	No	80	77	8.6 (17.6)	(44)		
$Zn_4O(T_2DC)(BTB)_{4/3}$	UMCM-2	5200	6060			No	46	7	6.9		6.4	[33]
COF												
COF-105		6636 (calc.)		5.22	No	100	77	4.67	18.05			[34]
COF-108		6298 (calc.)		5.59	No	100	77	4.51	17.80			[34]
COF-202			2690		1.09	No	100	298	1.52	8.08		[8a]
COF-202 Li			< 2690		< 1.09	Yes	100	298	4.39	25.86		[8a]

boil-off losses. The compression of natural gas (CNG) requires multistage compressors and heavy thick-walled cylindrical storage tanks. Still VED values of only 27% of that of gasoline are reached (9.2 MJ l^{-1} versus 34.2 MJ l^{-1}). These drawbacks render both LNG and CNG storage impractical for mobile applications. Storing natural gas in tanks filled with solid adsorbents represents a viable alternative. The lower pressures of 35–65 bar needed in such adsorbed natural gas (ANG) tanks and their operation at ambient temperatures eliminate the need for cryogenic or heavy tanks, making it safer and less energy intensive, and consequently also more cost-effective than CNG and LNG storage.

While hydrogen storage in MOFs suffers from intrinsically low isosteric heats of adsorption, the isosteric heats of methane adsorption are typically higher and often in the regime suitable for commercial applications. A high hydrogen storage capacity at ambient temperatures is only realized in the presence of open metal sites in MOFs (or intercalated metal ions in MOFs and COFs) that provide high binding energies (up to 13 kJ mol^{-1}). In contrast, methane typically adsorbs with higher binding energies even to nonpolar organic moieties (e.g. pore openings with multiple adsorptive sites). This adsorption behavior allows for a wider range of suitable materials compared to hydrogen storage. The high surface areas, regular pore shapes, and tunable pore sizes render MOFs superior to traditional porous materials such as zeolites or porous carbons. In the following, we will elucidate how these parameters can be used to design potent methane storage materials.

15.3.1 Optimizing MOFs for Methane Storage

A MOF must fulfill certain prerequisites to be suitable for application in ANG tanks and the most important are (i) the cost to synthesize a given material, (ii) the working capacity, (iii) the cycle stability, and (iv) the chemical stability toward the impurities present in natural gas. While the cost to synthesize a given material is mainly correlated to the demand, the latter three criteria represent intrinsic material properties that can be modified and optimized using the methods discussed in Chapters 1–6.

Methane storage is of particular interest for mobile applications. Here, the working capacity determines the possible driving range with a tank of a given volume. It is defined as the volume of methane stored per volume of material (in v v^{-1}) in a fully loaded tank (typically at 65 bar) minus the volume that remains in the tank when the depletion pressure (typically 5.8 bar) is reached. This implies that a given material must not only store a large quantity of methane at high pressures but also that a large fraction of this maximum capacity must be available within a pressure window ranging from 5 to 65 bar. Since the magnitude of the working capacity strongly depends on the structure and chemical nature of the MOF, the precise design of its structure and the underlying individual building units is crucial for the development of new high-performance materials. There are several structural factors such as (i) the surface area, (ii) the pore shape, diameter, or aperture, and (iii) the introduction of open metal sites or other strong adsorption sites that can be manipulated and optimized using reticular chemistry.

15.3.1.1 Optimization of the Pore Shape and Metrics

To derive appropriate design principles for MOFs with high methane uptake, it is instructive to first locate the preferred binding sites for methane within the pores of MOFs. Much research has been focused on identifying the primary adsorption sites in MOFs, and similar to other nonpolar gases methane is typically adsorbed close to the SBUs, arguably due to the more polar nature of the SBUs compared to the organic backbone. Additional binding sites are located on the faces and edges of the linker. Even though the interactions arising from these sites are much weaker than those arising from the SBUs, they play a key role in the design of materials with large gas uptake due to their great capacity for an increase in gas loading. Strategies to increase the number of adsorptive sites provided by linker design have been discussed earlier (see Chapter 2) [35]. For methane storage, however, a large surface area does not necessarily correspond to a large working capacity. To illustrate this, we analyze methane adsorption in two **fof** topology frameworks (NOTT-101 and NOTT-103) and a structurally related **stx** topology framework (NOTT-109). These MOFs are similar in terms of their structures but show significant differences in their pore sizes and surface areas (Figure 15.9). NOTT-109

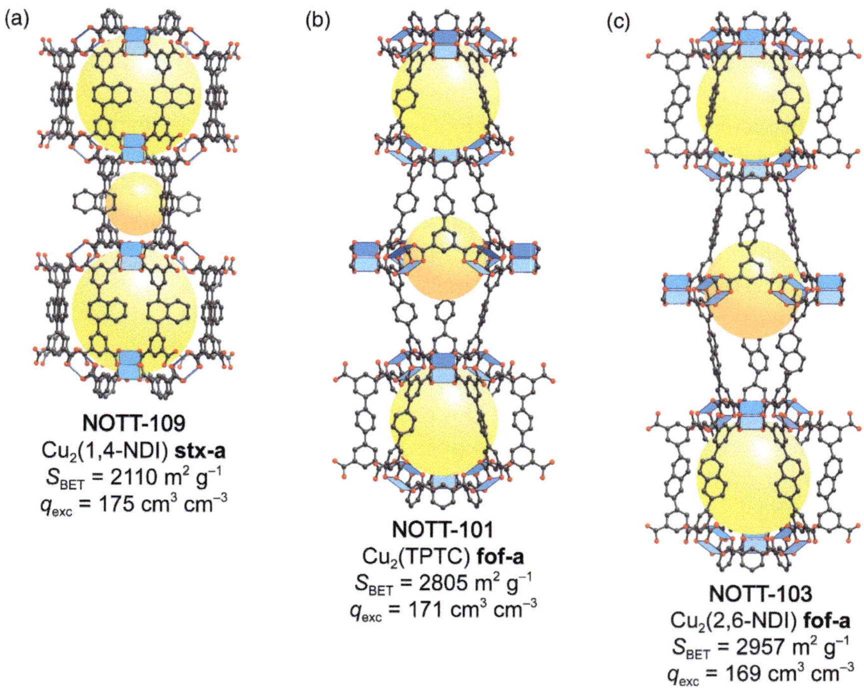

Figure 15.9 (a–c) Crystal structures of NOTT-109, NOTT-101, and NOTT-103 alongside the respective surface areas and excess methane uptake capacities. Although the structures are similar and the surface area increases from (a) to (c) the excess uptake decreases. This highlights the importance of the pore shape and size in realizing a high excess methane uptake. Color code: Cu, blue; C, gray; O, red.

has the lowest surface area (S_{BET} = 2110 m^2 g^{-1}), followed by NOTT-101 (S_{BET} = 2805 m^2 g^{-1}) and NOTT-103 (S_{BET} = 2958 m^2 g^{-1}). The comparison of the excess uptake at room temperature and 35 bar shows that the MOF with the lowest surface area has the highest excess uptake ($q_{exc(NOTT-109)}$ = 175 cm^3 cm^{-3}, $q_{exc(NOTT-103)}$ = 171 cm^3 cm^{-3}, and $q_{exc(NOTT-103)}$ = 169 cm^3 cm^{-3}) [36]. Based on this finding, an empirical formula providing a correlation between the gravimetric methane uptake and the pore volume can be established:

$$q_{exc} = -126.69 \times V_{pore}^2 + 381.62 \times V_{pore} - 12.57 \quad (15.1)$$

where q_{exc} is the excess gravimetric methane storage capacity (at 35 bar and 300 K in cm^3 g^{-1}) and V_{pore} is the free pore volume of the MOF (in cm^3 g^{-1}). Using this empirical equation, the methane storage performance of microporous MOF materials with a pore volume of less than 1.50 cm^3 g^{-1} can be predicted reasonably well. This example illustrates that not only a large surface area but also appropriately sized and shaped pores are of great importance in realizing a high methane working capacity.

Another important factor to consider when targeting high methane uptake is the optimization of the methane–framework interaction. We illustrate this by comparing the methane uptake characteristics of PCN-61 and PCN-68, two isoreticular frameworks with underlying **rht** topology. PCN-61 is constructed from H_6BTEI linkers and 4-c copper paddle wheel SBUs and its structure features pores of 12, 11.8, and 18.8 Å diameter and a surface area of 3000 m^2 g^{-1}. PCN-68 is the isoreticular expanded analog of PCN-61 and is built from 4-c paddle wheel SBUs and H_6PTEI linkers. Consequently, PCN-68 encompasses larger pores (12, 14.8, and 23.2 Å in diameter) and has a higher surface area (5109 m^2 g^{-1}) than PCN-61 [22]. Interestingly, PCN-61 still shows a significantly higher volumetric excess methane uptake (145 cm^3 cm^{-3}) than PCN-68 (99 cm^3 cm^{-3}), which is attributed to the stronger methane–framework interaction.

A strong interaction between methane and the framework is realized by the introduction of so-called "van der Waal pockets." Such pockets are cages where multiple aryl units are in close proximity, thereby enhancing the dispersive interactions between the adsorbate and the pore surface. Examples of MOFs with structures containing van der Waals pockets are PCN-14 and HKUST-1, two of the best methane storage materials today. PCN-14 is built from 4-c copper paddle wheel SBUs that are connected by ADIP linkers to form a framework of **fof** topology (Figure 15.10) [37]. The structure contains squashed cuboctahedral cages of approximately 12.5 Å diameter and has a comparatively low surface area of only 1753 m^2 g^{-1}. Nevertheless, the high density of anthracenyl units results in enhanced dispersive interactions with the methane molecules and consequently leads to a high gravimetric (15.3 wt%) and volumetric (220 cm^3 cm^{-3}) uptake.

The importance of appropriately sized pores becomes evident when comparing MOF-210 and Al-**soc**-MOF. The structures of both MOFs have been discussed previously (MOF-210, see Figure 5.9; Al-**soc**-MOF, see Figure 4.14). MOF-210 (**toz**) has two differently sized pores (22 × 22 Å and 22 × 51 Å) and a high surface area of 6240 m^2 g^{-1} [3]. The gravimetric working capacity of 0.376 g g^{-1} is one of the highest reported for a MOF. The volumetric capacity of 131 cm^3 cm^{-3} is

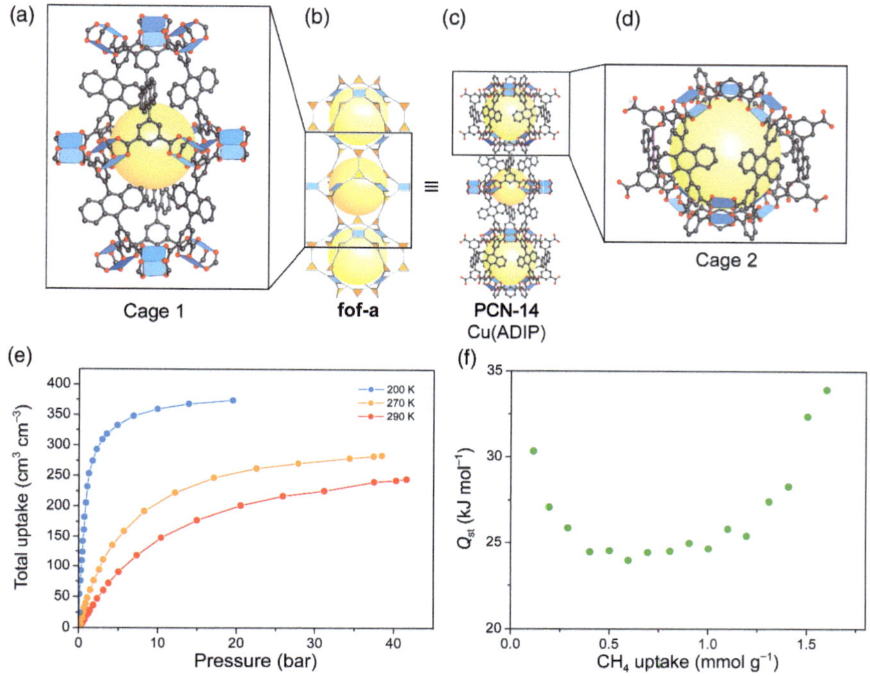

Figure 15.10 (a, d) The two distinct cages in the structure of PCN-14 are shaped in such a way as to maximaze dispersive interactions with methane molecules at the narrow pore apertures. This design feature leads to a high excess methane uptake despite the relatively low surface area of 1753 $m^2\ g^{-1}$. (b) Topology representation (**fof-a**) and (c) crystal structure of PCN-14. (e) Total methane uptake measured at 200 K (blue), 270 K (orange), and 290 K (red). (f) Isosteric heat of methane adsorption calculated for different loadings based on adsorption data collected at 270, 280, and 290 K indicating strong interactions of methane with the framework at low loadings [37]. Color code: Cu, blue; C, gray; O, red.

comparatively low due to the large pore size and pore volume of 3.6 $cm^3\ g^{-1}$.[1] The pores of Al-**soc**-MOF on the other hand are comparatively small (14.3 Å diameter), while a large surface area of 5585 $m^2\ g^{-1}$ is maintained. Therefore, it has a similar gravimetric (q_{grav} = 0.37 g g^{-1}) but significantly higher volumetric uptake (q_{vol} = 176 $cm^3\ cm^{-3}$) than MOF-210. The examples discussed so far highlight that small pores and narrow passages connecting them increase the strength of dispersion forces between the pore surface and incoming gas molecules and are thus beneficial for realizing high methane adsorption capacities. This can be achieved either by interpenetration or by designing MOFs with complex contorted pore systems. The latter principle is employed in the design of ST-1, ST-2, ST-3, and ST-4, a series of tertiary and quarternary MOFs with complex framework structures [38]. A flow diagram illustrating the synthesis and composition of these MOFs is given in Figure 15.11a. All structures are highly complex and are built from up to five different types of cage (ST-1, **muo**; ST-2, **umt**; ST-3, **ith-d**; ST-4, **ott**). The high methane storage capacity of these

1 The working capacity is measured between 65 and 5.8 bar.

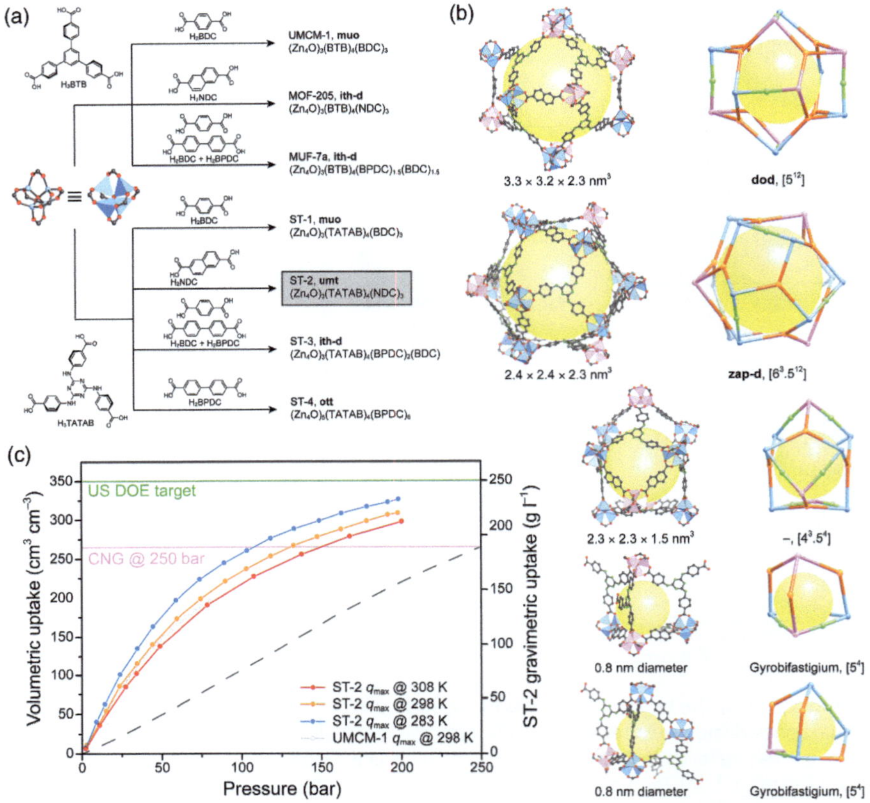

Figure 15.11 (a) Flow diagram visualizing the composition of tertiary and quaternary MOFs constructed from octahedral $Zn_4O(-COO)_6$ SBUs and a combination of linear ditopic and trigonal tritopic linkers. (b) Cages in the **umt** structure of ST-2. The two topologically distinct SBUs are shown in pink and blue. In the topological representation the linear linkers (commonly not show) are represented by green spheres for clarity; however, they do not represent vertices. All cages with exception of the [5^4] cages feature small "van der Waals" pockets and narrow pore openings. Additionally, the walls of the pores are lined with polar TATAB units. (c) Methane adsorption isotherms of ST-2 recorded at 283 K (blue), 298 K (orange), and 308 K (red) alongside a methane adsorption isotherm of UMCM-1 (gray) for comparison [38]. At a pressure of 135 bar ST-2 surpasses the capacity of compressed natural gas at 250 bar (light pink). All hydrogen atoms are omitted for clarity. Color code for crystal structure drawings: Zn, pink and blue; N, green; C, gray; O, red. Color code for the topology representation: 6-c SBUs, pink and blue; trigonal linker, orange; linear linker, green. The yellow spheres indicate the empty space within the respective cage.

compounds originates from the complex interconnected pore systems. Owing to a low methane uptake at pressures below the minimum delivery pressure, all four materials feature a high working capacity. Here, we discuss the best performing member of this series, ST-2. The **umt** net of ST-2 is constructed from four topologically distinguishable cages that are fused to form a highly complex pore system (Figure 15.11b). With the exception of the two topologically identical [5^4] cages, all cages feature polar TATAB units, small "pockets," and – with respect to the dimensions of the cages – relatively small apertures. These structural features help increase van der Waals interactions. Figure 15.11c shows methane adsorption isotherms of ST-2 measured at different temperatures. ST-2 clearly surpasses the storage capacity of compressed natural gas tanks at pressures >135 bar. The contorted pore system makes it possible to achieve both high volumetric (290 cm^3 cm^{-3}) and gravimetric (206 $g\,l^{-1}$) deliverable capacities between 5 and 200 bar.

15.3.1.2 Introduction of Polar Adsorption Sites

Similar to hydrogen, methane can interact with open metal sites, which leads to increased heats of adsorption. We introduced a selection of SBUs with potential open metal sites earlier in this chapter and the reader is referred to this section for further information. The plot of the total methane uptake against the pore volume of several MOFs shown in Figure 15.12 illustrates that as the pore volume increases, the total volumetric uptake first increases to reach a maximum and then decreases again. This implies that there is an upper limit to the methane storage capacity of framework materials at 65 bar and room temperature at a pore volume of about 1 cm^3 g^{-1}. Only three MOFs, namely HKUST-1, NiMOF-74, and UTSA-76, surpass this maximum. HKUST-1 and NiMOF-74 have some of the highest concentrations of open metal sites in MOFs, and UTSA-76 has a high density of functional pyrimidine groups.

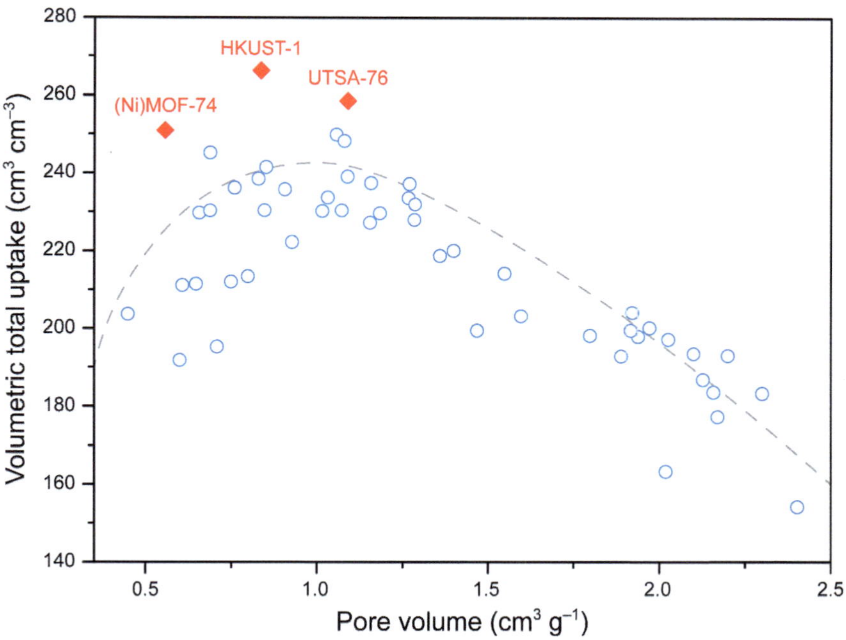

Figure 15.12 Correlation between the volumetric total methane uptake (cm^3 cm^{-3} at 65 bar and 298 K) and the pore volume, V_p (cm^3 g^{-1}). MOFs with open metal sites (HKUST-1 and (Ni)MOF-74) or high concentrations of polar entities dangling into the pores (UTSA-76) show higher volumetric capacities than MOFs without these features (blue circles).

Table 15.2 Compilation of the surface area, the pore size, and the gravimetric (wt%) and volumetric ($cm^3\ cm^{-3}$) methane uptake for a selection of promising MOFs and COFs.

Chemical formula	Common name	Surface area		Pore size (Å)	Pore volume ($cm^3\ g^{-1}$)	OMS	Methane storage capacity				References
		BET ($m^2\ g^{-1}$)	Langmuir ($m^2\ g^{-1}$)				p (bar)	T (K)	Uptake (wt%)	Uptake ($cm^3\ cm^{-3}$)	
MOF											
$Al_8(OH)_8(BTB)_4(H_2BTB)_4$	MOF-519	2400	2660	7.6 Å	0.938	No	35	298	—	200	[41]
$Al_8(OH)_8(BTB)_4(COO)_4$	MOF-520	3290	3930	16.2 × 9.9	1.277	No	35	298	—	162	[41]
$Al_3O(TCPT)_{1.5}(H_2O)_3]Cl^-$	Al-soc-MOF-1	5585	—	–	2.3	Yes	65	298	—	221	[42]
$Cu_2(ADIP)$	PCN-14	1753	2176	—	—	Yes	35	290	15.36	220	[37]
$Cu_3(BTC)_2$	HKUST-1	1850	—	—	0.780	Yes	35	303	—	267	[43]
$Cu_3(BTTTA)$	NU-125	3120	—	—	1.29	Yes	58	298	—	228	[44]
$Cu_2(TPTC)$	NOTT-101a	2805	—	13 × 24, 14	1.08	Yes	35	298	—	237	[36]
$Cu_2(PyDIA)$	UTSA-76a	2820	—	—	1.09	No	35	298	—	211	[45]
Cu(NDDI)	NOTT-103	2958	—	—	1.157	Yes	35	298	—	194	[36]
$Mg_2(DOBDC)$	MgMOF-74	—	1957	—	0.69	Yes	35	298	—	200	[46]
$Ni_2(DOBDC)$	Ni-MOF-74	—	1593	—	0.56	Yes	35	298	—	230	[46]
$Zn_4O(BDC)(BTAC)_{4/3}$	MOF-905	3490	3770	6.0, 18.0	1.34	No	1.1	298	7.7	—	[47]
$Zn_4O(BDC\text{-}Me_2)(BTAC)_{4/3}$	MOF-905-Me2	3640	3920	5.5, 17.6	1.39	No	1.1	298	11.0	—	[47]
$Zn_4O(BDC)_3$	MOF-5	3480	3860	12.8	1.39	No	80	298	22.6	274	[38, 47]
$Zn_4O(BTB)_3$	MOF-177	4700	5060	10.8	1.83	No	80	298	20.5	344	[38, 47]
$(Zn_4O)_3(TATAB)_4(BDC)_3$	ST-1	4412	—	1.3–3.4	2.39	No	80	298	37.3	—	[38]
$(Zn_4O)_3(TATAB)_4(NDC)_3$	ST-2	5172	—	1.3–3.0	2.44	No	80	298	40.1	—	[38]
$(Zn_4O)_3(TATAB)_4(BPDC)_2(BDC)$	ST-3	5660	—	1.3–3.6	2.67	No	80	298	41.0	—	[38]
COF											
$C_{25}H_{24}B_4O_8$	COF-102	3620	4650	12	1.55	No	35	298	17.72	127	[40, 48]
$C_{25}H_{24}B_4O_8 \times xLi$	COF-102 Li	<3620	<4650	<12	<1.55	Yes	35	298	33	327	[40]
$C_{25}H_{24}B_4O_8Si$	COF-103	3530	4630	12	1.54	No	35	298	16.69	108	[40, 48]
$C_{25}H_{24}B_4O_8Si \times xLi$	COF-103 Li	<3530	<4630	<12	<1.54	Yes	35	298	32.75	315	[40]

In a manner akin to that described for hydrogen storage, the methane storage capacity of MOFs can be enhanced by intercalation of Li^+ ions. This method is particularly interesting for COFs due to their intrinsic low affinity toward methane and an increase of the methane uptake by more than 200% is realized by Li^+ intercalation into 3D COFs (COF-102(Li) and COF-103(Li)) compared to the pristine materials [40].

15.3.2 Important MOFs for Methane Storage

MOFs must meet several prerequisites to be suitable candidates for implementation in methane storage applications (ANG storage): (i) a high working capacity (both gravimetric and volumetric), (ii) high cycle stability, and (iii) low cost of production on an industrial scale. Table 15.2 gives a compilation of MOFs with outstanding performance in methane storage as well as selected examples of COFs.

15.4 Summary

In this chapter, we discussed the design of MOFs for use in energy storage (methane and hydrogen) applications. We uncovered structural features that allow for a high uptake of nonpolar gases and ways to implement these features into extended structures. We saw that strongly polarizing structural fragments as well as van der Waals pockets help increase the uptake of methane and hydrogen and that the pore size of the storage materials must be designed with the size of the targeted adsorptive in mind. The importance of a high uptake between 5 and 65 bar due to practical reasons was highlighted and we gave examples of MOFs that show a good storage performance in this pressure range. We concluded each section with tables summarizing the best performing materials for both applications.

References

1 (a) He, Y., Zhou, W., Qian, G., and Chen, B. (2014). Methane storage in metal-organic frameworks. *Chemical Society Reviews* 43 (16): 5657–5678. (b) Simon, C.M., Kim, J., Gomez-Gualdron, D.A. et al. (2015). The materials genome in action: identifying the performance limits for methane storage. *Energy & Environmental Science* 8 (4): 1190–1199. (c) Gomez-Gualdron, D.A., Gutov, O.V., Krungleviciute, V. et al. (2014). Computational design of metal-organic frameworks based on stable zirconium building units for storage and delivery of methane. *Chemistry of Materials* 26 (19): 5632–5639.

2 Dresselhaus, M., Crabtree, G., Buchanan, M. et al. (2004). Basic Research Needs for the Hydrogen Economy: Report on the Basic Energy Sciences Workshop on Hydrogen Production, Storage, and Use. DOESC (USDOE Office of Science (SC)). Argonne, IL: Argonne National Laboratory (ANL).

3 Furukawa, H., Ko, N., Go, Y.B. et al. (2010). Ultrahigh porosity in metal-organic frameworks. *Science* 329 (5990): 424–428.

4 Wong-Foy, A.G., Matzger, A.J., and Yaghi, O.M. (2006). Exceptional H_2 saturation uptake in microporous metal-organic frameworks. *Journal of the American Chemical Society* 128 (11): 3494–3495.
5 Frost, H. and Snurr, R.Q. (2007). Design requirements for metal-organic frameworks as hydrogen storage materials. *The Journal of Physical Chemistry C* 111 (50): 18794–18803.
6 Bae, Y.-S. and Snurr, R.Q. (2010). Optimal isosteric heat of adsorption for hydrogen storage and delivery using metal-organic frameworks. *Microporous and Mesoporous Materials* 132 (1): 300–303.
7 Kubas, G.J. (2001). *Metal Dihydrogen and s-Bond Complexes: Structure, Theory, and Reactivity*. Springer Science & Business Media.
8 (a) Lan, J., Cao, D., and Wang, W. (2010). Li-doped and nondoped covalent organic borosilicate framework for hydrogen storage. *The Journal of Physical Chemistry C* 114 (7): 3108–3114. (b) Cao, D., Lan, J., Wang, W., and Smit, B. (2009). Lithium-doped 3D covalent organic frameworks: high-capacity hydrogen storage materials. *Angewandte Chemie* 121 (26): 4824–4827. (c) Klontzas, E., Tylianakis, E., and Froudakis, G.E. (2009). Hydrogen storage in lithium-functionalized 3-D covalent-organic framework materials. *The Journal of Physical Chemistry C* 113 (50): 21253–21257.
9 (a) Zou, X., Zhou, G., Duan, W. et al. (2010). A chemical modification strategy for hydrogen storage in covalent organic frameworks. *The Journal of Physical Chemistry C* 114 (31): 13402–13407. (b) Xiang, Z., Hu, Z., Yang, W., and Cao, D. (2012). Lithium doping on metal-organic frameworks for enhancing H_2 storage. *International Journal of Hydrogen Energy* 37 (1): 946–950.
10 (a) Chui, S.S.-Y., Lo, S.M.-F., Charmant, J.P.H. et al. (1999). A chemically functionalizable nanoporous material $[Cu_3(TMA)_2(H_2O)]_n$. *Science* 283 (5405): 1148–1150. (b) Dincă, M., Dailly, A., Liu, Y. et al. (2006). Hydrogen storage in a microporous metal-oganic framework with exposed Mn^{2+} coordination sites. *Journal of the American Chemical Society* 128 (51): 16876–16883.
11 Dincă, M. and Long, J.R. (2008). Hydrogen storage in microporous metal-organic frameworks with exposed metal sites. *Angewandte Chemie International Edition* 47 (36): 6766–6779.
12 (a) Sachdeva, S., Pustovarenko, A., Sudhölter, E.J. et al. (2016). Control of interpenetration of copper-based MOFs on supported surfaces by electrochemical synthesis. *CrystEngComm* 18 (22): 4018–4022. (b) Sun, D., Ma, S., Ke, Y. et al. (2006). An interweaving MOF with high hydrogen uptake. *Journal of the American Chemical Society* 128 (12): 3896–3897.
13 Ma, S., Sun, D., Ambrogio, M. et al. (2007). Framework-catenation isomerism in metal-organic frameworks and its impact on hydrogen uptake. *Journal of the American Chemical Society* 129 (7): 1858–1859.
14 (a) Li, Y., Yang, F.H., and Yang, R.T. (2007). Kinetics and mechanistic model for hydrogen spillover on bridged metal-organic frameworks. *The Journal of Physical Chemistry C* 111 (8): 3405–3411. (b) Li, Y. and Yang, R.T. (2006). Significantly enhanced hydrogen storage in metal-organic frameworks via spillover. *Journal of the American Chemical Society* 128 (3): 726–727. (c) Li, Y. and Yang, R.T. (2006). Hydrogen storage in metal-organic frameworks by bridged hydrogen spillover. *Journal of the American Chemical Society* 128

(25): 8136–8137. (d) Li, Y. and Yang, R.T. (2007). Gas adsorption and storage in metal-organic framework MOF-177. *Langmuir* 23 (26): 12937–12944. (e) Li, Y. and Yang, R.T. (2008). Hydrogen storage in metal-organic and covalent-organic frameworks by spillover. *AIChE Journal* 54 (1): 269–279. (f) Liu, Y.-Y., Zeng, J.-L., Zhang, J. et al. (2007). Improved hydrogen storage in the modified metal-organic frameworks by hydrogen spillover effect. *International Journal of Hydrogen Energy* 32 (16): 4005–4010.

15 Sumida, K., Hill, M.R., Horike, S. et al. (2009). Synthesis and hydrogen storage properties of $Be_{12}(OH)_{12}$(1,3,5-benzenetribenzoate)$_4$. *Journal of the American Chemical Society* 131 (42): 15120–15121.

16 (a) Bragg, W. (1923). Crystal structure of basic beryllium acetate. *Nature* 111: 532–532. (b) Pauling, L. and Sherman, J. (1934). The structure of the carboxyl group II. The crystal structure of basic beryllium acetate. *Proceedings of the National Academy of Sciences* 20: 340–345.

17 Latroche, M., Surblé, S., Serre, C. et al. (2006). Hydrogen storage in the giant-pore metal-organic frameworks MIL-100 and MIL-101. *Angewandte Chemie International Edition* 45 (48): 8227–8231.

18 Panella, B., Hirscher, M., Pütter, H., and Müller, U. (2006). Hydrogen adsorption in metal-organic frameworks: Cu-MOFs and Zn-MOFs compared. *Advanced Functional Materials* 16 (4): 520–524.

19 Lin, X., Telepeni, I., Blake, A.J. et al. (2009). High capacity hydrogen adsorption in Cu(II) tetracarboxylate framework materials: the role of pore size, ligand functionalization, and exposed metal sites. *Journal of the American Chemical Society* 131 (6): 2159–2171.

20 Lin, X., Jia, J., Zhao, X. et al. (2006). High H_2 adsorption by coordination-framework materials. *Angewandte Chemie International Edition* 45 (44): 7358–7364.

21 Wang, X.-S., Ma, S., Rauch, K. et al. (2008). Metal-organic frameworks based on double-bond-coupled di-isophthalate linkers with high hydrogen and methane uptakes. *Chemistry of Materials* 20 (9): 3145–3152.

22 Yuan, D., Zhao, D., Sun, D., and Zhou, H.C. (2010). An isoreticular series of metal-organic frameworks with dendritic hexacarboxylate ligands and exceptionally high gas-uptake capacity. *Angewandte Chemie International Edition* 49 (31): 5357–5361.

23 Farha, O.K., Yazaydın, A.Ö., Eryazici, I. et al. (2010). De novo synthesis of a metal-organic framework material featuring ultrahigh surface area and gas storage capacities. *Nature Chemistry* 2 (11): 944–948.

24 Guo, Z., Wu, H., Srinivas, G. et al. (2011). A metal-organic framework with optimized open metal sites and pore spaces for high methane storage at room temperature. *Angewandte Chemie International Edition* 50 (14): 3178–3181.

25 Liu, Y., Eubank, J.F., Cairns, A.J. et al. (2007). Assembly of metal-organic frameworks (MOFs) based on indium-trimer building blocks: a porous MOF with soc topology and high hydrogen storage. *Angewandte Chemie International Edition* 46 (18): 3278–3283.

26 Sumida, K., Brown, C.M., Herm, Z.R. et al. (2011). Hydrogen storage properties and neutron scattering studies of Mg_2(dobdc) – a metal-organic

framework with open Mg^{2+} adsorption sites. *Chemical Communications* 47 (4): 1157–1159.
27 Dinca, M., Dailly, A., Liu, Y. et al. (2006). Hydrogen storage in a microporous metal-organic framework with exposed Mn^{2+} coordination sites. *Journal of the American Chemical Society* 128 (51): 16876–16883.
28 Dinca, M., Dailly, A., Tsay, C., and Long, J.R. (2008). Expanded sodalite-type metal-organic frameworks: increased stability and H_2 adsorption through ligand-directed catenation. *Inorganic Chemistry* 47 (1): 11–13.
29 Ma, S., Wang, X.-S., Manis, E.S. et al. (2007). Metal-organic framework based on a trinickel secondary building unit exhibiting gas-sorption hysteresis. *Inorganic Chemistry* 46 (9): 3432–3434.
30 Park, K.S., Ni, Z., Côté, A.P. et al. (2006). Exceptional chemical and thermal stability of zeolitic imidazolate frameworks. *Proceedings of the National Academy of Sciences* 103 (27): 10186–10191.
31 Lee, Y.G., Moon, H.R., Cheon, Y.E., and Suh, M.P. (2008). A comparison of the H_2 sorption capacities of isostructural metal-organic frameworks with and without accessible metal sites: $[\{Zn_2(abtc)(dmf)_2\}_3]$ and $[\{Cu_2(abtc)(dmf)_2\}_3]$ versus $[\{Cu_2(abtc)\}_3]$. *Angewandte Chemie* 120 (40): 7855–7859.
32 Kim, H., Das, S., Kim, M.G. et al. (2011). Synthesis of phase-pure interpenetrated MOF-5 and its gas sorption properties. *Inorganic Chemistry* 50 (8): 3691–3696.
33 Koh, K., Wong-Foy, A.G., and Matzger, A.J. (2009). A porous coordination copolymer with over 5000 m_2/g BET surface area. *Journal of the American Chemical Society* 131 (12): 4184–4185.
34 (a) Babarao, R. and Jiang, J. (2008). Exceptionally high CO_2 storage in covalent-organic frameworks: atomistic simulation study. *Energy & Environmental Science* 1 (1): 139–143. (b) El-Kaderi, H.M., Hunt, J.R., Mendoza-Cortés, J.L. et al. (2007). Designed synthesis of 3D covalent organic frameworks. *Science* 316 (5822): 268–272.
35 Chae, H.K., Siberio-Pérez, D.Y., Kim, J. et al. (2004). A route to high surface area, porosity and inclusion of large molecules in crystals. *Nature* 427 (6974): 523–527.
36 He, Y., Zhou, W., Yildirim, T., and Chen, B. (2013). A series of metal-organic frameworks with high methane uptake and an empirical equation for predicting methane storage capacity. *Energy & Environmental Science* 6 (9): 2735–2744.
37 Ma, S., Sun, D., Simmons, J.M. et al. (2008). Metal-organic framework from an anthracene derivative containing nanoscopic cages exhibiting high methane uptake. *Journal of the American Chemical Society* 130 (3): 1012–1016.
38 Liang, C.-C., Shi, Z.-L., He, C.-T. et al. (2017). Engineering of pore geometry for ultrahigh capacity methane storage in mesoporous metal-organic frameworks. *Journal of the American Chemical Society* 139 (38): 13300–13303.
39 Li, B., Wen, H.-M., Zhou, W. et al. (2016). Porous metal-organic frameworks: promising materials for methane storage. *Chem* 1 (4): 557–580.
40 Lan, J., Cao, D., and Wang, W. (2009). High uptakes of methane in Li-doped 3D covalent organic frameworks. *Langmuir* 26 (1): 220–226.

41 Gándara, F., Furukawa, H., Lee, S., and Yaghi, O.M. (2014). High methane storage capacity in aluminum metal-organic frameworks. *Journal of the American Chemical Society* 136 (14): 5271–5274.

42 Alezi, D., Belmabkhout, Y., Suyetin, M. et al. (2015). MOF crystal chemistry paving the way to gas storage needs: aluminum-based soc-MOF for CH_4, O_2, and CO_2 storage. *Journal of the American Chemical Society* 137 (41): 13308–13318.

43 Wiersum, A.D., Chang, J.-S., Serre, C., and Llewellyn, P.L. (2013). An adsorbent performance indicator as a first step evaluation of novel sorbents for gas separations: application to metal-organic frameworks. *Langmuir* 29 (10): 3301–3309.

44 Wilmer, C.E., Farha, O.K., Yildirim, T. et al. (2013). Gram-scale, high-yield synthesis of a robust metal-organic framework for storing methane and other gases. *Energy & Environmental Science* 6 (4): 1158–1163.

45 Li, B., Wen, H.-M., Wang, H. et al. (2014). A porous metal-organic framework with dynamic pyrimidine groups exhibiting record high methane storage working capacity. *Journal of the American Chemical Society* 136 (17): 6207–6210.

46 Mason, J.A., Veenstra, M., and Long, J.R. (2014). Evaluating metal-organic frameworks for natural gas storage. *Chemical Science* 5 (1): 32–51.

47 Jiang, J., Furukawa, H., Zhang, Y.-B., and Yaghi, O.M. (2016). High methane storage working capacity in metal-organic frameworks with acrylate links. *Journal of the American Chemical Society* 138 (32): 10244–10251.

48 Furukawa, H. and Yaghi, O.M. (2009). Storage of hydrogen, methane, and carbon dioxide in highly porous covalent organic frameworks for clean energy applications. *Journal of the American Chemical Society* 131 (25): 8875–8883.

16

Liquid- and Gas-Phase Separation in MOFs

16.1 Introduction

Many industrial processes such as the purification of feedstock for the chemical industry, fuels, energy carriers, and exhaust gases as well as purification processes finding application in our daily life such as the purification of drinking water are based on separation. Most of these processes use porous solids for the selective separation of specific components from influent gaseous or liquid mixtures to give high-purity products.

Gases such as hydrogen, methane, and light hydrocarbons are sources of energy and are used in combustion engines or fuel cells [1]. This is, however, only possible if they are supplied in high purity. In Chapter 13 we discussed the problem of increasing CO_2 levels in the atmosphere and the application of metal-organic frameworks (MOFs) in CO_2 capture. This process is thermodynamically controlled (i.e. equilibrium adsorption), just like the purification of natural gas, syngas, and the separation of light hydrocarbons. In contrast, the separation of gaseous mixtures of different isomers of hydrocarbons is typically controlled by kinetics (Figure 16.1). A further development in terms of separation processes based on porous solids are mixed-matrix membranes (MMMs): composite membranes made from a polymer (membrane) and porous solid additives (filler). MMMs combine both the selectivity of porous adsorbents such as MOFs, zeolitic imidazolate frameworks (ZIFs), and covalent organic framework (COF) with the flux, facile preparation, and flexibility of polymers. They have been tested for many separation processes and their high performance illustrates how hybrid materials can surpass the performance of their individual components and in some cases also their sum [2].

MOF-based separation processes are not only applicable to gaseous mixtures but also to the separation of mixtures of liquids and the selective removal of specific components from them. In this context, the removal of biologically active molecules from aqueous solution is becoming increasingly important. Trace amounts of medical drugs and other biologically active molecules in rivers and drinking water worldwide pose a significant health concern [3]. Another example for a potential application of porous solids in the separation of liquids is the purification of liquid fuels, petroleum, and shale oil. They all contain cyclic amines that negatively influence their quality due to bad odor, acute

Introduction to Reticular Chemistry: Metal-Organic Frameworks and Covalent Organic Frameworks,
First Edition. Omar M. Yaghi, Markus J. Kalmutzki, and Christian S. Diercks.

Companion website: www.wiley.com/go/yaghi/reticular

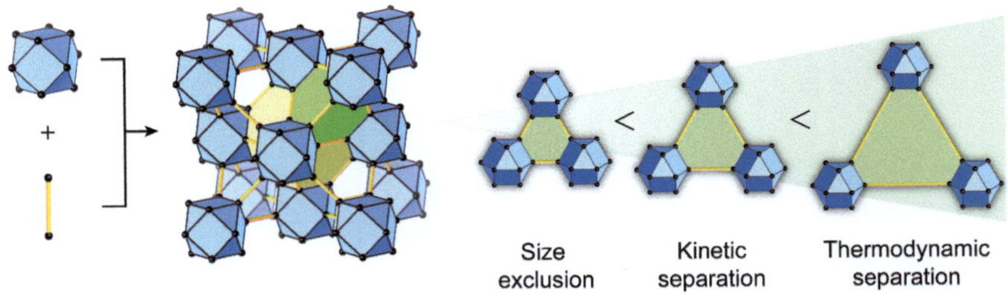

Figure 16.1 The predominant mechanism for separation is correlated to the ratio of the pore size and the kinetic diameter of the components in the mixture. While all three mechanisms can afford the separation of gas/liquid mixtures, only thermodynamic separation allows to selectively capture a specific component within the pores.

aquatic toxicity and carcinogenicity, and the increase formation of deposits. The combustion of fuels containing cyclic amines, or amines in general, leads to the formation of nitrous oxides (NO_x), which are the cause of acid rain [4].

To show high performance in the abovementioned processes, materials with high selectivity in the respective application must be developed. The rapidly growing number of compounds made by means of reticular chemistry show great promise for gas and liquid phase separation. Their facile synthesis in combination with unique structural and physical features compared to traditional porous materials (e.g. zeolites and porous carbons) makes them ideal for gas separation and the selective capture of molecules from the gas or liquid phase. These features include rational design approaches (see Chapters 4 and 5), the possibility to deliberately introduce functional sites (see Chapter 6), and ultra-high porosities with BET surface areas of over 6000 $m^2\ g^{-1}$ (see Chapter 2). The basic physical processes involved in the separation of gases and liquids in porous solids have been discussed in Chapter 13. Here, we will take a closer look at the separation of volatile organic molecules such as light hydrocarbons, aromatic compounds, and the adsorptive removal of bioactive molecules from water.

16.2 Separation of Hydrocarbons

Hydrocarbons are used as feedstock for the chemical industry and the separation of mixtures of hydrocarbon is one of the most important processes in the petrochemical industry [5]. Hydrocarbons are exclusively composed of carbon and hydrogen, and are categorized into alkanes (paraffins), alkenes (olefins), and aromatic hydrocarbons (naphthenic hydrocarbons). Many olefins (e.g. ethylene, propylene, and butadiene) and naphthenic hydrocarbons (e.g. benzene, toluene, and xylene) are important feedstocks for the chemical industry. *p*-Xylene is used as the starting material for the industrial synthesis of terephthalic acid (H_2BDC), which is an important component of many polymers such as PET (polyethyleneterephthalate) and also finds use in MOF chemistry as a linker [6]. The widespread industrial use of light hydrocarbons (C_1–C_4 fraction), isomers of alkanes, and especially C_8 naphthenic hydrocarbons (ethylbenzene, *o*-xylene, *m*-xylene, and *p*-xylene) highlights the importance of selective separation processes for the isolation of these compounds in pure phase.

The separation of alkane/alkene mixtures is commonly realized by cryogenic distillation. This process not only requires a large number of distillation stages and a high reflux ratio to obtain fractions of high purity but also operates at high pressures and cryogenic temperatures. These factors render cryogenic distillation less economical. In a similar manner, the separation of C_8 naphthenic hydrocarbons by distillation from an extracting solvent under reduced pressure is cost intensive, which renders the synthesis of these compounds (e.g. synthesis of ethylbenzene from ethylene and benzene) more profitable. This is also rooted in the fact that the separation of C_6 from C_7 and C_8 hydrocarbons is comparatively simple.

Natural gas contains different hydrocarbons such as methane (87–97%), ethane (1.5–9%), propane (1–1.5%), *iso*-butane (0.01–0.3%), and *n*-butane (0.01–0.3%) and the separation of methane from the other components is an essential industrial process. This is because all hydrocarbons present in natural gas have higher value as pure phase compounds that can be used as feedstock for the chemical industry (e.g. ethane, propane, *iso*-butane, etc.) or higher quality fuels (e.g. methane). C_2 and C_3 hydrocarbons are important raw materials for various products, such as acetic acid and polymers including rubbers and plastics. High-purity natural gas allows for a cleaner, more efficient combustion and consequently lower CO_2 emissions.

Separation using porous solids provides a more (cost) efficient alternative to the expensive processes outlined above. One key factor here is the structural tunability of MOFs, which enables the precise adjustment of a variety of parameters such as the shape, metrics, polarity, and functionality of the pores. Such control is not present in the inorganic and carbon-based materials that are currently employed in industrial separation processes. Aside from the tunability, MOFs can exhibit sensitivity toward external stimuli such as pressure or temperature that result in a gate-opening or breathing motion of the whole framework. This degree of flexibility can lead to an outstanding selectivity and performance in separation processes that is not possible with rigid porous materials (i.e. zeolites and porous carbon). In the following, we will develop principles for the design of potent MOFs and ZIFs with respect to the separation of light hydrocarbons (Section 16.2.1), light olefins and paraffins (Section 16.2.2), and aromatic C_8 hydrocarbons (Section 16.2.3).

16.2.1 C_1–C_5 Separation

The separation of C_1–C_5 hydrocarbons can be achieved in different ways. Here, we will focus on the adsorptive separation based on van der Waals interactions between the gas molecules and the pore surface. According to theoretical studies the enthalpy and entropy of adsorption become more negative with increasing chain length of the hydrocarbon [7]. Therefore, long alkane chains are adsorbed more strongly until the selectivity reaches a maximum at a certain pressure P/P_0 (Figure 16.2). When the pressure is further increased, the entropic cost of ordering the long alkane chains outweighs the enthalpic advantage. Similar results are

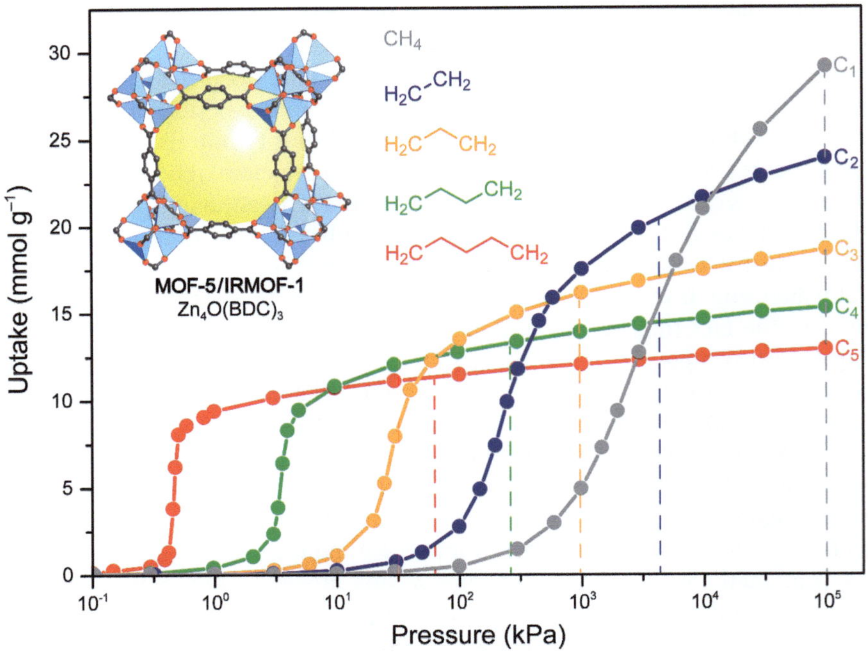

Figure 16.2 Calculated adsorption isotherms of linear C_1–C_5 alkanes in MOF-5 at 300 K [7]. In the low pressure region the high enthalpy of adsorption for long alkane chains outweighs the entropic cost of ordering them. At higher pressures the uptake of lower alkanes increases. The dotted lines represent the bulk saturation pressure of the corresponding component. Color code for isotherms: C_1, red; C_2, green; C_3, orange; C_4, blue; C_5, gray. The crystal structure of MOF-5 is shown in the insert. Color code: Zn, blue; C, gray; O, red.

observed in experimental studies for the adsorption of *n*-butane and methane in MOF-5 and HKUST-1 [8]. In both cases, the isosteric heat of adsorption is more than two times higher for *n*-butane (Q_{st} = −23.6 and −29.6 kJ mol^{-1}) than for methane (Q_{st} = −10.6 and −12.0 kJ mol^{-1}) [9].

Simulations reveal a dependence of the adsorption behavior on different structural factors [10]. The adsorption capacity in isoreticular MOFs is mainly correlated to the pore size, and a consequential decrease in the selectivity with increasing length of the linker is observed. For MOFs built from linkers of the same length, the strength of the interaction is correlated to the number of carbon atoms in the linker, which results in an increase in the selectivity for linkers with a larger aromatic backbone. This is illustrated by the example in Figure 16.3 showing the selectivity for methane in MOF-5 and the hypothetical isoreticular IRMOF-993 ($Zn_4O(ADC)_3$, where ADC = 9,10-anthracene dicarboxylate) [10, 11]. These results however cannot be confirmed experimentally because the reticulation of H_2ADC and Zn^{2+} yields PCN-13 rather than IRMOF-993. PCN-13 has a limited pore size of only 3.5 Å and therefore different gas adsorption properties (Figure 16.3a) [12].

Flexible MOFs commonly show steps in their adsorption isotherms due to swelling, gate-opening, or breathing effects. The MIL-53 family is a prominent

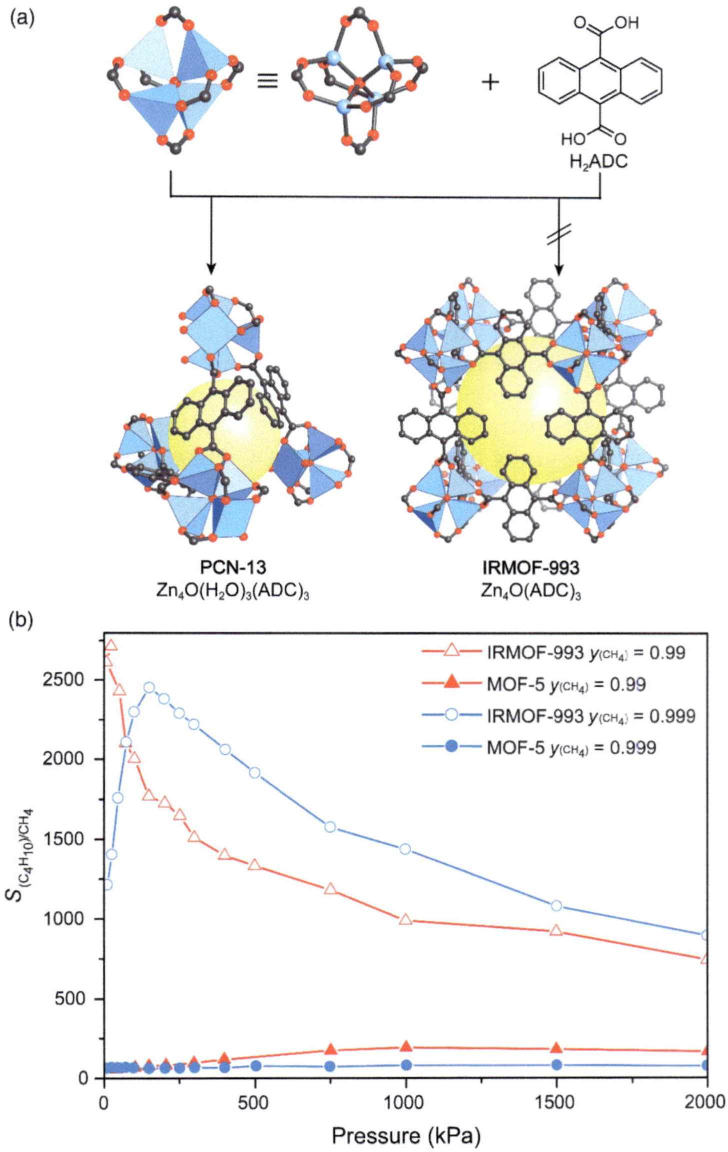

Figure 16.3 (a) Comparison of the crystal structures of PCN-13 and that of the hypothetical IRMOF-993. Only one pore is shown and all hydrogen atoms are omitted for clarity. IRMOF-993 cannot be prepared but serves as a good model for theoretical considerations regarding the correlation between the size of the aromatic backbone of the linker and the selectivity with respect to linear alkanes. (b) Comparison of the selectivity of MOF-5 (filled symbols) and IRMOF-933 (open symbols) in the removal of trace amounts of *n*-butane from methane as a function of the bulk pressure (triangles $y_{(CH_4)} = 0.99$; circles $y_{(CH_4)} = 0.999$).

example for this type of adsorption behavior. For the adsorption of *n*-propane through *n*-nonane the adsorption isotherms of MIL-53(Cr) (at 303 K) show an additional steep step at different pressures depending on the gas used, which is not observed for smaller hydrocarbons (methane and ethane) and is thus attributed to framework swelling [13]. MAMS-1 ($Ni_8(5\text{-BBDC})_6(\mu_3\text{-OH})_4$, where BBDC = 5-tert-butyl-1,3-benzenedicarboxylate), the first MOF-based mesh-adjustable molecular sieve, can discriminate methane/ethane and ethane/propane mixtures with high selectivity. Here, the selectivity is rooted in the capability of this framework to undergo a phase transition from a narrow pore (*np*) to a wide pore (*wp*) phase, also known as gate-opening [14]. The structure of MAMS-1 consists of hydrophobic gas storage chambers that are connected through hydrophilic channels. Influent gas molecules can enter the hydrophobic gas storage chambers only through their interface with the hydrophilic channels. Consequently, most gas is stored within these chambers. Each channel–chamber interface is framed by four BBDC linkers that act as a gate (Figure 16.4a,b). The thermally induced gate-opening effect in MAMS-1 is controlled by the amplitude of thermal vibration. The correlation between the width of the gate and the temperature gives rise to an equation that predicts the diameter of the gate for any given temperature. The diameter can vary between 2.9 and 5.0 Å, making it possible to separate methane and ethane from propane and butane, or methane from ethane, propane, and butane, depending on the temperature (Figure 16.4c).

In summary, for the adsorption of alkanes in MOFs, S-shaped Type-IV (and V) adsorption isotherms are often observed. Steep adsorption steps that occur at different pressures for different gases are utilized for their separation. A stronger adsorption is observed for longer alkane chains, and the adsorption capacity is usually higher for shorter alkanes because their smaller size increases the maximum loading. In contrast, the lower molecular surface decreases the strength of van der Waals interactions. These principles are used to facilitate the separation of a wide range of alkane mixtures. For more detailed information on specific separations, the reader is referred elsewhere [15].

16.2.2 Separation of Light Olefins and Paraffins

The separation of olefin/paraffin mixtures represents some of the most energy-intensive processes in the petrochemical industry [16]. Their similar molecular sizes, weights, and volatilities make these processes especially difficult. Alternative adsorption-based processes present the potential to significantly reduce operating expenses due to lower energy consumption. For this purpose, a number of adsorbents, mainly zeolites, have been evaluated, but only a few of them are capable of the kinetic separation of olefin/paraffin mixtures [17]. In contrast to most of the traditional adsorbents, MOFs often display a stronger affinity for saturated over unsaturated hydrocarbons in single component isotherms and their initial capacity is typically restored by regeneration using pressure-, vacuum-, and/or temperature-swing adsorption (PSA, VSA, and TSA). Employing materials that show this type of selectivity in the separation of olefin/paraffin mixtures leads to high-purity paraffins that

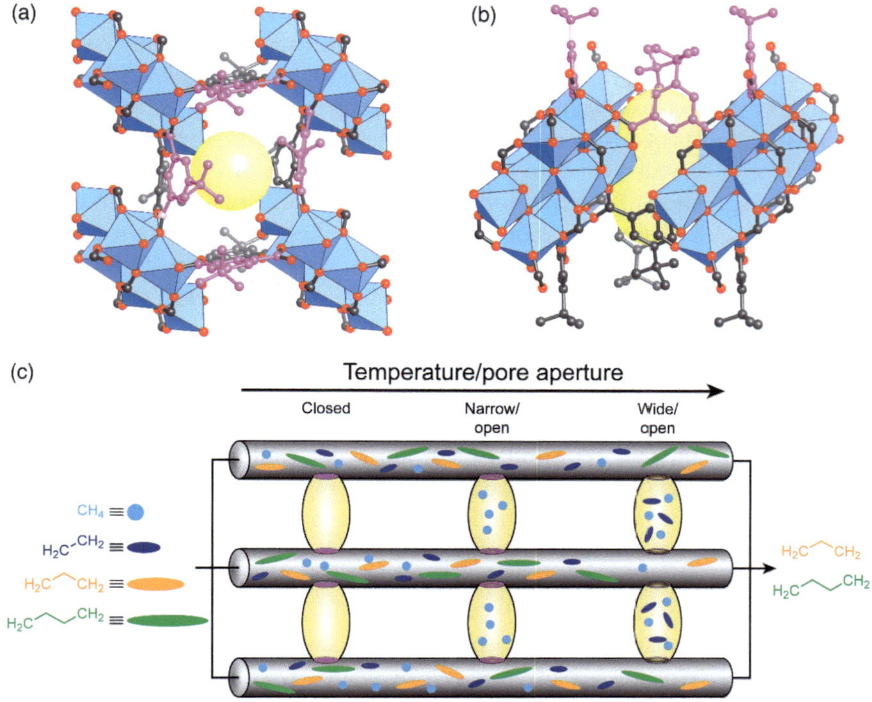

Figure 16.4 (a, b) Top and side view of the intersection connecting the hydrophilic 1D pores and the hydrophobic chambers. The BBDC linkers lining the opening of their intersections are highlighted in pink. (c) Schematic separation of C_1–C_4 alkanes in MAMS-1. The closed pores exclude any of the components from entering the hydrophobic chambers (left). Increasing the temperature results in a larger diameter of the intersection, allowing methane to pass into the hydrophobic chambers (center). A further increase in temperature results in intersection apertures large enough to allow for methane and ethane to enter the hydrophobic chambers (right). Propane and ethane are too large to enter the hydrophobic storage chambers and are excluded.

are required for polymerization processes. Such separation processes can be performed following four different mechanisms: (i) adsorptive or thermodynamic equilibrium separation, (ii) kinetic separation, (iii) separation based on gate-opening effects, and (iv) separation by molecular sieving. In the following, we will discuss these four mechanisms separately and highlight their material requirements.

16.2.2.1 Thermodynamic Separation of Olefin/Paraffin Mixtures

Thermodynamic separation relies on the selective adsorption of one component of the mixture over another. Open metal sites also play an important role in the separation of olefin/paraffin mixtures following a thermodynamic mechanism. This is because light olefins and paraffins such as ethane and ethylene are typically equally polarizable, possess no or only a small dipole moment, and generally have small quadrupole moments, and therefore strong adsorption sites (e.g. open metal sites) are required to increase the selectivity. This is however not always true, as illustrated by the fact that the preferential adsorption of ethylene over ethane in HKUST-1 is mainly correlated to stronger hydrogen bonding between ethylene and the basic oxygen atoms of the secondary building units (SBUs) and only partially to electrostatic interactions with copper open metal sites [18].

The isosteric heats of adsorption for ethane and ethylene differ only by approximately 3 kJ mol^{-1}, which leads to a low selectivity factor of 2 [19]. In a similar manner, the separation of propane and propylene reveals a preferential binding of propylene but with a larger difference in the isosteric heats of adsorption of about −13 kJ mol^{-1} [20]. Here, the stronger adsorption of propylene is attributed to the presence of open metal sites that give rise to an interaction between the bonding p-orbitals in propylene and the vacant copper s-orbitals [20, 21]. Evaluation of the potential of HKUST-1 for the separation of various olefin/paraffin mixtures (ethane/ethylene, propane/propylene, and *iso*-butane/*iso*-butylene) by simulated moving bed, PSA, and VSA reveals preferential adsorption of the unsaturated component [15b, 16, 21b, 22]. Similar behavior is observed for other MOFs with open metal sites. In the separation of olefin/paraffin mixtures using fully activated (Fe)MIL-100, adsorption of the olefin is typically favored [23].

From a practical point of view, alkane-selective adsorbents are favorable because it is hard to recover the desired alkene product from olefin-selective sorbents, and multiple separation cycles must be performed to obtain high-purity products such as those required to produce polymer grade polyethylene and polypropylene. The same separations can be realized within one cycle using alkane-selective adsorbents. There are, however, only very few MOFs that favor the adsorption of paraffins over olefins. One such MOF is MAF-49 [24]. The selectivity of MAF-49 for ethane over ethylene has its origin in the presence of multiple electronegative and electropositive groups that cover the inner pore surface. Ethane can form six C–H···N hydrogen bonds with these groups whereas ethylene can only form four hydrogen bonds. This means that the favored binding of ethane over ethylene is a result of the spatial arrangement of hydrogen bond acceptors in the pores of MAF-49.

To circumvent the adsorption of unsaturated over saturated species, the separation of olefin/paraffin mixtures in MOFs can also be realized under kinetic control, utilizing the gate-opening effect, or by taking advantage of shape selectivity and size exclusion. Separations following these mechanisms typically show lower selectivity. The highest selectivity is observed for adsorptive separation by MOFs with a high density of open metal sites.

16.2.2.2 Kinetic Separation of Olefin/Paraffin Mixtures

The selective separation of olefin/paraffin mixtures based on a kinetic mechanism makes use of the differences in diffusivities of the individual components of a mixture. Whether a separation process is governed by thermodynamics or kinetics can be assessed by analyzing single-component adsorption isotherms and conducting diffusion studies (see Chapter 13). A separation process is controlled by kinetic effects if the uptake and Q_{st} values for all components of the mixture are similar, but their diffusivities differ significantly.

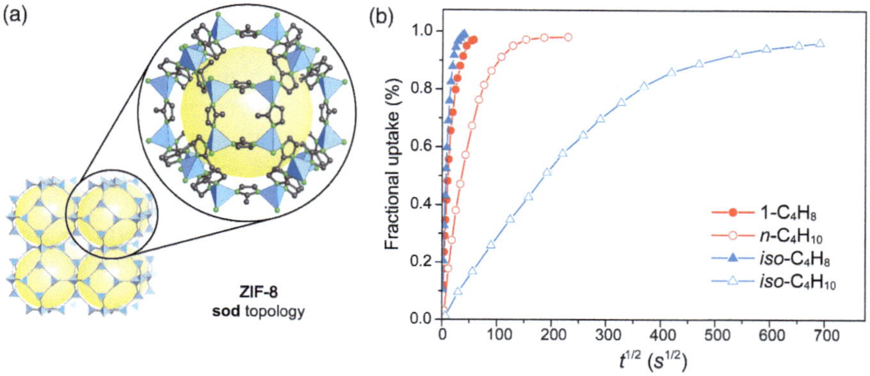

Figure 16.5 (a) Topological representation of the augmented sodalite (**sod**) net and one cage (**tro**) of the structure of ZIF-8. (b) Kinetic uptake curves in ZIF-8 recorded at 35 °C. The red curves represent the kinetic uptake of 1-C_4H_8 (closed symbols) and *n*-C_4H_{10}, and the blue curves represent the kinetic uptake curves of *iso*-C_4H_8 (closed symbols) and *iso*-C_4H_{10} (open symbols), respectively. A higher affinity toward the unsaturated components (1-propylene and *iso*-butylene) compared to the saturated counterparts (*n*-propane and *iso*-butane) is observed.

Kinetic separation is observed for propane/propylene mixtures using three isostructural ZIFs, $Zn(mIM)_2$ (ZIF-8, mIM = 2-methyle imidazolate), $Zn(cIM)_2$ (cIM = 2-chlorolimidazolate), and $Zn(bIM)_2$ (bIM = 2-bromolimidazolate) (Figure 16.5) [25]. Sieving materials such as ZIF-8 show a behavior similar to materials operation on an adsorptive mechanism where generally the unsaturated component is retained due to stronger interactions with its π-system. This finding is correlated to the effective size of the pore opening, giving rise to a significant difference in the diffusion rates for propylene and propane by a factor of 125. Interestingly, C_4 hydrocarbons (*n*-butane, *iso*-butane, *iso*-butene), which are significantly larger than the effective pore aperture for sieving (4.0–4.2 Å), can diffuse into the micropores of ZIF-8. The high kinetic selectivity of ZIF-8 in the separation of *iso*-butene/*iso*-butane and *n*-butane/*iso*-butane mixtures (at 308 K) of up to 180 and 2.5×10^6, respectively, has its origin in the combination of flexibility and dilation of the aperture [26]. Owing to its high performance in the separation of olefin/paraffin mixtures, ZIF-8 is one of the most popular MOFs for the fabrication of MMMs [27].

MOFs with pillared-layered structures are interesting for separation processes because here functionalization can give rise to anistropic diffusion properties. The stacking of 2D layers through pillars (typically N- or O-donor linkers) allows for the adjustment of the interlayer distance, which defines the pore diameter of the channels running parallel to the 2D layers. The pore diameter has a stark influence on the diffusivity of the permeating molecules, and its modification can therefore be utilized to adjust the selectivity in kinetic separation processes. Here we illustrate this concept for the separation of propane/propylene mixtures using a series of isostructural pillared-layered MOFs [28]. The structures of all

MOFs of this particular series are built from **sql** layers of $Zn_2(-COO_2)_4$ paddle wheel SBUs that are connected by BTEB (tetrakis(carboxyphenyl)benzene) linkers, and pillared along the *c*-direction by *R*-BPEE (dipyridylethene derivatives). The layered nature of these materials results in a plate-like crystal morphology (Figure 16.6a,b). The combination of different substitution patterns of both linkers gives rise to a series of MOFs with different pore apertures that can be controlled individually (Figure 16.6b). The modulation of the pore apertures has a significant effect on the kinetic selectivity in the separation of propane/propylene mixtures. The channels in the highly anisotropic structures have different pore

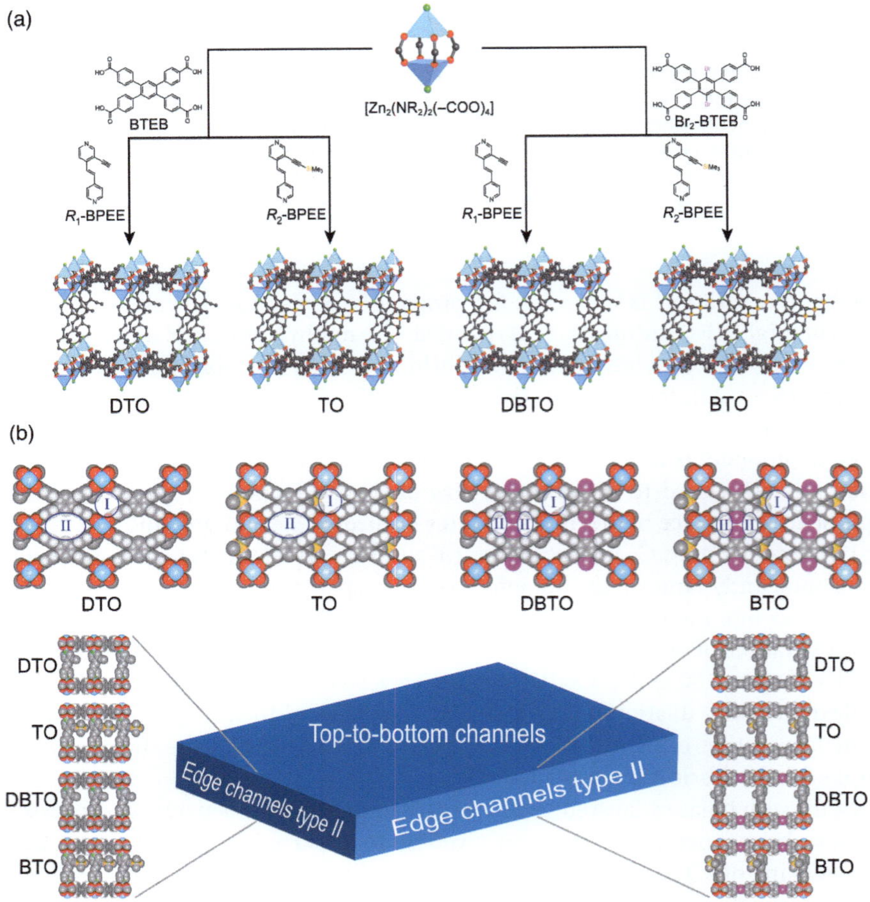

Figure 16.6 (a) Crystal structures of a series of isoreticular functionalized pillared-layered MOFs. The structures are built from **sql** layers of zinc paddle wheel SBUs that are connected by tetratopic BTEB or Br_2-BTEB linkers. These layers are pillared by pyridine-based struts (*R*-BPEE) to form the 3D frameworks DTO, TO, DBTO, and BTO (**fsc** topology). (b) All four compounds form crystals of plate-like morphology. Owing to the morphology of the crystals separation mainly occurs in the small channels (I) and (II), the larger edge-to-edge channels do not contribute to the separation. A top view of all types of channels is given and their orientation in the crystal is indicated.

diameters and run along all three directions of the plate-like crystals. Because the exposed surface of the different faces of the plate-like crystals are not identical, separation occurs mainly in the small channels I and II that run parallel to the BPEE linkers, from the top to the bottom face of the crystal (Figure 16.6b). The aperture of channel II is controlled by the substitution pattern of the BTEB linker while the apertures of channel I and the edge channels are controlled by substituents appended to the BPEE linker. Frameworks that are prepared using Br_2-BTEB display a higher selectivity in the separation of propane/propylene mixtures, which is attributed to the decreased pore aperture of channel II. This is supported by the fact that grinding the material leads to a significantly lower selectivity caused by the reduced "top-to-edges surface" ratio of the crystals.

16.2.2.3 Separation of Olefin/Paraffin Mixtures Utilizing the Gate-Opening Effect

Separation utilizing the gate-opening effect of framework structures also relies on differences in the diffusivity in a way similar to the kinetic separation mechanism outlined above. In contrast to this mechanism, the selectivity for paraffins is attributed to a gate-opening effect that controls the uptake and release of specific molecules at specific gate-opening pressures. In this context, ZIFs are interesting materials that can perform the separation of light hydrocarbons by gate-opening. ZIF-7 ($Zn(BIM)_2$, BIM = benzimidazolate) selectively adsorbs paraffins over olefins. This selectivity arises from interactions between the adsorbate molecules and the benzene rings of the BIM linkers that point into the narrow pore windows [29]. The gate-opening effect triggered by this interaction results in selective discrimination between molecules of similar size but (slightly) different shapes and facilitates the rapid desorption of the adsorbed species at relatively low temperatures. The difference in the adsorption behavior of different adsorbents is mainly correlated to their ability to form "adsorption complexes" at the external surface of the pore openings of ZIF-7 [29a, 30].

RPM3-Zn ($Zn_2(BPDC)_2(BPEE)$), a pillared-layered MOF, is another example for selective separation of ethane/ethylene mixtures based on the gate-opening effect. RPM3-Zn is built from dinulcear $Zn_2(-COO_2)_4$ paddle wheel SBUs that are connected by BPDC linkers to form **sql** layers, which are pillared by BPEE struts to form a framework with an overall **pcu** topology [31]. The adsorption isotherms for olefins and paraffins both show stepwise adsorption and pronounced hysteresis at specific gate-opening pressures and the gate-opening pressure itself strongly depends on the chain length of the adsorbate [32]. Gate-opening in RPM3-Zn is caused by hydrogen bonding between the methylene groups of ethylene and the end-on coordinated carboxylate oxygen of the BPDC linker as evidenced by Raman spectroscopy and density functional theory calculations.

16.2.2.4 Separation of Olefin/Paraffin Mixtures by Molecular Sieving

The presence of pores that are significantly smaller than the kinetic diameter of at least one of the components of the gas mixture can be used for its separation by molecular sieving. While ZIF-7 and ZIF-8 only show high selectivity up to a characteristic gate-opening pressure, materials with pore apertures that allow for the selective exclusion of molecules above a specific size limit (i.e. size exclusion) must be designed to facilitate this kind of selectivity over a larger pressure range. One such material is KAUST-7 (Ni(Pyr)$_2$(NbOF$_5$), also referred to as NbOFFIVE-1-Ni, where Pyr = pyrazine)) [33].[1] KAUST-7 has a pillared-layered structure built from **sql** layers of 6-coordinated nickel centers that are connected through Pyr linkers. These layers are pillared by inorganic NbOF$_5$-units to form a 3D framework of **pcu** topology and shows structural similarity to SIFSIX-3-Ni (Ni(Pyr)$_2$(SiF$_6$)) (Figure 16.7). The bulkier NbOF$_5$ units in the structure of KAUST-7 result in smaller pore apertures (3.047 Å) compared to that of SIFSIX-3-Ni (5.032 Å). The restricted pore size facilitates the exclusion of propane molecules whereas the slightly smaller propylene molecules can diffuse through the material. Consequently, KAUST-7 displays complete molecular exclusion of propane from a propane/propylene mixture at ambient temperature and atmospheric pressure.

16.2.3 Separation of Aromatic C$_8$ Isomers

Aromatic C$_8$ hydrocarbons such as xylene isomers and ethylbenzene are important feedstock chemicals for the synthesis of polymers and other value-added chemicals. Therefore, their selective and efficient separation is of great interest for the petrochemical industry. Figure 16.8 shows aromatic C$_8$ hydrocarbons and industrial products synthesized from them. Phthalic anhydride, a plasticizer in polymers, is synthesized from *o*-xylene. The oxidation of *m*-xylene yields isophthalic acid, more recently used in the synthesis of PET resin blends. *p*-Xylene (PX) is used to produce terephthalic acid, a basic component for the production of PET. Ethylbenzene is dehydrogenated to styrene and subsequently polymerized to give polystyrene.

The separation of aromatic C$_8$ hydrocarbons by distillation is energy intensive and requires enormous columns with about 150–200 plates and a high reflux ratio [34]. Research on adsorptive separation of aromatic C$_8$ hydrocarbons using zeolites indicates that this process may provide a more economical alternative. Consequently, MOFs and ZIFs have been intensely studied with respect to this application.

In the separation of a mixture of aromatic C$_8$ hydrocarbons ZIF-8 shows a selectivity of 4.0 and 2.4 for mixtures of *p*-/*o*-xylene and *p*-/*m*-xylene, respectively (see Figure 16.5) [35]. Considering that the pore aperture of ZIF-8 is approximately 3.4 Å in diameter, none of the aromatic C$_8$ hydrocarbons should be able to diffuse into the pores. However, the imidazolate linkers lining the 6-membered rings in the sodalite (**sod**) structure of ZIF-8 can act like saloon doors, resulting in an increase in the effective pore aperture to about 6.4 Å. This value is close to the kinetic diameter of *p*-xylene and thus enables the separation of xylene isomers owing to a significant difference in their diffusivities [36].

1 Strictly speaking, KAUST-7 is a coordination network, since its structure is built from single metal nodes that are connected through neutral pyrazine linkers to form **sql** layers. The pillar is an inorganic unit.

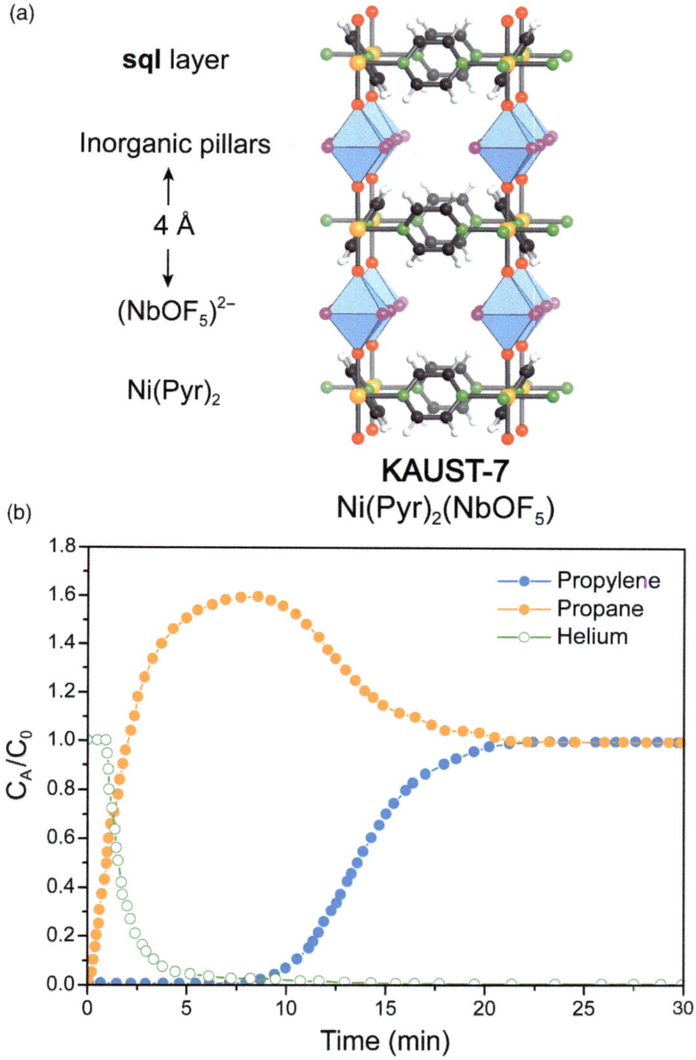

Figure 16.7 (a) Crystal structure of KAUST-7 highlighting the pillared-layered structure. The pyrazine linkers are tilted due to the use of the large $(NbOF_5)^{2-}$ leading to a short distance between neighboring two $(NbOF_5)^{2-}$ building units that belong to adjacent layers. (b) Breakthrough curve for the separation of a propane/propylene mixture illustrating the high selectivity of KAUST-7 for propylene over propane. Color code: Nb, blue; Ni, orange; F/O, red; O, pink; C, gray; N, green.

Compared to ZIFs, MOFs offer a higher degree of structural and functional tunability owing to the almost infinite number of possible SBU–linker combinations. This structural diversity offers great potential for the discovery of materials with high selectivity in the separation of aromatic C_8 hydrocarbons. To design para-selective MOFs, (computational) screening of MOF structures constructed from cages that have apertures in the size regime of the kinetic diameter of xylene

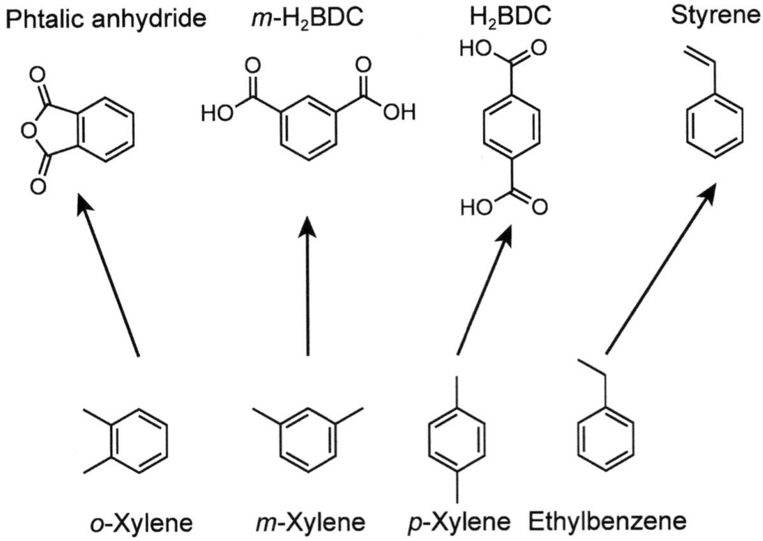

Figure 16.8 Aromatic C_8 hydrocarbons (bottom) alongside industrial products prepared therefrom (top). All compounds shown in the top row are important constitutents of polymers and thus produced in large quantities.

isomers is instructive. In the following, we will elaborate on selected MOFs that fulfill this requirement.

The 1D channels in JUC-77, (In(OH)(OBA) where H_2OBA = 4,4′-oxybis (benzoic acid)) have openings of 10.8 × 7.3 Å, which is close to the kinetic diameters of the xylene isomers. In the separation of xylene isomers, these channels act as molecular sieves that only allow the diffusion of *p*-xylene (the isomer with the smallest width) whereas *m*-xylene and *o*-xylene are excluded due to their larger size [37]. (Ti)MIL-125 and some of its derivatives also show potential for C_8 separation [38]. Here, the high para-selectivity originates from unique structural features that allow for more efficient packing of one component over the other. The structure of MIL-125 and its derivatives contains two types of cages; a large octahedral ($d = 12.5$ Å) and a smaller tetrahedral ($d \sim 6$ Å) cage (see Figure 4.32). Both cages are connected through narrow trigonal windows with an aperture of approximately 6 Å. (Ti)MIL-125-NH_2 displays a preference for *p*-xylene over the other xylene isomers, which decreases in the presence of ethylbenzene and strongly depends on the exact composition of the feed mixture [39]. Other examples relying on differences in the stacking efficiencies of different aromatic C_8 hydrocarbons are known; however, it is difficult to design MOFs showing this type of selectivity because it is highly dependent on the exact geometry and chemical nature of the pores. Among MOFs with the para-selectivity originating from differences in the packing efficiency, MAF-X8 is one of the best performers in the separation of *o*-xylene/*m*-xylene/*p*-xylene/ethylbenzene mixtures [40]. This is attributed to the geometry of the channels in the structure of MAF-X8 that allows for commensurate stacking of *p*-xylene. Calculations suggest that MAF-X8 can outperform state-of-the-art zeolites (e.g. BaX) currently used in industry for this type of separation, highlighting the potential of MOFs with respect to separation processes.

Another way to realize a high selectivity is to utilize shape-selective effects. Such effects are found for the C_8 aromatics separation using UiO-66 (see Figure 4.28). Molecular simulations and breakthrough experiments show that UiO-66 is highly ortho-selective, while the selectivity in the separation of *p*-xylene/*m*-xylene mixtures is modest [41]. The ortho-selectivity has its origin within the structure of UiO-66. Its **fcu** type structure contains octahedral ($d = 11$ Å) and tetrahedral cages ($d = 8$ Å) connected through narrow windows ($d = 5$–7 Å). The diameter of the tetrahedral cage is within the range of the kinetic diameter of ethylbenzene (6.7 Å), *o*-xylene (7.4 Å), *m*-xylene (7.1 Å), and *p*-xylene (6.7 Å) [42]. The stronger interaction with the bulkier ortho-isomer leads to its preferential adsorption and a selectivity pattern that is inverse to the molecular dimensions [43].

16.2.4 Mixed-Matrix Membranes

With respect to gas separation, membranes offer a number of benefits over packed beds of powders or shaped bodies and other gas separation technologies such as cryogenic distillation or selective condensation, requiring an energy-intensive gas–liquid phase change [44]. Gas separation using membranes does not require a phase change while also allowing for a more compact plant design and operation under continuous steady-state conditions, thus eliminating the need for regeneration of the separation medium. The general principle of membrane separation is shown in Figure 16.9. A feed-stream is passed along one side of the membrane and only one component permeates the membrane while all non-permeating components remain in the feed-stream. The separation process is driven by a pressure difference and the mechanism is strongly dependent on the nature of the material used to manufacture the membrane.

In separation processes using materials with a well-defined pore system such as MOFs and ZIFs adsorption, diffusion, and sometimes molecular sieving dominate the performance of the membrane. In contrast, the separation of gaseous mixtures using polymer membranes is governed by solution-diffusion processes. Early membrane systems include simple materials such as anisotropic cellulose acetate membranes that are applied in the separation of carbon dioxide from natural gas [45]. Today, polymeric membranes are used in many industrial processes, mainly in large-scale gas separation, owing to their facile processing and mechanical strength [46]. Even though polymeric membranes find widespread application they often suffer from low chemical and thermal stability. Additionally, their performance is limited by the Robeson upper boundary, which describes the trade-off between permeability and selectivity [47]. "Inorganic" membranes are made from materials such as zeolites, carbons, and more recently MOFs and ZIFs [48]. They are grouped into two categories: (i) porous inorganic membranes and (ii) dense (nonporous) inorganic membranes. On one hand, these membranes offer unique properties for gas separation such as their high thermal and chemical stability, high gas flux, and high selectivity.

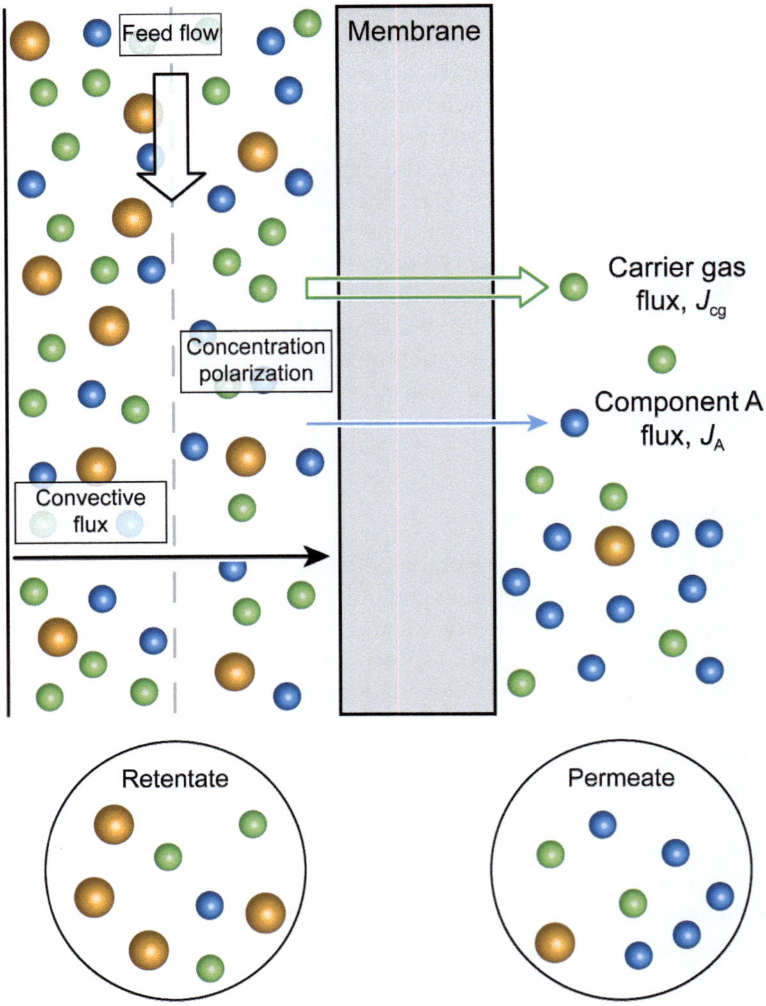

Figure 16.9 Schematic representation of the membrane separation process. A gas mixture that contains three components (gas A, blue; gas B, orange; carrier gas, green) is fed to the separation process. Concentration polarization leads to an increase of component A (c_A) close to the membrane. Gas A permeates the membrane with a larger flux than the carrier gas ($J_A > J_{cg}$) and the membrane is almost impermeable for gas B. The retentate is therefore mainly composed of gas B, whereas the permeate contains mainly gas A.

On the other hand, important factors such as reproducibility, long-term stability, and scalability require further development.

The limitations of polymeric and inorganic membranes are overcome by blending both components in so-called "mixed matrix membranes" (MMMs). These membranes consist of a blend of a polymeric matrix and filler particles (Figure 16.10). This allows them to overcome the Robeson upper boundary for polymeric membranes while avoiding inherent drawbacks common to inorganic membranes such as their brittleness. MMMs combine the facile

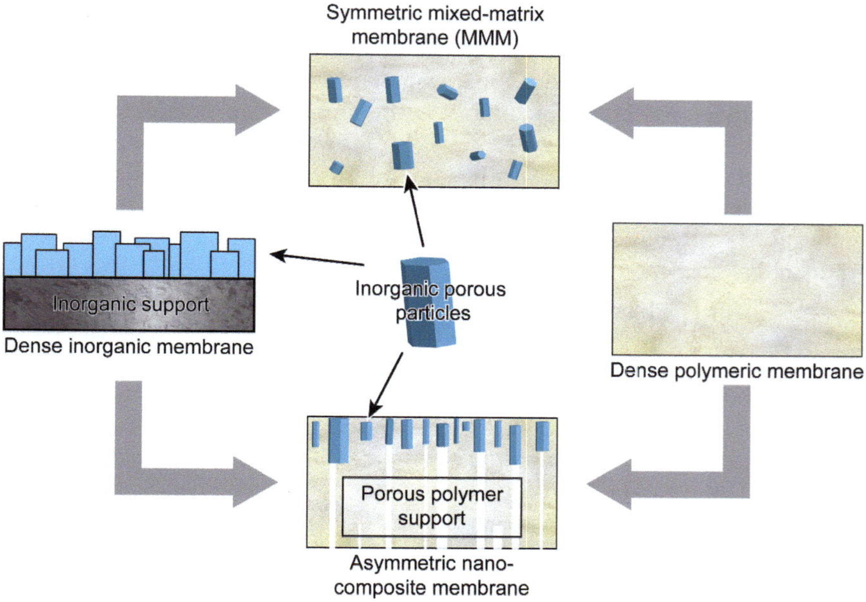

Figure 16.10 Schematic representation of different types of membranes. Both the asymmetric nanocomposite membrane, and the mixed-matrix membrane combine the advantages of polymeric membranes and inorganic porous materials, thereby eliminating the drawbacks presented by both individual methods.

processability and flexibility of polymeric membranes, while augmenting their limited permeability and selectivity with that of the inorganic component (e.g. MOF or ZIF) [45, 46, 47b, 48c, 49]. MOFs and ZIFs are both ideal materials for use as fillers in MMMs. Their synthesis in the form of porous nanoparticles is comparatively facile while the prospect of functionalization allows for the adjustment of structural and electronic features. Current research in this field is focused on enhancing the chemical compatibility between the polymer phase and the porous filler. This is typically realized by modulation of the particle size and size distribution, changes in morphology, judicious functionalization of the organic backbone, and surface modifications. The basic principles of separation are similar to those described "earlier" or "previously"; hence, we will discuss only one representative example. Many review articles addressing MMMs have been published, and the interested reader is referred to these articles [2a, 48c, 49a].

As we saw earlier, many ZIFs are promising candidates in diverse separation processes since they feature certain similarities to zeolites, which have been intensely studied as inorganic fillers for MMMs. ZIFs provide high chemical and thermal stability combined with high selectivity in many separation processes. MMMs of low thickness with enhanced performance and high selectivity compared to their individual components are prepared by embedding submicrometer ZIF-90 ($Zn(aIM)_2$) crystals into poly(imide) membranes [50]. While the adhesion of MOF and ZIF nanoparticles to the polymer blend is commonly challenging, in the case of poly(imide)-ZIF-90 MMMs good adhesion is achieved without any surface modifications.

16.3 Separation in Liquids

The application of MOFs and ZIFs in separation processes is not limited to gaseous mixtures. An increasing interest in the capture of molecules from liquid phase has emerged over the last decade. Two of the most promising applications in this regard are the adsorptive purification of waste water (removal of biologically active molecules and biomolecules) and the adsorptive purification of fuels (removal of cyclic amines), which are both currently performed on a large scale in outdated energy-intensive processes.

As a consequence of the rising world population and the concomitant growing use of medicinal drugs, the concentration of these bioactive molecules and their metabolic products in the water cycle steadily increases. This includes the contamination of surface water, wastewater, groundwater, and to a lesser extent also drinking water [3a]. This not only poses a direct health concern to humans but also has a dramatic impact on the ecosystem [3b–d]. Drinking water is typically purified by filtration, disinfection, and other treatments, such as the removal of organic biologically active molecules using O_3, H_2O_2, UV (ultraviolet) radiation, and their photocatalytic decomposition on TiO_2 or modified TiO_2 surfaces [51]. The more facile nature of adsorptive water purification can substantiate or even substitute these methods and presents the prospect of applying this type of treatment to river or sea water since it does not require a sophisticated infrastructure [52].

Liquid raw fuels typically contain cyclic amines that have adverse effects on the fuels' properties including bad smell, high toxicity, carcinogenicity, and their tendency to form deposits. The combustion of cyclic amines produces nitrous oxides (NO_x) that are one of the main causes of acid rain. Therefore, the removal of nitrogen-containing organic compounds from refinery streams is an important industrial process that is realized by catalytic hydrodenitrogenation (HDN) [53]. This process requires elevated temperatures and pressures, rendering it energy and cost intensive. Additionally, it is accompanied by an undesired decrease in the research octane number (RON) of the fuel. When combining these drawbacks with the fact that the heterocyclic compounds that are removed by HDN have a wide range of applications, it appears profitable to separate them in adsorption-based processes. Such processes not only provide a more energy-efficient alternative to HDM but they also do not result in a decreased RON and allow for the recovery of the cyclic amines.

In the following text we will outline the demands for porous absorbents in both processes regarding their structural features, toxicity, and performance, and give illustrative examples that elaborate the separation mechanisms.

16.3.1 Adsorption of Bioactive Molecules from Water

16.3.1.1 Toxicity of MOFs

For the application of MOFs in the purification of drinking water it is of utmost importance for the material to be hydrolytically stable to ensure that chemicals are not released into the drinking water by dissolution of the MOF. The SBUs in MOFs are often built from heavy metals with relatively high toxicity and many of the linkers used in MOF synthesis are harmful to humans and aquatic organisms. Therefore, the selection of MOFs constructed from building units that are non-hazardous to health and are stable under application conditions is as important as high selectivity and uptake.

Examples of linker molecules with low toxicity are H_2BDC and H_3BTC [54]. Similarly, metal ions with a high LD_{50} such as Fe, Al, Ti, Mg, and Ca must be selected when targeting MOFs with low toxicity [54b, 55]. Many nontoxic metals tend to form (purely) ionic bonds with carboxylates that render the corresponding MOFs hydrolytically unstable. A more detailed discussion of the hydrolytic stability of MOFs is given in Chapter 17. When nontoxic metal ions and linkers are combined to form framework structures, the resulting MOFs are also expected to display a low overall toxicity. This makes them promising materials for the adsorption-based separation of bioactive molecules in water.

16.3.1.2 Selective Adsorption of Drug Molecules from Water

The adsorption of medical drugs and their metabolic products from water using nontoxic MOFs is not only of interest for the purification of water but, in combination with the appropriate hydrolytic stability/instability, also for the controlled release of drugs in drug delivery systems [54b]. Many studies on adsorption-based separation in aqueous media have been performed using Fe-based MOFs that belong to the MIL series (e.g. MIL-53, MIL-88, MIL-100, and MIL-101) due to their low toxicity and high hydrolytic stability (Figure 16.11).

Figure 16.12 shows a range of drug molecules whose removal from aqueous solution using MOFs has been studied. Naproxene is a nonsteroidal anti-inflammatory drug available as over-the-counter medication, clofibric acid is an herbicide and plant growth regulator, furosemide is a diuretic medication with some veterinary use, sulfalazine is an anti-rheumatoid drug, doxorubine is an anticancer drug, and roxarsone and *p*-arsanilic acid have been used in growing poultry. Studies of the adsorption of these compounds show higher uptakes in MIL-100 and MIL-101 than for other common adsorbents such as activated carbon [56].

The uptake of drug molecules is often correlated to the pH of the solution. This is due to interactions arising from pH-dependent structural changes in both the

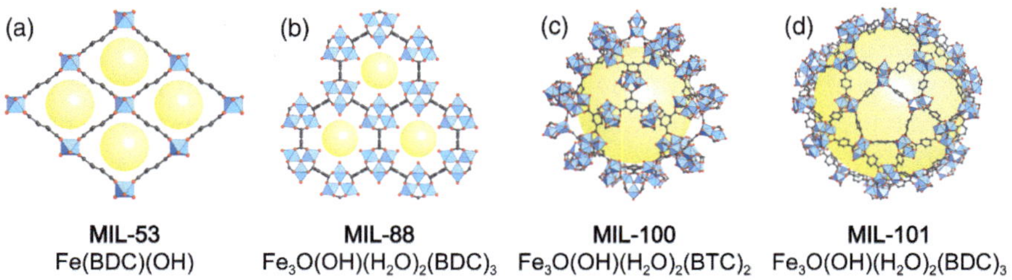

Figure 16.11 Biodegradable porous iron-based frameworks MIL-53, MIL-88, MIL-100 and, MIL-101 (a–d). All MOFs shown are built from iron-based SBUs and nontoxic organic linkers. The large cages of MIL-100 and MIL-101 make them especially interesting for the adsorptive removal of large drug molecules from water.

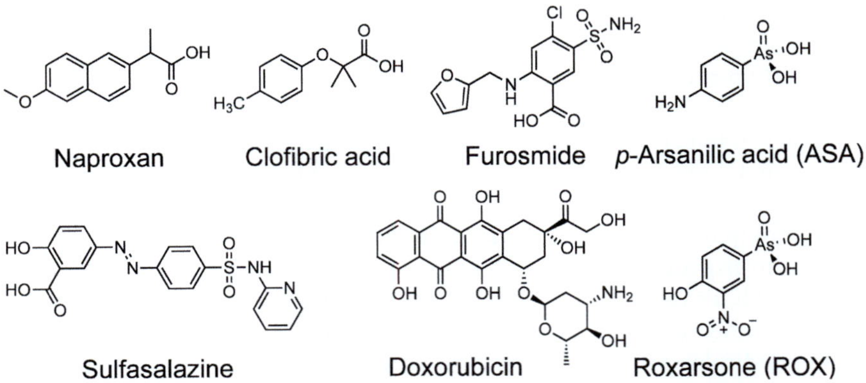

Figure 16.12 Compilation of drug molecules found in drinking water. These molecules are used in farming and human and veterinarian medication.

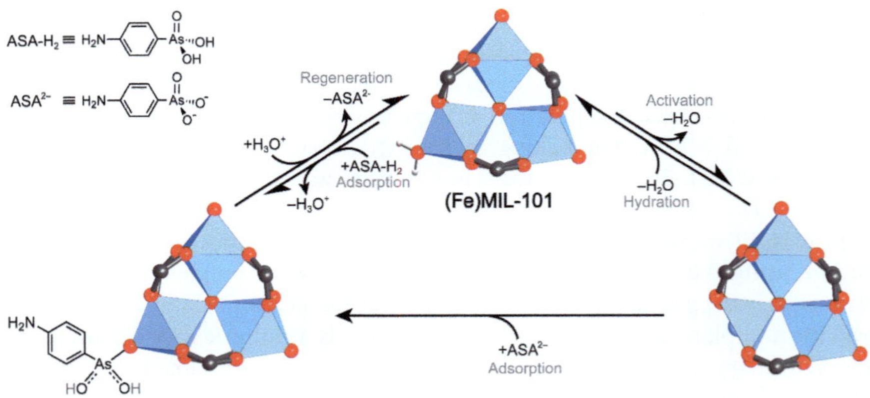

Figure 16.13 Proposed mechanism for the capture of *p*-arsanilic acid (ASA) by (Fe)MIL-101. ASA is adsorbed on the open metal sites of the trinuclear $Fe_3O(H_2O)_2(L)(-COO)_6$ SBU through an equilibrium reaction with water. Controlled release is realized by washing with acidic ethanol and affords the pristine hydrated form of the SBU.

adsorbent and the adsorbate. In most cases, the adsorption of these molecules takes place on the metal centers of the SBU. Figure 16.13 illustrates the mechanism for the adsorption of *p*-arsanilic acid on (Fe)MIL-100 [57].

Choosing a pH range that allows for the existence of a charged form of the adsorbate typically results in a higher uptake. Surface modifications can be used to modify the uptake and adsorption behavior. Such modifications are achieved in a manner akin to that described in Chapters 6, 13, and 14. MIL-101 with ethylenedamine grafted to its open metal sites shows an increased uptake of naproxen due to the stronger interaction of the deprotonated form of naproxen and the protonated amines dangling in the pores at pH of 3–4. In contrast, grafting aminomethanesulfonic acid (taurine) to the open metal sites in the same MOF leads to a significant decrease in the adsorption capacity due to the emergence of repulsive forces.

16.3.1.3 Selective Adsorption of Biomolecules from Water

MOFs can adsorb biomolecules from aqueous solutions, including compounds used in the diagnosis of diseases such as creatinine, a uremic toxin used as a probe for renal failure, and naturally occurring sugars. Of special interest is the separation of xylose. Xylose is a feedstock for the industrial production of furfural, which is further converted to furfuryl alcohol and subsequently used in the production of furan resins [58]. C_5 and C_6 sugars with *cis*-diols (glucose, mannose, and galactose) are known to interact strongly with boronic acid groups ($-BO_2H_2$) [59]. This principle can be implemented into MOF structures by partial replacement of carboxylate-based linkers with analogs bearing boronic acid groups. A MOF designed with this principle in mind is MIL-100(BO_2H_2), where the BTC linker is partially substituted by ditopic H_4BBDC (5-boronobenzene-1,3-dicarboxylate). MIL-100(BO_2H_2) is prepared from metallic chromium, H_4BBDC, and H_3BTC in water and in the presence of HF under hydrothermal conditions. At elevated pH (around 9) it selectively adsorbs galactose from a mixture of xylose, glucose, mannose, and galactose. Its crystallinity and 87% of its initial capacity are retained after regeneration under acidic conditions [60].

16.3.2 Adsorptive Purification of Fuels

HDN is used to remove cyclic amines from liquid fossil fuels [53]. In this process the cyclic amines are converted into ammonia and hydrocarbons. Since *N*-heterocyclic compounds find widespread application in industrial processes, their removal by adsorption and subsequent recovery by desorption appear promising. The adsorptive removal of aromatic *N*-heterocyclic compounds from liquid fossil fuels has been extensively studied since the early 2000s. More recently, the applicability of MOFs and ZIFs in this process has been explored [61]. Here, we will present selected examples illustrating ways to realize the removal of cyclic amines from liquid fuels using MOFs.

16.3.2.1 Aromatic *N*-Heterocyclic Compounds

Aromatic *N*-heterocycles (ANHs) are either basic or nonbasic. In basic ANHs, the lone pair of nitrogen is located in the plane of the molecule and has a basic nucleophilic character. In contrast, the lone pair in nonbasic ANHs is part of the heteroaromatic system and consequently perpendicular to the plane of the molecule. Figure 16.14 shows a selection of basic and nonbasic ANHs that are typically found in liquid fossil fuels.

16.3.2.2 Adsorptive Removal of Aromatic *N*-Heterocycles

Based on the nature of the ANH, different mechanisms predominate for the adsorption. Basic ANHs interact preferentially with Lewis acidic open metal sites. Therefore, the magnitude of the isosteric heat of adsorption in a series of isoreticular frameworks strongly depends on the metal constituting the SBUs. The number of adsorbed molecules is larger than that of the open metal sites. This indicates that there are additional secondary adsorption sites for basic ANHs. In contrast, nonbasic ANHs are adsorbed mainly through hydrogen bonding (polar interactions). This is illustrated by the fact that the uptake

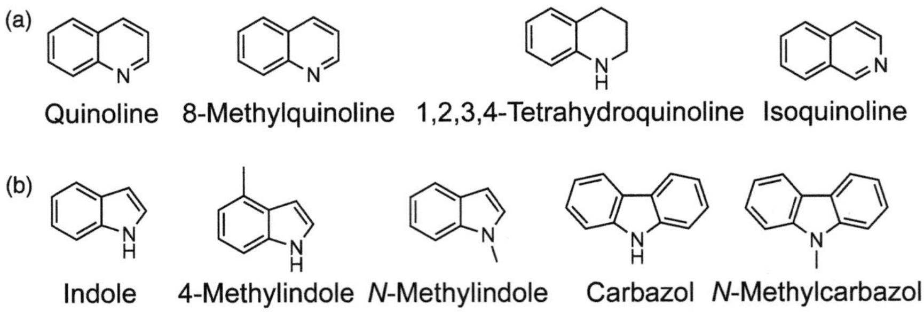

Figure 16.14 Basic (a) and nonbasic (b) aromatic *N*-heterocycles typically found in liquid fossil fuels. While the nitrogen lone pair is located in the plane of the molecule of basic heterocyclic amines, it is part of the aromatic system in nonbasic *N*-heterocycles.

capacity of nonbasic ANHs for a given MOF depends on the polarity of the solvent it is carried out in. The highest uptake is recorded in the presence of the least polar solvents (e.g. *n*-octane). MOFs that are impregnated with acidic molecules such as Keggin type polyoxometallates (e.g. PTA = phosphotungstic acid) display a higher uptake for basic ANH compounds than the pristine MOF. This is due to the stronger acid–base interactions compared to the non-functionalized MOF, whereas a decrease in uptake for nonbasic ANHs results from the smaller accessible pore space.

16.4 Summary

In this chapter, we discussed adsorptive separation processes of gaseous and liquid mixtures. We showed that three different mechanisms can afford the separation of gaseous mixtures (e.g. hydrocarbon mixtures, olefin/paraffin, aromatic C_8 hydrocarbons). The pore size can be adjusted to exclude certain molecules (size exclusion), lead to dissimilar diffusivities for different component of the mixture (kinetic separation), or allow all molecules to diffuse freely in which case the separation is based on adsorption–desorption equilibria (thermodynamic separation). This tailorability of the pore system can in certain structures be carried out anisotropically, which results in anistropic selectivity. The gate opening effect was in many cases shown to increase the selectivity or even enable separation processes that would not be possible in rigid materials. We outlined how the combination of a polymeric membrane and an MOF/ZIF filler can produce MMMs with properties surpassing those of the individual components. We illustrated how MOFs can be used to trap bioactive molecules from water and ANHs from liquid fuels and highlighted the importance of the structural and chemical tunability with respect to these applications.

References

1 (a) Lathia, R.V., Dobariya, K.S., and Patel, A. (2017). *Hydrogen Fuel Cells for Road Vehicles*. Elsevier. (b) Kramer, U., Lorenz, T., Hofmann, C. et al. (2017). Methane number effect on the efficiency of a downsized, dedicated, high performance compressed natural gas (CNG) direct injection engine. SAE Technical Paper 2017-01-0776.

2 (a) Dechnik, J., Gascon, J., Doonan, C. et al. (2017). New directions for mixed-matrix membranes. *Angewandte Chemie International Edition* 56 (32): 9292–9310. (b) Zornoza, B., Tellez, C., Coronas, J. et al. (2013). Metal organic framework based mixed matrix membranes: an increasingly important field of research with a large application potential. *Microporous and Mesoporous Materials* 166: 67–78. (c) Jeazet, H.B.T., Staudt, C., and Janiak, C. (2012). Metal-organic frameworks in mixed-matrix membranes for gas separation. *Dalton Transactions* 41 (46): 14003–14027.

3 (a) World Health Organization (2012). Pharmaceuticals in drinking water. http://www.who.int/water_sanitation_health/publications/pharmaceuticals-in-drinking-water/en/; (accessed 23 March 2018) (b) Cunningham, V.L., Buzby, M., Hutchinson, T. et al. (2006). *Effects of Human Pharmaceuticals on Aquatic Life: Next Steps*. ACS Publications. (c) Fent, K., Weston, A.A., and Caminada, D. (2006). Ecotoxicology of human pharmaceuticals. *Aquatic Toxicology* 76 (2): 122–159. (d) Kidd, K.A., Blanchfield, P.J., Mills, K.H. et al. (2007). Collapse of a fish population after exposure to a synthetic estrogen. *Proceedings of the National Academy of Sciences* 104 (21): 8897–8901.

4 Gaffney, J.S., Streit, G.E., Spall, W.D., and Hall, J.H. (1987). Beyond acid rain. Do soluble oxidants and organic toxinsinteract with SO_2 and NO_x to increase ecosystem effects? *Environmental Science & Technology* 21 (6): 519–524.

5 (a) Szmant, H.H. (1989). *Organic Building Blocks of the Chemical Industry*. Wiley. (b) Weissermel, K. (2008). *Industrial Organic Chemistry*. Wiley.

6 Tomás, R.A.F., Bordado, J.C.M., and Gomes, J.F.P. (2013). *p*-Xylene oxidation to terephthalic acid: a literature review oriented toward process optimization and development. *Chemical Reviews* 113 (10): 7421–7469.

7 Jiang, J. and Sandler, S.I. (2006). Monte Carlo simulation for the adsorption and separation of linear and branched alkanes in IRMOF-1. *Langmuir* 22 (13): 5702–5707.

8 Gutiérrez, I., Díaz, E., Vega, A., and Ordóñez, S. (2013). Consequences of cavity size and chemical environment on the adsorption properties of isoreticular metal-organic frameworks: an inverse gas chromatography study. *Journal of Chromatography A* 1274: 173–180.

9 Farrusseng, D., Daniel, C., Gaudillere, C. et al. (2009). Heats of adsorption for seven gases in three metal-organic frameworks: systematic comparison of experiment and simulation. *Langmuir* 25 (13): 7383–7388.

10 Düren, T. and Snurr, R.Q. (2004). Assessment of isoreticular metal-organic frameworks for adsorption separations: a molecular simulation study of methane/*n*-butane mixtures. *The Journal of Physical Chemistry B* 108 (40): 15703–15708.

11 Düren, T., Sarkisov, L., Yaghi, O.M., and Snurr, R.Q. (2004). Design of new materials for methane storage. *Langmuir* 20 (7): 2683–2689.

12 Ma, S., Wang, X.-S., Collier, C.D. et al. (2007). Ultramicroporous metal-organic framework based on 9,10-anthracenedicarboxylate for selective gas adsorption. *Inorganic Chemistry* 46 (21): 8499–8501.

13 (a) Trung, T.K., Trens, P., Tanchoux, N. et al. (2008). Hydrocarbon adsorption in the flexible metal organic frameworks MIL-53 (Al, Cr). *Journal of the American Chemical Society* 130 (50): 16926–16932. (b) Salles, F., Ghoufi, A., Maurin, G. et al. (2008). Molecular dynamics simulations of breathing MOFs: structural transformations of MIL-53(Cr) upon thermal activation and CO_2 adsorption. *Angewandte Chemie International Edition* 120 (44): 8615–8619.

14 Ma, S., Sun, D., Wang, X.S., and Zhou, H.C. (2007). A mesh-adjustable molecular sieve for general use in gas separation. *Angewandte Chemie International Edition* 46 (14): 2458–2462.

15 (a) Li, J.-R., Kuppler, R.J., and Zhou, H.-C. (2009). Selective gas adsorption and separation in metal-organic frameworks. *Chemical Society Reviews* 38 (5): 1477–1504. (b) Li, B., Wang, H., and Chen, B. (2014). Microporous metal-organic frameworks for gas separation. *Chemistry – An Asian Journal* 9 (6): 1474–1498. (c) Bao, Z., Chang, G., Xing, H. et al. (2016). Potential of microporous metal-organic frameworks for separation of hydrocarbon mixtures. *Energy & Environmental Science* 9 (12): 3612–3641. (d) Wang, Y. and Zhao, D. (2017). Beyond equilibrium: metal-organic frameworks for molecular sieving and kinetic gas separation. *Crystal Growth & Design* 17 (5): 2291–2308. (e) Herm, Z.R., Bloch, E.D., and Long, J.R. (2013). Hydrocarbon separations in metal-organic frameworks. *Chemistry of Materials* 26 (1): 323–338.

16 Eldridge, R.B. (1993). Olefin/paraffin separation technology: a review. *Industrial & Engineering Chemistry Research* 32 (10): 2208–2212.

17 (a) Padin, J., Rege, S.U., Yang, R.T., and Cheng, L.S. (2000). Molecular sieve sorbents for kinetic separation of propane/propylene. *Chemical Engineering Science* 55 (20): 4525–4535. (b) Da Silva, F.A. and Rodrigues, A.E. (1999). Adsorption equilibria and kinetics for propylene and propane over 13X and 4A zeolite pellets. *Industrial & Engineering Chemistry Research* 38 (5): 2051–2057. (c) Rege, S.U., Padin, J., and Yang, R.T. (1998). Olefin/paraffin separations by adsorption: π-complexation vs. kinetic separation. *AIChE Journal* 44 (4): 799–809. (d) Palomino, M., Cantín, A., Corma, A. et al. (2007). Pure silica ITQ-32 zeolite allows separation of linear olefins from paraffins. *Chemical Communications* (12): 1233–1235. (e) Da Silva, F. and Rodrigues, A. (2001). Propylene/propane separation by VSA using commercial 13X zeolite pellets. *AIChE Journal* 47 (2): 341–357. (f) Da Silva, F.A. and Rodrigues, A.E. (2001). Vacuum swing adsorption for propylene/propane separation with 4A zeolite. *Industrial & Engineering Chemistry Research* 40 (24):

5758–5774. (g) Takahashi, A., Yang, R.T., Munson, C.L., and Chinn, D. (2001). Cu(I)-Y-zeolite as a superior adsorbent for diene/olefin separation. *Langmuir* 17 (26): 8405–8413. (h) Bryan, P.F. (2004). Removal of propylene from fuel-grade propane. *Separation and Purification Reviews* 33 (2): 157–182. (i) Narin, G., Martins, V.F., Campo, M. et al. (2014). Light olefins/paraffins separation with 13X zeolite binderless beads. *Separation and Purification Technology* 133: 452–475.

18 (a) Wang, Q.M., Shen, D., Bülow, M. et al. (2002). Metallo-organic molecular sieve for gas separation and purification. *Microporous and Mesoporous Materials* 55 (2): 217–230. (b) Nicholson, T.M. and Bhatia, S.K. (2006). Electrostatically mediated specific adsorption of small molecules in metallo-organic frameworks. *The Journal of Physical Chemistry B* 110 (49): 24834–24836.

19 Nicholson, T.M. and Bhatia, S.K. (2007). Role of electrostatic effects in the pure component and binary adsorption of ethylene and ethane in Cu-tricarboxylate metal-organic frameworks. *Adsorption Science and Technology* 25 (8): 607–619.

20 Lamia, N., Jorge, M., Granato, M.A. et al. (2009). Adsorption of propane, propylene and isobutane on a metal-organic framework: molecular simulation and experiment. *Chemical Engineering Science* 64 (14): 3246–3259.

21 (a) Jorge, M., Lamia, N., and Rodrigues, A.E. (2010). Molecular simulation of propane/propylene separation on the metal-organic framework CuBTC. *Colloids and Surfaces A: Physicochemical and Engineering Aspects* 357 (1): 27–34. (b) Plaza, M., Ferreira, A., Santos, J. et al. (2012). Propane/propylene separation by adsorption using shaped copper trimesate MOF. *Microporous and Mesoporous Materials* 157: 101–111. (c) Rubeš, M., Wiersum, A.D., Llewellyn, P.L. et al. (2013). Adsorption of propane and propylene on CuBTC metal-organic framework: combined theoretical and experimental investigation. *The Journal of Physical Chemistry C* 117 (21): 11159–11167.

22 (a) Hartmann, M., Kunz, S., Himsl, D. et al. (2008). Adsorptive separation of isobutene and isobutane on $Cu_3(BTC)_2$. *Langmuir* 24 (16): 8634–8642. (b) Martins, V.F., Ribeiro, A.M., Ferreira, A. et al. (2015). Ethane/ethylene separation on a copper benzene-1,3,5-tricarboxylate MOF. *Separation and Purification Technology* 149: 445–456.

23 (a) Plaza, M., Ribeiro, A., Ferreira, A. et al. (2012). Separation of C_3/C_4 hydrocarbon mixtures by adsorption using a mesoporous iron MOF: MIL-100 (Fe). *Microporous and Mesoporous Materials* 153: 178–190. (b) Yoon, J.W., Seo, Y.K., Hwang, Y.K. et al. (2010). Controlled reducibility of a metal-organic framework with coordinatively unsaturated sites for preferential gas sorption. *Angewandte Chemie International Edition* 122 (34): 6085–6088. (c) Leclerc, H., Vimont, A., Lavalley, J.-C. et al. (2011). Infrared study of the influence of reducible iron(III) metal sites on the adsorption of CO, CO_2, propane, propene and propyne in the mesoporous metal-organic framework MIL-100. *Physical Chemistry Chemical Physics* 13 (24): 11748–11756. (d) Wuttke, S., Bazin, P., Vimont, A. et al. (2012). Discovering the active sites for C_3 separation in MIL-100(Fe) by using operando IR spectroscopy. *Chemistry – A European Journal* 18 (38): 11959–11967.

24 Liao, P.-Q., Zhang, W.-X., Zhang, J.-P., and Chen, X.-M. (2015). Efficient purification of ethene by an ethane-trapping metal-organic framework. *Nature Communications* 6: 8697.
25 Li, K., Olson, D.H., Seidel, J. et al. (2009). Zeolitic imidazolate frameworks for kinetic separation of propane and propene. *Journal of the American Chemical Society* 131 (30): 10368–10369.
26 Zhang, C., Lively, R.P., Zhang, K. et al. (2012). Unexpected molecular sieving properties of zeolitic imidazolate framework-8. *Journal of Physical Chemistry Letters* 3 (16): 2130–2134.
27 (a) Bux, H., Chmelik, C., Krishna, R., and Caro, J. (2011). Ethene/ethane separation by the MOF membrane ZIF-8: molecular correlation of permeation, adsorption, diffusion. *Journal of Membrane Science* 369 (1): 284–289. (b) Kwon, H.T. and Jeong, H.-K. (2013). Highly propylene-selective supported zeolite-imidazolate framework (ZIF-8) membranes synthesized by rapid microwave-assisted seeding and secondary growth. *Chemical Communications* 49 (37): 3854–3856. (c) Pan, Y., Liu, W., Zhao, Y. et al. (2015). Improved ZIF-8 membrane: effect of activation procedure and determination of diffusivities of light hydrocarbons. *Journal of Membrane Science* 493: 88–96. (d) Verploegh, R.J., Nair, S., and Sholl, D.S. (2015). Temperature and loading-dependent diffusion of light hydrocarbons in ZIF-8 as predicted through fully flexible molecular simulations. *Journal of the American Chemical Society* 137 (50): 15760–15771. (e) Benzaqui, M., Semino, R., Menguy, N. et al. (2016). Toward an understanding of the microstructure and interfacial properties of PIMs/ZIF-8 mixed matrix membranes. *ACS Applied Materials & Interfaces* 8 (40): 27311–27321.
28 Lee, C.Y., Bae, Y.-S., Jeong, N.C. et al. (2011). Kinetic separation of propene and propane in metal-organic frameworks: controlling diffusion rates in plate-shaped crystals via tuning of pore apertures and crystallite aspect ratios. *Journal of the American Chemical Society* 133 (14): 5228–5231.
29 (a) van den Bergh, J., Gücüyener, C., Pidko, E.A. et al. (2011). Understanding the anomalous alkane selectivity of ZIF-7 in the separation of light alkane/alkene mixtures. *Chemistry – A European Journal* 17 (32): 8832–8840. (b) Gücüyener, C., van den Bergh, J., Gascon, J., and Kapteijn, F. (2010). Ethane/ethene separation turned on its head: selective ethane adsorption on the metal-organic framework ZIF-7 through a gate-opening mechanism. *Journal of the American Chemical Society* 132 (50): 17704–17706.
30 Chen, D.-L., Wang, N., Xu, C. et al. (2015). A combined theoretical and experimental analysis on transient breakthroughs of C_2H_6/C_2H_4 in fixed beds packed with ZIF-7. *Microporous and Mesoporous Materials* 208: 55–65.
31 Lan, A., Li, K., Wu, H. et al. (2009). RPM3: a multifunctional microporous MOF with recyclable framework and high H_2 binding energy. *Inorganic Chemistry* 48 (15): 7165–7173.
32 Nijem, N., Wu, H., Canepa, P. et al. (2012). Tuning the gate opening pressure of metal-organic frameworks (MOFs) for the selective separation of hydrocarbons. *Journal of the American Chemical Society* 134 (37): 15201–15204.
33 Cadiau, A., Adil, K., Bhatt, P. et al. (2016). A metal-organic framework-based splitter for separating propylene from propane. *Science* 353 (6295): 137–140.

34 Moreira, M.A., Ferreira, A.F., Santos, J.C. et al. (2014). Hybrid process for *o*- and *p*-xylene production in aromatics plants. *Chemical Engineering and Technology* 37 (9): 1483–1492.

35 Zhang, K., Lively, R.P., Zhang, C. et al. (2013). Exploring the framework hydrophobicity and flexibility of ZIF-8: from biofuel recovery to hydrocarbon separations. *Journal of Physical Chemistry Letters* 4 (21): 3618–3622.

36 Peralta, D., Chaplais, G.r., Simon-Masseron, A.l. et al. (2012). Comparison of the behavior of metal-organic frameworks and zeolites for hydrocarbon separations. *Journal of the American Chemical Society* 134 (19): 8115–8126.

37 Jin, Z., Zhao, H.-Y., Zhao, X.-J. et al. (2010). A novel microporous MOF with the capability of selective adsorption of xylenes. *Chemical Communications* 46 (45): 8612–8614.

38 Vermoortele, F., Maes, M., Moghadam, P.Z. et al. (2011). *p*-Xylene-selective metal-organic frameworks: a case of topology-directed selectivity. *Journal of the American Chemical Society* 133 (46): 18526–18529.

39 Moreira, M.A., Santos, J.C., Ferreira, A.F. et al. (2012). Effect of ethylbenzene in *p*-xylene selectivity of the porous titanium amino terephthalate MIL-125(Ti)_NH_2. *Microporous and Mesoporous Materials* 158: 229–234.

40 Torres-Knoop, A., Krishna, R., and Dubbeldam, D. (2014). Separating xylene isomers by commensurate stacking of *p*-xylene within channels of MAF-X8. *Angewandte Chemie International Edition* 53 (30): 7774–7778.

41 (a) Chang, N. and Yan, X.-P. (2012). Exploring reverse shape selectivity and molecular sieving effect of metal-organic framework UIO-66 coated capillary column for gas chromatographic separation. *Journal of Chromatography A* 1257: 116–124. (b) Granato, M.A., Martins, V.D., Ferreira, A.F.P., and Rodrigues, A.E. (2014). Adsorption of xylene isomers in MOF UiO-66 by molecular simulation. *Microporous and Mesoporous Materials* 190: 165–170.

42 Cavka, J.H., Jakobsen, S., Olsbye, U. et al. (2008). A new zirconium inorganic building brick forming metal organic frameworks with exceptional stability. *Journal of the American Chemical Society* 130 (42): 13850–13851.

43 Bárcia, P.S., Guimarães, D., Mendes, P.A. et al. (2011). Reverse shape selectivity in the adsorption of hexane and xylene isomers in MOF UiO-66. *Microporous and Mesoporous Materials* 139 (1): 67–73.

44 Bernardo, P., Drioli, E., and Golemme, G. (2009). Membrane gas separation: a review/state of the art. *Industrial & Engineering Chemistry Research* 48: 4638–4663.

45 Baker, R.W. (2002). Future directions of membrane gas separation technology. *Industrial and Engineering Chemistry Research* 41 (6): 1393–1411.

46 (a) Aroon, M.A., Ismail, A.F., Matsuura, T., and Montazer-Rahmati, M.M. (2010). Performance studies of mixed matrix membranes for gas separation: a review. *Separation and Purification Technology* 75 (3): 229–242. (b) Ulbricht, M. (2006). Advanced functional polymer membranes. *Polymer* 47 (7): 2217–2262.

47 (a) Robeson, L.M. (1999). Polymer membranes for gas separation. *Current Opinion in Solid State and Materials Science* 4 (6): 549–552. (b) Robeson, L.M. (2008). The upper bound revisited. *Journal of Membrane Science* 320 (1): 390–400.

48 (a) Bastani, D., Esmaeili, N., and Asadollahi, M. (2013). Polymeric mixed matrix membranes containing zeolites as a filler for gas separation applications: a review. *Journal of Industrial and Engineering Chemistry* 19 (2): 375–393. (b) Saufi, S.M. and Ismail, A.F. (2002). Development and characterization of polyacrylonitrile (PAN) based carbon hollow fiber membrane. *Songklanakarin Journal of Science and Technology* 24: 843–854. (c) Seoane, B., Coronas, J., Gascon, I. et al. (2015). Metal-organic framework based mixed matrix membranes: a solution for highly efficient CO_2 capture? *Chemical Society Reviews* 44 (8): 2421–2454. (d) Perez, E.V., Balkus, K.J., Ferraris, J.P., and Musselman, I.H. (2009). Mixed-matrix membranes containing MOF-5 for gas separations. *Journal of Membrane Science* 328 (1): 165–173. (e) Bux, H., Liang, F., Li, Y. et al. (2009). Zeolitic imidazolate framework membrane with molecular sieving properties by microwave-assisted solvothermal synthesis. *Journal of the American Chemical Society* 131 (44): 16000–16001.

49 (a) Zhang, Y., Feng, X., Yuan, S. et al. (2016). Challenges and recent advances in MOF-polymer composite membranes for gas separation. *Inorganic Chemistry Frontiers* 3 (7): 896–909. (b) Chung, T.-S., Jiang, L.Y., Li, Y., and Kulprathipanja, S. (2007). Mixed matrix membranes (MMMs) comprising organic polymers with dispersed inorganic fillers for gas separation. *Progress in Polymer Science* 32 (4): 483–507.

50 Bae, T.H., Lee, J.S., Qiu, W. et al. (2010). A high-performance gas-separation membrane containing submicrometer-sized metal-organic framework crystals. *Angewandte Chemie International Edition* 49 (51): 9863–9866.

51 (a) Balcıoğlu, I.A. and Ötker, M. (2003). Treatment of pharmaceutical wastewater containing antibiotics by O_3 and O_3/H_2O_2 processes. *Chemosphere* 50 (1): 85–95. (b) Saud, P.S., Pant, B., Alam, A.-M. et al. (2015). Carbon quantum dots anchored TiO_2 nanofibers: effective photocatalyst for waste water treatment. *Ceramics International* 41 ((9): 11953–11959. (c) Asghar, A., Raman, A.A.A., and Daud, W.M.A.W. (2015). Advanced oxidation processes for in-situ production of hydrogen peroxide/hydroxyl radical for textile wastewater treatment: a review. *Journal of Cleaner Production* 87: 826–838.

52 Ternes, T.A., Meisenheimer, M., McDowell, D. et al. (2002). Removal of pharmaceuticals during drinking water treatment. *Environmental Science and Technology* 36 (17): 3855–3863.

53 Prins, R. (2001). Catalytic hydrodenitrogenation. *Advances in Catalysis* 46: 399–464.

54 (a) Dai, G., Cui, L., Song, L. et al. (2006). Metabolism of terephthalic acid and its effects on CYP4B1 induction. *Biomedical and Environmental Sciences* 19 (1): 8. (b) Horcajada, P., Chalati, T., Serre, C. et al. (2010). Porous metal-organic-framework nanoscale carriers as a potential platform for drug delivery and imaging. *Nature Materials* 9 (2): 172–178.

55 (a) Singh, R., Gautam, N., Mishra, A., and Gupta, R. (2011). Heavy metals and living systems: an overview. *Indian Journal of Pharmacology* 43 (3): 246. (b) Tchounwou, P.B., Yedjou, C.G., Patlolla, A.K., and Sutton, D.J. (2012). *Molecular, Clinical and Environmental Toxicology*, 133–164. Springer.

56 Cychosz, K.A. and Matzger, A.J. (2010). Water stability of microporous coordination polymers and the adsorption of pharmaceuticals from water. *Langmuir* 26 (22): 17198–17202.

57 Jun, J.W., Tong, M., Jung, B.K. et al. (2015). Effect of central metal ions of analogous metal-organic frameworks on adsorption of organoarsenic compounds from water: plausible mechanism of adsorption and water purification. *Chemistry – A European Journal* 21 (1): 347–354.

58 Kandola, B.K., Ebdon, J.R., and Chowdhury, K.P. (2015). Flame retardance and physical properties of novel cured blends of unsaturated polyester and furan resins. *Polymer* 7 (2): 298–315.

59 Lü, C., Li, H., Wang, H., and Liu, Z. (2013). Probing the interactions between boronic acids and cis-diol-containing biomolecules by affinity capillary electrophoresis. *Analytical Chemistry* 85 (4): 2361–2369.

60 Zhu, X., Gu, J., Zhu, J. et al. (2015). Metal-organic frameworks with boronic acid suspended and their implication for *cis*-diol moieties binding. *Advanced Functional Materials* 25 (25): 3847–3854.

61 (a) Hernández-Maldonado, A.J. and Yang, R.T. (2004). Denitrogenation of transportation fuels by zeolites at ambient temperature and pressure. *Angewandte Chemie International Edition* 116 (8): 1022–1024. (b) Almarri, M., Ma, X., and Song, C. (2009). Role of surface oxygen-containing functional groups in liquid-phase adsorption of nitrogen compounds on carbon-based adsorbents. *Energy & Fuels* 23 (8): 3940–3947.

17

Water Sorption Applications of MOFs

17.1 Introduction

The use of metal-organic frameworks (MOFs) in water sorption-based applications has attracted much interest. Such applications include adsorption-driven heat pumps (ADHPs), seawater desalination, drying of gas streams, humidity control in buildings, and water harvesting from air. All of these applications pose different requirements to the adsorbent in terms of its adsorption behavior [1]. The modular structure of MOFs makes them ideal candidates for such applications because it enables chemists to design their sorption properties with respect to parameters such as the total uptake, the position of the uptake on the P/P_0 scale, and the adsorption mechanism.

Even though all examples for processes involving the adsorption of water described above are fundamentally different and require significantly different adsorption properties, they all require the MOF adsorbent to be hydrolytically stable, have a high adsorption capacity, and have a high selectivity for water. This chapter will cover the principles of water vapor sorption in MOFs, provide guidelines for the design of next generation MOFs, and illustrate ways to optimize the water sorption properties of a given material. The adsorption of other vapors in MOFs have the same underlying general principles and are not discussed here.

17.2 Hydrolytic Stability of MOFs

Many MOFs, especially first-generation materials, undergo structural degradation in the presence of water. With respect to hydrolytic stability, we distinguish thermodynamic and kinetic factors (Table 17.1). It is important to note that with respect to thermodynamics the terms stable and unstable are used, while in the context of kinetics the corresponding terms are inert and labile. The fact that the hydrolytic stability of MOFs is related to specific structural allows to delineate strategies for the design of hydrolytically stable MOFs. In the following sections we will discuss the degradation of MOFs in the presence of water and derive principles for the design of hydrolytically stable MOFs.

Introduction to Reticular Chemistry: Metal-Organic Frameworks and Covalent Organic Frameworks,
First Edition. Omar M. Yaghi, Markus J. Kalmutzki, and Christian S. Diercks.

Companion website: www.wiley.com/go/yaghi/reticular

Table 17.1 Structural factors influencing the thermodynamic and kinetic stability of MOFs in the presence of water.

Thermodynamic stability/instability	Kinetic inertness/lability
• pK_a of the linker (basicity) • Oxidation state and ionic radius of the metal (acidity) • Reduction potential of the metal • Coordination geometry	• Steric hindrance (i.e. "coordination number") and rigidity of the constituents • Steric hindrance through interpenetration • Hydrophobicity • Electronic configuration of the metal ions

17.2.1 Experimental Assessment of the Hydrolytic Stability

The hydrolytic stability of MOFs is typically determined by comparing powder X-ray diffraction (PXRD) patterns before and after exposure to a specific amount of water (e.g. atmosphere with defined relative humidity, RH). While partial collapse of the pores only results in broadening of the reflection peaks due to partial loss of crystallinity and the formation of an amorphous side phase, the decomposition of MOFs by hydrolysis typically leads to complete amorphization and consequently the disappearance of all reflections in the PXRD pattern. To substantiate information gained from X-ray diffraction experiments it is instructive to compare surface areas calculated from gas adsorption measurements prior to and after exposure to humidity. Another decomposition pathway is the partial dissolution of MOFs. This inevitably leads to a mass loss, which is verified by weighing the MOF before and after it has been exposed to water.

To study the effects of water vapor on MOFs for specific applications, the gas composition should resemble that under application condition, e.g. $H_2O/CO_2/N_2$ 1 : 1.5 : 7.5 for carbon capture from post-combustion flue gas. In contrast, the isolated effect of water vapor on a MOF structure is studied using inert carrier gases such as N_2, He, and Ar and true water vapor adsorption isotherms can only be acquired when water vapor is dosed into the material in the absence of carrier gases.

Different applications require specific levels of hydrolytic stability. Minimum testing conditions for MOFs in "single-pass" and "cycled" applications, as well as liquid phase separation have been established and an overview is given in Table 17.2. While the abovementioned methods give information about the hydrolytic stability of MOFs, they provide little to no insight into the mechanism of degradation.

17.2.2 Degradation Mechanisms

The degradation of MOFs in the presence of water can occur either by hydrolysis or by linker replacement [2]. Hydrolysis is initiated by the cleavage of the metal–linker bond. Subsequently, a hydroxylated secondary building unit (SBU) is formed and the linker is released in its protonated (neutral) form.

$$M^{n+} - {}^{-}(O_2C)\text{-}R + H_2O \longrightarrow M^{n+} - (OH)^{-} + R\text{-}(CO_2H)$$

Table 17.2 Testing conditions to determine the water stability of a given material for different types of application.

Application	Testing conditions	Characterization methods
Gas or vapor phase		
Single pass/one-time use		
Single pass cartridge (e.g. gas mask or air filters)	Prolonged stability in ambient or humidified air	PXRD and sorption (BET)[a] measurements
Cycled/multiple use		
Gas separation packed bed (e.g. CO_2, NG, H_2)	Multiple adsorption desorption cycles with the relevant gas mixture (regeneration with the appropriate method; TSA[b], PSA[c], VSA[d])	PXRD, sorption (BET) measurements, adsorption capacity for each cycle, microscopy (e.g. SEM[e], TEM[f])
Gas separation membranes	Prolonged exposure to the relevant gas mixture under application conditions	PXRD, sorption (BET) measurements, microscopy (e.g. SEM, TEM, AFM[g])
Gas storage (e.g. H_2, CH_4, CO_2)	Prolonged storage of the relevant gas under application pressures	PXRD and sorption (BET) measurements, microscopy (e.g. SEM, TEM)
Aqueous phase		
Liquid phase catalysis, capture of molecules from aq. solutions, liquid phase separation	Immersion and prolonged stirring under relevant application conditions	PXRD and sorption (BET) measurements, solid-state mass loss, titration of the solution

a) Brunauer–Emmett–Teller surface area.
b) Temperature swing adsorption.
c) Pressure swing adsorption.
d) Vacuum swing adsorption.
e) Scanning electron microscope.
f) Transmission electron microscope.
g) Atomic force microscope.

The mechanism of linker displacement can be described as the insertion of water into the metal–linker bond. This leads to the formation of a hydrated SBU and the release of a deprotonated linker.

$$M^{n+} - {}^{-}(O_2C)\text{-}R + H_2O \longrightarrow M^{n+} - (OH_2) + R\text{-}(CO_2)^{-}$$

Ligand displacement is the most likely mechanism for the structural degradation of UiO-66 in neutral or basic aqueous conditions, whereas in the presence of acids another mechanism prevails [3]. Figure 17.1 illustrates the degradation of UiO-66 in the presence of water, base, and acid. Treatment with water leads to the hydration of the SBU and release of a deprotonated linker following the hydrolysis mechanism (Figure 17.1a). A similar mechanism (solvolysis) is suggested for the treatment of UiO-66 with alcohols such as methanol. Basic hydrolysis

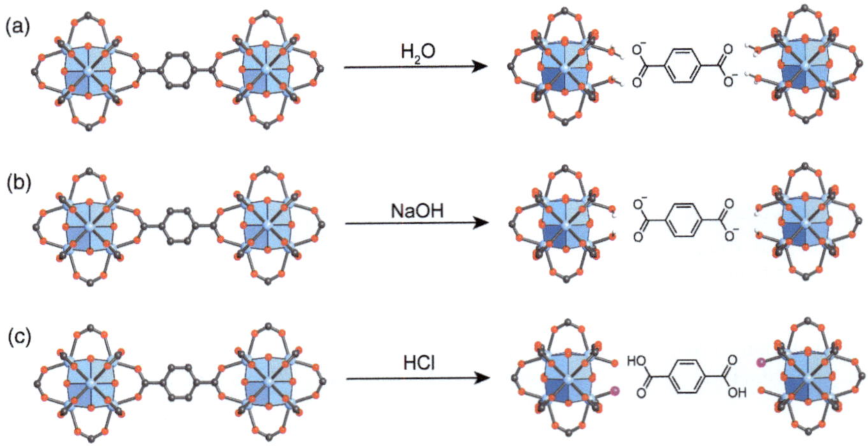

Figure 17.1 Degradation mechanisms of UiO-66 in the presence of water, base, and hydrochloric acid. (a) Hydrolysis in water affords the formation of a hydrated SBU, whereas (b) hydroxylation of the SBU affords the release of a deprotonated linker. (c) Treatment with HCl causes the chlorination of the SBU while the linker is released in its protonated neutral from. Color code: Zr, blue; C, gray; O, red; Cl, pink.

affords the hydroxylation of the SBU and the release of the deprotonated linker (Figure 17.1b), whereas acid treatment with hydrochloric acid gives the chlorinated SBU and the linker is released in its protonated form (Figure 17.1c). The chlorinated SBU may subsequently be hydrolyzed to give a neutral SBU bearing one addition –OH and one $-OH_2$ ligand.

17.2.3 Thermodynamic Stability

The thermodynamic stability is mainly defined by two factors: (i) the stability of the metal–linker bond and (ii) the energetic position of the frontier orbitals of the metal relative to those of water. It is easy to approximate the stability of the metal–linker bond, but it is more challenging to make a statement about the energetics of the frontier orbitals.

17.2.3.1 Strength of the Metal–Linker Bond

The covalent nature of the organic linkers renders them highly chemically stable entities and the molecular analogs of most SBUs are known to be stable in water. The weak point in the extended structure formed by their reticulation is the bond between them and its strength can (to some extent) be used as a measure of the hydrolytic stability of a MOF. The interaction between the metal centers in the SBU and the binding groups of the organic linker can be approximated as a Lewis acid–base interaction. This implies that the higher the pK_a of the protonated linker, the stronger the metal–linker bond as illustrated by the high hydrolytic stability of pyrazolate-based MOFs [4]. A comparison of di- and tritopic pyrazolate and carboxylate linkers based on a central aryl unit reveals dramatic differences in pK_a between these two binding groups (see Figure 17.2a). The higher pK_a values

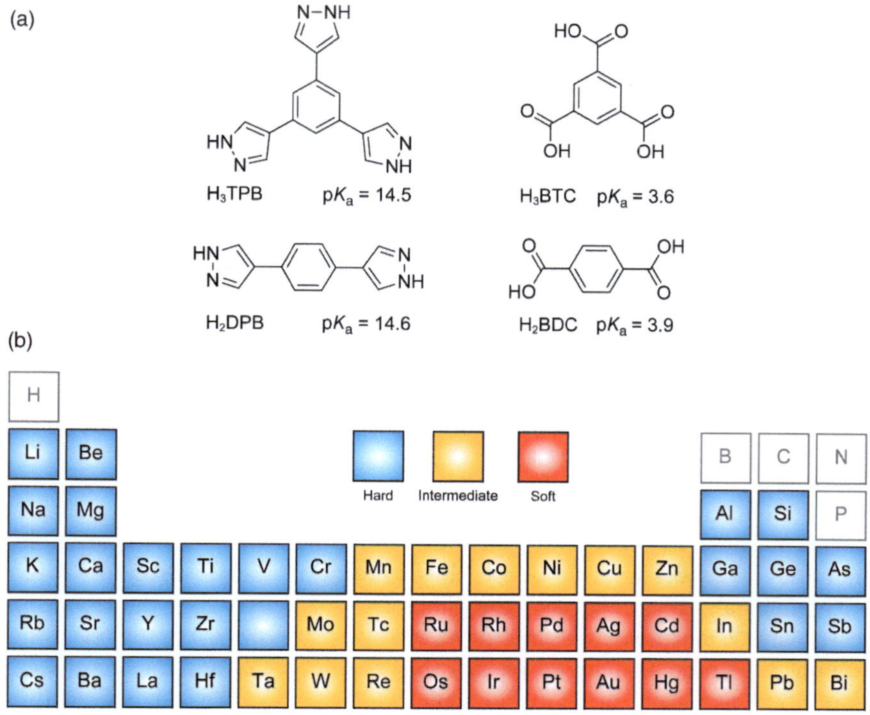

Figure 17.2 (a) Linker molecules with different binding groups and the corresponding pK_a values. Pyrazole is a weaker acid (the corresponding base is strong) and the resulting M—N bond is therefore thermodynamically more stable than the M—O bond in carboxylate-based MOFs. (b) Hard, intermediate, and soft acids and their position in the periodic table.

of the pyrazole-based linkers imply stronger bonding to the metal component of the framework and therefore a higher hydrolytic stability.

The chemical nature of the metal center determines the strength of the metal–linker bond in a similar way. The higher the charge on the metal and the smaller its radius, the more acidic it is, leading to stronger bonding with the organic linker. The radius and the precise electronic structure of the metal greatly influence the bond strength. This is best described by the hard–soft acid–base (HSAB) concept, which states that frontier orbitals of similar size and polarizability can overlap to form strong bonds. For MOFs, this means that the bond between hard carboxylate binding groups and hard Ti^{4+} centers (high charge/radius ratio) is stronger than that of hard carboxylate binding groups and soft Hg^{2+} centers (low charge/radius ratio). Figure 17.2b shows a periodic table where the metals have been divided into three categories: (i) hard acid, (ii) intermediate acid, and (iii) soft acid. The thermodynamic stability can correlate with the hydrolytic stability, but other factors must be taken into consideration as well.

17.2.3.2 Reactivity of Metals Toward Water

Considerations of the metal–linker bond strength based on pK_a values and the HSAB concept can serve as simple tools to approximate the stability of

MOFs. With respect to the hydrolytic stability, other thermodynamic factors such as the position of the metal frontier orbital with respect to those of water are important. The energetic position of the frontier orbitals of metal ions correlates with their reduction potential. The comparison of PXRD patterns of Co_2Cl_2(BTDD) (BTDD = (bis(1*H*-1,2,3-triazolo[4,5-*b*],[4′,5′-*i*])dibenzo[1,4] dioxin) and Mn_2Cl_2(BTDD) measured after exposure to moisture shows no changes for Co_2Cl_2(BTDD), whereas amorphization of Mn_2Cl_2(BTDD) is observed. When we consider the standard reduction potentials of Co^{2+} and Mn^{2+} we see that the reduction potential of Co^{2+} ($E^0 = -0.28$ V) is less negative than that of Mn^{2+} ($E^0 = -1.18$ V) [5]. As a rule of thumb, we can say that for a series of isoreticular MOFs constructed from different metals, those containing metals with lower (or more negative) reduction potentials are more likely to undergo hydrolysis [6].

17.2.4 Kinetic Inertness

A thermodynamically unstable MOF does not necessarily degrade in the presence of water. This is explained by the fact that thermodynamically unstable compounds can be kinetically inert, thus preventing their hydrolysis. Thermodynamic stability is related to the Gibbs free energy (ΔG) of the reaction, whereas kinetic inertness is related to the activation energy (E_a) of the reaction. Therefore, the products of hydrolysis may be of lower energy and thus represent the thermodynamically more stable state (ΔG more negative); however, the activation barrier may be too high to overcome under the chosen conditions (E_a too large). This is illustrated in Figure 17.3 where the reaction diagrams for

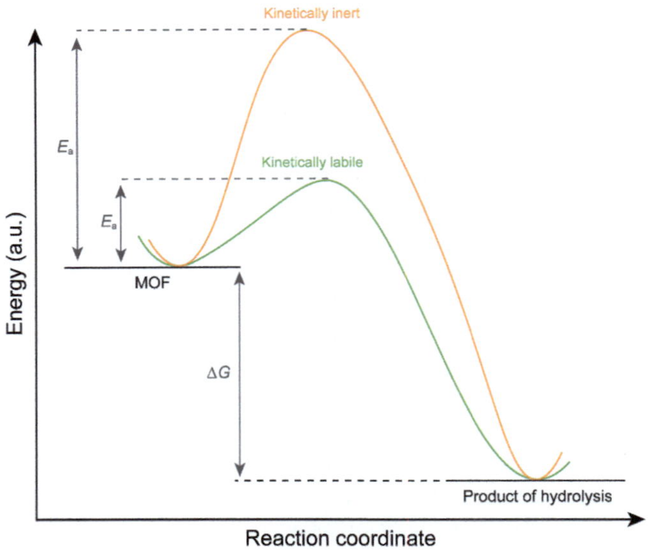

Figure 17.3 Comparison of the reaction diagrams of two thermodynamically unstable MOFs. In both cases the products of hydrolysis are thermodynamically favored (ΔG is negative). Kinetic inertness due to a high activation energy E_a (orange curve) results in an increased hydrolytic stability. Kinetic lability and the concomitant low activation energy for the hydrolysis renders the MOF unstable toward water (green curve).

the hydrolysis of two compounds with identical thermodynamic (in)stability but different kinetic lability are shown. For both compounds the hydrolysis is thermodynamically favored, but the difference in activation energy renders one of them hydrolytically more stable (orange curve) than the other (green curve).

There are multiple structural factors that can cause a MOF to be kinetically inert such as steric shielding of the SBU, rigidity of the SBU and/or the linker, the electronic configuration of the metal, and the hydrophobicity.

17.2.4.1 Steric Shielding

The most obvious way of introducing kinetic inertness is by steric shielding. This means that the attack of the thermodynamically unstable metal–linker bond by water is slowed down or prevented completely by the introduction of shielding moieties. This can be realized in three different ways: (i) SBUs with high connectivity, (ii) bulky linkers, and (iii) interpenetrating frameworks.

SBUs with high connectivity are frequently encountered in zirconium- and hafnium-based MOFs since these metals favor the formation of M_6O_8-core SBUs ($M = Zr^{4+}, Hf^{4+}$) with a maximum connectivity of 12 (Figure 17.4a). A prominent example is UiO-66 that is built from 12-c Zr_6O_8-core SBUs joined by ditopic BDC linkers to form a framework of **fcu** topology (see Figure 4.28) [7]. The high connectivity of the 12-c SBU endows UiO-66 with high hydrolytic stability by shielding through the binding groups of the linker. Other MOFs that are built from SBUs with high connectivity are hydrolytically stable, which confirms that steric shielding of the SBU by the binding groups is a viable approach for the design of water-stable MOFs [8]. Most of these SBUs are of the M_6O_8-type ($M = Zr^{4+}$, Hf^{4+}) or of rod-like nature and a selection is shown in Figure 17.4. The crowding

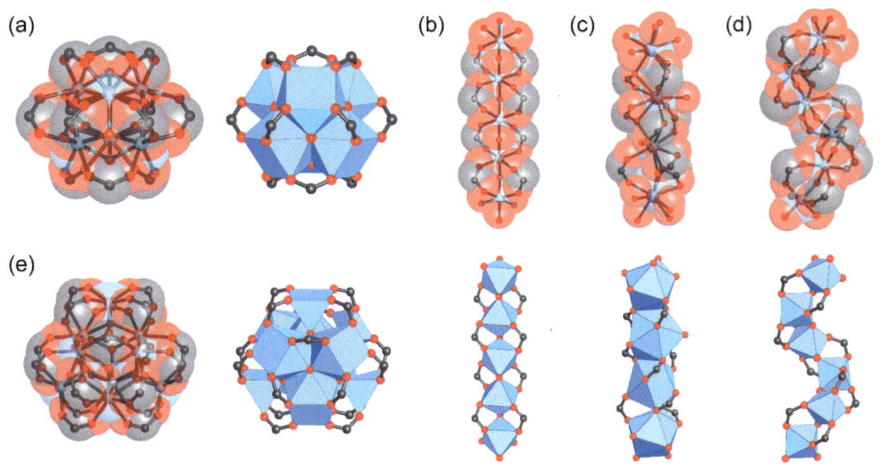

Figure 17.4 SBUs with high connectivity that endow MOFs with water stability due to steric shielding. Shown is a comparison of the polyhedral representation and the ball-and-stick representation overlaid with the corresponding space-filling model highlighting the steric shielding by the binding groups. (a) The most common SBU with high connectivity has the 12-c M_6O_8-core ($M = Zr^{4+}, Hf^{4+}$). Other SBUs include many rod-like SBUs such as that in (b) MIL-53, (c) La(BTB), or (d) CAU-10 as well as rare earth SBUs such as that of (e) **gea**-MOF-1. Color code: Metal, blue; C, gray; O, red.

of linkers around the SBU can prevent the formation of water clusters in close proximity and hinders water molecules from getting close to the metal centers as illustrated by the space-filling representation.

A second way to increase the kinetic inertness is the introduction of bulky substituents on the linker to sterically shield the SBUs. In contrast to the high connectivity of the SBUs discussed above, this is often possible by rational design of the linker. An example illustrating this approach is the functionalization of the BDC linker in DMOF ($Zn(BDC)(DABCO)_{0.5}$), a pillared-layered MOF built from 2D $Zn_2(BDC)_2$ **sql** layers that are pillared by DABCO linkers to form a **pcu** framework (Figure 17.5a) [9]. The zinc paddle wheel SBUs readily hydrolyze in the presence of water, leading to the structural degradation of DMOF [10]. An additional methyl substituents appended to the H_2BDC linker effectively shields the SBUs and the subsequent addition of further methyl groups permits systematic tuning of the kinetic water stability of DMOF derivatives [3]. While the additional methyl groups shield the SBUs, the increase in stability cannot be rationalized in terms of steric shielding alone but the increased hydrophobicity of DMOF derivatives has to be considered. Figure 17.5 gives a comparison of the pristine and modified DMOF, highlighting the shielding effect of the methyl substituents on the BDC linkers.

Interpenetration has as similar effect on the hydrolytic stability of MOFs, and the water stability of interpenetrated MOFs is consequently higher than that of their non-interpenetrated counterparts. Interpenetration of frameworks is

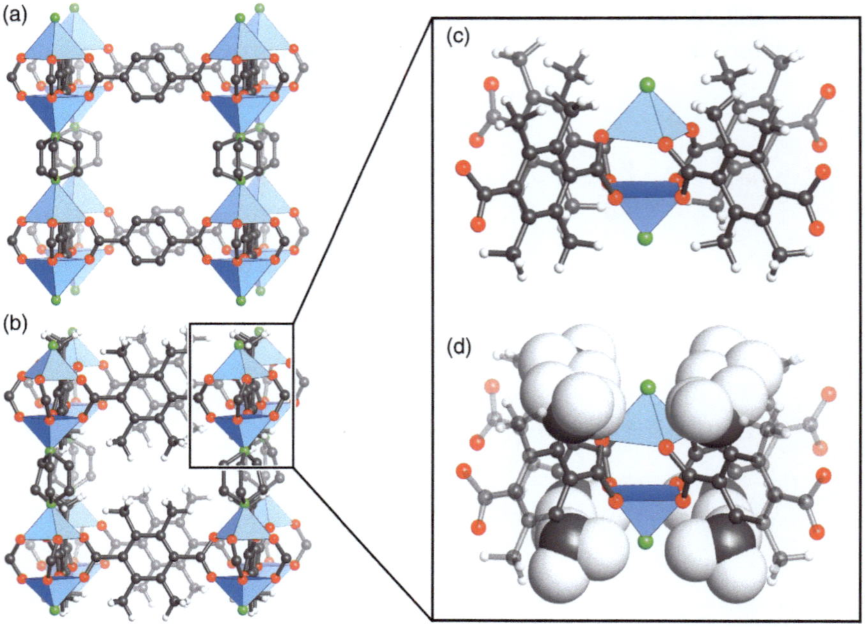

Figure 17.5 Crystal structure of (a) DMOF and (b) $(Me)_4$-DMOF (modeled based on DMOF). (c, d) Steric shielding effect of the methyl substituents on the $Zn_2(-COO)_4$ paddle wheel SBUs. Both the (c) ball-and-stick and (d) partial space-filling model are shown. The top and bottom of the paddle wheel SBUs are shielded by the DABCO ligands (omitted for clarity). Color code: Zn, blue; H, white; C, gray; N, green; O, red.

necessarily accompanied by a decrease in the pore size and pore volume, and thus results in lower uptake capacities. It is also important to remember that it is not necessarily possible to synthesize an interpenetrated counterpart for every MOF structure.

17.2.4.2 Hydrophobicity

Structural degradation by hydrolysis or linker displacement is a two-step process. In the first step, water clusters are formed near the SBU before the reaction of water and the metal centers in the SBU occurs and leads to the structural degradation of the MOF. The formation of water clusters near the SBU can be avoided in three ways: (i) Appending hydrophobic alkyl substituents to the linker to avoid the formation of water clusters close to the SBU through repulsive interactions. (ii) Introduction of polar functional groups such as amines and alkoxy groups that draw water away from the SBU and act as basins of attraction. (iii) Fluorination of the linker to give hydrophobic MOFs that repel water [11]. Even though the last approach is the most effective one, it eliminates the possibility for water adsorption to occur because water cannot permeate the surface of the MOF. Introducing alkyl chains or polar groups provides a more reasonable approach since both allow water to enter the pore system while minimizing the likelihood of hydrolysis. Introducing additional functional groups into the pores of the framework lowers the maximum capacity, and the nature of the functionalities has a dramatic impact on the position of the inflection point of the isotherm.

17.2.4.3 Electronic Configuration of the Metal Center

The electronic configuration of the metal centers in the SBUs can endow a thermodynamically unstable MOF with kinetic inertness. This concept is well established for transition metal complexes. Here we will illustrate this for octahedrally coordinated metal centers in $M_3OL_3(COO)_6$ SBU (Figure 17.6). Findings based on these considerations cannot be transferred directly to other structures where the coordination geometry of the metal is different which, according to the ligand field theory, leads to a different splitting of the d-orbitals. MIL-101 can be prepared from a variety of trivalent metals such as Cr^{3+}, Fe^{3+}, and Ti^{3+}. For coordination by carboxylate binding groups (i.e. $R-COO^-$) and

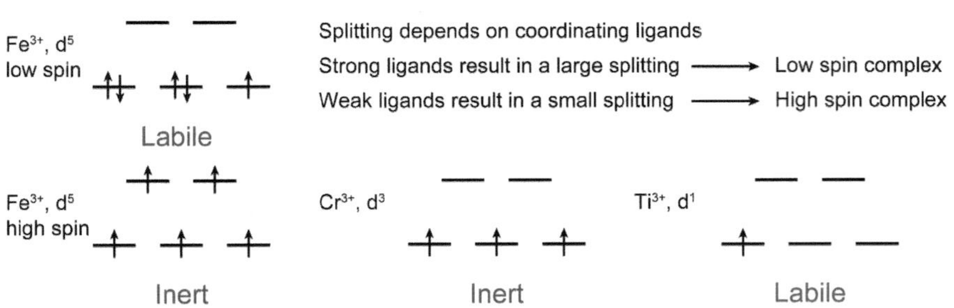

Figure 17.6 Electronic configuration of d-elements that are often used to construct the SBUs of MOFs. The weak field ligands typically present in MOFs (e.g. O^{2-}, OH^-, and $R-CO_2^-$) lead to a small ligand field splitting and high spin complexes. Asymmetric occupation of the orbitals (e.g. Ti^{3+} L.S.) renders the resulting complexes kinetically labile, whereas a symmetric occupation (e.g. Fe^{3+} and Cr^{3+} L.S.) results in kinetic inertness.

other weak field ligands (e.g. HO^-, H_2O) a high-spin configuration is expected. Cr^{3+} has a d^3 configuration, and according to Hund's rules all d-electrons are located in the t_{2g} orbitals. Therefore, (Cr)MIL-101 is kinetically inert. In the case of the d^5 ion Fe^{3+} every d-orbital is occupied by one electron and this symmetric occupation of d-orbitals renders Fe^{3+} kinetically inert. Things are different for Ti^{3+}. The d^1 configuration of Ti^{3+} results in a single d-electron occupying the degenerated t_{2g} orbitals. Such an asymmetric occupation of d-orbitals is kinetically labile. Even though this principle is very simple it can give qualitative information regarding the kinetic inertness of MOFs. A comparison of (Al, Cr)MIL-53, ((Al,Cr)(OH)(BDC)), and (V)MIL-47 (V(O)(BDC)), three isostructural **sra** frameworks, reveals that chemical stability follows the order $Cr^{+3} > Al^{+3} > V^{+4}$, a finding that is in good agreement with the decreasing inertness $Cr^{+3} > Al^{+3} > V^{+3} > V^{+4}$. In contrast, the bond strength (thermodynamic stability) of the M—O bond (calculated for the corresponding oxides) follows the reverse order V > Al > Cr [12]. This illustrates how kinetic inertness defines the stability toward water whereas the strength of the metal–linker bond is not always an appropriate indicator for the hydrolytic stability but always defines the thermal stability. The outstanding water stability of (Cr)MIL-53 originates not only from the kinetic inertness of Cr^{3+} but also from the large difference in the energy of the frontier orbitals of Cr^{3+} and water.

17.3 Water Adsorption in MOFs

17.3.1 Water Adsorption Isotherms

Vapor sorption is in many ways different from that of gases. The shape of water vapor adsorption isotherms gives direct information on the hydrophilicity or hydrophobicity of hydrolytically stable MOFs. Isotherms of MOFs that are not hydrolytically stable do not allow gaining such information because their shape is governed by the chemical nature of their decomposition products rather than that of the MOF itself. Hydrophilic materials show strong affinity to water molecules, a definition that cannot be directly applied to porous materials. Hydrophobic mesoporous silica (e.g. MCM-41 and SBA-15) adsorb more water than most more hydrophilic microporous zeolites. This finding is explained by the fact that the adsorption of water is largely dictated by the pore size. Therefore, the hydrophilicity of a given MOF is commonly determined by its selectivity for water over other components in a mixture. The position of the inflection point (α) on the water sorption isotherm is often used to determine the relative hydrophilicity. The inflection point is defined as the point at which half of the maximum uptake is reached. Figure 17.7 shows the isotherms of four materials ranging from strongly hydrophilic to strongly hydrophobic (a through d). Isotherm (a) has the shape of a typical Type I isotherm. Water is adsorbed at very low relative pressures and the maximum loading is therefore also reached at low relative pressures. The corresponding Henry's constant (slope at pressures close to $P/P_0 = 0$) is large and the inflection point is located at very low relative pressures. Isotherm (b) shows a smaller slope at low relative pressures and resembles a Type II or Type IV isotherm; hence the corresponding MOF is less hydrophilic than that leading to isotherm (a). Adsorbent (c) shows

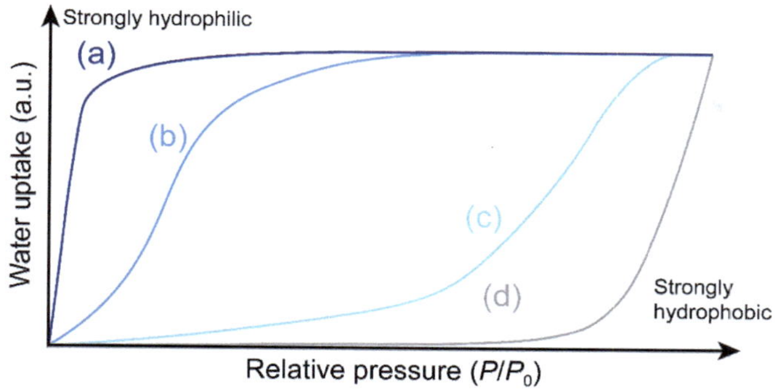

Figure 17.7 Schematic isotherms for (a) microporous strongly hydrophilic, (b) microporous less hydrophilic, (c) mesoporous "hydrophilic," and (d) strongly hydrophobic MOFs.

resembles a Type III isotherm. The uptake at low relative pressures is very low and is explained by the formation of clusters rather than a monolayer. The uptake eventually reaches the same maximum uptake as in cases (a) and (b) but at considerably higher relative pressures, which is indicative of a relatively hydrophobic pore system. The similar maximum capacity for (a)–(c) indicates similar pore volumes for all three materials while the different inflection points can be used as a relative measure of the hydrophilicity. A linear isotherm with a constant small slope like that shown in (b) is commonly observed for strongly hydrophobic materials, and in this case an uptake is only observed close to or at the saturation vapor pressure. It should be noted, that a comparison of the relative hydrophilicity/hydrophobicity of pore systems is strictly speaking only possible if the pore size and the dimensionality of all compared pore systems are identical (e.g. functionalized derivatives of the same MOF).

The adsorption of water in MOFs can occur following three main mechanisms. Structural factors such as the pore size and the nature of the primary adsorption sites influence which of these three mechanisms is observed and they are the subject of the Section 17.3.2.

17.3.2 Mechanisms of Water Adsorption in MOFs

17.3.2.1 Chemisorption on Open Metal Sites

Open metal sites are a frequently encountered structural motif in MOFs. They are formed by the removal of terminal neutral ligands (e.g. water, solvent) from the coordination sphere of the SBU in MOFs such as paddle wheel (e.g. HKUST-1) and many rod-like SBUs (e.g. MOF-74). In contrast to HKUST-1, which, similar to other MOFs constructed from copper paddle wheel SBUs, displays low stability in the presence of water, (M)MOF-74 (M = Zn, Ni, Co) shows a relatively high hydrolytic stability.[1] The structure of (M)MOF-74 is built from helical rod SBUs that are connected by H_4DOBDC linkers to form

1 Loss of terminal water from the SBUs is known for many rod-like and discrete SBUs (e.g. di- and trinuclear paddle wheel, $M_3O(H_2O)_2(COO)_6$, $Cu_3O(OH)(H_2O)_2(Py)_3$, $[M_3(-O)_3(H_2O)_3(COO)_3]_\infty$). The dehydration of the $Zr_6O_4(OH)_4(-COO)_{12}$ SBUs in UiO-66 follows a significantly different mechanism. Upon heating to temperatures above 250 °C the SBU loses four bridging –OH groups, which decreases the coordination number of Zirconium from 8 to 7 and results in a distortion of the cluster [13].

frameworks of **etb** topology with hexagonal 1D channels of approximately 11 Å that run along the crystallographic *c*-axis. Each pore is bordered by six SBUs with neutral water ligands bound to the metal centers that point into the center of the pore. Those terminal water ligands are removed by heating in dynamic vacuum, leaving behind open metal sites. Density functional theory (DFT) calculations for fully activated (Mg)MOF-74 show that a large unoccupied orbital is located at those open metal sites, making them strong Lewis acids [14]. Single-crystal X-ray diffraction studies on hydrated (Zn)MOF-74 show the different nature of the water species adsorbed in the framework. Figure 17.8a shows one pore of (Zn)MOF-74 with a view along the *c*-direction where all adsorbed water molecules are represented by their thermal ellipsoids [15]. Three different species can clearly be distinguished. The water coordinated directly to the SBU is bound strongly, which is reflected in small thermal ellipsoids. A second layer of water is adsorbed close to the SBUs. The strong physisorption interaction between these water molecules and the framework is weaker than that for chemisorption on the open metal sites and consequently the thermal ellipsoids for these water molecules are larger. Finally, weakly bound water molecules are located at the center of the pore. Figure 17.8c shows the different dehydration

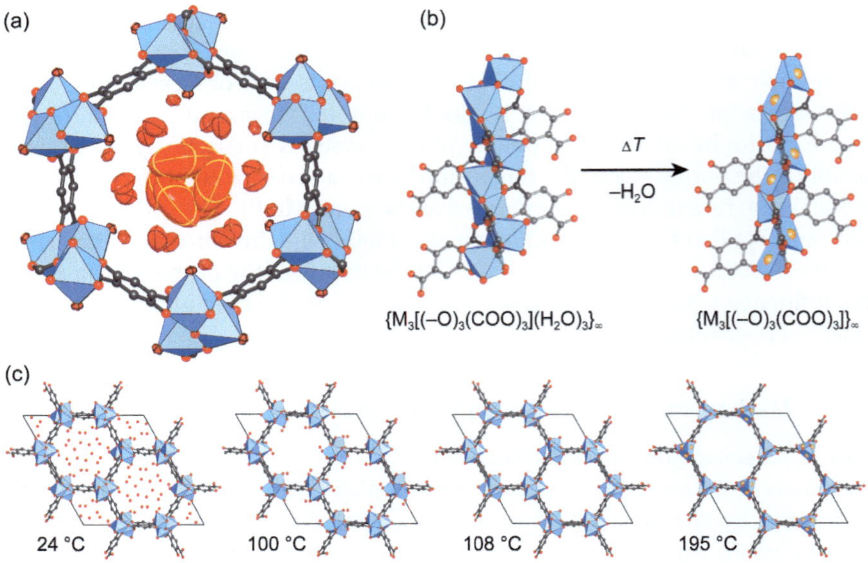

Figure 17.8 (a) One pore of hydrated (Zn)MOF-74 with view along the crystallographic *c*-axis. Three species of adsorbed water can be distinguished: (i) chemisorbed water on the open metal sites of the SBU, (ii) strongly physisorbed water close to the SBU, and (iii) loosely bound water at the center of the pore. All oxygen atoms of water are shown as their respective thermal ellipsoids (50%). (b) Segments of the SBU of (Zn)MOF-74 in its hydrated (left) and dehydrated (right) states; open metals sites are highlighted in orange. (c) Dehydration steps of (Zn)MOF-74 at different temperatures. Water located in the center of the pores is lost at 100 °C, water physisorbed close to the SBUs at 108 °C, and water chemisorbed on the open metal sites at 195 °C. All hydrogen atoms are omitted for clarity. Color code: octahedral Zn, blue; Zn open metal sites, orange; C, gray; O, red.

steps and Figure 17.8b shows fragments of the hydrated and dehydrated SBU of MOF-74. The high temperature required to regenerate MOF-74 to its full capacity is related to the strong binding of water to open metal sites, which is undesirable for most applications. It should also be pointed out that adsorption and desorption of water on open metal sites leads to global or local deformation of the structure, which can often result in gradual structural degradation.

The presence of open metal sites typically results in adsorption of water at low relative pressures (P/P_0) leading to Type I isotherms and a large Henrys constant K_H (slope of the isotherm at $P/P_0 \rightarrow 0$). A steep uptake at very low P/P_0 and a hysteresis (i.e. a discrepancy between the adsorption and desorption branch) in the low-pressure region of the isotherm is indicative of chemisorption. With respect to applications, this means that a high energy input is needed to restore the full capacity of such MOFs. Therefore, in many cases MOFs that adsorb water reversibly and require less energy for the complete regeneration of their capacity are more desirable.

17.3.2.2 Reversible Cluster Formation

The adsorption of water by reversible formation of water clusters is a mechanism commonly observed for porous carbons, and a similar mechanism has been observed for microporous MOFs such as MOF-801 ($Zr_6O_4(OH)_4(fumarate)_{12}$) [16]. MOF-801 is built from 12-c Zr_6O_8-core SBUs and fumarate linkers that are connected to form an **fcu** net that features two differently sized tetrahedral and one octahedral pore with diameters of 4.8, 5.6, and 7.4 Å, respectively (Figure 17.9a) [8e, 17]. The bridging –OH groups on the $Zr_6O_4(OH)_4(-COO)_{12}$ SBUs of MOF-801 are the primary adsorption sites. At low relative pressures water is adsorbed by hydrogen bonding to these –OH groups, which leads to the formation of tetrahedral water clusters within the small tetrahedral pores (Figure 17.9c). At higher relative pressures additional water molecules are confined within the tetrahedral pores and consequently a body-centered cubic water cluster is formed (Figure 17.9d). At even higher relative pressures, adsorption in a similar arrangement is observed for the octahedral pores (Figure 17.9b). The octahedral pores do not feature primary adsorption sites; however, water molecules adsorbed in the tetrahedral pores provide adsorption sites that facilitate the adsorption of water in the larger cavity of the framework. This leads to the formation of larger water clusters that eventually connect, which results in continuous pore filling. The different adsorption sites in the tetrahedral and octahedral pores of MOF-801 are shown in Figure 17.9.

Cluster-mediated adsorption has the advantage that a large working capacity in applications based on adsorption/desorption using mild temperature- and pressure swing adsorption (TSA and PSA) can be achieved. A cluster-mediated pore-filling mechanism can only occur for porous materials with pore diameters $<D_c$, the critical pore diameter for capillary condensation. The critical diameter for capillary condensation of water is 20.76 Å (at 25 °C) and can be calculated following Eq. (17.1):

$$D_c = \frac{4\sigma T_C}{(T_C - T)} \tag{17.1}$$

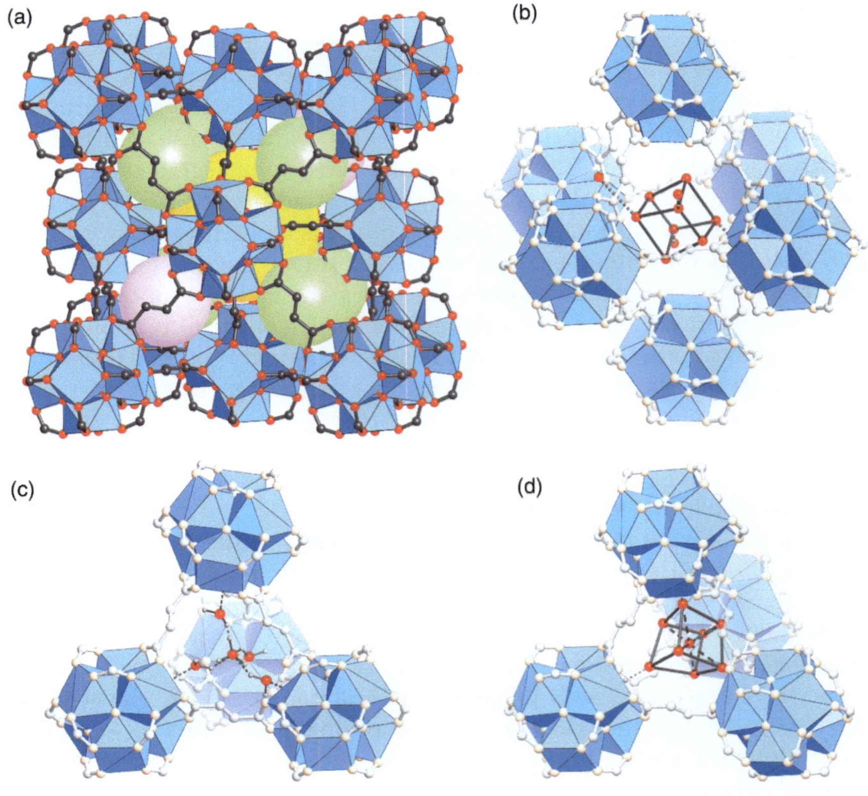

Figure 17.9 (a) Single-crystal structure of MOF-801. The pink, green, and yellow spheres highlight the small tetrahedral (4.8 Å), large tetrahedral (5.6 Å), and octahedral (7.4 Å) pores, respectively. (b)–(d) Three different water clusters are formed in the three distinct pores. (b) A cubic cluster is formed in the octahedral, (c) a tetrahedral cluster in the small tetrahedral, and (d) a body-centered cubic cluster in the large tetrahedral pore. All hydrogen atoms are omitted for clarity. Color code in (a): Zr, blue; C, gray; O, red. Color code in (b–d): Zr, blue; C, and O atoms that are part of the framework are shown in white and light orange, respectively, and the oxygen atoms of adsorbed water are shown in red.

where σ is the van der Waals diameter of the adsorbent and T_C and T are the critical temperature of the adsorbate and the adsorption temperature, respectively. Based on this, a large working capacity for microporous materials within the range of 10–30% RH is expected for materials with large pore volumes and pore sizes approaching this limit. The structure of Co_2Cl_2(BTDD), a MOF built from infinite helical SBUs that are connected through linear ditopic BTDD linkers, is composed of hexagonal channels with a diameter just above the critical pore diameter for capillary condensation of water (about 21 Å) for the fully activated MOF (Figure 17.10a) [5]. Hydration at low relative pressures results in occupation of the open metal sites by water (Figure 17.10c), which decreases the pore diameter to about 17 Å leading to a microporous material (Figure 17.10b). Therefore, the water adsorption isotherm of Co_2Cl_2(BTDD) is of Type IV and shows a steep uptake at low relative pressures

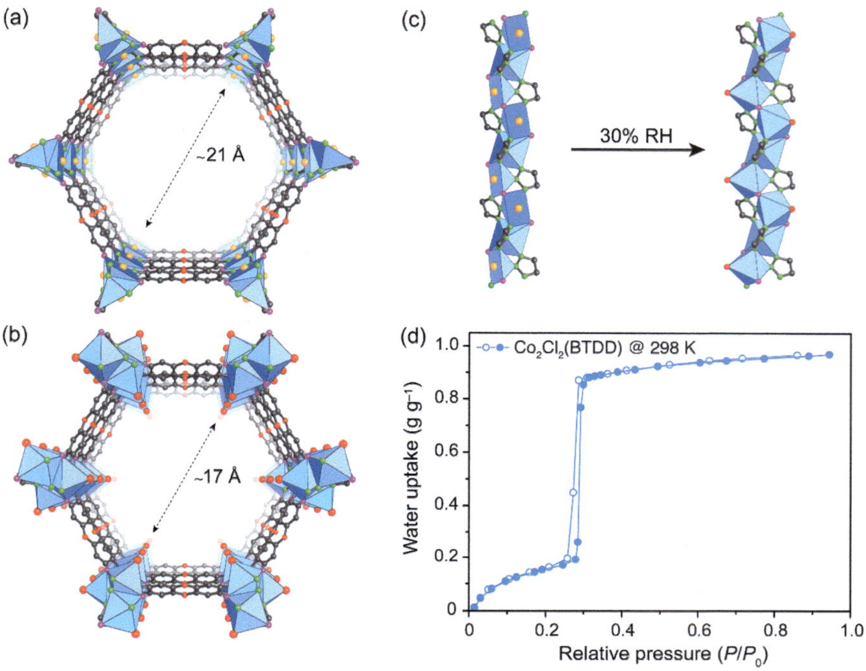

Figure 17.10 Crystal structure and water adsorption isotherm of Co_2Cl_2(BTDD). (a) One pore of fully activated Co_2Cl_2(BTDD) with view along the crystallographic *c*-axis. The fully activated structure features open metal sites pointed at the center of the pore. The pore diameter is about 21 Å, which is above the critical diameter for capillary condensation (D_c = 20.76 Å at 25 °C). (b) Adsorption of water on the open metal sites results in a decrease of the pore diameter to the microporous regime (about 17 Å). (c) Illustration of the hydration of the open metal sites on the SBUs of Co_2Cl_2(BTDD) at 30% RH. (d) Water adsorption isotherm of Co_2Cl_2BTDD recorded at 298 K. The low inflection point (α) at $P/P_0 = 0.29$ indicates a hydrophilic pore surrounding. Color code: octahedral Co, blue; Co open metal sites; orange; C, gray; N, green; O, red; Cl, pink.

(inflection point $\alpha = 0.29$), almost no hysteresis, and a maximum capacity of $q_{max} = 0.97\ g\,g^{-1}$ (Figure 17.10d). The large water uptake and the low inflection point are a result of the polar and hydrophilic nature of the BTDD linker, the large pore volume, and the pore diameter close to D_c. High initial Q_{st} values at low coverage indicate strong interactions between water and the framework, while Q_{st} values during pore filling (~44 kJ mol^{-1}) drop close to the evaporation enthalpy of water (−40.7 kJ mol^{-1}), indicating that in this pressure window water–water interactions are predominant.

17.3.2.3 Capillary Condensation

For MOFs with pore diameters larger than D_c the adsorption of water on primary adsorption sites or in the shape of mono-/multilayers is followed by capillary condensation. In contrast to the cluster-mediated filling in hydrophilic microporous MOFs, adsorption by capillary condensation in mesoporous MOFs is irreversible and results in Type IV and V isotherms that typically show a hysteresis loop (Figure 17.11). An example for a MOF showing this type of adsorption behavior

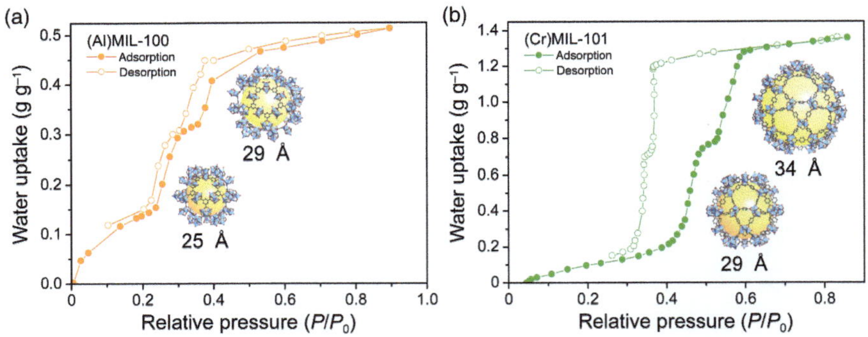

Figure 17.11 Typical water adsorption isotherms of mesoporous MOFs. An S-shaped isotherm and hysteresis are indicative of capillary condensation. The presence of two differently sized cages in the **mtn** type structures of (a) MIL-100 and (b) MIL-101 lead to two distinct steps in the corresponding adsorption isotherms. The adsorption is represented by filled symbols, the desorption by open symbols. All hydrogen atoms are omitted for clarity. Color code: Al and Cr, blue; C, gray; O, red.

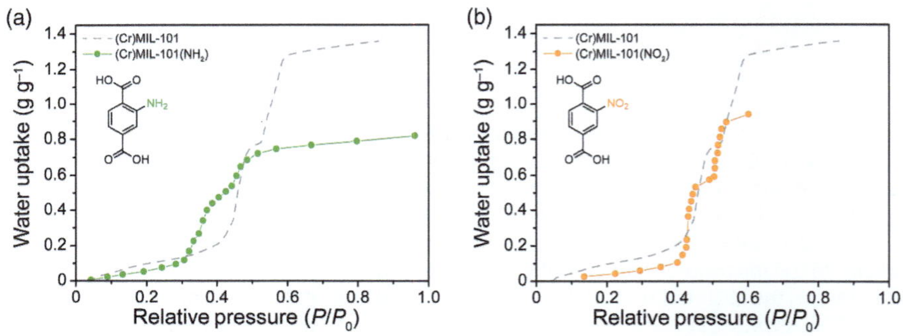

Figure 17.12 Influence of differently substituted BDC linkers in the structure of (Cr)MIL-101 on the water adsorption isotherm. (a) Functionalization with amine groups (–NH_2) renders the pores more hydrophilic and therefore shifts the inflection point to lower relative pressures. The decreased pore volume lowers the maximum adsorption capacity. (b) Hydroneutral nitro (–NO_2) substituents have a less pronounced effect on the inflection point; however, a decrease in the maximum capacity is observed. In both cases, the general shape of the isotherm is retained and the adsorption isotherm of pristine (Cr)MIL-101 is shown in gray.

is MIL-101. The **mtn** type structure of MIL-101 contains three differently sized cages, two of which are in the mesoporous regime and measure 29 and 34 Å in diameter (see Figure 4.12). The adsorption isotherm has the typical S-shape, showing low uptake at low relative pressures, a steep uptake at around $P/P_0 = 0.4$, and a second steep uptake at $P/P_0 = 0.5$ that are correlated to filling of the 29 and 34 Å large cages, respectively (Figure 17.11b) [18]. The adsorption is initiated by nucleation and growth on the inner surface, which is followed by capillary condensation at higher pressures that results in a hysteresis loop. This also manifests itself in the heats of adsorption determined at different coverage. The high heat of adsorption (about $-80\,\mathrm{kJ\,mol^{-1}}$) at low coverage is due to strong interactions of the open metal site with water, whereas a coverage of about 20%, Q_{st} decreases by almost 50% (about −45 to −50 kJ mol^{-1}), which is in the same range as the evaporation enthalpy of water (about −44 kJ mol^{-1}) [19].

17.4 Tuning the Adsorption Properties of MOFs by Introduction of Functional Groups

Functional groups can have a significant effect on the shape of the adsorption isotherm, the position of the inflection point, and the maximum uptake. The general shape of the isotherm is governed by the hydrophilicity/hydrophobicity (see Figure 17.7), pore diameter, and chemical nature of the adsorbent such as the presence of open metal sites. Hydrophilic MOFs with pore diameters <20 Å typically show S-shaped isotherms unless their structures contain open metal sites, in which case Type I isotherms are observed. Multiple adsorption steps common for structures with hierarchical pore systems and hysteresis are observed for mesoporous MOFs or MOFs with open metal sites. Aside from the presence of open metal sites, the abovementioned parameters cannot be altered for a given material. In contrast, this is possible for the hydrophilicity that can be altered by introducing functional groups onto the backbone of the linker. To illustrate the effect of different functional groups on the adsorption properties of MOFs we consider a series of functionalized (Cr)MIL-101 analogs. The water adsorption of pristine (Cr)MIL-101 has been discussed earlier and its adsorption isotherm is shown in Figure 17.11b. The influence of chemical modifications of the BDC linker by appending functional groups such as $-NO_2$ and $-NH_2$ groups reveals trends that are applicable to other MOFs [18, 20]. The relative pressure at which the inflection point of the isotherm is observed can act as an indicator for the hydrophobicity. The inflection point is shifted to lower relative pressures for $-NH_2$ functional groups, whereas $-NO_2$ groups seem to have a negligible effect on the position of the inflection point. In general, hydrophilic groups (e.g. $-NH_2$, −OH) shift the inflection point to lower pressures, hydroneutral groups (e.g. −COOR, −COR) have no effect on the position of the inflection point, and hydrophobic groups (e.g. −F, $-CH_3$) shift it to higher values. The classification of functional groups as hydrophilic, hydroneutral, and hydrophobic is based on their donor- and hydration number rather than the dipole moment [21]. In all three cases, the free pore volume is decreased by the additional substituent protruding into the pores, which leads to a lower maximum uptake. While the pressure at which capillary condensation occurs strongly depends on the chemical nature of the substituents appended to the linker, the desorption pressure is nearly unaltered. This is explained by the fact that the hydrophilic surface of (Cr)MIL-101(NH_2) is covered with adsorbed water at lower relative pressures, thus allowing for capillary condensation to take place at lower relative pressures. In the desorption step the pore size is the deciding factor, which is approximately the same for all three MOFs.

The principles discussed above are generally applicable and similar modifications carried out on other MOFs result in similar effects. Amino functionalization of the *m*-BDC linker in CAU-10 even results in a transition from a Type V to Type

I isotherms. In contrast, appending nonpolar hydrophobic groups such as $-CH_3$ result in a significant shift of the inflection point to higher relative pressures [22].

Functionalization of linkers can also influence the "breathing" behavior of MOFs. While breathing of MOFs in the presence of CO_2 is well studied only few examples of breathing initiated by the adsorption of water are known. Pristine (Al)MIL-53 shows no breathing in the presence of water, while (Al)MIL-53(OH) and (Al)MIL-53(NH_2) undergo structural change at high RH values [23]. The step around RH 80% in the water adsorption isotherm of (Al)MIL-53(OH) corresponds to a phase transition from the narrow pore (*np*) to the wide pore (*wp*) phase and results in a fivefold increase of the water uptake. This phenomenon may be explained by stronger interaction between –OH groups and water compared to the other isoreticular MOFs.

17.5 Adsorption-Driven Heat Pumps

Heating and cooling systems find widespread application in industry as well as in our daily lives. Such systems are mostly of mechanical nature due to comparatively low production costs and high coefficient of performance values [24]. Some major drawbacks of mechanical systems are that they consume large quantities of energy and rely on liquid refrigerants that have detrimental effects on the environment such as the depletion of the ozone layer and global warming [25]. While the use of chlorofluorocarbons and hydrochlorofluorocarbons as refrigerants has been prohibited, first by the Montreal agreement in 1988 and later in the Kyoto agreement in 1998, respectively, hydrofluorocarbons are still widely used [25c]. Therefore, the development of alternative technologies that operate with a higher energy efficiency and do not rely on the use of toxic refrigerant is sought after. For the evaluation of the efficiency for heat pump systems the primary energy efficiency must also be considered. Mechanical heat pumps operate using electrical power, which is generated in a separate process, thus lowering the overall efficiency. Today, absorption-driven heat pumps using salt solutions as the working liquid have found their way into many applications. Adsorption-driven heat pumps (ADHP) are based on the same working principle as their absorption-based counterparts but use a solid rather than a liquid adsorbent, making the devices more transportable and easier to handle. Furthermore, ADHPs can operate on different environmentally friendly primary energy sources such as solar energy, geothermal energy, or waste heat. Therefore, ADHPs represent an environmentally friendly alternative to mechanical heat pumps. In this section, the working principle of ADHPs is outlined and requirements regarding the vapor adsorption behavior of the adsorbent are discussed. ADHPs can operate on different working fluids (e.g. water, methanol, ethanol, and ammonia); here we will focus on water. A more detailed discussion of ADHPs is found in reference [1a].

17.5.1 Working Principles of Adsorption-Driven Heat Pumps

The concept of ADHPs may seem confusing at first since the overall process describes heat transfer from a reservoir of low temperature (T_1) to one of higher temperature (T_2). The second principle of thermodynamics seems to contradict this heat transfer, but in a closed cycle where multiple state functions change simultaneously, this is possible. The overall process in ADHPs is similar to that of their absorption-driven counterparts with the sole difference being that ADHPs rely on a solid adsorbent. In contrast to a liquid medium the solid adsorbent cannot be circulated, meaning that all four components ((i) adsorber, (ii) desorber, (iii) evaporator, and (iv) condenser) of the heat pump are of the same material. Therefore, an ADHP consists of only two compartments. It can be used for heating or cooling applications, depending on whether the released heat of adsorption or that consumed by evaporation of the working liquid is utilized. A simplified representation of the working cycle of an ADHP is given in Figure 17.13.

17.5.2 Thermodynamics of Adsorption-Driven Heat Pumps

The use of water as the working fluid in ADHPs has several advantages when compared to the working fluids used in mechanical heat pumps: (i) it is nontoxic, (ii) abundant and readily available, and (iii) has a high evaporation enthalpy (2500 kJ kg^{-1}). The only drawback is that a relatively low vapor pressure of 1.0–6.0 kPa must be used under cooling conditions. Other possible working fluids include short-chain alcohols (i.e. methanol or ethanol) or other small molecules (e.g. ammonia). While the Carnot cycle used in mechanical heat

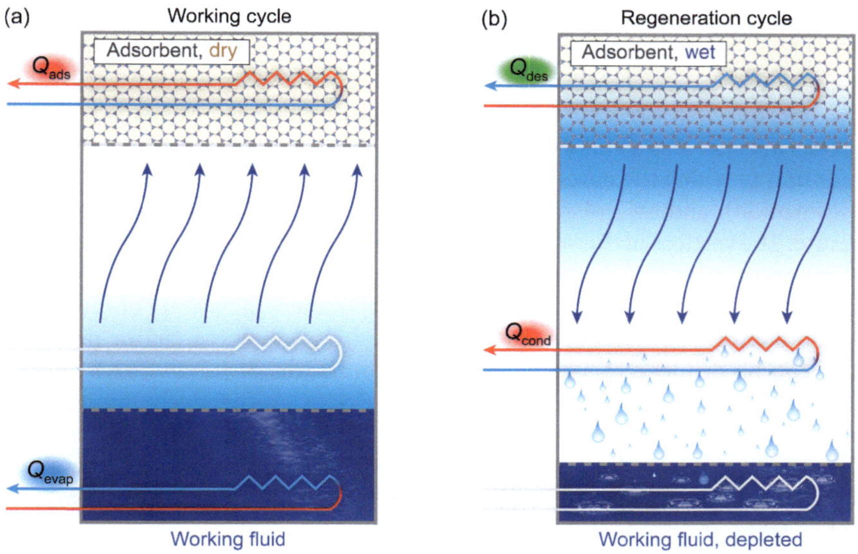

Figure 17.13 Working cycle and regeneration cycle of an adsorption-based heat pump. (a) Working cycle: the evaporation enthalpy $Q_{evap.}$ is consumed by the evaporation of the working fluid. The working fluid is then adsorbed by the adsorbent, leading to an increase in temperature due to the release of the heat of adsorption $Q_{ads.}$. (b) Regeneration cycle: desorption of the working fluid from the adsorbent is realized by heat transfer from an external source ($Q_{des.}$). The cycle is closed by the condensation of the liberated working fluid, which releases the heat of condensation ($Q_{cond.}$). Two applications are possible: (i) cooling ($Q_{evap.}$) or (ii) heating ($Q_{ads.} + Q_{cond.}$).

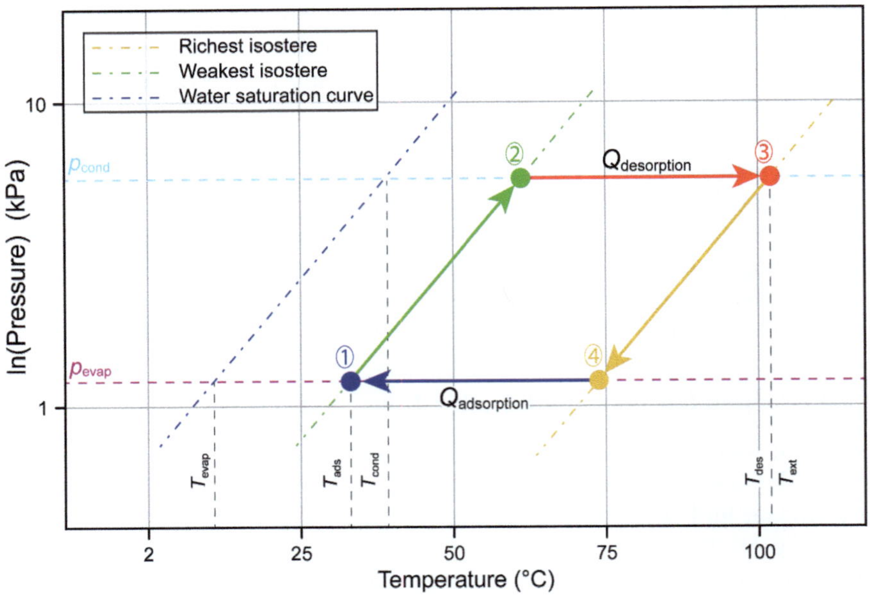

Figure 17.14 Arrhenius diagram for an adsorption-driven heat pump cycle. 4 → 1: Adsorption of the working fluid. 1 → 2: Isosteric compression. 2 → 3: Desorption by heat transfer from an external heat source. 3 → 4: Isosteric cooling.

pumps relies on two isothermal and two adiabatic compression and expansion steps, the four-step working cycle of ADHPs consists of two isosteres and two isobars. An isostere is a curve that describes a thermodynamic process involving adsorption in which the coverage does not change (i.e. constant loading, no ad- or desorption), and an isobar describes a thermodynamic process at constant pressure. The vertices of this cycle are defined by three temperatures: (i) the evaporator temperature (T_{evap}), the condenser temperature (T_{cond}), and the maximal desorption temperature (T_{des}, which is identical to the temperature of the external heat source T_{ext}). The condenser pressure (p_{cond}) and evaporator pressure (p_{evap}) can be deduced therefrom. For water, pressures of 1.2 and 5.6 kPa correspond to evaporation temperatures of 10 and 35 °C, respectively, appropriate temperature levels for cooling or low-temperature heating applications, respectively. Like any type of heat pump ADHPs operate in a discontinuous manner. The Arrhenius diagram of a full cycle is shown in Figure 17.14. It can be separated into two steps, the production and the regeneration step, that correspond to the adsorption and desorption of the working fluid, respectively. The working fluid exchange between these steps corresponds to the difference between the richest (4 → 1) and the weakest (2 → 3) isostere in the Arrhenius diagram.

The first step (4 → 1) is the production phase, where the working fluid is evaporated, cooling the system to T_{ads}. Isosteric heating (1 → 2) lifts the pressure level to the condenser pressure (p_{cond}). The desorption in the regeneration phase (2 → 3) is initiated by heat transfer from the external heat source until T_{ext}/T_{des} is reached. Different external heat sources that define the maximum desorption temperature are possible (e.g. solar thermal collector, geothermal, waste heat). Isosteric cooling (3 → 4) closes the cycle. The adsorption and desorption steps are isobaric processes; therefore, the suitability of MOFs for this application can be assessed by measuring the isotherms at p_{evap} and p_{cond} in the temperature intervals that correspond to steps 4 → 1 and 2 → 3. The temperature of the external heat source (T_{ext}) and the minimum adsorption temperature (T_{ads}) define the pressure window and thereby the reachable loading lift that is defined as the difference between the maximum loading (in $g_{water}\ g_{adsorbent}^{-1}$) for the weakest and the richest isotherm.

For applications of heat pumps for domestic heating, MOFs that adsorb water at comparatively low relative vapor pressure are favorable. This is the case for highly hydrophilic microporous MOFs that typically show a steep uptake at low relative pressures. An uptake between 10% and 30% RH is desirable to have a strong affinity for water while regeneration is still facile. MOFs that adsorb water at high relative pressures can be used in thermally driven room chillers. Since the desorption process in thermally driven chillers is fueled by an external heat source (e.g. district or solar heating) a higher evaporator temperature is perfectly acceptable and a material with comparatively low affinity toward water is in fact desirable. Interesting examples of such materials are members of MIL-100 and MIL-101 family [18, 26]. With 1.37 g g^{-1}, (Cr)MIL-101 has one of the highest water uptakes reported so far [27]. The areas of steep uptake are located between $P/P_0 = 0.3$ and 0.5. (Cr)MIL-101(NH_2) has a maximum lift of 0.55 g g^{-1}. Many other MOFs and different working fluids have been evaluated for use in adsorption-driven heat pumps [1a, 28].

17.6 Water Harvesting from Air

Today, a large proportion of the world is experiencing water stress caused by climate change and the increasing world population, and a growing portion of the world's population is expected to be affected by water scarcity in the future [29]. Only 2.5% of the water on earth is fresh water and a mere 0.3% of this fresh water are directly accessible in rivers and lakes, whereas 30.8% are groundwater and 68.9% are locked up in glaciers. The remaining 97.5% are seawater and bringing it to drinking water quality by desalination using current technologies is energy intensive. To provide sufficient amounts of water, especially in areas with poor infrastructure, new technologies must be developed. Earth-s atmosphere contains large quantities of water – equivalent to about 10% of all fresh water resources – that can potentially be used to address the global water problem. About 13 000 trillion liters of water are estimated to be contained within the atmosphere in the form of vapor and droplets and efforts have been made to collect this water using two types of water collection devices: fog collectors/dew nets and sorption-based systems [30] (Figure 17.15). Fog collectors rely on dewing on fine nets to wick water from air. Even though they can be used in dry regions that receive less than 1 mm of rain per year, they do require fog and light winds to operate, which severely limits their application. Sorption-based systems operate in adsorption–desorption cycles. Here, the desiccant is saturated

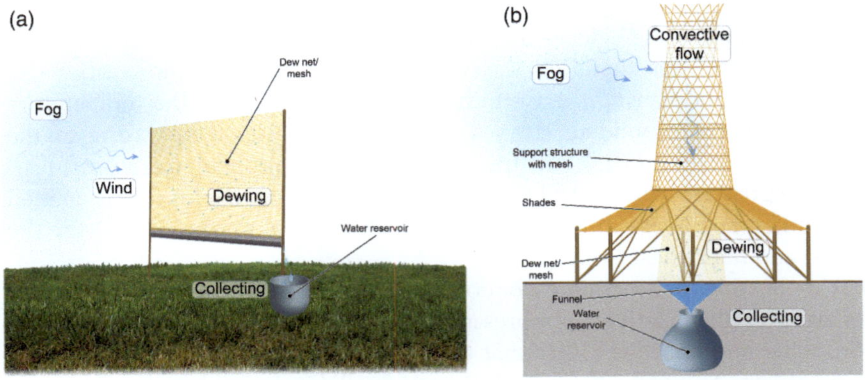

Figure 17.15 Schematics of two different concepts for fog collectors. (a) Illustration of a dew net. Air with 100% RH (fog) is transported to the dew net and consequently droplets form on the fine net. The droplets run down the net and are collected. (b) Tower for fog collection. In contrast to a dew net, the design of the tower sucks air to the mesh by convective air flow. Water droplets that form on the dew net run into a water reservoir.

with moisture under cold and humid conditions (during night) and desorption is achieved by heating (during the day). The desorbed moisture is then condensed, which requires a cold surface to reach the dew point for a given RH. Conventional desiccants such as $CaCl_2$ or silica gel show high water uptake at low RH, but due to the strong bonding of water to the desiccant a large amount of energy is required for its regeneration. This leads to low working capacities in autonomous devices.

Fog collectors require high humidity levels (close to 100% RH) to function, whereas the efficiency and viability of sorption-based water harvesting systems is limited by the adsorbent rather than the humidity levels. Therefore, the development of materials that can adsorb water at low RH and release it with minimum energy input is required. MOFs are interesting candidates for this application and we will illustrate the principles of water harvesting in autonomous low-energy devices and methods for the down-selection of promising materials.

17.6.1 Physical Background on Water Harvesting

To understand the requirements imposed on MOFs with respect to water harvesting applications it is instructive to consider the processes taking place within the water harvesting cycle. Figure 17.16a shows a schematic drawing of a water harvester [31]. The device consists of two main components: a case and a water sorption unit. The water sorption unit is a containment holding the MOF, and the case acts as an enclosure for the water sorption unit and as the condenser. To understand the processes within the device we need to consider the isotherms of the MOF and the psychometric chart for the given conditions. Figure 17.16b,c shows schematic isotherms at two different temperatures and a psychrometric chart between 0 and 65 °C at sea level, respectively. The water harvesting cycle starts with the saturation of the MOF at low temperatures and high RH during the night (I → ii, Figure 17.16b,c). Subsequent heating during the day leads to the

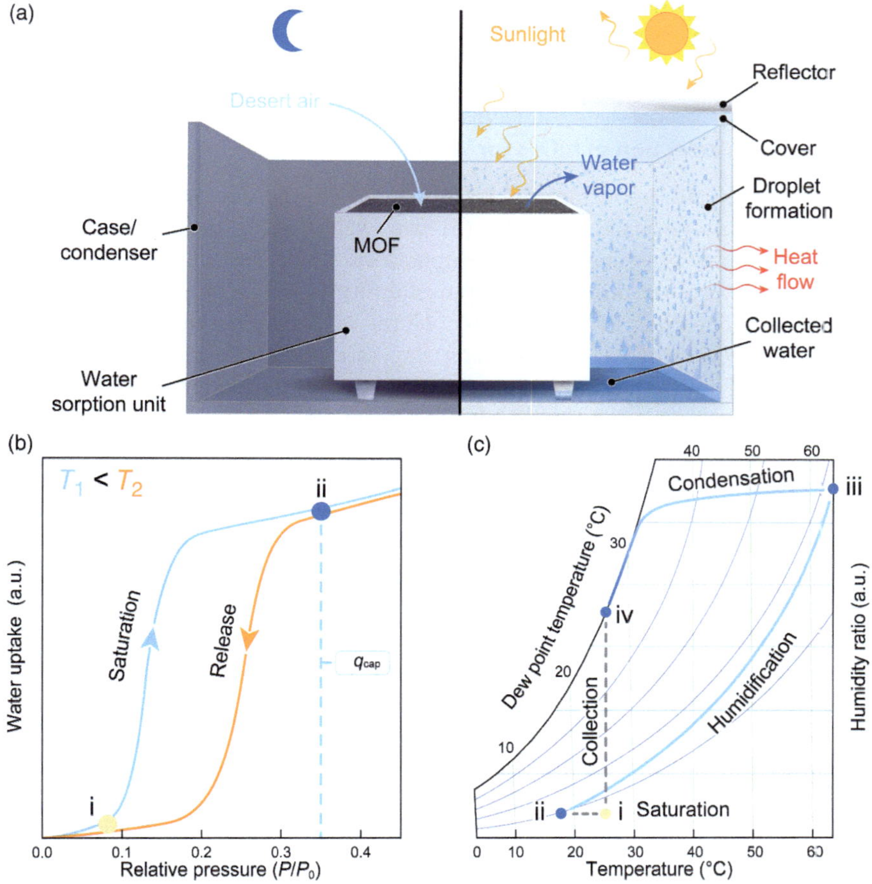

Figure 17.16 Working principles of a passive water harvester. (a) Schematic of a passive water harvester. At night the MOF is saturated (low temperatures, high RH). Solar thermal heating during the day leads to desorption of water and subsequent condensation on the cold surfaces of the case. (b) Schematic adsorption isotherms for two different temperatures ($T_1 < T_2$) illustrating the saturation with and release of water. (c) Schematic of the psychrometric chart at sea level. The temperature is plotted against the humidity ratio. An idealized water harvesting cycle is highlighted: starting at point i the MOF is saturated (overnight) until point ii is reached. A subsequent increase in temperature leads to desorption of water from the MOF that is accompanied by an increase in the humidity ratio until point iii is reached. When the humidified air is cooled at the condenser the dew point is reached and water vapor begins to condense (point iv).

desorption of water and a concomitant increase in the humidity ratio (ii → iii, Figure 17.16c). The water vapor is transported to the condenser by convection and cooled down to its dew point (iii → iv, Figure 17.16c). At the end of the day the condensed water is collected and the cycle starts over (iv → i, Figure 17.16c). To reach high efficiencies the MOF must fulfill the following requirements: high hydrolytic stability, adsorption of water at low RH, facile regeneration, high thermal conductivity, and high adsorption in the infrared range of the solar spectrum. The first three parameters can be adjusted by judicious choice of the building units and deliberate design of the MOF structure, whereas the latter two parameters can be enhanced by additives with good thermophysical properties (e.g. graphite) [32].

17.6.2 Down-selection of MOFs for Water Harvesting

Several factors are important in selecting a material for water harvesting applications: (i) the water stability, (ii) the cycling stability, (iii) the uptake capacity, (iv) the pressure and/or temperature swing response, and (v) the adsorption/desorption kinetics. Water stability and cycling stability have been discussed earlier in this chapter. Here, we will present a strategy for the down-selection of MOFs for water harvesting. This four-step process is illustrated in Figures 17.7 and 17.8. The first step is the determination of the water stability. For a water-stable MOF to be of interest for water harvesting applications, it should have a large uptake capacity. In addition to a large uptake the pressure window in which the uptake occurs and the location of the inflection point on the P/P_0 axis are important. A steep uptake in a narrow pressure region is advantageous since this will potentially allow for PSA within a narrow RH window. An inflection point at low pressures is favorable since it facilitates the saturation of the MOF at low RH. Figure 17.17a shows isotherms for three hypothetical materials. Only materials 1 and 2 are interesting candidates because they show a large uptake at low relative pressures within a narrow pressure window. Since the water harvesting process is a mixed PSA–TSA process, the temperature response of the material must be evaluated. For this purpose, isotherms at different temperature must be measured. An idealized set of isotherms for three different temperatures ($T_1 < T_2 < T_3$) is shown in Figure 17.17b. In this example, the inflection point is shifted significantly with temperature, which is desirable for efficient TSA cycles. The more pronounced

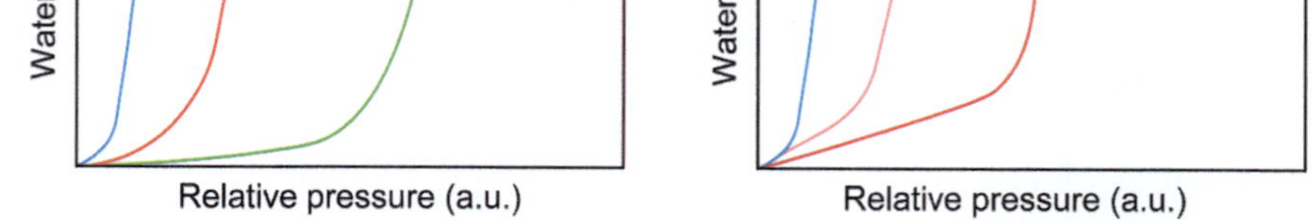

Figure 17.17 (a) Adsorption isotherms for materials of different hydrophilicity and pore dimensions. Material 1 is the most promising candidate because it shows a steep uptake at low relative pressures within a narrow pressure window. (b) Temperature-dependent water adsorption isotherms help analyze the temperature response and identify a reasonable temperature range for TSA.

the shift of the inflection point is at elevated temperatures, the more promising is the material.

The aforementioned measurements give idealized information on the adsorption properties under PSA and TSA conditions and represent equilibrium data. However, a kinetic component is always present, which leads to a lower usable working capacity for a defined cycle. Therefore, it is instructive to measure the kinetics of several adsorption–desorption cycles within different pressure and temperature intervals. Figure 17.18a shows an isotherm dissected into three pressure windows ($P_1 < P_2 < P_3$). These three segments have been chosen since it can be assumed that different adsorption mechanisms are predominant and therefore different kinetics are expected. Segment one is governed by monolayer adsorption, in segment two micropore filling dominates, and in segment three the diffusion is hindered and slows down the kinetics. The kinetics are determined by carrying out TGA (thermal gravimetric analysis) measurements either at constant temperature (Figure 17.18b, top right) or at constant RH (Figure 17.18b, bottom right). In both PSA and TSA experiments, material 1 shows faster kinetics and a larger maximum working capacity (MWC) for the selected TSA or PSA conditions. The working capacity within the water harvesting device is inevitably lower than the MWC determined in these experiments due to the influence of large-scale synthesis, processing and shaping of the material, limitations due to the design, and the specific cycling conditions. The kinetics determined in such measurements are strongly dependent on the synthesis and processing (e.g. powder, thin film, pressed pellets, or extrudates) of the MOF. Measurements of powders of different materials with the same packing density yield comparable results, and such measurements are inevitable for the proper evaluation of water adsorption materials.

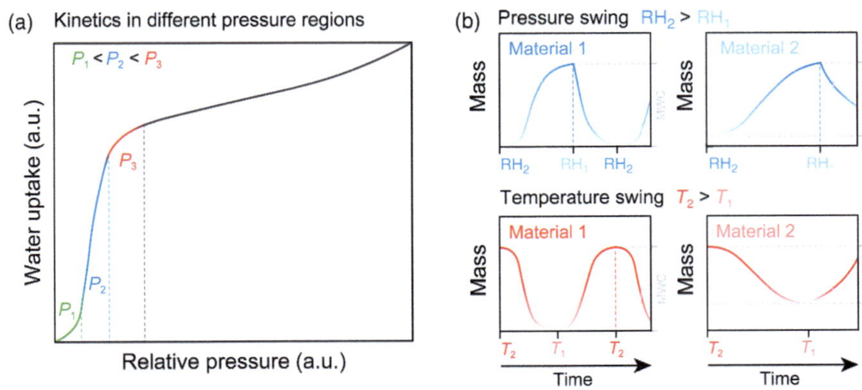

Figure 17.18 (a) Determination of the kinetics of water adsorption. Three segments of the isotherm are selected for TSA and PSA measurements. Those segments are selected based on the assumption that different mechanisms predominate in these pressure regions. (b) Pressure swing and temperature swing experiments help to further characterize the material and determine its suitability for water harvesting applications. Fast kinetics and a large working capacity are favorable.

17.7 Design of MOFs with Tailored Water Adsorption Properties

So far, we discussed different structural factors that have profound impact on the hydrolytic stability, the water adsorption mechanism, and the shape of the water sorption isotherm of MOFs [33]. In the following we will outline design principles for MOFs with tailored water adsorption properties. We will focus on influential structural parameters that are accessible by means of reticular chemistry and allow the adjustment of the water adsorption properties of MOFs.

17.7.1 Influence of the Linker Design

The hydrophilicity is the most influential physical parameter of the linker molecule with respect to the water adsorption properties of a MOF built therefrom. Hydrophilicity is a quantity that is not easily accessible. It is important to understand that the polarity of a given organic molecule does not necessarily correlate with its hydrophilicity. Nonpolar molecules are always hydrophobic, whereas polar molecules can be hydrophilic (e.g. R–NH_2, R–OH, and R–COOH), hydroneutral (e.g. R–NO_2, R–COR, R–COOR), or hydrophobic (e.g. R–Cl, R–F). As a rule of thumb, molecules that act as hydrogen bond donors or acceptors are hydrophilic. For potentially hydrophilic molecules the ratio between the overall exposed surface (van der Waals surface, A_{vdW}) and the polar surface (A_{polar}) helps rank their relative hydrophilicity as well as that of a pore system arising from its reticulation into a MOF. Introduction of heteroatoms capable of strong hydrogen bonding interactions render pore environments more hydrophilic. This principle is highlighted by the shift of the inflection point of the isotherm of CAU-10 analogs that are built from a variety of different linkers. Because all materials are of the same general structure, the shift of the inflection point is solely due to differences in the hydrophilicity of the linker (Figure 17.19). While appending hydrophilic functionalities to a linker molecule shifts the inflection point of the isotherm to lower relative pressures, in contrast to heterocyclic hydrophilic linker molecules, typically this is accompanied by a decrease of the maximum capacity, which is related to the decrease of the free pore volume.

17.7.2 Influence of the SBU

High valence metals lead to strong bonding with the organic linker and therefore the resulting frameworks are generally hydrolytically stable. Since a high atomic weight lowers the gravimetric capacity, the use of lighter elements such as aluminum or titanium, as compared to zirconium or hafnium, is favorable. In many MOF structures the bridging –OH groups in the SBUs act as primary adsorption sites by hydrogen bond formation with adsorbed water molecules [8e, 34]. Other primary adsorption sites with even stronger interaction (chemisorption) such as open metal sites result in Type I isotherms and hysteresis. In the context of large-scale applications, apart from the physical properties and the structural diversity associated with the use of certain metals their toxicity must also be considered. This is not only linked to health risks during the synthesis, but in the case of materials used for water harvesting applications, partial dissolution of the MOF results in the contamination of the harvested water. Therefore, when targeting materials for this application, the use of linkers and metals with a high LD_{50} is preferable (see Chapter 16) [35].

17.7.3 Influence of the Pore Size and Dimensionality of the Pore System

As illustrated in this chapter by several examples, different pore sizes lead to completely different adsorption mechanisms and isotherms. While microporous materials typically show S-shaped isotherms, an inflection point at comparatively low RH values, and no hysteresis, mesoporous materials with a pore diameter larger than $D_c = 20.76$ Å (at 25 °C) generally show pronounced hysteresis and an inflection point at high relative pressures. Consequently, microporous materials are favorable for water harvesting from air.

While it is difficult to quantify the influence of the dimensionality of the pore system, the magnitude of its effect on the adsorption properties is estimated to be rather small. To illustrate this hypothesis, we compare two structures with similar pore sizes and volumes but a different dimensionality of the pore system: (i) MOF-810 with a 3D pore system (**fcu** topology) and pore diameters between 4.8 and 7.4 Å and (ii) CAU-10 (Al(OH)(*m*-BDC)) with a 1D pore system (**yfm** topology) and pores of 7 Å diameter. Both structures have a similar pore volume ($V_p = 0.45$ and $0.43\ cm^3\ g^{-1}$, respectively), and bridging –OH groups of the SBUs are identified as the primary adsorption sites. The structural similarities outweigh the difference in the dimensionality of the pore system, as illustrated by similar inflection points (0.08 and 0.15, respectively) and maximum capacities ($q_{max} = 0.36\ g\ g^{-1}$) [8e, 36b].

17.7.4 Influence of Defects

In contrast to gas adsorption isotherms, water adsorption isotherms are very sensitive to the presence of defects in the crystal structure of the MOF. The shape of nitrogen and argon isotherms as well as the surface areas determined therefrom show only slight changes for high defect concentrations, whereas a significant shift of the inflection point as well as a higher adsorption capacity are commonly observed for the adsorption of water in such materials. The presence of defects in MOF-801 and UiO-66 (see Figure 4.28) renders their pores more hydrophilic and the larger pore volume leads to an increased maximum water uptake [8e, 37]. The presence of defects also manifests itself in the Q_{st} values calculated from water sorption isotherms. Q_{st} values at low coverage for defect-free UiO-66 are low ($Q_{st} = -15\ kJ\ mol^{-1}$) and increase with higher loading ($Q_{st} = -60\ kJ\ mol^{-1}$). In contrast, the Q_{st} values determined for defect-rich UiO-66 are significantly higher for low loading (−60 to $-70\ kJ\ mol^{-1}$). This is indicative of strong interactions between adsorbed water and the framework and consequently of hydrophilic pores. The presence and concentration of defects can significantly change the water adsorption behavior, which makes this method a valuable tool for the in-depth characterization of MOFs.

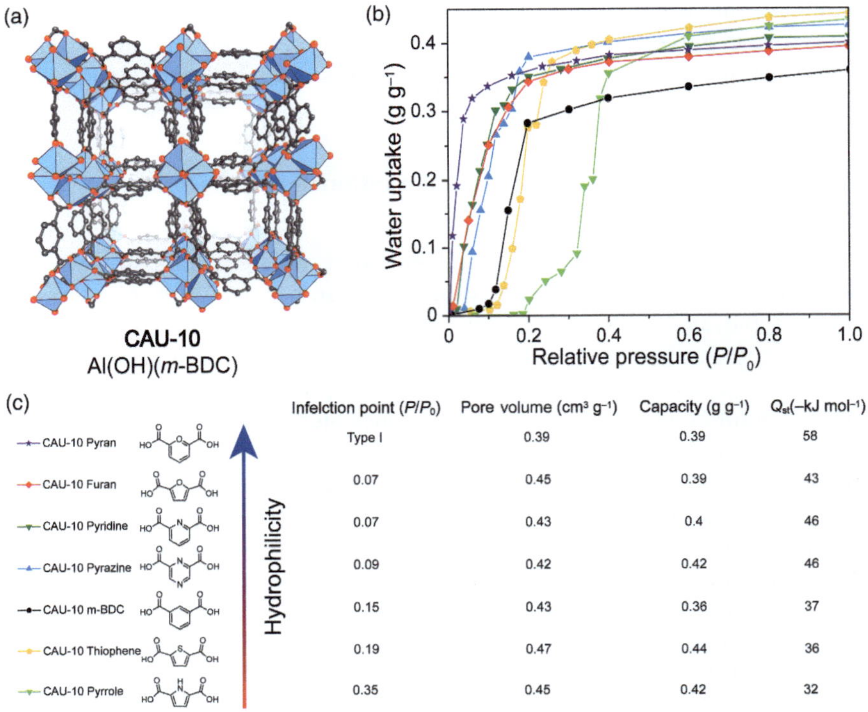

	Infelction point (P/P_0)	Pore volume ($cm^3\ g^{-1}$)	Capacity ($g\ g^{-1}$)	Q_{st}(–kJ mol^{-1})
CAU-10 Pyran	Type I	0.39	0.39	58
CAU-10 Furan	0.07	0.45	0.39	43
CAU-10 Pyridine	0.07	0.43	0.4	46
CAU-10 Pyrazine	0.09	0.42	0.42	46
CAU-10 m-BDC	0.15	0.43	0.36	37
CAU-10 Thiophene	0.19	0.47	0.44	36
CAU-10 Pyrrole	0.35	0.45	0.42	32

Figure 17.19 (a) Crystal structure of CAU-10 (**yfm**) with view along the *c*-direction. Source: Reinsch et al. 2012 [22]. Reproduced with permission of ACS.
(b) Water adsorption isotherms of CAU-10 analogs. All materials are isostructural and consequently the shift of the inflection point and the changes in the isosteric heat of adsorption are solely correlated to the hydrophilicity of the linker molecule.

17.8 Summary

In this chapter, we introduced the principles of water adsorption in MOFs. We first presented structural factors governing the thermodynamic and kinetic hydrolytic stability of MOFs and derived approaches to the design of hydrolytically stable MOFs therefrom. We discussed the possible water adsorption mechanisms and their correlation with the framework structure and specific structural features. We saw that the pore size has a strong effect on the shape of the isotherm and that large pores can even lead to irreversible water adsorption by capillary condensation. The effect of hydrophilic, hydroneutral, and hydrophobic functionalities on the shape of the water adsorption isotherm and the maximum capacity was discussed, revealing that hydrophilic groups shift the inflection point of the isotherm to lower relative pressures while hydrophobic groups shift it to higher relative pressure. This provides us with a tool to adjust the position of steep uptake by isoreticular functionalization. Subsequently, we discussed the two most studied applications with respect to water adsorption in MOFs: ADHPs and water harvesting from air. We illustrated the working principles of ADHPs and water harvesters, and conclude this chapter with considerations relevant to the design of next-generation materials for these applications.

References

1 (a) de Lange, M.F., Verouden, K.J., Vlugt, T.J. et al. (2015). Adsorption-driven heat pumps: the potential of metal-organic frameworks. *Chemical Reviews* 115 (22): 12205–12250. (b) Elsayed, E., Al-Dadah, R., Mahmoud, S. et al. (2017). CPO-27(Ni), aluminium fumarate and MIL-101(Cr) MOF materials for adsorption water desalination. *Desalination* 406: 25–36. (c) Ribeiro, A.M., Sauer, T.P., Grande, C.A. et al. (2008). Adsorption equilibrium and kinetics of water vapor on different adsorbents. *Industrial and Engineering Chemistry Research* 47 (18): 7019–7026. (d) Kanchanalai, P., Lively, R.P., Realff, M.J., and Kawajiri, Y. (2013). Cost and energy savings using an optimal design of reverse osmosis membrane pretreatment for dilute bioethanol purification. *Industrial and Engineering Chemistry Research* 52 (32): 11132–11141. (e) AbdulHalim, R.G., Bhatt, P.M., Belmabkhout, Y. et al. (2017). A fine-tuned metal-organic framework for autonomous indoor moisture control. *Journal of the American Chemical Society* 139 (31): 10715–10722. (f) Kim, H., Yang, S., Rao, S.R. et al. (2017). Water harvesting from air with metal-organic frameworks powered by natural sunlight. *Science* 356 (6336): 430–434.

2 Low, J.J., Benin, A.I., Jakubczak, P. et al. (2009). Virtual high throughput screening confirmed experimentally: porous coordination polymer hydration. *Journal of the American Chemical Society* 131 (43): 15834–15842.

3 DeCoste, J.B., Peterson, G.W., Jasuja, H. et al. (2013). Stability and degradation mechanisms of metal-organic frameworks containing the $Zr_6O_4(OH)_4$ secondary building unit. *Journal of Materials Chemistry A* 1 (18): 5642–5650.

4 (a) Choi, H.J., Dincă, M., Dailly, A., and Long, J.R. (2010). Hydrogen storage in water-stable metal-organic frameworks incorporating 1,3- and 1,4-benzenedipyrazolate. *Energy & Environmental Science* 3 (1): 117–123. (b) Colombo, V., Galli, S., Choi, H.J. et al. (2011). High thermal and chemical stability in pyrazolate-bridged metal-organic frameworks with exposed metal sites. *Chemical Science* 2 (7): 1311–1319.

5 Rieth, A.J., Yang, S., Wang, E.N., and Dincă, M. (2017). Record atmospheric fresh water capture and heat transfer with a material operating at the water uptake reversibility limit. *ACS Central Science* 3 (6): 668–672.

6 (a) Liu, J., Benin, A.I., Furtado, A.M. et al. (2011). Stability effects on CO_2 adsorption for the DOBDC series of metal-organic frameworks. *Langmuir* 27 (18): 11451–11456. (b) Kizzie, A.C., Wong-Foy, A.G., and Matzger, A.J. (2011). Effect of humidity on the performance of microporous coordination polymers as adsorbents for CO_2 capture. *Langmuir* 27 (10): 6368–6373.

7 (a) DeCoste, J.B., Peterson, G.W., Schindler, B.J. et al. (2013). The effect of water adsorption on the structure of the carboxylate containing metal-organic frameworks Cu-BTC, Mg-MOF-74, and UiO-66. *Journal of Materials Chemistry A* 1 (38): 11922–11932. (b) Jeremias, F., Lozan, V., Henninger, S.K., and Janiak, C. (2013). Programming MOFs for water sorption: amino-functionalized MIL-125 and UiO-66 for heat transformation and heat storage applications. *Dalton Transactions* 42 (45): 15967–15973.

8 (a) Guillerm, V., Ragon, F., Dan-Hardi, M. et al. (2012). A series of isoreticular, highly stable, porous zirconium oxide based metal-organic frameworks. *Angewandte Chemie International Edition* 51 (37): 9267–9271. (b) Bon, V., Senkovskyy, V., Senkovska, I., and Kaskel, S. (2012). Zr(IV) and Hf(IV) based metal-organic frameworks with reo-topology. *Chemical Communications* 48 (67): 8407–8409. (c) Bon, V., Senkovska, I., Baburin, I.A., and Kaskel, S. (2013). Zr-and Hf-based metal-organic frameworks: tracking down the polymorphism. *Crystal Growth and Design* 13 (3): 1231–1237. (d) Jiang, H.-L., Feng, D., Wang, K. et al. (2013). An exceptionally stable, porphyrinic Zr metal-organic framework exhibiting pH-dependent fluorescence. *Journal of the American Chemical Society* 135 (37): 13934–13938. (e) Furukawa, H., Gándara, F., Zhang, Y.-B. et al. (2014). Water adsorption in porous metal-organic frameworks and related materials. *Journal of the American Chemical Society* 136 (11): 4369–4381.

9 Dybtsev, D.N., Chun, H., and Kim, K. (2004). Rigid and flexible: a highly porous metal-organic framework with unusual guest-dependent dynamic behavior. *Angewandte Chemie International Edition* 116 (38): 5143–5146.

10 (a) Schoenecker, P.M., Carson, C.G., Jasuja, H. et al. (2012). Effect of water adsorption on retention of structure and surface area of metal-organic frameworks. *Industrial and Engineering Chemistry Research* 51 (18): 6513–6519. (b) Liang, Z., Marshall, M., and Chaffee, A.L. (2010). CO_2 adsorption, selectivity and water tolerance of pillared-layer metal organic frameworks. *Microporous and Mesoporous Materials* 132 (3): 305–310. (c) Tan, K., Nijem, N., Canepa, P. et al. (2012). Stability and hydrolyzation of metal organic frameworks with paddle-wheel SBUs upon hydration. *Chemistry of Materials* 24 (16): 3153–3167.

11 Bellarosa, L., Gutiérrez-Sevillano, J.J., Calero, S., and López, N. (2013). How ligands improve the hydrothermal stability and affect the adsorption in the IRMOF family. *Physical Chemistry Chemical Physics* 15 (40): 17696–17704.

12 Kang, I.J., Khan, N.A., Haque, E., and Jhung, S.H. (2011). Chemical and thermal stability of isotypic metal-organic frameworks: effect of metal ions. *Chemistry – A European Journal* 17 (23): 6437–6442.

13 (a) Cavka, J.H., Jakobsen, S., Olsbye, U. et al. (2008). A new zirconium inorganic building brick forming metal organic frameworks with exceptional stability. *Journal of the American Chemical Society* 130 (42): 13850–13851. (b) Valenzano, L., Civalleri, B., Chavan, S. et al. (2011). Disclosing the complex structure of UiO-66 metal organic framework: a synergic combination of experiment and theory. *Chemistry of Materials* 23 (7): 1700–1718.

(c) Wiersum, A.D., Soubeyrand-Lenoir, E., Yang, Q. et al. (2011). An evaluation of UiO-66 for gas-based applications. *Chemistry – An Asian Journal* 6 (12): 3270–3280.

14 Drisdell, W.S., Poloni, R., McDonald, T.M. et al. (2013). Probing adsorption interactions in metal-organic frameworks using X-ray spectroscopy. *Journal of the American Chemical Society* 135 (48): 18183–18190.

15 Dietzel, P.D., Johnsen, R.E., Blom, R., and Fjellvåg, H. (2008). Structural changes and coordinatively unsaturated metal atoms on dehydration of honeycomb analogous microporous metal-organic frameworks. *Chemistry – A European Journal* 14 (8): 2389–2397.

16 Do, D., Junpirom, S., and Do, H. (2009). A new adsorption–desorption model for water adsorption in activated carbon. *Carbon* 47 (6): 1466–1473.

17 Wißmann, G., Schaate, A., Lilienthal, S. et al. (2012). Modulated synthesis of Zr-fumarate MOF. *Microporous and Mesoporous Materials* 152: 64–70.

18 Akiyama, G., Matsuda, R., Sato, H. et al. (2012). Effect of functional groups in MIL-101 on water sorption behavior. *Microporous and Mesoporous Materials* 157: 89–93.

19 Jeremias, F., Khutia, A., Henninger, S.K., and Janiak, C. (2012). MIL-100(Al, Fe) as water adsorbents for heat transformation purposes – a promising application. *Journal of Materials Chemistry* 22 (20): 10148–10151.

20 Ko, N., Choi, P.G., Hong, J. et al. (2015). Tailoring the water adsorption properties of MIL-101 metal-organic frameworks by partial functionalization. *Journal of Materials Chemistry A* 3 (5): 2057–2064.

21 (a) Gutmann, V. (1976). Empirical parameters for donor and acceptor properties of solvents. *Electrochimica Acta* 21 (9): 661–670. (b) Gutmann, V. (1978). *Donor–Acceptor Approach to Molecular Interactions*. Plenum Press. (c) Sagawa, N. and Shikata, T. (2014). Are all polar molecules hydrophilic? Hydration numbers of nitro compounds and nitriles in aqueous solution. *Physical Chemistry Chemical Physics* 16 (26): 13262–13270.

22 Reinsch, H., van der Veen, M.A., Gil, B. et al. (2012). Structures, sorption characteristics, and nonlinear optical properties of a new series of highly stable aluminum MOFs. *Chemistry of Materials* 25 (1): 17–26.

23 Shigematsu, A., Yamada, T., and Kitagawa, H. (2011). Wide control of proton conductivity in porous coordination polymers. *Journal of the American Chemical Society* 133 (7): 2034–2036.

24 Chua, K., Chou, S., and Yang, W. (2010). Advances in heat pump systems: a review. *Applied Energy* 87 (12): 3611–3624.

25 (a) Metz, B., Solomon, S., Kuijpers, L. et al. (2005). *Safeguarding the Ozone Layer and the Global Climate System: Issues Related to Hydrofluorocarbons and Perfluorocarbons*. Cambridge University Press. (b) Velders, G.J., Andersen, S.O., Daniel, J.S. et al. (2007). The importance of the Montreal protocol in protecting climate. *Proceedings of the National Academy of Sciences* 104 (12): 4814–4819. (c) Oberthür, S. (2001). Linkages between the Montreal and Kyoto protocols – enhancing synergies between protecting the ozone layer and the global climate. *International Environmental Agreements: Politics, Law and Economics* 1 (3): 357–377.

26 Ehrenmann, J., Henninger, S.K., and Janiak, C. (2011). Water adsorption characteristics of MIL-101 for heat-transformation applications of MOFs. *European Journal of Inorganic Chemistry* 2011 (4): 471–474.

27 Küsgens, P., Rose, M., Senkovska, I. et al. (2009). Characterization of metal-organic frameworks by water adsorption. *Microporous and Mesoporous Materials* 120 (3): 325–330.

28 (a) Henninger, S.K., Jeremias, F., Kummer, H., and Janiak, C. (2012). MOFs for use in adsorption heat pump processes. *European Journal of Inorganic Chemistry* 2012 (16): 2625–2634. (b) Deria, P., Bury, W., Hod, I. et al. (2015). MOF functionalization via solvent-assisted ligand incorporation: phosphonates vs carboxylates. *Inorganic Chemistry* 54 (5): 2185–2192.

29 Vörösmarty, C.J., Green, P., Salisbury, J., and Lammers, R.B. (2000). Global water resources: vulnerability from climate change and population growth. *Science* 289 (5477): 284–288.

30 (a) Schemenauer, R.S. and Cereceda, P. (1994). A proposed standard fog collector for use in high-elevation regions. *Journal of Applied Meteorology* 33 (11): 1313–1322. (b) Klemm, O., Schemenauer, R.S., Lummerich, A. et al. (2012). Fog as a fresh-water resource: overview and perspectives. *Ambio* 41 (3): 221–234. (c) Park, K.-C., Chhatre, S.S., Srinivasan, S. et al. (2013). Optimal design of permeable fiber network structures for fog harvesting. *Langmuir* 29 (43): 13269–13277. (d) Wahlgren, R.V. (2001). Atmospheric water vapour processor designs for potable water production: a review. *Water Research* 35 (1): 1–22. (e) Muselli, M., Beysens, D., Marcillat, J. et al. (2002). Dew water collector for potable water in Ajaccio (Corsica Island, France). *Atmospheric Research* 64 (1): 297–312. (f) Clus, O., Ortega, P., Muselli, M. et al. (2008). Study of dew water collection in humid tropical islands. *Journal of Hydrology* 361 (1–2): 159–171. (g) Lee, A., Moon, M.-W., Lim, H. et al. (2012). Water harvest via dewing. *Langmuir* 28 (27): 10183–10191.

31 Fathieh, F., Kalmutzki, M.J., Kapustin, E.A. et al. (2018). Practical water production from desert air. *Science Advances* 4 (6): eaat3198.

32 Yang, S., Huang, X., Chen, G., and Wang, E.N. (2016). Three-dimensional graphene enhanced heat conduction of porous crystals. *Journal of Porous Materials* 23 (6): 1647–1652.

33 Canivet, J., Bonnefoy, J., Daniel, C. et al. (2014). Structure-property relationships of water adsorption in metal-organic frameworks. *New Journal of Chemistry* 38 (7): 3102–3111.

34 Cadiau, A., Lee, J.S., Damasceno Borges, D. et al. (2015). Design of hydrophilic metal organic framework water adsorbents for heat reallocation. *Advanced Materials* 27 (32): 4775–4780.

35 (a) Horcajada, P., Chalati, T., Serre, C. et al. (2010). Porous metal-organic-framework nanoscale carriers as a potential platform for drug delivery and imaging. *Nature Materials* 9 (2): 172–178. (b) Singh, R., Gautam, N., Mishra, A., and Gupta, R. (2011). Heavy metals and living systems: an overview. *Indian Journal of Pharmacology* 43 (3): 246. (c) Tchounwou, P.B., Yedjou, C.G., Patlolla, A.K., and Sutton, D.J. (2012). *Molecular, Clinical and Environmental Toxicology*, 133–164. Springer. (d) Venugopal, B. and Luckey, T.D. (1978). Metal toxicity in mammals. In: *Chemical Toxicity of Metals*

and Metalloids, vol. 2. Plenum Press. (e) Domingo, J. (1994). Metal-induced developmental toxicity in mammals: a review. *Journal of Toxicology and Environmental Health, Part A Current Issues* 42 (2): 123–141.

36 (a) Fröhlich, D., Henninger, S.K., and Janiak, C. (2014). Multicycle water vapour stability of microporous breathing MOF aluminium isophthalate CAU-10-H. *Dalton Transactions* 43 (41): 15300–15304. (b) Borges, D.D., Maurin, G., and Galvão, D.S. (2017). Design of porous metal-organic frameworks for adsorption driven thermal batteries. *MRS Advances* 2 (9): 519–524.

37 Ghosh, P., Colón, Y.J., and Snurr, R.Q. (2014). Water adsorption in UiO-66: the importance of defects. *Chemical Communications* 50 (77): 11329–11331.

Part IV

Special Topics

18

Topology

18.1 Introduction

Describing and understanding matter is at the heart of the chemical sciences. For finite units such as molecules this is generally trivial; however, the description of structures of (crystalline) solid-state materials is significantly more challenging and several methods can be used to describe such structures. The crystal structure is the most detailed description and includes information about the atomic composition, connectivity, spatial arrangement, and symmetry of the overall structure. This description can be very complicated and therefore, in the field of inorganic solid-state chemistry, a concept of describing crystal structures as a packing of one type of atom where other atoms occupy the voids of the structure is common practice. While this works well for ionic solids, this is less helpful for extended solids such as metal-organic frameworks (MOFs), zeolitic imidazolate frameworks (ZIFs), and covalent organic frameworks (COFs). The description of extended structures in terms of topology (from the Greek τόπος meaning "place," and λόγος meaning "study") is more frequently used [1]. This concept allows to simplify structures by only considering the connections between constituents, not their chemical nature. The differences in structure description are illustrated in Figure 18.1.

This approach significantly reduces the complexity of a given structure description [2]. Aside from the simplified description of crystal structures, this concept also allows for reverse engineering of crystal structures, or simply put, "to design solid-state materials" [3]. Before we consider different topologies we need to understand what "topology" is, how to determine the topology of a given structure, and introduce the terminology required for its description.

18.2 Graphs, Symmetry, and Topology

18.2.1 Graphs and Nets

Nets are a collection of nodes that are linked by connections (edges). They are a special kind of graph, an abstract mathematical object. For mathematical graphs we distinguish between infinite and finite graphs and classify them based on

Introduction to Reticular Chemistry: Metal-Organic Frameworks and Covalent Organic Frameworks, First Edition. Omar M. Yaghi, Markus J. Kalmutzki, and Christian S. Diercks.

Companion website: www.wiley.com/go/yaghi/reticular

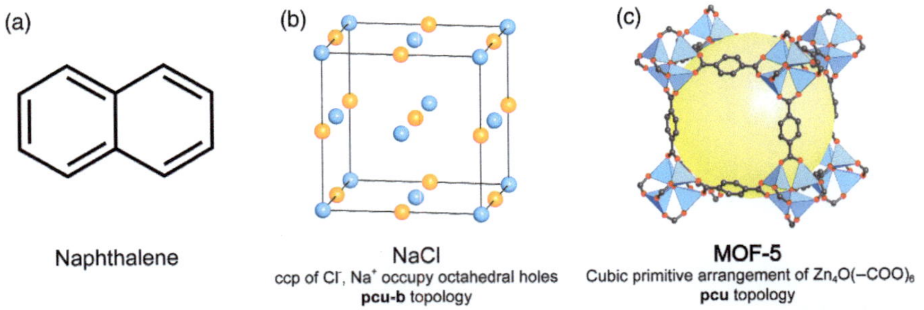

Figure 18.1 Comparison of different structure descriptions. (a) Molecular compounds are defined by the type of atoms and the chemical bonds between them, where specific units are given trivial names. (b) Ionic structures can be described by a packing of one type of ion (here cubic closest packing of Cl^-) where the other ions occupy the voids (here Na^+ occupies all octahedral holes). Alternatively, this arrangement can be described as a **pcu-b** net. (c) The crystal structure of MOF-5 can be simplified to a primitive cubic packing of $Zn_4O(–COO)_6$ SBUs that are connected by linear ditopic BDC linkers or as a **pcu** net.

their properties into (i) graphs that have loops, meaning vertices connected to themselves, (ii) graphs containing vertices that are connected by multiple links, (iii) graphs that are directional, and (iv) graphs that have loose ends.

With respect to the description of the crystal structures of MOFs in terms of nets there are only infinite 2- and 3-periodic graphs, meaning they do not possess any of the abovementioned properties (loops, multiple connections, directionality, and loose ends). It is important to note that periodicity is not synonymous with dimensionality. While every polyhedron is a 3D object, it is not periodic.

To understand these concepts and their applicability to the description of framework structures we consider a problem described by Leonhard Euler in 1735: "The seven bridges of Koenigsberg." A river flows through the city of Königsberg creating a central island and three riverbanks (bank A, B, and C) that are connected by seven bridges (Figure 18.2). Euler showed that to find

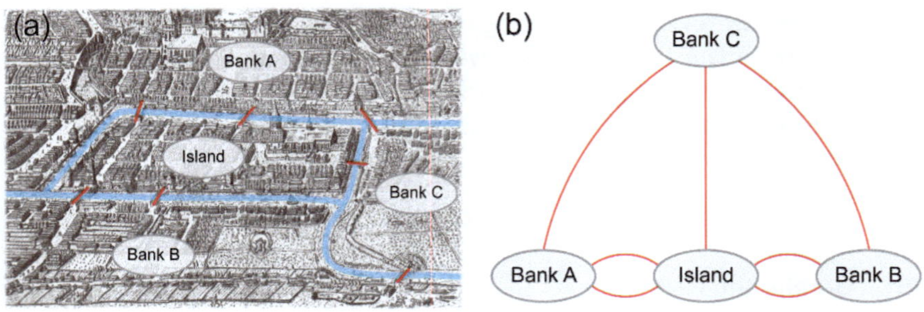

Figure 18.2 (a) The old city of Koenigsberg with the seven bridges highlighted in red. (b) Illustration of the graph describing this problem. Here, the riverbanks and the islands represent the nodes and the bridges represent the edges (red lines). From the analysis of this graph it becomes clear that it is impossible to walk through the city of Koenigsberg while passing every bridge only once.

out whether it is possible to cross the city while only crossing every bridge exactly once, neither the length of the bridges nor the distances between them is relevant. Solving this question only requires the knowledge over which bridges are connected to which islands or riverbanks.

18.2.2 Deconstruction of Crystal Structures into Their Underlying Nets

The above example highlights the need to identify the nodes and edges making up a crystal structure to establish a graph that describes it. This is referred to as the deconstruction of a crystal structure [4]. We will illustrate this process for a MOF but a similar procedure can be applied to other extended frameworks. The first step involves defining the building units that represent nodes and edges in the corresponding topology. For MOFs the building units are typically the inorganic secondary building units (SBUs) and the organic linkers. In special cases larger structural fragments can be chosen as the building units (see Section 5.2.3). To decide which building units are edges and which are nodes we need to define the number of "points of extensions" for each of them. The points of extension represent the number of connections a specific building unit makes to other building units in the structure. As an example, for carboxylate-based MOFs these points of extension are typically placed on the carboxylate carbon. Building units making only two connections represent edges, whereas all building units making three or more connections are nodes. In this chapter, we will refer to the connectivity using the abbreviation "X-c," indicating an X-connected vertex. To illustrate this process, we consider the structure of HKUST-1 and deconstruct it into its building units to derive the underlying topology. HKUST-1 is built from $Cu_2L_2(-COO)_4$ paddle wheel SBUs that are connected through BTC linkers (Figure 18.4). We separate the SBUs and the linker at the carboxylate carbon and are left with a paddle wheel, which can be reduced to a 4-c node, and the BTC linker, which can be reduced to a 3-c node. To derive the net formed by linking these nodes it is not sufficient to know their number of points of extension (here, three and four) and local symmetry (here, triangle and square) but also the connections between them and the spatial arrangement of the nodes with respect to each other. A net of nodes and edges is only unique if it cannot be transformed into another net by bending or stretching but only by breaking and making connections. This is illustrated in Figure 18.3. Connecting 3-c nodes to form 2D layers can lead to different structures, depending on the geometry of the 3-c nodes but all three structures have the same topology (Figure 18.3a). Figure 18.3b shows another case where even though all points are only connected to two other points and the order is identical, all three arrangements represent different topologies since they cannot be transformed into one another without breaking and making connections.

In our example, deconstructing the structure of HKUST-1 leads to a "twisted boracite" or **tbo** net (Figure 18.4). We will use the net identifiers of the reticular chemistry structure resource (RCSR) database [5]. According to the RCSR, topologies are denoted by three-letter identifiers (lower case, bold). These identifiers allow for the unambiguous naming of unique nets and we discuss the nomenclature of net topologies later in more detail.

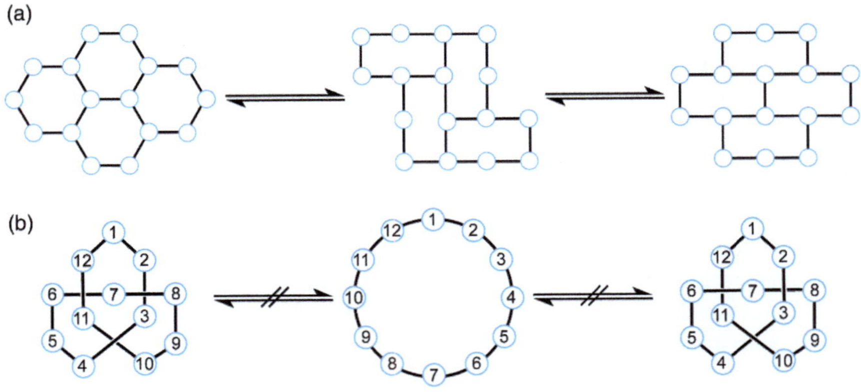

Figure 18.3 (a) Hexagonal layers of 3-c vertices can be distorted to form "brick-wall"-like layers. Since all three arrangements can be transformed into each other without the need to break and make connections they are "topologically identical" (**hcb**, honeycomb). (b) While the connectivity of all units is identical for all three arrangements, they cannot be transformed into one another without breaking and making of connections. Thus, they are topologically distinct even though the graphs are "isomorphic."

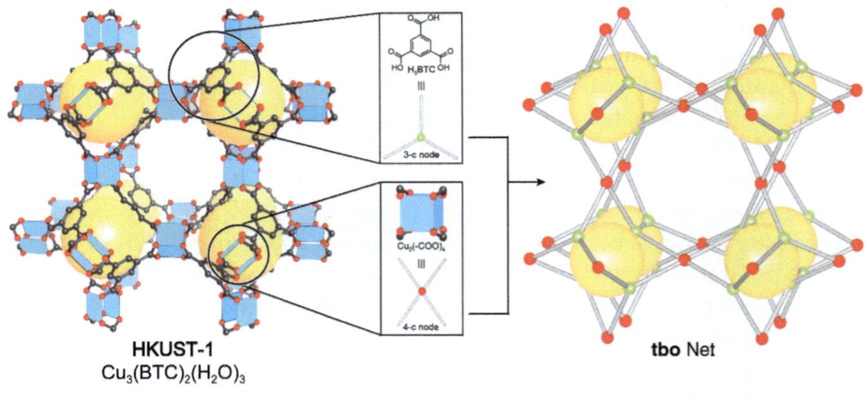

Figure 18.4 Deconstruction of the structure of HKUST-1 into its underlying building units to derive the topology. Dissecting the structure at the points of extension (here the carboxylate carbons) gives two different building units: the trigonal BTC linker and the square paddle wheel SBU. These units are connected to give an overall "twisted boracite" or **tbo** net (shown on the right).

Using this approach, crystal structures can be simplified and classified. The general principle used to deconstruct the structure of HKUST-1 can be transferred to other extended frameworks. For MOFs built from other constituents, appropriate points of extension must be chosen and for COF structures generally the structure is dissected into building units at the linkage (see Chapter 9). The simple net derived in this way is abstract and does not contain chemical information. Since chemists commonly use polyhedra to represent the shape of molecules, adding vertex figures to the vertices provides us with a more helpful

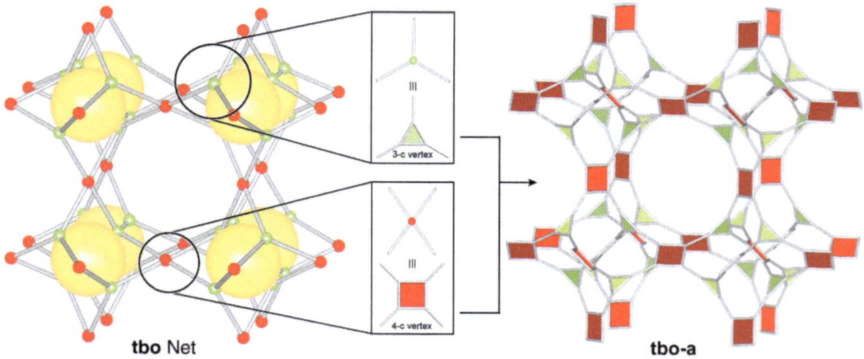

Figure 18.5 The simple net of HKUST-1 is transferred into its augmented form by replacing the nodes with their respective vertex figures. The augmented net is denoted by appending "-**a**" to the topology symbol. In comparison to the simple **tbo** net, the augmented **tbo-a** net shows more resemblance to the crystal structure of HKUST-1. Consequently, the augmented representation is more frequently used.

representation and we call this the augmented net, which is denoted by adding "-**a**" to the topology symbol. This process is illustrated in Figure 18.5 for the **tbo** net of HKUST-1.

18.2.3 Embeddings of Net Topologies

The previous discussion pointed out that a net is merely an arrangement of nodes that are connected by edges and does not include information about the specific symmetry of the structure it represents. However, the illustrations of the **tbo** and **tbo-a** net in Figure 18.5 are of high symmetry and resemble the structure of HKUST-1, which has that particular topology. This is because the location of the vertices and edges is typically chosen in such a way that the smallest number of different vertices and edges is required to realize the net while the connection scheme remains unaltered. Here, "different" means they are not related by symmetry. To do this the vertices and midpoints of the edges are positioned on coordinates with maximum site symmetry. This permits to create a real and simple representation of an abstract mathematical object and we refer to this as the embedding. In the embedding the abstract nodes and edges become real vertices and links and we will use this nomenclature when talking about the embedding of nets.

18.2.4 The Influence of Local Symmetry

As illustrated in Figure 18.3a, in terms of net topology we do not differentiate between vertices based on their geometry but rather on their connectivity. This means that both tetrahedra and squares are 4-c vertices regardless of the difference in their geometry. It is however possible that different arrangements of vertices of the same connectivity have different topologies. While the local

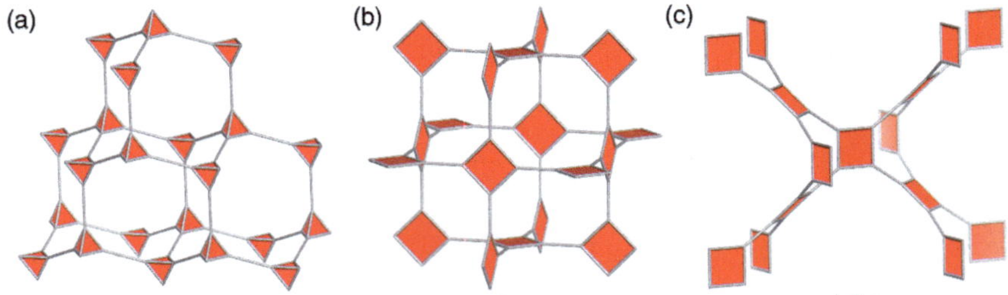

Figure 18.6 Augmented nets of (a) **dia**, (b) **nbo**, and (c) **lvt**. All three nets are 3-periodic and 4-c but differ in the local geometry of the vertices. These differences result in different arrangements of the vertices with respect to each other and thus different topologies. The three nets have different tilings that make their description unambiguous.

geometry of the building units is not a topological feature, it results in different arrangements in space when connecting these vertices. To illustrate this, we consider three 3-periodic 4-c uninodal nets: **dia**, **nbo**, and **lvt** (Figure 18.6). It is clear that linking tetrahedra or squares results in two different arrangements and thus two different topologies. Smaller changes in the local symmetry of the vertex or more precisely the angles between them when they are linked into an extended structure can also lead to different arrangements. The influence of the angles between square vertices on the resulting topology and ways to influence them have been discussed in Section 4.2.2 and we showed that it is not only possible to form different 3-periodic 4-c nets but also 0-periodic, 1-periodic, and 2-periodic nets (see Figure 4.3). This highlights that the local symmetry of the vertices is an important factor for both the design of new materials based on topological considerations and the deduction of topologies. To render the assignment of topologies unambiguous, further parameters are needed to characterize nets. Unfortunately, many different terms or symbols are used to characterize nets and their constituents. Here, we will discuss the concepts of tilings, transitivity, and the face symbols as these are most frequently used in the context of reticular materials [6]. Further methods are discussed in the literature [7].

18.2.5 Vertex Symbols

2-periodic nets consist of cycles and the vertex symbols of 2-periodic nets give all possible "strong rings" within the net. Only those cycles of a graph that are not the sum of smaller cycles are strong rings. This is illustrated in Figure 18.7 for an **fxt** net. The strong rings in an **fxt** net are 4-, 6-, and 12-membered (in blue, red, and yellow respectively, Figure 18.7a), whereas the 8-membered cycle (green, Figure 18.7b) is the sum of the 4- and 6-membered ring and therefore does not constitute a strong ring. The vertex symbol of a 2-periodic net is denoted as [*q*], where *q* is the size of the strong ring. The **fxt** net is consequently denoted as [4.6.12].

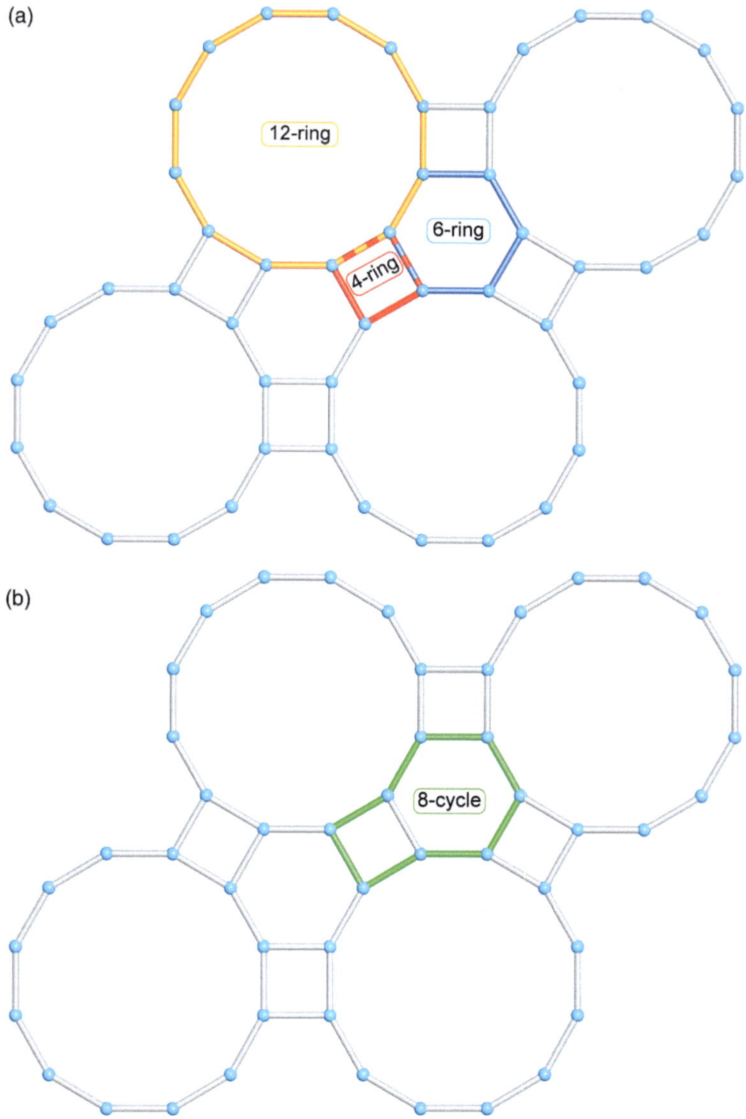

Figure 18.7 Illustration of a fragment of the **fxt** net. (a) The smallest possible rings are strong rings. The **fxt** net has three different types of strong ring: 12-, 6-, and 4-membered (yellow, blue, and red, respectively). (b) Closed cycles that are a combination of smaller rings (8-membered ring, green) are not strong rings. Here the ring could be closed by taking a shortcut.

18.2.6 Tilings and Face Symbols

In a tiling, the space is divided into generalized face-sharing polyhedra or polygons for 3-periodic or 2-periodic nets, respectively. These geometric constructs fill the open space encompassed by the net. Tilings may have curved faces and are

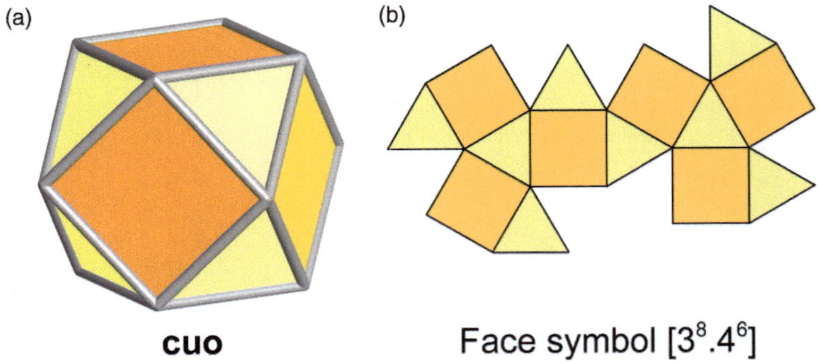

Figure 18.8 Determination of the face symbol of a polyhedron of **cuo** topology (cuboctahedron). (a) The polyhedron is composed of eight 3-membered rings and six 4-membered rings. (b) This is more easily seen by unfolding the polyhedron. Counting the number of faces facilitates the deduction of the face symbol $[3^8.4^6]$.

not necessarily convex but are completely space filling. In terms of porous frameworks, tilings are an idealized representation of the pores of the open extended structures, which renders this approach particularly useful for the description of porous materials.

Before we discuss tilings of 3-periodic nets we will consider a 0-periodic net: the **cuo** (cuboctahedron). The cuboctahedron has 3- and 4-membered rings, and again the same rules apply; only those rings that are not the sum of shorter rings are counted. The polyhedron consists of eight 3-membered and six 4-membered rings. In a tiling, we define the individual tiles with their face symbol $[p^q]$, where q is the number of p-sided faces. Accordingly, the face symbol of the cuboctahedron is $[3^8.4^6]$ (Figure 18.8).

The tiling of 3-periodic structures is determined in a similar way and we will show this for the **dia** and the **lon** net. Both nets are composed of 4-c vertices that are connected by edges. The **dia** net has one kind of tile with four identical faces of 6-membered rings. The natural tiling of the diamond net is the adamantane polyhedron with the face symbol $[6^4]$. In contrast, the **lon** net, often referred to as hexagonal diamond, has a natural tiling with equal numbers of two distinct tiles – one with three identical faces of 6-membered rings and one with five identical faces of 6-membered rings. The face symbols of 3-periodic structures with multiple tiles are denoted as $[q_1^p.q_2^p\ldots]$; therefore the face symbol of **lon** is $[6^3.6^5]$ [3a] (Figure 18.9).

The vertices of the net constitute the vertices of the tiles and the edges of the net constitute their borders – we say the tiling carries the net. The concept of tilings is not only important for the unambiguous assignment of a net topology but is also a helpful tool in reticular chemistry because they allow for the facile deduction and visualization of the different pores present in framework structures. For example, when comparing the tilings of the **dia** and the **lon** net it is immediately evident that there is just one kind of cage in frameworks with **dia** topology but two distinct cages in frameworks with **lon** topology.

Tilings can further be used to derive the transitivity of a net, which is helpful in enumerating and classifying them. The transitivity is described by a set of

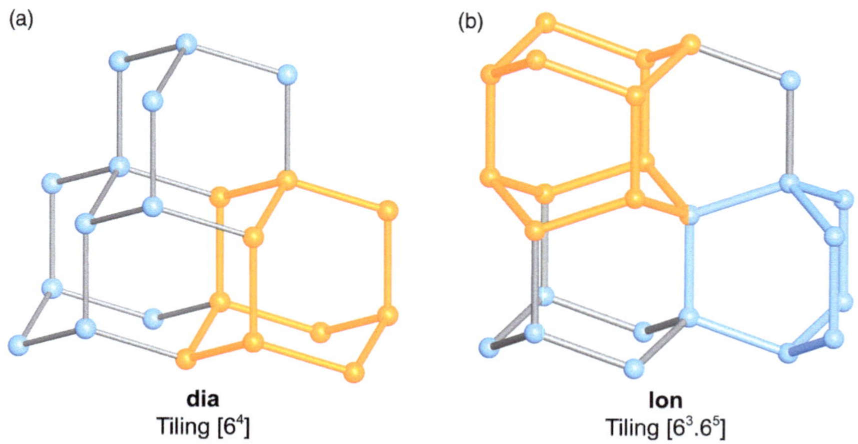

Figure 18.9 Comparison of the simple nets of (a) **dia** and (b) **lon**. Both nets are built from 4-c tetrahedral vertices but they differ in their tiling. (a) While **dia** has only one kind of tile with the face symbol [6^4], (b) **lon** has two types of tile and the face symbol [$6^3.6^5$].

four parameters, *pqrs* [8]. These four parameters represent the number of topologically distinct vertices (*p*), the number of topologically distinct edges (*q*), the number of distinct faces (or rings) (*r*), and the number of different types of tiles (*s*). We will use the transitivity of nets later when we consider specific 3-, 2-, and 0- periodic nets (polyhedra).

18.3 Nomenclature

In the context of reticular chemistry net topologies are denoted by three-letter (lower case, bold) identifiers as specified in the RCSR database [5]. Every net topology in the RCSR database is assigned a unique symbol. Many net topologies are named after minerals or other naturally occurring compounds such as sodalite (**sod**), a naturally occurring zeolitic mineral, or diamond (**dia**). New topologies are assigned new 3-letter acronyms that may be related to the name of a compound or a specific connectivity but can generally be chosen arbitrarily.

We already discussed the simple and augmented nets earlier in this chapter. There are further derived nets that are signified by a suffix appended to the basic symbol. In the following, derived nets that are of relevance in the context of this book will be introduced. These are (i) augmented nets (suffix "**-a**"), (ii) binary nets (suffix "**-b**"), (iii) dual nets (suffix "**-d**"), (iv) interpenetrated/catenated nets (suffix "**-c**"), (v) cross-linked, (vi) weaving (suffix "**-w**"), and (vii) interlocking (suffix "**-y**").

18.3.1 Augmented Nets

Replacing the nodes in a net with the corresponding polygons or polyhedra, which we refer to as vertex figures, gives the augmented net. The process of

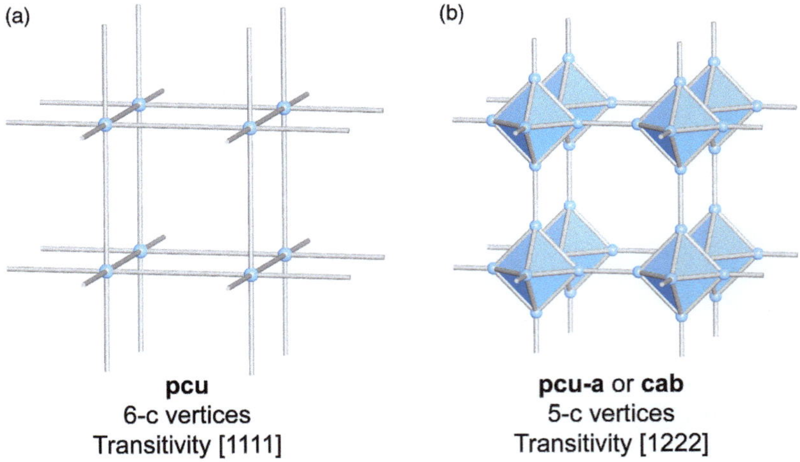

Figure 18.10 Comparison of the **pcu** and the augmented **pcu** (**pcu-a** or **cab**) net. (a) In the **pcu** net all vertices are 6-connected and the transitivity is [1111]. (b) In contrast, the new vertices in the **pcu-a** (**cab**) net are 5-connected and the transitivity is [1222].

augmentation is not based on graph theory and the new set of vertices and edges created in this process do not necessarily form a closed graph. In the augmented net the connectivity of the nodes as a whole does not change, but since each node is composed of multiple new vertices the connectivity of those fragments changes. The augmented representation of a net is helpful in that is includes the local geometry of the vertices, which makes it easier to design building units of a specific geometry to target a framework of a particular topology. Augmented nets are denoted with a suffix "**-a**." In special cases, the augmented net has its own name since the arrangement and connectivity of the new vertices are identical to another simple net. The augmented net of **pcu** (primitive cubic) is not termed **pcu-a** but **cab** (calcium boride, CaB_6) to reflect that the connectivity and arrangement of new vertices is identical to that in the **cab** net (Figure 18.10).

18.3.2 Binary Nets

In the topological description only the connectivity of the vertices, their spatial arrangement, and the connections between them are taken into account but not their chemical nature. In reticular chemistry, however, structures can be composed of two different building units of the same connectivity and coordination geometry, such as two chemically distinct tetrahedral building units within one framework of **dia** topology. This aspect can be added to the three-letter identifier of the net by appending the suffix "**-b**", as in "binary net." An example for a binary net is sphalerite, ZnS (**dia-b**), the binary version of diamond (**dia**) (Figure 18.11).

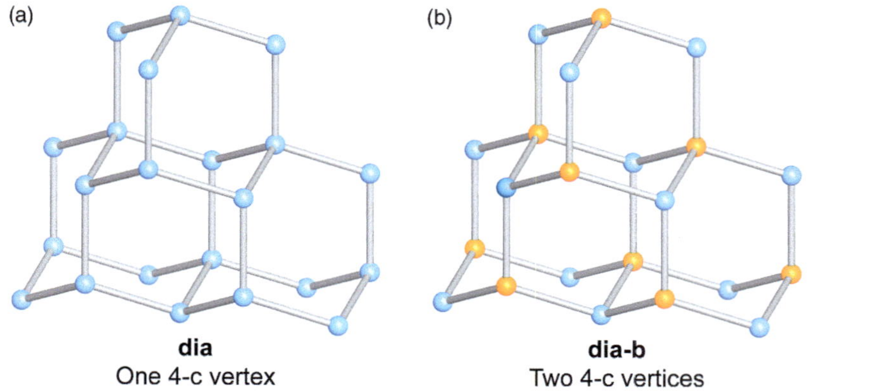

Figure 18.11 Comparison of the uninodal **dia** net and its derived binary **dia-b** net. (a) In the **dia** net all vertices are not only topologically but also chemically identical, whereas in (b) the binary **dia-b** net chemically distinct vertices are arranged in an alternating manner.

18.3.3 Dual Nets

In porous framework structures, the open space encompassed by the framework is not necessarily empty but it can be occupied by guest molecules or even a second framework structure. Such frameworks that extend throughout the pore system of another framework are described by "dual nets." In Section 18.2.5 we illustrated an approach to creating a tiling for a given net. The tiling can be used to derive the dual net for a given net topology. To do so a new vertex is placed at the center of each tile and these new vertices are consequently connected though new edges that pass through the faces of the tiles. This gives a new net within the first net, termed the dual net, and the suffix "**-d**" is appended to the RCSR identifier. There are two scenarios regarding the nature of the dual net. (i) The dual net is of a different topology than the first net and is therefore referred to as a "hetero-dual." (ii) The dual net has the same topology as the first net and is thus referred to as a "self-dual" [8]. The concept of the dual net is illustrated in Figure 18.12 for the **dia** net (a self-dual net) and the **lon** net (a hetero-dual net).

The first net and its dual net are connected by mechanical, not chemical bonds. This type of entanglement is also referred to as "interpenetration." Owing to the difference in topology between the original net and its hetero-dual, interpenetration is commonly not observed for these nets.

18.3.4 Interpenetrated/Catenated Nets

Open space is energetically not favored and therefore it is not surprising that open framework structures that encompass a large amount of open space tend to interpenetrate. In such structures, the empty space is partially occupied by one or more additional frameworks. These additional frameworks are not connected among each other by chemical bonds but are mechanically entangled.

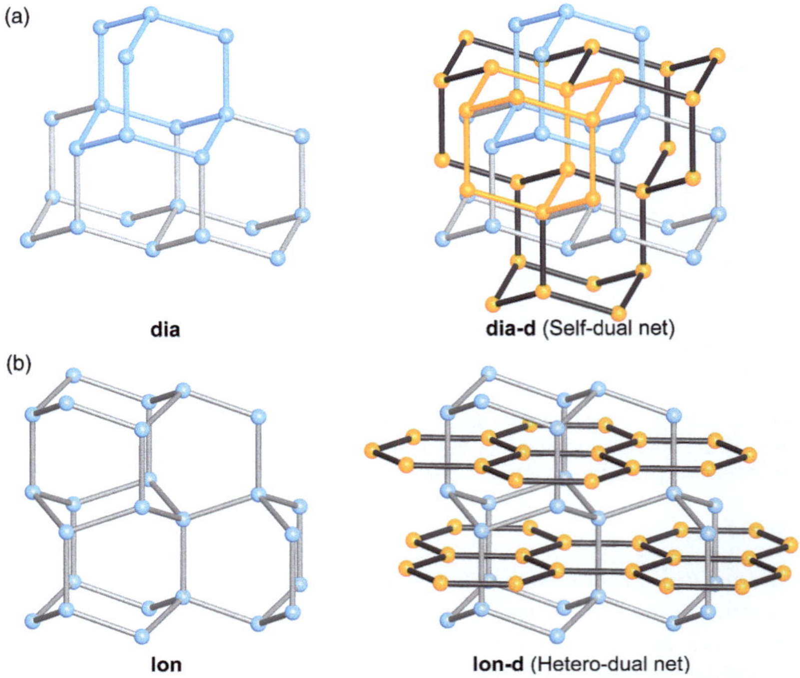

Figure 18.12 (a) Comparison of the simple **dia** net (left) and its dual **dia-d** (right). The dual net of **dia** is also a **dia** net, and thus it is self-dual. Self-dual nets are also often highlighted by a suffix "**-c**" (for catenated). One cage of the simple net (blue) and the self-dual net (orange) are highlighted for clarity. (b) Comparison of the simple **lon** net (left) and its dual **lon-d** (right). The dual of the **lon** net is a **gra** net (offset **hcb** layers); thus, it is a hetero-dual.

Interpenetrated structures are given the suffix "**-c**," for "catenated." The presence of multiple interpenetrating frameworks is highlighted by adding their number to the suffix "**-c**" (e.g. a triply interpenetrated **dia** net is denoted as **dia-c3**). The example in Figure 18.13 shows the non-interpenetrated and interpenetrated **pcu** net. This example highlights that in general, MOFs with self-dual nets, large pores, and large pore apertures are likely to form interpenetrated structures to avoid the unfavorable open space. This concept is helpful in the design of MOFs since it allows to target structures that are unlikely to interpenetrate (hetero-dual nets).

18.3.5 Cross-Linked Nets

Since nets cannot contain multiple connections between the same vertices as in "double bonds," the term "cross-linking" of nets is used to describe structures where the same vertices are connected by more than one edge. One example of a structure is bio-MOF-100. The structure has an **lcs** net; however, every vertex is connected to each adjacent vertex by three linkers, and thus the structure is referred to as a "triply cross-linked **lcs** net" (Figure 18.14).

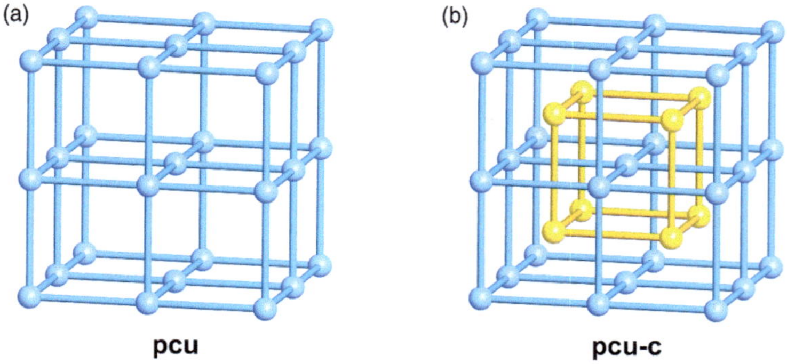

Figure 18.13 (a) Non-interpenetrated and (b) interpenetrated **pcu** net. Depending on the pore size it is possible that more than one interpenetrating framework forms. The vertices that mark the origin of the interpenetrating frameworks lie along the diagonal of the cubic unit cell.

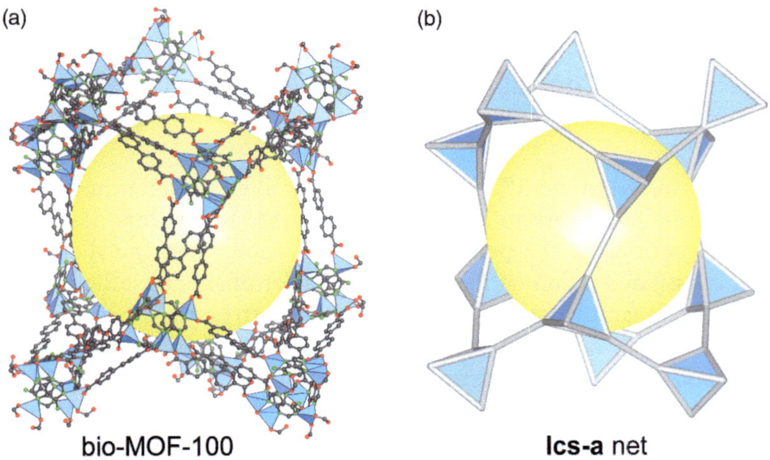

Figure 18.14 (a) One cage in the structure of bio-MOF-100 and (b) the corresponding topology representation as an **lcs-a** net. Since the tetrahedral nodes are connected to each other by three linkers, this is referred to as triply cross-linked. In the augmented net each link represents three linkers in the crystal structure and therefore, the **lcs** net remains unaltered.

18.3.6 Weaving and Interlocking Nets

In addition to the nets discussed thus far, recently structures with weaving topologies have been reported. In such structures infinite elements are entangled into extended structures in a manner akin to the weaving of fabrics from threads. The identifier of woven nets is denoted by appending the suffix "**-w**." A COF that has such a woven net (**dia-w**) is described in more detail in 21.3.2 (see Figure 21.9). It is also possible (even though thus far not reported) to form nets of interlocked rings, which is highlighted by appending the suffix "**-y**" to the RCSR identifier. Figure 18.15 shows an example of a woven (**dia-w**) and an interlocking (**sod-y**) net.

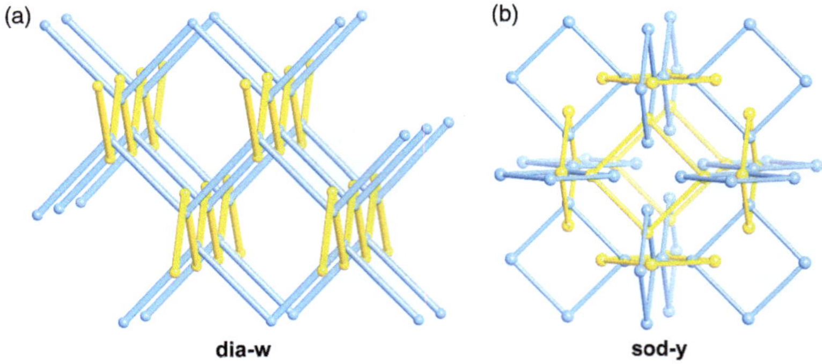

Figure 18.15 (a) Woven diamond net (**dia-w**). In the woven **dia-w** net independent 1D struts are woven in an up–down–up pattern. The points at which the struts cross (points of registry) correspond to the tetrahedral nodes in the **dia** net. (b) Interlocking sodalite net (**sod-y**). The interlocking **sod-y** net consists of 4-membered ring that are interlocking at each corner of the ring. The points at which the rings cross (points of registry) correspond to the tetrahedral nodes in the **sod** net.

18.4 The Reticular Chemistry Structure Resource (RCSR) Database

We have already referred to the RCSR database earlier in this chapter and have used the three-letter identifiers to name topologies [5]. Today, this database comprises 2803 3-periodic, 300 2-periodic, and 78 0-periodic nets (polyhedral), each of which is assigned a unique identifier. These identifiers are sufficient to define a certain net unambiguously. The database lists attributes for every net entry such as the topological density, the number of distinct vertices and edges, the tiling, the topology of the dual net, and the transitivity (Figure 18.16) [6]. The transitivity and the topological density can be used to predict thermodynamic and kinetic products of reactions. Nature prefers the formation of highly symmetric and dense structures, and consequently those are thermodynamically favored. To illustrate this, we consider the **ctn** and the **bor** net. Both consist of square 4-c and trigonal 3-c vertices and in both arrangements each 4-c vertex is connected to four 3-c vertices, and each 3-c vertex is connected to three 4-c vertices. Both nets have the transitivity [2122] but due to a difference in their tiling, they differ in their topological density. The **ctn** net has a higher topological density ($d_{top} = 0.5513$) than the **bor** net ($d_{top} = 0.4763$), and this is reflected in the finding that in COF chemistry the **ctn** net is more likely to be formed from the combination of square 4-c and trigonal 3-c building units than the **bor** net.

To allow for the visualization of the nets in the RCSR database structural data (space groups and fractional coordinates of all vertices and edges) for their highest symmetry embedding are given. In the highest symmetry embedding, where the vertices and centers of the edges occupy positions of maximum site symmetry, the lengths of the edges are constrained to be equal to unity, and the volume is subject to that constraint.

Dia

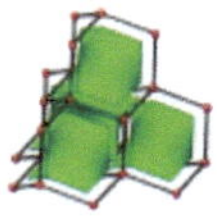

RCSR reference: http://rcsr.net/nets/dia

Names: diamond, sqc6, 4/6/c1
Keywords: regular net, uniform net, self dual net, quasisimple tiling, good
References: Acta Cryst. A59, 22-27 (2003), Acta Cryst. A60, 517-520 (2004)

Embed type	Space group	Volume	Density	Genus	td10	Deg freedom
1a	Fd-3m	12.3168	0.6495	3	981	1

A	B	C	α	β	γ
2.3094	2.3094	2.3094	90.0	90.0	90.0

Vertices: 1

Vertex	cn	x	y	z	Symbolic	Wyckoff	Symmetry	Order
V1	4	0.1250	0.1250	0.1250	1/8, 1/8, 1/8	8 a	–43m	24

Vertex	cs_1	cs_2	cs_3	cs_4	cs_5	cs_6	cs_7	cs_8	cs_9	cs_{10}	cum_{10}	Vertex symbol
V1	4	12	24	42	64	92	124	162	204	252	981	6(2).6(2).6(2).6(2).6(2).6(2)

Edges: 1

Edge	x	y	z	Symbolic	Wyckoff	Symmetry
E1	0.0000	0.0000	0.0000	0, 0, 0	16 c	–3m

Tiling:

Tiling	Dual	Vertices	Edges	Faces	Tiles	*D*-symbol
[6^4]	dia	1	1	1	1	2

Figure 18.16 Entry for the **dia** net in the RCSR database. A representation of the topology is given at the top (here, the tiling is also shown). The space group, unit cell parameters, and fractional coordinates for the vertices and edges represent the highest symmetry embedding of the net and give information about their local site symmetry. The transitivity (see tiling: vertices, edges, faces, tiles) and the topological density can help to predict the outcome of a reticulation.

18.5 Important 3-Periodic Nets

We discussed the concept of transitivity in Section 18.2.5. The transitivity is described by the four parameter "*pqrs*" of the tiling carrying that particular net where *p* is the number of topologically distinct vertices, *q* is the number of topologically distinct edges, *r* is the number of distinct faces or rings in the tiling, and *s* is the number of different types of tiles [8]. 3-periodic nets with transitivity 1111 are referred to as "regular nets" (**bcu**, **dia**, **nbo**, **pcu**, and **srs**), 3-periodic nets with transitivity 1112 are referred to as "quasiregular nets" (**fcu**), and those

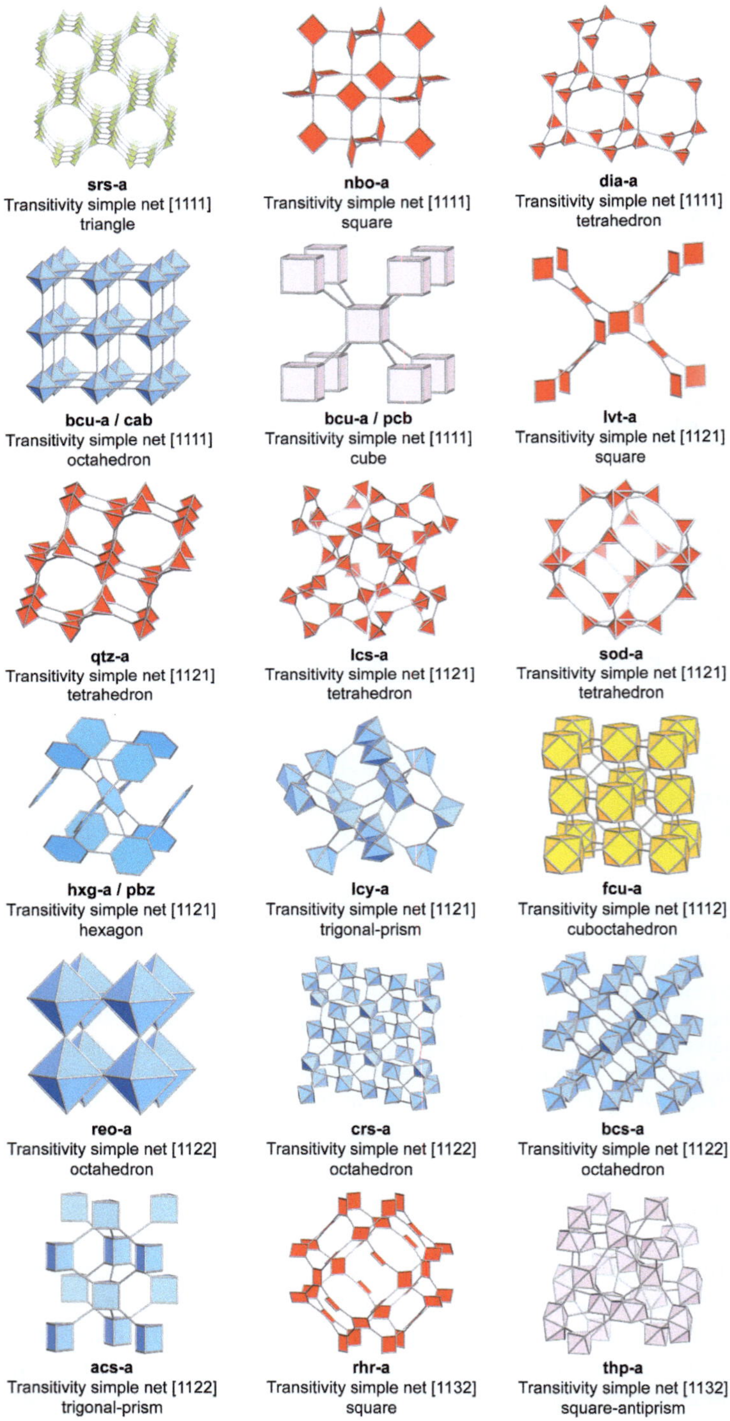

Figure 18.17 3-periodic "regular," "quasiregular," and "semiregular" nets sorted by the digit sum of the transitivity for the corresponding simple net. All nets are shown in their augmented form, with their respective RCSR identifier, and the transitivity of the corresponding simple net. 3-c vertices are shown in green, 4-c vertices in red, 6-c vertices in blue, 8-c vertices in pink, and 12-c vertices in orange.

with transitivity 11*rs* are referred to as "semiregular" (**lvt**, **sod**, **lcs**, **lcv**, **qtz**, **hxq**, **lcy**, **crs**, **bcs**, **acs**, **reo**, **thp**, **rhr**, and **ana**). These three groups of 3-periodic nets are the ones that are most likely to form when one type of vertex is linked by one type of edge, or in terms of MOF chemistry, one type of SBU is reticulated with a ditopic linker [3c, 6]. All regular, quasiregular, and semiregular nets are shown in Figure 18.17 in their augmented form. As outlined before, for the combination of two building units of similar (or identical) geometry, those nets with the lowest transitivity, the most symmetric highest symmetry embedding, and the highest topological density are the most likely to form. This holds especially true for combinations of single metal ions linked by flexible linkers. We refer to these topologies as the default topologies for a given combination of vertex geometries. In reticular chemistry, geometrically well-defined, rigid building units are used to form extended framework structures, and by their judicious choice and under appropriate reaction conditions, topologies other than the default ones can be targeted.

All nets shown in Figure 18.17 consists of one type of vertex and one type of edge and are thus referred to as "uninodal edge-transitive nets." Edge-transitive nets are the most likely products of reticular synthesis. Therefore, nets that are edge-transitive but built from two different types of vertices ("binodal"), where each type of vertex is only connected to the second type of vertex ("bipartite"), are of great importance for reticular chemistry. All binodal, bipartite edge-transitive nets are given in Figure 18.18.

Aside from the nets discussed above, where discrete vertices are connected by edges, 1D vertices can be linked into nets. Such structures are found in MOF chemistry, where rod-like SBUs can form to give 3D frameworks. These structures are not as common as those discussed above and for a more detail discussion of rod-MOFs and their topological description, the reader is referred elsewhere [9].

18.6 Important 2-Periodic Nets

Many frameworks have 2D layered structures. This is especially true for COFs. Such structures can either be described as 3-periodic (stacked 2-periodic nets) or 2-periodic nets. An example for this is the description of staggered **hcb** layers by the **gra** net (see Figure 18.12). This description is confusing since the 2D layers are only held together by weak interactions rather than actual "links," which makes the definition of edges between vertices of neighboring layers difficult. Consequently, the description of such structures as 3-periodic nets is generally not helpful for the overall understanding and the topological description of layered structures as 2-periodic nets is preferred. In analogy to the transitivity of 3-periodic nets, the transitivity of the tiling of a plane with p kinds of vertex, q kinds of edge, and r kinds of ring is given as [pqr] [6]. There are a total of five possible edge-transitive 2-periodic nets: (i) three "regular nets" with transitivity 111 (**hcb**, **sql**, and **hxl**), (ii) one "quasiregular net" with transitivity 112 (**kgm**), and (iii) and one binodal net with transitivity 211 (**kgd**). These nets are shown in Figure 18.19 alongside their augmented versions.

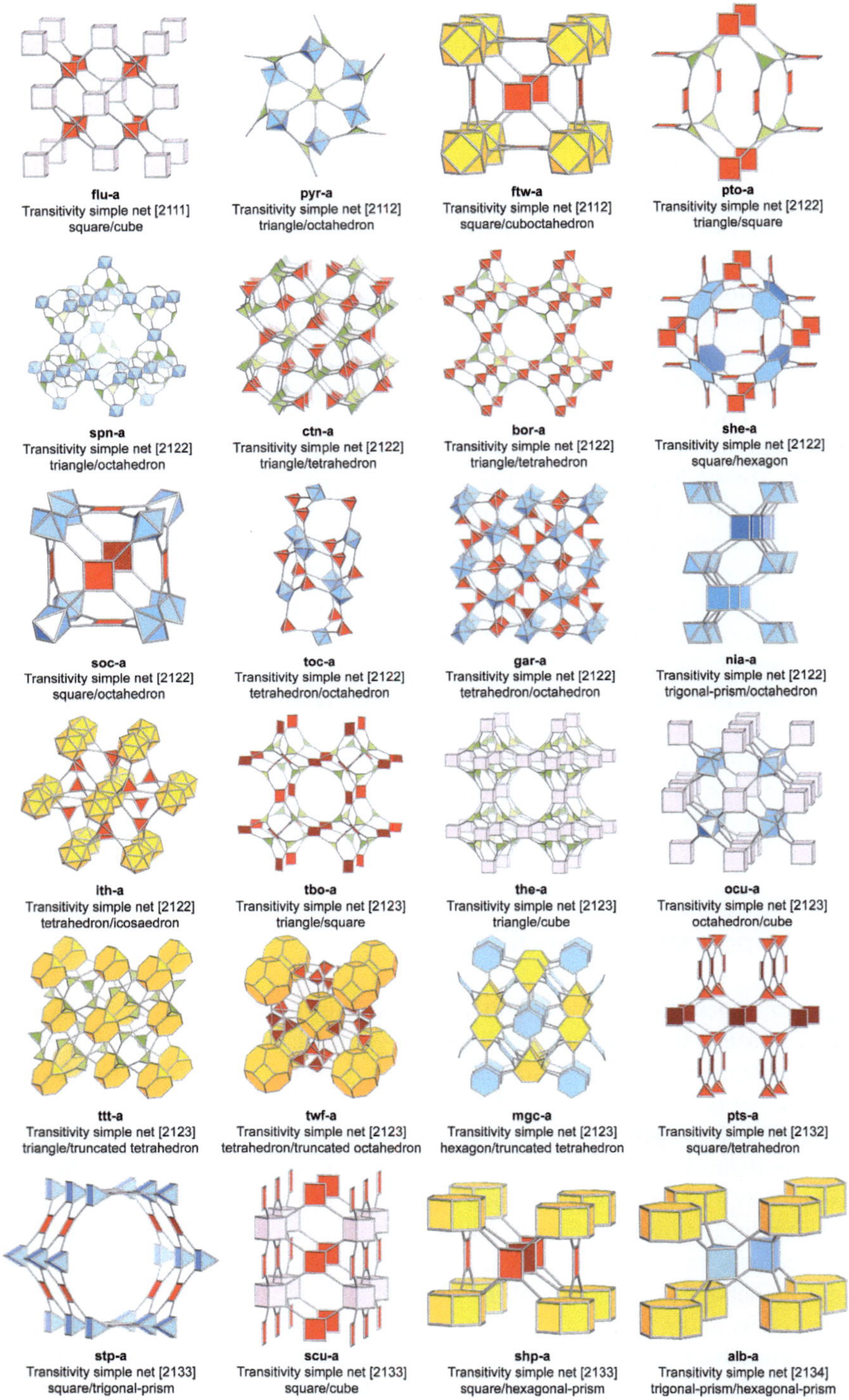

Figure 18.18 3-periodic binodal edge-transitive nets sorted by the digit sum of the transitivity for the corresponding simple net. All nets are shown in their augmented form, with their respective RCSR identifier and the transitivity of the corresponding simple net. 3-c vertices are shown in green, 4-c vertices in red, 6-c vertices in blue, 8-c vertices in pink, and 12-c as well as 24-c vertices in orange.

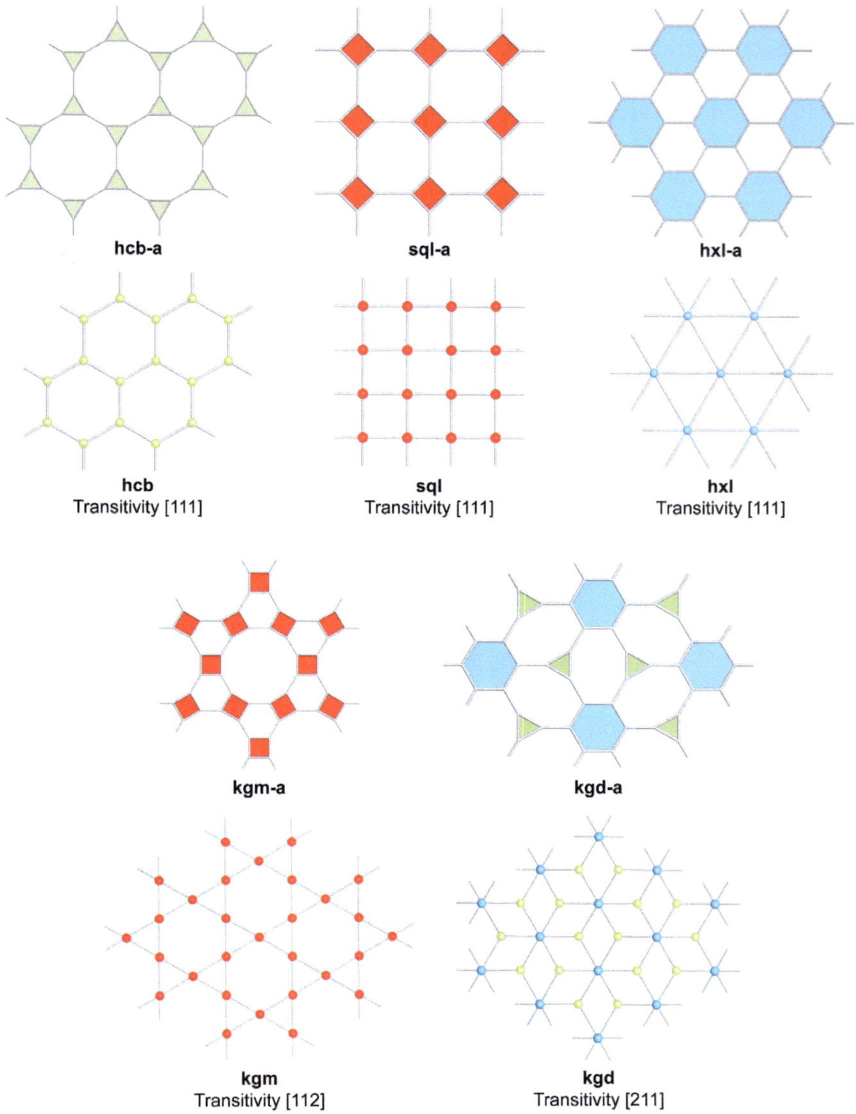

Figure 18.19 The five edge-transitive 2-periodic nets. Both the simple and augmented nets are shown and the transitivity for the simple net is given. 3-c vertices are shown in green, 4-c vertices in red, 6-c vertices in blue.

18.7 Important 0-Periodic Nets/Polyhedra

Reticular chemistry does not only involve the making of 2D and 3D extended structures but also that of complex molecular moieties (see Chapter 19). So far, we discussed the topological description of framework structures (e.g. MOFs, ZIFs, and COFs) as 2- or 3-periodic nets, and discrete polyhedra are described in a similar way. This is important not only for the description of molecular entities such as metal-organic and covalent organic polyhedra (MOPs and COPs), but also for the topological description of individual cages within framework structures [10]. As in the case of 3-periodic and 2-periodic nets, the most frequently encountered 0-periodic polyhedra in reticular chemistry are edge transitive and among these, the most commonly observed ones are "regular convex polyhedra" or "Platonic solids." Such polyhedra have one kind of vertex, one kind of edge, and one kind of face and consequently have a transitivity of [111] (**tet**, **oct**, **cub**, **ico**, and **dod**). The second most common polyhedra are "quasiregular" polyhedra, which have one kind of vertex and edge, but two kinds of face and therefore a transitivity of [112] (**rdo** and **cuo**). There are two more edge-transitive polyhedra with a transitivity [211], meaning they have two kinds of vertex, one kind

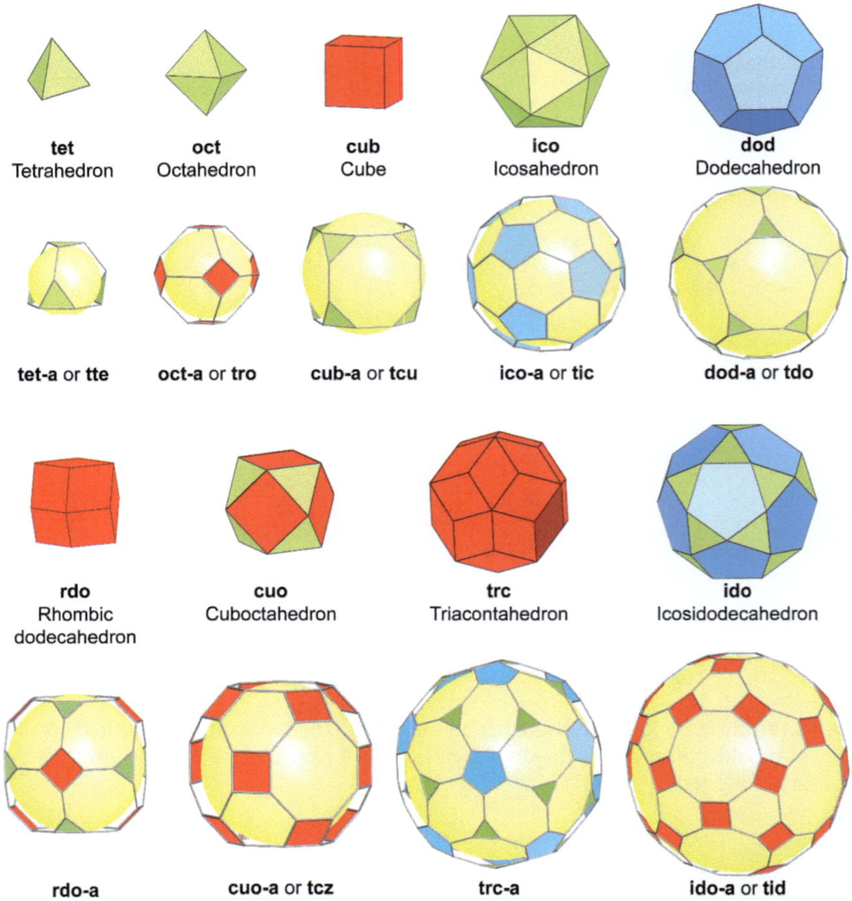

Figure 18.20 0-periodic "regular," "quasiregular," and "semiregular" tilings. Both the Platonic solids and their augmented counterparts are shown. 3-c vertices and trigonal faces are shown in green, 4-c vertices and quadrangular faces in red, and 5-c vertices and pentagonal faces in blue. The yellow sphere in the augmented representation is added for clarity and represents the potential voids within compounds of this topology.

of edge, and one kind of face (**trc** and **ido**). These polyhedra are the duals of the "quasiregular" polyhedra.

Edge transitive polyhedra are the most common polyhedra in reticular chemistry. The reason therefore is that the vertex figure of their augmented versions (in their highest symmetry embedding) are regular polygons (e.g. triangle, square, and pentagon), shapes that are synthetically accessible molecular building units [6] (Figure 18.20).

18.8 Summary

In this chapter, we introduced the concept of "topology" as a way to describe, simplify, and classify structures of extended solids based on the connectivity of their building units, their spatial arrangement, and the resulting pattern of connections arising therefrom. We introduced the nomenclature used by the RCSR to unambiguously identify net topologies. After introducing the concept of transitivity, we took a closer look at 3-periodic, 2-periodic, and 0-periodic edge-transitive nets, which are the most frequently encountered nets in reticular chemistry. The concept of topology is not only helpful in simplifying crystal structures and classifying them, but also for the deliberate and rational design of new materials, and this aspect is discussed in more detail in other chapters of this book (for MOFs (Chapters 1–6), ZIFs (Chapter 20), COFs (Chapters 7–11), and MOPs and COPs (Chapter 19)).

References

1 Öhrström, L. (2016). Designing, describing and disseminating new materials by using the network topology approach. *Chemistry: A European Journal* 22 (39): 13758–13763.

2 (a) Wells, A. (1954). The geometrical basis of crystal chemistry. Part 1. *Acta Crystallographica* 7 (8–9): 535–544. (b) Wells, A.F. (1984). *Structural Inorganic Chemistry*, 5e. Oxford University Press.

3 (a) Yaghi, O.M., O'Keeffe, M., Ockwig, N.W. et al. (2003). Reticular synthesis and the design of new materials. *Nature* 423 (6941): 705–714. (b) Euler, L. (1953). Leonhard Euler and the Königsberg bridges. *Scientific American* 189 (1): 66–70. (c) Ockwig, N.W., Delgado-Friedrichs, O., O'Keeffe, M., and Yaghi, O.M. (2005). Reticular chemistry: occurrence and taxonomy of nets and grammar for the design of frameworks. *Accounts of Chemical Research* 38 (3): 176–182.

4 (a) O'Keeffe, M. and Yaghi, O.M. (2012). Deconstructing the crystal structures of metal-organic frameworks and related materials into their underlying nets. *Chemical Reviews* 112 (2): 675–702. (b) Li, M., Li, D., O'Keeffe, M., and Yaghi, O.M. (2013). Topological analysis of metal-organic frameworks with polytopic linkers and/or multiple building units and the minimal transitivity principle. *Chemical Reviews* 114 (2): 1343–1370.

5 O'Keeffe, M., Peskov, M.A., Ramsden, S.J., and Yaghi, O.M. (2008). The reticular chemistry structure resource (RCSR) database of, and symbols for, crystal nets. *Accounts of Chemical Research* 41 (12): 1782–1789.
6 Delgado-Friedrichs, O., O'Keeffe, M., and Yaghi, O.M. (2007). Taxonomy of periodic nets and the design of materials. *Physical Chemistry Chemical Physics* 9 (9): 1035–1043.
7 Hoffmann, F. and Föba, M. (2016). *The Chemistry of Metal-Organic Frameworks: Synthesis, Characterization, and Applications*, Chapter 2 (ed. S. Kaskel), 5–40. Wiley.
8 Delgado-Friedrichs, O. and O'Keeffe, M. (2005). Crystal nets as graphs: terminology and definitions. *Journal of Solid State Chemistry* 178 (8): 2480–2485.
9 (a) Schoedel, A., Li, M., Li, D. et al. (2016). Structures of metal-organic frameworks with rod secondary building units. *Chemical Reviews* 116 (19): 12466–12535. (b) Rosi, N.L., Kim, J., Eddaoudi, M. et al. (2005). Rod packings and metal-organic frameworks constructed from rod-shaped secondary building units. *Journal of the American Chemical Society* 127 (5): 1504–1518.
10 (a) Lu, Z., Knobler, C.B., Furukawa, H. et al. (2009). Synthesis and structure of chemically stable metal-organic polyhedra. *Journal of the American Chemical Society* 131 (35): 12532–12533. (b) Tranchemontagne, D.J., Ni, Z., O'Keeffe, M., and Yaghi, O.M. (2008). Reticular chemistry of metal-organic polyhedra. *Angewandte Chemie International Edition* 47 (28): 5136–5147.

19

Metal-Organic Polyhedra and Covalent Organic Polyhedra

19.1 Introduction

Reticular chemistry is concerned with linking molecular building units through strong bonds into predefined structures [1]. Most products of reticular synthesis are 2D or 3D extended frameworks; however, this chemistry can further be applied to target the formation of discrete 0D compounds. In the following text, we describe how the design principles that we have covered in the context of metal-organic frameworks (MOFs) and covalent organic frameworks (COFs) can be applied to the synthesis of metal-organic polyhedra (MOPs) and covalent organic polyhedra (COPs)–discrete intrinsically porous cage compounds [2]. We limit the discussion to porous cage compounds that sustain permanent porosity and whose structures are unambiguously confirmed by X-ray diffraction. With respect to MOPs only structures based on secondary building units (SBUs) are discussed. For further information on coordination cages based on single metal nodes the reader is referred to the literature [3]. For COPs the discussion is limited to structures based on linkage chemistries utilized in COF chemistry and for other shape-persistent organic cages we refer the reader to the literature [4].

19.2 General Considerations for the Design of MOPs and COPs

The nine different shapes of polyhedra that are most likely to form as products of reticular synthesis of MOPs and COPs are the edge-transitive convex polyhedra that have already been described in Chapter 18. They can be divided into (i) Vertex-, face-, and edge-transitive regular polyhedra (tetrahedron, octahedron, cube, icosahedron, and dodecahedron); (ii) quasiregular polyhedra, which are edge and vertex transitive but have two kinds of face (triacontahedron and icosidodecahedron), and (iii) the duals of these two quasiregular polyhedra, which are edge and face transitive but have two kinds of vertex (rhombic dodecahedron and octahedron) [5]. The topology can give an idea about the connectivity the building units must have to target polyhedra carrying their respective nets. However this knowledge alone does not suffice to specifically target MOPs and

Introduction to Reticular Chemistry: Metal-Organic Frameworks and Covalent Organic Frameworks, First Edition. Omar M. Yaghi, Markus J. Kalmutzki, and Christian S. Diercks.

Companion website: www.wiley.com/go/yaghi/reticular

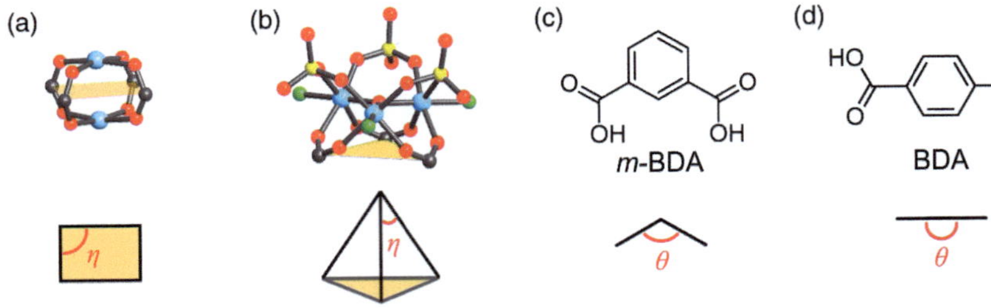

Figure 19.1 Illustration of the angles η in the molecular building units that form the vertices and the angle θ of the linkers representing the edges of the targeted topology. (a) The angle η in a $Cu_2(-COO)_4$ copper paddle wheel SBU. (b) The angle η in an $Fe_3OL_3(-COO)_3(SO_4)_3$ SBU. (c) The angle θ in *m*-BDC, and in BDC (d). All hydrogen atoms are omitted for clarity. Color code for (a, b): M, blue; C, gray; O, red; S, yellow; N, green.

COPs because the competitive formation of extended 2D and 3D frameworks with the same connectivity is generally more favorable. Hence, the formation of MOPs and COPs requires the judicious choice of both the angle θ between the points of extension of the molecules that represent the edges (linkers with two points of extension) and the angle η between the respective points of extension of the vertex (SBU or linker with more than two points of extension) as illustrated in Figure 19.1 [2].

In this chapter, selected examples for the synthesis of MOPs and COPs representing different polyhedra are highlighted and geometric requirements for the building units deciphered.

19.3 MOPs and COPs Based on the Tetrahedron

To target MOPs or COPs with an underlying tetrahedron (**tet**) topology, four 3-c SBUs with angle η need to be joined by six ditopic linkers with angle θ. In Figure 19.2 the two extreme scenarios of configurations that can lead to a tetrahedron are illustrated: (i) $\eta = 60°$, $\theta = 180°$ and (ii) $\eta = 120°$, $\theta = 70.5°$. These are the angles for the two idealized extreme conformations that a tetrahedron can assume and, of course, in reality the angles found in structures of this topology are typically somewhere between these values. Nonetheless, the idealized structures serve as a valuable reference for targeting MOPs and COPs of **tet** topology.

An isoreticular series of MOPs assuming a conformation close to the first extreme case of a tetrahedron (i) is constructed from partially sulfate-capped $Fe_3OL_3(-COO)_3(SO_4)_3$ SBUs connected by linear ditopic BDC, BPDC, and BPTC to yield IRMOP-50, IRMOP-51, and IRMOP-53, respectively (Figure 19.3) [6]. In this series, the pore size in the center of the tetrahedron is varied systematically: 7.3, 10.4, and 13.3 Å for IRMOP-50, IRMOP-51, and IRMOP-53, respectively. The surface area of IRMOP-51 determined by nitrogen adsorption is calculated to be 480 $m^2\ g^{-1}$. In contrast, the isoreticular expanded IRMOP-53 has a lower surface area of just 387 $m^2\ g^{-1}$. This appears counterintuitive at first but can be rationalized by the fact that for the case of discrete 0D polyhedra not only the intrinsic pore of the cage is relevant to the surface area but also the

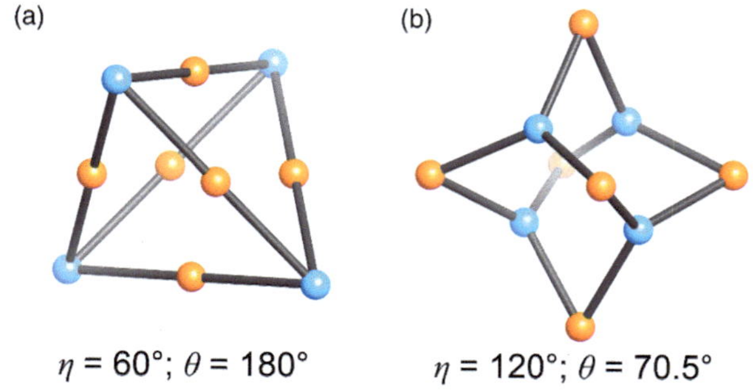

Figure 19.2 Two extreme cases of four 3-c vertices linked by edges to form a tetrahedron. (a) One extreme case is the linking of vertices with the minimum angle of $\eta = 60°$ with linear edges of $\theta = 180°$. (b) In the second extreme case the vertices have the maximum angle of $\eta = 120°$ and an angle of $\theta = 70.5°$. In reality, the angles are found somewhere between these values. Color code: vertices, blue; edges, orange.

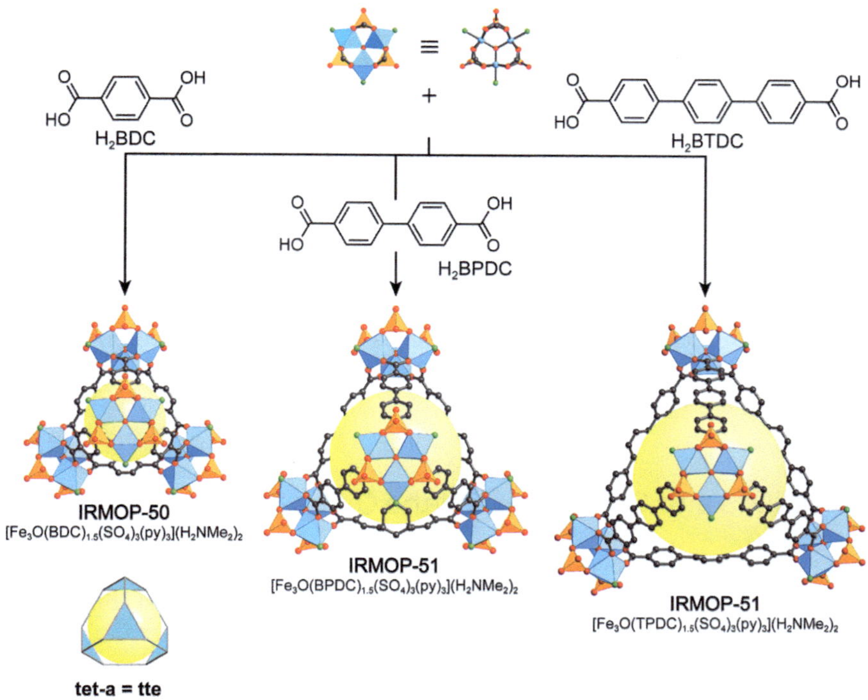

Figure 19.3 MOPs assuming a tetrahedron topology. Partially sulfate capped $Fe_3OL_3(–COO)_3(SO_4)_3$ SBUs are linked by linear ditopic BDC, BPDC, and TPDC to yield an isoreticular series of MOPs termed IRMOP-50, IRMOP-51, and IRMOP-53, respectively. All hydrogen atoms are omitted for clarity. Color code: Fe, blue; S, orange; C, gray; N, green; O, red.

extrinsic pores resulting from how the molecular cages are packed in the solid state. In IRMOP-51, 76% of the crystal structure is free volume as opposed to only 70.5% in IRMOP-53, and consequently the higher surface area of IRMOP-51 is rationalized by the additional uptake in the extrinsic pores of the crystal.

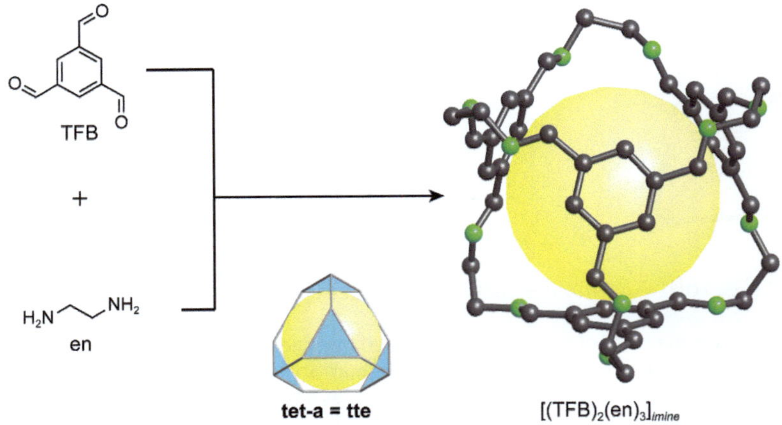

Figure 19.4 Reticulation of TFB with en yields the **tet** topology COP $[(TFB)_2(en)_3]_{imine}$, which resembles the second extreme case for the tetrahedron with angles $\eta = 118°$, $\theta = 74.5°$. In the crystal structure the cages pack in a face-sharing manner to yield a structure of **dia** topology held together by weak interactions. All hydrogen atoms are omitted for clarity. Color code: C, gray; N, green.

The other extreme scenario is exemplified by a COP constructed from tritopic TFB with ditopic en (1,2-ethylene diamine) of molecular formula $[(TFB)_2(en)_3]_{imine}$ (Figure 19.4) [7]. In this structure, the angles, $\eta = 118°$, $\theta = 74.5°$ are close to the values in the second extreme conformation of the tetrahedron (Figure 19.2b). The pore size of the cage has a diameter of 7.8 Å and in the crystal structure its trigonal faces are packed in a face-to-face manner to yield channels of 5.8 Å. The packing resembles a diamond-like arrangement of the COPs with an interconnected pore system that endows the material with a BET surface area of $624\,m^2\,g^{-1}$.

19.4 MOPs and COPs Based on the Octahedron

MOPs and COPs based on the octahedron are composed of six 4-c vertices with angle η joined by links with angle θ. There are two extreme cases of configurations: (i) $\eta = 60°$, $\theta = 180°$ and (ii) $\eta = 90°$, $\theta = 90°$ (Figure 19.5).

As of now, no MOP or COP resembling the extreme case (i) is known in the literature. An example for the second type of configuration is the structure of MOP-28 (Figure 19.6) [8]. The cage is constructed from $Cu_2(-COO)_4$ paddle wheel SBUs linked by bent H_2TTC (2,2′:5′,2″-terthiophene-5,5″-dicarboxylic acid) linkers. In the crystal structure the copper paddle wheels are slightly distorted but maintain an average angle of $\eta = 90°$. The TTC linkers also have the ideal angle of $\theta = 90°$ found in the second extreme case of an octahedron. The pore size of MOP-28 is 27.0 Å with large, 9 Å wide pore openings. The structure is architecturally stable and sustains permanent porosity with a surface area of $1100\,m^2\,g^{-1}$.[1] It is important to consider that discrete porous cages often lose

1 This is the first MOP that was proved to sustain permanent porosity.

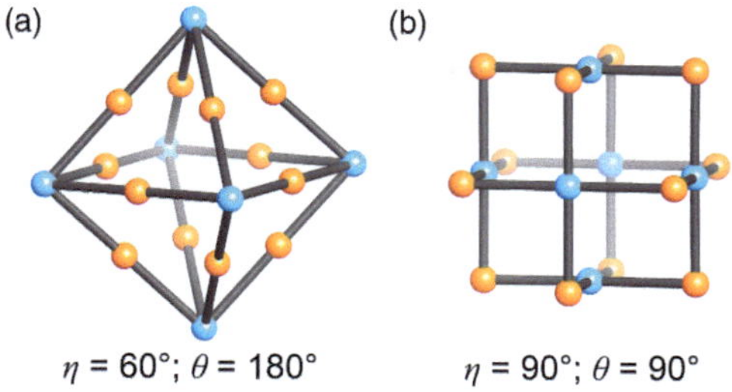

Figure 19.5 Two extreme cases of 6 4-c vertices linked by 12 edges to form an octahedron. (a) One extreme case is the linking of vertices with the minimum angle of $\eta = 60°$ with linear edges of $\theta = 180°$. (b) In the second extreme case the vertices have the maximum angle of $\eta = 90°$ and an angle of $\theta = 90°$. The angles in synthetic structures are found within these boundaries. Color code: vertices, blue; edges, orange.

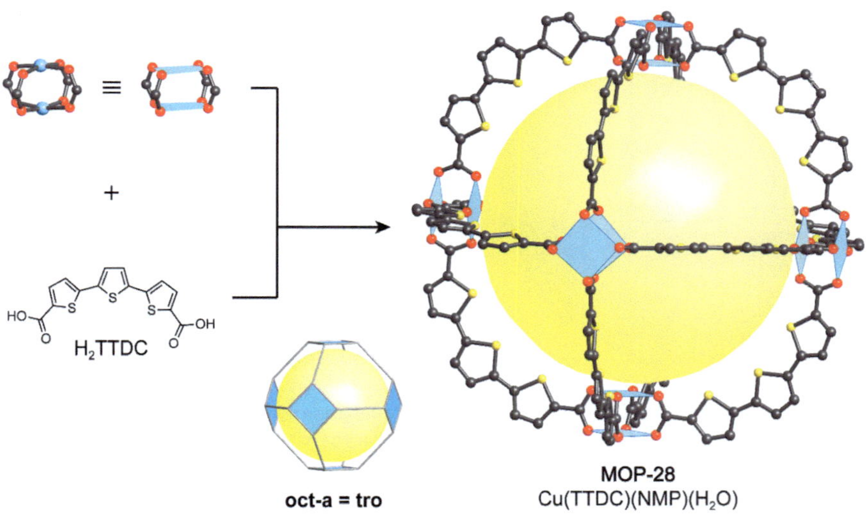

Figure 19.6 The structure of MOP-28, an example of a MOP in the shape of an octahedron. The averaged angles within the $Cu_2(-COO)_4$ paddle wheel of $\eta = 90°$ and the $\theta = 90°$ angles between the points of extension of the bent ditopic TTDC linker perfectly match the angles of the second extreme case of a cube (Figure 19.5b). All hydrogen atoms are omitted for clarity. Color code: Cu, blue; C, gray; O, red; S, yellow.

porosity upon cycling due to changes in the crystal packing but in the case of MOP-28 no decrease in nitrogen uptake or in surface area is detected.

19.5 MOPs and COPs Based on Cubes and Heterocubes

In cubes eight 3-c vertices with angle η are joined by ditopic edges with an angle θ. The two extreme cases are formed with angles (i) $\eta = 90°$, $\theta = 180°$ and (ii) $\eta = 120°$, $\theta = 109.5°$ (Figure 19.7).

While there are several coordination cages (polyhedra based on single-metal nodes) that assume the first of the two extreme cases of an octahedron, no MOPs

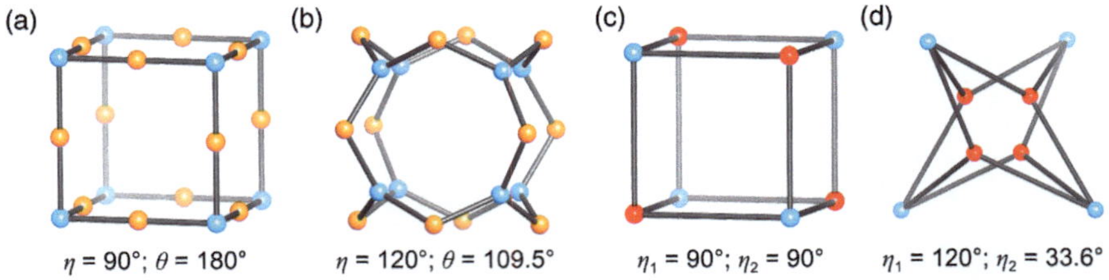

Figure 19.7 (a, b) Two extreme cases to construct cubes from 8 3-c vertices and 12 edges: (a) $\eta = 90°$ and $\theta = 180°$, (b) $\eta = 120°$ and edges $\theta = 109.5°$. (c, d) Construction of heterocubes each built from two sets of four respective 3-c vertices. There are two extreme cases: (c) $\eta = 90°$ and $\theta = 90°$, (d) $\eta = 120°$ and $\theta = 33.6°$. Color code: vertices, blue or red; edges, orange.

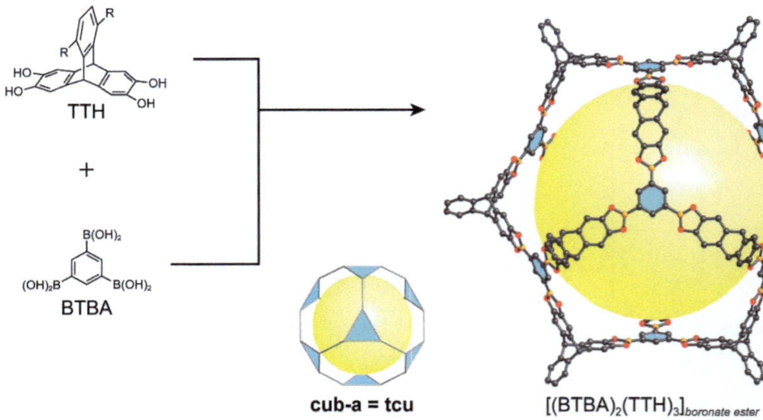

Figure 19.8 Synthesis of $[(BTBA)_2(TTH)_3]_{boronate\ ester}$, a mesoporous COP. Reticulation of trigonal tritopic BTBA with linear ditopic TTH yields a cage similar to that of the second extreme case for the conformation of cubes. The angles $\eta = 118°$ and edges $\theta = 110°$ are very close to the idealized angles. All hydrogen atoms are omitted for clarity. Center of the tritopic linker, blue; B, orange; C, gray; O, red.

are known. For examples on single-metal coordination cages we refer the reader to the literature [9].

An example for the second extreme case of a cube is found in the boronate ester-linked COP $[(BTBA)_2(TTH)_3]_{boronate\ ester}$ (BTBA = benzene-1,3,5-triyltriboronic acid, TTH = ((9s,10s)-13,16-diethyl-9,10-dihydro-9,10-[1,2]benzenoanthracene-2,3,6,7-tetraol). In its structure, 8 BTBA and 12 TTH are linked by boronate ester bonds (Figure 19.8) [10]. The resulting COP has a large pore of 2.4 nm diameter and can be activated to yield a permanently mesoporous cage with an exceptionally high surface area of $3758\ m^2\ g^{-1}$. Depending on the functionalization pattern of the TTH building unit, a catenated version of $[(BTBA)_2(TTH)_3]_{boronate\ ester}$ is isolated where two cages quadruply interlock to yield a catenane in high yield (62%). This is interesting because in general the formation of catenanes or related interlocking molecules requires a template to obtain them in high yield. Multicomponent reactions such as the 96-fold

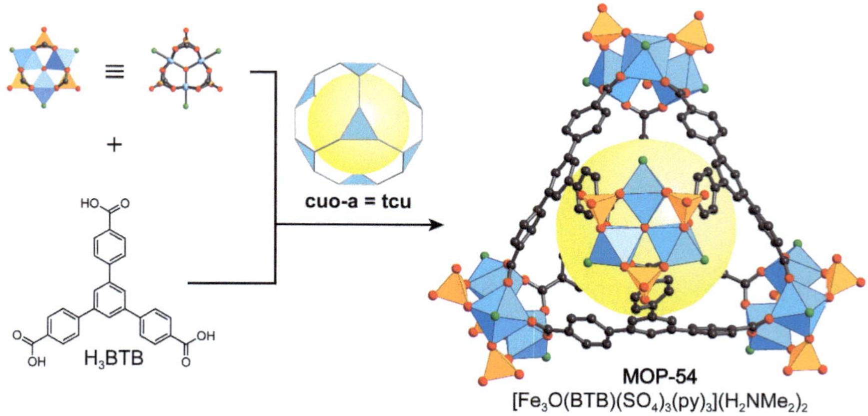

Figure 19.9 Synthesis of a MOP heterocube. Reticulation of H_3BTB with Fe^{2+} yields MOP-54. The angle $\eta_1 = 113°$ in the BTB linker and $\eta_2 = 67.1°$ in the SBU deviate substantially from the values of the idealized two extreme cases for a heterocube. All hydrogen atoms are omitted for clarity. Color code: Fe, blue; S, orange; C, gray; N, green; O, red.

boronate ester bond formation in this example further complicate their formation. This phenomenon is reminiscent of the interpenetration of 2D and 3D frameworks and is likely thermodynamically favorable as no large open pores are generated during the synthesis [11].

Another way to construct cube-shaped MOPs is by making so-called heterocubes, where two different trivalent SBUs with links of angles η_1 and η_2 are joined together. Here again, there are two extreme scenarios: (i) $\eta_1 = \eta_2 = 90°$ and (ii) $\eta_1 = 120°$ and $\eta_2 = 33.6°$ (Figure 19.7c,d).

An example for the second extreme case of a heterocube is MOP-54, constructed from partially sulfate-capped $Fe_3OL_3(-COO)_3(SO_4)_3$ SBUs and trigonal tritopic BTB linkers (Figure 19.9). The structure of MOP-54 is related to the structures of IRMOP-50, IRMOP-51, and IRMOP-53, yet topologically they are fundamentally different. While in the case of the IRMOP series six linear ditopic linkers are lying on the edges of a tetrahedron formed by four SBUs, in the case of MOP-54 four trigonal tritopic linkers occupy the faces of the tetrahedron. Topologically the IRMOP series are tetrahedra, whereas MOP-54 is classified as a heterocube. The structure resembles more the second extreme case of a heterocube despite significant deviations from the idealized angles ($\eta_1 = 113°$; $\eta_2 = 67.1°$) (see Figure 19.7d) [6]. This example highlights that in some cases strong structural similarities may be lost in the topological description of reticular materials and, as such, topology should only serve as a tool and by no means as a replacement for a deep understanding of the chemical nature of a given material.

19.6 MOPs Based on the Cuboctahedron

In a cuboctahedron, 12 4-c vertices are joined by 20 ditopic edges. The two extreme cases are shown in Figure 19.10. In the first case the edges are straight

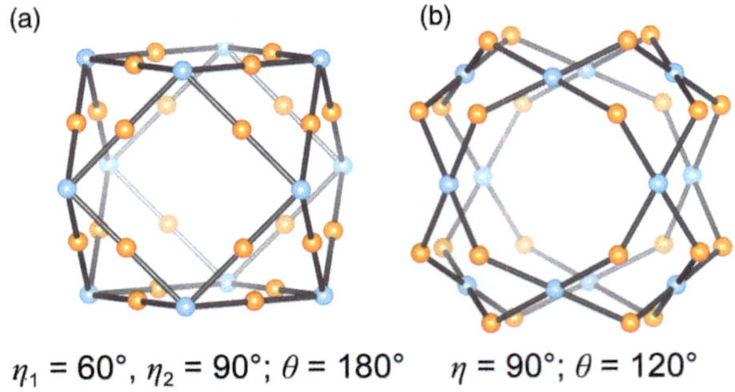

Figure 19.10 Two extreme cases of 12 4-c vertices linked by 20 edges to form a cuboctahedron. (a) One extreme case is the linking of vertices with the minimum angle of $\eta_1 = 60°$ and $\eta_2 = 90°$ with linear edges of $\theta = 180°$. (b) In the second extreme case the vertices have the maximum angle of $\eta = 90°$ and an angle of $\theta = 117°$. Color code: vertices, blue; edges, orange.

($\theta = 180°$), and the tetravalent vertices are not at sites of fourfold symmetry and thus it is no longer possible to give a single value of η to relate the angles. Consequently, there are two kinds of angle between the edges from the center of the vertices of $\eta_1 = 60°$ and $\eta_2 = 90°$ (Figure 19.10a). In the second case, the angle between the square planar vertices are $\eta_1 = 90°$ and the edges are bent to an angle of $\theta = 117°$ (Figure 19.10a).

MOP-1 is constructed from $Cu_2(-COO)_4$ paddle wheel SBUs linked by bent m-H_2BDC linkers. The 90° angles in the paddle wheel SBUs and the 120° angles in the m-BDC linker are almost identical to the idealized second extreme case with 90° angles between the points of extension of the paddle wheel SBU and 120° angles between the carboxylate carbons of the linker (Figure 19.10b). MOP-1 has already been discussed in Section 4.2.2 in the context of the influence of linker geometry on the linking of paddle wheel SBUs (Figure 19.11). It is worthwhile

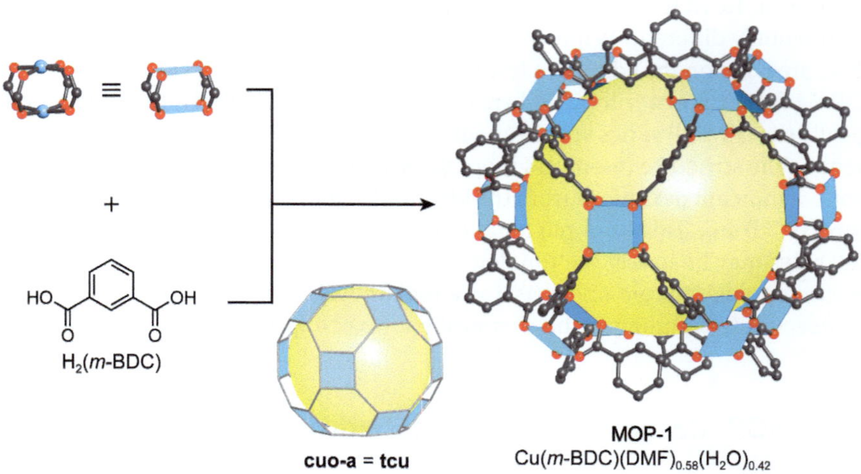

Figure 19.11 The cuboctahedral MOP-1 is constructed from $Cu_2(-COO)_4$ paddle wheel SBUs and bent m-H_2BDC linkers. The angles in the SBU and the linker are identical to those in the idealized extreme case 2 with $\eta = 90°$ and $\theta = 120°$. All hydrogen atoms are omitted for clarity. Color code: Cu, blue; C, gray; O, red.

mentioning that MOPs such as MOP-1 are used as tertiary building units in the synthesis of complex hierarchical MOFs.

19.7 Summary

In this chapter, we covered the reticular synthesis of MOPs and COPs and outlined the necessary structural requirements imposed on the building units to selectively target the formation of these discrete compounds over their extended counterparts, MOFs and COFs. We have discussed the angles η and θ between the points of extension of the building units representing the vertex and the edge of the underlying topology, respectively. The two extreme cases that these two angles can assume in MOPs and COPs of **tet**, **oct**, **cub**, and **cuo** topologies were shown and illustrative examples of polyhedra of these structure types were given and their angular information examined. Finally, important considerations with respect to the surface area of discrete polyhedra were covered and the interplay between intrinsic surface area of the polyhedra and the extrinsic surface area imposed by the packing of these discrete molecules in the crystal structure were discussed.

References

1 (a) Diercks, C.S. and Yaghi, O.M. (2017). The atom, the molecule, and the covalent organic framework. *Science* 355 (6328): eaal1585. (b) Yaghi, O.M., O'Keeffe, M., Ockwig, N.W. et al. (2003). Reticular synthesis and the design of new materials. *Nature* 423 (6941): 705–714. (c) Furukawa, H., Cordova, K.E., O'Keeffe, M., and Yaghi, O.M. (2013). The chemistry and applications of metal-organic frameworks. *Science* 341 (6149): 1230444.

2 Tranchemontagne, D.J., Ni, Z., O'Keeffe, M., and Yaghi, O.M. (2008). Reticular chemistry of metal-organic polyhedra. *Angewandte Chemie International Edition* 47 (28): 5136–5147.

3 (a) Fujita, M. (1998). Metal-directed self-assembly of two-and three-dimensional synthetic receptors. *Chemical Society Reviews* 27 (6): 417–425. (b) Seidel, S.R. and Stang, P.J. (2002). High-symmetry coordination cages via self-assembly. *Accounts of Chemical Research* 35 (11): 972–983. (c) Caulder, D.L. and Raymond, K.N. (1999). Supermolecules by design. *Accounts of Chemical Research* 32 (11): 975–982. (d) Han, M., Engelhard, D.M., and Clever, G.H. (2014). Self-assembled coordination cages based on banana-shaped ligands. *Chemical Society Reviews* 43 (6): 1848–1860.

4 (a) Mastalerz, M. (2010). Shape-persistent organic cage compounds by dynamic covalent bond formation. *Angewandte Chemie International Edition* 49 (30): 5042–5053. (b) Zhang, G. and Mastalerz, M. (2014). Organic cage compounds – from shape-persistency to function. *Chemical Society Reviews* 43 (6): 1934–1947. (c) Hasell, T. and Cooper, A.I. (2016). Porous organic cages: soluble, modular and molecular pores. *Nature Reviews Materials* 1 (9): 16053.

5 (a) Delgado-Friedrichs, O., O'Keeffe, M., and Yaghi, O.M. (2007). Taxonomy of periodic nets and the design of materials. *Physical Chemistry Chemical*

Physics 9 (9): 1035–1043. (b) Ockwig, N.W., Delgado-Friedrichs, O., O'Keeffe, M., and Yaghi, O.M. (2005). Reticular chemistry: occurrence and taxonomy of nets and grammar for the design of frameworks. *Accounts of Chemical Research* 38 (3): 176–182.

6 Sudik, A.C., Millward, A.R., Ockwig, N.W. et al. (2005). Design, synthesis, structure, and gas (N_2, Ar, CO_2, CH_4, and H_2) sorption properties of porous metal-organic tetrahedral and heterocuboidal polyhedra. *Journal of the American Chemical Society* 127 (19): 7110–7118.

7 Tozawa, T., Jones, J.T., Swamy, S.I. et al. (2009). Porous organic cages. *Nature Materials* 8 (12): 973–978.

8 Ni, Z., Yassar, A., Antoun, T., and Yaghi, O.M. (2005). Porous metal-organic truncated octahedron constructed from paddle-wheel squares and terthiophene links. *Journal of the American Chemical Society* 127 (37): 12752–12753.

9 Liu, Y., Kravtsov, V., Walsh, R.D. et al. (2004). Directed assembly of metal-organic cubes from deliberately predesigned molecular building blocks. *Chemical Communications* (24): 2806–2807.

10 Zhang, G., Presly, O., White, F. et al. (2014). A permanent mesoporous organic cage with an exceptionally high surface area. *Angewandte Chemie International Edition* 53 (6): 1516–1520.

11 Zhang, G., Presly, O., White, F. et al. (2014). A shape-persistent quadruply interlocked giant cage catenane with two distinct pores in the solid state. *Angewandte Chemie International Edition* 53 (20): 5126–5130.

20

Zeolitic Imidazolate Frameworks

20.1 Introduction

Zeolitic imidazolate frameworks (ZIFs) are a subclass of metal-organic frameworks (MOFs). Similar to MOFs, their structures are built from organic linkers and inorganic nodes. However, whereas MOF structures typically consist of linkers bearing chelating functionalities (e.g. carboxylate, pyrazolate) favoring the formation of polynuclear secondary building units (SBUs), the imidazole linkers used to construct ZIFs induce the formation of tetrahedral single transition metal nodes, and therefore the formation of tetrahedral topologies as in zeolites is favored (Figure 20.1). The strong interaction between the charged imidazolate linkers and the metal ions, in combination with the preference for the formation of rigid cages, renders ZIFs highly robust porous materials, thus setting them apart from classic coordination networks (see Chapter 1). As the name suggests, the structures of ZIFs are closely related to those of zeolites, which is ascribed to the combination of tetrahedral nodes and bent linkers mimicking the structural features of zeolites. In ZIFs, the angle between two metal centers bridged by an imidazolate unit is typically ∠(M–imidazolate–M) = 145°, a value close or even identical to that found for the ∠(Si–O–Si) angle in zeolites. To understand the development of porous metal-organic materials with zeolitic structures it is instructive to discuss the development of zeolites and their underlying structural chemistry.

The name zeolite is derived from the Greek ζέω (*zéō*), meaning "to boil" and λίθος (*líthos*), meaning "stone." Zeolites are crystalline aluminosilicates of group I and II elements (Na, K, Mg, Ca) with the general chemical formula $M_{2/n}O \cdot Al_2O_3 \cdot ySiO_2 \cdot wH_2O$, where n is the cation valence, y is 2–200, and w is the amount of water contained in the pores. They can be described as complex crystalline purely inorganic extended structures. The primary structural units (AlO_4 or SiO_4) are linked into larger so-called secondary building units that are linked to form extended 3D framework structures with pore sizes typically ranging from 0.3 to 1.0 nm and pore volumes between 0.10 and 0.35 $cm^3\ g^{-1}$. The presence of AlO_4 units in the silicate structures leads to charged cationic frameworks with counter ions residing within their pores. The term "zeolite" was coined by the Swedish mineralogist Alex F. Cronstedt in 1756 who discovered the first zeolite mineral (Stilbite) and observed that the material liberated large

Introduction to Reticular Chemistry: Metal-Organic Frameworks and Covalent Organic Frameworks, First Edition. Omar M. Yaghi, Markus J. Kalmutzki, and Christian S. Diercks.

Companion website: www.wiley.com/go/yaghi/reticular

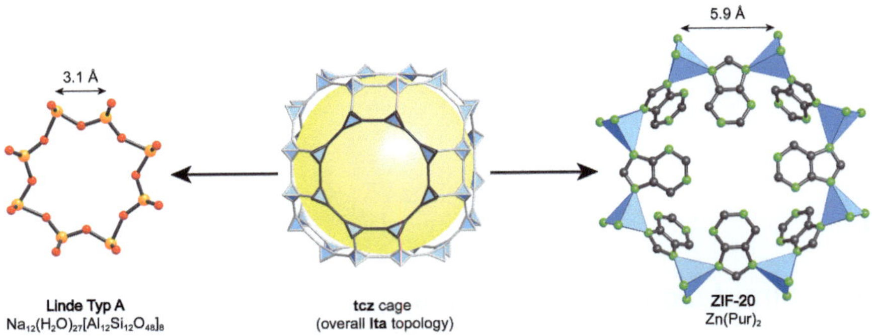

Figure 20.1 Zeolite versus ZIF: comparison of the 8-membered rings constituting the **tcz** cages in frameworks of **lta** topology. The silicon/aluminum nodes of the zeolite structure are replaced by tetrahedral metal ions (here Zn) and the oxygen links are replaced by imidazole-based linkers (here Pur = purine). Color code: Zn, blue; C, gray; N, green; O, red; Al and Si, orange.

quantities of "steam" when heated [1]. In 1840 Augustin A. Damour reported that zeolites are indeed capable of reversible uptake of water with no apparent change in morphology [2]. Twenty-two years later Henri É. Saint Claire-Deville described the first hydrothermal synthesis of a zeolite: levynite, a naturally occurring mineral [3]. At this point, only little was known about the properties of zeolites and it was not until 1896 that Georges Friedel developed the idea that the structure of dehydrated zeolites is similar to an open sponge-like framework, after having observed the occlusion of various liquids such as alcohol, benzene, and chloroform by dehydrated zeolites [4]. In the mid-1930s Richard M. Barrer began his pioneering work on zeolites, studying their synthesis and adsorption properties, in particular the diffusion of ions in open structures and subsequently classified them according to their pore sizes [5]. In 1948, Barrer reported the first definitive synthesis of zeolites, including both a synthetic analog of a zeolite mineral (Mordenite) and a novel, synthetic zeolite (KFI) [6]. With a steadily increasing number of synthetically accessible zeolites and a growing interest in their properties, their properties were studied in the search for new approaches to the separation and purification of air. In the mid- to late-1940s, Robert M. Milton confirmed that zeolites are indeed capable of reversible gas adsorption, rendering them highly interesting materials for industrial applications. The 1980s was a period of discovery of new zeolites. Extensive work on the synthesis and applications of ZSM-5 ($Na_nAl_nSi_{96-n}O_{192}·16H_2O$, where $0 < n < 27$) and a growing number of other members of the high silica zeolite family led to the discovery of microporous crystalline aluminophosphate molecular sieves in 1982 by Stephen T. Wilson et al. at Union Carbide [7]. Soon after, many more members of the family of aluminophosphates, SAPO, MeAPO, MeAPSO, ElAPO, and ElAPSO, were discovered and efforts to synthesize metallosilicate molecular sieves containing elements other than silicon and aluminum, such as titanium, iron, gallium, or germanium, were made [8]. Until today, the combination of different tetrahedral MO_4 units linked to form larger 3D entities gave rise to about 180 different structure types [9]. The definite geometry of the tetrahedral

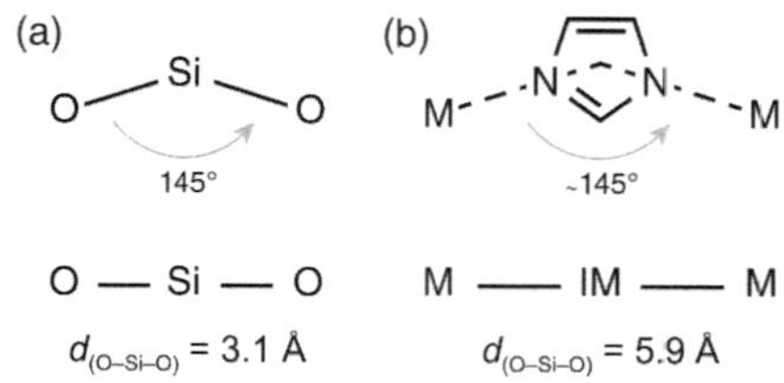

Figure 20.2 (a) Comparison of ∠(O–Si–O) in zeolites and (b) the ∠(M–IM–M) angle in ZIFs. Both are approximately 145°, but the O–Si–O-distance is 3.1 Å compared to the M–IM–M distance of 5.9 Å, leading to expanded pore sizes for identical topologies in ZIFs.

building units provides zeolites with rigid architecturally and mechanically stable structures. Even though a wide variety of different structures is accessible and certain methods for their functionalization have been established, zeolites suffer from the inherent limitation imparted by their inorganic building units. In a manner akin to that described for carboxylate MOFs, extended metal-organic materials with zeolitic structures can be prepared by judicious choice of appropriate starting materials. In order to prepare extended framework structures with tetrahedral (zeolitic) topologies, tetrahedral nodes have to be linked by building units with an angle close to, if not identical to, that of the ∠(Si–O–Si) = 145° fragment found in zeolites [10]. With an average ∠(M–imidazolate–M) angle of around 145°, imidazoles satisfy this prerequisite (Figure 20.2).[1]

Their reticulation with metal ions leads to the formation of structures with tetrahedrally coordinated metal cations playing the role of the tetrahedral silicon/aluminum atoms and the imidazolate anions forming bridges that mimic the role of oxygen atoms in zeolites [11]. Today, ZIFs with more than 100 different topologies, often analogous, but not limited to those known in zeolite chemistry, have been synthesized and structurally characterized. Their open permanently porous structures initiated intense research in the field of ZIFs, mostly driven by the prospect of expanding their pore sizes, functionalizing their pores, and exploring new sorption, separation, and catalytic properties [10c, 11].

20.2 Zeolitic Framework Structures

Extended metal-organic materials with zeolitic structures are divided into two groups: ZIFs and Z-MOFs. In analogy to MOFs, the topologies of Z-MOFs are commonly referred to by a lower case (bold) three-letter code (e.g. **rho**, the codes are identical to the RCSR codes) whereas zeolites and ZIFs having zeolite topologies are often given an uppercase three-letter code (e.g. RHO, as implemented by the Structure Commission of the International Zeolite Association, "IZA"). Throughout this chapter all topologies will be referred to with their three-letter codes according to the RCSR.

20.2.1 Zeolite-Like Metal-Organic Frameworks (Z-MOFs)

Z-MOFs are typically constructed from carboxylate functionalized imidazoles or pyrimidine units and tetrahedrally coordinated single-metal nodes linked together into tetrahedral structures. In contrast to the formation reactions of

1 Some MOFs fulfill these requirements as well and form tetrahedral structures. Owing to this they are commonly referred to as Z-MOFs. For more details, the reader is referred elsewhere [10a].

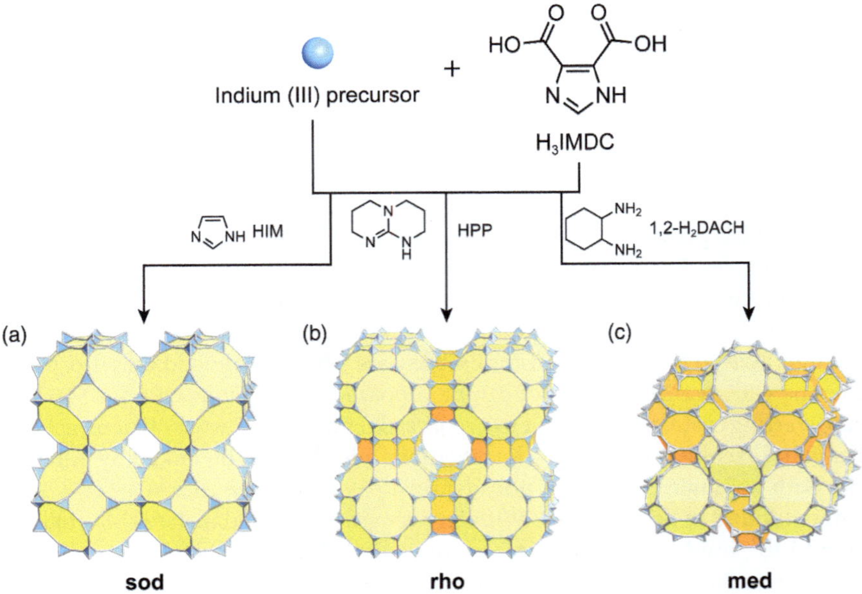

Figure 20.3 Synthesis of Z-MOFs by reticulating indium ions and H_3IMDC (imidazole dicarboxylic acid) in the presence of different SDAs. Three different framework topologies are realized. (a) A framework of **sod** topology forms in the presence of HIM, (b) the **rho** topology is favored in the presence of HHP, (c) and a framework of **med** topology is obtained when 1,2-H_2DACH is used [13].

MOFs based on polynuclear SBUs, the formation of Z-MOFs often requires the presence of a structure-directing agent (SDA) and a mixture of the same components can result in different topologies depending on the SDA used. Reticulation of In^{3+} ions and H_3IMDC (1*H*-imidazole-4,5-dicarboxylic acid) with three different SDAs yields three Z-MOFs of different topologies [12]. The presence of imidazole (HIM) affords the formation of a Z-MOF with **sod** topology (Figure 20.3a), HPP (1,3,4,6,7,8-hexahydro-2*H*-pyrimido[1,2-*a*]pyrimidine) affords the formation of an **rho** topology (Figure 20.3b), and the presence of 1,2-H_2DACH (1,2-diaminocyclohexane) affords the formation of an **med** net (Figure 20.3c), a topology not known for zeolites. In the **sod** Z-MOF, each indium center is 6-coordinate by two hetero-chelating HIMDC linkers and two HIMDC nitrogen donors, leading to the desired tetrahedral $InN_4(-COO)_2$ building unit. In contrast, in the **rho** Z-MOF each indium center is 8-coordinate by four hetero-chelating HIMDC linkers to give tetrahedral $InN_4(-COO)_4$ building units. Both materials possess pore volumes that are up to eight times as large than those of their purely inorganic counterparts. In the structure of **usf**-Z-MOF (**med** topology), the indium single-metal nodes are coordinated in a tetrahedral manner by four HIMDC linkers forming 8-c $InN_4(COO)_4$ building units [13]. A comparison of all three topologies is given in Figure 20.3.

Since, in terms of topology, the vertices and nodes can be chosen in such way that a significant simplification of the overall structure is achieved, tetrahedral subunits assembled from SBUs and carboxylate linkers can also be described by tetrahedral, zeolite-like topologies [10a, 14]. We already encountered such MOFs with zeolite-like structures in earlier chapters exemplified by MIL-100 and MIL-101, both of which are built from vertex-sharing tetrahedral tertiary building units (TBUs) resulting in an overall tetrahedral **mtn** topology (see Figure 2.13) [14].

20.2.2 Zeolitic Imidazolate Frameworks (ZIFs)

As outlined earlier, imidazolates can link tetrahedral metal centers with an M–IM–M angle close to 145°, which is not the case for triazolates or tetrazolates, making imidazolates ideal building units for extended structures with tetrahedral, zeolite-like topologies. The combination of imidazole derivatives and tetrahedral metal nodes allows for the formation of a wide range of different ZIF structures. The strong bond between the charged imidazolate linker and the metal center along with the hydrophobic nature of ZIFs gives rise to high chemical and thermal stability, whereas their tetrahedral structures provide for high mechanical and architectural stability. These properties, in combination with the chemical tunability inherent to reticular materials, make ZIFs interesting in the context of gas adsorption, gas separation, and catalysis.

Early examples of ZIFs were limited to non-porous, dense structures. This can be ascribed to the fact that the energetically most favorable spatial arrangement for the combination of tetrahedral metal nodes and (unsubstituted) imidazolate units is also the most dense one [15]. The development of porous ZIFs featuring characteristics of both zeolites and MOFs, combining uniform cage-like pores with high crystallinity and permanent porosity, initiated intense research in this field. In contrast to the pores without walls commonly observed in carboxylate-based MOFs (see Chapters 1–6), the structures of ZIFs are typically built form cages composed of multiple fused rings of metal nodes connected by imidazolate linkers. Therefore, the pore opening is dictated by the size of the rings constituting these cages and is thus often relatively small. The specific architecture of these cages in combination with the relatively short imidazole linker endow ZIFs with mechanical and architectural stability and allows them to support permanent porosity. Indeed, ZIFs are often more stable than other extended metal-organic materials as highlighted by the fact that ZIF-8 retains both crystallinity and porosity when refluxed in water, alkaline solutions, or organic solvents over an extended period of time.

The functionalization of the imidazole linker can be achieved in manners akin to those used to functionalize MOFs (see Chapter 6). Pre-synthetic functionalization of the imidazole linker not only imparts functionality but the steric demand of the functionality appended to the linker as well as its position also has an impact on the structure formed upon reticulation. Design principles making use of this finding allow for the design and synthesis of ZIFs built from cages of specific size and we will discuss them in more detail later in this chapter. Some examples of imidazole derivatives used in the synthesis of ZIFs are given in Figure 20.4. Before elucidating the principles underlying the design of new ZIFs, it is instructive to take a closer look at typical synthetic approaches.

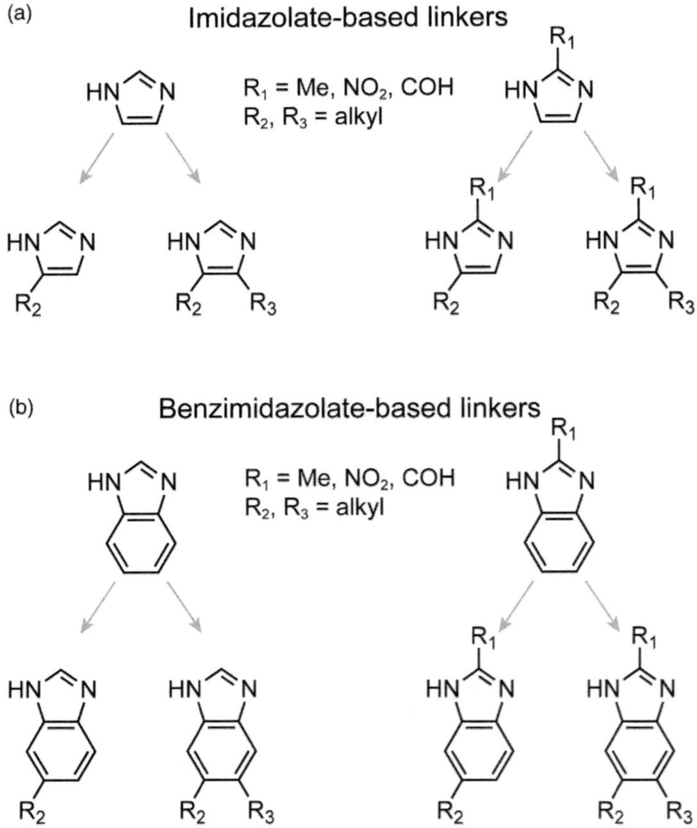

Figure 20.4 (a) Imidazolate and (b) benzimidazolate linkers used to construct ZIFs. Different substitution patterns are shown and typical substituents are given. Functionalization can be performed pre- or post-synthetically allowing for the synthesis of a wide range of structures with different functionalities.

20.3 Synthesis of ZIFs

While SDAs, such as templates, are commonly used in the synthesis of zeolites and Z-MOFs, the formation of ZIFs does not rely on such additives. ZIFs are commonly prepared by reacting the appropriate hydrated metal salts, most often zinc and other metals that prefer a tetrahedral coordination, and an imidazole (or functionalized derivatives) in an amide solution (e.g. DMF, DEF) at temperatures ranging from 85 to 150 °C. Similar to the synthesis of carboxylate-based frameworks, the slow release of amine base upon decomposition of the amide at elevated temperatures deprotonates the linker and thereby initiates framework formation. A similarly slow deprotonation is achieved by layering of solutions. This method is used to synthesize ZIF-8. Here, a solution of HmIM (2-methylimidazole) and 2,2′-bipyridine in ethanol is layered onto a solution of $[Zn(OAc)_2]{\cdot}2H_2O$ dissolved in concentrated aqueous ammonia [16].

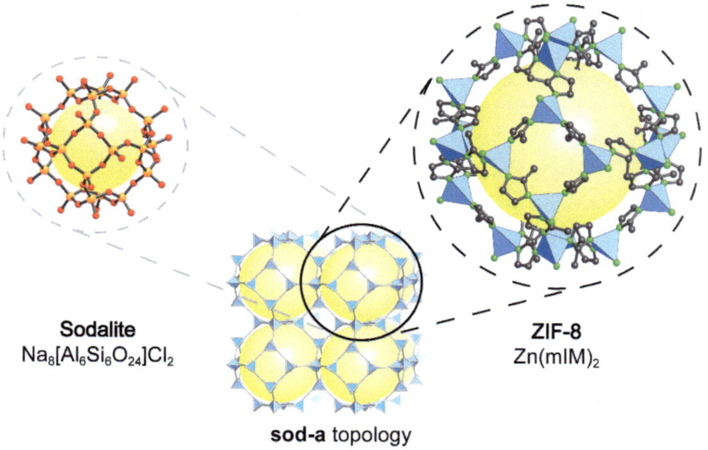

Figure 20.5 The augmented **sod** net is named after the natural mineral Sodalite. It is built from cages of **tro** topology that are connected through 4-membered rings. Each **tro** cage is built from 24 tetrahedral nodes. Owing to the increased distance of the tetrahedral nodes in ZIFs compared to zeolites, their pores are significantly larger. The **tro** cages in ZIF-8 ($Zn(mIM)_2$) have a diameter of 11.6 Å and a pore apertures of 3.4 Å. Color code: Zn, blue; C, gray; N, green; O, red; Al and Si, orange.

The slow diffusion between the layers provides for low reaction rates and thus for the reversibility required to afford a crystalline product of chemical formula $Zn(mIM)_2$, termed ZIF-8. The structure has a sodalite topology (SOD or **sod**), named after the naturally occurring mineral sodalite. The augmented **sod** net alongside the fundamental **tro** cages of sodalite and ZIF-8 are depicted in Figure 20.5. The **tro** cages in the structure of ZIF-8 have a diameter of 11.6 Å and are constructed from 4- to 6-membered rings that are fused along common edges $[4^6.6^8]$. Connecting these cages through the 4-membered rings results in a 3D pore system with channels of 3.4 Å.

The formation of frameworks with a wide range of topologies can be realized by employing substituted imidazole linkers in the synthesis. The specific substitution pattern directs the formation of different topologies and a precise design of the linker is therefore inevitable.

20.4 Prominent ZIF Structures

The structures of ZIFs are built from tetrahedral nodes that are geometrically similar to the MO_4 units in zeolites. All zeolite nets found in ZIFs are uninodal, whereas for zeolites, the majority of their nets is not. The number of possible structures built from tetrahedral units is expected to increase exponentially with the number of chemically different vertices and edges, which illustrates the vast structural possibilities of ZIF chemistry. Table 20.1 shows a compilation of frequently encountered topologies alongside representative ZIF structures. A selection of zeolite topologies found in ZIF chemistry is shown in Figure 20.6.

Table 20.1 Compilation of ZIFs, their chemical formulae, and topologies.

Common name	Chemical composition	RCSR code	Zeolite code	References
ZIF-14	$Zn(eIM)_2$	**ana**	ANA	[17]
ZIF-386	$Zn(nBIM)_{0.85}(nIM)_{0.70}(IM)_{0.45}$	—	AFX	[18]
ZIF-725	$Zn(bBIM)_{1.35}(nIM)_{0.40}(IM)_{0.25}$	**bam**	—	[18]
ZIF-62	$Co(IM)_2$	**cag**	—	[17b]
ZIF-303	$Zn(cBIM)_{0.70}(nIM)_{0.30}(IM)_{1.00}$	—	CHA	[18]
TIF-4	$Zn(IM)_{1.5}(mBIM)_{0.5}$	**coi**	—	[19]
ZIF-64	$Zn(IM)_2$	**crb**	BCT	[17b]
—	$Pr(IM)_5$	**crs**	—	[10c]
ZIF-3	$Zn_2(IM)_4$	**dft**	—	[20]
ZIF-23	$Zn(4ab\backslash BIM)_2$	**dia**	—	[21]
BIF-6	$CuBH(IM)_3$	**fes**	—	[22]
ZIF-73	$Zn(nIM)_{1.74}(mBIM)_{0.26}$	**frl**	—	[17b]
ZIF-5	$Zn_3In_2(IM)_{12}$	**gar**	—	[20]
ZIF-615	$Zn(cBIM)_{1.05}(4\text{-}nIM)_{0.95}$	**gcc**	—	[18]
ZIF-6	$Zn(IM)_2$	**gis**	GIS	[20]
ZIF-486	$Zn(nBIM)_{0.20}(mIM)_{0.65}(IM)_{1.15}$	**gme**	GME	[18]
ZIF-360	$Zn(bBIM)_{1.00}(nIM)_{0.70}(IM)_{0.30}$	**kfi**	KFI	[18]
ZIF-72	$Zn(dcIM)_2$	**lcs**	—	[17b]
ZIF-376	$Zn(nBIM)_{0.25}(mIM)_{0.25}(IM)_{1.50}$	**lta**	LTA	[18]
—	$Cd(Im)_2bipy$	**mab**	—	[23]
ZIF-60	$Zn_2(IM)_3(mIM)$	**mer**	MER	[17b]
—	$Cu(IM)_2$	**mog**	—	[24]
ZIF-100	$Zn_{20}(cBIM)_{39}(OH)$	**moz**	—	[10b]
—	$Co(IM)_2$	**neb**	—	[25]
—	$Co_2(IM)_4$	**nog**	—	[25]
TIF-3	Zn(IM)(mBIM)	**pcb**	ACO	[26]
ZIF-95	$Zn(cBIM)_2$	**poz**	—	[10b]
—	$Fe(mIM)_2$	**qtz**	—	[27]
ZIF-11	$Zn(BIM)_2$	**rho**	RHO	[20]
ZIF-8	$Zn(mIM)_2$	**sod**	SOD	[16]
BIF-8	$CuBH(eIM)_3$	**srs-c-b**	—	[22]
BIF-7	CuBH(mIM)3	**ths-c-b**	—	[22]
ZIF-412	$Zn(BIM)_{1.13}(nIM)_{0.62}(IM)_{0.25}$	**ucb**	—	[18]
ZIF-516	$Zn(mBIM)_{1.23}(bBIM)_{0.77}$	**ykh**	—	[18]
TIF-1Zn	$Zn(dmBIM)_2$	**zea**	—	[28]
TIF-2	$Zn(IM)_{1.1}(mBIM)_{0.9}$	**zeb**	—	[26]
ZIF-61	Zn(IM)(mIM)	**zni**	—	[17b]

Topologies are given both as the notation referring to zeolite topologies and those found in the RCSR database.

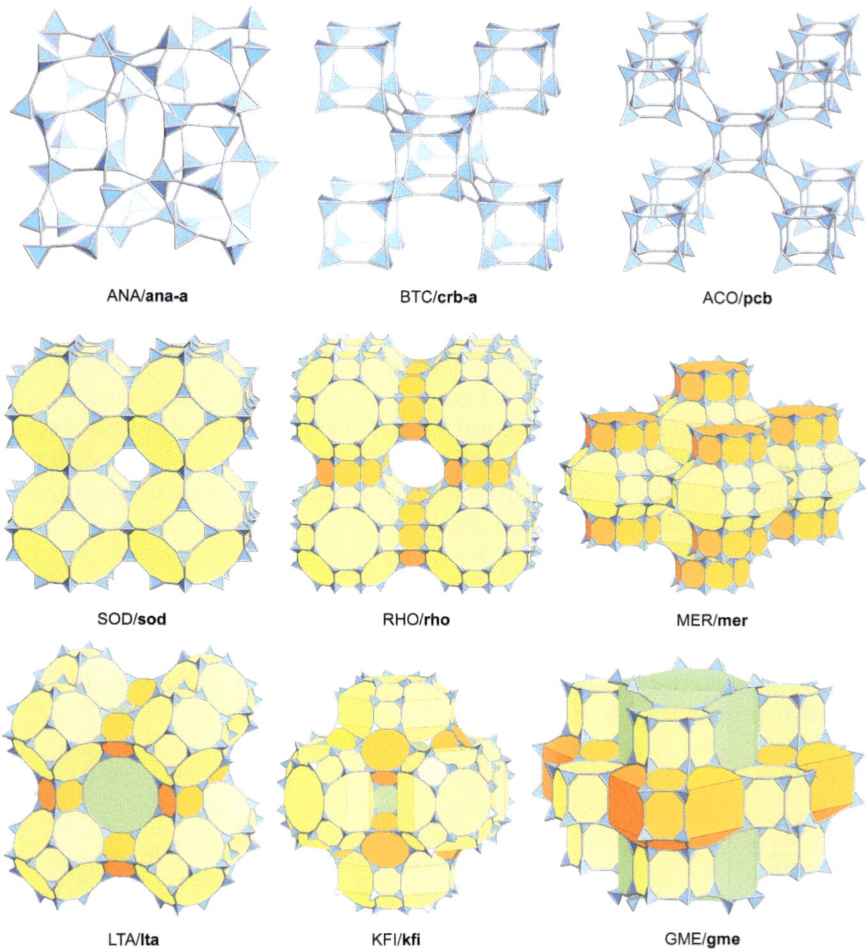

Figure 20.6 Zeolite topologies commonly found in ZIF chemistry. All frameworks are constructed from tetrahedral nodes but differ in the types of rings formed by linking these nodes. Each blue tetrahedron represents a metal node, and the differently colored polyhedra indicate different types of cage. The following frameworks are shown: ANA (**ana**), BTC (**crb**), ACO (**pcb**), SOD (**sod**), RHO (**rho**), MER (**mer**), LTA (**lta**), KFO (**kfi**), and GME (**gme**). Capitalized three-letter codes are used to describe zeolite topologies, while MOF and ZIF topologies are referred to by lowercase bolded three-letter codes.

20.5 Design of ZIFs

Even though it is known that experimental parameters such as the temperature, the concentration of starting materials, and the solvents employed can influence the formation of ZIFs in terms of topology, there are no general principles for their rationalization. In MOF chemistry, the precise geometry of the linker allows to direct the synthesis toward specific topologies. Structural consideration of the imidazole linker in general and the steric demand of its substituents in particular form the basis of the three general design principles that allow for the rational design of ZIFs by linker-directed synthesis: (i) the maximum size of the pore opening is determined by the size and shape of the imidazole linker, which is described by a steric index (δ); (ii) the combination of linkers with large and small δ is required for the formation of large cages; and (iii) changing the ratio of a given set of imidazole linkers facilitates the formation of cages with different topologies and metrics.

20.5.1 The Steric Index δ as a Design Tool

The framework formation of ZIFs is mainly guided by the substitution pattern of the imidazole linkers employed in their synthesis. Imidazole can be substituted in two distinguishable positions: (i) the 2-position and (ii) the 4- and 5-position. We define distances in the molecular structure of the imidazole linker that are correlated to the size of appended functionalities in these positions. The size of substituents in the 2-position defines l_2 whereas substituents in the 4- or 5-positions influence the value of $l_{4,5}$ (Figure 20.7). Based on these two distances we can calculate the steric index δ according to Eq. (20.1):

$$\delta = V \times l \tag{20.1}$$

where V is the van der Waals volume of the linker and l the longer distance (l_2 or $l_{4,5}$). The steric index gives a measure of the size and shape of a given imidazole linker. Figure 20.7 gives examples of different imidazole linkers and their respective steric indices δ.

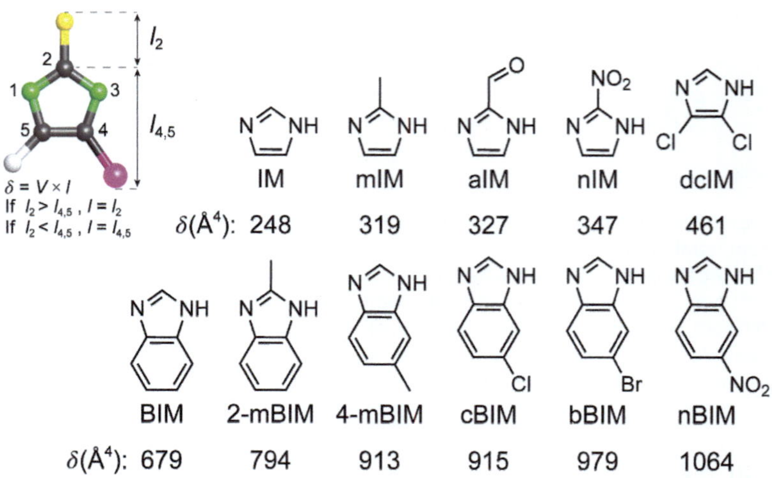

Figure 20.7 Definition of the steric index (δ). Two lengths are defined, l_2 and $l_{4,5}$. For the calculation of the steric index only the larger value is used and multiplied with the van der Waals volume of the linker. Several examples of substituted imidazoles and benzimidazoles are shown and their respective values of δ are given. Color code: N, green; C, gray; substituents in the 2-, 4-, and 5-position are shown in yellow, pink, and white, respectively.

20.5.1.1 Principle I: Control over the Maximum Pore Opening

The size of the largest ring (pore opening) in ZIFs is correlated to the steric index of the imidazole linker constituting this particular ring. To take advantage of this principle it is instructive to search for regularities in the orientation of differently substituted imidazolate linkers with respect to the ring they are forming. In general, substituents in the 2-position tend to point into the small (mostly 4-membered) rings, whereas substituents in the 4- and 5-position are pointing toward the center of 8-membered and larger rings, and both the 2- and 4-/5-positions are found to point into 6-membered rings. In structures built from cages that are constructed from fused 8-membered rings, substituents in the 4- and 5-positions are necessarily forced to point toward the center of the 4-membered rings. Thus, such structures can only form if the substituents on the 4- and 5-positions are comparatively small. This entails that increasing the bulkiness of substituents in the 4- and 5-positions, which equals a longer distance $l_{4,5}$ and therefore a large steric index δ, will inevitably lead to the formation of large rings. Since the 2-position points toward the center of the small rings due to its low steric demand, the formation of small rings is not precluded when linkers with large δ arising from large values of $l_{4,5}$ are used. The same holds true for structures built from a combination of linkers with large and small δ. In contrast, large rings cannot be obtained when only linkers with small δ are used. The first principle can be summarized as follows: the steric index δ of the imidazole linker dictates the maximum ring size and therefore the size of the pore opening of the resulting ZIF structure.

20.5.1.2 Principle II: Control over the Maximum Cage Size

The presence of large rings within a ZIF structure realized by employing imidazole linkers with large δ is not necessarily synonymous to the presence of large cages. The formation of large cages (or large internal pores) in tetrahedral structures relies on a combination of both large and small rings. Hence, to design a ZIF structure composed of large cages, it is necessary to mix both linkers with small and large δ in the appropriate ratio. The importance of balancing the ratio of imidazole linkers with large and small δ becomes clear when comparing the structures of ZIF-412 ($Zn(BIM)_{1.13}(nIM)_{0.62}(IM)_{0.25}$) and ZIF-68 (Zn(BIM)(nIM), where BIM = benzimidazolate, nIM = 2-nitroimiazolate, and IM = imidazolate) (Figure 20.8). Both structures contain 8- and 12-membered rings facilitated by sterically demanding BIM and nIM linkers with large δ of 679 and 347 $Å^4$, respectively. However, the largest cage in ZIF-412 is twice the size of that in ZIF-68 due to the additional comparatively small IM linker ($\delta = 248$ $Å^4$). This can be explained by the fact that the small IM linker enables the formation of 4- and 6-membered rings that can combine with the larger rings built from the bulkier BIM and nIM linkers to form large cages. In general, the largest cages are found in structures containing 75–90% of bulky (large δ) and 10–25% of small linkers (small δ). Thus far, ZIF-412 marks the largest cage reported for ZIFs.

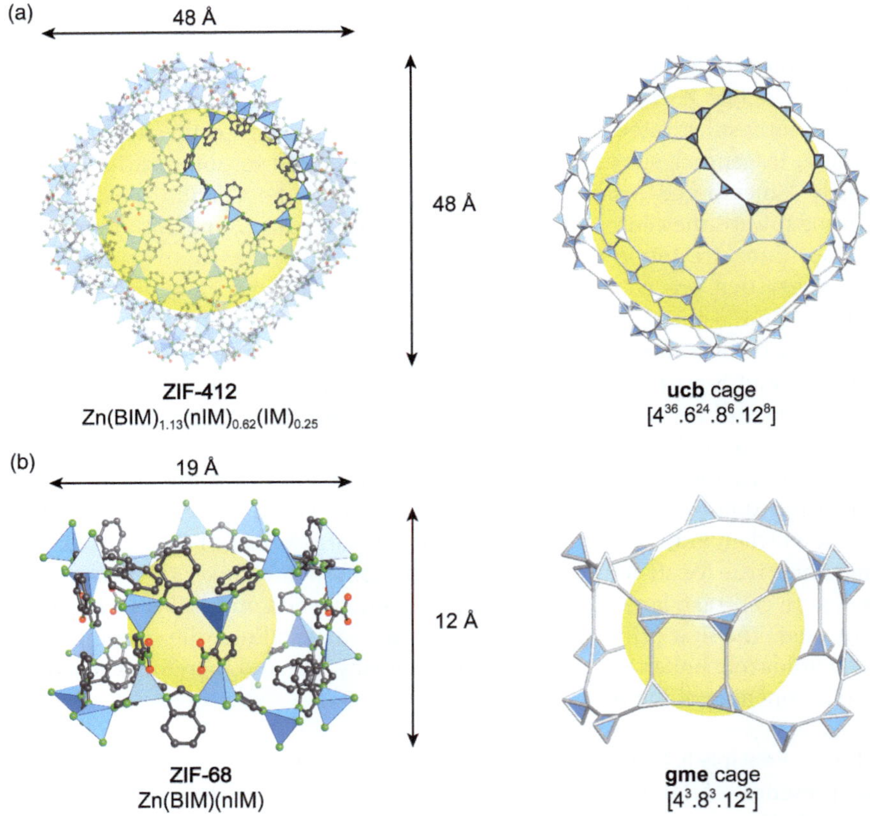

Figure 20.8 Comparison of the largest cages found in the crystal structures of ZIF-412 and ZIF-68 [17b, 18]. (a) The largest cage in ZIF-412 is built from 36 4-membered, 24 6-membered, 6 8-membered, and 8 12-membered rings and measures 48 Å in diameter. (b) The largest cage in ZIF-68 is significantly smaller due to the lack of IM linkers (small δ). It consists of only three 4-membered, three 8-membered, and two 12-membered rings. In both cases, the largest cage is shown as a crystal structure representation alongside the corresponding topology. This comparison highlights the importance of linkers with small values of δ for the formation of large cages. Color code: Zn, blue; C, gray; N, green; O, red.

20.5.1.3 Principle III: Control over the Structural Tunability

The two principles discussed above provide a rationale for the design of ZIF structures with large rings and large cages. The ratio of imidazole linkers with different steric indices provides an additional tool to access a variety of structures built from the same general components. Employing linkers with different δ therefore not only allows for the formation of structures with the maximum pore opening or cage size, but the variation of the ratio of the linkers further facilitates the formation of structures with ring and pore sizes of any value up to that maximum. The more complex the composition, meaning the larger the number of differently substituted imidazole linkers used, the more potent this principle gets, providing access to an almost infinite number of structures. This can be illustrated by the finding that three ZIFs with different underlying topologies are formed when Zn^{2+} ions are reticulated with a mixture of different ratios of nbIM ($\delta = 1064\,Å^4$), mIM ($\delta = 319\,Å^4$), and IM linkers ($\delta = 248\,Å^4$); ZIF-486 ($Zn(nbIM)_{0.20}(mIM)_{0.65}(IM)_{1.15}$, **gme**), ZIF-376 ($Zn(nbIM)_{0.25}(mIM)_{0.25}(IM)_{1.5}$, **lta**), and ZIF-414 ($Zn(nbIM)_{0.91}(mIM)_{0.62}(IM)_{0.47}$, **ucb**). The largest cages in the structures vary between 22.6, 27.5, and 45.8 Å, respectively.

20.5.2 Functionalization of ZIFs

Applying methods similar or identical to those described in Chapter 6, ZIFs can be functionalized pre-synthetically, *in situ*, or by post-synthetic modification (PSM) – a broadness of possible functionalization not known for their inorganic zeolite counterparts. Here, we will discuss a few selected examples of the functionalization and modification of ZIFs.

To introduce functionality by performing reactions on the organic backbone of a ZIF, functional groups facilitating these reactions must be installed pre-synthetically or by linker exchange. An imidazole linker that carries a suitable functional group is aIM (1*H*-imidazole-2-carbaldehyde). The aldehyde group in the 2-posititon can be reacted with amines to form an imine bond or it can be reduced to give an alcohol group using reducing agents such as $NaBH_4$. The reaction of ZIF-90 ($Zn(aIM)_2$), a framework of **sod** topology built from aIM linkers and Zn^{2+} nodes, with ethanolamine yields ZIF-92 and reduction of ZIF-90 with $NaBH_4$ yields ZIF-91 (Figure 20.9) [29]. In both cases the crystallinity and porosity of the framework is retained.

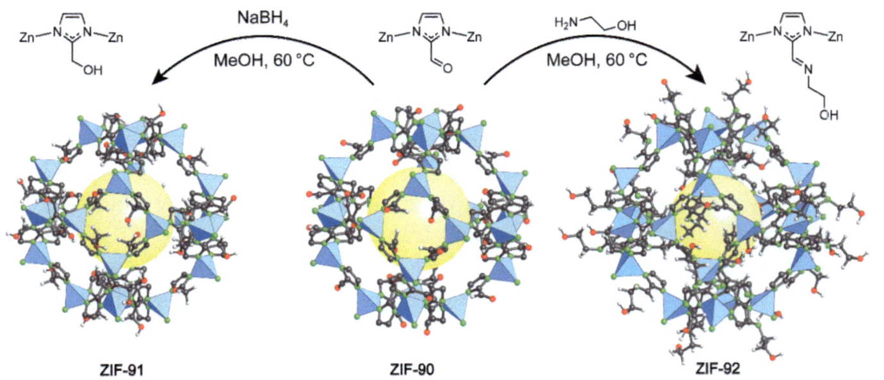

Figure 20.9 Post-synthetic functionalization of ZIF-90. The aIM linker in ZIF-90 (center) can either be reduced to yield alcohols (left, ZIF-91) or reacted with amines to yield the corresponding imines (right, ZIF-92). In both cases, the porosity and crystallinity of the material are retained. Only one cage is shown and all hydrogen atoms on the imidazolate are omitted for clarity. Color code: Zn, blue; C, gray; N, green; O, red.

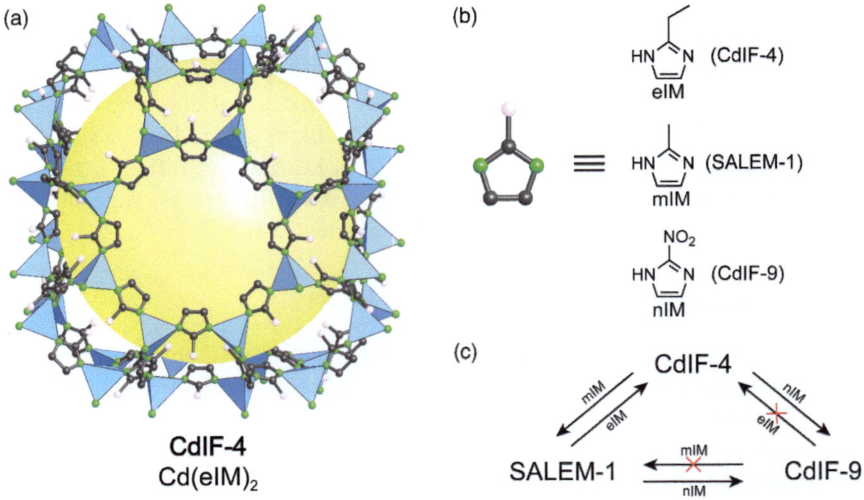

Figure 20.10 Linker exchange reactions in CdIF-4 [32]. (a) The **rho** net of CdIF-4 is built from eIM linkers connecting Cd^{2+} centers. (b) Linker exchange reactions using three differently functionalized imidazole linkers have been studied, yielding three ZIFs in single-crystal to single-crystal transformations. (c) Possible linker exchange reactions between CdIF-4, CdIF-9, and SALEM-1.

Such modifications find use in the preparation of mixed-matrix membranes (MMMs). MMMs are an intensely studied area of ZIF research with focus on applications such as gas separation. Many methods used to prepare MMMs take advantage of PSMs. One example is the deposition of ZIF-90 on an alumina surfaces. This is achieved by modifying an alumina surface with 3-aminopropyltriethoxysilane (APTES) to provide free amines for consecutive imine condensation with the aIM linkers and subsequent *in situ* crystallization of ZIF-90 on the surface [30]. The remaining unmodified aIM linkers can then be functionalized in order to tune the properties of the membranes [31]. ZIFs have also been shown to be good candidates for linker exchange reactions [32]. The 2-ethylimidazole (eIM) linker in CdIF-4, a cadmium-based ZIF with underlying **rho** topology, can be exchanged by nIM or mIM under relatively mild conditions in a single-crystal to single-crystal transformation affording CdIF-9 and SALEM-1, respectively (Figure 20.10). The linker exchange in MMMs is another tool to fine-tune the properties of the membrane as demonstrated for the adjustment of the interplay between selectivity and permeability of ZIF-7 MMMs [33].

20.6 Summary

In this chapter, we introduced ZIFs, a special subclass of MOFs. We saw that similarly to inorganic zeolites, the structures of ZIFs are constructed from tetrahedral nodes connected by linkers. The combination of tetrahedral single-metal nodes and imidazole-based linkers results in tetrahedral structures that feature large cages that are connected through narrow windows. Even though ZIFs are only composed of linked tetrahedra, the structural diversity seems limitless. We discussed the structural features of ZIFs, general synthetic approaches, and highlighted some frequently encountered topologies. The steric index was introduced as a means to target the formation of ZIFs with large cages and we showed that analogous to MOFs, the organic imidazolate linkers can be functionalized using PSM.

References

1 Cronstedt, A.F. (1756). *Rön och beskrifning om en obekant bärg art, som kallas Zeolites*. Stockholm: Svenska Vetenskaps akademiens Handlingar (trans. J.L. Schlenker and G.H. Kühl. Proceedings of the 9th International Conference on Zeolites. 1993).

2 Damour, A. (1840). Über das Bleigummi und thonerdhaltiges phosphorsaures Bleioxyd von Huelgoat. *Annales des Mines* 17: 191.

3 de St. Claire-Deville, H. (1862). Reproduction de la levyne. *Comptes Rendus* 54 (1862): 324–327.

4 (a) Friedel, G. (1896). New experiments on zeolites. *Bulletin de la Société Française de Minéralogie* 19: 363–390. (b) Friedel, G. (1896). Sur quelques propriétés nouvelles des zéolithes. *Bulletin de la Société Française de Minéralogie, Paris* 19: 94–118.

5 (a) Sherman, J.D. (1999). Synthetic zeolites and other microporous oxide molecular sieves. *Proceedings of the National Academy of Sciences* 96 (7): 3471–3478. (b) Barrer, R.M. (1945). Separation of mixtures using zeolites as molecular sieves. I. Three classes of molecular-sieve zeolite. *Journal of the Society of Chemical Industry* 41 (12): 130–133.

6 (a) Barrer, R.M. (1981). Zeolites and their synthesis. *Zeolites* 1 (3): 130–140. (b) Barrer, R.M. (1948). 33. Synthesis of a zeolitic mineral with chabazite-like sorptive properties. *Journal of the Chemical Society (Resumed)* 127–132.

7 Wilson, S.T., Lok, B.M., Messina, C.A. et al. (1982). Aluminophosphate molecular sieves: a new class of microporous crystalline inorganic solids. *Journal of the American Chemical Society* 104 (4): 1146–1147.

8 Flanigen, E.M., Lok, B.M., Patton, R.L., and Wilson, S.T. (1986). Aluminophosphate molecular sieves and the periodic table. *Pure and Applied Chemistry* 58 (10): 1351–1358.

9 (a) Baerlocher, C., McCusker, L.B., and Olson, D.H. (2007). *Atlas of Zeolite Framework Types*. Elsevier. (b) Corma, A., Díaz-Cabañas, M.J., Jiang, J. et al. (2010). Extra-large pore zeolite (ITQ-40) with the lowest framework density containing double four- and double three-rings. *Proceedings of the National Academy of Sciences of the United States of America* 107 (32): 13997–14002. (c) Flanigen, E.M., Broach, R.W., and Wilson, S.T. (2010). Introduction. S. Kulprathipanja In: *Zeolites in Industrial Separation and Catalysis*, Wiley-VCH 1–26.

10 (a) Eddaoudi, M., Sava, D.F., Eubank, J.F. et al. (2015). Zeolite-like metal-organic frameworks (ZMOFs): design, synthesis, and properties. *Chemical Society Reviews* 44 (1): 228–249. (b) Wang, B., Cote, A.P., Furukawa, H.

et al. (2008). Colossal cages in zeolitic imidazolate frameworks as selective carbon dioxide reservoirs. *Nature* 453 (7192): 207–211. (c) Phan, A., Doonan, C.J., Uribe-Romo, F.J. et al. (2010). Synthesis, structure, and carbon dioxide capture properties of zeolitic imidazolate frameworks. *Accounts of Chemical Research* 43 (1): 58–67.

11 Chen, B., Yang, Z., Zhu, Y., and Xia, Y. (2014). Zeolitic imidazolate framework materials: recent progress in synthesis and applications. *Journal of Materials Chemistry A* 2 (40): 16811–16831.

12 Liu, Y., Kravtsov, V.C., Larsen, R., and Eddaoudi, M. (2006). Molecular building blocks approach to the assembly of zeolite-like metal-organic frameworks (ZMOFs) with extra-large cavities. *Chemical Communications* (14): 1488–1490.

13 Liu, Y., Kravtsov, V.C., and Eddaoudi, M. (2008). Template-directed assembly of zeolite-like metal-organic frameworks (ZMOFs): a usf-ZMOF with an unprecedented zeolite topology. *Angewandte Chemie International Edition* 47 (44): 8446–8449.

14 (a) Férey, G., Mellot-Draznieks, C., Serre, C., and Millange, F. (2005). Crystallized frameworks with giant pores: are there limits to the possible? *Accounts of Chemical Research* 38 (4): 217–225. (b) Férey, G., Mellot-Draznieks, C., Serre, C. et al. (2005). A chromium terephthalate-based solid with unusually large pore volumes and surface area. *Science* 309 (5743): 2040–2042. (c) Férey, G., Serre, C., Mellot-Draznieks, C. et al. (2004). A hybrid solid with giant pores prepared by a combination of targeted chemistry, simulation, and powder diffraction. *Angewandte Chemie International Edition* 116 (46): 6456–6461.

15 Baburin, I., Leoni, S., and Seifert, G. (2008). Enumeration of not-yet-synthesized zeolitic zinc imidazolate MOF networks: a topological and DFT approach. *The Journal of Physical Chemistry B* 112 (31): 9437–9443.

16 Huang, X., Zhang, J., and Chen, X. (2003). $[Zn(bim)_2]\cdot(H_2O)_{1.67}$: a metal-organic open-framework with sodalite topology. *Chinese Science Bulletin* 48 (15): 1531–1534.

17 (a) Huang, X.-C., Lin, Y.-Y., Zhang, J.-P., and Chen, X.-M. (2006). Ligand-directed strategy for zeolite-type metal-organic frameworks: zinc(II) imidazolates with unusual zeolitic topologies. *Angewandte Chemie International Edition* 45 (10): 1557–1559. (b) Banerjee, R., Phan, A., Wang, B. et al. (2008). High-throughput synthesis of zeolitic imidazolate frameworks and application to CO_2 capture. *Science* 319 (5865): 939–943.

18 Yang, J., Zhang, Y.-B., Liu, Q. et al. (2017). Principles of designing extra-large pore openings and cages in zeolitic imidazolate frameworks. *Journal of the American Chemical Society* 139 (18): 6448–6455.

19 Fu, Y.-M., Zhao, Y.-H., Lan, Y.-Q. et al. (2007). A chiral 3D polymer with right- and left-helices based on 2,2′-biimidazole: synthesis, crystal structure and fluorescent property. *Inorganic Chemistry Communications* 10 (6): 720–723.

20 Park, K.S., Ni, Z., Côté, A.P. et al. (2006). Exceptional chemical and thermal stability of zeolitic imidazolate frameworks. *Proceedings of the National Academy of Sciences of the United States of America* 103 (27): 10186–10191.

21 Hayashi, H., Cote, A.P., Furukawa, H. et al. (2007). Zeolite A imidazolate frameworks. *Nature Materials* 6 (7): 501–506.

22 Zhang, J., Wu, T., Zhou, C. et al. (2009). Zeolitic boron imidazolate frameworks. *Angewandte Chemie International Edition* 48 (14): 2542–2545.

23 Chen, W.-T., Fang, X.-N., Luo, Q.-Y. et al. (2007). Poly[μ_2-4,4′-bipyridine-di-μ_2-bromido-cadmium(II)], with novel colour-tunable fluorescence. *Acta Crystallographica Section C: Crystal Structure Communications* 63 (9): 398–400.

24 Masciocchi, N., Bruni, S., Cariati, E. et al. (2001). Extended polymorphism in copper(II) imidazolate polymers: a spectroscopic and XRPD structural study. *Inorganic Chemistry* 40 (23): 5897–5905.

25 Tian, Y.-Q., Cai, C.-X., Ren, X.-M. et al. (2003). The silica-like extended polymorphism of cobalt(II) imidazolate three-dimensional frameworks: X-ray single-crystal structures and magnetic properties. *Chemistry – A European Journal* 9 (22): 5673–5685.

26 Wu, T., Bu, X., Zhang, J., and Feng, P. (2008). New zeolitic imidazolate frameworks: from unprecedented assembly of cubic clusters to ordered cooperative organization of complementary ligands. *Chemistry of Materials* 20 (24): 7377–7382.

27 Spek, A., Duisenberg, A., and Feiters, M. (1983). The structure of the three-dimensional polymer poly [μ-hexakis(2-methylimidazolato-*N,N′*)-triiron(II)],$[Fe_3(C_4H_5N_2)_6]_n$. *Acta Crystallographica Section C: Crystal Structure Communications* 39 (9): 1212–1214.

28 Wu, T., Bu, X., Liu, R. et al. (2008). A new zeolitic topology with sixteen-membered ring and multidimensional large pore channels. *Chemistry – A European Journal* 14 (26): 7771–7773.

29 Morris, W., Doonan, C.J., Furukawa, H. et al. (2008). Crystals as molecules: postsynthesis covalent functionalization of zeolitic imidazolate frameworks. *Journal of the American Chemical Society* 130 (38): 12626–12627.

30 (a) Huang, A., Bux, H., Steinbach, F., and Caro, J. (2010). Molecular-sieve membrane with hydrogen permselectivity: ZIF-22 in LTA topology prepared with 3-aminopropyltriethoxysilane as covalent linker. *Angewandte Chemie International Edition* 122 (29): 5078–5081. (b) Huang, A., Dou, W., and Caro, J.r. (2010). Steam-stable zeolitic imidazolate framework ZIF-90 membrane with hydrogen selectivity through covalent functionalization. *Journal of the American Chemical Society* 132 (44): 15562–15564.

31 Huang, A. and Caro, J. (2011). Covalent post-functionalization of zeolitic imidazolate framework ZIF-90 membrane for enhanced hydrogen selectivity. *Angewandte Chemie International Edition* 50 (21): 4979–4982.

32 (a) Karagiaridi, O., Bury, W., Sarjeant, A.A. et al. (2012). Synthesis and characterization of isostructural cadmium zeolitic imidazolate frameworks via solvent-assisted linker exchange. *Chemical Science* 3 (11): 3256–3260.

33 Al-Maythalony, B.A., Alloush, A.M., Faizan, M. et al. (2017). Tuning the interplay between selectivity and permeability of ZIF-7 mixed matrix membranes. *Matrix* 13: 25.

21

Dynamic Frameworks

21.1 Introduction

Reticular chemistry is concerned with linking molecular building units through strong bonds into well-defined extended structures such as metal-organic frameworks (MOFs) and covalent organic frameworks (COFs) [1]. The resulting materials are known to be architecturally stable and to sustain permanent porosity. This is generally attributed to the strong directional bonding making up the backbone of these materials. The large accessible internal voids within the structures of MOFs and COFs further endow them with the seemingly contradictory prospect of large amplitude motion of their constituents in the solid state, an aspect rarely achieved in conventional extended structures. To understand this, it is instructive to consider what criteria need to be met to create extended solids capable of motion of their constituents without collapse or deterioration of the overall structure. Two general points need to be considered: (i) the constituents must be able to move about without interfering with each other which makes the use of open porous frameworks a necessary requirement, and (ii) specific weak points must be introduced in the structure to control where the motion is to take place. Both aspects can be addressed in reticular chemistry. The construction from molecular building units affords porous reticular frameworks with pre-determined composition and ensures the prospect of distinctly different kind of bonds within one single framework. Different strategies can be applied to target MOFs/COFs capable of large amplitude motion in the solid state and they can be categorized based on the prevalent modes of framework dynamics. In general, we distinguish four distinct cases: (i) Synchronized global dynamics where two or more discrete configurations of the framework backbone exist and can be interconverted by an external stimulus such as gas pressure or temperature. (ii) Synchronized local dynamics where the backbone remains unaffected but a synchronized motion of functionality appended to the backbone can be triggered by an external stimulus. (iii) Independent local dynamics where the backbone of the framework is held together by mechanical rather than chemical bonds thus allowing for motion of the framework constituents without the need for making or breaking of covalent bonds and no need for an external stimulus, and (iv) independent local dynamics where mechanically entangled species are appended to the backbone

Introduction to Reticular Chemistry: Metal-Organic Frameworks and Covalent Organic Frameworks,
First Edition. Omar M. Yaghi, Markus J. Kalmutzki, and Christian S. Diercks.

Companion website: www.wiley.com/go/yaghi/reticular

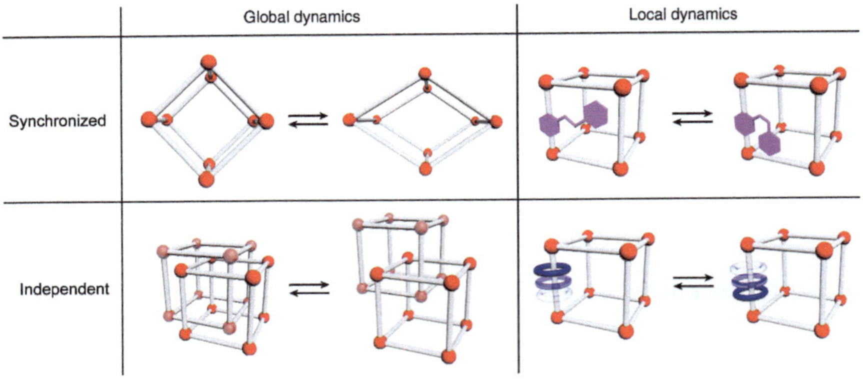

Figure 21.1 Representative modes of dynamics in extended framework structures. The dynamics can be either global and affect the entire backbone of the framework or local and independent of the framework backbone. Additionally, we distinguish synchronized dynamics, where distinct states are accessed that are structurally well-defined from independent motion in frameworks where the motion is not synchronized throughout the entire framework. Global synchronized dynamics are found in so-called "breathing" MOFs. Local dynamics are found in frameworks with molecular switches appended to their backbone. Global independent dynamics are found in interpenetrated or woven frameworks where the mechanically entangled substituents show large degrees of freedom of motion without the need for making or breaking of bonds. Similarly, local independent dynamics are found in frameworks with mechanically entangled rings appended to their backbone.

and move without affecting it (Figure 21.1). In this chapter we conceptualize the different modes of dynamics in extended framework structures and highlight their respective underlying design principles.

21.2 Flexibility in Synchronized Dynamics

In "flexible" MOFs and COFs the framework gets flexed meaning that a force is exerted on the material to bring about the structural change. On a molecular level this translates into making and breaking of bonds or distortions in bond lengths and angles. In this context the motion of the framework can be global when the entire backbone of the framework is dynamic, or local when the motion is independent of the backbone. Global flexibility in MOFs is found in "breathing MOFs" which, when triggered by an external stimulus, show substantial changes in their internal void space. Local flexibility can be achieved by decorating the backbone of a framework with molecular switches that can accommodate two (or more) distinct conformations without affecting the integrity of the framework. In both cases the dynamic motion of the constituents is ordered and gets triggered by an external stimulus which supplies the necessary energy for the structural change to occur.

21.2.1 Synchronized Global Dynamics

It was found that certain MOFs undergo reversible structural phase transitions in response to external stimuli (e.g. guest inclusion, heat, gas pressure, etc.), often accompanied by drastic changes in pore volume resulting from the expansion or contraction. In general, such materials cooperatively switch between two or more distinct states with full retention of long-range order, a phenomenon commonly referred to as "breathing." The resulting large amplitude structural motion is in stark contrast to what is observed in traditional crystalline solids, where such motion would result in structure collapse. The crystallographically well-defined transitions of breathing frameworks not only allow for the unambiguous elucidation of the individual states but furthermore provide a handle for the identification of the inherent structural features that endow these otherwise rigid frameworks with flexibility. Upon exposure to external stimuli flexible frameworks distort at their weakest points, and in this regard several structural components feature prominently: (i) MOFs comprising rod secondary building units (SBUs) and square-shaped 1D pore systems, (ii) discrete SBUs with flexible coordination geometry around the metal centers, and (iii) MOFs comprising inherently flexible linkers [2].

21.2.1.1 Breathing in MOFs Built from Rod SBUs

The first example of a MOF exhibiting large amplitude structural flexibility was MIL-53, a framework of general composition (M^{3+}(OH)(BDC), where M = Al, Fe, Cr, Sc, Ga) [3]. In the crystal structure, one-dimensional rod SBUs are linked by BDC linkers to afford a three-dimensional **sra** topology extended structure featuring square shaped 1D channels along the crystallographic *c*-axis (Figure 21.2). Upon exposure to external stimuli of different nature, three distinct phases of (Cr)MIL-53 are observed; as-synthesized (*as*), narrow pore (*np*), and wide pore (*wp*). In (Cr)MIL-53-*as* (V = 1440 Å^3) the channels are occupied by disordered H_2BDC molecules. Upon heating to 573 K these unbound moieties are removed resulting in (Cr)MIL-53-*wp* (V = 1486 Å^3) and subsequent cooling in air leads to (Cr)MIL-53-*np* with significantly decreased pore volume (V = 1012 Å^3). In these fully reversible phase transitions, both, the inorganic SBU, as well as the entirely sp^2 hybridized linker remain unchanged. Consequently, the origin of flexibility must be due to the junction between these two constituents. Indeed, crystallographic data corroborates that the weak point in the structure is in fact the coordination geometry around the octahedral Cr^{3+} ions. This manifests itself in rotation of the linker around the carboxylate O–O axis during the phase transitions, resulting in the corresponding dihedral angles between the O–Cr–Cr–O and the O–C–O planes of 177.51° for (Cr)MIL-53-*as*, 180.1° for (Cr)MIL-53-*wp*, and 139.1° for (Cr)MIL-53-*np*, respectively. More generally it is observed that in the absence of guests and at low temperature the framework has narrow pores and in the presence of guest molecules or at high temperature the pore opens up to yield (Cr)MIL-53-*wp* (Figure 21.2) [4]. It must be noted that in this case, the presence of square-shaped tetragonal pores is crucial in that it allows for the linker to rotate around all SBUs in a synchronized manner. During the framework distortion the carboxylate binding groups rotate around the O–O axis in a "knee cap" like motion.

The importance of this finding is illustrated by comparison to MIL-68 (M^{3+}(BDC)(OH), where M = V, Fe, Al, In), a MOF of the same chemical composition and with the same SBU as MIL-53, but of different structure type.

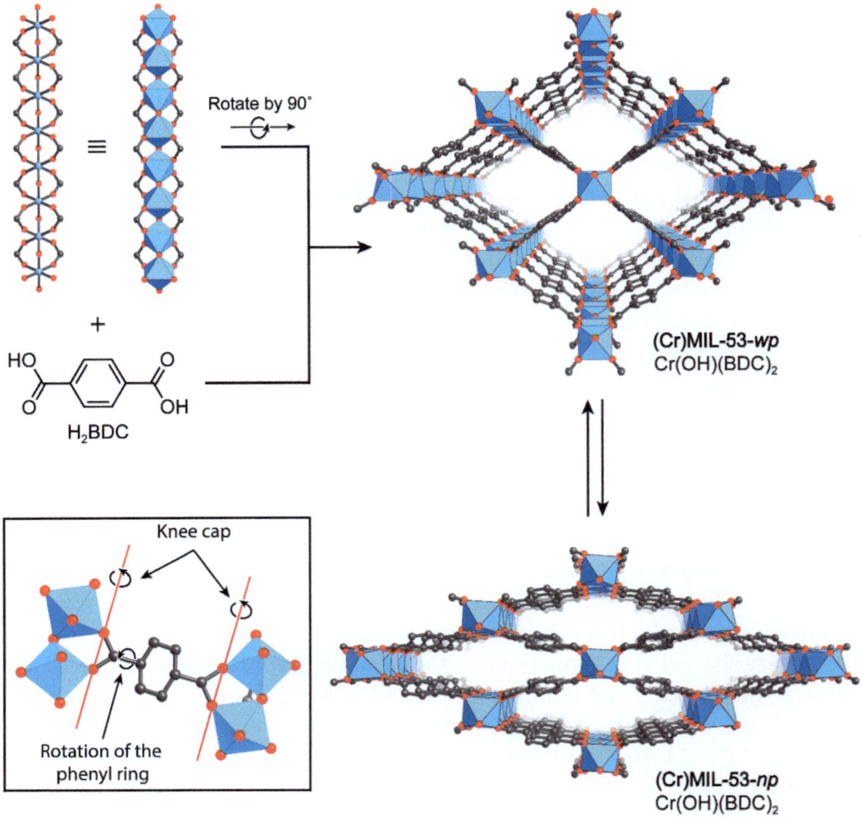

Figure 21.2 Breathing behavior of (Cr)MIL-53, constructed from linear rod SBUs and BDC linkers. In the presence of guests and at elevated temperatures the framework features wide trapezoidal channels along the crystallographic *c*-axis. Upon cooling to room temperature and/or removal of guest molecules the framework distorts by rotation around the carboxylate O–O axis in a "knee cap" fashion assisted by rotation of the central phenyl ring of the BDC linker, thus resulting in narrow trapezoidal pores. The framework distortion is fully reversible. All hydrogen atoms are omitted for clarity. Color code: Cr, blue; C, gray; O, red.

In contrast to the square-shaped pores of MIL-53, MIL-68 has a **rad** net with parallel hexagonal and triangular channels running along the crystallographic *c*-axis. As a result, synchronized rotation of the linker around the SBUs is prohibited and consequentially MIL-68 does not display a breathing behavior [5].

21.2.1.2 Breathing in MOFs Built from Discrete SBUs

Discrete SBUs can also lead to flexible frameworks. The fact that the inorganic SBUs are 0D not 1D enables a higher degree of flexibility, the main source of which still comes from the rotation of chelating linkers around the metal centers. However, in contrast to 1D rod SBUs the expansion/contraction motion in the case of discrete SBUs is not necessarily restricted to two dimensions.

Intuitively flexible structures expand upon inclusion and contract when the guest is expelled. Contrarily, the opposite is observed in the breathing behavior

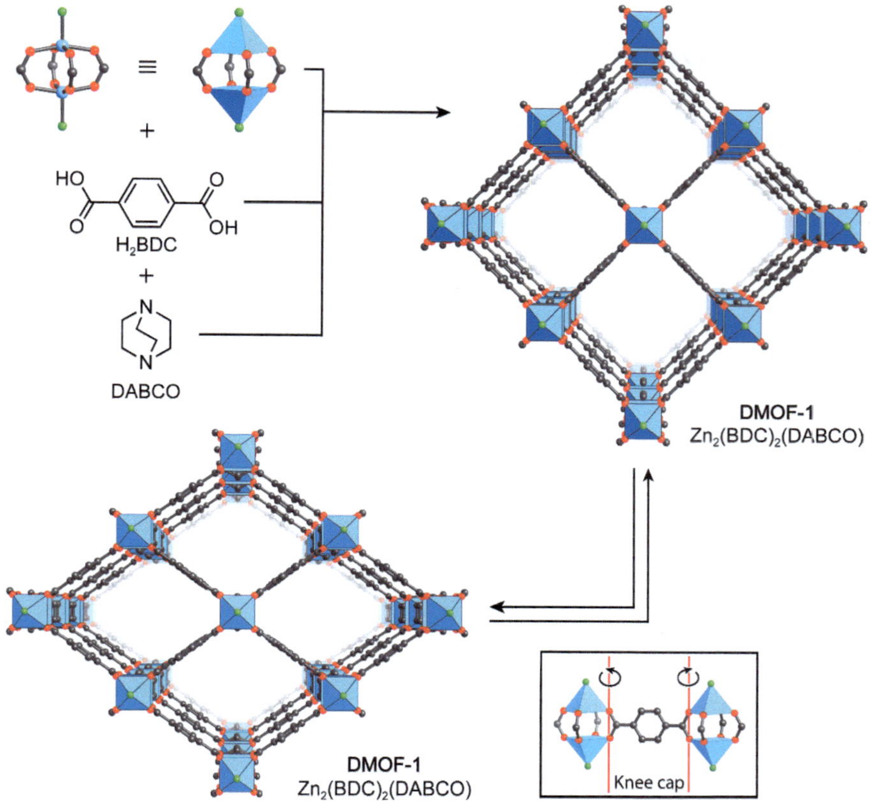

Figure 21.3 DMOF-1 is constructed from copper paddle wheel SBUs that are connected by BDC linkers to yield $Zn_2(BDC)_2$ square grid layers. These layers are pillared by DABCO molecules to yield a framework with underlying **pcu** topology. The empty framework features wide tetragonal channels along the crystallographic *c*-axis. Upon inclusion of benzene molecules, the pore aperture adopts a narrow trapezoidal shape. The weak point in the structure is the carboxylate metal coordination which can rotate along the carboxylate O—O bond. All hydrogen atoms are omitted for clarity. Color code: Zn, blue; C, gray; N, green; O, red.

of DMOF-1 ($Zn_2(BDC)_2(DABCO){\cdot}4DMF$, where DABCO = 1,4-diazabicyclo [2.2.2]octane). DMOF-1 is composed of dinuclear paddle wheel SBUs, which are bridged by linear ditopic BDC linkers to form distorted 2D square-grid ($Zn_2(BDC)_2$) layers (Figure 21.3). The axial sites of the paddle wheels are occupied by DABCO pillars to extend the 2D layers into a 3D framework of **pcu** topology. The evacuated open framework with tetragonal square-shaped channels shrinks upon inclusion of benzene into a trapezoidal conformation. Again, the weak point of the structure is the coordination environment around the inorganic SBU. A decrease in volume accompanies this distortion (1147.6 to 1114.2 $Å^3$) with the thermodynamic driving force being favorable host–guest interactions [6].

Reticulation of iron acetate with H_2NDC (naphthalene-2,6-dicarboxylic acid) yields (Fe)MIL-88(C) ($Fe_3(O)(OH)(H_2O)_2(NDC)_3$) of **asc** topology (Figure 21.4).

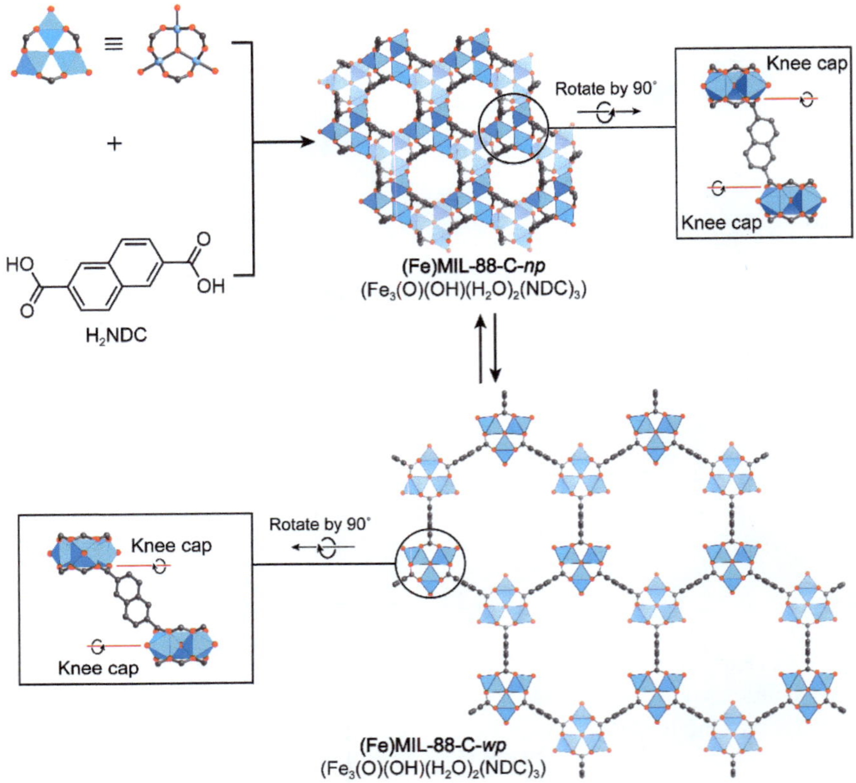

Figure 21.4 Flexibility in the 3D **acs** topology framework (Fe)MIL-88(C) constructed from linear ditopic BDC linkers and discrete $Fe_3(O)(OH)(H_2O)_2$ SBUs. Distortion of the O–O axes of the carboxylates coordinated to the SBU by 30° in a "knee cap" kind of fashion due to favorable interactions with DMF translates into an increased SBU–SBU distance and a concomitant increase in the unit cell volume of 230%. All hydrogen atoms are omitted for clarity. Color code: Zn, blue; C, gray; N, green; O, red.

The framework shows a breathing behavior with a difference in unit cell volume between the *np* and the *wp* form of up to 230%. This is remarkable, especially when compared to MOFs based on rod SBUs which display only up to 40% difference in unit cell volume between their respective *np* and *wp* phases. In the case of (Fe)MIL-88(C) the closed form minimizes lattice energy and exists in the absence of guests. To accommodate guest inclusion, the carboxylates coordinated to the trimeric metal SBUs can rotate around the O–O axis by up to 30°, thus extending the SBU–SBU distance which in turn leads to an expansion in all three lattice dimensions in a manner akin to swelling. Selective adsorption toward guests of different chemical nature is observed. Specifically, (Fe)MIL-88(C) expands from a unit cell volume of 2120 $Å^3$ in the dry *np* state to 5695 $Å^3$ upon exposure to DMF in the *wp* state. In contrast, in the presence of water, methanol, and lutidine, the unit cell remains largely unaltered (2270 $Å^3$) [7].

21.2.1.3 Flexibility Through Distorted Organic Linkers

Rigidity of organic linkers is considered a necessary requirement for the construction of architecturally stable frameworks. In rare cases, bending of these presumably inflexible units can however be induced by external stimuli to provide for flexibility, as illustrated by DUT-49 (Cu_2(BBCDC), where BBCDC = 9*H*-carbazole-3,6-dicarboxylate) [8]. DUT-49 is constructed from copper paddle wheel SBUs and tetratopic H_4BBCDC linkers. In the structure, the paddle wheel SBUs are connected by carbazole-3,6-dicarboxylate moieties to form cuboctahedral MOPs (see Chapter 21) [9]. These 12-c tertiary building units (TBUs) are connected by linear biphenyl links to yield a framework of overall **fcu** topology (Figure 21.5). The framework has a hierarchical pore structure with three distinct kinds of pore: 12 Å large cuboctahedral cages, 18 Å large tetrahedral cages, and 26 Å large octahedral cages [8]. In stark contrast to commonly observed gas adsorption behavior of microporous solids, which display an increase of gas uptake with increasing external gas pressure, DUT-49 exhibits an unusual negative gas adsorption isotherm [10]. Spontaneous desorption of gas is observed (10 and 30 kPa for methane and *n*-butane, respectively) upon increasing the gas pressure. Careful examination of single crystal X-ray diffraction data reveals that the framework undergoes a dramatic contraction (61% decrease in pore volume) caused by rotation of the TBUs induced by a severe distortion of the linker backbone (Figure 21.5). Theoretical calculations suggest that the high level of strain on the organic molecule can be compensated for by the energy benefits provided by greater methane affinity of the contracted/smaller pores. Such framework deformation results in decreased porosity and release of previously adsorbed gas [11].

The aforementioned three modes of flexibility only represent the most common scenarios reported for breathing MOFs. In these cases, the entire framework undergoes synchronized phase transitions induced by external stimuli. The weak points in the structure can either be the coordination junction between chelating linkers and SBUs or the organic linkers themselves.

21.2.2 Synchronized Local Dynamics

In breathing MOFs the backbone of the framework is distorted significantly upon framework expansion/contraction. Over time this can lead to structure deterioration. To circumvent the need for the frameworks' backbone to distort, molecular photoswitches (i.e. organic molecules which change their conformation or structure upon interaction with light) are implemented into MOFs by appending them to the organic linkers. In this way the pore size, shape, and hence the ability to adsorb certain guests, are directly be controlled by light irradiation. The most common way to introduce photoswitches is their addition as a functional side group to the linker. Even though there is a wide range of different molecules that show switching upon irradiation with light, thus far only the prototypical azobenzene photoswitch has been incorporated into MOFs. In most cases the azobenzene photoswitch dangles into the pore and after light

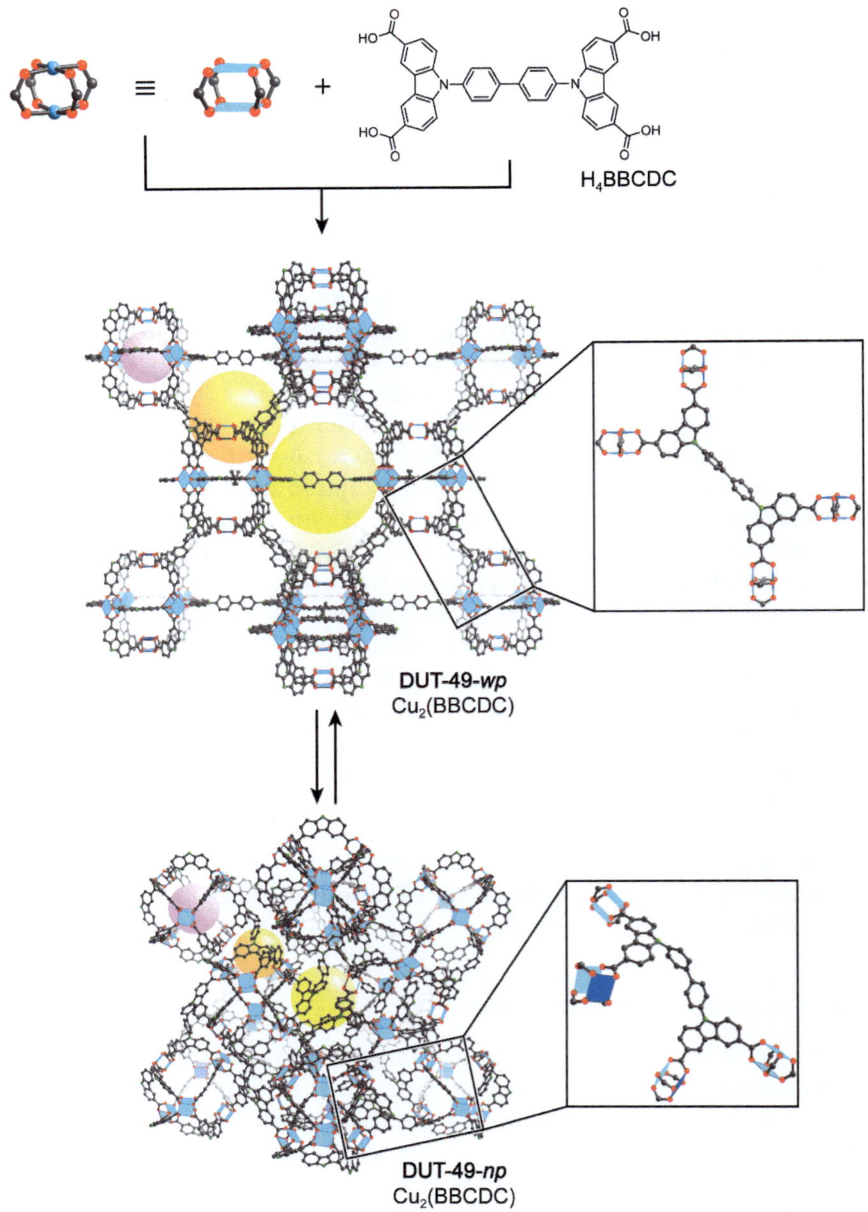

Figure 21.5 DUT-49 is constructed from cuboctahedral 12-c TBUs formed from paddle wheel SBUs and carbazole-3,6-dicarboxylate moieties of the linker. These TBUs are connected by biphenyl bridges to yield a framework of **fcu** topology. Upon adsorption of methane (10 kPa) and butane (30 kPa) the framework shows negative gas adsorption due to structural rearrangement induced by bending of the organic linker which decreases the internal pore volume of the structure. All hydrogen atoms are omitted for clarity. Color code: Zn, blue; C, gray; N, green; O, red.

exposure (365 nm) changes from a *trans* configuration to a *cis* configuration. This is usually accompanied by a substantial change in the pore aperture and thus the accessibility of the internal void space. The switching back to the initial state can be triggered by either absorption of light of a higher wavelength (440 nm) or by heating thus rendering the whole process reversible. IRMOF-74-III(azo) (Mg(azo-DOT-III)$_2$) is an isoreticular analog of the prototypical **etb** topology MOF-74. It is constructed. from 1D SBUs that are connected by azo-DOT-III linkers. The structure of IRMOF-74-III(azo) features mesoporous hexagonal 1D channels running along the crystallographic *c*-axis. (Figure 21.6) [12]. Each azo-DOT-III linker in IRMOF-74-III(azo) is decorated with a photoswitchable azobenzene unit, which can be converted between its *cis* and *trans* conformation

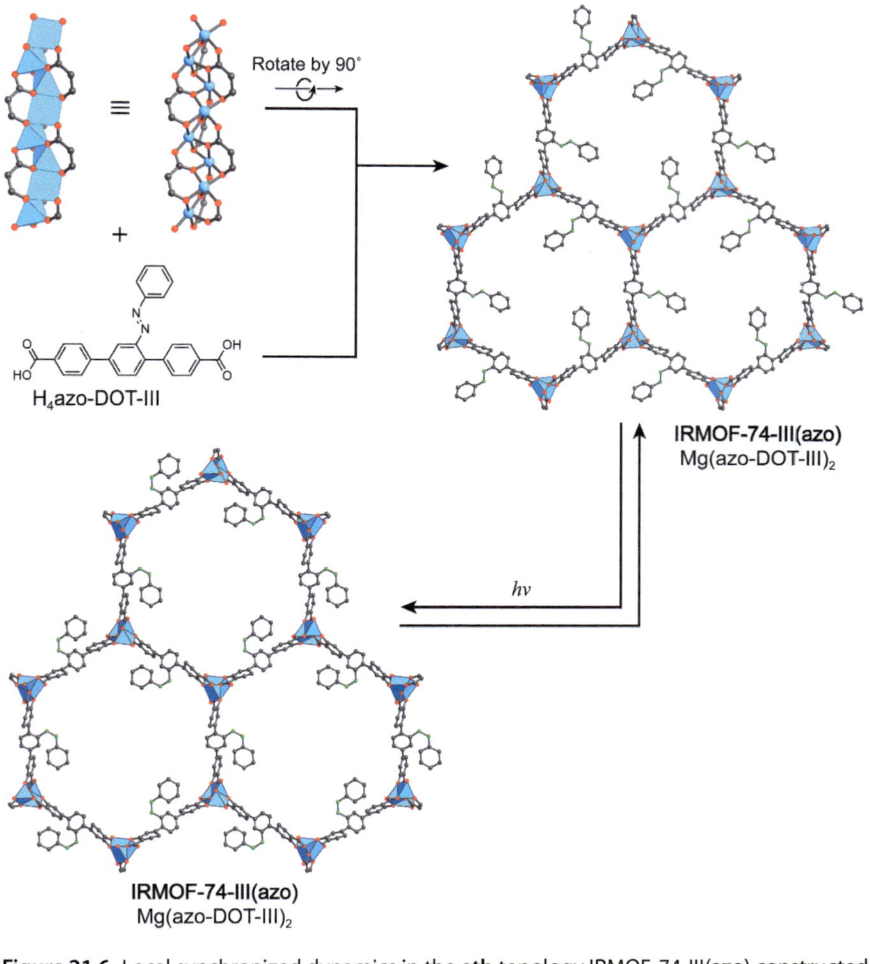

Figure 21.6 Local synchronized dynamics in the **etb** topology IRMOF-74-III(azo) constructed from rod SBUs and azo-DOT-III linkers. Isomerization of the appended photoswitches between the *cis* and the *trans* conformation is affected by irradiation with light of 408 nm in wavelength. This motion occurs uniformly throughout the framework and is independent of the framework backbone. All hydrogen atoms are omitted for clarity. Color code: Mg, blue; C, gray; N, green; O, red.

by excitation at 408 nm. This switching is independent of the framework backbone but directly affects the pore aperture of the material which expands from 8.3 Å in the *trans* conformation to 10.3 Å in the *cis* conformation. The utility of this switching is highlighted by the controlled loading and release of a propidium iodide dye. Spectroscopic studies show no release of the luminescent dye under ambient conditions but upon irradiation of the MOF at 408 nm, the rapid wagging motion inherent to the repetitive isomerization of the azobenzene functionality triggers the release of the dye from the pores.

While in photoswitchable MOFs the backbone of the framework remains unaffected of the motion of the switch, the strategy still relies on the isomerization (i.e. flexing) of covalent bonds, which is not desirable in the context of longevity and endurance of materials.

21.3 Independent Dynamics in Frameworks

A fundamentally different approach toward dynamics in extended structures are adaptive frameworks; materials where the framework dynamics are not triggered by a specific external stimulus but that can adapt, locally or globally, to various external influences. To realize such frameworks, the constituents need to be able to move freely and thus without the need for flexing of bonds. This is achieved by employing mechanical instead of chemical bonds. Mechanical bonds hold molecular entities together by physical entanglement rather than actual chemical bonding. This is advantageous in the context of dynamics because the constituents of such frameworks can move independently without parting company. Again, we distinguish global dynamics from local dynamics. The introduction of mechanically bonded rings onto the struts of an extended backbone is referred to as "robust dynamics." In this scenario, the motion of the ring is independent of the framework backbone and can occur in the absence of an external stimulus. Global dynamics are observed in extended interpenetrated structures where individual frameworks can move with respect to each other, and in 3D frameworks formed by weaving of molecular threads. It is important to note that in contrast to flexible frameworks where an external stimulus affects synchronized motions of the respective building units, adaptive dynamics based on mechanical bonding can occur in the absence of external stimuli and are therefore independent as opposed to synchronized.

21.3.1 Independent Local Dynamics

To overcome the inherent limitations of flexible frameworks with respect to longevity and endurance, mechanically interlocked molecules are inserted into the rigid framework backbone in such way that they are capable of dynamics without compromising the fidelity of the entire system (robust dynamics) [13]. The fact that in such systems the constituents are linked mechanically, not chemically, allows for large amplitude molecular motion of the units with respect to each other. On the molecular level this approach has proved fruitful for the development of molecular machines, muscles, and rotors where during a switching process only weak non-covalent bonds are broken and reformed again in a fully reversible and highly controllable fashion [14]. In the structure of MOF-1040 ($Zn_4O(BPCu)_3$, where BPCu = Cu^{2+}-4,7,10,13,16,19,22, 25-octaoxa-2(2,9)-phenanthrolina-1,3(1,4)-dibenzenacyclohexacosaphane@4,4′-(1,10-phenanthroline-3,8-diyl)dibenzoate), macrocycles are mechanically appended to its backbone. MOF-1040 is prepared by reticulating copper(I)-bisphenanthroline complex-based pseudorotaxane linker H_2BPCu with Zn^{2+} ions into a framework of **pcu** topology, isoreticular to MOF-5.The single crystal structure of the twofold interpenetrated framework confirms that the chemical integrity of the copper(I)-complexed pseudorotaxane is maintained under the MOF synthesis condition. The oxidation of Cu^{1+} to Cu^{2+} within the framework can be carried out under ambient conditions without affecting the integrity of the MOF backbone, corroborating the chemical accessibility of the copper centers in the framework. In theory, upon demetallation, the discrete macrocycle that is connected to the framework only by means of mechanical bonding can move about freely on the strut between neighboring SBUs, however, in this case the demetallation of the framework does not go to completion, presumably due to steric shielding of the Cu^{1+} centers by the framework (Figure 21.7).

In contrast a MOF constructed from $Zn_4O(\text{-COO})_6$ SBUs and tetratopic H_4BITC (18-crown-6 @ 4,4′,4″,4‴-(1,4-phenylenebis(1*H*-benzo[*d*]imidazole-2,4,7-triyl))tetrabenzoic acid) linkers, termed UWDM-4 ($(Zn_4O)_2(BITC)_3$), contains 18-crown-6 ethers on the crossbars between two triphenyl-dicarboxylic

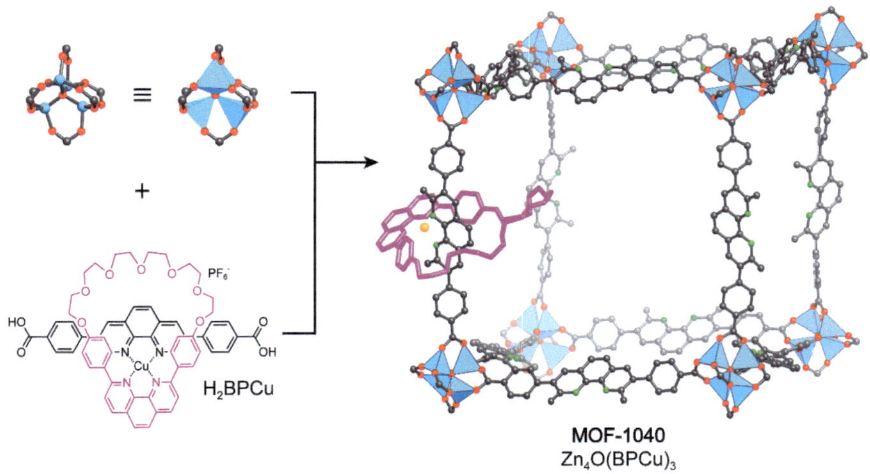

Figure 21.7 Mechanically bonded macrocycles appended to the backbone of MOF-1040. MOF-1040 is constructed from $Zn_4O(-COO)_6$ SBUs and ditopic BPCu linkers. The linker is a pseudorotaxane where the linear strut and the macrocycle are held together by coordination to a Cu(I) template. Upon reticulation the ring is trapped on the strut between two neighboring SBUs. Demetallation of the framework proves difficult due to steric shielding of the Cu(I) centers in the twofold interpenetrated structure. All hydrogen atoms and interpenetrating nets are omitted for clarity. Only one macrocycle is shown for clarity and highlighted in pink. Color code: Zn, blue; Cu, orange; C, gray; N, green; O, red.

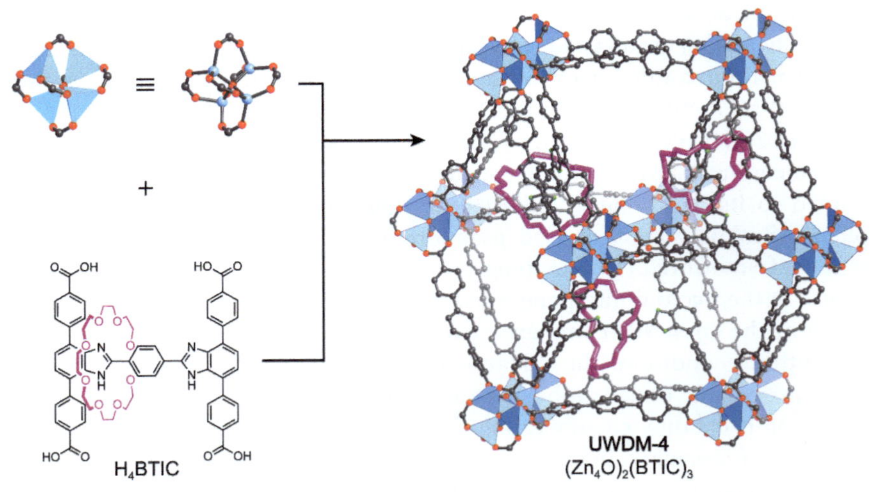

Figure 21.8 Local independent dynamics of 18-crown-6 ethers mechanically bonded to the backbone of UWDM-4. The framework is constructed from $Zn_4O(-COO)_6$ SBUs and tetratopic BITC linkers and assumes an overall **acs** topology. The 18-crown-6 ether is appended to a crossbar between two triphenyldicarboxylic acids on the linker and does not detach from the linker during the MOF synthesis despite the lack of strong chemical interactions. At room temperature the linker rapidly shuffles between two imidazole moieties on the crossbar. Only upon cooling this motion is slowed down and the macrocycle is confined to either of the two imidazole moieties on the linkers' crossbar. All hydrogen atoms and interpenetrating frameworks are omitted for clarity. Color code: Zn, blue; C, gray; N, green; O, red.

acids. In the linker the macrocycle is not bound through metal coordination but held in place by two bulky stoppers. The molecular ring can move freely but interacts preferably with the two electron deficient imidazole moieties on opposing sites of the crossbar (Figure 21.8) [15]. The translational "shuttling" of the crown ether is observed by variable-temperature solid-state nuclear magnetic resonance (NMR). At room temperature, only one resonance structure is detected for occupied and unoccupied imidazole moieties on the linker, indicating rapid shuttling of the ring between the two sites beyond the NMR timescale. Upon a decrease in temperature the shuttling motion slows down which allows for the peak to be resolved into two distinct resonance states.

21.3.2 Independent Global Dynamics

The use of mechanical bonds is not limited to the local motion of molecular fragments on a rigid backbone, but it is also possible to synthesize frameworks that are entirely constructed from mechanical bonds which allows for global dynamics and rearrangement of the whole structure without the making or breaking of chemical bonds. In principal there are three distinct cases of how a framework can be composed of mechanically linked building units. These scenarios are: (i) interpenetration of extended 3D and 2D frameworks, (ii) weaving of 1D molecular threads held together by mechanical entanglement, and (iii) a structure built up entirely from discrete (0D) interlocking rings [1]. Interpenetration of 3D and

2D frameworks is well established in solid-state materials such as MOFs, coordination networks, or inorganic extended structures. However, in the case of 3D and 2D interpenetration, the potential for motion of the fragments with respect to each other is limited because the entire extended 3D or 2D frameworks would have to be dislocated.

In contrast, the formation of extended structures based on 1D weaving or 0D interlocking offers many more degrees of freedom of the building units and thus allows for enormous spatial deviations to take place without deteriorating the overall structure. The synthesis of such structures is challenging but the reticular synthesis of COFs provides a directed and general approach to access materials with these modes of entanglement. Woven COFs have been accessed synthetically as illustrated by the 3D framework COF-505-Cu ((Cu)(BF_4)[(PDB)(BZ)]$_{imine}$, where PDB = 0,13,16,19,22,25-octaoxa-2(2,9)-phenanthrolina-1,3(1,4)-dibenzenacyclohexacosaphane). COF-505-Cu is built from distorted tetrahedral tetratopic functionalized copper bis-phenanthroline complexes, (Cu$(PDB)_2$(BF_4)), and linear ditopic benzidine linkers through imine bond formation to give a framework of **dia** topology (Figure 21.9) [16]. In the structure

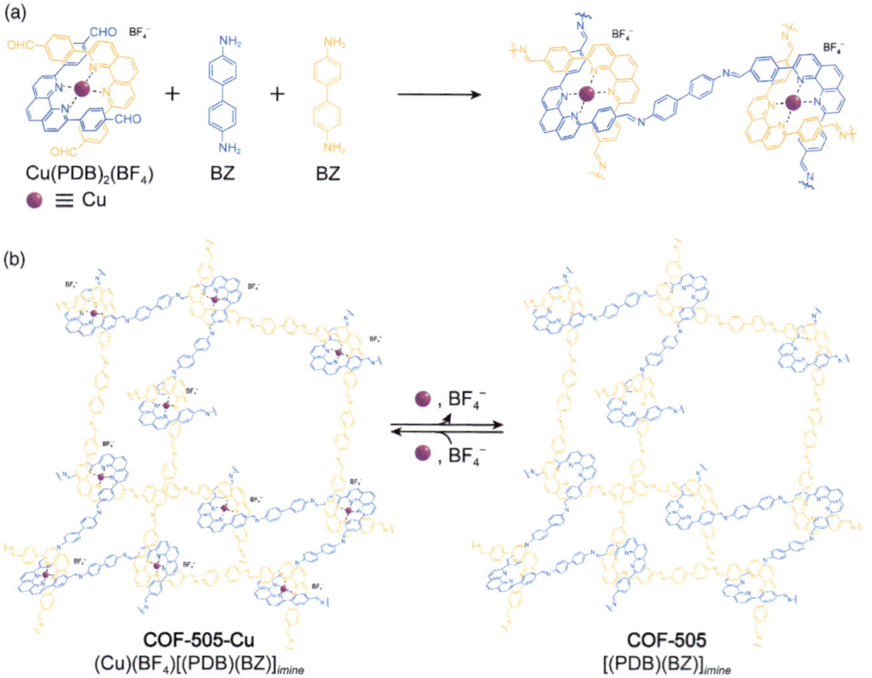

Figure 21.9 (a) Reticulation of Cu$(PDB)_2$(BF_4) with BZ by imine bond formation. (b) Illustration of one adamantane cage of the resulting COF-505-Cu and the demetallated woven framework COF-505. The reversible demetallation of COF-505-Cu results in a change of the topology of the framework from **dia** to **dia-w**. The backbone of the **dia-w** COF-505 is comprised of linear interwoven threads that are held together only by mechanical but not by chemical bonding. As a consequence the framework is endowed with many additional degrees of freedom which allow for large deviations to take place. This is evidenced in a 10-fold increase in elasticity as determined by nano-indentation by atomic force microscopy.

the copper centers only serve as templates for bringing the threads into a woven pattern instead of the more commonly observed parallel arrangement. Post-synthetically these Cu^{1+} ions can be reversibly removed and re-added without loss of the overall structure. Upon demetallation the threads in the resulting **dia-w** topology framework COF-505 ($[(PDB)(BZ)]_{imine}$) are endowed with many additional degrees of freedom for enormous deviations to take place between them, without unzipping the weaving of the overall structure. This translates into a 10-fold increase in elasticity of the material as determined by nano-indentation by atomic force microscope (AFM).

The largest degrees of freedom that can be achieved in extended frameworks based on mechanical bonds are expected in structures composed of discrete rings. The synthesis of extended structures composed of 0D rings that are held together exclusively by mechanical bonds. This synthetic target is a long-standing challenge. We expect that such structures can be achieved with a strategy akin to what has been used for the generation of woven structures [17].

21.4 Summary

In this chapter we covered the modes of dynamics in extended framework materials. The importance of both the porosity and the presence of distinctly different kinds of bonding within one single structure with respect to large amplitude motion of their constituents were examined. Modes of framework dynamics were categorized into global and local, depending on if the motion affects the entire backbone of the material or is independent of it. Such dynamic processes can be affected by a specific external stimulus which generally results in synchronized motion. Global synchronized motion is observed in flexible breathing frameworks and the weak point in the framework that distorts in response to the stimuli can be both, the inorganic SBU or the organic linker. Local synchronized motion can be achieved by appending photoswitches to the backbone of the MOF or COF. Synchronized framework dynamics rely on the making and breaking or on the flexing of covalent bonds which is detrimental in the context of material longevity. In contrast independent framework dynamics are based on mechanical bonding and only weak, non-covalent interactions need to be overcome to affect the motion. On a local level this can be achieved by appending mechanically bonded rings onto the struts of the framework and on a global level such dynamics are observed in woven materials where the 1D framework constituents are held together by mechanical bonds. It is worthwhile noting that in independent framework dynamics the motion can be autonomous as opposed to synchronized and can happen in the absence of specific external stimuli.

References

1 (a) Furukawa, H., Cordova, K.E., O'Keeffe, M., and Yaghi, O.M. (2013). The chemistry and applications of metal-organic frameworks. *Science* 341 (6149): 1230444. (b) Diercks, C.S. and Yaghi, O.M. (2017). The atom, the molecule,

and the covalent organic framework. *Science* 355 (6328): eaal1585. (c) Yaghi, O.M., O'Keeffe, M., Ockwig, N.W. et al. (2003). Reticular synthesis and the design of new materials. *Nature* 423 (6941): 705–714.

2 (a) Schneemann, A., Bon, V., Schwedler, I. et al. (2014). Flexible metal-organic frameworks. *Chemical Society Reviews* 43 (16): 6062–6096. (b) Fletcher, A.J., Thomas, K.M., and Rosseinsky, M.J. (2005). Flexibility in metal-organic framework materials: impact on sorption properties. *Journal of Solid State Chemistry* 178 (8): 2491–2510. (c) Sarkisov, L., Martin, R.L., Haranczyk, M., and Smit, B. (2014). On the flexibility of metal-organic frameworks. *Journal of the American Chemical Society* 136 (6): 2228–2231. (d) Bennett, T.D., Cheetham, A.K., Fuchs, A.H., and Coudert, F.-X. (2017). Interplay between defects, disorder and flexibility in metal-organic frameworks. *Nature Chemistry* 9: 11–16.

3 (a) Loiseau, T., Serre, C., Huguenard, C. et al. (2004). A rationale for the large breathing of the porous aluminum terephthalate (MIL-53) upon hydration. *Chemistry – A European Journal* 10 (6): 1373–1382. (b) Hamon, L., Serre, C., Devic, T. et al. (2009). Comparative study of hydrogen sulfide adsorption in the MIL-53 (Al, Cr, Fe), MIL-47 (V), MIL-100 (Cr), and MIL-101 (Cr) metal-organic frameworks at room temperature. *Journal of the American Chemical Society* 131 (25): 8775–8777. (c) Mowat, J.P., Seymour, V.R., Griffin, J.M. et al. (2012). A novel structural form of MIL-53 observed for the scandium analogue and its response to temperature variation and CO_2 adsorption. *Dalton Transactions* 41 (14): 3937–3941. (d) Volkringer, C., Loiseau, T., Guillou, N. et al. (2009). XRD and IR structural investigations of a particular breathing effect in the MOF-type gallium terephthalate MIL-53 (Ga). *Dalton Transactions* (12): 2241–2249.

4 Serre, C., Millange, F., Thouvenot, C. et al. (2002). Very large breathing effect in the first nanoporous chromium(III)-based solids: MIL-53 or $Cr^{III}(OH)\cdot\{O_2C{-}C_6H_4{-}CO_2\}\cdot\{HO_2C{-}C_6H_4{-}CO_2H\}_x\cdot H_2O_y$. *Journal of the American Chemical Society* 124 (45): 13519–13526.

5 (a) Volkringer, C., Meddouri, M., Loiseau, T. et al. (2008). The Kagomé topology of the gallium and indium metal-organic framework types with a MIL-68 structure: synthesis, XRD, solid-state NMR characterizations, and hydrogen adsorption. *Inorganic Chemistry* 47 (24): 11892–11901. (b) Barthelet, K., Marrot, J., Férey, G., and Riou, D. (2004). $V^{III}(OH)\{O_2C{-}C_6H_4{-}CO_2\}\cdot(HO_2C{-}C_6H_4{-}CO_2H)_x(DMF)_y(H_2O)_z$ (or MIL-68), a new vanadocarboxylate with a large pore hybrid topology: reticular synthesis with infinite inorganic building blocks? *Chemical Communications* (5): 520–521.

6 Dybtsev, D.N., Chun, H., and Kim, K. (2004). Rigid and flexible: a highly porous metal-organic framework with unusual guest-dependent dynamic behavior. *Angewandte Chemie International Edition* 116 (38): 5143–5146.

7 Surblé, S., Serre, C., Mellot-Draznieks, C. et al. (2006). A new isoreticular class of metal-organic-frameworks with the MIL-88 topology. *Chemical Communications* (3): 284–286.

8 Stoeck, U., Krause, S., Bon, V. et al. (2012). A highly porous metal-organic framework, constructed from a cuboctahedral super-molecular building

block, with exceptionally high methane uptake. *Chemical Communications* 48 (88): 10841–10843.

9 Tranchemontagne, D.J., Ni, Z., O'Keeffe, M., and Yaghi, O.M. (2008). Reticular chemistry of metal-organic polyhedra. *Angewandte Chemie International Edition* 47 (28): 5136–5147.

10 Krause, S., Bon, V., Senkovska, I. et al. (2016). A pressure-amplifying framework material with negative gas adsorption transitions. *Nature* 532 (7599): 348–352.

11 Schaber, J., Krause, S., Paasch, S. et al. (2017). *In situ* monitoring of unique switching transitions in the pressure-amplifying flexible framework material DUT-49 by high-pressure ^{129}Xe NMR spectroscopy. *The Journal of Physical Chemistry C* 121 (9): 5195–5200.

12 Brown, J.W., Henderson, B.L., Kiesz, M.D. et al. (2013). Photophysical pore control in an azobenzene-containing metal-organic framework. *Chemical Science* 4 (7): 2858–2864.

13 (a) Deng, H., Olson, M.A., Stoddart, J.F., and Yaghi, O.M. (2010). Robust dynamics. *Nature Chemistry* 2 (6): 439–443. (b) Fan, C.B., Liu, Z.Q., Le Gong, L. et al. (2017). Photoswitching adsorption selectivity in a diarylethene–azobenzene MOF. *Chemical Communications* 53 (4): 763–766. (c) Müller, K., Knebel, A., Zhao, F. et al. (2017). Switching thin films of azobenzene-containing metal-organic frameworks with visible light. *Chemistry – A European Journal* 23 (23): 5434–5438. (d) Kanj, A.B., Müller, K., and Heinke, L. (2017). Stimuli-responsive metal-organic frameworks with photoswitchable azobenzene side groups. *Macromolecular Rapid Communications* 39 (1): 1700239. (e) Müller, K., Helfferich, J., Zhao, F. et al. (2018). Switching the proton conduction in nanoporous, crystalline materials by light. *Advanced Materials* 30 (8): 1706551.

14 (a) Sauvage, J.-P. (1998). Transition metal-containing rotaxanes and catenanes in motion: toward molecular machines and motors. *Accounts of Chemical Research* 31 (10): 611–619. (b) Collin, J.-P., Dietrich-Buchecker, C., Gaviña, P. et al. (2001). Shuttles and muscles: Linear molecular machines based on transition metals. *Accounts of Chemical Research* 34 (6): 477–487. (c) Balzani, V., Credi, A., Raymo, F.M., and Stoddart, J.F. (2000). Artificial molecular machines. *Angewandte Chemie International Edition* 39 (19): 3348–3391. (d) Balzani, V., Gómez-López, M., and Stoddart, J.F. (1998). Molecular machines. *Accounts of Chemical Research* 31 (7): 405–414.

15 Zhu, K., O'keefe, C.A., Vukotic, V.N. et al. (2015). A molecular shuttle that operates inside a metal-organic framework. *Nature Chemistry* 7 (6): 514–519.

16 Liu, Y., Ma, Y., Zhao, Y. et al. (2016). Weaving of organic threads into a crystalline covalent organic framework. *Science* 351 (6271): 365–369.

17 Zhao, Y., Guo, L., Gándara, F. et al. (2017). A synthetic route for crystals of woven structures, uniform nanocrystals, and thin films of imine covalent organic frameworks. *Journal of the American Chemical Society* 139 (37): 13166–13172.

Index

Introduction to Reticular Chemistry: Metal-Organic Frameworks and Covalent Organic Frameworks, First Edition. Omar M. Yaghi, Markus J. Kalmutzki, and Christian S. Diercks.

Companion website: www.wiley.com/go/yaghi/reticular

n

o

p

W

X

Z